T0399450

Microbiome-Host Interactions

Microbiome-Host Interactions

Edited by
D. Dhanasekaran
Dhiraj Paul
N. Amaresan
A. Sankaranarayanan
Yogesh S. Shouche

CRC Press
Taylor & Francis Group
Boca Raton London New York

CRC Press is an imprint of the
Taylor & Francis Group, an **informa** business

First edition published 2021
by CRC Press
6000 Broken Sound Parkway NW, Suite 300, Boca Raton, FL 33487-2742

and by CRC Press
2 Park Square, Milton Park, Abingdon, Oxon, OX14 4RN

CRC Press is an imprint of Taylor & Francis Group, LLC

Library of Congress Cataloging-in-Publication Data

Names: Dhanasekaran, Dharumadurai, editor. | Paul, Dhiraj, editor. | Amaresan, N., editor. | Sankaranarayanan, A., editor. | Shouche, Yogesh, editor.
Title: Microbiome-host interactions / edited by D. Dhanasekaran, Dhiraj Paul, N. Amaresan, A. Sankaranarayanan, Yogesh Shouche.
Description: First edition. | Boca Raton : CRC Press, 2021. | Includes bibliographical references and index.
Identifiers: LCCN 2020044928 | ISBN 9780367479909 (hardback) | ISBN 9781003037521 (ebook)
Subjects: LCSH: Microbial ecology. | Microorganisms—Behavior.
Classification: LCC QR100 .M5315 2021 | DDC 579/.17—dc23
LC record available at https://lccn.loc.gov/2020044928

ISBN: 978-0-367-47990-9 (hbk)
ISBN: 978-0-367-71619-6 (pbk)
ISBN: 978-1-003-03752-1 (ebk)

Typeset in Times
by codeMantra

Contents

Section III Animal Microbiome

Section IV Plant Microbiome

Section V Environmental Microbiome

UNIVERSIDADE ESTADUAL DE CAMPINAS
COLÉGIO TÉCNICO DE CAMPINAS – COTUCA
Rua Jorge de Figueiredo Correa, 735 –Parque Taquaral– Campinas / SP – CEP: 13.087-261 – Caixa Postal 6139
Fone: (19) 3521-9904 – CNPJ: 46.068.425/0001-33 www.cotuca.unicamp.br

Foreword

The term microbiome has become more and more frequent in the everyday world of academic research, and it is also increasingly becoming known to the general public. The main reason for the popularity of this terminology is certainly associated with the numerous discoveries relating it to health and well-being. Studies of the human microbiome reveal that the microorganisms that develop in our body, especially in the intestine, have a systemic effect on our entire organism, that is, our microbial population affects positively or negatively all the organs and systems of our body. Since these microorganisms are present in almost all parts of our body, they are decisive for the establishment of health and well-being or the development of diseases.

Much is already known about the effects of the microbiome in the prevention or prevalence of chronic nontransmissible diseases, such as diabetes, obesity, cancer, cardiovascular diseases among others. In addition, there is also a growing interest in knowing the effects of microbial populations on the development of other animals, plants and the environmental balance of the planet, as the discoveries and applications appear to be almost inexhaustible.

Given the importance and breadth of the theme, this book provides a broad and multidisciplinary overview of the most recent issues related to different microbiomes. It presents the specificities, and the most current topics of this theme applied to human beings, other animals, plants, the environment and it also presents and discusses the newest and most advanced microbial analysis and characterisation techniques.

I am sure that this book is an important contribution to compile, enlighten and deepen the discussions about the different microbiomes and their roles in building a more sustainable world and a healthier life. I congratulate the editorial team and all the researchers who worked conjointly to elaborate such a wide and high-quality work, which contributes greatly to the discussions in this important area of scientific knowledge.

Rafael Resende Maldonado.
Prof. Dr.Rafael Resende Maldonado
Food Department, Technical College of Campinas
University of Campinas

Preface

Over millions of years, there has been a strong association between microbial communities and hosts under diverse environmental conditions. Microbiome is the combined genetic material of the microorganisms in a particular environment. Microbiome is an integral part of both biotic environment and abiotic environment. Microbiome is a popular term that is nowadays used among the researchers in the field of biology due to its influential role, and it is sprawling its wings in different fields like environment, plant, animal, and human biology and its related subjects. Microbiome research is a rolling field of science, and hence, the US National Microbiome Initiative reported the increasing trends of microbiome science in the past decade.

This book comprises five sections: omic and computational techniques used for microbiome analysis with the narrative beneath the focus of human microbiome, animal microbiome, plant microbiome, and environmental microbiome. Moreover, this book also provides a comprehensive account on microbiome role in health and disease, gut, skin microbiota, oral and gastrointestinal microbiome in chronic inflammatory disease, reproductive tract microbiome in humans and animals, mycobiome in agroecosystem, rice root and actinorhizal root nodule microbiome, speleothems, and microbiome of marine shallow water hydrothermal vents.

This book comprises a total of 28 chapters from multiple contributors around the world, including Brazil, China, Egypt, India, Italy, Kingdom of Saudi Arabia, Portugal, and the United States. This book will be valuable to the postgraduate students, teachers, researchers, microbiologists, computational biologist, and other professionals who interested to fortify and expand their knowledge about microbiome and its role in different fields.

We are extremely thankful to all the authors who contributed chapters for their prompt and timely responses. We extend our earnest appreciation to Renu Upadhyay and Jyotsna Jangra of Taylor &Francis group and their team for their constant encouragement and help in bringing out the volume in the present form. We also express our gratitude to Prof. Rafael Resende Maldonado, Food Department, Technical College of Campinas, University of Campinas, Brazil, for his support.

We are also indebted to CRC Press, Taylor & Francis Florida, USA; the Authorities of Bharathidasan University, Tiruchirappalli, Tamil Nadu, India; UkaTarsadia University, Surat, Gujarat, India; and National Centre for Cell Science, Pune, India, for their support in the task of publishing this book.

D. Dhanasekaran
Dhiraj Paul
N. Amaresan
A. Sankaranarayanan
Yogesh S. Shouche

Editors

D. Dhanasekaran is an Associate Professor in the Department of Microbiology, School of Life Sciences, Bharathidasan University, Tiruchirappalli, India. He has experience in the fields of actinobacteriology and mycology. His current research focuses on microbiome profiling of actinorhizal root nodules, lichen, poultry gut, and cattle reproductive system. He was awarded UGC-Raman Postdoctoral Fellowship and worked in the Department of Molecular, Cellular, and Biomedical Sciences at the University of New Hampshire, Durham, USA. He has qualified the Tamil Nadu State Eligibility Test (SET) for Lectureship in Life Science. He has deposited around 106 nucleotide sequences and 7 metagenome sequences; drafted genome sequence of *Blastococcus* sp. CT_GayMR20 in GenBank and five bioactive compounds in PubChem; and published 105 research and review articles—including one paper in *Nature Group Journal Scientific Report*—and 23 book chapters. He has an h-index of 25 with total citations of 1995 as per Google Scholar. He has edited seven books on *Antimicrobial Compounds: Synthetic and Natural Compounds, Microbial Control of Vector Borne Diseases, Fermented Food Products*, CRC Press, Taylor & Francis Group, New York, *Fungicides for Plant and Animal Diseases, Actinobacteria: Basics and Biotechnological Applications, Algae-Organisms for Imminent Biotechnology, Microbial Biofilms – Importance and Applications* under in-tech open access publisher Eastern Europe. He has guided 12PhD candidates and organized several national-level symposia, conference, and workshop programs. He has received research funding from the Department of Biotechnology, University Grant Commission, Indian Council for Medical Research and International Foundation for Science, Sweden; International Society for Microbial Ecology, The Netherlands; and Tamil Nadu State Council for Science and Technology. He is a member of the American Society for Microbiology, North American Mycology Association, Mycological Society of India, National Academy of Biological Sciences, Society for Alternatives to Animal Experiments, and a member of several editorial boards in national, international journals, doctoral committee member, board of study member in microbiology, and reviewer in the scientific journals and research grants. As per the reports of *Indian Journal of Experimental Biology*, 51, 2013, Dr. Dhanasekaran is rated second among the top five institutions in the field of actinobacteria research in India.

Dhiraj Paul is a Scientist at the National Center for Microbial Resources (NCMR-NCCS), Pune. Previously, he was a scientist (DST, Govt. of India Fast-track fellowship) at the Microbial Ecology Laboratory, NCCS, Pune, India. He earned a PhD at the Environmental Microbiology Laboratory, Department of Biotechnology, IIT Kharagpur, India. He has published his findings in reputed journals such as *mSystems, AEM, Bioresource Technology, PLOS One, Geomicrobiology, Water research, Frontiers in Microbiology,* and *Microbiology*. He also edited the books and authors of a number of book chapters of CRC Press and Springer. He has expertise in microbial ecology, geomicrobiology, microbial diversity, and metagenomics. He has received an AMI Young Scientist Award, DST Young Scientist, and a number of fellowships (CSIR, GATE, DBT) in life sciences. Presently, he is working on microbial ecology of hypersaline Lonar Lake ecosystem, deep biosphere, and human microbiome using multiomics approaches.

N. Amaresan is an Assistant Professor at C.G. Bhakta Institute of Biotechnology, Uka Tarsadia University, Gujarat. He has over 15 years of experience in teaching and research, and made several original and novel discoveries, especially in various allied fields of microbiology, mainly plant–microbe interactions, bioremediation, plant pathology, and others. For his original discoveries on agriculturally important microorganisms, he has received Young Scientist Awards by Association of Microbiologists of India and National Academy of Biological Sciences. He was also awarded a visiting scientist fellowship from National Academy of India to learn advanced techniques and Early Career Research Award by Department of Science and Technology, Government of India. He has handled many funded projects sponsored by DST, DBT, GEMI, etc. He has published more than 70 research articles and books of national and international repute. He also deposited over 500 bacterial 16S rDNA and fungal ITS rDNA sequences in the GenBank

(NCBI, EMBL, and DDBJ), and also preserved over 150 microbial germplasms in various culture collection centers of India.

A. Sankaranarayanan has been associated since 2015 with C.G. Bhakta Institute of Biotechnology, Uka Tarsadia University, Surat of Gujarat State of India. He has experience in the fields of fermented food products and antimicrobial activity of herbal and nanoparticles against pathogens. His research focuses on microbes in fermented food products and removal of bacteria from food by dielectrophoresis. He has published 18 chapters in books and 50 research articles in international and national journals of repute, and he has authored 6 books which published by international publishers, guided 5 PhDs and 16 MPhil scholars, and operated five external funded projects and two institute-funded projects. From 2002 to 2015, he worked as an Assistant Professor and Head of the Department of Microbiology, K.S.R. College of Arts and Sciences, Tiruchengode, Tamil Nadu. He was awarded by the Indian Academy of Sciences (IASc), the National Academy of Sciences (NAS), and the National Academy of Sciences (TNAS)-sponsored summer research fellowship for young teachers consecutively for three years. His name is included as a mentor in DST-mentors/resource persons for summer/winter camps and other INSPIRE initiatives, Department of Science and Technology, Government of India, New Delhi. He is a grant reviewer in British Society of Antimicrobial Chemotherapy (BSAC), UK. He has involved himself in the organization of various national and international seminars and symposia. He is actively involved as an editor or an editorial board member in journals and reviewer in various international national journals, and acted as an external examiner to adjudicate PhD theses of various universities in India.

Yogesh S. Shouche is a Senior Scientist and Head of the National Centre for Microbial Resource (NCMR) at the National Centre for Cell Science (NCCS), Pune, India. He has 28 years of research experience in the field of microbial ecology, microbial molecular taxonomy, and biodiversity. He has written more than 330 publications in reputed journals of relevant area and has been the editor and reviewer of publications in journals such as *FEMS Ecology, International Journal of Systematic and Evolutionary Microbiology,* and *Microbial Ecology*. He and his team are actively working on the human microbiome, the role of gut microbial community in healthy individuals, and Indian population.

Contributors

B. Abirami
Centre for Drug Discovery and Development
Sathyabama Institute of Science and Technology
Chennai, India

Elizama Aguiar-Oliveira
Exact Science and Technology Department
State Univeristy of Santa Cruz (UESC)
Bahia, Brazil

Pavithra Amritkumar
Department of Biotechnology
Women's Christian College
Chennai, India

M. Angappan
Division of Animal Health
ICAR Research Complex for NEH Region
Meghalaya, India
and
ICAR – National Research Centre
Medziphema, India

G. Archunan
Department of Animal Science
Bharathidasan University
Tiruchirappalli, India

Sunil Banskar
University of Arizona
Tucson, Arizona

Kiron Bhakat
Department of Microbiology
University of Kalyani
Kalyani, India

P.V. Bhaskar
National Centre for Polar and Ocean Research
Ministry of Earth Sciences
Vasco da Gama, India

Shrikant Bhute
University of Nevada
Las Vegas, Nevada

K. P. Biam
ICAR – National Research Centre
Medziphema, India

Federico Bussolino
Department of Oncology
Candiolo Cancer Institute – IRCCS
Torino, Italy

Ana Lúcia Alves Caram
Nutrition Course
UNIMOGI
São Paulo, Brazil

J. Helan Chandra
Department of Biodivision
Ampersand Academy
Chennai, India

Nikeeta S. Chavan
National Centre for Microbial Resource
National Centre for Cell Science
Pune, India

Ana Colaço
Instituto do MarandInstituto de Investigação em Ciências do Mar
Okeanos da Universidade dos Açores
Horta, Portugal

Diana R. Cundell
Department of Biology
Jefferson University
Philadelphia, Pennsylvania

Somarani Dash
Department of Biotechnology
Gauhati University
Jalukbari, India

Syed G. Dastager
National Collection of Industrial Microorganisms
CSIR – National Chemical Laboratory
Pune, India

C. Anchana Devi
Department of Biotechnology
Women's Christian College
Chennai, India

T. Ramani Devi
Ramakrishna Medical Center
Tiruchirappalli, India

Sahabram Dewala
National Centre for Cell Science
Pune, India

D. Dhanasekaran
Department of Microbiology, School of Life Sciences
Bharathidasan University
Tiruchirappalli, India

Mitesh Dwivedi
Faculty of Science
C. G. Bhakta Institute of Biotechnology
Uka Tarsadia University
Tarsadi, India

S. Ghatak
Division of Animal Health
ICAR Research Complex for NEH Region
Meghalaya, India
and
ICAR – National Research Centre
Medziphema, India

Prashant S. Giri
Faculty of Science
C. G. Bhakta Institute of Biotechnology
Uka Tarsadia University
Tarsadi, India

V. Gopikrishnan
Centre for Drug Discovery and Development
Sathyabama Institute of Science and Technology
Chennai, India

Shanmugaraj Gowrishankar
Department of Biotechnology
Alagappa University
Karaikudi, India

Mohamed Yousuff Mohamed Imran
Department of Biotechnology and Microbiology
National College (Autonomous)
Tiruchirappalli, India

Ekramul Islam
Department of Microbiology
University of Kalyani
Kalyani, India

Rathinam Arthur James
Department of Marine Science
Bharathidasan University
Trichy, India

S. Kalaiselvi
Department of Microbiology
Government Arts and Science College (W)
Orathanadu, India

Arumugam Kamaladevi
Department of Animal Science
Bharathidasan University
Tiruchirappalli, India

Eliana Setsuko Kamimura
Food Engineering Department
Faculty of Animal Science and Food Engineering
University of São Paulo (USP)
São Paulo, Brazil

Haresh Kumar Keharia
Department of Biosciences
UGC – Centre of Advanced Study
Sardar Patel University
Anand, India

Ashraf Y. Z. Khalifa
Biological Sciences Department
King Faisal University
Hofuf, Kingdom of Saudi Arabia
and
Botany and Microbiology Department
University of Beni-Suef
Beni-Suef, Egypt

Ahmed Abdul Haleem Khan
Department of Botany
Telangana University
Nizamabad, India

M. H. Khan
ICAR – National Research Centre
Medziphema, India

Mojibur R. Khan
Institute of Advanced Study in Science and Technology
Vigyan Path
Guwahati, India

Snehal O. Kulkarni
National Centre for Cell Science
Savitribai Phule Pune University Campus
Pune, India

Arun Kumar
Institute of Advanced Study in Science and Technology
Paschim Boragaon
Guwahati, India

Selvanathan Latha
Department of Microbiology, School of Life Sciences
Bharathidasan University
Tiruchirappalli, India

Wen Jun-Li
State Key Laboratory of Desert and Oasis Ecology
Xinjiang Institute of Ecology and Geography
Chinese Academy of Sciences
Urumqi, China

Rafael Resende Maldonado
Food Department
Technical College of Campinas (COTUCA)
University of Campinas
R. Jorge Figueiredo Correa
São Paulo, Brazil

K. Manigundan
Centre for Drug Discovery and Development
Sathyabama Institute of Science and Technology
Chennai, India

C. Manikkaraja
Department of Animal Science
Bharathidasan University
Tiruchirappalli, India

Natesan Manoharan
Department of Marine Science
Bharathidasan University
Tiruchirappalli, India

Narayanasamy Marappa
Department of Microbiology
School of Life Sciences
Bharathidasan University
Tiruchirappalli, India

Mônica Roberta Mazalli
Food Engineering Department
Faculty of Animal Science and Food Engineering
University of São Paulo (USP)
São Paulo, Brazil

A. A. P. Milton
Division of Animal Health
ICAR Research Complex for NEH Region
Meghalaya, India
and
ICAR – National Research Centre
Medziphema, India

Chellandi Mohandass
CSIR – National Institute of Oceanography
Regional Centre
Mumbai, India

Dattatray S. Mongad
National Centre for Microbial Resource
National Centre for Cell Science
Pune, India

D. Mudgil
Department of Environmental Science and Engineering
Guru Jambheshwar University of Science and Technology
Hisar, India
and
Zakir Husain Delhi College
University of Delhi
New Delhi, India

M.S. Murugan
Veterinary University Training and Research Centre
Tamil Nadu Veterinary and Animal Sciences University (TANUVAS)
Rajapalayam, India

Sandhya Nanjani
Department of Biosciences
UGC Centre of Advanced Study
Sardar Patel University
Gujarat, India

Daniela Soares de Oliveira
Nutrition Course
Municipal College Professor Franco Montoro (FMPFM)
São Paulo, Brazil

Shunmugiah Karutha Pandian
Department of Biotechnology
Alagappa University
Karaikudi, India

A. Panneerselvam
PG & Research Department of Botany and Microbiology
AVVM Sri Pushpam College
Poondi, India

Sushant Parab
Universita degli Studi di Torino
Turin, Italy

Murali Sankar Perumal
Department of Plant Molecular Biology and Bioinformatics
CPMB&B
Tamil Nadu Agricultural University
Coimbatore, India

Vivek Joshi
Division of Animal Health
ICAR Research Complex for NEH Region
Meghalaya, India
and
ICAR – National Research Centre
Medziphema, India

G. Bhuvana Priya
Division of Animal Health
ICAR Research Complex for NEH Region
Meghalaya, India
and
ICAR – National Research Centre
Medziphema, India

M. Radhakrishnan
Centre for Drug Discovery and Development
Sathyabama Institute of Science and Technology
Chennai, India

Raju Rajasabapathy
Department of Marine Science
Bharathidasan University
Trichy, India

Govindan Nadar Rajivgandhi
State Key Laboratory of Biocontrol
Guangdong Provincial Key Laboratory of Plant Resources and Southern Marine Science and Engineering
Guangdong Laboratory (Zhuhai)
School of Life Sciences

Sun Yat-Sen University
Guangzhou, China

Govindan Ramachandran
Department of Marine Science
Bharathidasan University
Tiruchirappalli, India

Nazar Reehana
Department of Microbiology
Jamal Mohamed College (Autonomous)
Tiruchirappalli, India

Shreedevasena Sakthibalan
Department of Plant Pathology
CPGSAS
Umiam, India

A. Sankaranarayanan
Faculty of Science
C. G. Bhakta Institute of Biotechnology
Uka Tarsadia University
Tarsadi, India

Priyanka Sarkar
Wellcome-DBT (Indian Alliance) Lab, Asian Healthcare Foundation
Asian Institute of Gastroenterology
Hyderabad, India

T. Savitha
Department of Microbiology
Tiruppur Kumaran College for Women
Tiruppur, India

Firdosh Shah
Faculty of Science
C. G. Bhakta Institute of Biotechnology
Uka Tarsadia University
Tarsadi, India

T. Shanmugasundaram
DRDO – Centre for Life Science
Bharathiar University Campus
Coimbatore, India

Yogesh S. Shouche
National Centre for Microbial Resource
National Centre for Cell Science
Pune, India

M. Srinivasan
Department of Animal Science
Bharathidasan University
Tiruchirappalli, India

Nooruddin Thajuddin
Department of Microbiology, School of Life Sciences
Bharathidasan University
Tiruchirappalli, India

Manuela Tripepi
Department of Biology
Jefferson University
Philadelphia, Pennsylvania

R. Vikram
ICAR – National Research Centre
Medziphema, India

R.T.V. Vimala
Department of Biotechnology
Bharathidasan University
Tiruchirappalli, India

1

An Insight of Microbiome Science

T. Savitha
Tiruppur Kumaran College for Women

A. Sankaranarayanan
Uka Tarsadia University

Ashraf Y. Z. Khalifa
King Faisal University
University of Beni-Suef

CONTENTS

1.1 Introduction

Microbiome is a popular and repeatedly used term due to its widespread activities in various scientific platforms and due to its influential roles. It is sprawling its wings in the diversified fields of biology, and the microbes have a symbiotic and pathogenic association with various biotic organisms, including all vertebrates and plants. It is a great promising tool and that's the reason in the past decade more than 1.7 billion US $ spent toward human microbiome research (Proctor, 2019). The microbiome science nowadays attracts more researchers due to its importance in diversified fields. Further, the new discoveries and concepts have been emerging from this field changing the perception of the importance of this field among the researchers (Walter and Ley, 2011; O'Callaghan et al., 2016).

Microbiome is defined as a "collective genes and genomes" of all the microorganisms dwelling in a particular environment (Marchesi and Ravel, 2015). Another short definition is, "Microbiome means genes associated with the microbiota" (Amato, et al., 2016). Diversified microbiomes are present in the different parts of body, environment, and other habitats (Gilbert and Stephens, 2018). In addition, their presence depends on the surrounding conditions (Derrien et al., 2019; de Steenhuijsen Piters et al., 2015). There are many factors that are involved in cohabitation of microbiome in biota.

Human microbiome was considered as one among the organs of human body (Baquero and Nombela, 2012), merely a decade ago. The relationship between microbiome and humans starts within 20 minutes after the birth of the infant. The microbiome of the infant resembles the microbiome of mother's vagina (i.e., infants born through vagina) or that of skin (i.e., infants born through cesarean) (Dominguez-Bello et al., 2010; Rasmussen et al., 2020). Microorganisms are dwelling on the surface and in the tissues of small-sized marine biotic organisms as well as in larger-sized animals. There are possible chances of incorporating microorganisms as a sentinel in the host due to the rapid change in the marine environment by pollution (Ainsworth and Gates, 2016). It is not a surprise to state that around 40 trillion microbial cells reside in, and on the surface of, the human body—especially, more microbial flora are present in the gut (Nagasaka et al., 2020). Microbiomes sprawl their wings and are not restricted only to the gut, skin (Bangert et al., 2011), lower respiratory tract (Charlson et al., 2011), nasal cavity, and oropharynx (Beck et al., 2012; Dickson et al., 2013). As per the report, the highest microbial diversity and density were reported in the gastrointestinal tract followed by the skin.

More than a decade ago, NIH (National Institute of Health) has initiated the human microbiome project and reported the presence of microbiota in different parts of the body. Further, the genome sequence technologies and metagenomic analysis have enhanced the interest of scientists to know the function and various research aspects of microbiome (Amon and Sanderson, 2017). In addition, sequencing techniques, genetic fingerprinting techniques, quantitative polymerase chain reaction (PCR)approach, multilocus gene typing, fluorescent-tagged oligonucleotide probes, and metagenomic gene sequencing approaches are additional techniques to concentrate more on the microbiome-based research studies (O'Callaghan et al., 2016).

Microbiota present in different parts of the plants is called as "plant holobiont/plant microbiome" (Hassani et al., 2018). It plays a pivotal role in supporting the growth and maintenance of plant health (Lindow and Brandl, 2003; Buee et al., 2009; Berendsen et al., 2012; Vorholt, 2012; Hassani et al., 2018) due to its presence in phyllosphere, rhizosphere, endosphere, and episphere (outside) (Dastogeer et al., 2020). The association between plants and microbiome was developed 700 million years ago (Heckman et al., 2001). The various functions rendered by plant microbiome are presented in Figure 1.1. In this juncture, it is a need to mention that the metabolites secreted by bacterial endophytes present in plants showed more efficiency in the growth of the plant than the metabolites secreted by the plants in association with the microbes in open environments (like rhizosphere microbes). The reason is that the metabolites are secreted in open environment, and there are more chances of becoming affected by various extraneous factors such as biotic and abiotic (Santoyo et al., 2017).

For the maintenance of animal health, including humans, the resident microbial communities exert a prominent role. These communities alleviate various chronic diseases, and metabolic and neurobiological diseases; thus, the host microbial traits play a predominant role (Douglas, 2019). A new concept has recently emerged and named as "one health concept," which insists the relationship between humans and animal environment, and it envisages the shifting of pathogens from animals and environment to human populations (Trinh et al., 2018). Like humans, animal microbiomes influence the health of livestock, disease vectors, and pets. In addition, pet animals share their microbiome diversity with pet owners' skin (Ross et al., 2017; Song et al., 2013). There exists a strong association between host—especially animals—and microbes; hence, the bonding may be known as "biocenosis" or "living community" (Bosch and Miller, 2016).

The resident microbial communities influence the health and fitness of animals and humans by their presence (Douglas, 2019). In animals, the various functions of gut microbiome were listed elsewhere (O'Callaghan et al., 2016). In animals,

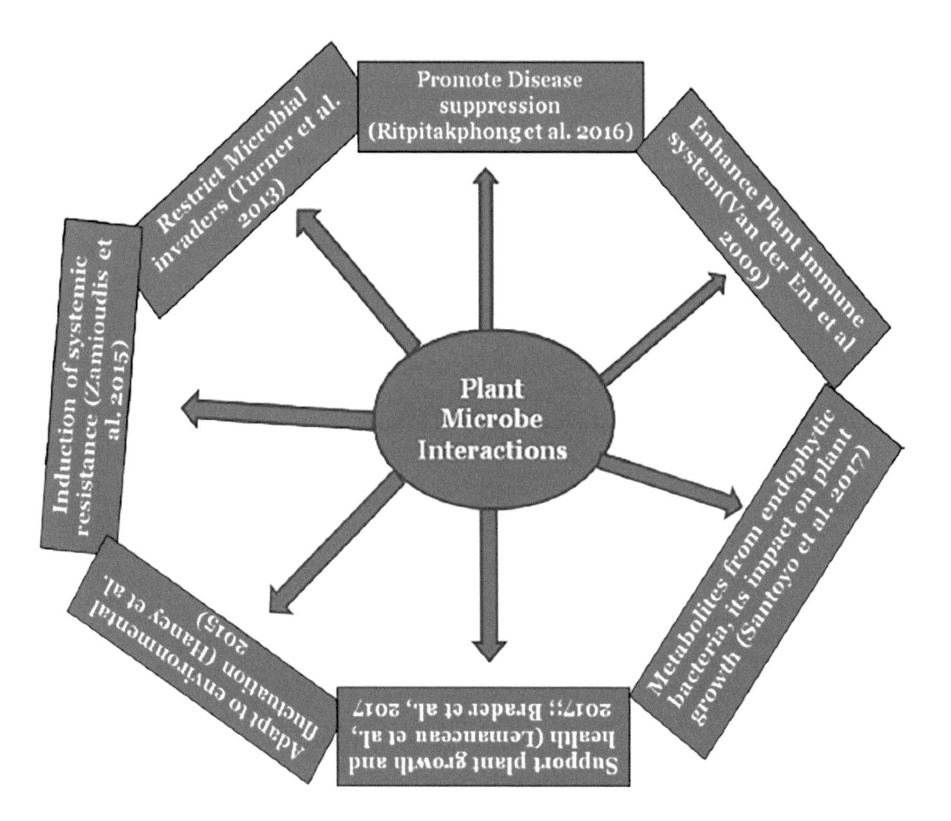

FIGURE 1.1 Various functions exerted by microbiome in plants.

the predominant microbial group is bacteria—especially strict anaerobes—followed by also protozoa, viruses, and fungi. The microbiota varies in animals based on the food habit—especially, the rumen of cow and sheep is dominated by fiber-degrading bacteria (Simpson et al., 2002) followed by the methanogenic bacteria (Boomker and Cronje, 2000). In another interesting observation, the specific members of microbiome determined the milk-yielding capacity of animals (Jewell et al., 2015), especially cows.

With the increasing trends of industrialization and urbanization, environmental microbiome plays an important role in the Universe. Due to increasing trends of pollutants, especially air pollutants (Dujardin et al., 2020; Rajagopalan et al., 2018; Huang et al., 2018), fine (Zhang et al., 2017) and particulate matter (Belis et al., 2013), and industrial emissions (Taiwo et al., 2014), the microbiome plays a major role in preventing the deterioration of the health of the human being. In light of the above, this chapter concentrates on the ecological attributes in understanding of the microbiome; classification of microbiomes; various microbiomes in gut, genital tract, skin, oral cavity, and respiratory organs; and future perspectives of microbiome.

1.2 Ecological Theory in Understanding of the Microbiome

Ecological understanding plays a vital role in the microbial assemblages of microbiome. The microbial assemblages formed as community in any biotic/abiotic sources based on dispersal, local diversification, environmental selection, and ecological drift (Costello et al., 2012). For example, the association between the host and gut microbiota depends on host genome, nutrition, and lifestyle of the particular host (Nicholson et al., 2012). The host and the conditions determine what type of microbiota exists in the particular environment. In case of vaginal microbiome, the microbiome compositional changes related to pregnancy and menstruation can lead to diseases like bacterial vaginosis (Greenbaum et al., 2018). In another study about the gut microbiota, the stability of the microbiome members depends on immunosuppression, spatial structuring, and other allied factors related to them (Coyte et al., 2015). The diet and body size affect the microbiome diversity in animals; especially, the physiology of the gut also plays a vital role in microbiome diversity (Reese and Dunn, 2018). Various soil abiotic parameters and competition among microbes play a role in the diversity of microbiome in soil and soil-associated plant rhizospheric environment—especially physical interaction, secretion, and plant immunity (Tkacz et al., 2020).

1.3 Different Types of Microbiome

1.3.1 Microbiome

The first metaphors of human-associated microbiota arose back from the 1670s to the 1680s, when Antonie van Leeuwenhoek started with his newly developed hand-crafted microscopes. In a letter written to the Royal Society of London in 1683, he described and illustrated five different kinds of bacteria (called by him as "animalcules") present in his own mouth and that of others, and afterward, he also compared his own oral and fecal microbiota, determining that there are differences between body sites as well as between health and disease. Some of the first direct observations of bacteria were of human-associated microbiota. In 1853, Joseph Leidy published a book entitled *A flora and fauna within living animals*, which was considered to be the origin of microbiota research. Then, the works of Pasteur, Metchnikoff, Koch, Escherich, Kendall, and few others pave a path for the study of host–microbe interactions. The above legends formed a platform for the human–microbiome research.

Scientists defined the word "micro" to indicate an ecosystem which is invisible to the human eyes, made up of mostly bacteria, viruses, archaea, and also fungi, and exerts an important role in the safeguarding the stability to the specific environment. The phrase "biome" denotes an ecosystem made up of flora and fauna. Microbes are predominantly ubiquitous in nature in all forms of life such as human body, earth's soils and sediments, oceans, and freshwater systems; in atmosphere; and also in extreme environments like hydrothermal vents and subglacial lakes. Hence, scientists coined the term "microbiome" to refer all these genes associated with those life forms. Microbiomes possess many universal aspects to all categories. Their population are plenteous and dissimilar, differing from place to place and even individuals. They also vibrantly modified their response to factors such as diet or climate. Finally, they are interrelated with their host and engaged in mutual relationship, which is both essential and beneficial in terms of the host and their resident microorganisms.

1.3.2 Types of Microbiome

The present scenario of research studies focuses on understanding of the healthy microbiome establishment and in turn balanced communities in various ecosystems along with their impact on human and environmental health. In addition, microbiome research is mandatory for space exploration and planetary science. Exhaustive knowledge about the microbes on earth brings out better understanding toward alternate forms of life and also aids for the searching of life forms in other planets, which helps to establish earth-like, human-friendly environments faraway also. Knowledge on the role of microbiota, including their dynamic interactions with their hosts and other microbes, can enable the engineering of new diagnostic techniques and interventional strategies that can be used in a diverse spectrum of fields (Figure 1.2), spanning from ecology and agriculture to medicine and from forensics to exobiology (Cullen et al., 2020). Microbiota plays a deep-seated role in the large expansion and defense mechanism of the human body. The majority of indigenous microbiota exist in a mutually beneficial relationship with their hosts, while few of these are opportunistic pathogens that can lead to life-threatening diseases and chronic infections. These microbial communities act as the primary defense against infections induced by nonindigenous invasive organisms.

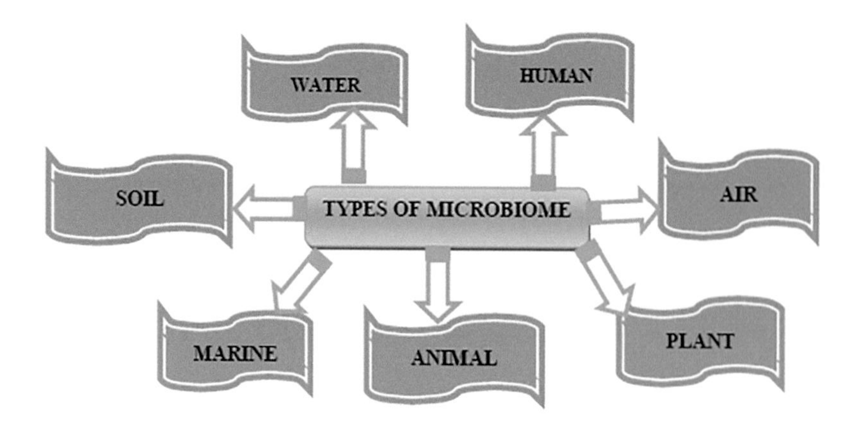

FIGURE 1.2 Types of microbiome existed in the environment.

Microbes are highly abundant and diverse, and play an important role in the ecological system. For example, the ocean contains the estimated 1.3×1,028 archaea cells, 3.1×1,028 bacterial cells, and 1×1,030 viral particles. The bacterial diversity is estimated to be about 160 for a ml of ocean water, 6,400–38,000 for a gram of soil, and 70 for a ml of sewage water (Suttle, 2007; Curtis et al., 2002).

1.3.2.1 Soil Microbiome

The soil microbiome governs biogeochemical cycling of macronutrients, micronutrients, and other elements that are vital for the growth of plants and animal life. Understanding and predicting the impact of climate change on soil microbiomes and the ecosystem services they afford is a grand challenge and major opportunity towards emerging problem facing our planet. Jansson and Hofmocke (2020) reported the knowledge about the impacts of climate change on soil microbes in different climate-sensitive soil ecosystems, as well as the potential ways that soil microbes can be harnessed to help mitigate the negative consequences of climate change.

1.3.2.2 Marine Microbiome

The marine microbiome is one of the largest microbiomes on the planet. Microbes (<3μm) dominate the marine environment with 10^4–10^6 cells in each millimeter of sea water (Sunagawa et al., 2015). Microbial communities are the dominant drivers of biogeochemical processes and form the basis of a complex food web which is fueled by the fixation of carbon dioxide by oxygenic photosynthesis. The microbes play a central role in the energy generation of the earth—through photosynthesis at the surface and chemosynthesis in darker ocean depths. The ocean harbors hugely diverse, but largely unknown microbes. Unlocking the genetic code of marine microbes can have an enormous functional potential for human, animal, and plant disease treatment and the understanding of climate change. Even though the tremendous explosion of technology arises, a complete global picture of the distribution, diversity, function, and understanding of the ecological determinants of their composition and their response to environmental change remains a challengeable area of research.

1.3.2.3 Animal Microbiome

Emerging area of thrust is to explore about the factors that could affect the microbiome of animals about their sustainability in the ecosystem, species, and/or populations. The compositions of the bacterial communities of animals, including invertebrates and vertebrates, are shaped by multiple factors such as host genotype (Wong et al., 2013), diet (Staubach et al., 2013), life stage (Wang et al., 2011), laboratory rearing (Morrow et al., 2015), and their ecological and physiological conditions. In addition, these microbiomes are proposed to exhibit the impacts on the nutritional supplementation, tolerance to environmental perturbations, and maintenance and/or development of the immune system.

1.3.2.4 Plant Microbiome

Microbes that are having an intact association with certain plant species or genotype, independent of soil and other environmental conditions, are collectively termed as "core plant microbiome" (Toju et al., 2018). Plants are naturally enriched with their surroundings with numerous microbes in their specific organs. The microbial communities of the plant are termed as "plant microbiota" (all microbes) or the "plant microbiome" (all microbial genomes) in the rhizosphere, phyllosphere, and endosphere, which supports for the plant sustainability and growth in an efficient way (Brader et al., 2017; Lemanceau et al., 2017).

1.3.2.5 Human Microbiome

The **human microbiome** encompasses bacteria, archaea, viruses, and eukaryotes, which inhabit interiorly and exteriorly our bodies. These microbes impact human physiology, both in health and in disease, contributing to the enhancement or impairment of metabolic and immune functions. Microbes colonize various sites, where they may adapt to specific features of each niche. Facultative anaerobes are more dominant in the gastrointestinal tract, whereas strict aerobes inhabit the respiratory tract, nasal cavity, and skin surface. The indigenous organisms in the human body are well adapted to the immune system due to the biological interaction of the organisms with the immune system. An alteration in the intestinal

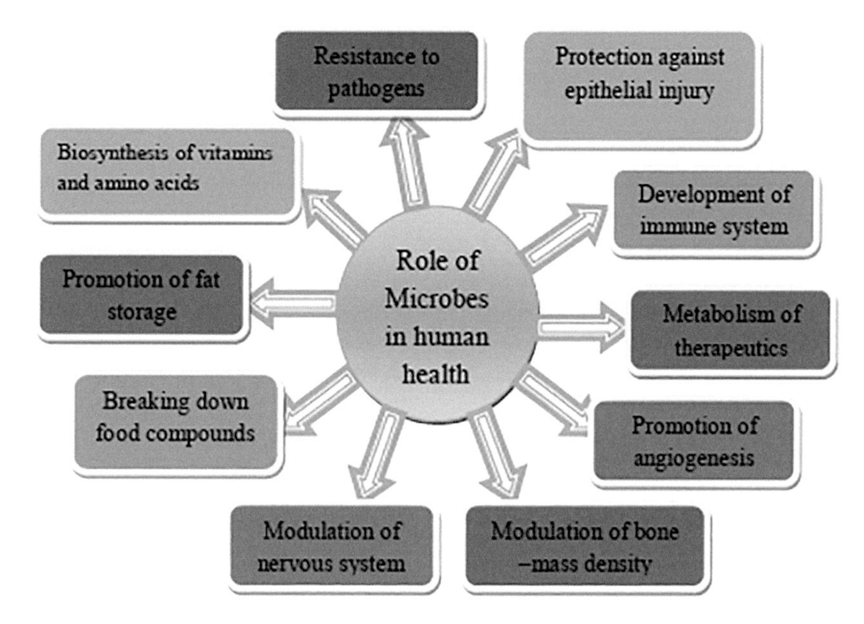

FIGURE 1.3 Role of microbiomes in human healthcare.

microbial community plays a major role in human health and establishment of pathogenesis. The unique diversity of the human microbiota accounts for the specific metabolic activities and functions of these microbes within each body site. Hence, it is more important to understand the microbial composition and activities of the human microbiome as they are involved in maintaining health (Figure 1.3) and also in establishing disease states (Ogunrinola et al., 2020).

1.4 Classification of Microbiomes in Human Health

Different sites of the human body can be seen as unique biomes, with drastically different environments and nutrient availability, which in turn can promote the existence of different communities, again which reflects the health condition, lifestyle, and also other factors (Integrative HMP Research Network Consortium, 2014). The human microbiome is an aggregate of all microbiota (Figure 1.4) that reside on or within human tissues and biofluids along with the corresponding anatomical sites in which they reside, including the skin, mammary glands, placenta, seminal fluids, uterus, ovarian follicles, lungs, saliva, oral mucosa, conjunctiva, biliary tract, and gastrointestinal tract (Marchesi and Ravel, 2015). These microbiomes exhibit their variety of roles not only in the improvement of health status but also in the establishment of disease conditions through different mechanisms (Figure 1.5).

1.4.1 The Skin Microbiome

The skin is colonized by numerous, diverse microbial phyla that function to protect the host in healthy individuals. Namely, *Firmicutes*, *Actinobacteria*, *Bacteroides*, and Proteobacteria comprise the largest bacterial populations on healthy human skin (Thio, 2018). The skin microbiome confers numerous benefits to the host, including resisting pathogen colonization, maintaining the skin barrier, and modulating the inflammatory response (Sanford and Gallo, 2013). The species *Staphylococcus hominis* frequently produces antimicrobial substance, micrococcin P (MP1), which shows (Liu et al., 2020) the strongest and broadest antimicrobial activity in probiotic types of topical application to reduce infection and accelerate healing of infected wounds.

1.4.2 Oral Microbiome

The mouth is an open system. Microbes are inhaled with every breath; ingested with every meal or drink; and introduced by close contact with other humans, animals, or our physical surroundings. As a warm, moist, and nutrient-rich environment, the mouth presents a good shelter for microbes (Welch et al., 2019). The mouth represents a multiplicity of local environments in communication with each other via saliva. The spatial organization of microbes within the mouth is shaped by opposing forces in dynamic equilibrium in flow of salivary and adhesion, shedding, and colonization, and also interactions among and between microbes and the hosts (Welch et al., 2020). An oral cavity contains one of the most diverse and unique communities of microbes in the human body.

1.4.3 The Respiratory Microbiome

The lower respiratory tract is sterile in healthy humans. Hilty et al. (2010) showed that the lower respiratory tract in health adults consists of mainly anaerobes such as genus *Prevotella* using NGS (next-generation sequencing). Not only *Bacteroides* phylum—including *Prevotella, Firmicutes*, Proteobacteria genera, and *Veillonella, Fusobacterium, Streptococcus*, *Pseudomonas* genera—are the main bacteria, but also *Neisseria* and *Haemophilus* genera are less common in the lower respiratory tract in healthy subjects (Charlson et al., 2011). The composition of the respiratory microbiome in the first few months of life is likely influenced by external

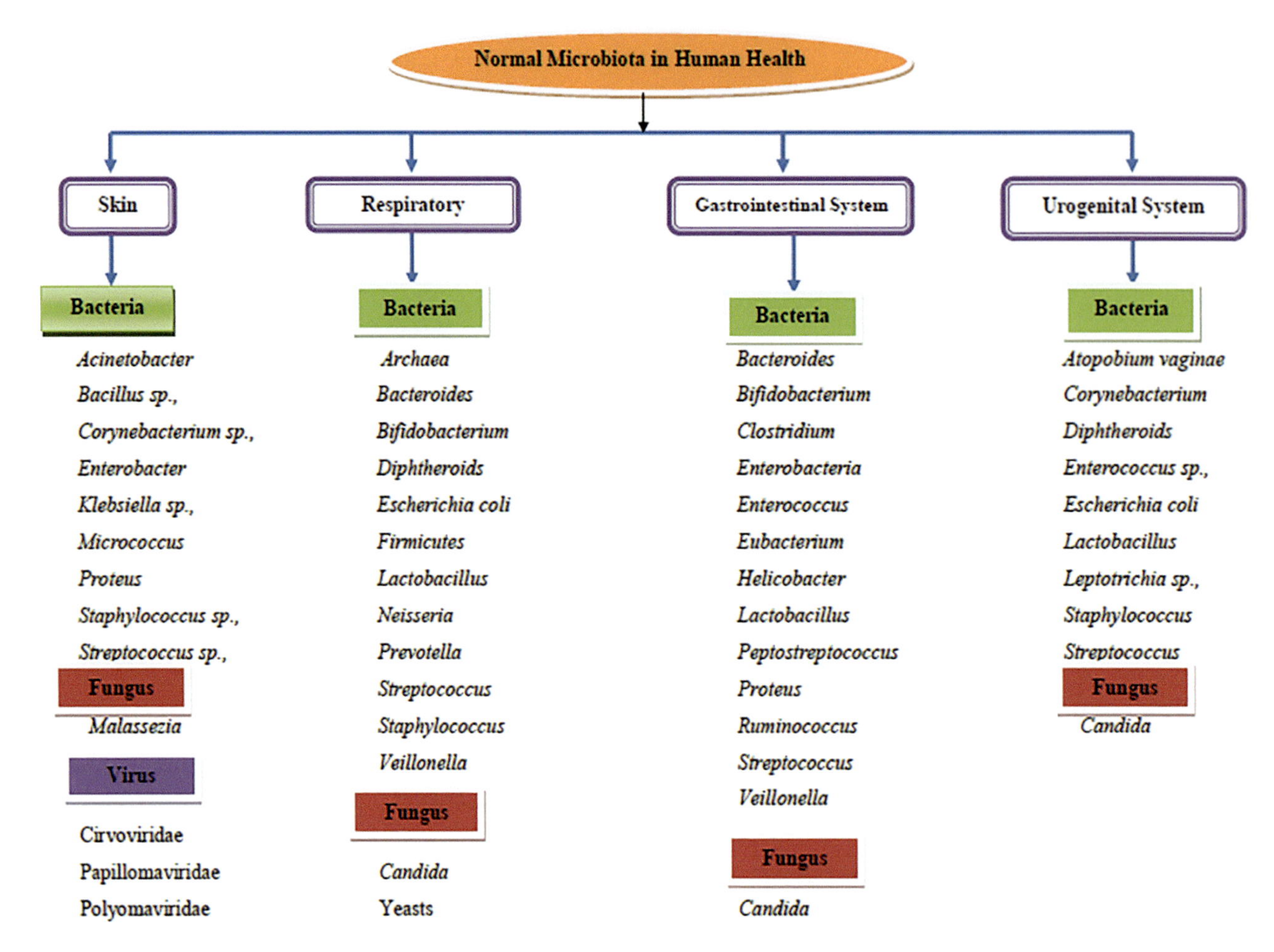

FIGURE 1.4 Normal microbiota of human health.

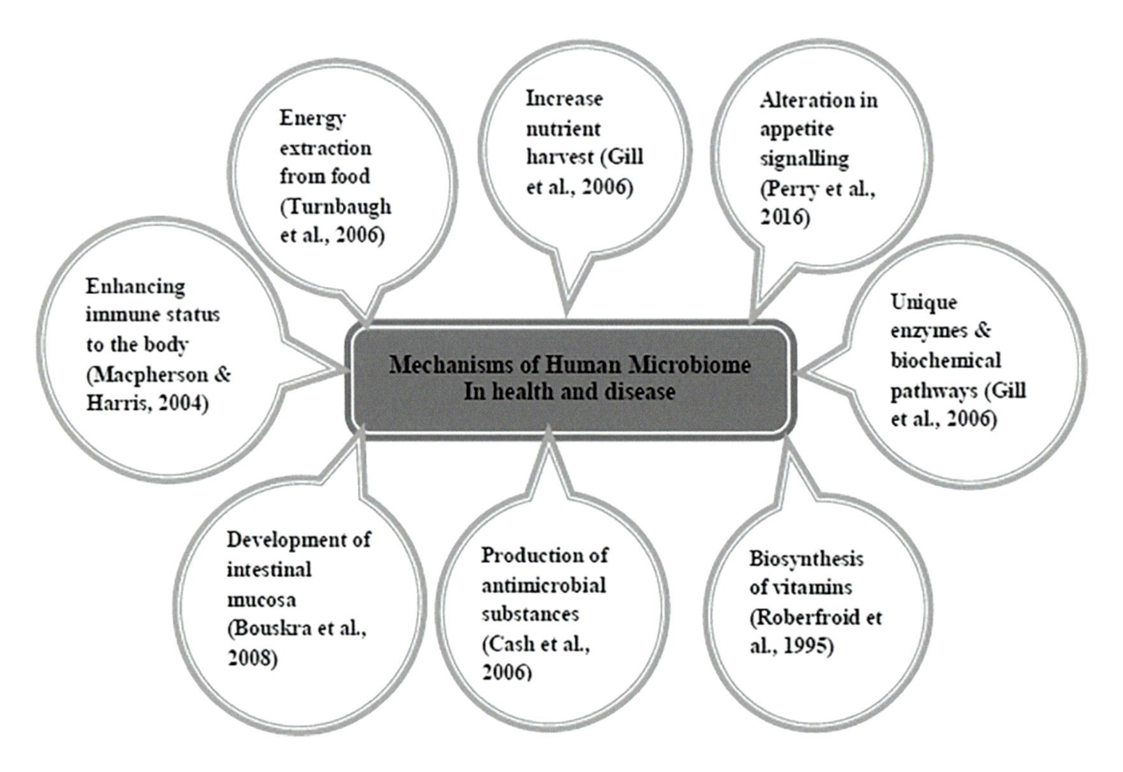

FIGURE 1.5 Mechanisms of human microbiome in health and disease.

factors such as environment, mode of delivery, and infant feeding practices, which are also associated with susceptibility to respiratory microbiota profiles early in life and are associated with an increased risk and frequency of subsequent respiratory tract infections, disease severity, and occurrence of wheeze in later childhood. Early interactions between infectious agents such as viruses and the respiratory microbiome have shown to modulate host immune responses potentially affecting the course of the disease and future respiratory health (Unger and Bogaert, 2017).

1.4.4 Gut Microbiome

The **gut microbiome** is an assortment of dissimilar bacteria that usually reside within the gastrointestinal tract. In the present arena, the relationship between the gut microbiome and the health status is emerging area of interest in medical research. It not only imparts barrier role in the gastrointestinal tract but also contributes to active immune function of gut microbiota with the relationship between dysbiosis of gut microbiota and certain inflammatory and malignant disease states of the gastrointestinal system (Nagasaka et al., 2020). The gastrointestinal tract contains several microbiomes, with the largest diversity and abundance of microbes residing in the colon, hereafter referred to as the "gut microbiome." Development of the human gut microbiome starts immediately afterbirth. After profound changes during the first few years of infancy, the mature gut microbiome stays relatively stable within a certain range but continues to vary as it is prone to changes made by different external environments (Derrien et al., 2019). The composition of the respiratory microbiome in the first few months of life is likely influenced by external factors such as environment, mode of delivery, and infant feeding practices, which are associated with susceptibility to respiratory tract infections and wheezing illness/asthma.

Many research findings evidenced that respiratory microbiota profiles early in life are associated with an increased risk and frequency of subsequent respiratory tract infections, disease severity, and occurrence of wheeze in later childhood. Early interactions between infectious agents such as viruses and the respiratory microbiome have shown to modulate host immune responses potentially affecting the course of the disease and future respiratory health. In-depth understanding of these interactions will help in the development of new therapeutic agents or preventive measures that may modify respiratory health outcomes and help us to stratify at-risk populations to better target our current interventional approaches (Unger and Bogaert, 2017).

1.4.5 Genital Microbiome

Female vaginal ecosystem is thought to have been shaped over the years by coevolutionary processes occurring between the particular microbial partners and the human host. Vaginal secretions contain numerous microbes, and the host provides them nutrients for their growth and development. Disruptions in vaginal association with the microbiome lead to the change in the vaginal environment, which enhanced the risk of acquiring diseases, including sexually transmitted infections, bacterial vaginosis, fungal infections, and preterm birth (Gupta et al., 2019). Characterization of the male genital tract microbiota has always holdup behind investigations in other body sites while comparing the female genital system. Mandar (2013) reviewed genome search of male genital system; only limited publications are emerged: penis microbiome—five publications; urethra microbiome—three publications; and male genital tract microbiome—one publication, whereas there are no publications in semen microbiome and coronal sulcus microbiome. This is quite interesting area of research because commensal bacteria may mediate the male reproductive tract homeostasis.

1.5 Future Perspectives

The microbiome-based research is sprawling its wings in the diversified fields, especially using more modern molecular gadgets, gene technology, and omics technology. More methodological advances have already propelled and revealed new interesting findings. Still more methodological advances have yet to develop, which will generate new ideas and new hypothesis in the microbiome domains. The bidirectional approach between host and microbiome understanding and integration of ecological and host approaches may bring more novel findings in the field of microbiome domain in the near future.

REFERENCES

Ainsworth, T.D. and Gates, R.D. 2016. Corals' microbial sentinels. *Science* 352; 1518–1519. doi: 10.1126/science.aad9957.

Amato, K.R., Metcalf J.L., Song, S.J., Hale, V.L., Clayton, J., Ackermann, G., Humphrey, G., Niu, K., Cui, D., Zhao, H., Schrenzel, M.D., Tan, C.L., Knight, R. and Braun J. 2016. Using the gut microbiota as a novel tool for examining colobine primate GI health. *Glob Ecol Conserv* 7; 225–237.

Amon, P. and Sanderson, I. 2017. What is the microbiome? *Arch Dis Child Educ Pract Ed* 102; 258–261.

Bangert, C., Brunner, P.M. and Stingl, G. 2011. Immune functions of the skin. *Clin Dermatol* 29(4); 360–376.

Baquero, F. and Nombela, C. 2012. The microbiome as a human organ. *Clin Microbiol Infect* 18; 2–4.

Beck, J.M., Young, V.B. and Huffnagle, G.B. 2012. The microbiome of the lung. *Transl Res* 160; 258–66.

Belis, C.A., Karagulian, F., Larsen, B.R. and Hopke, B.K. 2013. Critical review and meta-analysis of ambient particulate matter source apportionment using receptor models in Europe. *Atmos Environ* 69; 94–108.

Berendsen, R.L., Pieterse, C.M.J. and Bakker, P.A.H.M. 2012. The rhizosphere microbiome and plant health. *Trends Plant Sci* 17; 478–86.

Boomker, E. and Cronje, P. 2000. *Ruminant Physiology: Digestion, Metabolism, Growth, and Reproduction.* Wallingford, Oxon, UK: CABI.

Bosch, T.C. and Miller, D.J. 2016. *The Holobiont Imperative: Perspectives from Early Emerging Animals.* Wien: Springer.

Brader, G., Compant, S., Vescio, K., Mitter, B., Trognitz, F., Ma, L.J. and Sessitsch, A. 2017. Ecology and genomic insights into plant-pathogenic and plant – non pathogenic endophytes. *Ann Rev Phytopathol* 55; 61–83.

Buee, M., De Boer, W., Martin, F., van Overbeek, L. and Jurkevitch, E. 2009. The rhizosphere zoo: an overview of plant-associated communities of microorganisms, including phages, bacteria, archaea, and fungi, and of some of their structuring factors. *Plant Soil* 321; 189–212.

Charlson, E.S., Bittinger, K., Haas, A.R., et al. 2011. Topographical continuity of bacterial populations in the healthy human respiratory tract. *Am J Respir Crit Care Med* 184; 957–963.

Costello, E.K., Stagman, K., Dethlefsen, L., Bohannan, B.J.M. and Relman, D.A. 2012. The application of ecological theory toward an understanding of the Human Microbiome. *Science* 336; 1255–1262.

Coyte, K.Z., Schluter, J. and Foster, K.R. 2015. The ecology of the microbiome: networks, competition and Stability. *Science* 350(6261); 663–666. doi: 10.1126/science.aad2602.

Cullen, C.M., Aneja, K.K. and Beyhan, S., et al. 2020. Emerging priorities for microbiome research. *Front Microbiol* 11. Article 136. doi: 10.3389/fmicb.2020.00136.

Curtis, T.P., Sloan, W.T. and Scannell, J.W. 2002. Estimating prokaryotic diversity and its limits. *Proc Natl Acad Sci* 99(16); 10494–10499. doi: 10.1073/pnas.142680199.

Dastogeer, K.M.G., Tumpa, F.H., Sultana, A., Akter, M.A. and Chakraborty, A. 2020. Plant microbiome–an account of the factors that shape community composition and diversity. *Current Plant Biology*. doi: 10.1016/j.cpb.2020. 100161.

de Steenhuijsen Piters, W.A., Sanders, E.A. and Bogaert, D. 2015. The role of the local microbial ecosystem in respiratory health and disease. *Philos Trans R Soc Lond B Biol Sci* 370; 1675.

Derrien, M., Alvarez, A.S. and de Vos, W.M. 2019. The gut microbiota in the first decade of life. *Trends Microbiol* 27(12); 997–1010.

Dickson, R.P., Erb-Downward, J.R. and Huffnagle, G.B. 2013. The role of the bacterial microbiome in lung disease. *Expert Rev Respir Med* 7; 245–257.

Dominguez-Bello, M.G., Costello, E.K., Contreras, M., Magris, M., Hidalgo, G., Fierer, N. and Knight, R. 2010. Delivery mode shapes the acquisition and structure of the initial microbiota across multiple body habitats in newborns. *Proc Natl Acad Sci USA* 107(26); 11971–11975.

Douglas, A.E. 2019. Simple animal models for microbiome research. *Nat Rev Microbiol* 17; 764–775.

Dujardin, C.E., Mars, R.A.T., Manemann, S.M., Kashyap, P.C., Clements, N.S., Hassett, L.C. and Roger, V.L. 2020. Impact of air quality on the gastrointestinal microbiome: a review. *Environ Res* 109485. doi:10.1016/j.envres.2020.109485.

Gilbert, J.A. and Stephens, B. 2018. Microbiology of the built environment. *Nat RevMicrobiol* 16(11); 661–670.

Greenbaum, S., Greenbaum, G., Moran-Gilad, J. and Weintruab, A.Y. 2018. Ecogoical dynamics of the vaginal microbiome in relation to health and disease. *Am J Obstet Gynecol*. doi:10.1016/j.ajog.2018.11.1089.

Gupta, S., Kakkar, V. and Bhushan, I. 2019. Crosstalk between vaginal microbiome and female health: a review. *Microb Pathog* 136; 103696. doi: 10.1016/j.micpath.2019.103696.

Hassani, M.A., Duran, P. and Hacquard, S. 2018. Microbial interactions within the plant holobiont. *Microbiome* 6; 58. doi: 10.1186/s40168-018-0445-0.

Heckman, D.S., Geiser, D.M., Eidell, B.R., Stauffer, R.L, Kardos, N.L. and Hedges, S.B. 2001. Molecular evidence for the early colonization of land by fungi and plants. *Science* 293; 1129–1133.

Hilty, M., Burke, C. and Pedro, H., et al. 2010. Disordered microbial communities in asthmatic airways. *PLoS One* 5; e8578.

Huang, J., Pan, X., Guo, X. and Li, G. 2018. Impacts of air pollution wave on years of life lost: a crucial way to communicate the health risks of air pollution to the public. *Environ Int* 113; 42–49.

IntegrativeHMP (iHMP) Research Network Consortium. 2014. The Integrative Human Microbiome Project: dynamic analysis of microbiome – host omics profiles during periods of human health and disease. *Cell Host Microbe* 16; 276–298.

Jansson, J.K. and Hofmocke, K.S. 2020. Soil microbiomes and climate change. *Nat Rev Microbiol* 18; 35–46. doi: 10.1038/s41579-019-0265-7.

Jewell, K.A., McCormick, C.A., Odt, C.L., Weimer, P.J. and Suen, G. 2015. Ruminal bacterial community composition in dairy cows is dynamic over the course of two lactations and correlates with feed efficiency. *Appl Environ Microbiol* 81; 4697–710.

Lemanceau, P., Blouin, M., Muller, D. and Moenne-Loccoz, Y. 2017. Let the core microbiota be functional. *Trends Plant Sci* 22; 583–595.

Lindow, S.E. and Brandl, M.T. 2003. Microbiology of the phyllosphere. *Appl Environ Microbiol* 69; 1875–1883.

Liu, Y., Liu, Y., Du, Z., et al. Skin microbiota analysis-inspired development of novel anti-infectives. *Microbiome* 8; 85. doi: 10.1186/s40168-020-00866-1.

Mandar, R. 2013. Microbiota of male genital tract: impact on the health of man and his partner. *Pharmacol Res* 69; 32–41. doi: 10.1016/j.phrs.2012.10.019.

Marchesi, J.R. and Ravel, J. 2015. The vocabulary of microbiome research: a proposal. *Microbiome* 3; 31. doi:10.1186/s40168-015-0094-5.

Morrow, J.L., Frommer, M., Shearman, D.C.A. and Riegler, M. 2015. The microbiome of field-caught and laboratory adapted Australian tephritid fruit fly species with different host plant use and specialization. *Microb Ecol* 70(2); 498–508.

Nagasaka, M., Sexton, R., Alhasan, R., Rahman, S., Asfar, A.S. and Sukari, A. 2020. Gut microbiome and response to checkpoint inhibitors in non-small cell lung cancer – a review. *Crit Rev Oncol Hematol*. doi: 10.1016/j.critrevonc.2019.102841.

Nicholson, J.K., Holmes, E., Kinross, J., Burecelin, R., Gibson, G., Jia, W. and Pettersson, S. 2012. Host-gut microbiota metabolic interactions. *Science* 336(6086); 1262–1267. doi: 10.1126/science.1223813.

O'Callaghan, T.F., Ross, R.P., Stanton, C. and Clarke, G. 2016. The gut microbiome as a virtual endocrine organ with implications for farm and domestic animal endocrinology. *Domes Anim Endocrinol* 56; S44–S55. doi:10.1016/j.domaniend.2016.05.003.

Ogunrinola, G.A., Oyewale, J.O., Oshamika, O.O. and Olasehinde, G.I. 2020. The human microbiome and its impacts on health. *Int J Microbiol* 2020. Article ID 8045646. doi: 10.1155/2020/8045646.

Proctor, L. 2019. Priorities for the next 10 years of human microbiome research. *Nature* 569; 623–625.

Rajagopalan, S., Al-Kindi, S.G. and Brook, R.D. 2018. Air pollution and cardiovascular disease: JACC state-of-the-art review. *J Am Coll Cardiol* 72(17); 2054–2070.

Rasmussen, M.A., Thorsen, J., Dominguez-Bello, M.G., Blaser, M.J., Mortensen, M.S., Brejnrod, A.D., Shah, S.A., Hjelmso, M.H., Lehtimaki, J., Trivedi, U., Bisgaard, H., Sorensen, S.J. and Stokholm, J. 2020. Ecological succession in the vaginal microbiota during pregnancy and birth. *ISME J.* doi: 10.1038/s41396-020-0686-3.

Reese, A.T. and Dunn, R.R. 2018. Drivers of microbiome biodiversity: a review of general rules, feces, and ignorance. *mBio* 9; e01294–18. doi: 10.1128/mBio.01294-18.

Ross, A.A., Doxey, A.C. and Neufeld, J.D. 2017.The skin microbiome of cohabiting couples. *mSystems* 2; e00043–17. doi: 10.1128/mSystems.00043-17.

Sanford, J.A. and Gallo, R.L. 2013. Functions of the skin microbiota in health and disease. *SeminImmunol* 25(5); 370–377.

Santoyo, G., Hernandez-Pacheco, C., Hernandex-Salmeron, J. and Hernandez-Leon, R. 2017. The role of abiotic factors modulating the plant-microbe soil interactions: toward sustainable agriculture. A review. *Span J Agric Res* 15.doi: 10.5424/sjar/2017151-9990.

Simpson, J.M., Kocherginskaya, S.A., Aminov, R.I., Skerlos, L.T., Bradley, T.M., Mackie, R.I. and White, B.A. 2002. Comparative microbial diversity in the gastrointestinal tracts of food animal species. *Integr Comp Biol* 42; 327–331.

Song, S.J., Lauber, C., Costello, E.K., et al. 2013. Cohabiting family members share microbiota with one another and with their dogs. *Elife* 2; e00458. doi: 10.7554/eLife.00458.

Staubach, F., Baines, J.F., Kunzel, S., Bik, E.M. and Petrov, D.A. 2013. Host species and environmental effects on bacterial communities associated with Drosophila in the laboratory and in the natural environment. *PLoS One* 8(8). Article ID e70749.

Sunagawa, S., Coelho, L.P., Chaffron, S., et al. 2015. Structure and function of the global ocean microbiome. *Science* 348(6237); 1261359. doi: 10.1126/science 1261359.

Suttle, C.A. 2007. Marine viruses – major players in the global ecosystem. *Nat Rev Microbiol* 5(10); 801–812. doi: 10.1038/nrmicro1750.

Taiwo, A.M., Harrison, R.M. and Shi, Z. 2014. A review of receptor modelling of industrially emitted particulate matter. *Atmos Environ* 97; 109–120.

Thio, H.B. 2018. The microbiome in psoriasis and Psoriatic arthritis: the skin perspective. *JRheumatolSuppl* 94; 30–31. doi:10.3899/jrheum.180133.

Tkacz, A., Bestion, E., Bo, Z., Hortala, M. and Poole, P.S. 2020. Influence of plant fraction, soil, and plant species on microbiota: a multi kingdom comparison. *mBio* 11; e02785–19. doi: 10.1128/mBio.02785-19.

Toju, H., Peay, K.G., Yamamichi, M., et al. 2018. Core microbiomes for sustainable agrosystems. *Nat Plants* 4; 247–257.

Trinh, P., Zaneveld, J.R., Safranek, S., Rabinowitz, P.M. 2018. One health relationships betweenhuman, animal and environmental microbiomes: a mini-review. *Front Public Health* 6; 325. doi: 10.3389/fpubh.2018.00235.

Turner, T.R., James, E.K. and Poole, P.S. 2013. The plant microbiome. *Genome Biol* 14; 209.

Unger, S.A. and Bogaert, D. 2017. The respiratory microbiome and respiratory infections. *J Infect* 74; S84–S88.

Vorholt, J.A. 2012. Microbial life in the phyllosphere. *Nat Rev Microbiol* 10; 828–840.

Walter, J. and Ley, R. 2011. The human gut microbiome: ecology and recent evolutionary changes. *Ann Rev Microbiol* 65; 411–429.

Wang, Y., Gilbreath, T.M., Yan, K.G. and Xu, J. 2011. Dynamic gut microbiome across life history of the malaria mosquito *Anopheles gambiae* in Kenya. *PLoS ONE* 6(9). Article ID e24767.

Welch, J.L.M., Dewhirst, F.E. and Borisy, G.G. 2019. Biogeography of the oral microbiome. the site-specialist hypothesis. *AnnRevMicrobiol* 73; 335–358.

Welch, J.L.M., Ramirez-Puebla, S.T. and Borisy, G.G. 2020. Oral microbiome geography: microscale habitat and niche. *Cell Host & Microbe* 28(2); 160–168. doi: 10.1016/j.chrom.2020.07.009.

Wong, A.C.N., Chaston, J.M. and Douglas, A.E. 2013. The inconstant gut microbiota of Drosophila species revealed by 16S rRNA gene analysis. *ISME J* 7(10); 1922–1932.

Zhang, Y., Cai, J., Wang, S., He, K. and Zheng, M. 2017. Review of receptor based source apportionment research of fine particulate matter and its challenges in China. *Sci Total Environ* 586; 917–929.

Section I

Omics and Computational Techniques Used for Microbiome Analysis

2

Multi-Omics: Overview, Challenges, and Applications

Sushant Parab and Federico Bussolino
University of Turin
Candiolo Cancer Institute – IRCCS

CONTENTS

2.1 Introduction

Microbiome plays a very important role in host health, disease, and the environment. In the past few decades, the study of microbial communities has been accelerated by the evolution of sequencing technologies, which has made easier to sequence the whole DNA or RNA content within a day. Even though targeted approaches like PCR (polymerase chain reaction)-amplicon sequencing of 16S rRNA gene analysis were mainly in focus, researchers are taking a shift towards studying the whole microbial community by a metagenomics (entirety of genomes/genes) and metatranscriptomics (entirety of transcriptome/transcripts) approach (Aguiar-Pulido et al. 2016). The advantages of these two approaches are that they can answer these two basic questions quite nicely: "Who all are there?" and "What are they doing?"

"Who all are there?": To date, molecular-based taxonomic analysis of the microbial data highly relies on the amplicon sequencing (e.g., 16S rDNA/rRNA for bacteria, 18S rDNA/rRNA for protozoa). Even though this approach is rapid and low cost, the taxonomic assessment can be misleading due to the PCR biases from the primer selection (Hong et al. 2009) and amplification cycling conditions (Huber et al. 2009). Hence, a metagenomic and metatranscriptomic approach can be less-biased and more quantitative. Metagenomics is packed with information about the present taxonomies/genes defining which all features are present or absent in the microbial data. But it thus fails to tell much about their functions—this is where the metatranscriptomics play a key role.

"What are they doing?": A metatranscriptomic analysis allows understanding of how the microbiome responds to a particular environment by studying the genes expressed in that microbiome. It can estimate the taxonomic composition; predict open-reading frames (ORFs); and sometimes also identify the novel sites of transcription and/or translation. Metatranscriptomics can enable more complete protein sequence databases for metaproteomics or other downstream analysis.

In this chapter, we have summarized some frequently used and recently updated tools/algorithms which are used in metagenomic and metatranscriptomic data analyses.

2.2 Assemblers/Tools Used for Metagenome Analysis

2.2.1 AbySS 2.0

For large and complex genomes, de novo assembly remains a challenging problem, which reconstructs the chromosome sequences from sequencing reads in an order of magnitude shorter than the target genome (Nagarajan and Pop 2013). The de novo assembly is a more valuable approach as compared to the reference sequence assembly in detecting variants between individuals/genomes.

AbySS 1.0 was the first scalable de novo assembly algorithm that could assemble a human genome, using short reads from a high-throughput sequencing platform (Simpson et al. 2009). However, the algorithm required a large amount of memory distributed across a number of computer nodes (through message-passing interface protocol); still, it has found applications in many large cancer cohort studies (Yip et al. 2011; Roberts et al. 2012; Ley et al. 2013; Morin et al. 2013; Pugh et al. 2013). ABySS 1.0 is a multistage de novo pipeline consisting of unitig, contig, and scaffold stages. The most time-consuming stage of ABySS 1.0 is the unitig (the initial assembly of sequences according to the de Bruijn graph) assembly stage and is also its peak memory requirement. The main focus of AbySS 2.0 has been at this unitig assembly stage, where the introduction of Bloom filter-based (Bloom 1970) implementation reduces the overall memory requirements enabling assembly of large genomes on a single machine. ABySS 2.0 is built on the aspects of Minia (Chikhi and Rizk 2013), where the very first use of Bloom filters for performing a de novo assembly was demonstrated.

The parts of ABySS 2.0 algorithm that are new with respect to Minia are as follows:

1. The use of solid reads to seed contig traversals,
2. The use of look-ahead for error correction and elimination of Bloom filter false positives rather than a separate data structure,
3. The use of a new hashing algorithm, ntHash (Mohamadi et al. 2016), designed for processing DNA/RNA sequences efficiently.

2.2.2 MEGAHIT v1.0

The Metagenome assembler MEGAHIT v0.1 (Li et al. 2014) was first released in 2014, which can assemble large and complex metagenomic datasets in time- and memory-efficient manner. It makes use of a compressed data structure called "succinct de Bruijn graphs" (Bowe 2012) for the reduction of memory usage. The new version MEGAHIT v1.0 (Li et al. 2016) has few updates as compared to the older one:

1. New modules for local assembly and long-bubble merging to refine assembly,
2. A new CPU-based algorithm to speed up the time-consuming construction of succinct de Bruijn graph (SdBG), and
3. A revamp to the software architecture and data structures to reduce memory usage. Along with this, there are also some more new features added such as memory capacity, different preset combinations, and visualization of contig graphs.

SdBG in MEGAHIT is directly defined on the edges of its corresponding DBG (an edge-based de Bruijn graph). The k-mer size k is considered as the most important parameter for all the DBG-based assemblers, and MEGAHIT adopts a multiple k-mer size strategy (Peng et al. 2010), similar to other metagenome assemblers like IDBA-UD (Peng et al. 2012) and SPAdes (Bankevich et al. 2012). The new updates from MEGAHIT v1.0 such as "Local assembly" were first used in IBDA-UD, and it makes use of contigs as "anchor" and assembles a set of local reads. Bubbles in a DBG group are usually introduced by errors/SNPs/INDELs. In MEGAHIT v1.0, these bubbles are merged by a long-bubble merging component, but in a mild manner, which is a modified version of that used in IDBA-UD. MEGAHIT v1.0 improves the older version more significantly in assembly quality, including completeness and contiguity; it is more efficient and faster than its older version. The output of contigs in FASTG format can be visualized in contig graph using Bandage (Wick et al. 2015). MEGAHIT assemblies of public datasets can be openly accessed at MEGABOX.

2.2.3 SGA

Most of the short-read assemblers rely on the de Bruijn graph model of sequence assembly, which requires breaking the reads into k-mers (Pevzner et al. 2001). These de Bruijn graph assemblers attempt to recover the information lost from the breaking down of the reads, and attempt to resolve small repeats, using complicated read threading algorithms. SGA (string graph assembler, Simpson and Durbin 2012), on the other hand, avoids this problem by using an overlap-based string graph model of assembly.

Keeping all the reads intact, the string graph model creates a graph from overlaps between reads. SGA is among the first assemblers to implement a string graph approach for assembling short reads and also the very first to exploit a compressed index of a set of sequence reads. The SGA pipeline has three main stages: error correction, contig assembly, and scaffolding.

1. The first error correction stage begins by filtering or trimming the reads with multiple low-quality base calls. The FM index (full-text minute-space index, Ferragina and Manzini 2000) is constructed for the filtered set of reads (SGA index) and base calling errors are detected/corrected using k-mer frequencies (SGA correct).
2. The second contig assembly stage takes the corrected reads as input, re-indexes them, and removes the duplicated and low-quality reads, and then, a string graph is built constructing contigs.
3. The final scaffolding stage realigns the original reads to the contigs using BWA, constructs a scaffold graph using these alignments, and outputs a final set of scaffold in FASTA format.

SGA is an expansion of an algorithm to construct an assembly string graph (Myers 2005) for a set of error-free sequence reads using the FM index (Simpson and Durbin 2010). SGA acts as a modular pipeline, which allows it to be easily extended as improved algorithms are developed or sequencing technology changes.

2.2.4 Kraken

Kraken (Wood et al. 2014) program is fast and highly accurate for assigning taxonomic labels to metagenomics sequences. It is able to achieve genus-level sensitivity and precision that are very similar to those obtained by the fastest BLAST program, Megablast (Morgulis et al. 2008). At the core of Kraken, there is a database consisting of a k-merrecords and the lowest common ancestor (LCA) of all organisms whose genomes contain that k-mer. This database was built using a user-specified library of genomes which allows a quick lookup of the most specific node in the taxonomic tree that is associated with a given k-mer. Sequences are classified by querying the database for each k-mer in a sequence and then using the resulting set of LCA taxa to determine an appropriate label for the sequence. Sequences that have no k-mers in the databases are left unclassified by Kraken. Kraken by default builds the database with $k=31$, but this value can be modified by the user. Since most of the true species in metagenomics datasets today are unknown, two simulated metagenomes were created by combining real sequences obtained from projects that sequenced isolated microbial genomes. Data sequenced by the Illumina HiSeq and MiSeq sequencing platforms was used for creating these simulated metagenomes. In addition to the two metagenomes constructed, a third metagenomics sample (simBA-5) was created by covering a much broader range of the sequence phylogeny. This sample featured simulated bacterial and archaeal reads from RefSeq. Kraken's accuracy might be comparable to that of Megablast for classifying short reads. An important constrain for Kraken is its memory usage: At present, the default database requires 70 GB, a value that will grow in linear proportion to the number of distinct k-mers in the genomic library. The Kraken database which is tuned to query overlapping k-mers rapidly enables kraken to produce results faster.

2.2.5 SUPER-FOCUS

SUPER-FOCUS (Silva et al. 2016), Subsystems Profile by database Reduction using FOCUS, is a homology-based approach to report the subsystems present in metagenomics databases and to profile their abundance.

SUPER-FOCUS aligns all the input data against a reduced database which contains only the subsystems present in the organisms in that metagenome. The speed-up drives from three improvements compared with the standard metagenome annotation pipelines.

1. The SEED database was clustered using CD-HIT (Huang et al. 2010) using a similar approach as previously discussed and applied to the GenBank NR database (Li et al. 2012);
2. The metagenomics query sequences are profiled using FOCUS (Silva et al. 2014), an ultra-fast tool that identifies the organisms in the metagenome;
3. Align input data against the reduced dataset using RAPSearch2, DIAMOND, or blastx (Berendzen et al. 2012).

SUPER-FOCUS has few additional advantages in the functional profiling as compared with other tools:

(1) It uses a fast aligner, (2) it uses clustered databases in order to obtain a fast profile with little loss of sensitivity, and (3) it focuses on the microbes present in the input data to attain a more microbial profile.

2.2.6 StrainPhlAn

It is a novel method and implementation to profile microbial strains from metagenomes at a resolution comparable with that of isolate sequencing, and apply it to thousands of gut samples spanning multiple host populations (Truong et al. 2017). The method is based on reconstructing consensus sequence variants within species-specific marker genes and using them to estimate strain-level phylogenies.

The StrainPhlAn workflow: Metagenomics reads in each sample are first mapped against the species-specific MetaPhlAn2 markers using Bowtie2. The resulting alignments are processed with BAMtools (Li et al. 2009) to estimate the consensus sequence of each detected species-specific marker. This is performed using a simple majority rule to infer each nucleotide of the markers. Strain-specific markers can also be extracted from the available reference genomes (using BLASTN) (Altschul et al. 1990) to include them in the downstream analysis. The port-processing operations include multiple sequence alignment (MSA) on high-quality consensus sequences and concatenate them in larger alignments for each species. Reconstructed markers with the lack of coverage >20% are discarded. Strain profiling in a sample is only provided for species where reconstructed markers exceed 80% of the total number of markers available for that species in MetaPhlAn2. The reconstructed markers are then aligned using MUSCLE (Table 2.1).

2.3 Assemblers/Tools Used for Metatranscriptome Analysis

2.3.1 Leimena-2013

Leimena-2013 (the pipeline does not have a specific name, and hence, we use the first author's name followed by the publishing year to represent it), a reliable metatranscriptome data analysis pipeline, was developed. The setup of this pipeline is very generic and can be applied for bacterial metatranscriptome analysis in any chosen niche.

The pipeline includes the following four stages:

1. Removal of ribosomal RNA sequences,
2. Taxonomic identity and functional assignment of mRNA reads,

TABLE 2.1

Assemblers/Tools Used for Metagenomic Analysis

Tool	AbySS 2.0	MEGAHITv1.0	SGA	Kraken	SUPER-FOCUS	StrainPhlAn
PubMed link	https://pubmed.ncbi.nlm.nih.gov/28232478/	https://pubmed.ncbi.nlm.nih.gov/27012178/	https://pubmed.ncbi.nlm.nih.gov/22156294/	https://pubmed.ncbi.nlm.nih.gov/24580807/	https://pubmed.ncbi.nlm.nih.gov/26454280/	https://pubmed.ncbi.nlm.nih.gov/28167665/
Short intro	ABySS 2.0 is built on the aspects of Minia, where the very first use of Bloom filters for performing a de novo assembly was demonstrated.	MEGAHIT uses a succinct de Bruijn graph algorithm along with some updates on local assembly, reduce memory usage.	SGA uses an overlap-based string graph model of assembly instead of de Bruijn graph model.	Kraken is software used in metagenomics study for assigning taxonomic labels to short DNA sequences.	SUPER-FOCUS uses a homology-based approach adopting a reduced SEED database to report the subsystems present in metagenomics samples.	StrainPhlAn is a strain-level metagenome profiling tool by profiling microbes from known species with strain-level resolution.
Operating system	Linux	Linux	Linux	Linux	–	MacOS, Linux
Algorithm used	Bloom filter	Succinct de Bruijn graph	Overlap-based string graph model	Exact alignment of k-mers	BLASTX, RAPSearch2, DIAMOND	MetaPhlAn2
Databases	-	-	-	Kraken database, RefSeq, NCBI taxonomy database	Reduced SEED database	MetaPhlAn2
Web address	https://github.com/bcgsc/abyss-2.0-giab	https://hku-bal.github.io/megabox	https://github.com/jts/sga	https://ccb.jhu.edu/software/kraken/	http://edwards.sds.edu/superfocus/	http://segatalab.cibio.unitn.it/tools/strainphlan/
Publication date and citation counts	February 14, 2017; 95	June 1, 2016; 138	March 22, 2012; 258	March 3, 2014; 959	October 9, 2015; 39	February 2, 2017; 121

3. Classification of mRNA in gene and intergenic reads,
4. Functional assignment.

Workflow: The rRNA/tRNA reads were removed from the unique Illumina reads using SortMeRNA software (Kopylova et al. 2012) followed by BLASTN alignment to NCBI and SILVA ribosomal databases (Pruesse et al. 2007). The mRNA reads were assigned to the prokaryote genomes of NCBI using MegaBLAST (Morgulis et al. 2008) followed by BLASTN, followed by classification according to alignment bit scores using a minimum bit score of 148 and 110 for the prediction of phylogenetic origin at the genus and family level, respectively. The genome-assigned reads were classified into protein-encoding or non-coding reads, followed by COG (cluster of orthologous groups of proteins) (Tatusov et al. 2000) and KEGG (Kanehisa et al. 2000) functional annotation and metabolic mapping. Additional functional assignment was made for evaluation purposes by assigning 10% of randomly selected unassigned reads (bit score ≤74) to the NCBI protein database followed by MetaHIT (Qin et al. 2010) and SI metagenome databases using BLASTX.

2.3.2 MetaTrans

MetaTrans (Martinez et al. 2016) is a downloadable, open-source, efficient metatranscriptomic pipeline developed for a paired-end RNA-Seq analysis. The pipeline is designed to perform two types of RNA-Seq analyses: taxonomic and gene expression.

Workflow: The raw paired-end reads were subjected to quality control and adjustment using the FastQC tool and Kraken pipeline (turquoise boxes, Wood et al. 2014). The rRNA/tRNA reads were then separated from the non-rRNA/tRNA reads using SortMeRNA software (green boxes), for taxonomic (clear blue boxes) and functional analyses (pink boxes), respectively. For the taxonomic analysis, the reads were mapped against the 16S rRNA Greengenes v13.5 database (McDonald et al. 2012) using SOAP2 (Li et al. 2009). For the functional analysis, the reads were subjected to the FragGeneScan (Rho et al. 2010) to predict putative genes before being mapped against a functional database (MetaHIT-2014 or M5nr) also using the SOAP2 tool.

A metatranscriptomic pipeline is implemented by making use of the multi-threading capacity of modern computers and then validated its functionality by comparing different methods for taxonomy profiling. The pipeline is implemented on the basis of a constantly changing environment, thus offering the possibility to easily integrate third-party tools that can improve parts of the pipeline or change entire modules as long as the input/output folder structure is preserved.

2.3.3 SAMSA

SAMSA (Simple Analysis of Metatranscriptome Sequence Annotations) pipeline is designed to fully analyze and characterize bacterial transcription of gut microbiome data (Westreich et al. 2016). The pipeline works in four stages:

1. The preprocessing phase,
2. The annotation phase,
3. The aggregation phase,
4. The analysis phase.

Workflow: During the preprocessing phase, the raw sequences are trimmed to remove low-quality reads and adapters using Trimmomatic (Bolger et al. 2014). Then, the remaining good-quality reads are aligned to each other using FLASh, a short-read aligner (Magoc and Salzberg 2011). In the annotation phase, these sequences are submitted to Metagenomic Rapid Annotations using Subsystems Technology (MG-RAST, Meyer et al. 2008a). MG-RAST includes several steps, in which for each sequence cluster MG-RAST selects the best match through the sBLAT similarity search. All annotations with an acceptable best match to the NCBI Reference Database (ReFSeq, Tatusova et al. 2014) are downloaded from MG-RAST in a tab-delimited form using RESTful API interface (Wilke et al. 2015). In the aggregation phase, a custom Python program parses each annotated output, storing each unique annotation match and also maintaining counts of each unique annotation. In the final phase, i.e., the analysis phase, all the information from the aggregate phase is passed into custom R scripts to generate barplots and dendrograms. To find differentially expressed transcripts, DESeq2 (Love et al. 2014) package is used to test the output files (Table 2.2).

- The algorithms such as ABySS 2.0, MEGAHIT, and SGA can also be used for metatranscriptomic data analysis.
- They are already explained in the previous section.

2.4 Functional Analysis Using Different Databases

2.4.1 MG-RAST

Over the past few years, the major challenge in metagenomics has been in analyzing sequences rather than generating new metagenomes/samples. MG-RAST (Keegan et al. 2016) is an open-source pipeline for metagenomes that constructs functional assignments of sequences by comparing nucleotide and protein databases. The pipeline is implemented in Perl by using a number of open-source components, including SEED framework (Overbeek et al. 2005), NCBI BLAST (Altschul et al. 1997), SQLite, and Sun Grid (http://gridengine.sunsource.net/) as components. The pipeline also uses publically available SEED subsystems.

Overview of the workflow: Once the metagenome is uploaded, a normalization step is executed generating unique internal IDs and removing duplicates. In the second step, BLASTX (Altschul et al. 1997) is used for screening the sequences for potential protein-encoding genes (PEGs) against the SEED nr database and other sources (Overbeek et al. 2005). A threshold/cutoff of 0.01 for an expect value (E) is used. In parallel with the BLASTX search, the sequences are compared to all accessory databases such as rDNA databases, GREENGENES (DeSantis et al. 2006), RDP-II (Cole et al. 2007), European 16S RNA database (Wuyts et al. 2002), and ACLAME database of mobile elements (Leplae et al. 2004). In the last step, a

TABLE 2.2
Assemblers/Tools Used for Metatranscriptomic Analysis

Tool	Leimena-2013	MetaTrans	SAMSA
PubMed link	https://pubmed.ncbi.nlm.nih.gov/23915218/	https://pubmed.ncbi.nlm.nih.gov/27211518/	https://pubmed.ncbi.nlm.nih.gov/27687690/
Short intro	This comprehensive metatranscriptome analysis pipeline is designed for function assignment and mapping based on given RNA-seq data. The performance of this pipeline has been evaluated using human small intestine microbiota data.	MetaTrans is an open-source pipeline that integrates quality control, rRNA removal, and read mapping for taxonomic and gene expression analysis.	SAMSA is a comprehensive pipeline for metatranscriptome analyses by working with MG-RAST, providing an ability to fully analyze the expression activity within microbial communities.
Operating system		Linux	Unix/MacOS/Windows
Algorithm/tools used	SortMeRNA, BLASTN, MegaBLAST, KAAS	Kraken, SortMeRNA, UCLUST, SOAP2, FragGeneScan, DESeq2	Trimmomatic, FLASH, MG-RAST, DESeq2
Databases	SILVA, COG, MetaHIT, human small intestinal metagenome database, KEGG	SILVA-115, Greengenes-13.5, Rfam-11.0, tRNA-all, PhiX genome, MetaHIT-2014, M5nr-20130801	NCBI database, SEED subsystems reference database
Web address		http://www.metatrans.org/	https://github.com/transcript/SAMSA
Publication date and citation counts	August 2, 2013; 50	May 23, 2016; 18	March 29, 2016; 19

phylogenetic tree is constructed by using the information from the SEED nr database and the similarities to ribosomal RNA database. Functional classifications of the PEGs are computed against SEED FIGfams (Meyer et al. 2008b).

The central utility of the MG-RAST platform is the abundance of comparative metagenomics tools. Since MG-RAST handles both assembled and unassembled data, while comparing metagenomes the advantages of each approach should be considered. MG-RAST thus provides an integration of metagenome data, microbial genomics, and manually curated annotations.

2.4.2 MEGAN

MEGAN (Huson et al. 2007) allows laptop analysis of large metagenomics datasets. MEGAN can be applied to DNA reads of any metagenomics project, regardless of the sequencing technology used.

MEGAN workflow is as follows:

1. Metagenome reads are collected from the sample.
2. The resulting reads are then compared with one or more different databases (nr, nt, env-nr, env-nt, etc) using BLAST (Altschul et al. 1990) or a similar comparison tool.
3. The results obtained by the comparison are collected and processed by MEGAN to assign a taxon ID to each sequence based on the NCBI taxonomy.
4. Finally, the program assigns each read to the LCA of the set of taxa that it hits in the comparison.

MEGAN diverges from the previous metagenomics pipelines and builds on the statistical power of comparing random sequences against databases of known sequences. It demonstrates that even given the current incomplete and biased databases, a meaningful categorization of reads is possible as the first phylogenetic analysis in metagenome data.

2.4.3 COG

Orthologs are direct evolutionary counterparts as opposed to paralogs which are genes within the same genome by duplication (Fitch 1970, 1995). The COGs database has been designed as an effort to classify proteins from completely sequenced genomes on the basis of the orthology concept (Tatusov et al. 1997). The COGs reflect one-to-one, one-to-many, and many-to-many orthologous relationships.

The construction of COGs includes the following steps:

1. Perform all-against-all protein sequence comparison.
2. Detect and discard the obvious paralogs, i.e., proteins from the same genome that are more similar to each other than to any proteins from other species.
3. Taking into account the paralogous groups from the previous step, detect triangle of mutually consistent, genome-specific best hits (BeTs).
4. Merge triangle with a common side to form COGs.
5. To eliminate false positives and to identify groups of multidomain proteins, a case-by-case analysis of each COG is done. These multidomain proteins are split into single-domain segments and all the above steps are then repeated, resulting in assigning of individual domains of COGs in accordance with their distinct evolutionary affinities.
6. Examinations of large COGs that include multiple members which are split into two or more smaller ones using phylogenetic trees, cluster analysis, etc.

The most straightforward application of the COGs is for the prediction of functions of individual proteins or protein sets, and those from newly completed genomes. This is done by using the COGNITOR program. The COG WWW site (http://www.ncbi.nlm.nih.gov/COG) contains the following types of data:

1. BLAST search outputs for each member of the COG along with their respective links to GenBank and Entrez-Genome entries,
2. MSA of the COG members using ClustalW program (Thompson et al. 1994),
3. A cluster dendrogram generated using BLAST scores.

2.5 Multi-Omics—A Future Full of Promises and Challenges

This new field of studying genomes and transcriptomes has seen a rapid increase in the number of metagenomics and metatranscriptomics projects. To realize the full potential of meta-analysis, the deposition of adequate sample metadata should become important focus of future efforts along with the timely standardization. The rise of long-read technologies shows great promise; long reads will be able to help with all aspects of analysis (assembly, taxonomy determination, functional analysis) and will provide better resolution of transcript isoforms on different genes with high similarity. Thus, one additional area that opens for more research is the benchmarking of the performance and accuracy of bioinformatics tools and pipelines for meta-analysis. The complexity of microbiomes presents a great challenge in benchmarking these algorithms. Sequencing technologies will continue to push the development of new algorithms and will help the field progress, and possibly coalesce towards accurate and appropriate workflows. Despite some of the issues, it is clear that next generation of meta-analysis tools hold great promise in facilitating our understanding of the biologically active fraction of microbiomes and the relevant pathways involved.

REFERENCES

Aguiar-Pulido, V, Huang, W, Suarez-Ulloa, V, Cickovski, T, Mathee, K, and Narasimhan, G. 2016. Metagenomics, metatranscriptomics, and metabolomics approaches for microbiome analysis. *Evol Bioinform* Online 12(Suppl. 1):5–16.

Altschul, SF, Gish, W, Miller, W, Myers, EW, and Lipman, DJ. 1990. Basic local alignment search tool. *J Mol Biol* 215:403–410.

Altschul, SF, Madden, TL, Schaffer, AA, Zhang, J, Zhang, Z, Miller, W, and Lipman, DJ. 1997. Gapped BLAST and PSI-BLAST: a new generation ofprotein database search programs. *Nucleic Acids Res* 25:3389–3402.

Bankevich, A, et al. 2012. SPAdes: a new genome assembly algorithm and its applications to single-cell sequencing. *J Comput Biol* 19(5):455–477.

Berendzen, J, et al. 2012. Rapid phylogenetic and functional classification of short genomic fragments with signature peptides. *BMC Res Notes* 5:460.

Bloom, BH. 1970. Space/time trade-offs in hash coding with allowable errors. *Commun ACM* 13:422–426.

Bolger, AM, Lohse, M, and Usadel, B. 2014. Trimmomatic: a flexible trimmer for Illumina sequence data. *Bioinformatics* 30(15):2114–2120.

Bowe. 2012. Succinct de Bruijn Graphs, in: B Raphael, J Tang (Eds.), *Algorithms in Bioinformatics*, Springer, Berlin Heidelberg, pp. 225–235.

Chikhi, R, and Rizk, G. 2013. Space-efficient and exact de Bruijn graph representation based on a Bloom filter. *Algorithms MolBiol* 8:1.

Cole, JR, Chai, B, Farris, RJ, Wang, Q, Kulam-Syed-Mohideen, AS, McGarrell, DM, Bandela, AM, Cardenas, E, Garrity, GM, and Tiedje, JM. 2007. The ribosomal database project (RDP-II): introducing myRDPspace and quality controlled public data. *Nucleic Acids Res*:D169–D172.

DeSantis, TZ, Hugenholtz, P, Larsen, N, Rojas, M, Brodie, EL, Keller, K, Huber, T, Dalevil, D, Hu, P, and Andersen, GL. 2006. Green genes, a chimerachecked16S rRNA gene database and workbench compatible with ARB. *Appl Environ Microbiol* 72(7):5069–5072.

Ferragina, P, and Manzini, G. 2000. Opportunistic data structures with applications. In *Proceedings of the 41st Annual Symposium on Foundations of Computer Science*, pp. 390–398. IEEE Computer Society, Washington, DC. doi:10.1109/SFCS.2000.892127.

Fitch, WM. 1970. System. *Zool* 19:99–106.

Fitch, WM. 1995. Uses for evolutionary trees. *Phil Trans R Soc Lond B Biol Sci* 349:93–102.

Forouzan, E, Shariati, P, Maleki, MSM, Karkhane, AA, and Yakhchali, B. 2018. Practical evaluation of 11 de novo assemblers in metagenome assembly. *J Microbiol Methods* 151:99–105, ISSN 0167-7012. doi:10.1016/j.mimet.2018.06.007.

gridengine – Project home [http://gridengine.sunsource.net/].

Hong, S, Bunge, J, Leslin, C, Jeon, S, and Epstein, SS. 2009. Polymerase chain reaction primers miss half of rRNA microbial diversity. *ISME J* 3:1365–1373.

Huang, Y, et al. 2010. CD-HIT suite: a web server for clustering and comparing biological sequences. *Bioinformatics* 26:680–682.

Huber, JA, Morrison, HG, Huse, SM, Neal, PR, Sogin, ML, and Mark Welch, DB. 2009. Effect of PCR amplicon size on assessments of clone library microbial diversity and community structure. *Environ Microbiol* 11:1292–1302.

Huson, DH, Auch, AF, Qi, J, and Schuster, SC. 2007. MEGAN analysis of metagenomic data. *Genome Res* 17(3):377–386. doi:10.1101/gr.5969107.

Jackman, SD, et al. 2017. ABySS 2.0: resource-efficient assembly of large genomes using a Bloom filter. *Genome Res* 27(5):768–777. doi:10.1101/gr.214346.116.

Kanehisa, M, and Goto, S. 2000. KEGG: kyoto encyclopedia of genes and genomes. *Nucleic Acids Res* 28:27–30.

Keegan KP, Glass EM, Meyer F. 2016. MG-RAST, a Metagenomics service for analysis of microbial community structure and function. *Methods Mol Biol.* 1399: 207–233. doi: 10.1007/978-1-4939-3369-3_13.

Kopylova, E, Noé, L, and Touzet, H. 2012. SortMeRNA: fast and accurate filtering of ribosomal RNAs in metatranscriptomic data. *Bioinformatics* 28(24):3211–3217. doi:10.1093/bioinformatics/bts611.

Leimena, MM, Ramiro-Garcia, J, Davids, M, et al. 2013. A comprehensive metatranscriptome analysis pipeline and its validation using human small intestine microbiota datasets. *BMC Genomics* 14:530. doi:10.1186/1471-2164-14-530.

Leplae, R, Hebrant, A, Wodak, SJ, and Toussaint, A. 2004. ACLAME: a CLAssification of Mobile genetic Elements. *Nucleic Acids Res*:D45–D49.

Ley, T, et al. 2013. Genomic and epigenomic landscapes of adult de novo acute myeloid leukemia. *N Engl J Med* 368:2059–2074.

Li, R, et al. 2009. SOAP2: an improved ultrafast tool for short read alignment. *Bioinformatics* 25:1966–1967.

Li, W, et al. 2012. Ultrafast clustering algorithms for metagenomic sequence analysis. *Brief Bioinform* 13:656–668.

Li, D, et al. 2016. MEGAHIT v1.0: a fast and scalable metagenome assembler driven by advanced methodologies and community practices. *Elsevier* 102:3–11.

Li, D, Liu, C-M, Luo, R, Sadakane, K, and Lam, T-W. 2014. MEGAHIT: an ultra-fast single-node solution for large and complex metagenomics assembly via succinct *de Bruijn* graph. *Bioinformatics* 31(10):1674–1676.

Li, F, Neves, ALA, Ghoshal, B, and Guan, LL. 2018. Symposium review: mining metagenomic and metatranscriptomic data for clues about microbial metabolic functions in ruminants. *J Dairy Sci* 101(6):5605–5618, ISSN 0022-0302. doi:10.3168/jds.2017-13356.

Love, MI, Huber, W, and Anders, S. 2014. Moderated estimation of fold change and dispersion for RNA-seq data with DESeq2. *Genome Biol* 15(12):550.

Magoc, T, and Salzberg, SL. 2011. FLASH: fast length adjustment of short reads to improve genome assemblies. *Bioinformatics* 27(21):2957–2963.

Martinez, X, et al. 2016. MetaTrans: an open-source pipeline for metatranscriptomics. *Sci Rep* 6:26447. doi:10.1038/srep26447.

McDonald, D, et al. 2012. An improved Greengenes taxonomy with explicit ranks for ecological and evolutionary analyses of bacteria and archaea. *ISME J* 6:610–618.

Meyer, F, et al. 2008a. The metagenomics RAST server – a public resource for the automatic phylogenetic and functional analysis of metagenomes. *BMC Bioinformatics* 9:386. doi:10.1186/1471-2105-9-386.

Meyer, F, Overbeek, R, and Rodriquez, A. 2008b. FIGfams – Yet another protein family collection. in press.

Mohamadi, H, Chu, J, Vandervalk, BP, and Birol, I. 2016. ntHash: recursive nucleotide hashing. *Bioinformatics* 32:3492–3494.

Morgulis, A, et al. 2008. Database indexing for production Mega BLAST searches. *Bioinformatics* 24:1757–1764.

Morin, RD, et al. 2013. Mutational and structural analysis of diffuse large B-cell lymphoma using whole-genome sequencing. *Blood* 122:1256–1265.

Myers, EW. 2005. The fragment assembly string graph. *Bioinformatics* 21(Suppl 2):ii79–ii85.

Nagarajan, N, and Pop, M. 2013. Sequence assembly demystified. *Nat Rev Genet* 14:157–167.

Niu, S-Y, Yang, J, McDermaid, A, Zhao, J, Kang, Y, and Ma, Q. 2018. Bioinformatics tools for quantitative and functional metagenome and metatranscriptome data analysis in microbes. *Brief Bioinform* 19(6):1415–1429. doi:10.1093/bib/bbx051.

Overbeek, R, et al. 2005. The subsystems approach to genome annotation and its use in the project to annotate 1000 genomes. *Nucleic Acids Res* 33(17):5691–5702.

Peng, Y, Leung, H, Yiu, SM, and Chin, F. 2010. IDBA – A Practical Iterative de Bruijn Graph De Novo Assembler, in: B Berger (Ed.), *Research in Computational Molecular Biology*, Springer, Berlin Heidelberg, pp. 426–440.

Peng, Y, Leung, HCM, Yiu, SM, and Chin, FYL. 2012. IDBA-UD: a *de novo* assembler for single-cell and metagenomic sequencing data with highly uneven depth. *Bioinformatics* 28(11):1420–1428.

Pevzner, PA, Tang, H, and Waterman, MS. 2001. An Eulerian path approach to DNA fragment assembly. *Proc Natl Acad Sci* 98:9748–9753.

Pruesse, E, et al. 2007. SILVA: a comprehensive online resource for quality checked and aligned ribosomal RNA sequence data compatible with ARB. *Nucleic Acids Res* 35:7188–7196.

Pugh, TJ, et al. 2013. The genetic landscape of high-risk neuroblastoma. *Nat Genet* 45:279–284.

Qin, J, et al. 2010. A human gut microbial gene catalogue established by metagenomic sequencing. *Nature* 464:59–65.

Rho, M, Tang, H, and Ye, Y. 2010. FragGeneScan: predicting genes in short and error-prone reads. *Nucleic Acids Res* 38:e191.

Roberts, KG, et al. 2012. Genetic alterations activating kinase and cytokine receptor signaling in high-risk acute lymphoblastic leukemia. *Cancer Cell* 22:153–166.

Shakya, M, Lo, CC, and Chain, PSG. 2019. Advances and challenges in metatranscriptomic analysis. *Front Genet* 10:904. doi:10.3389/fgene.2019.00904.

Silva GGZ, Cuevas DA, Dutilh BE, and Edwards RA. 2014. FOCUS: an alignment-free model to identify organisms in metagenomes using non-negative least squares. *PeerJ* 2:e425 doi:10.7717/peerj.425

Silva, GG, Green, KT, Dutilh, BE, and Edwards, RA. 2016. SUPER-FOCUS: a tool for agile functional analysis of shotgun metagenomic data. *Bioinformatics* 32(3):354–361. doi:10.1093/bioinformatics/btv584.

Simpson, JT, and Durbin, R. 2010. Efficient construction of an assembly string graph using the FM-index. *Bioinformatics* 26:i367–i373.

Simpson, JT, and Durbin, R. 2012. Efficient de novo assembly of large genomes using compressed data structures. *Genome Res* 22(3):549–556.

Simpson, JT, Wong, K, Jackman, SD, Schein, JE, Jones, SJ, and Birol, I. 2009. ABySS: a parallel assembler for short read sequence data *Genome Res* 19:1117–1123.

Tatusova, T, et al. 2014. RefSeq microbial genomes database: new representation and annotation strategy. *Nucleic Acids Res* 42(Database issue):D553–D559.

Tatusov, RL, Galperin, MY, Natale, DA, and Koonin, EV. 2000. The COG database: a tool for genome-scale analysis of protein functions and evolution. *Nucleic Acids Res* 28:33–36.

Tatusov, RL, Koonin, EV, and Lipman, DJ. 1997. A genomic perspective on protein families. *Science* 278(5338):631–637. doi: 10.1126/science.278.5338.631

Thompson, JD, Higgins, DG, and Gibson, TJ. 1994. CLUSTAL W: improving the sensitivity of progressive multiple sequence alignment through sequence weighting, position-specific gap penalties and weight matrix choice. *Nucleic Acids Res* 22(22):4673–4680. doi: 10.1093/nar/22.22.4673

Truong, DT, Tett, A, Pasolli, E, Huttenhower, C, and Segata, N. 2017. Microbial strain-level population structure and genetic diversity from metagenomes. *Genome Res* 27(4):626–638. doi:10.1101/gr.216242.116.

Westreich, ST, Korf, I, Mills, DA, and Lemay, DG. 2016. SAMSA: a comprehensive metatranscriptome analysis pipeline. *BMC Bioinformat* 17(1):399. doi:10.1186/s12859-016-1270-8.

Wick, RR, et al. 2015. Bandage: interactive visualisation of de novo genome assemblies. *Bioinformatics* 31(20):3350–3352.

Wilke, A, et al. 2015. A RESTful API for accessing microbial community data for MG-RAST. *PLoS Comput Biol* 11(1):e1004008.

Wood, DE, and Salzberg, SL. 2014. Kraken: ultrafast metagenomic sequence classification using exact alignments. *Genome Biol* 15(3):R46. doi:10.1186/gb-2014-15-3-r46.

Wuyts, J, Peer, YV de, Winkelmans, T, and De Wachter, R. 2002. The European database on small subunit ribosomal RNA. *Nucleic Acids Res* 30(1):183–185.

Yip, S, et al. 2011. Concurrent CIC mutations, IDH mutations, and 1p/19q loss distinguish oligodendrogliomas from other cancers. *J Pathol* 226:7–16.

3

Computational Techniques Used for Microbial Diversity Analysis

Dattatray S. Mongad, Nikeeta S. Chavan, and Yogesh S. Shouche
National Centre for Cell Science

CONTENTS

3.1 Introduction

Microbiota is a collective term for microbial taxa associated within an ecosystem, while the catalog of these microbes and their genes can be termed as "microbiome" (Ursell et al. 2012). Microorganisms play a vital role in ecosystem's or host's health like soil, rivers, lakes, humans, and mammals. Any change in abundance of these microbes as compared to healthy state is called as "dysbiosis," which can be correlated with a number of human diseases like diabetes (Upadhyaya and Banerjee 2015), cancer (Osman et al. 2018), and celiac disease (Bodkhe et al. 2019). Microbiome is considered as virtual organ of human body, which plays a vital role in human's biology, immune system, and physiology (Evans, Morris, and Marchesi 2013). In recent decades, microbiome studies have been routinely carried out to discover the potential of microbiome living inside the human gut environment. Traditionally, microbiome studies were carried out using culture-dependent methods which can be time-consuming and laborious, and cannot provide the full picture about the interactions between microbes in an ecosystem. Due to a decrease in sequencing cost in recent years, culture-independent approaches like 16S rRNA gene amplicon sequencing and metagenome sequencing are being routinely used in microbiome studies.

3.1.1 16S rRNA Amplicon Sequencing

16S rRNA gene is widely accepted as a phylogenetic marker for taxonomic identification, classification, and phylogenetic analysis of novel microorganisms (Srinivasan et al. 2015). The 16S rRNA gene has highly conserved sequences separated by nine hypervariable regions (V1–V9), which makes it a reliable phylogenetic marker. In 16S rRNA amplicon sequencing, one or multiple variable regions of 16S rRNA gene (V4, V3–V5, etc.) are targeted, amplified, and sequenced, while in metagenome sequencing, all the DNA contents from each sample are sequenced. Amplification is done using a set of universal primer pairs available for variable regions along with heterogeneity spacers (0–7 nucleotides) to increase effectiveness of sequencing reaction. Hence, the amplified product will have sequencing primer and heterogeneity spacer on both sides of targeted sequence. In barcoding step, 5' & 3' index (5–7 nucleotides) and flow cell linker adapters are added on both sides of amplified product, which are essential for the attachment of DNA fragments to flow cell. Hence, the final product that has biological (targeted variable region) and nonbiological (adapters, index, sequencing primer, and heterogeneity spacers) sequences is proceeded for sequencing with different chemistry of sequencers depending on its length (Figure 3.1).

3.1.2 Approaches for Data Analysis of 16S rRNA Amplicon Sequences

Briefly, 16S rRNA gene amplicon sequencing data analysis involves steps like demultiplexing, quality check, removal of chimeric sequences, OTU (operational taxonomic unit) picking or sample inference, taxonomy assignment, construction of abundance table, and downstream analysis. Figure 3.2 shows two broad categories of approaches used for the analysis of amplicon data, viz., OTU- and ASV (amplicon sequence variant)-based. OTUs can be defined as group of sequences (based on similarity cutoff, e.g., 97%)—which represents a specific taxa—while ASVs can detect one nucleotide difference in sequence and attempt to recover the exact biological sequence from samples which can be referred to as "zero noise OTUs" or "zOTUs" (Prodan et al. 2020). Both the approaches (discussed in detail in the respective sections) are different but have huge impact on results. The preprocessing of sequences (quality analysis, filtering and trimming of primers, and low-quality tails) and downstream analysis using abundance table remain the same in both approaches.

3.2 Materials

The online version of this tutorial, including QIIME1, DADA2, and downstream analysis, can be found at https://github.com/nikeetaC/amplicon_data_analysis.

3.2.1 Software

For the microbiome data analysis using OTU approach, we will use QIIME1 (v1.9.1) (Caporaso et al. 2010); the installation guide is available at http://qiime.org/install/install.html. QIIME1 uses Greengenes, a chimera-checked 16S rRNA gene database (DeSantis et al. 2006) as default reference database. For ASV approach, we will use DADA2 (v1.14.1); the installation instruction is available at https://benjjneb.github.io/dada2/dada-installation.html. Along with these two software packages, we will be using Microbiome Helper (https://github.com/LangilleLab/microbiome_helper/wiki), which has several scripts to automate QIIME1 workflow (Comeau, Douglas,

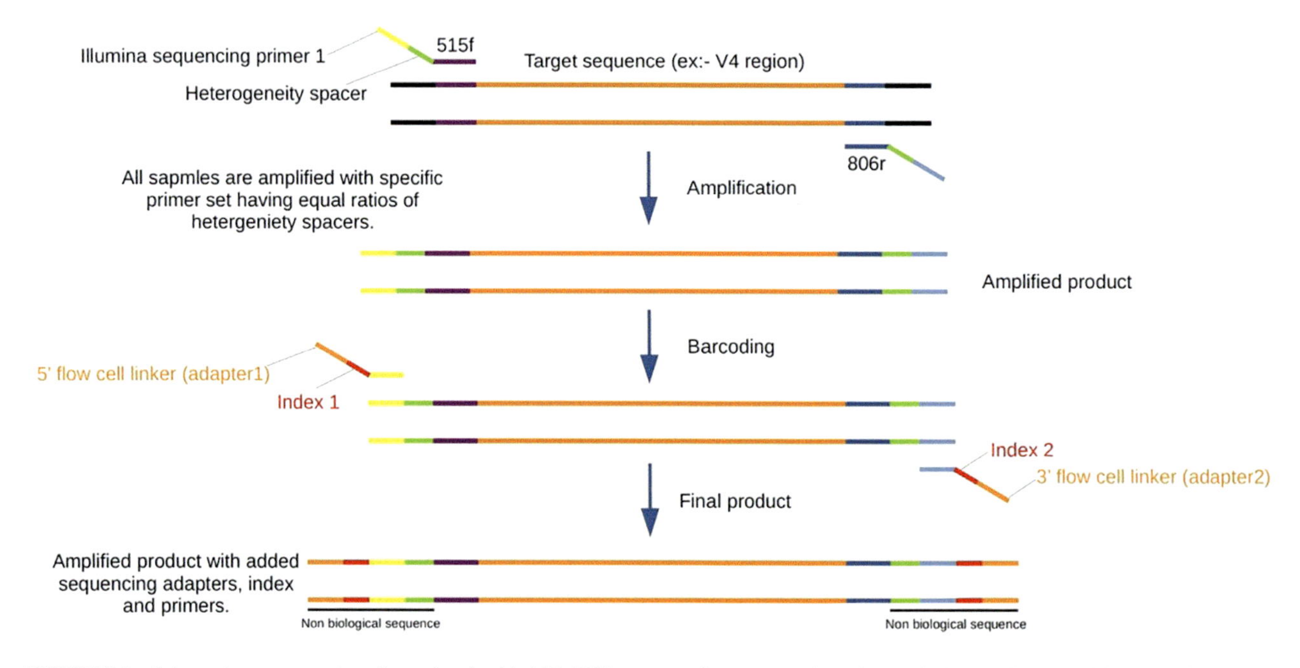

FIGURE 3.1 Schematic representation of steps involved in 16S rRNA gene amplicon sequencing after the isolation of DNA. Initial steps of amplification add heterogeneity spacers and sequencing primers in the amplified product, while the addition of adapter and index sequences is done in barcoding step. These nonbiological sequences are shown by black line in the final product. (Adapted from Holm et al. (2019).)

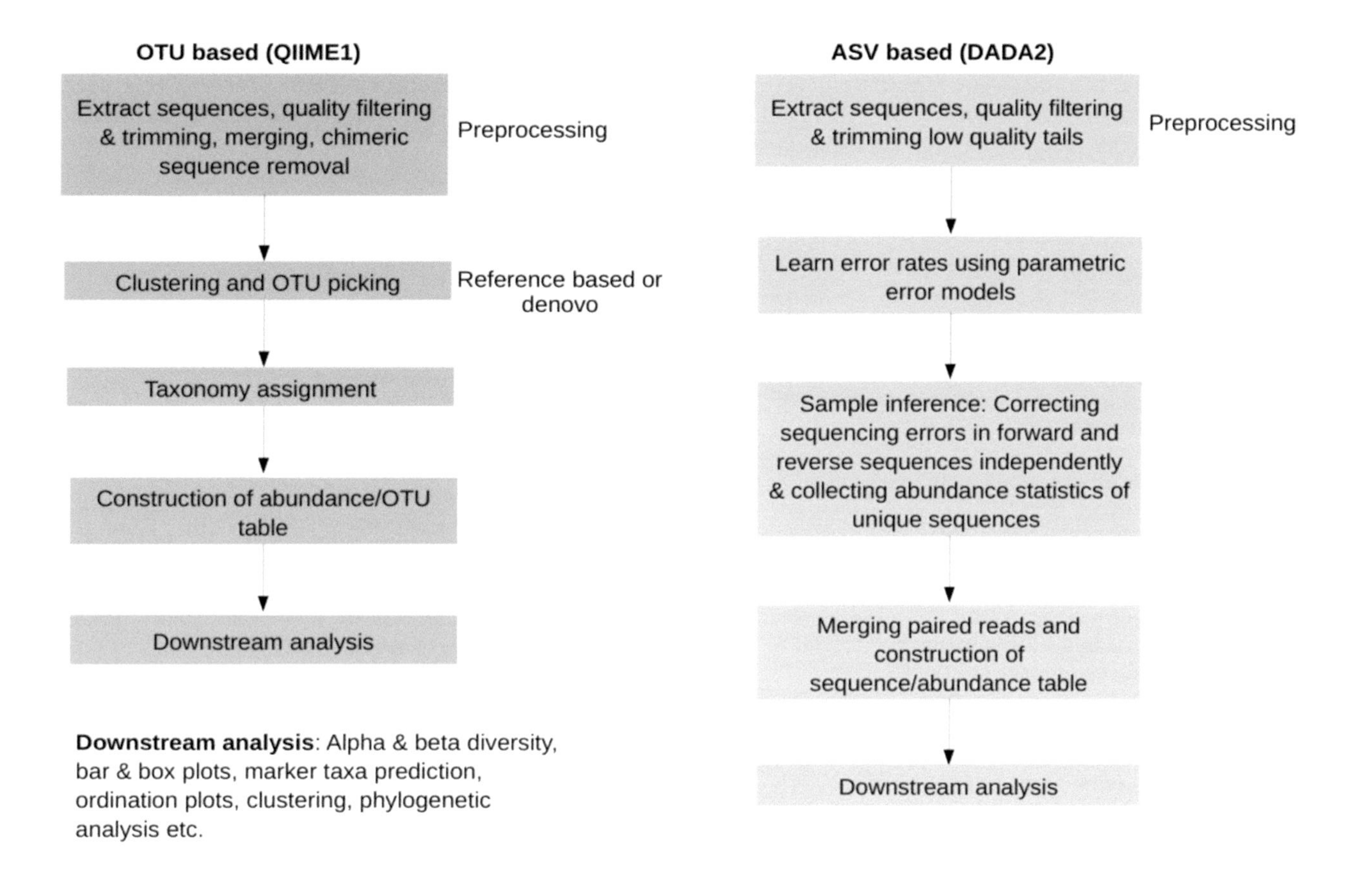

FIGURE 3.2 Overview of OTU- and ASV-based approaches for 16S rRNA amplicon sequence data using QIIME1 and DADA2, respectively.

and Langille 2017), FASTQC for quality analysis, R (v3.6.3), phyloseq (v1.30.0) (McMurdie and Holmes 2013), and other R packages for downstream analysis.

3.2.2 Sequence and Sample Metadata

For tutorial purpose using QIIME1 and DADA2, we will work on the same data used for mothur MiSeq SOP (Kozich et al. 2013) available at http://www.mothur.org/w/images/d/d6/MiSeqSOPData.zip. QIIME1 and DADA2 both can work on data generated from Ion-Torrent, Illumina, or 454 pyrosequencing, which can be paired or single-end sequences. These tools require demultiplexed FASTQ files of each sample; i.e., each sample is split into individual FASTQ file. In some cases, sequences from different samples can be lumped together in single FASTQ file, and each sample is labeled using barcodes (unique sequences for each sample). To show how demultiplexing works using QIIME1, we have used sequences (*illumina/forward_reads.fastq.gz*), barcodes (*illumina/barcodes.fastq.gz*), and mapping file (*illumina/map.tsv*) available at ftp://ftp.microbio.me/qiime/tutorial_files/moving_pictures_tutorial-1.9.0.tgz derived from Moving Pictures of Human Microbiome study (Caporaso et al. 2011). Mapping file is important to tell which barcodes correspond to which samples. The first line (header) of mapping file starts with "#," which has mandatory fields line "*SampleID*," "*BarcodeSequence*," "*LinkerPrimerSequence*," and "*Description*" with some optional metadata about each sample. The "*SampleID*" column must be unique to each sample and will be used to add labels in each sequence while demultiplexing. After generating mapping file, prior to demultiplexing, one should validate it by using *validate_mapping_file.py* command.

```
#SampleID     BarcodeSequence
LinkerPrimerSequence SampleType     Description
L1S8    AGCTGACTAGTC  GTGCCAGCMGCCGCGGTAA  gut
1_Fece_10_28_2008
L1S140 ATGGCAGCTCTA  GTGCCAGCMGCCGCGGTAA  gut
2_Fece_10_28_2008
L1S57   ACACACTATGGC  GTGCCAGCMGCCGCGGTAA  gut
1_Fece_1_20_2009
L1S208 CTGAGATACGCG  GTGCCAGCMGCCGCGGTAA  gut
2_Fece_1_20_2009
L1S76   ACTACGTGTGGT  GTGCCAGCMGCCGCGGTAA  gut
1_Fece_2_17_2009
L1S105 AGTGCGATGCGT  GTGCCAGCMGCCGCGGTAA  gut
1_Fece_3_17_2009
L1S257 CCGACTGAGATG  GTGCCAGCMGCCGCGGTAA  gut
2_Fece_3_17_2009
L1S281 CCTCTCGTGATC  GTGCCAGCMGCCGCGGTAA  gut
2_Fece_4_14_2009
```

QIIME1 provides a script (*split_libraries_fastq.py*) for demultiplexing and quality filtering purpose, which gives filtered sequences from each sample into single FASTA file, which can be proceeded for chimera removal and directly to OTU picking. The following command shows demultiplexing with a minimum quality cutoff 20: *split_libraries_fastq.py -i forward_reads.fastq.gz -m map.tsv -b barcodes.fastq.gz -o slout -q 20*

If you want manual inspection of quality of sequences from each sample, you can use fastq-multx (https://github.com/brwnj/fastq-multx) to write sequences in separate FASTQ files. It requires same mapping file but with mandatory headers "*id*" and "*seq*." Following is a snapshot of mapping file and fastq-multx command for demultiplexing

```
id      seq     LinkerPrimerSequence SampleType
Description
L1S8    AGCTGACTAGTC GTGCCAGCMGCCGCGGTAA    gut
1_Fece_10_28_2008
L1S140 ATGGCAGCTCTA GTGCCAGCMGCCGCGGTAA    gut
2_Fece_10_28_2008
L1S57   ACACACTATGGC GTGCCAGCMGCCGCGGTAA    gut
1_Fece_1_20_2009
L1S208 CTGAGATACGCG GTGCCAGCMGCCGCGGTAA    gut
2_Fece_1_20_2009
L1S76   ACTACGTGTGGT GTGCCAGCMGCCGCGGTAA    gut
1_Fece_2_17_2009
L1S105 AGTGCGATGCGT GTGCCAGCMGCCGCGGTAA    gut
1_Fece_3_17_2009
L1S257 CCGACTGAGATG GTGCCAGCMGCCGCGGTAA    gut
2_Fece_3_17_2009
L1S281 CCTCTCGTGATC GTGCCAGCMGCCGCGGTAA
gut     2_Fece_4_14_2009
# fastq-multx
fastq-multx -B map.tsv barcodes.fastq.gz
forward_reads.fastq.gz -o n/a -o %.fastq
# for paired end data
fastq-multx -B map.tsv barcodes.fastq.gz
forward_reads.fastq.gz reverse_reads.fastq.gz
-o n/a -o %_R1.fastq -o %_R2.fastq
```

3.3 Methods

3.3.1 Merging Paired-End Reads, Quality Analysis, and Filtering

At this point, it can be assumed that you have demultiplexed paired-end reads. Hence, we have used MiSeq SOP paired-end data for subsequent analysis onward. First, we have to merge/stitch paired-end reads. We will run PEAR: a fast and accurate paired-end read merger for stitching paired-end reads (Zhang et al. 2014) using "*run_pear.pl*" script from microbiome helper. This script assumes that paired-end files have the same named with "_R1" in forward and "_R2" in reverse read filename.

```
# change directory to path where fastq
files are located
cd MiSeq_SOP
run_pear.pl -p 4 -o stitched_reads
*.fastq
```

Now we will inspect the quality of each assembled read using FASTQC, which is the simplest tool to visualize the quality of each sequence through a variety of charts. We will run FASTQC on every FASTQ file; we have shown per-base quality of forward and reverse reads of sample "F3D0_S188_L001" and "F3D1_S189_L001" (Figure 3.3). FASTQC color-codes the background of per-base quality plot in three different colors: very good-quality calls (green), reasonable quality (yellow), and poor quality (red). Our both samples show a sudden drop of quality score beyond ~260 base pairs. It suggests us that we should trim bad-quality bases (< Q20) and set the minimum filtering length cutoff to 250 bp. There are tools available to automate the trimming and filtering process like *fastq_quality_filter* from FASTX-Toolkit (http://hannonlab.cshl.edu/fastx_toolkit/), which can trim sequences on the basis of quality cutoff and filter them using length or percent good-quality bases.

Microbiome helper provides a Perl script (read_filter.pl) to trim and filter multiple samples at once. We will give a list of stitched files to be filtered, generated by running PEAR.

```
read_filter.pl -q 20 -p 90 -l 250
--primer_check none --thread 4 stitched_
reads/*.assembled.* -b /path_to_bbmap
```

Here, we have used Q20 quality cutoff with a minimum length of reads to be 250. "–p 90" specifies that 90% of bases from reads should have a minimum quality cutoff (> = Q20) to qualify the read.

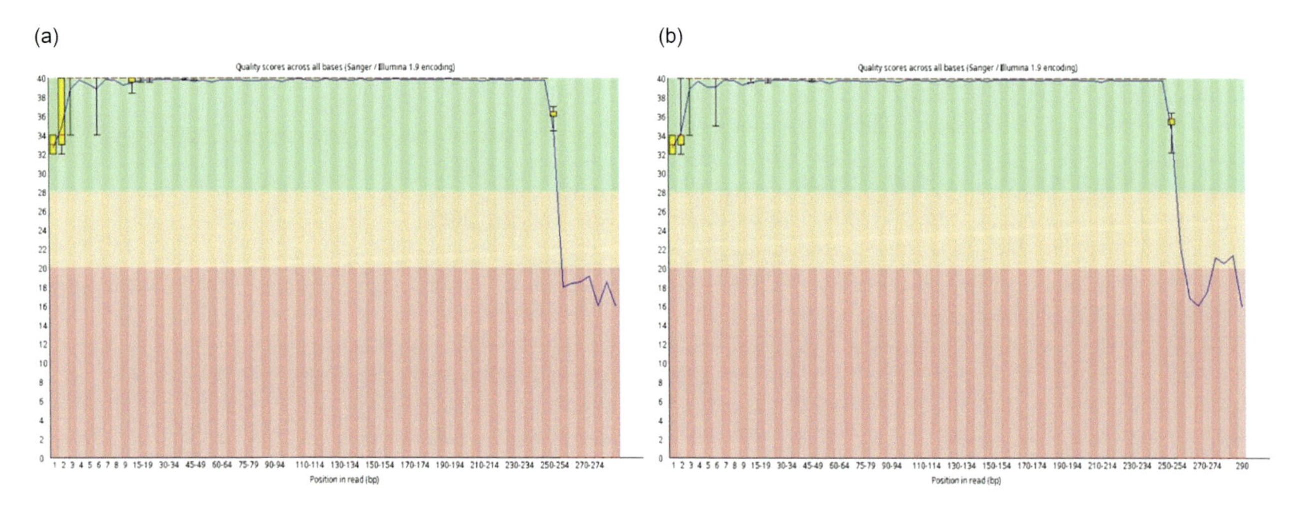

FIGURE 3.3 Box–whisker plot of per-base quality scores for samples "F3D0_S188_L001" (a) and "F3D1_S189_L001" (b). Central red line is the median of phred scores, yellow box represents the interquartile range (25%–75%), and the blue line is the mean phred score at each base pair position.

3.3.2 Chimera Removal

Chimeras are sequences formed from two or more biological sequences due to incomplete extension in PCR (polymerase chain reaction) and are very common in amplicon sequencing.

```
# convert FASTQ files to FASTA
run_fastq_to_fasta.pl -p 4 -o fasta_files
filtered_reads/*fastq
# Remove chimeric sequences from each
samples
chimera_filter.pl --thread 4 -type 1 -db
gg_13_8_otus/rep_set/97_otus.fasta
fasta_files/*
```

These heterogeneous sequences can be falsely considered as novel sequences and hence must be removed from subsequent analysis. We will use "*run_fastq_to_fasta.pl*" to convert FASTQ files to FASTA format and "*chimera_filter.pl*" from microbiome helper package which uses VSEARCH (Rognes et al. 2016) with UCHIME (Edgar et al. 2011) reference-based algorithm to detect and remove chimeric sequences from input FASTA files.

Here, we have used greengenes database (https://greengenes.secondgenome.com/?prefix=downloads/greengenes_database/gg_13_5/) but you can use any reference data provided FASTA file. Now, we have to merge all sequences from different samples into single FASTA files with sampleIDs present in headers of each sequence. "*add_qiime_labels.py*" requires a mapping file to specify sampleIDs to each FASTA file. "*create_qiime_map.pl*" from microbiome helper package can create a mapping file with the first column as sampleIDs and another "FileInput" column containing filenames.

```
create_qiime_map.pl non_chimeras/*.fasta >map.
tsv
# print mapping file (map.tsv)
head map.tsv

#SampleID  BarcodeSequence
LinkerPrimerSequence  FileInput
F3D0    F3D0_S188_L001.assembled_filtered.
nonchimera.fasta
F3D141 F3D141_S207_L001.assembled_filtered.
nonchimera.fasta
F3D142 F3D142_S208_L001.assembled_filtered.
nonchimera.fasta

# create combined fasta file with sequences
from each sample
add_qiime_labels.py -m map.tsv -i non_
chimeras/ -c FileInput -o combined_seqs
```

This will generate a single FASTA file (*combined_seqs.fna*) inside the output directory. This FASTA file format is similar to one generated using "*split_libraries_fastq.py*" during demultiplexing. If you have proceeded with demultiplexing step with this command, then you can use "*seqs.fna*" (which have sample Ids in headers already) file as it is for chimera removal by using the following command:

```
vsearch --uchime_ref seqs.fna --db
gg_13_8_otus/rep_set/97_otus.fasta -
nonchimeras \combined_fasta.fna
```

3.3.3 OTU-Picking Strategies

OTU picking is the process of clustering similar sequences (using percent identity cutoff) in a single unit called as "operational taxonomic unit." For clustering, QIIME1 wraps external tools, including uclust (Edgar 2010), usearch, and SortMeRNA (Kopylova, Noé, and Touzet 2012). QIIME1 provides three different OTU-picking strategies which are demonstrated below. User can choose any of the three OTU-picking strategies depending on the aim of study. We have demonstrated and discussed each of them briefly using the default percent identity cutoff (97%); however, one can find more information at http://qiime.org/tutorials/otu_picking.html.

3.3.3.1 Closed Reference

In closed-reference OTU-picking method, clustering is done against reference dataset. Reads which are not similar to any reference sequence will be excluded from further analysis. This method can be used for large datasets as it can be run in parallel manner; this speeds up clustering process. "*pick_closed_reference_otus.py*" script is used for closed-reference OTU picking as follows:

```
pick_closed_reference_otus.py -i
combined_seqs/combined_seqs.fna -o
closed_ref_otus -s
```

The advantage of using this method is reliable taxonomic assignments as OTUs are already defined in the reference dataset collection. It can only be used for analyzing samples with well studied microbial taxa sequences which are present in reference database. The unknown sequences may fail to align with reference sequence set hence get discarded.

3.3.3.2 De novo

In de novo OTU picking, all sequences are compared with one another without any external reference database and clustered using the given percent identity cutoff. "*pick_de_novo_otus.py*" script is used for de novo OTU picking as follows:

```
pick_de_novo_otus.py -i combined_seqs/
combined_seqs.fna -o denovo_otus -O 4 -a
```

This method is not preferred for large data as it can be time-consuming. It will be used when there is no reference clustering dataset available. The advantage of using this method is that all reads from the dataset will be clustered and used in further analysis.

3.3.3.3 *Open Reference*

Open-reference OTU-picking method is a combination of both reference-based and de novo approaches for clustering. It uses a closed-reference approach for clustering of reads which map to reference dataset and de novo clustering method for reads which do not hit reference datasets. "pick_open_reference_otus.py" script is used for open-reference OTU picking as follows:

```
pick_open_reference_otus.py -i combined_
seqs/combined_seqs.fna -o open_ref_otus
-m uclust
```

This method cannot be used with nonoverlapping amplicons, and the presence of reference dataset is required. With this method, partial parallel processing can be done, which speeds up process as compared to de novo approach.

3.3.4 OTU Table

OTU-picking commands will generate OTU table file named "*otu_table.biom*" or "*otu_table_w_tax.biom*" (in case of open-reference OTU picking). This file contains the abundance values of each OTU from each sample. You can read more about BIOM format at https://biom-format.org/. We will add sample metadata (map.tsv) to biom file generated by closed-reference OTU-picking method with some additional information.

head map.tsv
#SampleID BarcodeSequence LinkerPrimerSequence FileInput Days Time
F3D0 F3D0_S188_L001.assembled_filtered.nonchimera.fasta 0 Early
F3D141 F3D141_S207_L001.assembled_filtered.nonchimera.fasta 141 Late
F3D142 F3D142_S208_L001.assembled_filtered.nonchimera.fasta 142 Late
F3D143 F3D143_S209_L001.assembled_filtered.nonchimera.fasta 143 Late
F3D144 F3D144_S210_L001.assembled_filtered.nonchimera.fasta 144 Late
F3D145 F3D145_S211_L001.assembled_filtered.nonchimera.fasta 145 Late
F3D146 F3D146_S212_L001.assembled_filtered.nonchimera.fasta 146 Late
F3D147 F3D147_S213_L001.assembled_filtered.nonchimera.fasta 147 Late
F3D148 F3D148_S214_L001.assembled_filtered.nonchimera.fasta 148 Late

 # add sample metadata from map.tsv to closed_ref_otus/otu_table.biom
biom add-metadata -i closed_ref_otus/otu_table.biom -m map.tsv -o closed_ref_otus/otu_table.biom
 # summarize the OTU table
biom summarize-table -i closed_ref_otus/otu_table.biom
Num samples: 20
Num observations: 779
Total count: 1,30,832
Table density (fraction of non-zero values): 0.287

Counts/sample summary:
Min: 2,671.000
Max: 17,389.000
Median: 5,066.000
Mean: 6,541.600
Std. dev.: 3,731.746
Sample Metadata Categories: BarcodeSequence; LinkerPrimerSequence; Group; Days; FileInput
Observation Metadata Categories: taxonomy

3.3.5 Converting BIOM Table into Phyloseq Object

Phyloseq is an R package to store, analyze, and generate reproducible interactive graphics of microbiome data (McMurdie and Holmes 2013). A phyloseq class object has different levels of data and functions to access OTU table (*otu_table*), sample metadata (*sample_data*), taxonomic table (*tax_table*), tree (*phy_tree*), and representative sequences (*BioStrings::DNAStringSet*). We will build a phyloseq object using OTU table generated from closed-reference OTU-picking approach and save it in "*ps_qiime1.rds*" file for downstream analysis.

```
# set working directory to store files and
RData
setwd('../qiime1/')
# Check/install and load dada2 package
if(!requireNamespace("phyloseq",quietly = T)){
    if(!requireNamespace("BiocManager",quietly
    = T)){
         install.packages("BiocManager")
    }
    BiocManager::install('phyloseq',version
    ='1.30.0')
}
library(phyloseq)
packageVersion('phyloseq')
## [1] '1.30.0'
# import QIIME1 data
ps <-import_biom(BIOMfilename ='../qiime1/
MiSeq_SOP/closed_ref_otus/otu_table_w_
metadata.biom')
## Warning in strsplit(conditionMessage(e),
"\n"): input string 1 is invalid in
## this locale
# add phylogenetic tree
phy_tree(ps) <-read_tree(treefile ='../qiime1/
MiSeq_SOP/closed_ref_otus/97_otus.tree')

# add OTU sequences in phyloseq object
library(Biostrings)
seqs <-readDNAStringSet(filepath ='../qiime1/
MiSeq_SOP/closed_ref_otus/rep_set/combined_
seqs_rep_set.fasta')
names(seqs) <-gsub(names(seqs),pattern
='\\s.*',replacement ='',perl = T)
ps <-merge_phyloseq(ps,seqs)
# print phyloseq object
ps
```

```
## phyloseq-class experiment-level object
## otu_table() OTU Table:    [ 779 taxa and 20
samples ]
## sample_data() Sample Data:   [ 20 samples by
5 sample variables ]
## tax_table() Taxonomy Table: [ 779 taxa by 7
taxonomic ranks ]
## phy_tree() Phylogenetic Tree: [ 779 tips
and 778 internal nodes ]
## refseq()   DNAStringSet:   [ 779 reference
sequences ]

# save phyloseq object
saveRDS(ps,file='ps_qiime1.rds')
```

3.4 DADA2 Tutorial

By clustering filtered sequences into OTUs with some percent identity threshold, QIIME1 ignores the sequencing errors by allowing dissimilarity within the sequences in an OTU. On the other hand, DADA2 works on philosophy that a single nucleotide variation within the sequences can provide information about fine scales of temporal variations or ecological niche (Callahan et al. 2016). DADA2 implements Divisive Amplicon Denoising Algorithm (DADA), which uses parametric models to learn the error rates in sequencing data and systematically corrects the sequencing errors forming partitions of unique sequences which can be called as "ASVs" (a substitute to OTU). DADA2 pipeline tutorial is available at https://benjjneb.github.io/dada2/tutorial.html. The following tutorial will use same data and give more insights to each and every step. DADA2 is built in R; hence, some basic knowledge of R is must in order to understand the following tutorial.

3.4.1 Preparation of Data

```
# Check/install and load dada2 package
if(!requireNamespace("dada2",quietly=T)){
    if(!requireNamespace("BiocManager",quietl
    y=T)){
        install.packages("BiocManager")
    }
    BiocManager::install('dada2',vers
    ion='1.14.1')
}
library(dada2)
## Loading required package: Rcpp
packageVersion('dada2')
## [1] '1.14.1'
# Set working directory. All files generated
will be stored at following path
setwd('../data/dada2/')

# Store path of sequence data. If forward and
reverse reads are at different locations
please use pathF and pathR variables at
subsequent steps
path <-'../data/dada2/MiSeq_SOP/'

# MiSeq SOP data filenames have format
SAMPLENAME_R1_001.fastq and SAMPLENAME_R2_001.
fastq
# Here we will make two vectors for storing
list of absolute paths of forward and reverse
files.
forward.files <-sort(list.files(path,
pattern="_R1_001.fastq", full.names =TRUE))

reverse.files <-sort(list.files(path,
pattern="_R2_001.fastq", full.names =TRUE))

# We can make a vector of sample names by
extracting from filenames, assuming filenames
have format: SAMPLENAME_XXX.fastq
sample.names <-sapply(strsplit(basename(forwa
rd.files), "_"), `[`, 1)
```

3.4.2 Sequence Quality Check

Every sequencing run can have unique quality profiles. Assuming that all samples in MiSeq SOP data are sequenced in single run, we can get snapshot of quality profiles of all samples by inspecting one or two samples only. Figure 3.4 shows that quality profiles for forward and reverse reads start dropping by 240 and 160 bp, respectively.

```
plotQualityProfile(forward.files[1:2])
plotQualityProfile(reverse.files[1:2])
```

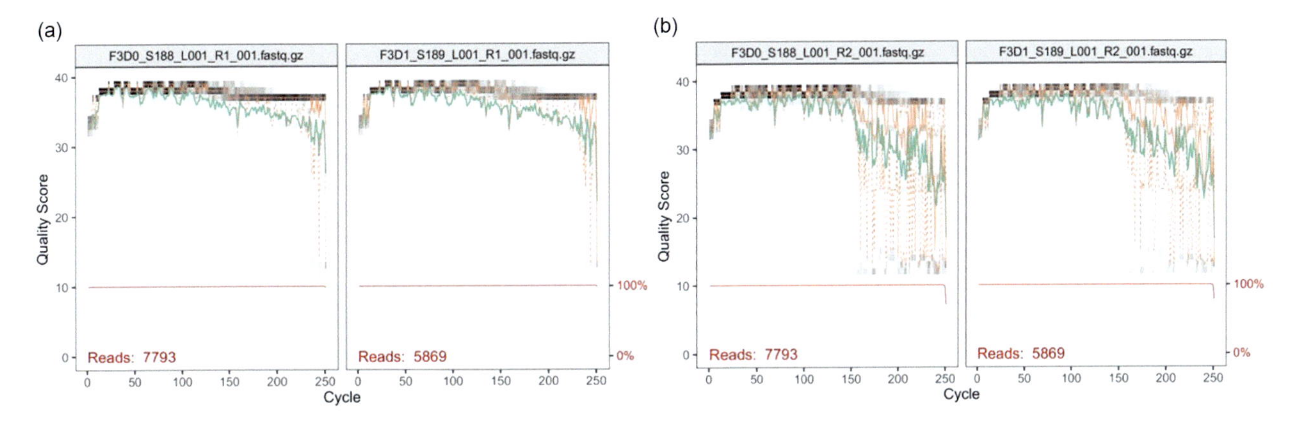

FIGURE 3.4 Quality profiles for forward (a) and reverse (b) reads of samples *F3D0_S188* and *F3D0_S188*. Positions in bp and phred scores are represented on x-axis and y-axis, respectively.

3.4.3 Filtering and Trimming

"*filterAndTrim*" function from DADA2 provides different ways to trim and filter reads. The following commands will truncate forward and reverse reads after 240 and 160 bp or at first instance where quality score is less than truncQ. This function will discard reads which have "*N*" character or matches with phiX genome. As we are using 2X250 V4 sequencing data and length of V4 region of 16S rRNA gene is approximately 254 bp, the forward and reverse reads are almost overlapping. Hence, we need not worry too much about trimming criteria. But with less overlapping regions like V2–V4 or V3–V5, one must consider the length of overlapping region to merge these reads later.

```
# Create folders to store the filtered reads
filt.forward <-file.path(path, "filteredF",
paste0(sample.names, "_F_filt.fastq.gz"))
filt.reverse <-file.path(path, "filteredR",
paste0(sample.names, "_R_filt.fastq.gz"))
names(filt.forward) <-sample.names
names(filt.reverse) <-sample.names

# If your reads contain any non-biological
sequence (primers), add trimLeft=c(15,15).
This will trim 15 bp from 5' & 3' end of
forward and reverse reads.
out <-filterAndTrim(forward.files, filt.
forward, reverse.files, filt.reverse,
truncLen=c(240,160), maxN=0, maxEE=c(2,2),
truncQ=20, rm.phix=TRUE, compress=TRUE,
multithread=TRUE)

head(out)
##          reads.in reads.out
## F3D0_S188_L001_R1_001.fastq.gz  7793  7113
## F3D1_S189_L001_R1_001.fastq.gz  5869  5299
## F3D141_S207_L001_R1_001.fastq.gz  5958
5463
## F3D142_S208_L001_R1_001.fastq.gz  3183
2914
## F3D143_S209_L001_R1_001.fastq.gz  3178  2941
## F3D144_S210_L001_R1_001.fastq.gz  4827
4312
```

3.4.4 Learning Error Rates

As every amplicon dataset/sequencing run has different error rates, the parametric error model has to build separately on each dataset. DADA2 uses subset of reads from input data to build a model, which quantifies the rate at which sequence(i) is produced from sequence(j) as a function of composition and quality values of sequences. This model keeps estimating the errors until these estimates converge with the observed error rates.

```
errF <-learnErrors(filt.forward,
multithread=TRUE)
## 33514080 total bases in 139642 reads from
20 samples will be used for learning the error
rates.
errR <-learnErrors(filt.reverse,
multithread=TRUE)
## 22342720 total bases in 139642 reads from
20 samples will be used for learning the error
rates.
# plot error model (here forward plot is shown
only)
plotErrors(errF, nominalQ=TRUE)
```

Figure 3.5 shows the estimated and observed error for forward reads. The black line represents the estimated error rates, which converge with the observed error rate (black points) for all possible 16 transitions. With an increase in quality scores, the error frequency also drops, which shows that the error model is good fit, and we can proceed with confidence.

3.4.5 Sample Inference

The "*dada*" functions apply the core "Divisive Amplicon Denoising Algorithm" on forward and reverse reads separately using the learned error model. This step does depreplication

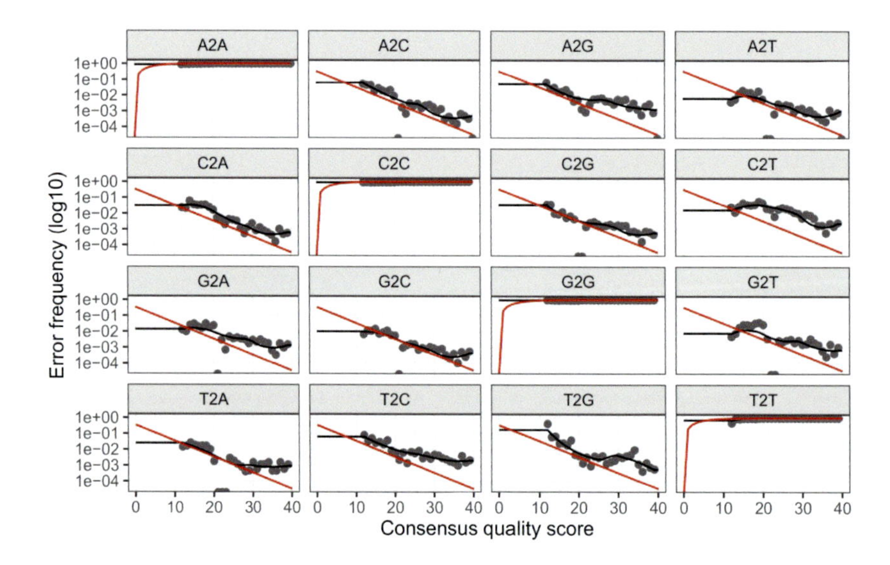

FIGURE 3.5 Error model for forward reads. X-axis and y-axis represent the quality scores and respective error frequency, respectively. The red line shows the nominal error rates which can be expected.

(picking unique sequences), and then run divisive partitioning algorithm which starts by placing all unique sequences in single partition, and depending on the calculated abundance p-value (if<user cutoff), a new partition is formed with new unique sequences. Basically, this step tries to identify the variants in sequences and classify them in different partitions after corrections. This step is carried out in an iterative manner for each sample until no new partition of sequences is possible.

```
dadaFs <-dada(filt.forward, err=errF,
multithread=TRUE)
dadaRs <-dada(filt.reverse, err=errR,
multithread=TRUE)
dadaFs[[1]]
## dada-class: object describing DADA2
denoising results
## 128 sequence variants were inferred from
1979 input unique sequences.
## Key parameters: OMEGA_A = 1e-40, OMEGA_C =
1e-40, BAND_SIZE = 16
```

3.4.6 Merging Paired-End Reads

"*mergePairs*" aligns forward and reverse sequences using global ends-free alignment and merges them if two sequences overlap perfectly. The number of mismatch and overlap length can be adjusted using "*maxMismatch*" and "*minOverlap*" arguments, which have default values of 0 and 20. Again, as this was 2X250 V4 sequencing data, the overlap length can be high as 148 (see "*nmatch*" column in the following output).

```
mergers <-mergePairs(dadaFs, filt.forward,
dadaRs, filt.reverse, verbose=TRUE)
# Inspect the merger data.frame from the first
sample
head(mergers[[1]], n=4)

## sequence
## 1 TACGGAGGATGCGAGCGTTATCCGGATTTATTGGGTTTA-
AAGGGTGCGCAGGCGGAAGATCAAGTCAGCGGTAAAATTGAGAG-
GCTCAACCTCTTCGAGCCGTTGAAACTGGTTTTCTTGAGTGAGC-
GAGAAGTATGCGGAATGCGTGGTGTAGCGGTGAAATGCATAG-
ATATCACGCAGAACTCCGATTGCGAAGGCAGCATACCGGCGCT-
CAACTGACGCTCATGCACGAAAGTGTGGGTATCGAACAGG
##2
TACGGAGGATGCGAGCGTTATCCGGATTTATTGGGTTTAAAGGGT-
GCGTAGGCGG
GGCGGCCTGCCAAGTCAGCGGTAAAATTGCGGGGCTCAACCCC-
GTACAGCCGTTGAAACTGCCGGGCTCGAGTGGGCGAGAAGTAT-
GCGGAATGCGTGGTGTAGCGGTGAAATGCATAGATATCACG-
CAGAACCCCGATTGCGAAGGCAGCATACCGGCGCCCTACTGAC-
GCTGAGGCACGAAAGTGCGGGGATCAAACAGG
## 3 TACGGAGGATGCGAGCGTTATCCGGATTTATTGGGTT-
TAAAGGGTGCGTAGGCGGGCTGTTAAGTCAGCGGTCAAAT-
GTCGGGGCTCAACCCCGGCCTGCCGTTGAAACTGGCGGCCTC-
GAGTGGGCGAGAAGTATGCGGAATGCGTGGTGTAGCGGTGAAATG-
CATAGATATCACGCAGAACTCCGATTGCGAAGGCAGCATACCGGC-
GCCCGACTGACGCTGAGGCACGAAAGCGTGGGTATCGAACAGG
## 4 TACGGAGGATGCGAGCGTTATCCGGATTTATTGGGTT-
TAAAGGGTGCGTAGGCGGGCTTTTAAGTCAGCGGTAAAAATTC-
GGGGCTCAACCCCGTCCGGCCGTTGAAACTGGGGGCCTT-
GAGTGGGCGAGAAGAAGGCGGAATGCGTGGTGTAGCGGTGAAATG-
CATAGATATCACGCAGAACCCCGATTGCGAAGGCAGCCTTCCGGC-
GCCCTACTGACGCTGAGGCACGAAAGTGCGGGGATCGAACAGG
## abundance forward reverse nmatch nmismatch
nindel prefer accept
## 1  579  1  1 148   0  0  1 TRUE
## 2  470  2  2 148   0  0  2 TRUE
## 3  449  3  4 148   0  0  1 TRUE
## 4  430  4  3 148   0  0  2 TRUE
```

3.4.7 Construct Sequence Table and Remove Chimeras

The sequence table is just the abundance matrix with rows as samples and columns as ASV sequences. Here, we got 293 ASV sequences in a total of 20 samples. DADA2 uses de novo method to identify chimeras by performing Needleman and Wunsch global alignment of each sequence with all more abundant sequences to find the combination of left and right parent of child sequences without any mismatch.

```
seqtab <-makeSequenceTable(mergers)
dim(seqtab)
## [1] 20 293

# Identify and remove chimeric sequences
seqtab.nochim <-removeBimeraDenovo(seqtab,
method="consensus", multithread=TRUE,
verbose=TRUE)
## Identified 61 bimeras out of 293 input
sequences.
dim(seqtab.nochim)
## [1] 20 232
sum(seqtab.nochim)/sum(seqtab)
## [1] 0.964064
# Hence, 96.40% of our merged sequences are
non-chimeric
```

3.4.8 Assign Taxonomy

DADA2 implements Ribosomal Database Project (RDP) naive Bayesian classifier, which can rapidly and accurately classify 16S rRNA gene sequences into higher-order taxonomy (Wang et al. 2007). Here, we have used SILVA (v138) database (Pruesse et al. 2007) but you can use any DADA2 formatted database available at https://benjjneb.github.io/dada2/training.html. DADA2 also allows assigning species to ASV sequences which exactly match with reference database. "*addSpecies*" function adds species column to taxonomy table.

```
taxa <-assignTaxonomy(seqtab.nochim, "silva_
nr_v138_train_set.fa.gz", multithread=TRUE)
taxa <-addSpecies(taxa, "silva_species_
assignment_v138.fa.gz")
```

3.4.9 Constructing Phyloseq Object

Now, we will construct phyloseq object using variable seqtab, taxa, and sample metadata file (*map.tsv*). We have constructed neighbor-joining tree using phangorn R package (Schliep 2011) and then fit a GTR+G+I (generalized time-reversible with gamma rate variation) maximum-likelihood tree.

```
if(!requireNamespace("phyloseq",quietly = T)){
    if(!requireNamespace("BiocManager",quietly
    = T)){
          install.packages("BiocManager")
    }
    BiocManager::install('phyloseq',version
    ='1.30.0')
}
library(phyloseq)
packageVersion('phyloseq')
## [1] '1.30.0'
# read sample metadata
samdf <-read.table(file ='map.tsv',header =
T,sep ='\t',row.names =1)
ps <-phyloseq(otu_table(seqtab.nochim,
taxa_are_rows=FALSE),
sample_data(samdf),
tax_table(taxa))
# Remove mock sample
ps <-prune_samples(sample_names(ps) !="Mock",
ps)

# Add ASV sequences in refseq slot
dna <-Biostrings::DNAStringSet(taxa_names(ps))
names(dna) <-taxa_names(ps)
ps <-merge_phyloseq(ps, dna)
ps
## phyloseq-class experiment-level object
## otu_table() OTU Table:   [ 232 taxa and 19
samples ]
## sample_data() Sample Data:   [ 19 samples by
2 sample variables ]
## tax_table() Taxonomy Table: [ 232 taxa by 7
taxonomic ranks ]
## refseq()  DNAStringSet:   [ 232 reference
sequences ]
# Create phylogenetic tree by aligning ASV
sequences
library(DECIPHER)
library(phangorn)
seqs <-getSequences(seqtab)
names(seqs) <-seqs

# perform alignment
alignment <-AlignSeqs(DNAStringSet(seqs),
anchor=NA)
phang.align <-phyDat(as(alignment, "matrix"),
type="DNA")
dm <-dist.ml(phang.align)
treeNJ <-NJ(dm) # Note, tip order != sequence
order
fit =pml(treeNJ, data=phang.align)
## negative edges length changed to 0!

# compute likelihood
fitGTR <-update(fit, k=4, inv=0.2)
fitGTR <-optim.pml(fitGTR, model="GTR",
optInv=TRUE, optGamma=TRUE,
rearrangement ="stochastic", control =pml.
control(trace =0))
detach("package:phangorn", unload=TRUE)

# add tree to phyloseq object
phy_tree(ps) <-fitGTR$tree
ps
## phyloseq-class experiment-level object
## otu_table() OTU Table:    [ 232 taxa and 19
samples ]
## sample_data() Sample Data:   [ 19 samples by
2 sample variables ]
## tax_table() Taxonomy Table: [ 232 taxa by 7
taxonomic ranks ]
## phy_tree() Phylogenetic Tree: [ 232 tips
and 230 internal nodes ]
## refseq()  DNAStringSet:   [ 232 reference
sequences ]
# change ASV names and save phyloseq object
taxa_names(ps) <-paste0("ASV", seq(ntaxa(ps)))
saveRDS(ps,file ='ps_dada2.rds')
```

3.5 Downstream Analysis

3.5.1 Alpha Diversity

Determining alpha diversity indices is the first step in microbiome data analysis to access the differences between samples with respect to richness and diversity. Here, we have plotted observed (number of unique OTUs/ASVs), Chao1 (richness), and Shannon (diversity) index for samples from QIIME1 and DADA2 approaches (Figure 3.6).

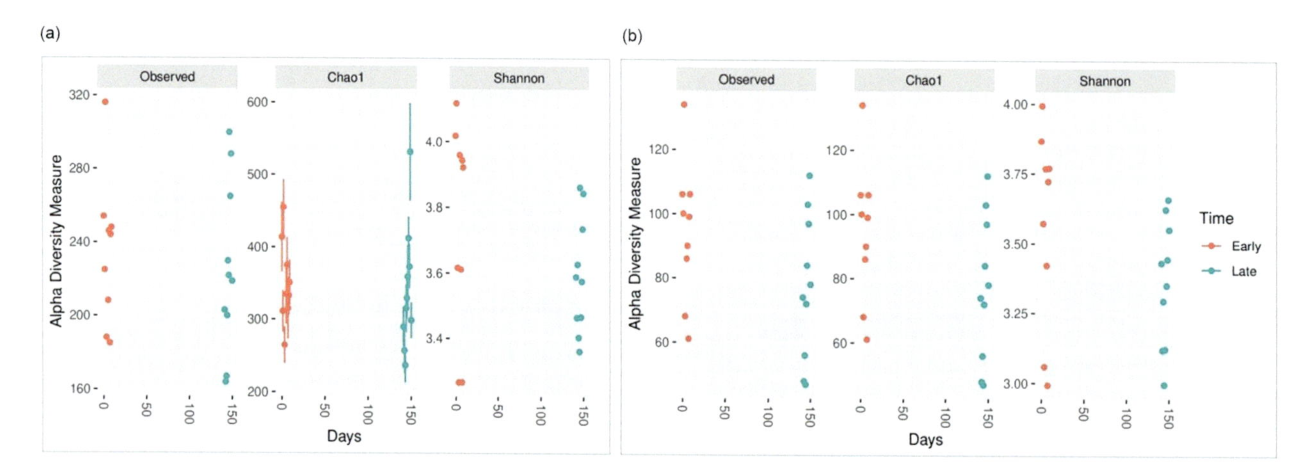

FIGURE 3.6 Alpha diversity indices using QIIME1 (a) and DADA2 (b) phyloseq objects. Chao1 alpha diversity index is plotted with standard errors.

```
setwd('../data/analysis/')
library(phyloseq)
# read saved phyloseq object from DADA2
ps.dada2 <-readRDS(file ='../dada2/ps_dada2.
rds')
ps.dada2
## phyloseq-class experiment-level object
## otu_table() OTU Table:   [ 232 taxa and 19
samples ]
## sample_data() Sample Data:   [ 19 samples by
2 sample variables ]
## tax_table() Taxonomy Table: [ 232 taxa by 7
taxonomic ranks ]
## phy_tree() Phylogenetic Tree: [ 232 tips
and 230 internal nodes ]
## refseq()  DNAStringSet:   [ 232 reference
sequences ]

# read saved phyloseq object from QIIME1
ps.qiime1 <-readRDS(file ='../qiime1/ps_
qiime1.rds')
ps.qiime1
## phyloseq-class experiment-level object
## otu_table() OTU Table:   [ 779 taxa and 19
samples ]
## sample_data() Sample Data:   [ 19 samples by
5 sample variables ]
## tax_table() Taxonomy Table: [ 779 taxa by 7
taxonomic ranks ]
## phy_tree() Phylogenetic Tree: [ 779 tips
and 778 internal nodes ]
## refseq()  DNAStringSet:   [ 779 reference
sequences ]

# Plot alpha diversity indices
p1 <-plot_richness(physeq = ps.dada2,x
='Days',color ='Time',measures =c('Observed','
Chao1','Shannon'))
print(p1)
p2 <-plot_richness(physeq = ps.qiime1,x
='Days',color ='Time',measures =c('Observed','
Chao1','Shannon'))
print(p2)
```

3.5.2 Beta Diversity

Beta diversity measures the differences between samples from the same or different group. Briefly, beta diversity measures can be quantitative (e.g., Bray–Curtis), qualitative based on presence/absence data (e.g., Jaccard index), or phylogeny-based (e.g., Unifrac) (Goodrich et al. 2014). Here, we have used Bray–Curtis metric, and ordination was plotted using nonmetric multidimensional scaling (NMDS). However, phyloseq "*ordinate*" function provides different distance metrics with different ordination methods. To account for differences in library size between samples, we have to transform data prior to calculation distances. Here, we have transformed the abundance values to relative proportions using "*transform_sample_counts*" functions, which can take any defined function to transform data (Figure 3.7).

```
# Transform data to relative proportions
ps.dada2.prop <-transform_sample_counts(physeq
= ps.dada2,function(x) x/sum(x))
ps.qiime1.prop <-transform_sample_counts(ps.
qiime1,function(x) x/sum(x))

# Ordination using bray-curtis distance metric
ps.dada2.ord <-ordinate(physeq = ps.dada2.
prop,method ='NMDS',distance ='bray')
ps.qiime1.ord <-ordinate(physeq = ps.qiime1.
prop,method ='NMDS',distance ='bray')
# ordination plots
p1 <-plot_ordination(physeq = ps.dada2.
prop,ordination = ps.dada2.ord,color ='Time')
print(p1)
p2 <-plot_ordination(physeq = ps.qiime1.
prop,ordination = ps.qiime1.ord,color ='Time')
print(p2)
```

3.5.3 Bar Charts

Bar charts are useful to investigate the abundance of different taxa or OTUs/ASVs in different samples. Here, we have selected and plotted the abundance constituting top-20 OTUs/

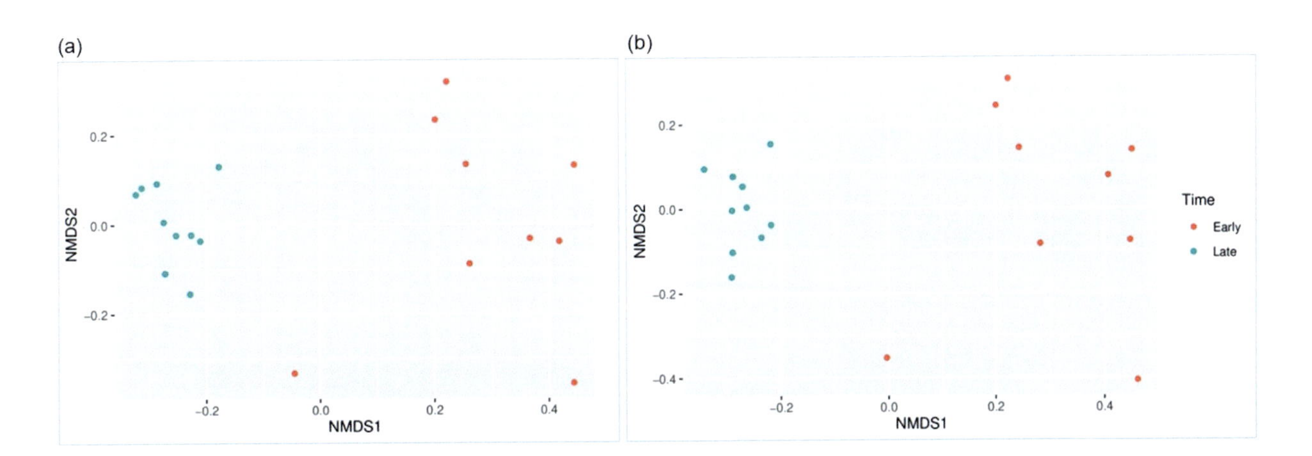

FIGURE 3.7 Beta diversity plots using QIIME1 (a) and DADA2 (b) phyloseq objects. Each point represents a sample from early (red) or late (blue) category.

ASVs from QIIME1 and DADA2 phyloseq objects which belong to different families (Figure 3.8).

```
# Change type of data and levels for Days
column in sample metadata# This step is just
to change the order of Days in sample
metadata. If you don't want to change the
order, you can skip this step
sample_data(ps.dada2.prop)$Days <- factor(as.
character(sample_data(ps.dada2.
prop)$Days),levels =c("0","1","2","3","5","6",
"7","8","9","141","142","143","144","145","146
","147","148","149","150"))
sample_data(ps.qiime1.prop)$Days <- factor(as.
character(sample_data(ps.qiime1.
prop)$Days),levels = c("0","1","2","3","5","6"
,"7","8","9","141","142","143","144","145","14
6","147","148","149","150"))

# plotting abundance of top20 OTUs/ASVs
# DADA2
library(ggplot2)top20 <- names(sort(taxa_
sums(ps.dada2.prop),decreasing = T))[1:20]
ps.dada2.top20 <- prune_taxa(top20,ps.dada2.
prop)p1 <- plot_bar(physeq = ps.dada2.top20,x
= 'Days',fill = 'Family') + facet_
wrap(~Time,scales = 'free_x') + geom_
bar(stat='identity') +
scale_fill_brewer(palette = 'Paired')print(p1)
# QIIME1
top20 <- names(sort(taxa_sums(ps.qiime1.
prop),decreasing = T))[1:20]ps.qiime1.top20
<- prune_taxa(top20,ps.qiime1.prop)
#sample_data(ps.qiime1.top20)$Days <-
as.character(sample_data(ps.qiime1.
top20)$Days)
p2 <- plot_bar(physeq = ps.qiime1.top20,x =
'Days',fill = 'Rank5') + facet_wrap(~Time,
scales="free_x") + geom_bar(stat = 'identity')
+ scale_fill_brewer(palette = 'Paired')
print(p2)
```

3.5.3.1 *Tax_glom*

It is sometimes required to do analysis at taxa level rather than at OTU/ASV level. Here, we have consolidated the abundance table at "*family*" level (sum up abundance of OTUs/ASVs, which belong to the same family) (Figure 3.9).

```
# Here we will consolidate the abundances of
each OTUs/ASVs at Family level and then plot
top10 Families.
# DADA2
ps.dada2.prop.glom <-tax_glom(ps.dada2.
prop,taxrank ='Family')
top10 <-names(sort(taxa_sums(ps.dada2.prop.
glom),decreasing = T))[1:10]
ps.dada2.top10 <-prune_taxa(top10,ps.dada2.
prop.glom)
p1 <-plot_bar(physeq = ps.dada2.top10,x
='Days',fill ='Family') +facet_
wrap(~Time,scales ='free_x') +geom_
bar(stat='identity')
+scale_fill_brewer(palette ='Paired')
print(p1)

# QIIME1
ps.qiime1.prop.glom <-tax_glom(ps.qiime1.
prop,taxrank ='Rank5')
top10 <-names(sort(taxa_sums(ps.qiime1.prop.
glom),decreasing = T))[1:10]
ps.qiime.top10 <-prune_taxa(top10,ps.qiime1.
prop.glom)
p2 <-plot_bar(physeq = ps.qiime.top10,x
='Days',fill ='Rank5') +facet_
wrap(~Time,scales ='free_x') +geom_
bar(stat='identity')
+scale_fill_brewer(palette ='Paired')
print(p2)
```

3.5.4 Differential Abundance Testing

Testing the abundance of different OTUs/ASVs for differential abundance in different groups is one of the expected outcomes from clinical microbiome studies. A lot of tools like DESeq (Love, Huber, and Anders 2014) and metagenomeSeq (Goodrich et al. 2014) have been developed for this purpose. We have used ALDEx2 (Fernandes et al. 2013) R package, which takes sample variations within the same group into consideration while testing for differential abundance (Figure 3.10).

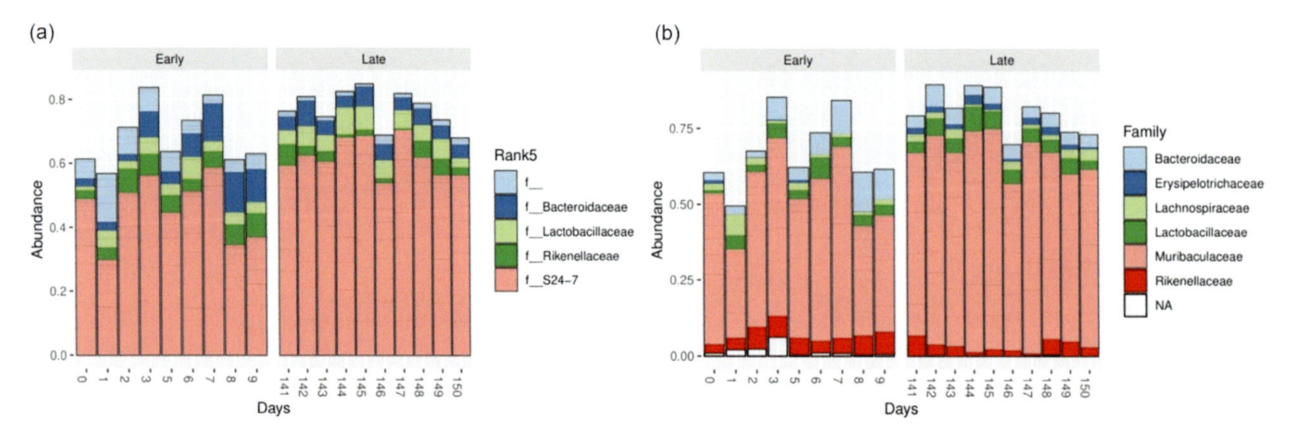

FIGURE 3.8 Bar charts showing abundance profiles at the family level, constituting top-20 OTUs (a) and ASVs (b). Abundance of OTUs/ASVs belonging to the same family is filled with the same color.

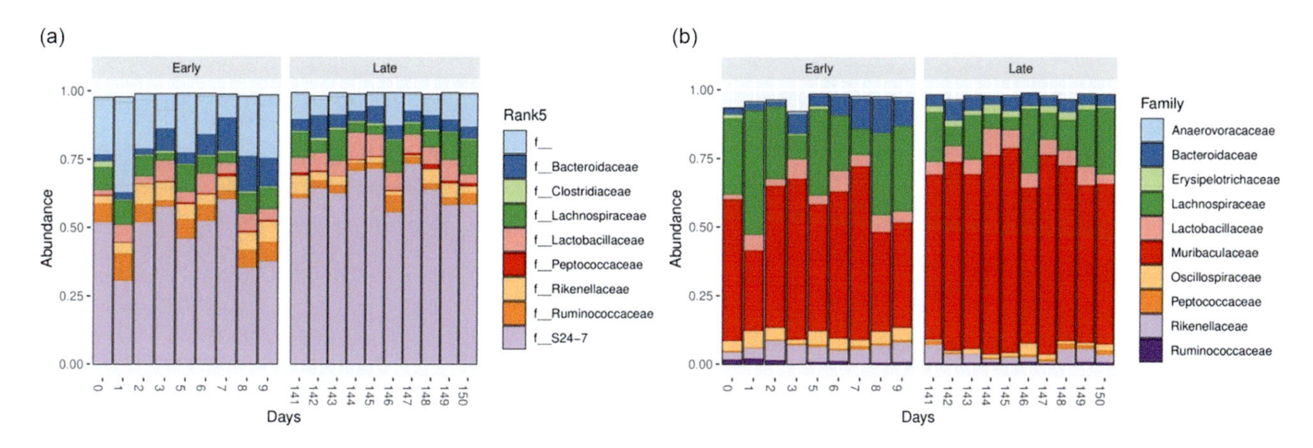

FIGURE 3.9 Distribution of relative abundance of different taxa (family) in QIIME1 (a)- and DADA2 (b)-derived phyloseq objects.

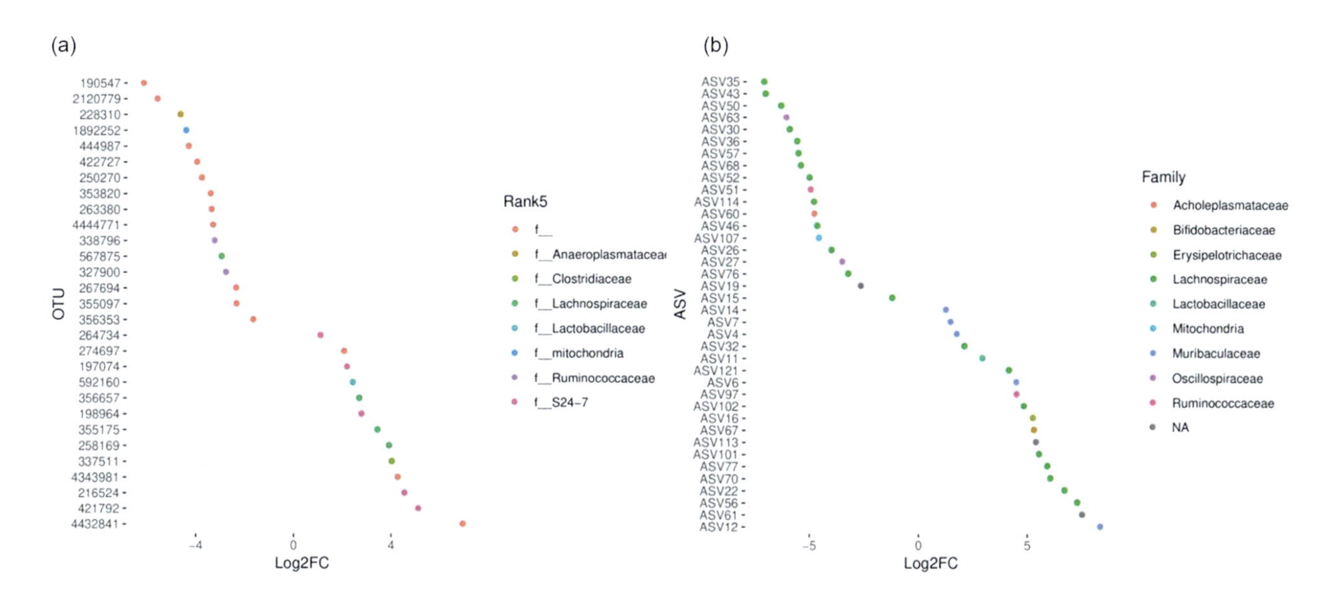

FIGURE 3.10 Differential abundance of different OTUs (a) and ASVs (b) within "*Time*" group. Points are colored using the family-level taxonomy.

```
library(ALDEx2)
# DADA2
otu.tab <-t(otu_table(ps.dada2))
conds <-sample_data(ps.dada2)$Time
# Run aldex function
x.all <-aldex(reads = otu.tab,conditions =
conds)
## aldex.clr: generating Monte-Carlo instances
and clr values
## operating in serial mode
## computing center with all features
## aldex.ttest: doing t-test
## aldex.effect: calculating effect sizes
x.all <-x.all[x.all$wi.eBH<=0.05,]
# filter significant ASVs
# add taxonomy to result table
x.all <-merge(x.all,tax_table(ps.dada2),by=0)
x.all <-x.all[order(x.all$diff.btw,decreasing
= T),] # sort table by log2FC
x.all$Row.names <-factor(x = x.all$Row.
names,levels = x.all$Row.names)
#plot
p1 <-ggplot(data = x.all,mapping =aes(x =
diff.btw,y = Row.names, color=Family)) +geom_
point() +xlab('Log2FC') +ylab('ASV')
print(p1)

# QIIME1
otu.tab <-otu_table(ps.qiime1)
conds <-sample_data(ps.qiime1)$Time
# Run aldex function
x.all <-aldex(reads = otu.tab,conditions =
conds)
## aldex.clr: generating Monte-Carlo instances
and clr values
## operating in serial mode
## computing center with all features
## aldex.ttest: doing t-test
## aldex.effect: calculating effect sizes
x.all <-x.all[x.all$wi.eBH<=0.05,]
# filter significant ASVs
# add taxonomy to result table
x.all <-merge(x.all,tax_table(ps.
qiime1),by=0)
```

```
x.all <-x.all[order(x.all$diff.btw,decreasing
= T),] # sort table by log2FC
x.all$Row.names <-factor(x = x.all$Row.
names,levels = x.all$Row.names)
#plot
p2 <-ggplot(data = x.all,mapping =aes(x =
diff.btw,y = Row.names, color=Rank5)) +geom_
point() +xlab('Log2FC') +ylab('OTU')
print(p2)
```

3.6 Conclusions

The above tutorial gives insights into some basic steps and strategies to analyze microbiome data. However, there are plenty of tools to analyze the microbiome data with ease and with different aspects. For the beginners, we will recommend using MicrobiomeAnalyst, an interactive web server for statistical and visual analyses of microbiome data (Dhariwal et al. 2017) using abundance table.

Exact sequence variants (or ASVs) can detect single nucleotide variation in data and should replace OTUs concept (Callahan, McMurdie, and Holmes 2017). There are several studies which reveal that OTU- and ASV-based approaches show quantitative differences in abundance of microbiome but the alpha and beta diversity indices are highly correlated using these two approaches (Glassman and Martiny 2018). ASV-based approaches are also shown to perform better in case of complex diversity and presence of contamination (Caruso et al. 2019). The choice of 16S rRNA database and taxonomic assignment methods is also known to impact the abundance of microbiome in downstream analysis (Almeida et al. 2018). We will strictly recommend the user to choose analysis approach, 16S rRNA database and taxonomic assignment method depending on your data and ecosystem.

REFERENCES

Almeida, A., Mitchell, A. L., Tarkowska, A., & Finn, R. D. (2018). Benchmarking taxonomic assignments based on 16S rRNA gene profiling of the microbiota from commonly sampled environments. *GigaScience*, 7(5), giy054.

Bodkhe, R., Shetty, S. A., Dhotre, D. P., Verma, A. K., Bhatia, K., Mishra, A., … & Perumal, R. C. (2019). Comparison of small gut and whole gut microbiota of first-degree relatives with adult celiac disease patients and controls. *Frontiers in Microbiology*, 10, 164.

Callahan, B. J., McMurdie, P. J., & Holmes, S. P. (2017). Exact sequence variants should replace operational taxonomic units in marker-gene data analysis. *The ISME Journal*, 11(12), 2639.

Callahan, B. J., McMurdie, P. J., Rosen, M. J., Han, A. W., Johnson, A. J. A., & Holmes, S. P. (2016). DADA2: high-resolution sample inference from Illumina amplicon data. *Nature Methods*, 13(7), 581.

Caporaso, J. G., Kuczynski, J., Stombaugh, J., Bittinger, K., Bushman, F. D., Costello, E. K., … & Huttley, G. A. (2010). QIIME allows analysis of high-throughput community sequencing data. *Nature Methods*, 7(5), 335.

Caporaso, J. G., Lauber, C. L., Costello, E. K., Berg-Lyons, D., Gonzalez, A., Stombaugh, J., … & Gordon, J. I. (2011). Moving pictures of the human microbiome. *Genome Biology*, 12(5), R50.

Caruso, V., Song, X., Asquith, M., & Karstens, L. (2019). Performance of microbiome sequence inference methods in environments with varying biomass. *MSystems*, 4(1), e00163–18.

Comeau, A. M., Douglas, G. M., & Langille, M. G. (2017). Microbiome helper: a custom and streamlined workflow for microbiome research. *mSystems*, 2(1), e00127-16.

DeSantis, T. Z., Hugenholtz, P., Larsen, N., Rojas, M., Brodie, E. L., Keller, K., … & Andersen, G. L. (2006). Greengenes, a chimera-checked 16S rRNA gene database and workbench compatible with ARB. *Applied Environmental Microbiology*, 72(7), 5069–5072.

Dhariwal, A., Chong, J., Habib, S., King, I. L., Agellon, L. B., & Xia, J. (2017). MicrobiomeAnalyst: a web-based tool for comprehensive statistical, visual and meta-analysis of microbiome data. *Nucleic Acids Research*, 45(W1), W180–W188.

Edgar, R. C. (2010). Search and clustering orders of magnitude faster than BLAST. *Bioinformatics*, 26(19), 2460–2461.

Edgar, R. C., Haas, B. J., Clemente, J. C., Quince, C., & Knight, R. (2011). UCHIME improves sensitivity and speed of chimera detection. *Bioinformatics*, 27(16), 2194–2200.

Evans, J. M., Morris, L. S., & Marchesi, J. R. (2013). The gut microbiome: the role of a virtual organ in the endocrinology of the host. *Journal of Endocrinology*, 218(3), R37–R47.

Fernandes, A. D., Macklaim, J. M., Linn, T. G., Reid, G., & Gloor, G. B. (2013). ANOVA-like differential expression (ALDEx) analysis for mixed population RNA-Seq. *PLoS ONE*, 8(7), e67019.

Glassman, S. I., & Martiny, J. B. (2018). Broadscale ecological patterns are robust to use of exact sequence variants versus operational taxonomic units. *MSphere*, 3(4), e00148–18.

Goodrich, J. K., Di Rienzi, S. C., Poole, A. C., Koren, O., Walters, W. A., Caporaso, J. G., … & Ley, R. E. (2014). Conducting a microbiome study. *Cell*, 158(2), 250–262.

Holm, J. B., Humphrys, M. S., Robinson, C. K., Settles, M. L., Ott, S., Fu, L., … & Gravitt, P. E. (2019). Ultrahigh-throughput multiplexing and sequencing of>500-base-pair amplicon regions on the Illumina HiSeq 2500 Platform. *MSystems*, 4(1), e00029–19.

Kopylova, E., Noé, L., & Touzet, H. (2012). SortMeRNA: fast and accurate filtering of ribosomal RNAs in metatranscriptomic data. *Bioinformatics*, 28(24), 3211–3217.

Kozich, J. J., Westcott, S. L., Baxter, N. T., Highlander, S. K., & Schloss, P. D. (2013). Development of a dual-index sequencing strategy and curation pipeline for analyzing amplicon sequence data on the MiSeq Illumina sequencing platform. *Applied Environmental Microbiology*, 79(17), 5112–5120.

Love, M. I., Huber, W., & Anders, S. (2014). Moderated estimation of fold change and dispersion for RNA-seq data with DESeq2. *Genome Biology*, 15(12), 550.

McMurdie, P. J., & Holmes, S. (2013). phyloseq: an R package for reproducible interactive analysis and graphics of microbiome census data. *PLoS ONE*, 8(4), e61217.

Osman, M. A., Neoh, H. M., Ab Mutalib, N. S., Chin, S. F., & Jamal, R. (2018). 16S rRNA gene sequencing for deciphering the colorectal Cancer gut microbiome: current protocols and workflows. *Frontiers in Microbiology*, 9, 767.

Prodan, A., Tremaroli, V., Brolin, H., Zwinderman, A. H., Nieuwdorp, M., & Levin, E. (2020). Comparing bioinformatic pipelines for microbial 16S rRNA amplicon sequencing. PLoS ONE, 15(1), e0227434

Pruesse, E., Quast, C., Knittel, K., Fuchs, B. M., Ludwig, W., Peplies, J., & Glöckner, F. O. (2007). SILVA: a comprehensive online resource for quality checked and aligned ribosomal RNA sequence data compatible with ARB. *Nucleic Acids Research*, 35(21), 7188–7196.

Rognes, T., Flouri, T., Nichols, B., Quince, C., & Mahé, F. (2016). VSEARCH: a versatile open source tool for metagenomics. *PeerJ*, 4, e2584.

Schliep, K. P. (2011). phangorn: phylogenetic analysis in R. *Bioinformatics*, 27(4), 592–593.

Srinivasan, R., Karaoz, U., Volegova, M., MacKichan, J., Kato-Maeda, M., Miller, S., … & Lynch, S. V. (2015). Use of 16S rRNA gene for identification of a broad range of clinically relevant bacterial pathogens. *PLoS One*, 10(2), e0117617.

Upadhyaya, S., & Banerjee, G. (2015). Type 2 diabetes and gut microbiome: at the intersection of known and unknown. *Gut Microbes*, 6(2), 85–92.

Ursell, L. K., Metcalf, J. L., Parfrey, L. W., & Knight, R. (2012). Defining the human microbiome. *Nutrition Reviews*, 70(suppl_1), S38–S44.

Wang, Q., Garrity, G. M., Tiedje, J. M., & Cole, J. R. (2007). Naive Bayesian classifier for rapid assignment of rRNA sequences into the new bacterial taxonomy. *Applied Environmental Microbiology*, 73(16), 5261–5267.

Zhang, J., Kobert, K., Flouri, T., & Stamatakis, A. (2014). PEAR: a fast and accurate Illumina Paired-End reAd mergeR. *Bioinformatics*, 30(5), 614–620.

4

Downstream Analysis and Visualization-Knowledge Discovery – Alpha and Beta Diversity

Murali Sankar Perumal
CPMB&B, TNAU

Shreedevasena Sakthibalan
School of Crop Protection, CPGSAS

CONTENTS

4.1 Introduction

Microorganisms are the basic and biological active mass communities in the earth for all biological systems for better livelihood of all living things in the world (Dominguez-Bello & Blaser 2015). They are key drivers for the life of all organisms on the earth with their biomutual relationships and synthesis of beneficial activities (Whitman et al. 1998). Naturally, the microbial communities are found in all the sources, *viz*., soil, water, and air with heavy bonding relationship to them. In soil, they are decision-makers for promotion or suffering of plant growth, for cycling of nutrients, in soil ecosystem, and in the modulation of nutritional status (Falkowski et al. 2008). The recycling of nutrients and exploration of their effects differ due to the presence of microbial communities in soil around the plant (Filzmoser et al. 2009). So the unique diversity is formed in the nature of ecosystem, which reflects the genomic nature of microbial communities (Dhariwal et al. 2017). The higher level of variance in the beneficial and harmful effects against all living things such as plants, animals, and humans is exploited (Wilson et al. 2000). The unpredicted positive and negative correlation effects of microorganisms on ecosystem were extensively studied through their functional and genetical nature of approaches (Figure 4.1) (Greene et al. 2016).

Microbiome is a highly diversified zone for the presence of beneficial and nonbeneficial organisms and eluted with diversified microbial bioactive compounds for better life of other organisms due to accumulation in the ecosystems (Li 2015). Studying the microbial diversity is an enormous scientific approach for largely explored activities that are represented through genetic and biological pool of novel genes, biosynthetic pathways, and metabolic products (Mardis 2008). Mostly, 90% of microbial consortium cannot be cultured

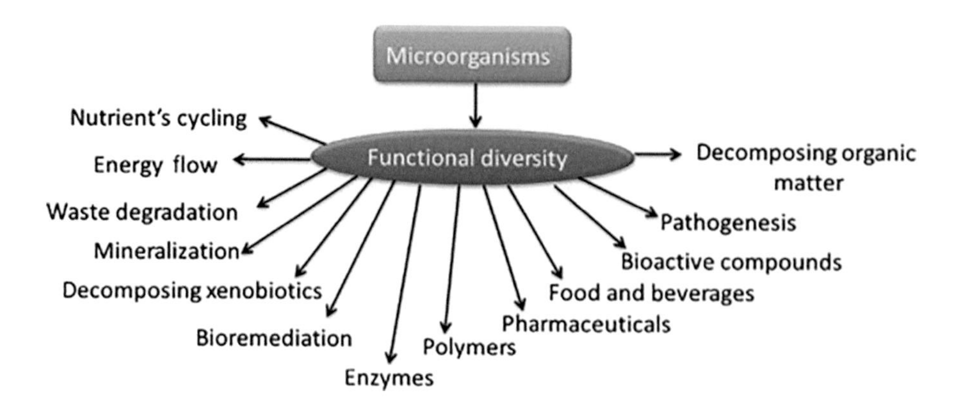

FIGURE 4.1 Basic functional diversity of microorganisms in the earth.

under laboratory conditions. Hence, there is a requirement of another option for culture-independent methods with the well-established genetic and ecostatic tools for the identification of microbiota (White et al. 2009).

Up to the last decade, microorganisms and their grouping nature were studied and differentiated through their cultural, morphological, phenotypical, physiological, biochemical and molecular characteristics via platform of DNA (Turnbaugh et al. 2007). They act as elemental basis, drive the speciation, and underpin other state of biodiversity—including functional traits, species, and ecosystems (McGill 2010). Due to high consumption of time and cost and being essential for high-throughput capacity, NGS (next-generation sequencing), gene profiling, whole-genome sequencing concepts, and marker genes have preferred to the huge number of datasets and libraries for microbial communities (Meyer et al. 2008). So advanced strategies are needed for studying the microbial diversity in all ways because they are easy to handle and time-adaptable and show higher efficacy.

These choices provide a better knowledge in bioinformatics for analyzing the microbial communities before isolation, after isolation, and in the absence of isolation conditions. An extensive approach was developed based on environmentally, genomically and functionally adapted traits of microbial communities (Mendes et al. 2018). This prospective was given a large size of big data focused through environmental changes in global-wide use to humans and animals for the analysis of diversity (Gilbert et al. 2014). These diversity analyses or technologies were applied in the extreme environmental zones with changes through downstream analysis approaches with visual basis.

4.2 Genesis of Downstream Analysis and Visualization Technologies

Naturally occurring changes in the environment, which reflect in all living organism's behavioral nature, and phenotypic, morphological, and molecular characterizations with functional abilities, were explored in different manner, including also humans. This fact generated the evolutionary biology (Koonin & Wolf 2012). During 1977, applications of culturing and microscopic methods attained several limitations and showed variabilities in microbiological studies, thus moving towards the other step of molecular fingerprinting (Lagier et al. 2015). Small subunits and large subunits, *viz.*, 16S (SSU) 5S, 5.8S, 23S, 28S (LSU) rRNA, and ITS (internal transcribed spacer)/IGS (intergenic spacer) were found in prokaryotes and eukaryotes (Pace et al. 1985). In the 1990s, several approaches were implemented for diversity analysis; *viz.*, denaturing gradient gel electrophoresis (DGGE) (Muyzer et al. 1993), terminal restriction fragment length polymorphism (T-RFLP) (Liu et al. 1997), random-amplified polymorphic DNA (RAPD) (Dubey et al. 2010), and automated ribosomal intergenic space analysis (ARISA) (Fisher & Triplett 1999) all are based on traveling characteristics of amplicon size, and this banding pattern is called as "operational taxonomic unit" (OTU) (Figure 4.2) (Tikhonov et al. 2015). These OTUs mostly focused on the environmental and clinical-based discrimination on big group of species (Humbert et al. 2010). During 2006, 454 pyrosequencing is used for studying the diversity profile in marine water based on two characteristics, namely, rarefaction curves very far from reaching the saturation point—which indicates a much larger microbial diversity suspected—and a highly uneven community, with abundant tags. These spaced tags provide a duplicated region called as "rare biosphere" (Lynch & Neufeld 2015). Later, sample-specific barcode sequences (SSBs) were widely used to introduce numerous samples in large comparative microbiome analysis (Anderson et al. 2008).

Currently, microbiome diversity analysis is a web-based platform through comprehensive approaches with highly significant output. It was formatted a single unique platform with researchers, clinicians, biologists, and statisticians to provide a well-established tool for microbiome data processing, statistical analysis, functional profiling, and comparison with public data. It contains four modules, *viz.*, marker-gene data profiling (MDP), shotgun data profiling (SDP), projection with public data (PPD), taxon set enrichment analysis (TESA), VAMPS (visualization and analysis of microbial population structures—http:// vamps.mbl.edu), and web-based shiny diversity (Figure 4.3): (1) how to prepare, process, and normalize data; (2). community profiling and interpretation; (3). meta-analysis; and (4). visual exploration of results with the highest significance within ~70 minutes (Chong et al. 2020).

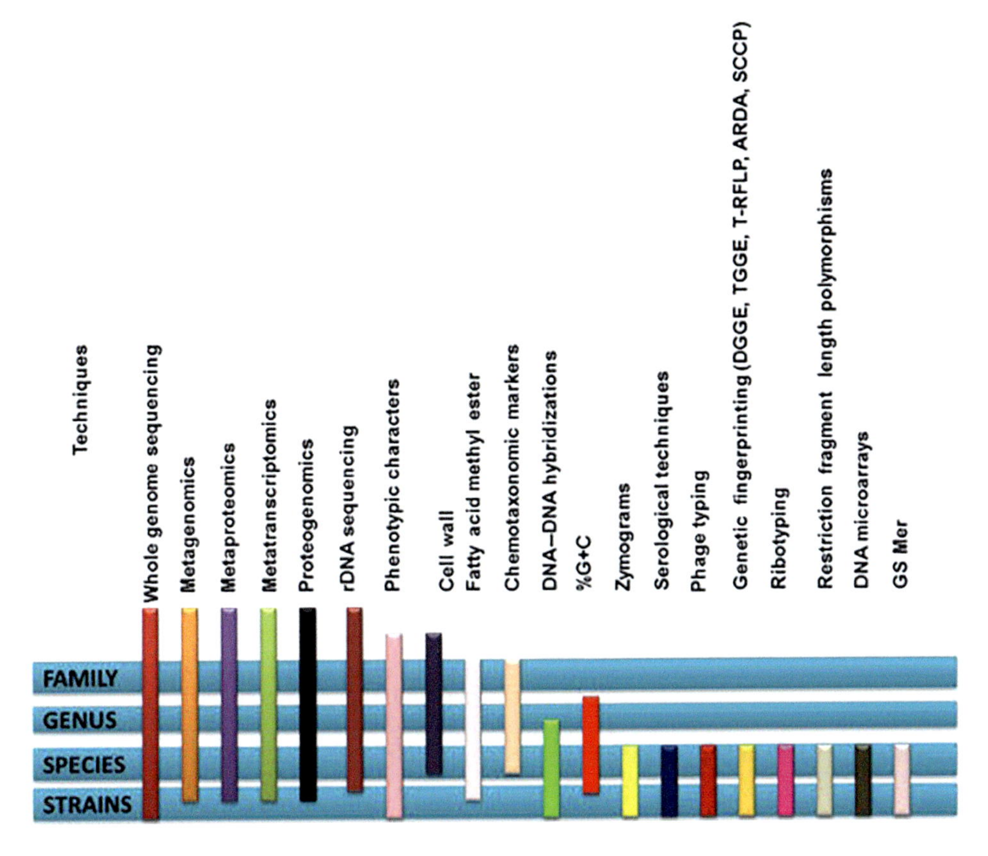

FIGURE 4.2 Downstream analysis and visualization techniques for microbiome diversity studies.

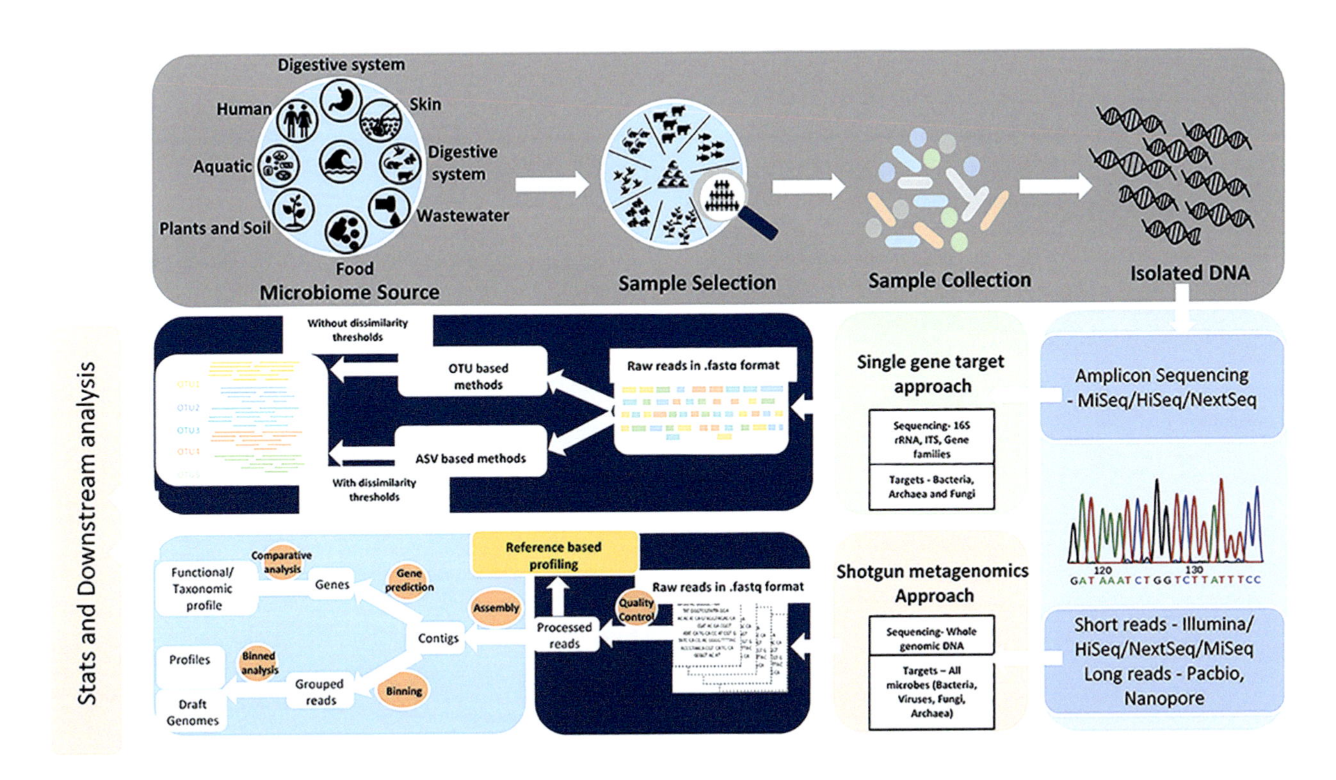

FIGURE 4.3 Multidimensional approaches of downstream analysis.

4.3 Challenges

Naturally, there are some unwanted issues or disturbances (minor) to promote a big problem in the study of microbiome diversity analysis, *viz*., environmental changes, human nature (handling and preservation), and technical (lack of knowledge sources for interpretation). They are classified as experimental and computational challenges (Bharti & Grimm 2019). When we correct it or carried out it, a better result will be obtained in a progressive manner (Figure 4.4).

4.4 Experimental Bias (Narrow Spectrum)

4.4.1 Sample Collection and Handling

Handling of environmental samples after collection is a crucial aspect in nucleic acid-based methods for diversity analysis studies. Collection of samples from an extensive ecosystem is a common problem; i.e., it is difficult to determine the amount of microbial DNA present in different samples (Thomas et al. 2012). Additionally, some parameters play a vital role in making error in sample collection and handling by *contamination* (different agroclimatic conditions such as various crops, temperature, humidity, water source, rainfall distribution, and soil physiochemical properties), *transportation* (transit time requirement), and *storage and safety* (compositional and biosynthesis variation occurred during the long-term storage and make a change [16S rRNA–bacteria stored at >14 days on −20°C]—play a major role (Salter et al. 2014; Choo et al. 2015).

4.4.2 Nucleic Acid Extraction and Preparation

The choice of nucleic acid (DNA / RNA) isolation protocols could cause some bias during sequencing and affects the downstream analysis. For example, isolation of DNA from gram^{+ve} bacteria is harder due to the presence of thick peptidoglycan cell walls (Cuthbertson et al. 2014). Also, bead beating or chemical lysis affects the amount of nucleic acid load during PCR (polymerase chain reaction). As single marker / target gene NGS approaches, amplification using barcode primers, purification, and preparation of purified DNA libraries are done before sequencing. So, quantification of DNA is essential for avoiding some basic drawbacks in the sequencing, and it provides a well-predicted DNA library (Ardui et al. 2018).

4.5 Computational Bias (Broad Spectrum)

In recent years, sequencing technologies were implemented with new methods, algorithms, and computational tools for functional analyses and annotations (Treangen & Salzberg 2011). They met with several challenges like complexity of nonsignificant biological data, lack of metadata, and unavailability of standard data computational sources (Fricke & Rasko 2014). They were expressed in interpretations and poor outcome of results with noncorrelated ones (Cole et al. 2014).

4.5.1 Amplicon Sequencing Analysis

Gene-specific marker-based analysis is to differentiate an error from the real nucleotides; for this purpose, two major tool groups are used: (1). OTU-based and (2). amplicon

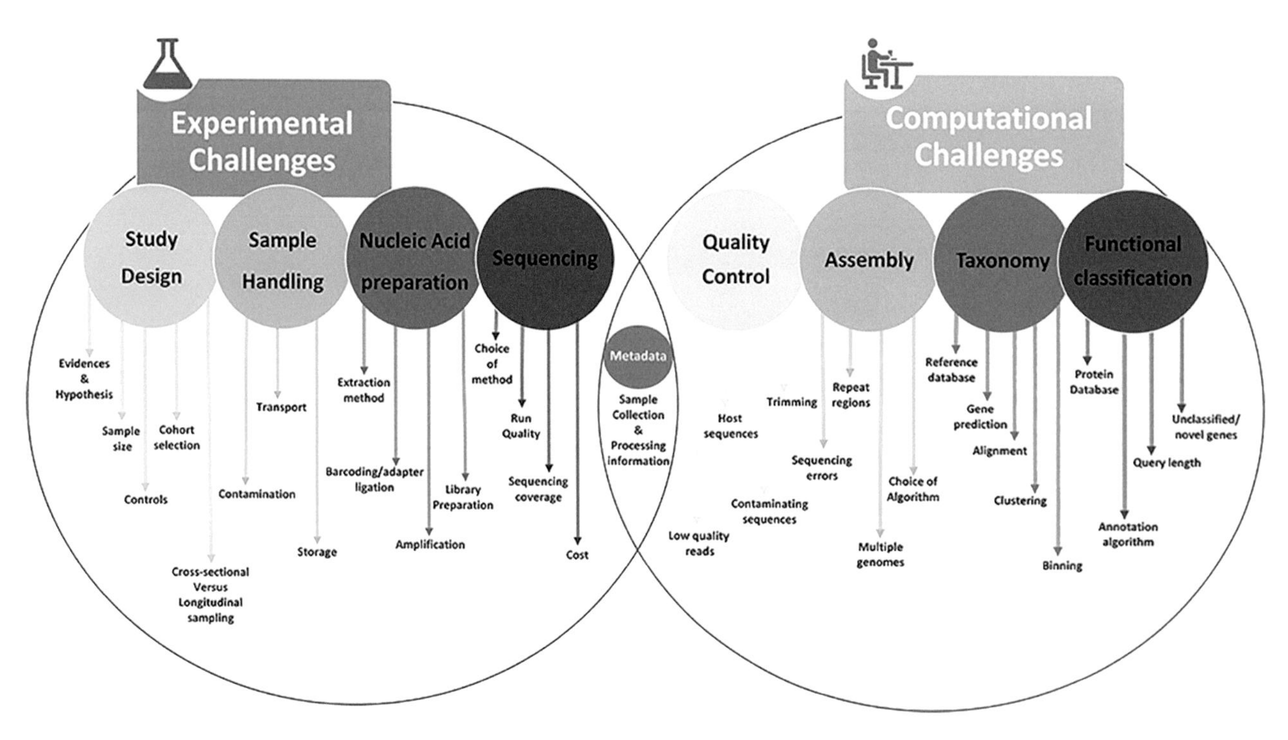

FIGURE 4.4 Classification of challenges in microbiome diversity studies.

sequence variant (ASV) (Sczyrba et al. 2017). OTU-based tool group resolves sequencing errors through clustering reads on prenoted identical threshold OTUs, whereas ASV is used as denoising approach on bio sequences before the initiation of amplification and sequence errors with a vital impact on alpha diversity analysis (Westcott & Schloss 2015).

4.5.2 Metagenomic Sequencing Analysis

A number of tools and algorithms are available for high-throughput metagenomic diversity analysis. In this method, data analysis, assembly, binning followed by taxonomic and functional profiling with the most comprehensive approaches are involved (Bolger et al. 2014)

4.6 Quality of Nucleic Acids

A quality control was mainly focused on the nucleic acids (DNA/RNA) for avoiding the mixture and removal of low-quality raw reads through trimming and the identification of contaminants and removal tools (Trimmomatic, sickle, BBTools and DeconSeq) (Eisen 2007). Basically, two types of read lengths (short-read/long-read) were processed in environmental samples depending upon the composites design (Figure 4.5) (Handelsman et al. 1998).

4.6.1 Short-Read Metagenomic Analysis (NGS)

It's a basic sequencing for random DNA strands of 100–1,000 base pairs. Based on the size limit, the longer sequences were subdivided into smaller fragments, which were sequenced separately with assembling (Junemann 2013). Two principle methods are involved in fragmentation and sequencing: (a). primer walking or chromosome walking progresses through the entire strand piece by piece, whereas shotgun sequencing does random reading of short-read fragments (Figure 4.4) (Metzker 2010). Then, it is comprised of several methods, *viz.*, massively parallel signature sequencing (MPSS) (Brenner et al. 2000), polony sequencing (Margulies & Egholm 2005), 454 pyrosequencing (Shendure et al. 2005), Illumina sequencing (Bentley & Balasubramanian 2008), combinatorial probe anchor synthesis (cPAS) (Drmanac et al. 2010), SOLiD sequencing (Huang et al. 2017), iron torrent semiconductor sequencing (Rusk 2011), DNA nanoball sequencing (Porreca 2010), heliscope single-molecule sequencing (El-Metwally et al. 2014), and microfluidic systems (Abate et al. 2013).

4.6.2 Long-Read Metagenomic Analysis (Third-Gen.)

This is a high-throughput sequencing (HTS) (~5,00,000 seq.) run in parallel, which applies to exome sequencing, whole-genome sequencing, resequencing, transcriptome profiling

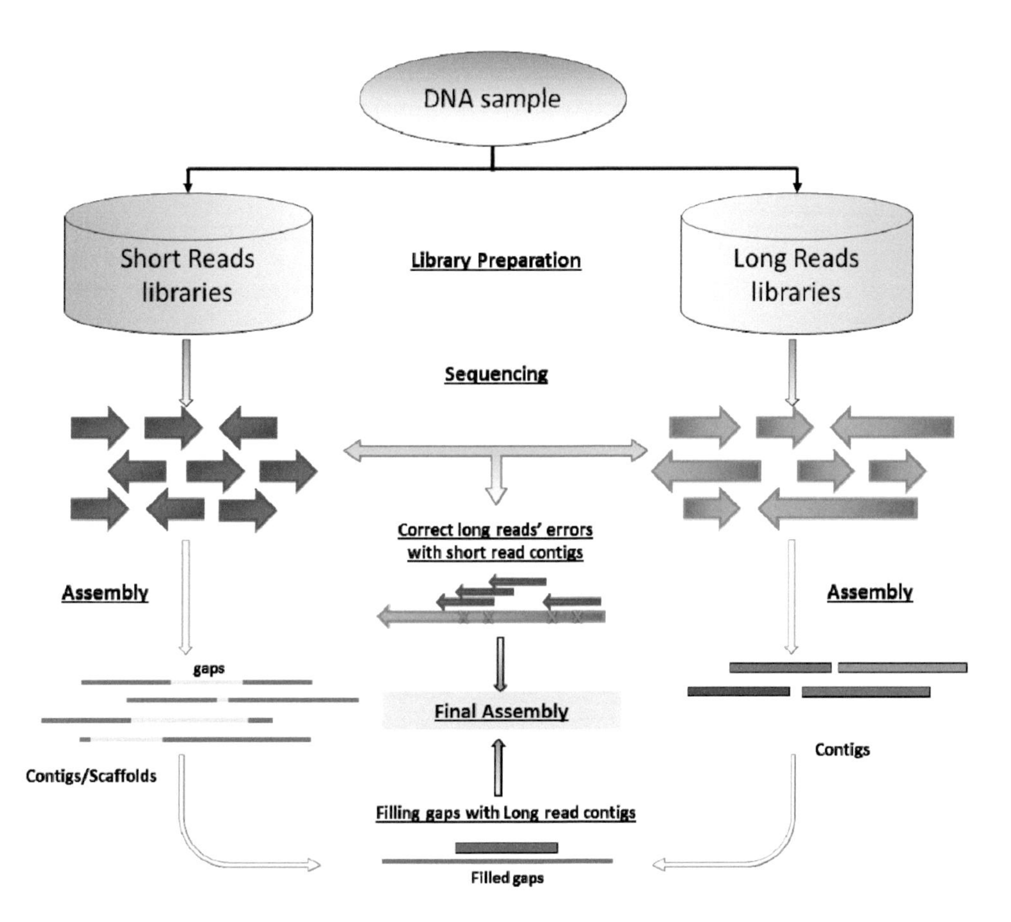

FIGURE 4.5 Protocol for short-and long-read sequencing libraries preparation.

(mRNA–tRNA), nucleic acids (DNA/RNA)–protein interactions (ChiP sequencing), and epigenome characterization with inter- and intraspecific variations in the studied species (Hall 2007). At present, it was done by two tools, *viz.*, single-molecule real-time sequencing (SMRT) based on zero-mode wave guides (ZMW) equipped with small well and sequencing with unmodified polymerase fluorescent-labelled nucleotides (Clarke et al. 2009); nanopore DNA sequencing by DNA passing *via* nanopore changes its ion electronic diode flow and time gradient. Both are referred to as "third-generation sequencing or high-throughput sequencing (HTS)" (dela Torre et al. 2012).

4.6.2.1 Applications of HTS

It offers several advantages compared with DNA microarrays—especially it is more precise and is not subjected to cross-hybridization, and provides higher accuracy and large-dimensional range (10^5)/DNA sequencing × (10^2)/microarrays (Wang et al. 2009). It is more applicable in large-scale projects, *viz.*, human genome project; analyzes gene expression (GTEx); and discovers the molecular underpinnings of human diseases (Alzheimer, diabetics, cancer and autism) (Landt et al. 2012; Lawrence et al. 2014).

4.6.3 Limitations

From this approach, accuracy and coverage across the genome are still problematic, especially for (GC)-rich regions with homopolymer stretches (Ross et al. 2013). The short-read sequences have a limit of large repeat regions and structural variations and leaving of significant portions of the genome at inaccurate and CLARITY issue also occurred (Brownstein et al. 2014). In complete transcriptomes, individual allelic and spliced isoforms hindered by these short-reads were manipulated by the synthetic long-read methods (Tilgner et al. 2015).

The aforementioned technologies are highly employed in analyzing the diversity from big zone of different environmental sources in all aspects, *viz.*, big data formatting, assembling and analysis through visual basis also.

4.7 Knowledge Discovery

Due to increasing of new trends in agriculture, medicine, biotechnology and nanoscience, the personalized, predictive and participatory data are essential for day-to-day science and technology (Hood & Friend 2011). It reflects on various essential needs on omics world and interpretation with emerging diseases and diagnosis with heterogeneous nature of big data (Barrera et al. 2004). It's a comprehensive nature, assembling of numerous data (not only) and most essential for analysis with computational approaching knowledge (Holzinger et al. 2014). In past decades, several approaches, *viz.*, statistical, graph theoretical methods, data mining, and machine learning, have been applied in diversity analysis, but they give a result of unsatisfactory and confused one (Holzinger & Zupan 2013). So, a big challenge to work towards implementing human control over machine intelligence, integration and visual analytic methods is to only provide better performance in biotechnological studies (Beslon et al. 2010). Nowadays, only hand-operating apps and software packages are based on human–computer interaction (HCI) and knowledge discovery from data (KDD), thus enabling the work easily (Shneiderman 2002).

4.8 Knowledge Discovery Protocol

This protocol is categorized into four different coordinating phases (Figure 4.6)

1a. *Interactive data integration, data fusion, and preselection of datasets:*

Collection of wide data (it is heterogeneous, complexity, noisy/dirty data) after fusion with manually through biological, physiological and phenotypical grouping (data integration).

2b. *Interactive sampling, cleansing, preprocessing, and mapping:*

Merging or joining the data in nature manner; it creates impurities and identifies the differed one and the accuracy is maintained by a broad range of including only and making a high-dimensional data.

3c. *Interactive advanced data mining methods and pattern discovery:*

Nonstructural polydimensional data were found to be accumulated from data mining processes, *viz.*, graph, entropy, and topological-based. We concluded that information and graph theory generated a reproducible pattern form (human easy handle).

4d. *Interactive visualization, HCI analytics and decision supportiveness:*

Finally, the results were obtained through the man-favor algorithms with high-dimensional and better-visualized optimistic standard data (avoidance of inaccuracy, dirty data, unwanted data) and saved as good practice of protection with safety (Inselberg 2005).

4.9 Alpha and Beta Diversities

Multiple samples were summarized as a table of number of reads count per OTU sample. These tables are sparsely scattered, huge communities with long OTU tail under rare biosphere. The true absence/presence of OTU is referred from counts of 0 to forward move. It results in different sequence yields, unequal pooled sequencing libraries and very short detection limit. Due to the lack of clarity, arbitrary approaches (MOTHUR and QIIME) were adopted with discard of OTU counts (less) oriented with environment. Lack or less number of OTU counts synthesis has great effect on diversity and sequencing. So, alternate approach has been introduced to diversity analyses (alpha and beta) (McMurdie & Holmes 2014).

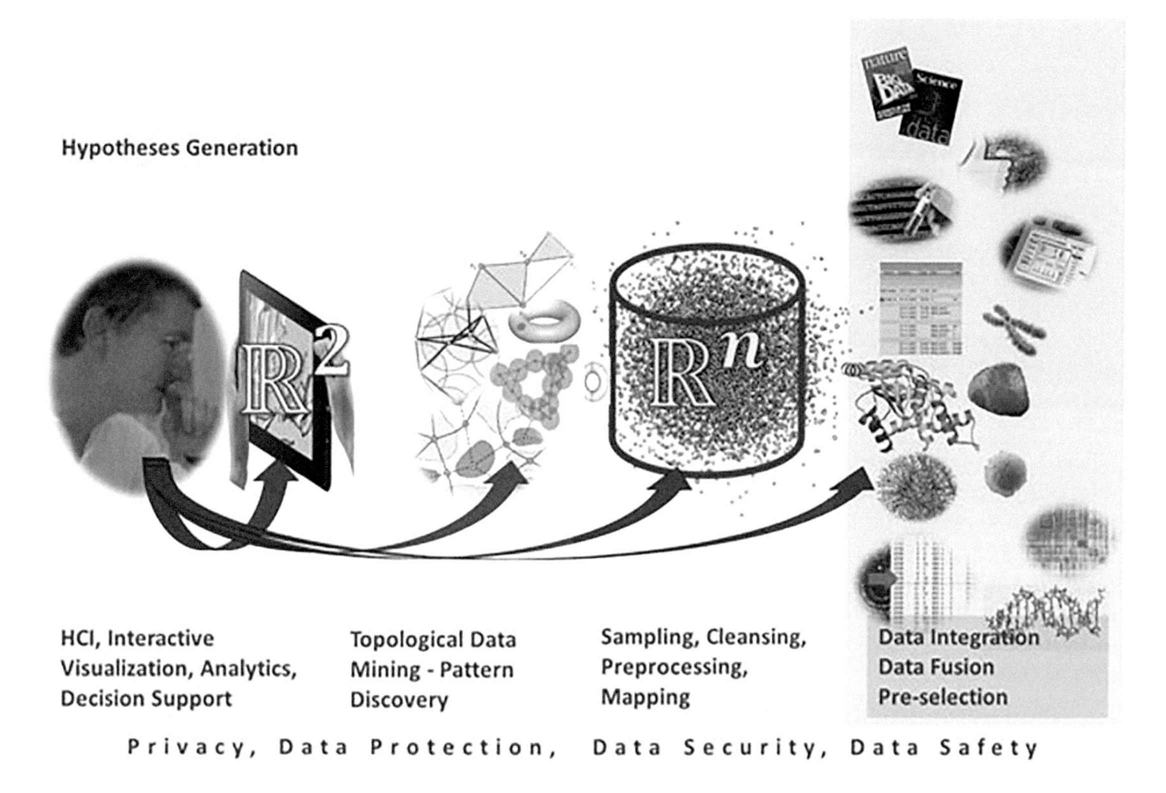

FIGURE 4.6 Image of knowledge discovery process.

4.9.1 α-Diversity

Alpha diversity means the richness in species of particular stand or community or a given stratum or group of organisms in a group (Whittaker et al. 1960). In ecology, the diversity is assumed within a single sample or set of replicates through counting how many different OTUs are in the sample, and it does not identify every single taxon in a sample. To calculate the tail of the species, OTUs abundance distribution and observations of singletons (*sp.* once) and doubletons (*sp.* twice) are taken into account:

$$S_{\text{est}} = S_{\text{obs}} + \frac{f_1^2}{2f_2},$$

where

S_{est} is the estimated species richness, S_{obs} is the observed species richness (f_1 is the number of singletons and f_2 is the number of doubletons). Related to Chao1 is ACE (abundance-based coverage estimator), which considers the ratio not only of singletons and doubletons, but also of all OTUs observed up to an arbitrary count, most usually 10.

$$S_{\text{ace}} = F_{\text{abund}} + \frac{f_{\text{rare}}}{C_{\text{ace}}} + \gamma^2_{\text{ace}} \frac{f_1}{C_{\text{ace}}}$$

$$C_{\text{ace}} = 1 - \frac{f_1}{n_{\text{rare}}} \quad ,$$

$$\gamma^2_{\text{ace}} = \max\left[0, \frac{f_{\text{rare}} \sum_{i=1}^{10} i(i-1) f_i}{C_{\text{ace}} n_{\text{rare}} (n_{\text{rare}} - 1)} - 1\right]$$

where

f_{abund} is the number of OTU above the abundance threshold, f_{rare} is the number of OTU at or below the threshold (f_i is the number of OTU observed i times, n_{rare} is the total number of individuals in rare OTU, C_{ace} is the sample coverage estimator, and γ^2_{ace} is the estimated coefficient of variation for rare OTU). Finally, the variation was documented by richness of relativeness of OTUs read within the samples (10 *sp.*) in single genus. Further deep statistics is described by Hugerth and Andersson (2017).

4.9.2 β-Diversity

Beta diversity is the true diversity, which is defined as a ratio between regional and local species variability. The total species diversity in a landscape (γ) is determined by two different things, α-species diversity at the habitat level and β-differentiation among habitats. This tool is used to study the compositional heterogeneity in species.

The true diversity is measured by

$$d(S_1, S_2) = \sqrt{\sum (S_{1i} - S_{2i})^2},$$

where S_1 and S_2 are two samples, and S_{1i} and S_{2i} are the abundance of OTU i in samples S_1 and S_2, respectively.

The alternative of OTU-based distance progresses with quantitative and qualitative trait metrics to promote the lead of differentiation by dissimilarity relativeness (PCoA).

4.9.3 Limitations

Alpha diversity is applicable in both extinct and extant landscapes. There is an understanding of the change in species composition from local to regional as working fundamental patterns and spatial changes relationship between diversity and latitude (β-diversity) (Jost 2007).

4.10 Conclusions

Human life on earth was mostly depending upon the microorganisms, and it highly influenced the biome zone through their beneficial activities in evolutionary biology on single-celled organisms to huge organisms. Microbiologists have been striving to study the ecosystem with biologists for obtaining the sources from microbiome for use in agriculture (varieties and crop management), medicine (curing new diseases and pharmaceuticals), and nanotechnology for better livelihood on earth. From historical, microbial ecology diversity studies were handled with microscopic visualization to move forward third-generation sequencing (omics era) and high-throughput technologies were coupled with statistics and visual science by HCIs. In the future, it yielded better performance by advanced technologies, which were associated with contribution of sciences; *viz*., microbiology, bioinformatics and statistics and computer–human intelligence are at play when using a single platform.

REFERENCES

Abate, A. R., Hung, T., Sperling, R. A., Mary, P., Rotem, A. & Agresti, J. J. (2013). DNA sequence analysis with droplet-based microfluidics. *Lab on a Chip*, 13(24), 4864–4869.

Andersson, A. F., Lindberg, M., Jakbosson, H., Bäckhed, F., Nyrén, P. & Engstrand, L. (2008). Comparative analysis of human gut microbiota by barcoded pyrosequencing. *PLoS ONE*, 3, e2836.

Ardui, S., Ameur, A. & Vermeesch, J. R. (2018). Single molecule real time (SMRT) sequencing comes of age: applications and utilities for medical diagnostics. *Nucleic Acids Research*, 46, 2159–2168.

Barrera, J., Cesar-Jr, R. M., Ferreira, J. E. & Gubitoso, M. D. (2004). An environment for knowledge discovery in biology. *Computers in Biology and Medicine*, 34(5), 427–447.

Bentley, D. R. & Balasubramanian, S. (2008). Accurate whole human genome sequencing using reversible terminator chemistry. *Nature*, 456(7218), 53–59.

Beslon, G., Parsons, D. P., Peña, J.-M., Rigotti, C. & Sanchez-Dehesa, Y. (2010). From digital genetics to knowledge discovery: perspectives in genetic network understanding. *Intelligent Data Analysis*, 14(2), 173–191.

Bharti, R. & Grimm, D. G. (2019). Current challenges and best-practice protocols for microbiome analysis. *Briefings in Bioinformatics*, 00(00), 1–16.

Bolger, A., Lohse, M. & Usadel, B. (2014). Trimmomatic: a flexible trimmer for illumina sequence data. *Bioinformatics*, 30, 2114–2120.

Brenner, S., Johnson, M., Bridgham, J., Golda, G., Lloyd, D. H., Johnson, D., Luo, S., McCurdy, S., Foy, M., Ewan, M., Roth, R., George, D., Eletr, S., Albrecht, G., Vermaas, E., Williams, S. R., Moon, K., Burcham, T., Pallas, M., DuBridge, R. B., Kirchner, J., Fearon, K., Mao, J. & Corcoran, K. (2000). Gene expression analysis by massively parallel signature sequencing (MPSS) on microbead arrays. *Nature Biotechnology*, 18(6), 630–634.

Brownstein, Ca., Beggs, A. H., Homer, N., Merriman, B., Yu, T. W., Flannery, K. C., DeChene, E. T., Towne, M. C., Savage, S. K. & Price, E. N. (2014). An international effort towards developing standards for best practices in analysis, interpretation and reporting of clinical genome sequencing results in the CLARITY challenge. *Genome Biology*, 15, R53.

Chong, J., Liu, P., Zhou, G. & Xia, J. (2020). Using microbiome analyst for comprehensive statistical, functional and meta- analysis of microbiome data. *Nature Protocols*, 15, 799–821.

Choo, J. M., Leong, L. E. & Rogers, G. B. (2015). Sample storage conditions significantly influence faecal microbiome profiles. *Scientific Reports*, 5, 16350.

Clarke, J., Wu, H. C., Jayasinghe, L., Patel, A., Reid, S. & Bayley, H. (2009). Continuous base identification for single-molecule nanopore DNA sequencing. *Nature Nanotechnology*, 4(4), 265–270.

Cole, J. R., Wang, Q. & Fish, J. A. (2014). Ribosomal database project: data and tools for high throughput rRNA analysis. *Nucleic Acids Research*, 42, D633–D642.

Cuthbertson, L., Rogers, G. B. & Walker, A. W. (2014). Time between collection and storage significantly influences bacterial sequence composition in sputum samples from cystic fibrosis respiratory infections. *Journal of Clinical Microbiology*, 52, 3011–3016.

dela Torre, R., Larkin, J., Singer, A. & Meller, A. (2012). Fabrication and characterization of solid-state nanopore arrays for high-throughput DNA sequencing. *Nanotechnology*, 23(38), 385308.

Dhariwal, A., Chong, J., Habib, S., King, I. L., Agellon, L. B. & Xia, J. (2017). Microbiome analyst: a web-based tool for comprehensive statistical, visual and meta-analysis of microbiome. *Nucleic Acids Research*, 45, W180–W188.

Dominguez-Bello, M. G. & Blaser, M. J. (2015). Asthma: undoing millions of years of coevolution in early life? *Science Translational Medicine*, 7, (307), f39.

Drmanac, R., Sparks, A. B., Callow, M. J., Halpern, A. L., Burns, N. L. & Kermani, B. G. (2010). Human genome sequencing using unchained base reads on self-assembling DNA nanoarrays. *Science*, 327(5961), 78–81.

Dubey, S. C., Tripathi, A. & Singh, S. R. (2010). ITS-RFLP fingerprinting and molecular marker for detection of *Fusarium oxysporum* f. sp. *ciceris*. *Folia Microbiologica*, 55(6), 629–634.

Eisen, J. A. (2007). Environmental shotgun sequencing: its potential and challenges for studying the hidden world of microbes. *PLoS Biology*, 5(3), e82.

El-Metwally, S., Ouda, O. M. & Helmy, M. (2014). Next generation sequencing technologies and challenges in sequence assembly. Springer *Briefs in Systems Biology*, 7, 51–59.

Falkowski, P. G., Fenchel, T. & Delong, E. F. (2008). The microbial engines that drive Earth's biogeochemical cycles. *Science*, 320, 1034–1039.

Filzmoser, P., Hron, K. & Reimann, C. (2009). Univariate statistical analysis of environmental (compositional) data: problems and possibilities. *Science of the Total Environment*, 407, 6100–6108.

Fisher, M. M. & Triplett, E. W. (1999). Automated approach for ribosomal intergenic spacer analysis of microbial diversity and its application to freshwater bacterial communities. *Applied Environmental Microbiology*, 65, 4630–4636.

Fricke, W. F. & Rasko, D. A. (2014). Bacterial genome sequencing in the clinic: bioinformatic challenges and solutions. *Nature Reviews Genetics*, 15, 49–55.

Gilbert, J. A., Jansson, J. K. & Knight, R. (2014). The Earth Microbiome project: successes and aspirations. *BMC Biology*, 12, 69.

Greene, C. S., Stanton, B. A., Hogan, D. A. & Bromberg, Y. (2016). Computational approaches to study microbes and microbiomes. *Pacific Symposium on Biocomputing*, 21, 557–567.

Hall, N. (2007). Advanced sequencing technologies and their wider impact in microbiology. *Journal of Experimental Biology*, 210(9), 1518–1525.

Handelsman, J., Rondon, M. R., Brady, S. F., Clardy, J. & Goodman, R. M. (1998). Molecular biological access to the chemistry of unknown soil microbes: a new frontier for natural products. *Chemistry & Biology*, 5(10), R245–R249.

Holzinger, A., Dehmer, M. & Jurisica, I. (2014). Knowledge discovery and interactive data mining in bioinformatics – state-of-the-art, future challenges and research directions. *BMC Bioinformatics*, 15(6), 11.

Holzinger, A. & Zupan, M. (2013). KNODWAT: a scientific framework application for testing knowledge discovery methods for the biomedical domain. *BMC Bioinformatics*, 14(1), 191.

Hood, L. & Friend S. H. (2011). Predictive, personalized, preventive, participatory (P4) cancer medicine. *Nature Reviews Clinical Oncology*, 8(3), 184–187.

Huang, J., Liang, X., Xuan, Y., Geng, C., Li, Y. & Lu, H. (2017). A reference human genome dataset of the BGISEQ-500 sequencer. *Giga Science*, 6(5), 1–9.

Hugerth, L. W. & Andersson, A. F. (2017). Analysing microbial community composition through amplicon sequencing: from sampling to hypothesis testing. *Frontiers in Microbiology*, 8, 1561.

Humbert, J. F., Quiblier, C. & Gugger, M. (2010). Molecular approaches for monitoring potentially toxic marine and freshwater phytoplankton species. *Analytical and Bioanalytical Chemistry*, 397, 1723–1732.

Inselberg, A. (2005). Visualization of concept formation and learning. *Kybernetes*, 34 (1/2), 151-166.

Jost, L. (2007). Partitioning diversity into independent alpha and beta components. *Ecology*, 88, 2427–2439.

Junemann, S. (2013). Updating benchtop sequencing performance comparison. *Nature Biotechnology*, 31(4), 294–296.

Koonin, E. V. & Wolf, Y. I. (2012). Evolution of microbes and viruses: a paradigm shift in evolutionary biology? *Frontiers in Cellular and Infection Microbiology*, 2, 119.

Lagier, J. C., Edouard, S., Pagnier, I., Mediannikov, O., Drancourt, M. & Raoult, D. (2015). Current and past strategies for bacterial culture in clinical microbiology. *Clinical Microbiology Reviews*, 28, 208–236.

Landt, S. G., Marinov, G. K., Kundaje, A., Kheradpour, P., Pauli, F., Batzoglou, S., Bernstein, B. E., Bickel, P., Brown, J. B. & Cayting, P. (2012). ChIP-seq guidelines and practices of the ENCODE and modENCODE consortia. *Genome Research*, 22, 1813–1831.

Lawrence, M. S., Stojanov, P., Mermel, C. H., Robinson, J. T., Garraway, La., Golub, T. R., Meyerson, M., Gabriel, S. B., Lander, E. S. & Getz, G. (2014). Discovery and saturation analysis of cancer genes across 21 tumour types. *Nature*, 505, 495–501.

Li, H. Z. (2015). Microbiome, metagenomics, and high-dimensional compositional data analysis. *Annual Review of Statistics and its Application*, 2, 73–94.

Liu, W. T., Marsh, T. L., Cheng, H. & Forney, L. J. (1997). Characterization of microbial diversity by determining terminal restriction fragment length polymorphisms of genes encoding 16S rRNA. *Applied and Environmental Microbiology*, 63, 4516–4522.

Lynch, M. D. J. & Neufeld, J. D. (2015). Ecology and exploration of the rare biosphere. *Nature Reviews Microbiology*, 13, 217–229.

Mardis, E. R. (2008). The impact of next-generation sequencing technology on genetics. *Trends in Genetics*, 24, 133–141.

Margulies, M. & Egholm, M. (2005). Genome sequencing in open microfabricated high density picoliter reactors. *Nature*, 437(7057), 376–380.

McGill, B. J. (2010). Ecology matters of scale. *Science*, 328, 575–576.

McMurdie, P. J. & Holmes, S. (2014). Waste not, want not: why rarefying microbiome data is inadmissible. *PLoS Computational Biology*, 10, e1003531.

Mendes, L. W., Mendes, R., Raaijmakers, J. M. & Tsai, S. M. (2018). Breeding for soil-borne pathogen resistance impacts active rhizosphere microbiome of common bean. *ISME Journal*, 12, 3038–3042.

Metzker, M. L. (2010). Sequencing technologies – the next generation. *Nature Reviews Genetics*, 11(1), 31–46.

Meyer, F., Paarmann, D., D'Souza, M., Olson, R., Glass, E. M., Kubal, M., Paczian, T., Rodriguez, A., Stevens, R. & Wilke, A. (2008). The metagenomics RAST server-a public resource for the automatic phylogenetic and functional analysis of metagenomes. *BMC Bioinformatics*, 9, 386.

Muyzer, G., de Waal, E. C. & Uitterlinden, A. G. (1993). Profiling of complex microbial populations by denaturing gradient gel electrophoresis analysis of polymerase chain reaction-amplified genes coding for 16S rRNA. *Applied and Environmental Microbiology*, 59, 695–700.

Pace, N. R., Stahl, D. A., Lane, D. J., and Olsen, G. J. (1985). Analyzing natural microbial populations by rRNA sequences. *ASM News*, 51, 4–12.

Porreca, G. J. (2010). Genome sequencing on nanoballs. *Nature Biotechnology*, 28(1), 43–44.

Ross, M. G., Russ, C., Costello, M., Hollinger, A., Lennon, N. J., Hegarty, R., Nusbaum, C. & Jaffe, D. B. (2013). Characterizing and measuring bias in sequence data. *Genome Biology*, 14, R51.

Rusk, N. (2011). Torrents of sequence. *Nature Methods*, 8(1), 44.

Salter, S. J., Cox, M. J. & Turek, E. M. (2014). Reagent and laboratory contamination can critically impact sequence-based microbiome analyses. *BMC Biology*, 12, 87.

Sczyrba, A., Hofmann, P. & Belmann, P. (2017). Critical assessment of metagenome interpretation-a benchmark of metagenomics software. *Nature Methods*, 14, 1063–1071.

Shendure, J., Porreca, G. J., Reppas, N. B., Lin, X., McCutcheon, J. P., Rosenbaum, A. M., Wang, M. D., Zhang, K., Mitra, R. D. & Church, G. M. (2005). Accurate multiplex polony sequencing of an evolved bacterial genome. *Science*, 309(5741), 1728–1732.

Shneiderman, B. (2002). Inventing discovery tools: combining information visualization with data mining. *Information Visualization*, 1(1), 5–12.

Thomas, T., Gilbert, J. & Meyer, F. (2012). Metagenomics – a guide from sampling to data analysis. *Microbial Informatics and Experimentation*, 2, 3–50.

Tikhonov, M., Leach, R. W. & Wingreen, N. S. (2015). Interpreting 16S metagenomic data without clustering to achieve sub-OTU resolution. *ISME Journal*, 9, 68–80.

Tilgner, H., Jahanbani, F., Blauwkamp, T., Moshrefi, A., Jaeger, E., Chen, F., Harel, I., Bustamante, C., Rasmussen, M. & Snyder, M. (2015). Comprehensive transcriptome analysis using synthetic long read sequencing reveals molecular co-association of distant splicing events. *Nature Biotechnology*, 33 (7), 736-742.

Treangen, T. J. & Salzberg, S. L. (2011). Repetitive DNA and next generation sequencing: computational challenges and solutions. *Nature Reviews Genetics*, 13, 36–46.

Turnbaugh, P. J., Ley, R. E., Hamady, M., Fraser-Liggett, C. M., Knight, R. & Gordon, J. I. (2007). The human microbiome project. *Nature*, 449, 804–810.

Wang, Z., Gerstein, M. & Snyder, M. (2009). RNA-Seq: a revolutionary tool for transcriptomics. *Nature Reviews Genetics*, 10, 57–63.

Westcott, S. L. & Schloss, P. D. (2015). De novo clustering methods outperform reference- based methods for assigning 16S rRNA gene sequences to operational taxonomic units. *Peer Journal*, 3, e1487.

White, J. R., Nagarajan, N. & Pop, M. (2009). Statistical methods for detecting differentially abundant features in clinical metagenomic samples. *PLoS Computational Biology*, 5, e1000352.

Whitman, W. B., Coleman, D. C. & Wiebe, W. J. (1998). Prokaryotes: the unseen majority. *Proceedings of the National Academy of Sciences*, 95, 6578–6583.

Whittaker, R. H. (1960). Vegetation of the Siskiyou mountains, Oregon and California. *Ecological Monographs*, 30, 279–338.

Wilson, C. A., Kreychman, J. & Gerstein, M. (2000). Assessing annotation transfer for genomics: quantifying the relations between protein sequence, structure and function through traditional and probabilistic scores. *Journal of Molecular Biology*, 297, 233–249.

5

Biostatistics Including Multivariate Analysis Commonly Used for Microbiome Analysis/Study

Priyanka Sarkar
Asian Institute of Gastroenterology

CONTENTS

5.1 Introduction

> "Death sits in the bowels" and "bad digestion is the root of all evil".
>
> *Hippocrates*

Humans have been harboring a vast diversity of microbes since the origin (Ley et al., 2008). Every nook and corner of the human body is inhabited by these microorganisms. Recently, human gut microbiota has been under the limelight due to its impactful role in diverse host functions such as metabolism, physiology, nutrition, and immune functions (Guinane and Cotter, 2013). Disruption of the gut microbiota or dysbiosis can lead to significant disease development such as diabetes and inflammatory bowel disease (IBD) (Qin et al., 2012; Nishida et al., 2018). The relationship between microbes and their host is multidimensional. Factors of a host such as diet, genetics, mode of delivery, exposure to medicines and other xenobiotics can greatly influence the shape and size of the host gut microbiome (Wen and Duffy, 2017). Thus, it can be concluded that adaptations acquired by humans with respect to the changes faced by an individual or a population leave an impact on the microbial makeup leading to compositional variations. However, a dilemma faced during microbiome analysis is the selection of relevant methods to precisely depict the age-old relationship between the host and its microbiome. Therefore, several statistical and computational tools are being developed to have a better resolution on the data. The inception of the Human Microbiome Project in 2008 also marked the revolutionary phase for the development of bioinformatics tools and computational methods (Peterson et al., 2009). However, the crucial drawback that is halting the progress is the huge variability existing in the cohorts and, bioinformatics workflows along with experimental designs. These significantly affect the computational analysis along with statistical assessment. The variability also has an impact on the standardization and validation of the statistical and computational tools.

This chapter summarizes the hypotheses, methods, and statistical analyses involved in depicting the role of microbiome in the host. Additionally, the chapter will also summarize the challenges and limitations associated with the usage of biostatistical tools.

5.2 Hypothesis of Microbiome Study

Microbiome test first begins with the ethical approval that includes detailed protocols of screening the volunteers, inclusion/exclusion criteria, hypotheses, methodologies, techniques involved, and predicted outcome followed by the principles depicted in the Declaration of Helsinki (Figure 5.1). Microbiome studies in the current scenario deal with two main aims: *first*, to profile the microbiome of the host in association with the other parameters such as genetic, environmental, diet, health/disease status of the host and *second*, to identify the possible causal factor impacting the host–microbiome profiles leading to microbial dysbiosis and even diseases. The focus of these schemes is to determine the mechanisms of host–microbial dysbiosis that leads to the onset of various diseases. This will further help in developing therapeutic interventions in disease management (Virgin and Todd, 2011; Spor et al., 2011). Over the decades, three types of hypotheses have been developed to capture the dynamic association between the host microbiome and host health.

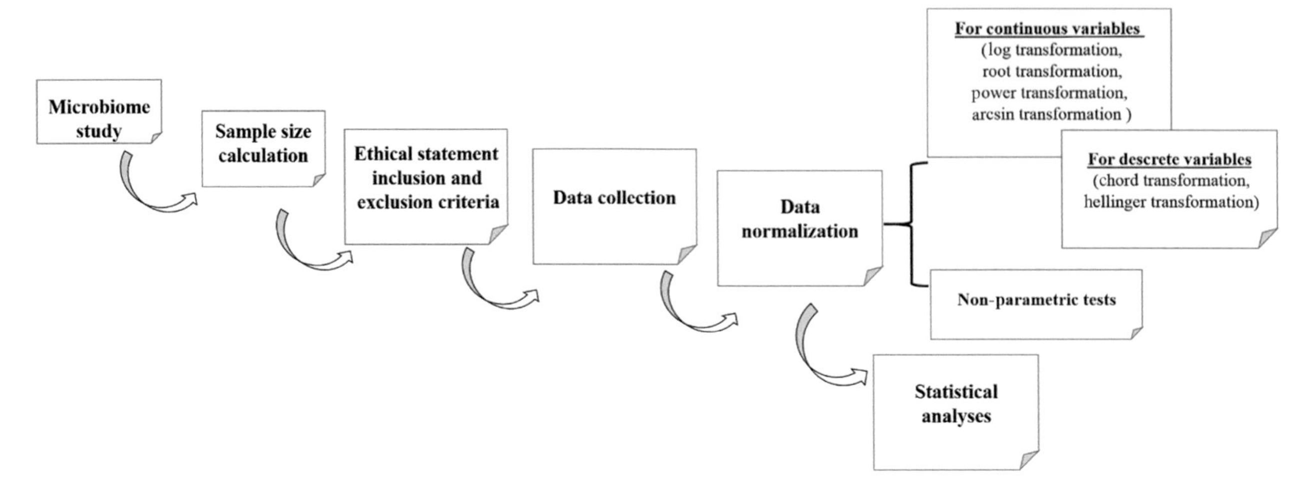

FIGURE 5.1 Flowchart of the microbiome analyses.

Hypothesis 1 evaluates the linkage between the host and its environment. Hypothesis 2 deals with the evaluation of the co-association between host microbiome and its host, following the linkage of the dysbiotic microbiome with host health and disease development. Hypothesis 3 deals with the co-association of the host microbiome with its biological covariates or surrounding environments which possibly lead to the disease development. Besides these three, scientists have also developed their own statistical hypotheses according to their respective research goals. Moreover, all these hypotheses can be categorized into two broad categories, viz. null (H_0) and alternate (H_1) hypotheses. The null hypothesis is employed if there are no differences between the microbiome makeups of two cohorts or there is no impact of the environment or other biological factors on host gut microbiome. On the contrary, the H_1 is applied if there are differences exist between the two groups. Briefly, all hypotheses run around a common core theme that deals with the impacts of environment or other factors on bacterial diversity in terms of bacterial richness and evenness, phylogenetic divergences, common bacterial strains, and unique bacterial populations in the host.

5.3 Statistical Methods Commonly Used for Microbiome Studies

The establishment of hypothesis requires appropriate and relevant testing methods. In this context, we have reviewed the classical statistical tools, multivariate statistical analysis methods, and recently developed models and tools that play a vital role in analyzing the microbiome data. Microbial studies are concerned with two types of diversities: α and β diversities. The α-diversity generally refers to the diversity within a particular individual or a group and is usually expressed by the species richness (i.e. the number of species in the individual or group) and species evenness. On the other hand, the β diversity deals with the diversity between two or more groups. To calculate the α diversity, Shannon, Simpson, Observed OTUs (Operational Taxonomic Units), PD (Phylogenetic diversity) whole tree, Chao1 indices (indicates species richness), etc. are commonly used. However, for the selection of relevant statistical tools, the factors that need to be considered are the heterogeneity of the microbial data, samples size of the experiment, and data distribution (i.e. normal, non-normal). Broadly, the statistical tools are two types, parametric and non-parametric, depending on the data distribution, which is discussed in the following.

Statistical tests are the mathematical tools that aid in assessing the quantitative data generated via a research study. Several statistical tools are available to analyze the gut microbiome-related data. However, the multitude of these tests often makes it difficult to remember and decide the most suitable test as per the need of the situation. Hence, it is essential to ponder upon diverse points such as study design, types of data, and groups to be compared, while selecting the relevant statistical test. For instance, for carrying out hypothesis testing in microbial taxa, it is crucial to compare the α and β diversity indices. Further, for the utilization of diverse tests such as *t*-test (for two groups), analysis of variance (ANOVA, for three or more groups), or the corresponding non-parametric statistical tests, it is necessary to consider the number of experimental groups, condition, and even the distribution of the data. To begin with, standard *t*-test finds its applicability for comparing the α diversity that prevails between the two cohorts (La Rosa et al., 2015). Additionally, it can also be applied for comparing the microbial abundance data in a population (Chen et al., 2012a; Kim et al., 2013; Iwai et al., 2012; Hsiao et al., 2013). Another widely used test in microbiome analysis is the Wilcoxon Rank-Sum test. It is the counterpart of *Z* test. In this test, the comparison is done between two population median values of the variables. However, the test could be applied under certain circumstances. First, the two groups that are being tested must be independent of one another. Further, it is essential that the groups must have similar distributions. Additionally, it considers numeric and ordinal data. The Mann Whitney *U* test, which is the non-parametric substitute for Student's *t*-test, is also widely used for microbiome data analysis. The test has been used to compare the α diversity in the rat model of NAFLD to know the impact of herbal formulation on diseased animals' gut microbiome (Yin et al., 2013). However, in case of more than two groups, the most used statistical tool is either the one-way ANOVA or its relevant non-parametric Kruskal–Wallis *H* test. The selection of the test depends on the type of distribution of the variables. This statistical tool had been used for varied purposes such as to determine the significance of the microbial taxon identified in a sample or group of samples with the aid of the bioinformatic pipelines such as QIIME and MG-RAST. Further, ANOVA has been used for comparing the proportional microbial abundance in the groups (Alekseyenko et al., 2013). Moreover, the assessment of the risk model performance in the context of comparing gut microbiome and its functional capacity in the intestinal locations has also been performed with ANOVA (Yang et al., 2016). On the other hand, the Kruskal–Wallis *H* test is also used to determine the prevalence of the statistically significant differences between three or more groups of an independent variable (Jandhyala et al., 2017). Additionally, it takes into consideration the groups that possess diverse standard deviations and have different distributions. This statistical tool has been used to develop linear discriminant analysis effect size (LEfSe) analysis, which is predominantly being used in microbiome studies to measure the microbial richness of the groups (Carmody et al., 2015; Muniz et al., 2018; Chumpitazi et al., 2015; Zhang et al., 2016). Followed by this, another statistical analysis referred to as Post Hoc test has been readily used for exploring the significant differences prevailing between three or more groups. It is commonly performed along with ANOVA to provide specific information about the groups that possess the significant differences.

Additionally, the exploration of the microbiome for analyzing their role in maintaining and disrupting the health of their host species has resulted in the interrogation of thousands of taxa simultaneously. In this course, it is essential to use the statistical tool for obtaining microbiome read counts (Hu et al., 2018). However, taxa count data retrieved as a result of the RNA-Seq (16S) or shotgun metagenomic sequencing or gene sequencing-based experiments are often over-dispersed and coupled with several zero. To deal with such data, negative binomial models along with **zero-inflated models** have been considered (Zhang et al., 2017).

The negative binomial model has been used for testing the differences that existed in the sequence tag abundance (Zhang et al., 2017). Moreover, it is also used for detecting the differentially profuse characteristics prevailing in the clinically metagenomic samples (White et al., 2009). The count response follows the following negative binomial distribution. The differential analysis is quite complex owing to the microbiome count data characteristics like the overdispersion along with the fluctuating library size. In order to cope with these challenges, analytical tools like the R Bioconductor packages, viz. DESeq2 and edgeR, have been integrated in the classical **binomial model** (Robinson et al., 2010; Anders and Huber, 2010). However, these models are restricted to be used for a limited number of samples, which often leads to a significant loss of information. To overcome this, **Bayesian hierarchical negative binomial model** has been developed, which can be used for hundreds of host characteristics. Moreover, the model can offer a comprehensive solution considering the coefficient estimation without compromising the type I error. The model was applied to the subset of the American Gut Project to know the healthy microbiome composition and effect of certain diet, environment, and genotypes of the host on its microbiome (Pendegraft et al., 2019). Further, **negative binomial mixed models** have also been developed for detecting the linkage of the host microbiome with its environmental or clinical factors. This tool has been used for the correlation of microbiome count considering heterogeneity dependence in microbiome dimensions. This model was applied to analyze the mouse gut microbiome to evaluate the response of high-fat diet in its composition by evaluating the key altered microbial taxa (Zhang et al., 2017). The outcome of the application revealed that the model could outperform the previously designed model with respect to both empirical powers along with a type I error. As the designed model was used for assessing the clustered microbiome data, in the subsequent model, its applicability was extended to the longitudinal microbiome count data. This extended version can include diverse forms of fixed and random effects. Along with this, it is capable of incorporating diverse correlational structures within the same subjects amid observations. Henceforth, using this model, addressing the longitudinal microbiome count data makes possible.

Apart from this, various zero-inflated models have also been proposed with amalgamated characteristics of diverse models. One such model is the **zero-inflated beta-binomial model** (ZIBB). The proposed model is a mixture of two components (Zhou and Wright, 2015; Hu et al., 2018). The first one is a zero model that accounts for excess zero in the read count data, while the second component is a count model for capturing the remaining components with the utilization of the beta-binomial regression. The second component allows for the overdispersion effect. This model has the ability to control the type I error. Moreover, it also has higher power for detecting the taxa linked with a phenotype. However, for a smaller proportion of OTUs, it is likely that this model might fail to converge. The reason that had been attributed for this is the prevalence of fewer non-zero counts. In such a case, the power to detect the alliance with the experimental variable is quite low. To overcome the challenge, the permutation values had been coded in place of the significance value by a recent study to depict the microbial changes (Hu et al., 2018).

5.4 Multivariate Statistical Tools Used in Microbiome Studies

Multivariate statistics is used for assessing more than two variables simultaneously. This statistical model is used for analyzing the microbial communities with respect to the environmental covariates along with the microbial data. Establishment of the relationship between the microbiome composition and the environmental factors is difficult with the aid of OTUs or taxa abundances as the data are non-normal, having multiple dimensions along with the phylogenetic structures. In such a case, multivariate tools provide an advantage in testing the association by allowing to choose the relevant distance measure matrices. Thereafter, estimated distances are analyzed, where the distance matrices are equated between two microbial samples. Multivariate models

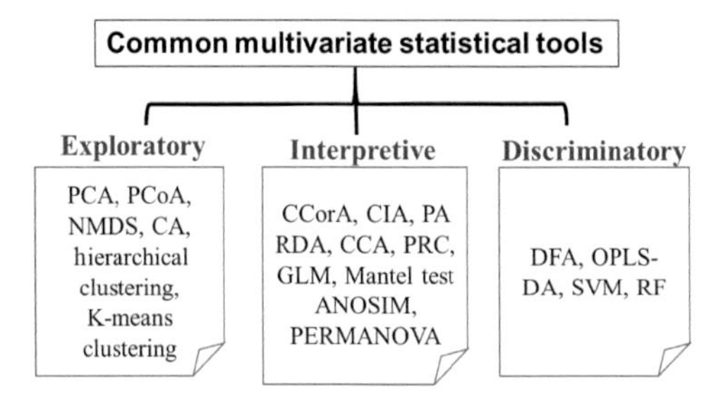

FIGURE 5.2 Different categories of multivariate statistical analysis plots.

have been undergoing diverse modifications from its classical forms. Based on the primary study goal of a multivariate analysis, the diverse techniques can be classified into three broad categories: exploratory, interpretive, and discriminatory methods (Figure 5.2). With the aid of the exploratory methods, it becomes possible to investigate the relationships that exist among the objects depending on the measured values of the variable. On the other hand, interpretive methods along with a measured set of variables take into consideration explanatory variables as supplementary for finding the axes in the multidimensional dataset. Additionally, discriminatory methods are the extension of interpretive techniques. With the aid of this method, it becomes possible to describe the synthetic variables of the hyperspace planes.

One of the classical examples of multivariate analysis that fits into the microbiome data analysis is the principal components analysis (PCA). It is the simplest approach that aids to combine multiple variables into one plot. This method can be used for accomplishing two goals. First, it becomes possible to achieve the visualized form of the relationship existing between the samples based on all the features with the principal component scores; second, it provides the opportunity to relate the features within and across the diverse variables via principal component directions. This multivariate tool also suffers a couple of limitations. First, it is not possible to retrieve the relationship existing among the sets of the variables that define the table. Second, in case if some variables have more expression or values than others, it is likely those will dominate the subsequent ordination. These limitations could be overcome by the application of CCA (canonical correspondence analysis) and MFA (multiple factor analysis). The statistical tool CCA finds its application in comparing the sets of features across the samples; similar to PCA, it offers low-dimensional representation of the data. However, it permits the comparison of the observations at the variable level for finding the low-dimensional representations with a maximum covariance. This process is analogous to the process of interpretation of the maximum variance. In the study conducted by Chen et al. (2013), a modified version of CCA, ssCCA (structure-constrained sparse canonical correlation analysis), has been successfully evaluated to identify the key microbial variation in the human gut microbiome dataset in association with the nutrient intake. On the other hand, MFA is a factorial method in which the group of individuals are represented as structured variables either in the qualitative or quantitative form. The most crucial advantage of this statistical tool is that it supports the integration of the structured group of variables. This tool has been used to comprehend the interaction occurring between the microbiome with that of the antibiotic intake. Moreover, it permits the integration of the structured groups of variables (Raymond et al., 2016).

The other statistical tool that has emerged as one of the prominent analytical methods for facilitating variation analysis in the species abundance associating environmental conditions, is Co-inertia Analysis (CIA). It is slightly different from CCA as it maximizes the covariance rather than correlation between the scores for finding its first directions. This method has been applied in several studies, especially for revealing the relationship that exists between the human-associated microbiome and metabolome datasets in the elderly gut (Claesson et al., 2012).

Additionally, principal coordinate analysis (PCoA) is another tool that finds its applicability in exploring and visualizing the similarities and dissimilarities prevailing among the microbiome data. The initiation of this analysis takes place either with a similarity or dissimilarity indices. Further, the ordination technique also helps to reduce the dimensionality of the microbiome datasets. This helps to visualize the summary of the β diversity relationship either in the two- or three-dimensional scatterplots. In these plots, each of the principal coordinates explains inertia (fraction of the variability). This provides the visual representation of the microbial community with respect to the compositional differences prevailing among the samples (Goodrich et al., 2014). Further, the quantification of the dissimilarity prevailing between the microbial compositions between two diverse samples or environments is performed with PCoA coupled with Bray–Curtis distance matrix. This algorithm was developed by J. Roger Bray and John T. Curtis (Bray and Curtis, 1957) based on the abundance or read count, and is bounded from 0 to 1. The value 0 indicates that both the samples have similar types of species, while the value 1 indicates that the samples or particular microbial variable are separate in between two samples or groups. This statistical tool has been used by Li et al. as a metric of similarity between the gut bacterial communities prevailing in the captive and musk deer depending on the OTUs abundance in the sample (Li et al., 2017b). Moreover, the Euclidean matrix has also been used in constructing PCoA plots depicting the microbial profiles of the groups (Hawinkel et al., 2019).

The other algorithm that is used to measure the differences existing between the sequences retrieved from the 16S rRNA of the two microbial samples is UniFrac. The difference is measured as the amount of the evolutionary history which is unique to either of the two. Further, it is measured as the fraction of the branch length in a phylogenetic tree. This distance measuring tool was classified as weighted and unweighted in 2007 by Lozupone et al. (2007, 2011). Unweighted UniFrac takes into consideration the information related to the presence and absence of the microbial species in the samples. Further, it counts the fraction of the branch length that is unique to each microbial community. On the other hand, the species abundance is included in the weighted one. Despite popularity, the major limitation associated with UniFrac is that both the measures assign excessive weight either to the abundant or rare lineages, respectively. Therefore, it would not be possible to detect the changes in the moderately abundant lineages or moderately abundant microbial taxa. However, in order to overcome this limitation, Chen et al. (2012a) have developed a generalized UniFrac distance to measure the changes in microbial composition in a wide range of microbial abundance data in the samples. This tool has been developed based on the variance-adjusted weighted UniFrac distance, commonly called as VA-WUniFrac (Chang et al., 2011).

5.5 Statistical Tests to Signify the Multivariate Plots

In this section, we will discuss the commonly used tests to signify the output generated in the multivariate analysis.

Permutational multivariate ANOVA, abbreviated as PERMANOVA, first described by Anderson, is a geometric partitioning of variation and semi-parametric multivariate statistical test (Anderson, 2014). It finds its utilization in comparing the groups of objects along with testing the null hypothesis. The analysis is carried out considering either Euclidean or non-Euclidean-embeddable dissimilarity matrices. This statistical tool has been improved by Tang et al. (2016) and has been named as PERMANOVA-S, to test the association of microbiome composition with the covariates of interest impacting the microbial compositional variations. The advantage of this modified tool over the original is that it allows adjustments of the flexible cofounder along with assemblage of multiple distances.

Parallel to PERMANOVA, the other most widely used tool in microbiology is ANOSIM (Analysis Of SIMilarities). It was developed by K. R. Clarke (Clarke, 1993) for comparing the similarities within and between the groups via distance matrix. This is also performed to evaluate the null hypothesis that indicates that the average rank similarity is the same for samples belonging to either similar or different groups. Li et al. (2017b) have used ANOSIM to demonstrate the differences prevailing in the gut microbiota between the captive and the wild forest musk deer. Further, this tool has also been used to depict tribe-specific gut microbial composition (Dehingia et al., 2015). The study conducted by Sharma et al. (2019) has reported the significance of convergence using ANOSIM between the microbiome profiles of US Airforce cadets that lived in neighboring squadrons. Additionally, ANOSIM has also been applied for evaluating the dissimilarities present in the microbial composition between the ADHD (attention-deficit hyperactivity disorder) patients and healthy controls (Prehn-Kristensen et al., 2018).

In addition, multi-response permutation procedures (MRPP) are also commonly used to confirm whether the groups have a significant different variables expression between them. This test has been used to test the significant difference of the β diversity of the microbes residing in the soil sample based on both geographic and disease factors (Liu et al., 2016). Further, it is also used to compare the community dissimilarities coupled with the Bray–Curtis distances matrix (Yan et al., 2016). This statistical tool calculates the distances between the observations and generates a weighted average of the distance. On the other hand, for testing alliance/co-association between the host microbiome and the environmental factor, Mantel's test has been used; this test has been developed by Nathan Mantel to explore the co-relation between two variables in biological data (Mantel, 1967). Recently, it has been used to depict the association among the host–microbial profile, host phylogeny, and metabolite profiles (Li et al., 2017a). Further, the test has been used to assess the statistical significance of the dissimilarities prevailing in the bacterial communities of Chimpanzee along with Sooty mangabey microbiomes (Gogarten et al., 2018).

Moreover, the average percent contribution of individual variables to the dissimilarity between subjects in a Bray–Curtis dissimilarity matrix could be done with the use of the similarity percentages breakdown (SIMPER) procedure (Clarke, 1993). It helps to identify the variables that are significantly responsible for the differences of the group of samples. This method is widely used in the ecology to determine which group has the most evident differences. It has also been used to identify 16S and 18S rRNA gene amplicon sequence variants rather than OTUs that were contributing to the differences (Wright et al., 2019). However, this test has limited capacity in detecting the key different taxa in between two groups which may result in failure to accurately account for the mean–variance relationship.

Apart from the above-mentioned statistical tools, Mandal et al. (2015) have introduced a novel statistical tool, termed as analysis of composition of microbiomes (ANCOM). This statistical tool helps to compare the compositional data of microbiomes in two or more samples or populations. Additionally, the framework is free from distributional assumptions; therefore, it can be applied with a linear model framework for adjusting the covariates along with model longitudinal data. Furthermore, the statistical framework is also efficient in scaling the comparison of the samples that include thousands of taxa. In addition, as compared to the performance of the standard t-test and a recently published methodology called zero-inflated Gaussian (ZIG) methodology, ANCOM has the ability to control the false discovery rate (FDR). This statistical tool has been used to collect the information about the taxa abundance related to the gut bacteria composition and metabolites in 1-month-old infants, in association with the concentration of the environmental toxicants in mothers' breast milk (Iszatt et al., 2019).

The other vital tool that has been developed for carrying out microbiome compositional data analysis is ALDEX (ANOVA-like differential expression tool). This tool is featured to exploit the log-ratio transformations. It is applied for analyzing the metagenomic data and distinguishing the abundant features in three or more groups (Fernandes et al., 2014). At present, the second version of this compositional analytical tool is being readily used, which uses Bayesian methods for inferring the technical and statistical errors. This tool has been used to identify the key differential microbial taxa that creates the distinction between tongue dorsum and buccal mucosal microbiome studied in the Human Microbiome Project dataset (Fernandes et al., 2014). Moreover, $ALDEX_2$ is capable of reducing the false-positive results. As compared to ALDEX that used effect size and post hoc for significance test, $ALDEX_2$ uses Welch's *t* or Wilcoxon tests. Further, the minimum dataset required for ALDEX is two samples per group, while for $ALDEX_2$, it is three samples per group.

The dynamicity of the microbes coupled with interaction with the host and the environment varies over time. Therefore, it is crucial to understand the microbial profile as per the interactive fluctuation. This can be achieved with the aid of the longitudinal microbiome data analysis. In this regard, diverse computation methods have been developed and applied in the microbiome-based studies. One such model is the regression-based time series model. This model helps to analyze the dependent variables along with relative abundance of OTUs. It has also been applied for assessing the function of time with diverse independent covariates. The method has been used for the evaluation of the human vaginal microbiome during the menstrual period along with other covariates (Gajer et al., 2012; Gerber, 2015). However, it is crucial to modify the existing longitudinal approaches for achieving the proper application of the statistical tools while dealing with dynamic and complicated data of microbiome. This is because microbiomes not only have causative effects on the host but the composition is also personalized. Moreover, the mutual relationship that is being shared between the host and the microbiome favors the applicability of either causal inference model, or mediation analysis and longitudinal analysis.

In addition to evaluating the performance properties of these tests that is type I and II errors, P values, power and sample size calculations, and other results, Monte Carlo simulation study is commonly applied, which further predicts the probability of different outcomes in a process/analysis that cannot easily be predicted due to the intervention of random variables. Additionally, the Dirichlet-multinomial distribution has been used in recent studies for depicting the microbiome data. This tool is used for power calculation along with sample sizes for experimental designs (Holmes et al., 2012). It has also been used for power calculation along with sample sizes for experimental designs and to compare microbiome across the groups (La Rosa et al., 2012a). Additionally, it has also been applied for estimating the key parameters that describe the properties of the microbiome (La Rosa et al., 2012a; Holmes et al., 2012). Moreover, it can retain ample amounts of information available in the data; therefore, it is considered to have an advantage over non-parametric models like bootstrapping and permutation testing.

5.6 Limitations and Future Direction

Multivariate statistical tools have rendered a multitude of benefits in the microbiome-based studies. However, the outputs generated by these statistical tools are far more difficult to interpret than its univariate counterpart. Moreover, many of the multivariate methods demand computational expertise and computing resources for dealing with large datasets. Further, the crucial limitation associated with the ordination techniques is that distance matrices often do not correctly account for the mean–variance relationship while combining the data across the variables. Statistical tools and models are being readily developed and used to analyze the association of the microbial community with other environmental covariates. Therefore, with the further development of either the novel statistical tools or modifying the existing ones, it could be possible to understand the outcomes of this interactions significantly. One such model is **null model** which tends to satisfy the collection of the constraints that is the contrarily considered to be an unbiased random structure. β Mean-nearest taxon distance (β-MNTD; Fine and Kembel, 2011; Stegen et al., 2013) is used in the model to measure the degree of certain variables impacting the host–microbiome composition. Recently, this approach has been used to calculate the degree to which the fish microbial composition is determined by environmental selection (Zha et al., 2020).

The other model that holds future promise in the field of microbiome research is **Source Tracker Analysis**, which is a Bayesian approach that has the ability to forecast the source of microbial communities present within the input samples (Knights et al., 2011). Moreover, contamination is one of the most significant issues associated with metagenomic studies, and with the application of this model, it becomes possible to measure the extent of contamination. **Partial redundancy (pRDA) analysis** is another analytical tool that intends to remove the consequence of one or more explanatory variables, on a set of response variables. This can be readily used when well-featured variables obscure the effects of other valuable explanatory ones. Followed by this, the other key statistical test is **partial least squares (PLS) regression** analysis. It is associated to certain extent with principal components regression rather than finding or discovering the hyperplanes that exist between response and independent variables. The model finds its application in finding the relationship that exists between the two matrices. Thus, it takes a latent variable approach in which the covariance structures are modeled in these two spaces. This model was initially developed by Swedish statistician Herman O. A. Woldand Svante Wold, as a statistical tool for social sciences, which is also used in microbiome data analysis (Wold et al., 2001).

The multivariate analytical technique provides the several choices for selecting the algorithm to analyze the large-scale dataset. We have tabulated a few appropriate choices based on the research goal, data input structure, and expected relationships among variables in Table 5.1. The utilization of these techniques will help in the development of the linkage or association of the microbial communities with its ecology, function, environmental gradients, and variables associated with time and space. Further, with the utility of the multivariate tests, it could be possible to discriminate between the sites, diseases, and environments based on the microbial community (Lozupone et al., 2013; Shankar et al., 2015). Additionally,

TABLE 5.1

Potential Choices of Multivariate Plots Based on the Research Aims

Research Aim	Analysis Tool
To explore the relationships that exist among the measured variables	Principal component analysis (PCA) Principal coordinate analysis (PCoA) Non-metric multidimensional scaling (NMDS) Correspondence analysis (CA) Cluster analysis (CLA): Hierarchical clustering (HCA) and Disjoint clustering (K-means clustering)
For the exploration of key similarities in the variables of the samples	Canonical correlation analysis (CCorA) Co-inertia analysis (CIA) Procrustes analysis (PA) Redundancy analysis (RDA) Canonical correspondence analysis (CCA) Principal response curves (PRC) Generalized linear modeling (GLM)
To explore the variations or key discriminant variables in the samples	Discriminant function analysis (DFA) Orthogonal projections to latent structures discriminant analysis (OPLS-DA) Support vector machine (SVM) Random forest (RF) distribution analysis

it will also help to gain a deeper insight into the mechanistic interactions between the complex microbial communities along with their host and environment, and can also be useful in providing the personalized solution to the host.

Acknowledgments

Priyanka gratefully acknowledges the guidance of Dr. Rupjyoti Talukdar, Dr. Hilloljyoti Singha, and Dr. Mojibur R. Khan, especially Dr. Singha for teaching her Biostatistics. The author is thankful to Dr. Talukdar for providing the input towards the manuscript preparation and fellowship to the author under the Wellcome/DBT India Alliance project grant (No. IA/CPHS/17/1/503358). Asian Healthcare Foundation, Asian Institute of Gastroenterology is also thankfully accredited for providing the essential research infrastructural facilities to the author.

REFERENCES

Alekseyenko, A.V., Perez-Perez, G.I., De Souza, A., Strober, B., Gao, Z., Bihan, M., Li, K., Methé, B. A. and Blaser, M. J. 2013. Community differentiation of the cutaneous microbiota in psoriasis. *Microbiome* 1(1), 31.

Anders, S. and Huber, W. 2010. Differential expression analysis for sequence count data. *Nature Precedings* 30, 1.

Anderson, M.J. 2014. Permutational multivariate analysis of variance (PERMANOVA). *Wiley Stats Ref: Statistics Reference Online*, 1–15.

Bray, J.R. and Curtis, J.T. 1957. An ordination of upland forest communities of southern Wisconsin. *Ecological Monographs* 27(4), 325–349. Change in Marine Communities: An Approach to Statistical Analysis and Interpretation.

Carmody, R.N., Gerber, G.K., Luevano Jr, J.M., Gatti, D.M., Somes, L., Svenson, K.L. and Turnbaugh, P.J. 2015. Diet dominates host genotype in shaping the murine gut microbiota. *Cell Host & Microbe* 17(1), 72–84.

Chang, Q., Luan, Y. and Sun, F. 2011. Variance adjusted weighted UniFrac: a powerful beta diversity measure for comparing communities based on phylogeny. *BMC Bioinformatics* 12(1), 118.

Chen, J., Bittinger, K., Charlson, E.S., Hoffmann, C., Lewis, J., Wu, G.D., Collman, R.G., Bushman, F.D. and Li, H. 2012a. Associating microbiome composition with environmental covariates using generalized UniFrac distances. *Bioinformatics* 28(16), 2106–2113.

Chen, J., Bushman, F.D., Lewis, J.D., Wu, G.D. and Li, H. 2013. Structure-constrained sparse canonical correlation analysis with an application to microbiome data analysis. *Biostatistics* 14(2), 244–258.

Chen, W., Liu, F., Ling, Z., Tong, X. and Xiang, C. 2012b. Human intestinal lumen and mucosa-associated microbiota in patients with colorectal cancer. *PLoS One* 7(6), e39743.

Chumpitazi, B.P., Cope, J.L., Hollister, E.B., Tsai, C.M., McMeans, A.R., Luna, R.A., Versalovic, J. and Shulman, R.J. 2015. Randomised clinical trial: gut microbiome biomarkers are associated with clinical response to a low FODMAP diet in children with the irritable bowel syndrome. *Alimentary Pharmacology Therapeutics* 42(4), 418–427.

Claesson, M.J., Jeffery, I.B., Conde, S., Power, S.E., O'connor, E.M., Cusack, S., Harris, H.M., Coakley, M., Lakshminarayanan, B., O'Sullivan, O. and Fitzgerald, G.F. 2012. Gut microbiota composition correlates with diet and health in the elderly. *Nature* 488(7410), 178–184.

Clarke, K.R. 1993. Non-parametric multivariate analyses of changes in community structure. *Austral Ecology* 18(1), 117–143.

Dehingia, M., Talukdar, N. C., Talukdar, R., Reddy, N., Mande, S. S., Deka, M. and Khan, M. R. 2015. Gut bacterial diversity of the tribes of India and comparison with the worldwide data. *Scientific Reports* 5, 18563.

Fernandes, A.D., Reid, J.N., Macklaim, J.M., McMurrough, T.A., Edgell, D.R. and Gloor, G.B. 2014. Unifying the analysis of high-throughput sequencing datasets: characterizing RNA-seq, 16S rRNA gene sequencing and selective growth experiments by compositional data analysis. *Microbiome* 2(1), 15.

Fine, P.V. and Kembel, S.W. 2011. Phylogenetic community structure and phylogenetic turnover across space and edaphic gradients in western Amazonian tree communities. *Ecography* 34(4), 552–565.

Gajer, P., Brotman, R.M., Bai, G., Sakamoto, J., Schütte, U.M., Zhong, X., Koenig, S.S., Fu, L., Ma, Z.S., Zhou, X. and Abdo, Z. 2012. Temporal dynamics of the human vaginal microbiota. *Science Translational Medicine* 4(132), 132ra52–132ra52.

Gerber, G.K. 2015. Longitudinal microbiome data analysis. *Metagenomics for Microbiology*, 97–111. Academic Press.

Gogarten, J.F., Davies, T.J., Benjamino, J., Gogarten, J.P., Graf, J., Mielke, A., Mundry, R., Nelson, M.C., Wittig, R.M., Leendertz, F.H. and Calvignac-Spencer, S. 2018. Factors influencing bacterial microbiome composition in a wild non-human primate community in Taï National Park, Côte d'Ivoire. *ISME* 12(10), 2559–2574.

Goodrich, J.K., Di Rienzi, S.C., Poole, A.C., Koren, O., Walters, W.A., Caporaso, J.G., Knight, R. and Ley, R.E. 2014. Conducting a microbiome study. *Cell* 158(2), 250–262.

Guinane, C.M. and Cotter, P.D. 2013. Role of the gut microbiota in health and chronic gastrointestinal disease: understanding a hidden metabolic organ. *Therapeutic Advances in Gastroenterology* 6(4), 295–308.

Hawinkel, S., Kerckhof, F.M., Bijnens, L. and Thas, O. 2019. A unified framework for unconstrained and constrained ordination of microbiome read count data. *PLoS One* 14(2), e0205474.

Holmes, I., Harris, K. and Quince, C. 2012. Dirichlet multinomial mixtures: generative models for microbial metagenomics. *PLoS One* 7(2), e30126.

Hsiao, E.Y., McBride, S.W., Hsien, S., Sharon, G., Hyde, E.R., McCue, T., Codelli, J.A., Chow, J., Reisman, S.E., Petrosino, J.F. and Patterson, P.H. 2013. The microbiota modulates gut physiology and behavioral abnormalities associated with autism. *Cell* 155(7):1451.

Hu, T., Gallins, P. and Zhou, Y.H. 2018. A zero-inflated beta-binomial model for microbiome data analysis. *Stat* 7(1), e185.

Iszatt, N., Janssen, S., Lenters, V., Dahl, C., Stigum, H., Knight, R., Mandal, S., Peddada, S., González, A., Midtvedt, T. and Eggesbø, M. 2019. Environmental toxicants in breast milk of Norwegian mothers and gut bacteria composition and metabolites in their infants at 1 month. *Microbiome* 7(1), 34.

Iwai, S., Fei, M., Huang, D., Fong, S., Subramanian, A., Grieco, K., Lynch, S.V. and Huang, L. 2012. Oral and airway microbiota in HIV-infected pneumonia patients. *Journal of Clinical Microbiology* 50(9), 2995–3002.

Jandhyala, S.M., Madhulika, A., Deepika, G., Rao, G.V., Reddy, D.N., Subramanyam, C., Sasikala, M. and Talukdar, R. 2017. Altered intestinal microbiota in patients with chronic pancreatitis: implications in diabetes and metabolic abnormalities. *Scientific Reports* 7, 43640.

Kim, K.A., Jung, I.H., Park, S.H., Ahn, Y.T., Huh, C.S. and Kim, D.H. 2013. Comparative analysis of the gut microbiota in people with different levels of ginsenoside Rb1 degradation to compound K. *PLoS One* 8(4), e62409.

Knights, D., Kuczynski, J., Charlson, E.S., Zaneveld, J., Mozer, M.C., Collman, R.G., Bushman, F.D., Knight, R. and Kelley, S.T. 2011. Bayesian community-wide culture-independent microbial source tracking. *Nature Methods* 8(9), 761–763.

La Rosa, P.S., Brooks, J.P., Deych, E., Boone, E.L., Edwards, D.J., Wang, Q., Sodergren, E., Weinstock, G. and Shannon, W.D. 2012a. Hypothesis testing and power calculations for taxonomic-based human microbiome data. *PLoS One* 7(12), e52078.

La Rosa, P.S., Shands, B., Deych, E., Zhou, Y., Sodergren, E., Weinstock, G. and Shannon, W.D. 2012b. Statistical object data analysis of taxonomic trees from human microbiome data. *PLoS One* 7(11), e48996.

La Rosa, P.S., Zhou, Y., Sodergren, E., Weinstock, G. and Shannon, W.D. 2015. Hypothesis testing of metagenomic data. *Metagenomics for Microbiology*, 81–96. Academic Press.

Ley, R.E., Hamady, M., Lozupone, C., Turnbaugh, P.J., Ramey, R.R., Bircher, J.S., Schlegel, M.L., Tucker, T.A., Schrenzel, M.D., Knight, R. and Gordon, J.I. 2008. Evolution of mammals and their gut microbes. *Science* 320(5883), 1647–1651.

Li, Y., Hu, X., Yang, S., Zhou, J., Zhang, T., Qi, L., Sun, X., Fan, M., Xu, S., Cha, M. and Zhang, M. 2017b. Comparative analysis of the gut microbiota composition between captive and wild forest musk deer. *Frontiers in Microbiology* 8, 1705.

Li, T., Long, M., Li, H., Gatesoupe, F.J., Zhang, X., Zhang, Q., Feng, D. and Li, A. 2017a. Multi-omics analysis reveals a correlation between the host phylogeny, gut microbiota and metabolite profiles in cyprinid fishes. *Frontiers in Microbiology* 8, 454.

Liu, X., Zhang, S., Jiang, Q., Bai, Y., Shen, G., Li, S. and Ding, W. 2016. Using community analysis to explore bacterial indicators for disease suppression of tobacco bacterial wilt. *Scientific Reports* 6(1), 1–11.

Lozupone, C.A., Hamady, M., Kelley, S.T. and Knight, R. 2007. Quantitative and qualitative β diversity measures lead to different insights into factors that structure microbial communities. *Applied and Environmental Microbiology* 73(5), 1576–1585.

Lozupone, C.A., Li, M., Campbell, T.B., Flores, S.C., Linderman, D., Gebert, M.J., Knight, R., Fontenot, A.P. and Palmer, B.E. 2013. Alterations in the gut microbiota associated with HIV-1 infection. *Cell Host & Microbe* 14(3), 329–339.

Lozupone, C., Lladser, M.E., Knights, D., Stombaugh, J. and Knight, R. 2011. UniFrac: an effective distance metric for microbial community comparison. *ISME* 5(2), 169–172.

Mandal, S., Van Treuren, W., White, R.A., Eggesbø, M., Knight, R. and Peddada, S.D. 2015. Analysis of composition of microbiomes: a novel method for studying microbial composition. *Microbial Ecology in Health and Disease* 26(1), 27663.

Mantel, N. 1967. The detection of disease clustering and a generalized regression approach. *Cancer Research* 27(2 Part 1), 209–220.

Nishida, A., Inoue, R., Inatomi, O., Bamba, S., Naito, Y. and Andoh, A. 2018. Gut microbiota in the pathogenesis of inflammatory bowel disease. *Clinical Journal of Gastroenterology* 11(1), 1–10.

Pendegraft, A. H., Guo, B. and Yi, N. 2019. Bayesian hierarchical negative binomial models for multivariable analyses with applications to human microbiome count data. PLoS ONE. 14(8), e0220961.

Peterson, J., Garges, S., Giovanni, M., McInnes, P., Wang, L., Schloss, J.A., Bonazzi, V., McEwen, J.E., Wetterstrand, K.A., Deal, C. and Baker, C.C. 2009. The NIH human microbiome project. *Genome Research* 19(12), 2317–2323.

Prehn-Kristensen, A., Zimmermann, A., Tittmann, L., Lieb, W., Schreiber, S., Baving, L. and Fischer, A. 2018. Reduced microbiome alpha diversity in young patients with ADHD. *PLoS One* 13(7), e0200728.

Qin, J., Li, Y., Cai, Z., Li, S., Zhu, J., Zhang, F., Liang, S., Zhang, W., Guan, Y., Shen, D. and Peng, Y. 2012. A metagenome-wide association study of gut microbiota in type 2 diabetes. *Nature* 490(7418), 55–60.

Raymond, F., Ouameur, A.A., Déraspe, M., Iqbal, N., Gingras, H., Dridi, B., Leprohon, P., Plante, P.L., Giroux, R., Bérubé, È. and Frenette, J. 2016. The initial state of the human gut microbiome determines its reshaping by antibiotics. *ISME* 10(3), 707–720.

Robinson, M. D., McCarthy, D. J., Smyth, G. K. 2010. edgeR: a Bioconductor package for differential expression analysis of digital gene expression data. *Bioinformatics* 26(1), 139–140.

Shankar, V., Homer, D., Rigsbee, L., Khamis, H. J., Michail, S., Raymer, M., Reo, N. V. and Paliy, O. 2015. The networks of human gut microbe–metabolite associations are different between health and irritable bowel syndrome. *The ISME Journal* 9(8), 1899–1903.

Sharma, A., Richardson, M., Cralle, L., Stamper, C.E., Maestre, J.P., Stearns-Yoder, K.A., Postolache, T.T., Bates, K.L., Kinney, K.A., Brenner, L.A. and Lowry, C.A. 2019. Longitudinal homogenization of the microbiome between both occupants and the built environment in a cohort of United States Air Force Cadets. *Microbiome* 7(1), 1–17.

Spor, A., Koren, O. and Ley, R. 2011. Unravelling the effects of the environment and host genotype on the gut microbiome. *Nature Reviews Microbiology* 9(4), 279–290.

Stegen, J.C., Lin, X., Fredrickson, J.K., Chen, X., Kennedy, D.W., Murray, C.J., Rockhold, M.L. and Konopka, A. 2013. Quantifying community assembly processes and identifying features that impose them. *ISME* 7(11), 2069–2079.

Tang, Z.Z., Chen, G. and Alekseyenko, A.V. 2016. PERMANOVA-S: association test for microbial community composition that accommodates confounders and multiple distances. *Bioinformatics* 32(17), 2618–2625.

Virgin, H.W. and Todd, J.A. 2011. Metagenomics and personalized medicine. *Cell* 147(1), 44–56.

Wen, L. and Duffy, A. 2017. Factors influencing the gut microbiota, inflammation, and type 2 diabetes. *The Journal of Nutrition* 147(7), 1468S–1475S.

White, J.R., Nagarajan, N. and Pop, M. 2009. Statistical methods for detecting differentially abundant features in clinical metagenomic samples. *PLoS Computational Biology* 5(4), e1000352.

Wold, S., Sjöström, M. and Eriksson, L. 2001. PLS-regression: a basic tool of chemometrics. *Chemometrics and Intelligent Laboratory Systems* 58(2), 109–130.

Wright, R.J., Gibson, M.I. and Christie-Oleza, J.A. 2019. Understanding microbial community dynamics to improve optimal microbiome selection. *Microbiome* 7(1), 1–14.

Yan, Q., Li, J., Yu, Y., Wang, J., He, Z., Van Nostrand, J.D., Kempher, M.L., Wu, L., Wang, Y., Liao, L. and Li, X. 2016. Environmental filtering decreases with fish development for the assembly of gut microbiota. *Environmental Microbiology* 18(12), 4739–4754.

Yang, H., Huang, X., Fang, S., Xin, W., Huang, L. and Chen, C. 2016. Uncovering the composition of microbial community structure and metagenomics among three gut locations in pigs with distinct fatness. *Scientific Reports* 6, 27427.

Yin, X., Peng, J., Zhao, L., Yu, Y., Zhang, X., Liu, P., Feng, Q., Hu, Y. and Pang, X. 2013. Structural changes of gut microbiota in a rat non-alcoholic fatty liver disease model treated with a Chinese herbal formula. *Systematic and Applied Microbiology*. 36(3), 188–196.

Zha, Y., Lindström, E.S., Eiler, A. and Svanbäck, R. 2020. Different roles of environmental selection, dispersal, and drift in the assembly of intestinal microbial communities of freshwater fish with and without a stomach. *Frontiers in Ecology and Evolution* 8, 152.

Zhang, C., Derrien, M., Levenez, F., Brazeilles, R., Ballal, S.A., Kim, J., Degivry, M.C., Quéré, G., Garault, P., van HylckamaVlieg, J.E. and Garrett, W.S. 2016. Ecological robustness of the gut microbiota in response to ingestion of transient food-borne microbes. *ISME* 10(9), 2235–2245.

Zhang, X., Mallick, H., Tang, Z., Zhang, L., Cui, X., Benson, A.K. and Yi, N. 2017. Negative binomial mixed models for analyzing microbiome count data. *BMC Bioinformatics* 18(1), 4.

Zhou, Y.H. and Wright, F.A. 2015. Hypothesis testing at the extremes: fast and robust association for high-throughput data. *Biostatistics* 16(3), 611–625.

Section II

Human Microbiome

6

Structure and Functional Role of Microbiome Associated with Specific Organs of Healthy Individuals

Shanmugaraj Gowrishankar
Alagappa University

Arumugam Kamaladevi
Bharathidasan University

Shunmugiah Karutha Pandian
Alagappa University

CONTENTS

6.1 Introduction

Until before a decade, microbiology associated with human infection had the perception of identifying and characterizing single microbe either bacteria/fungi or viruses. Later on, scientists started to work and understand the role of microbes that inhabit diverse parts and organs of human body (Marchesi and Ravel 2015). Today, with the advent of recent improvements in sequencing technologies, scientists delineated the way in which these microbes assemble into communities of different stages of complexity, and they coined the term as human 'microbiome'. Human microbiome is defined as the total communities of microorganisms residing in every parts/organs (*viz.* skin, respiratory tract including the nose, gastrointestinal (GI) tract including the mouth, gut, genitalia, ocular surface, etc.) and biofluids of human body (Backhed et al. 2012). Through a deeper sense of microbial ecology of human, we presently believe that humans encompass microbes toting up to three times of total human cells (37 trillion). Notably, the research outcomes on microbe–human (mammalian) interaction have significantly widened our knowledge on how the inter-kingdom crosstalk influences acute and chronic health illnesses (Davenport et al. 2017).

In the contemporary concept of human health, these native microbiota are being recognized as a crucial component and has acquired noteworthy attention. However, the structural composition and metabolic functional characteristics of a healthy microbiome still have huge gap to define precisely (Backhed et al. 2012; Marchesi and Ravel 2015; Davenport et al. 2017). On the other hand, the prototypes of microbial association in diversified infection states have been well demonstrated, which further necessitates the dire need of precise understanding about the structure and functional role of microbiome associated with specific organs of healthy individuals (Marchesi and Ravel 2015). Since insights into the human healthy microbiome in stipulations of public well-being and medicine persist to be mysterious, there exists several regional microbiome projects under progress as a linked global network, *viz.* International Human Microbiome Consortium, the US National Institutes of Health's Human Microbiome Project, the Canadian Microbiome Initiative, the European Commission's Metagenomics of the Human Intestinal Tract project, etc. (The Human Microbiome Project 2012; Backhed et al. 2012; Davenport et al. 2017).

6.2 Microbiome of the Gastrointestinal Tract

Intestinal microbiome in healthy individual commences to colonize at birth and is influenced chiefly by the model of delivery and feeding during early life (Moreno-Indias et al. 2014).

Later, the introduction of solid foods reassembles the infant to adult-like microbiota, after which it remains moderately stable throughout adulthood. The microbiome is also temporarily affected by diets, infections, use of antibiotics, etc., yet these disruptions are able to regain its former state shortly. The human gut is a densely populated and well-characterized microbial ecosystem found in the body. It is estimated that more than 1,000–1,500 species of microbes are dwelling in the adult intestinal tract (DiBaise et al. 2008). Every individual harbors about 160 bacterial species and 10 million microbial genes which attributes to a unique individual signatured microbial make-up (Li et al. 2014). In addition, the host genetics may also influence the variations in intestinal microbiota that affect the function and composition of indigenous bacterial community (Moreno-Indias et al. 2014).

The human gut microbiome is majorly occupied by two phyla such as *Bacteroidetes* and *Firmicutes* (Mariat et al. 2009). The phylum *Bacteroidetes* belongs to the genera *Bacteroides* and *Prevotella*, and *Firmicutes* belongs to *Clostridium*, *Eubacterium*, and *Ruminococcus*. In addition, *Proteobacteria*, *Lentisphaerase*, *Archea* (methanogens), and fungi are also a part of healthy human microbiome. The metagenomics data imply that individual disturbance in microbiome composition does not alter the metabolic pathways in the GI tract of healthy subjects (Lewis et al. 2012; Rup 2012). The microbial community found in the GI tract forms a dynamic ecosystem and exerts critical metabolic, immunological, and physiological functions in host. The concrete ecosystem of the GI tract depends on the physicochemical conditions such as pH, redox potential, motility, nutrients, mucus, and digestive enzymes. These factors alone or together create a unique environment that favors the growth of particular bacteria at the site. The human GI tract includes upper GI tract, small intestine, and large intestine, and the microbial abundance varies upon the function of the site. The microbial community present at each site of an organ is identified by the metatranscriptomics, metabolomics, and metaproteomics approaches. These approaches provide insights on microbial gene regulation, gene expression, and synthesis of metabolites including proteins, vitamins, and other regulatory elements.

The functional potential of the gut microbiome is well explored by employing metagenomic studies. The GI tract of each person harbors about ten million bacterial genes that are majorly involved in bacterial metabolism (Li et al. 2014; Turnbaugh et al. 2009). Furthermore, the recent techniques such as metabolomics, metatranscriptomics, and metaproteomics provide insights on microbial gene regulation, expression, and regulatory elements. Alike compositional diversity, the functional variation of conserved and stable microbial ecosystem encodes similar metabolic and functional pathways (Zoetendal et al. 2008; Lewis et al. 2012; Rup 2012).

6.3 Microbiome of the Upper Gastrointestinal Tract of Healthy Individuals

The composition of microbial ecosystem in the different parts of the GI tract such as esophagus, stomach, and duodenum are still scary owing to its poor accessibility and invasive procedure to obtain samples. Available resources on esophagus microbiota demonstrate the conserved microbial community to chiefly include *Firmicutes*, *Bacteroides*, *Actinobacteria*, *Proteobacteria*, *Fusobacteria*, *TM7*, *Streptococcus*, *Prevotella*, and *Veillonella* (Pei et al. 2004; Fillon et al. 2012). Like oral cavity, ingested food does not reside in the esophagus for long period to permit the establishment of resident microbiome. Notably, the stomach is the initial part of the GI tract that resides the food for extensive time. Hence, the stomach and descend regions of the GI tract accommodate spatially specific microbial distributions. The microbiota of the GI tract includes a population of microbes associated with gastric content and mucosal layer (Wang and Yang 2013). The gastric juice in the stomach provides a low pH that allows only the growth of acid-resistant bacteria. Hence, the total microbial population in the stomach is estimated about 10^3–10^4 bacterial cells mL^{-1} (Tlaskalova-Hogenova et al. 2011). Diet and influx of bacteria from the mouth, esophagus, and duodenum are the critical factors that affect the microbiota of gastric layer. However, these factors at its lesser extend influence the microbial population of mucosa, and maintain the stability of mucosa-associated microbiome (Wang and Yang 2013). An approach by a culture-dependent method revealed the abundance of the phyla *Firmicutes*, *Proteobacteria*, *Bacteroidetes*, and *Fusobacteria* and common genera such as *Streptococcus*, *Prevotella*, *Porphyromonas*, *Neisseria*, *Haemophilus*, and *Veillonella*. However, the distribution of the taxa is noted to be highly variable at its genus level among the individuals (Stearns et al. 2011; Bik et al. 2006). Paradoxically, the GI tract of most of the healthy population contains about 50% of *Helicobacter pylori*, which is associated with gastric diseases, *viz.* gastritis and cancer (Wang and Yang 2013).

Till date, the microbiota function of the upper GI tract is not well studied. Though the ecology of the esophagus and stomach is minimally explored, its role in protection against pathogens is well recognized. The microenvironment of normal microbiota prohibits the growth of various etiological agents by competing for nutrients and binding sites, triggering immune responses, and synthesizing antimicrobial peptides (AMPs). An *in vitro* and *in vivo* paradigm deciphers the colonization resistance potential of stomach commensal microbiota against *H. pylori*. The probiotic bacteria such as *Bifidobacterium*, *Lactobacillus*, and *Saccharomyces* involve chiefly in hampering the growth of pathogens and provide a persistent pathogen resistance in the stomach (Wang and Yang 2013). Importantly, microbes-associated pathogenic substances, *viz.* lipopolysaccharide, lipotichoeic acid, lipoproteins, flagellins, and nucleic acids, could also interact with the common microbiota of mucosa and induce chronic inflammation, alter mucin production, and cause metaplasia that eventually leads to illness (Tlaskalova-Hogenova et al. 2011; Cheng et al. 2013). A comparative study of microbial composition between a celiac disease patient and a healthy individual reinforces the role of microbiota in immune response, inflammation, and in maintaining gut homeostasis (Cheng et al. 2013; Wacklin et al. 2013). The degree of gut homeostasis relies on the adequate activation of toll-like receptors (TLRs) by microbes-associated motifs, regulation of pathogen-specific immune response, and the potential to regulate tight junction proteins such as mucin and AMPs (Cheng et al. 2013).

6.4 Microbiome of the Small Intestine of Healthy Individuals

The small intestine is a site where the enzymatic digestion of food and absorption of nutrients take place. Investigations on healthy GI microbial diversity is limited because the majority of the reported findings are only based on the biopsy specimens associated with GI disorders. In addition, there is a lack of focus on culture-independent investigations on the human duodenal resident microbiota. A recent study on duodenal biopsies from children by employing 16S rRNA gene-targeted HITChip analysis revealed the unique and different individual microbial profiles containing 13 bacteria groups including *Proteobacteria*, *Bacilli*, and *Bacteroidetes* (Cheng et al. 2013). It also documented the microbes that are predominantly present, *viz. Sutterella wadsworthensis et rel.*, *Streptococcus mitis et rel.*, *Aquabacterium*, *Streptococcus intermedius et rel.*, and *Prevotella melaninogenica et rel.* (Cheng et al. 2013). Nistal et al. (2012) conducted another study by 16S rRNA sequencing on biopsies from children and adults, and detected the microbial population belonging to the phyla *Firmicutes*, *Proteobacteria*, and *Bacteroidetes*. They also found *Actinobacteria*, *Fusobacteria*, and *Deinococcus-Thermus* to be dominant. Interestingly, most of the identified sequences in both age groups are categorized into the species of *Streptococcus* and *Prevotella*, and 5% of the identified sequences from the healthy individuals are not assigned to any known genus. Moreover, as anticipated, the microbial richness in adult group is higher than that in the children group, with the microbial species of *Veillonella*, *Neisseria*, *Haemophilus*, *Methylobacterium*, and *Mycobacterium*. It is also remarkable to note that the microbial community present in the duodenum is similar to the microbial diversity of oral cavity and esophagus, but different from the diversity of lower GI tract (Wacklin et al. 2013).

Along the intestine, the number of bacterial cells is found to be increased, and it is estimated that the jejunum harbors 10^5–10^6 bacterial cells mL^{-1} (Tlaskalova-Hogenova et al. 2011). The mucosa biopsies of human jejunum showed the abundance of *Streptococcus* and *Proteobacteria* to about 68% and 13%, respectively (Wang et al. 2005). Additionally, the ileostoma effluent and jejunum resides a microbial population dominated by *Bacilli* (*Streptococcus* spp.), *Clostridium* cluster IX (*Veillonella* spp.), *Clostridium* cluster XIVa, and *Gammaproteobacteria* (Zoetendal et al. 2012). The another study by Booijink et al. (2010) on ileostoma effluent documented the members of rich species such as *Lactobacillales* and *Clostridiales*, *Streptococcus bovisrelated* and *Veillonella* group, as well as species belonging to *Clostridium* cluster I and *Enterococcus*. However, the ileum-associated *Bacteroidetes* and *Clostridium* clusters III, IV, and XIVa are thinly populated in ileostoma effluent samples. The ileal digesta contains 10^8–10^9 cells mL^{-1}. The bacterial community in the lumen is found to be dominated by species belonging to *Bacteroidetes* and *Clostridium* clusters IV and XIVa (Tlaskalova-Hogenova et al. 2011; Wang et al. 2005). In parallel, microbial characterization of ileostoma effluent and ileum samples revealed the short- and long-term variations in the microbial profile within individuals. Also, there is a large inter-individual variability observed between the healthy individuals (Booijink et al. 2010).

The enzymatic digestion of ingested food and absorption of nutrients take place in the small intestine. Therefore, diet plays a pivotal role in modulating the microbial population by selectively shifting the group of bacteria equipped to degrade different dietary substances (Moreno-Indias et al. 2014). For instance, certain *Lactobacillus* spp. that reside in duodenum and jejunum determine the body weight by altering the rate of breakdown of dietary carbohydrate and fat in the host (Moreno-Indias et al. 2014). The predominant member of *Streptococcus* and *Veillonella* spp. in jejunum and ileum aides the rapid metabolism of lactate to acetate and propionate (Booijink et al. 2010). A metatranscriptomic analysis on ileostoma effluent demonstrated the abundance of *Streptococcus* bacteria that facilitate transport and metabolism of diet-derived carbohydrates (El Aidy et al. 2015). Alongside, the microbiota of the small intestine plays a vital role in the development of immune system and homeostasis. The gut-associated lymphoid tissue (GALT) and payer's patches-associated commensal bacteria induce specific immune response in host and help to eliminate the pathogenic bacteria (El Aidy et al. 2015). The microbes-derived metabolites or toxins may alter gene expressions through the gut–brain axis. This consequently influences the endocrine function (e.g., secretion of glucagon, serotonin, and incretins) and affects the mood or behavior of the host (Moreno-Indias et al. 2014; El Aidy et al. 2015a).

6.5 Microbiome of the Large Intestine of Healthy Individuals

The colon microbial diversity is stabilized and restored by the projection-like rudimentary structure called appendix (Bollinger et al. 2007). Unlike the small intestine, the structure and function of microbial diversity of the large intestine is studied to a greater extent. The ease of fecal collection made it possible to explore and enumerate the microbial diversity of the large intestine. The large intestine is found to harbor a high density of microbial population of about 10^{11}–10^{12} bacterial cells mL^{-1} (Tlaskalova-Hogenova et al. 2011). The predominant and versatile microbial population in the human large intestine includes *Bacteroides*, *Clostridium* clusters, *Bifidobacterium*, *Enterobacteriaceae*, and *Eubacterium*. Although the large intestine is very long and divided into five anatomical regions, the microbial diversity is found to be uniform throughout the entire length (Gerritsen et al. 2011). Nevertheless, similar to other parts of the GI tract, the large intestine also shows wide difference in the microbial ecosystem that prevails in the lumen and mucosal layer. The fecal samples help to characterize the microbial ecosystem of luminal fraction, whereas mucosal layer requires intense invasive methods in biopsy collections. The microbial ecosystem in the large intestine is relatively stable and highly unique (Lahti et al. 2014). Generally, the stable microbial ecosystem in the intestine is chiefly affected by age, diseases, and use of antibiotics (Lahti et al. 2014). Recent cohort studies on healthy individuals suggested that the fecal microbiota majorly contains three enterotypes of bacterial population, *viz. Bacteroides*, *Prevotella*, and *Ruminococcus* (Arumugam et al. 2011; Benson et al. 2010). The presence of these enterotypes is independent of age, ethnicity, gender, and body mass. Nevertheless, the abundance of these microbial populations is not found in both samples collected from elders (Claesson et al. 2012) and adults (Huse et al. 2012). Later, in 2014, a study by Lathti et al. suggested an alternative to enterotype theory defining the presence of different groups of microbial population including *Dialister* spp., *Bacteroides fragilis*, *Prevotella melaninogenica*, *P. oralis*, and two groups of uncultured *Clostridiales* clusters I and II in the healthy adult western human population. This bimodal distribution is represented as tipping element (Lahti et al. 2014). These bacterial groups are either abundant or absent and in unstable levels. Moreover, the population of *Bacteroides* and *Prevotella* have been believed to shift other bacteria, which results in maintaining the microbial composition of colonic ecosystem with specific enterotypes (Lahti et al. 2014).

The tremendous role of common microbiota of the large intestine is to accelerate the rate of digestion and metabolism, and provide an accessible source of energy like short-chain fatty acid (SCFA) (Leser and Molbak 2009). In addition, the colonic microbiome is a key source of vitamins B12 and K, which inhibit the colonization of pathogenic bacteria by regulating host immune responses (Moreno-Indias et al. 2014; Leser and Molbak 2009). Analyzing the colon ecosystem of Japanese healthy individuals revealed the enrichment of genes encoded for carbohydrate metabolism and transport, peptidase activity and anaerobic pyruvate metabolism, AMP transport, and multidrug efflux pump peptides. However, a minimum number of genes that encode for fatty acid metabolism is noted (Kurokawa et al. 2007). Amusingly, the observed pattern of gene distribution is not found in breast-fed infants, which clearly demonstrates that the infant microbiota remains less complex, less stable, and highly adaptable. However, the adult microbiota is highly diverse and stable, and it exerts large variability in inter-individual microbial composition among individuals (Turnbaugh et al. 2009; Kurokawa et al. 2007). An insight analysis on functional differences associated with different enterotypes suggests that *Bacteroides* and *Prevotella* produce vitamins to accelerate metabolism of carbohydrate (Arumugam et al. 2011). A decade ago, *Clostridium* clusters IV and XIVa attracted huge attention for their potential to produce SCFA and butyrate. *Eubacterium rectale* and *Faecalibacterium prausnitzii* are the major butyrate-producing bacterial species (Louis and Flint 2009). The presence of butyrate producers is believed to inhibit functional dysbiosis and pathogenic infection by attenuating the oxidative stress. The butyrate producers influence the bifidobacteria that assist the breakdown of polysaccharides (Moreno-Indias et al. 2014).

6.6 Microbiome of the Oral Cavity of Healthy Individuals

The oral cavity includes several different niches that provide a platform for the growth of unique microbial population. These microbes are commonly associated with teeth, braces, and mucosal surfaces of cheeks and tongue. However, no bacteria reside in the lumen because of the short food passage time. The diverse microbial ecosystem of the oral cavity contains 10^{12} bacterial cells, which include *Actinobacteria*, *Firmicutes*, *Spirochaetes*, *Tenericutes*, *Bacteroidetes*, *Proteobacteria*, and *Synergistetes* (Wade 2013; Tlaskalova-Hogenova et al. 2011; Simon-Soro et al. 2013; He et al. 2015). The genera of *Actinomyces*, *Selenomonas*, *Streptococcus*, *Neisseria*, *Porphyromonas*, and *Veillonella* are predominantly observed. In addition, fungi, viruses, protozoa, and a minimum number of methanogenic *Archaea* are also found in the microbiota of the oral cavity. Notably, the distribution of the microbial composition differs among the individuals based on the factors such as age, diet, oral health, and hygiene (Wade 2013).

The oral cavity initiates the interface between diet, host, and microbiota. In addition to the food ingested by the host, the oral microbiome derive their nutrients from glycoproteins found in saliva and gingival crevicular fluid (Homer et al. 1990). The complete catabolism of these glycoproteins requires the collective action of different species of oral bacteria. The sequential catabolic process includes the breakdown of oligosaccharides and proteins by a combined proteolytic, endopeptidase, and glycosidic activity of *Porphyromonas gingivalis*, *Prevotella nigrescens*, *Prevotella intermedia*, *Peptostreptoccus micros*, and *Streptococci* species (Wickstrom et al. 2009). Amino acids are the reduced outcome of proteins which are further fermented to SCFA by methanogenic *Archaea* and other bacteria (Wade 2013). In addition, the process and end-product of degradation of gluten and nitrate by microbial enzymatic activity reinforces the health and well-being of the host. Concomitantly, any intervention in these processes or functions disturbs the normal mechanism of metabolism and causes diseases in the host (Hezel and Weitzberg 2015; Helmerhorst et al. 2010; Zamakhchari et al. 2011).

Intriguingly, the mouth is an open environment that acts as a barrier against the colonization of any opportunistic pathogen that enters with the food. The oral microbiota from mice leverages a community defense response against the invasion of pathogens. *S. saprophyticus* present in oral microbiota senses the invading pathogens and initiates the defense pathway by triggering the mediator *S. infantis*, and finally kills the invader by producing hydrogen peroxide by *S. sanguinis* (He et al. 2015). Besides their barrier activity against pathogens, they play a crucial role in maintaining the host–microbiome homeostasis. The oral microbiome interacts with the mucosal cells and trains the host defense system to differentiate between the invaders and mouth commensals by regulating the proinflammatory cytokines (Srinivasan et al. 2010).

6.7 Microbiome of the Nasal Region of Healthy Individuals

The advent of next-generation sequencing, metaproteomics, and metagenomics approaches has gained growing attention in the past few years (Cho and Blaser 2012). The culture- and sequencing-based approaches revealed that major population of the nasal microbiome consists of *Actinobacteria*, *Firmicutes*, and other bacteria in the skin (Huse et al. 2012; Lemon et al. 2010). The 16S rRNA sequencing in different nasal sites like middle meatus and sphenoethmoidal recess showed a microbial population of *Actinobacteria*, *Firmicutes*, and *Proteobacteria* (Yan et al. 2013). However, the nasal vestibule is enriched with *Firmicutes* and *S. aureus* (Yan et al. 2013). Another study employing 16S rRNA sequencing identified 140 different taxa from the samples obtained from anterior and posterior vestibules and middle and inferior meatuses (Kaspar et al. 2016). A similar core profile is identified in anterior nasal cavity and nasopharynx (De Boeck et al. 2017). The lower respiratory tract is dominated by *Streptococcus*, *Haemophilus*, *Corynebacterium*, *Staphylococcus*, and *Dolosigranulum* (De Boeck et al. 2017; Einarsson et al. 2016). Humans have different compositions of microbiome in the nasal region in different life stages (Langevin et al. 2017). Adult nasal microbiome is over-represented with *Proteobacteria* (*Moraxella*, *Haemophilus*, and *Neisseria*), *Firmicutes* (*Streptococcus*, *Dolosigranulum*, *Gemella*, and *Granulicatella*), *Actinobacteria* (*Corynebacterium*, *Propionibacterium*, and *Turicella*), *Simonsiella*, and *Streptococcus* (Hofstra et al. 2015).

Any alteration in the nasal microbiome leads to acute respiratory tract infection caused by diverse pathogenic viruses, virome, bacteriome, etc. A mouse model of influenza virus respiratory infection displayed that super-infection by *S. aureus* is generally enhanced by the production of interferons I and III (Tarabichi et al. 2015; De Boeck et al. 2017). *S. aureus* present in the nasal cavity combined with *Chlamydia* and other *Staphylococcus* species causes a severe respiratory infection (Salzano et al. 1992). The potentially disturbed microbiota in the nasal region favors the colonization of *S. pneumonia*, *H. influenzae*, and *S. aureus* (Chonmaitree et al. 2017; Pettigrew et al. 2012). Infants with severe *S. pneumoniae* infection showed lower abundances of *Corynebacterium* and *Dolosigranulum* than their healthy counterparts (Laufer et al. 2011; Bomar et al. 2016). In addition, the level of *Lactococcus* is negatively regulated by the *Streptococcus* genus (Laufer et al. 2011). A study investigating the critical role of nasal microbiota with the severity or susceptibility of bronchiolitis documented that 18% of the hospitalized infants are at the risk of development of asthma (Hasegawa et al. 2013). The nasopharyngeal microbiota triggers the host immune response through enhanced secretion of CCL5 and serum LL-37, and protects the host against infection (Salzano et al. 1992; Hasegawa et al. 2017).

Chronic rhinosinusitis (CRS) is a chronic inflammatory disorder of paranasal sinuses. Recently, the polymicrobial mechanism behind CRS has been demonstrated (Cope et al. 2017; Choi et al. 2014). In CRS patients, the level of bacterial colonization and phylum level of abundance remain stable, whereas specific bacterial genera are found to be relatively increase (Ramakrishnan et al. 2015; Psaltis and Wormald 2017). In addition, *Anaerococcus*, *Corynebacterium*, *Finegoldia*, *Peptoniphilus*, and *Propionibacterium*, which are all health-associated bacteria in the upper respiratory tract, are found to be depleted in CRS patients (Hoggard et al. 2017; Wagner Mackenzie et al. 2017). This deleterious shift away from the healthy microbial population may eventually increase the inflammatory response and clinical severity (Hoggard et al. 2018).

6.8 Microbiome of the Skin of Healthy Individuals

The microbial niche of the skin is influenced by humidity, temperature, pH of sweat, hair follicle, sebaceous secretion, and apocrine gland which create a unique microbiota (Grice et al. 2008). The 16S rRNA phylotyping deciphered the abundance of 19 phyla which majorly include *Actinobacteria*, *Firmicutes*, *Proteobacteria*, and *Bacteroidetes*, and genera *Corynebacterium*, *Propionibacterium*, and *Staphylococcus* (Grice et al. 2009). The abundance of microbial population is specific to the site of the niche. For instance, sebaceous site is majorly occupied by *Propionibacterium* and *Staphylococcus* species. The species of *Corynebacterium* along with *Staphylococcus* is predominant in the moist cells like axilla. On the contrary, the dry site is occupied by the mixed bacterial population rich in *β-Proteobacteria* and *Flavobacteriales* species (Grice et al. 2009; Fitz-Gibbon et al. 2013). Analyzing the bacterial niches of the skin by conventional culture methods and sequencing technologies revealed the predominant distribution of *Staphylococcus epidermidis* than *S. aureus* and other bacteria (Kuehnert et al. 2006; Human Microbiome Project 2012). In addition, the genomic analysis of samples collected from different sites of body from 129 males and 113 females deciphered a moderate alpha-diversity (diversity within single samples) and a high beta-diversity (comparisons between samples from the same habitat among subjects) in the skin. Though researchers accelerate the research on microbiome, the composition of microbial skin inhabitants is still not clear. According to the 18S rRNA analysis, skin microbiome contains the highest population of *Malassezia* species including the most frequent isolates of *Malassezia globosa*, *Malassezia restricta*, and *Malassezia sympodialis* (Paulino et al. 2008; Gioti et al. 2013; Findley et al. 2013). The richness of the microbial population is very specific to the site, and it is majorly determined by the microbial lipid requirement (Saunders et al. 2012). For instance, *M. globosa* dominates in the back, occiput and inguinal crease, while *M. restricta* is predominant in the scalp (Clavaud et al. 2013), auditory canal, retroauricular crease, and glabella (Findley et al. 2013). The hair follicles and rim of eyelids are occupied by *Demodex folluculorum* and *Demodex brevis*, respectively (Lacey et al. 2009). In spite of several practical difficulties in culturing and sequencing viruses, a recent high-throughput metagenomic sequencing investigation on skin microbiome revealed the high diversity of viral DNA on a Merkel cell carcinoma patient (Foulongne et al. 2012). Since the study is focused on different aspects of the essential role of viruses on human skin microbiota, their mutualism to the host is not clearly demonstrated. Interestingly, it has been hypothesized that pathogenic human papillomavirus is also a common component of skin microbiome (Antonsson et al. 2003; Delwart 2007; Singh et al. 2009; Rosenthal et al. 2011; Foulongne et al. 2012).

In adults, the individual variation in the microbial composition of skin microbiota remains stable. Paradoxically, shift in the microbial communities alters the host–microbiome interface, which is always associated with diseases (Kong et al. 2012; Fry et al. 2013; Srinivas et al. 2013). The microbiome provides an immune shield to the skin by triggering TLRs. Activating TLRs triggers a distinct pattern of gene expression that eventually provokes a variety of immune defense responses. Conventionally, these immune responses are proinflammatory and considered to fight against pathogenic infection. For instance, *S. epidermidis*, a skin commensal, stimulates TLR3-dependent inflammation by initiating TLR2-mediated crosstalk mechanism to diminish inflammations (Lai et al. 2009). In addition, it also stimulates keratinocytes to enhance the synthesis of AMPs through a TLR2-dependent immune response (Lai et al. 2010). In a comparative study, the T-cells from germ-free (GF) mice produced less inflammatory molecules such as interferon-γ (IFN- γ) and interleukin-17A (IL-17A). By contrast, the introduction of *S. epidermidis* alone is sufficient enough to restore the production of IL-17A through the activation of T-cells in the skin. However, this kind of immune response is not supported in the gut. *S. epidermidis* is also proven to inhibit the infection by protozoan parasite *Leishmania major* in GF mice (Naik et al. 2012). The microbiome of the skin plays a crucial role in reducing various skin diseases such as acne (Iinuma et al. 2009; McDowell et al. 2011; Williams et al. 2012), atopic dermatitis (Saunders et al. 2012; Gioti et al. 2013), psoriasis (Gao et al. 2008; Cho and Blaser 2012), rosacea (Da Silva et al. 2008; Iram et al. 2012), and seborrheic dermatitis (Gupta et al. 2004).

6.9 Microbiome of Urogenital Organs of Healthy Individuals

The microbiome of male and female documented a distinct pattern (Marchesi and Ravel 2015). Analysis through culture-independent and molecular approaches assessed the microbial composition of urogenital organs (van de Wijgert et al. 2014). The 16S rRNA analyses deciphered that females have more heterogeneous bacterial population than males. The urine samples from healthy females showed a high number of *E. coli* than the male counterparts (Lewis et al. 2013). The other microbial populations predominantly identified are *Actinobacteria*, *Lactobacillus*, *Corynebacterium*, *Streptococcus*, and *Bacteroidetes* (Lewis et al. 2013; Nelson et al. 2012). However, few studies have reported the urinary microbiome in men. The 16S rRNA analysis in adolescent males found that *Streptococcus*, *Gardnerella*, *Lactobacillus*, and *Veillonella* are predominantly distributed (Nelson et al. 2012). The 16S rRNA coupled with metaproteomics analyses in both males and females reported that males have a higher distribution of *Corynebacterium* (Fouts et al. 2012). In addition, *Enterococcus*, *Proteus*, *Klebsiella*, and *Aerococcus* are also predominant in urine of males. These classes of bacteria are also found to be abundant in proximal urethra, bladder, and distal urethra. The composition of the urine determines the variation in the urinary microbiome. Women produce an excess amount of citrate but less calcium and oxalate, whereas male excretes high creatinine (Ipe et al. 2016). These compositional variations influence the niche, and eventually contribute to the health status of the host.

A plethora of studies have demonstrated the beneficial roles of urinary tract microbiota. Variations in the common components of urinary microbiota induce emergence of several chronic urinary tract symptoms (Wolfe et al. 2012; Khasriya et al. 2013; Hilt et al. 2014). An altered gut and urinary microbiome was observed in patients who tend to kidney stones than the healthy controls (Mehta et al. 2016). Contribution of urinary tract microbiota is crucial in bladder cancer. Indeed, *Mycobacterium bovis* is used to treat cancer in urothelial bladder (Wein 2015). A double-blind, placebo-controlled randomized trial suggested that the oral administration of *Lactobacillus casei* alleviates the recurrence of superficial bladder cancer (Aso and Akazan 1992). In addition, it acts as a barrier for uropathogens by secreting bacteriostatic and bactericidal molecules (Addis 1926). Most of the human bacterial pathogens are believed to be a common component of normal microflora that cause infection during various adverse host factors (Vayssier-Taussat et al. 2014; Lagier et al. 2018; Dubourg et al. 2018). The occurrence of microbiome dysbiosis due to diet and external factors affects the urogenital microbiome and causes symptoms (Abat et al. 2016). Moreover, the rapid growth of certain virulent bacteria provides an important inoculum to cause urinary tract infections (UTIs) in the host (Forsyth et al. 2018; Pompilio et al. 2018). Recently, fecal microbiota transplantation is considered as a potential strategy to combat recurrent UTIs by antibiotic-resistant *Clostridium difficile* (Tariq et al. 2017). The frequent intake of fermented milk products reduced the risk of recurrent UTIs (Makino et al. 2010). Similarly, decreased abundance of lactobacillus in the urinary microbiome is linked to interstitial cystitis (Lagier et al. 2016). The 16s rRNA analysis revealed the reduced levels of *Eggerthella sinensis*, *Collinsella aerofaciens*, *Faecalibacterium prausnitzii*, *Odoribacter splanchnicus*, and *Lactonifractor longoviformis* in interstitial cystitis patients (Braundmeier-Fleming et al. 2016). The prebiotics bacteria in the microbiome can act as a prophylaxis of UTIs, vulvovaginal candidiasis, human papillomavirus, and vaginosis (Hiergeist and Gessner 2017).

This chapter has summarized the structural compositions and functional roles of microbiome in healthy individuals in terms of public health concern.

Acknowledgments

The authors sincerely acknowledge UGC-SAP [Grant No. F.5-1/2018/DRS-II (SAP-II)], DST-FIST [Grant No. SR/FST/LSI-639/2015 (C)], and DST PURSE [Grant No. SR/PURSE Phase 2/38 (G)] for rendering instrumentation and infrastructure facilities. SG gratefully acknowledges UGC for Start-Up

Grant (Grant No. F.30–381/2017(BSR)/F.D. Diary No. 2892), Alagappa University for AURF (Ref. ALU:AURF Start-Up Grant: 2018), and RUSA 2.0 [F.24–51/2014-U, Policy (TN Multi-Gen), Department of Education, Government of India].

REFERENCES

Abat, C., M. Huart, V. Garcia, G. Dubourg, and D. Raoult. 2016. Enterococcus faecalis urinary-tract infections: do they have a zoonotic origin? *J Infect* 73, 4: 305–13.

Addis, T. 1926. The number of formed elements in the urinary sediment of normal individuals. *J Clin Invest* 2: 409–15.

Antonsson, A., C. Erfurt, K. Hazard, V. Holmgren, M. Simon, A. Kataoka, S. Hossain, et al. 2003. Prevalence and type spectrum of human papillomaviruses in healthy skin samples collected in three continents. *J Gen Virol* 84, 7: 1881–6.

Arumugam, M., J. Raes, E. Pelletier, D. Le Paslier, T. Yamada, D. R. Mende, G. R. Fernandes, et al. 2011. Enterotypes of the human gut microbiome. *Nature* 473, 7346: 174–80.

Aso, Y., and H. Akazan. 1992. Prophylactic effect of a Lactobacillus casei preparation on the recurrence of superficial bladder cancer. Blp Study Group. *Urol Int* 49, 3: 125–9.

Backhed, F., C. M. Fraser, Y. Ringel, M. E. Sanders, R. B. Sartor, P. M. Sherman, J. Versalovic, et al. 2012. Defining a healthy human gut microbiome: current concepts, future directions, and clinical applications. *Cell Host Microbe* 12, 5: 611–22.

Benson, A. K., S. A. Kelly, R. Legge, F. Ma, S. J. Low, J. Kim, M. Zhang, et al. 2010. Individuality in gut microbiota composition is a complex polygenic trait shaped by multiple environmental and host genetic factors. *Proc Natl Acad Sci U S A* 107, 44: 18933–8.

Bik, E. M., P. B. Eckburg, S. R. Gill, K. E. Nelson, E. A. Purdom, F. Francois, G. Perez-Perez, et al. 2006. Molecular analysis of the bacterial microbiota in the human stomach. *Proc Natl Acad Sci U S A* 103, 3: 732–7.

Bollinger, R. R., A. S. Barbas, E. L. Bush, S. S. Lin, and W. Parker. 2007. Biofilms in the normal human large bowel: fact rather than fiction. *Gut* 56, 10: 1481–2.

Bomar, L., S. D. Brugger, B. H. Yost, S. S. Davies, and K. P. Lemon. 2016. *Corynebacterium accolens* releases antipneumococcal free fatty acids from human nostril and skin surface triacylglycerols. *mBio* 7, 1: e01725–15.

Booijink, C. C., S. El-Aidy, M. Rajilic-Stojanovic, H. G. Heilig, F. J. Troost, H. Smidt, M. Kleerebezem, et al. 2010. High temporal and inter-individual variation detected in the human ileal microbiota. *Environ Microbiol* 12, 12: 3213–27.

Braundmeier-Fleming, A., N. T. Russell, W. Yang, M. Y. Nas, R. E. Yaggie, M. Berry, L. Bachrach, et al. 2016. Stool-based biomarkers of interstitial cystitis/bladder pain syndrome. *Sci Rep* 6: 26083.

Cheng, J., A. M. Palva, W. M. de Vos, and R. Satokari. 2013. Contribution of the intestinal microbiota to human health: from birth to 100 years of age. *Curr Top Microbiol Immunol* 358: 323–46.

Cho, I., and M. J. Blaser. 2012. The human microbiome: at the interface of health and disease. *Nat Rev Genet* 13, 4: 260–70.

Choi, E. B., S. W. Hong, D. K. Kim, S. G. Jeon, K. R. Kim, S. H. Cho, Y. S. Gho, et al. 2014. Decreased diversity of nasal microbiota and their secreted extracellular vesicles in patients with chronic rhinosinusitis based on a metagenomic analysis. *Allergy* 69, 4: 517–26.

Chonmaitree, T., K. Jennings, G. Golovko, K. Khanipov, M. Pimenova, J. A. Patel, D. P. McCormick, et al. 2017. Nasopharyngeal microbiota in infants and changes during viral upper respiratory tract infection and acute otitis media. *PLoS One* 12, 7: e0180630.

Claesson, M. J., I. B. Jeffery, S. Conde, S. E. Power, E. M. O'Connor, S. Cusack, H. M. Harris, et al. 2012. Gut microbiota composition correlates with diet and health in the elderly. *Nature* 488, 7410: 178–84.

Clavaud, C., R. Jourdain, A. Bar-Hen, M. Tichit, C. Bouchier, F. Pouradier, C. El Rawadi, et al. 2013. Dandruff is associated with disequilibrium in the proportion of the major bacterial and fungal populations colonizing the scalp. *PLoS One* 8, 3: e58203.

Cope, E. K., A. N. Goldberg, S. D. Pletcher, and S. V. Lynch. 2017. Compositionally and functionally distinct sinus microbiota in chronic rhinosinusitis patients have immunological and clinically divergent consequences. *Microbiome* 5, 1: 53.

Da Silva, C. A., D. Hartl, W. Liu, C. G. Lee, and J. A. Elias. 2008. Tlr-2 and Il-17a in chitin-induced macrophage activation and acute inflammation. *J Immunol* 181, 6: 4279–86.

Davenport, E. R., J. G. Sanders, S. J. Song, K. R. Amato, A. G. Clark, and R. Knight. 2017. The human microbiome in evolution. *BMC Biol* 15, 1: 127.

De Boeck, I., S. Wittouck, S. Wuyts, E. F. M. Oerlemans, M. F. L. van den Broek, D. Vandenheuvel, O. Vanderveken, et al. 2017. Comparing the healthy nose and nasopharynx microbiota reveals continuity as well as niche-specificity. *Front Microbiol* 8: 2372.

Delwart, E. L. 2007. Viral metagenomics. *Rev Med Virol* 17, 2: 115–31.

DiBaise, J. K., H. Zhang, M. D. Crowell, R. Krajmalnik-Brown, G. A. Decker, and B. E. Rittmann. 2008. Gut microbiota and its possible relationship with obesity. *Mayo Clin Proc* 83, 4: 460–9.

Dubourg, G., S. Baron, F. Cadoret, C. Couderc, P. E. Fournier, J. C. Lagier, and D. Raoult. 2018. From culturomics to clinical microbiology and forward. *Emerg Infect Dis* 24, 9: 1683–90.

Einarsson, G. G., D. M. Comer, L. McIlreavey, J. Parkhill, M. Ennis, M. M. Tunney, and J. S. Elborn. 2016. Community dynamics and the lower airway microbiota in stable chronic obstructive pulmonary disease, smokers and healthy non-smokers. *Thorax* 71, 9: 795–803.

El Aidy, S., B. van den Bogert, and M. Kleerebezem. 2015. The small intestine microbiota, nutritional modulation and relevance for health. *Curr Opin Biotechnol* 32: 14–20.

El Aidy, S., T. G. Dinan, and J. F. Cryan. 2015a. Gut microbiota: the conductor in the orchestra of immune-neuroendocrine communication. *Clin Ther* 37, 5: 954–67.

Fillon, S. A., J. K. Harris, B. D. Wagner, C. J. Kelly, M. J. Stevens, W. Moore, R. Fang, et al. 2012. Novel device to sample the esophageal microbiome--the esophageal string test. *PLoS One* 7, 9: e42938.

Findley, K., J. Oh, J. Yang, S. Conlan, C. Deming, J. A. Meyer, D. Schoenfeld, et al. 2013. Topographic diversity of fungal and bacterial communities in human skin. *Nature* 498, 7454: 367–70.

Fitz-Gibbon, S., S. Tomida, B. H. Chiu, L. Nguyen, C. Du, M. Liu, D. Elashoff, et al. 2013. Propionibacterium acnes strain populations in the human skin microbiome associated with acne. *J Invest Dermatol* 133, 9: 2152–60.

Forsyth, V. S., C. E. Armbruster, S. N. Smith, A. Pirani, A. C. Springman, M. S. Walters, G. R. Nielubowicz, et al. 2018. Rapid growth of uropathogenic escherichia coli during human urinary tract infection. *mBio* 9, 2: e00186–18.

Foulongne, V., V. Sauvage, C. Hebert, O. Dereure, J. Cheval, M. A. Gouilh, K. Pariente, et al. 2012. Human skin microbiota: high diversity of DNA viruses identified on the human skin by high throughput sequencing. *PLoS One* 7, 6: e38499.

Fouts, D.E., R. Pieper, S. Szpakowski, H. Pohl, S. Knoblach, M.-J. Suh, S.-T. Huang, et al. 2012. Integrated next-generation sequencing for 16S rDNA and metaproteomics differentiate the healthy urine microbiome from asymptomatic bacteriuria in neuropathic bladder associated with spinal cord injury. *J Trans Med* 10: 174.

Fry, L., B. S. Baker, A. V. Powles, A. Fahlen, and L. Engstrand. 2013. Is chronic plaque psoriasis triggered by microbiota in the skin? *Br J Dermatol* 169, 1: 47–52.

Gao, Z., C. H. Tseng, B. E. Strober, Z. Pei, and M. J. Blaser. 2008. Substantial alterations of the cutaneous bacterial biota in psoriatic lesions. *PLoS One* 3, 7: e2719.

Gerritsen, J., H. Smidt, G. T. Rijkers, and W. M. de Vos. 2011. Intestinal microbiota in human health and disease: the impact of probiotics. *Genes Nutr* 6, 3: 209–40.

Gioti, A., B. Nystedt, W. Li, J. Xu, A. Andersson, A. F. Averette, K. Munch, et al. 2013. genomic insights into the atopic eczema-associated skin commensal yeast *Malassezia Sympodialis*. *mBio* 4, 1: e00572–12.

Grice, E. A., H. H. Kong, G. Renaud, A. C. Young, G. G. Bouffard, R. W. Blakesley, T. G. Wolfsberg, et al. 2008. A diversity profile of the human skin microbiota. *Genome Res* 18, 7: 1043–50.

Grice, E. A., H. H. Kong, S. Conlan, C. B. Deming, J. Davis, A. C. Young, G. G. Bouffard, et al. 2009. Topographical and temporal diversity of the human skin microbiome. *Science* 324, 5931: 1190–2.

Gupta, A. K., R. Batra, R. Bluhm, T. Boekhout, and T. L. Dawson, Jr. 2004. Skin diseases associated with *Malassezia* species. *J Am Acad Dermatol* 51, 5: 785–98.

Hasegawa, K., J. M. Mansbach, N. J. Ajami, J. F. Petrosino, R. J. Freishtat, S. J. Teach, P. A. Piedra, et al. 2017. Serum cathelicidin, nasopharyngeal microbiota, and disease severity among infants hospitalized with bronchiolitis. *J Allergy Clin Immunol* 139, 4: 1383–86 e6.

Hasegawa, K., Y. Tsugawa, D. F. Brown, J. M. Mansbach, and C. A. Camargo, Jr. 2013. Trends in bronchiolitis hospitalizations in the United States, 2000–2009. *Pediatrics* 132, 1: 28–36.

He, J., Y. Li, Y. Cao, J. Xue, and X. Zhou. 2015. The oral microbiome diversity and its relation to human diseases. *Folia Microbiol* 60, 1: 69–80.

Helmerhorst, E. J., M. Zamakhchari, D. Schuppan, and F. G. Oppenheim. 2010. Discovery of a novel and rich source of gluten-degrading microbial enzymes in the oral cavity. *PLoS One* 5, 10: e13264.

Hezel, M. P., and E. Weitzberg. 2015. The oral microbiome and nitric oxide homoeostasis. *Oral Dis* 21, 1: 7–16.

Hiergeist, A., and A. Gessner. 2017. Clinical implications of the microbiome in urinary tract diseases. *Curr Opin Urol* 27, 2: 93–8.

Hilt, E. E., K. McKinley, M. M. Pearce, A. B. Rosenfeld, M. J. Zilliox, E. R. Mueller, L. Brubaker, et al. 2014. Urine is not sterile: use of enhanced urine culture techniques to detect resident bacterial flora in the adult female bladder. *J Clin Microbiol* 52, 3: 871–6.

Hofstra, J. J., S. Matamoros, M. A. van de Pol, B. de Wever, M. W. Tanck, H. Wendt-Knol, M. Deijs, et al. 2015. Changes in microbiota during experimental human rhinovirus infection. *BMC Infect Dis* 15: 336.

Hoggard, M., K. Biswas, M. Zoing, B. Wagner Mackenzie, M. W. Taylor, and R. G. Douglas. 2017. Evidence of microbiota dysbiosis in chronic rhinosinusitis. *Int Forum Allergy Rhinol* 7, 3: 230–9.

Hoggard, M., S. Waldvogel-Thurlow, M. Zoing, K. Chang, F. J. Radcliff, B. Wagner Mackenzie, K. Biswas, et al. 2018. Inflammatory endotypes and microbial associations in chronic rhinosinusitis. *Front Immunol* 9: 2065.

Homer, K. A., R. A. Whiley, and D. Beighton. 1990. Proteolytic activity of oral *Streptococci*. *FEMS Microbiol Lett* 55, 3: 257–60.

Huse, S. M., Y. Ye, Y. Zhou, and A. A. Fodor. 2012. A core human microbiome as viewed through 16s rRNA sequence clusters. *PLoS One* 7, 6: e34242.

Iinuma, K., T. Sato, N. Akimoto, N. Noguchi, M. Sasatsu, S. Nishijima, I. Kurokawa, and A. Ito. 2009. Involvement of propionibacterium acnes in the augmentation of lipogenesis in hamster sebaceous glands *in Vivo* and *in Vitro*. *J Invest Dermatol* 129, 9: 2113–9.

Ipe, D. S., E. Horton, and G. C. Ulett. 2016. The basics of bacteriuria: strategies of microbes for persistence in urine. *Front Cell Infect Microbiol* 6: 14.

Iram, N., M. Mildner, M. Prior, P. Petzelbauer, C. Fiala, S. Hacker, A. Schoppl, et al. 2012. Age-related changes in expression and function of toll-like receptors in human skin. *Development* 139, 22: 4210–9.

Kaspar, U., A. Kriegeskorte, T. Schubert, G. Peters, C. Rudack, D. H. Pieper, M. Wos-Oxley, et al. 2016. The culturome of the human nose habitats reveals individual bacterial fingerprint patterns. *Environ Microbiol* 18, 7: 2130–42.

Khasriya, R., S. Sathiananthamoorthy, S. Ismail, M. Kelsey, M. Wilson, J. L. Rohn, and J. Malone-Lee. 2013. Spectrum of bacterial colonization associated with urothelial cells from patients with chronic lower urinary tract symptoms. *J Clin Microbiol* 51, 7: 2054–62.

Kong, H. H., J. Oh, C. Deming, S. Conlan, E. A. Grice, M. A. Beatson, E. Nomicos, et al. 2012. Temporal shifts in the skin microbiome associated with disease flares and treatment in children with atopic dermatitis. *Genome Res* 22, 5: 850–9.

Kuehnert, M. J., D. Kruszon-Moran, H. A. Hill, G. McQuillan, S. K. McAllister, G. Fosheim, L. K. McDougal, et al. 2006. Prevalence of staphylococcus aureus nasal colonization in the United States, 2001–2002. *J Infect Dis* 193, 2: 172–9.

Kurokawa, K., T. Itoh, T. Kuwahara, K. Oshima, H. Toh, A. Toyoda, H. Takami, et al. 2007. Comparative metagenomics revealed commonly enriched gene sets in human gut microbiomes. *DNA Res* 14, 4: 169–81.

Lacey, N., K. Kavanagh, and S. C. Tseng. 2009. Under the lash: demodex mites in human diseases. *Biochem (Lond)* 31, 4: 2–6.

Lagier, J. C., G. Dubourg, M. Million, F. Cadoret, M. Bilen, F. Fenollar, A. Levasseur, et al. 2018. Culturing the human microbiota and culturomics. *Nat Rev Microbiol* 16: 540–50.

Lagier, J. C., S. Khelaifia, M. T. Alou, S. Ndongo, N. Dione, P. Hugon, A. Caputo, et al. 2016. Culture of previously uncultured members of the human gut microbiota by culturomics. *Nat Microbiol* 1: 16203.

Lahti, L., J. Salojarvi, A. Salonen, M. Scheffer, and W. M. de Vos. 2014. Tipping elements in the human intestinal ecosystem. *Nat Commun* 5: 4344.

Lai, Y., A. Di Nardo, T. Nakatsuji, A. Leichtle, Y. Yang, A. L. Cogen, Z. R. Wu, et al. 2009. Commensal bacteria regulate toll-like receptor 3-dependent inflammation after skin injury. *Nat Med* 15, 12: 1377–82.

Lai, Y., A. L. Cogen, K. A. Radek, H. J. Park, D. T. Macleod, A. Leichtle, A. F. Ryan, et al. 2010. Activation of tlr2 by a small molecule produced by staphylococcus epidermidis increases antimicrobial defense against bacterial skin infections. *J Invest Dermatol* 130, 9: 2211–21.

Langevin, S., M. Pichon, E. Smith, J. Morrison, Z. Bent, R. Green, K. Barker, et al. 2017. Early nasopharyngeal microbial signature associated with severe influenza in children: a retrospective pilot study. *J Gen Virol* 98, 10: 2425–37.

Laufer, A. S., J. P. Metlay, J. F. Gent, K. P. Fennie, Y. Kong, and M. M. Pettigrew. 2011. Microbial communities of the upper respiratory tract and otitis media in children. *mBio* 2, 1: e00245–10.

Lemon, K. P., V. Klepac-Ceraj, H. K. Schiffer, E. L. Brodie, S. V. Lynch, and R. Kolter. 2010. Comparative analyses of the bacterial microbiota of the human nostril and oropharynx. *mBio* 1, 3: e00129–10.

Leser, T. D., and L. Molbak. 2009. mBetter living through microbial action: the benefits of the mammalian gastrointestinal microbiota on the host. *Environ Microbiol* 11, 9: 2194–206.

Lewis, C. M., Jr, A. Obregon-Tito, R. Y. Tito, M. W. Foster, and P. G. Spicer. 2012. The human microbiome project: lessons from human genomics. *Trends Microbiol* 20, 1: 1–4.

Lewis, D. A., R. Brown, J. Williams, P. White, S. K. Jacobson, J. R. Marchesi, and M. J. Drake. 2013. The human urinary microbiome; bacterial DNA in voided urine of asymptomatic adults. *Front Cell Infect Microbiol* 3: 41.

Li, J., H. Jia, X. Cai, H. Zhong, Q. Feng, S. Sunagawa, M. Arumugam, et al. 2014. An integrated catalog of reference genes in the human gut microbiome. *Nat Biotechnol* 32, 8: 834–41.

Louis, P., and H. J. Flint. 2009. Diversity, metabolism and microbial ecology of butyrate-producing bacteria from the human large intestine. *FEMS Microbiol Lett* 294, 1: 1–8.

Makino, S., S. Ikegami, A. Kume, H. Horiuchi, H. Sasaki, and N. Orii. 2010. Reducing the risk of infection in the elderly by dietary intake of yoghurt fermented with lactobacillus delbrueckii ssp. bulgaricus Oll1073r-1. *Br J Nutr* 104, 7: 998–1006.

Marchesi, J. R., and J. Ravel. 2015. The vocabulary of microbiome research: a proposal. *Microbiome* 3: 31.

Mariat, D., O. Firmesse, F. Levenez, V. Guimaraes, H. Sokol, J. Dore, G. Corthier, et al. 2009. The firmicutes/bacteroidetes ratio of the human microbiota changes with age. *BMC Microbiol* 9: 123.

McDowell, A., A. Gao, E. Barnard, C. Fink, P. I. Murray, C. G. Dowson, I. Nagy, et al. 2011. A novel multilocus sequence typing scheme for the opportunistic pathogen propionibacterium acnes and characterization of type I cell surface-associated antigens. *Microbiology* 157, 7: 1990–2003.

Mehta, M., D.S. Goldfarb., and L. Nazzal. 2016. The role of the microbiome in kidney stone formation. *Int J Surg* 36: 607–12.

Moreno-Indias, I., F. Cardona, F. J. Tinahones, and M. I. Queipo-Ortuno. 2014. Impact of the gut microbiota on the development of obesity and type 2 diabetes mellitus. *Front Microbiol* 5: 190.

Naik, S., N. Bouladoux, C. Wilhelm, M. J. Molloy, R. Salcedo, W. Kastenmuller, C. Deming, et al. 2012. Compartmentalized control of skin immunity by resident commensals. *Science* 337, 6098: 1115–9.

Nelson, D. E., Q. Dong, B. Van der Pol, E. Toh, B. Fan, B. P. Katz, D. Mi, et al. 2012. Bacterial communities of the coronal sulcus and distal urethra of adolescent males. *PLoS One* 7, 5: e36298.

Nistal, E., A. Caminero, A. R. Herran, L. Arias, S. Vivas, J. M. de Morales, S. Calleja, et al. 2012. Differences of small intestinal bacteria populations in adults and children with/without celiac disease: effect of age, gluten diet, and disease. *Inflamm Bowel Dis* 18, 4: 649–56.

Paulino, L. C., C. H. Tseng, and M. J. Blaser. 2008. Analysis of malassezia microbiota in healthy superficial human skin and in psoriatic lesions by multiplex real-time pcr. *FEMS Yeast Res* 8, 3: 460–71.

Pei, Z., E. J. Bini, L. Yang, M. Zhou, F. Francois, and M. J. Blaser. 2004. Bacterial biota in the human distal esophagus. *Proc Natl Acad Sci U S A* 101, 12: 4250–5.

Pettigrew, M. M., A. S. Laufer, J. F. Gent, Y. Kong, K. P. Fennie, and J. P. Metlay. 2012. Upper respiratory tract microbial communities, acute otitis media pathogens, and antibiotic use in healthy and sick children. *Appl Environ Microbiol* 78, 17: 6262–70.

Pompilio, A., V. Crocetta, V. Savini, D. Petrelli, M. Di Nicola, S. Bucco, L. Amoroso, et al. 2018. Phylogenetic relationships, biofilm formation, motility, antibiotic resistance and extended virulence genotypes among escherichia coli strains from women with community-onset primitive acute pyelonephritis. *PLoS One* 13, 5: e0196260.

Psaltis, A. J., and P. J. Wormald. 2017. Therapy of sinonasal microbiome in crs: a critical approach. *Curr Allergy Asthma Rep* 17, 9: 59.

Ramakrishnan, V. R., L. J. Hauser, L. M. Feazel, D. Jr, C. E. Robertson, and D. N. Frank. 2015. Sinus microbiota varies among chronic rhinosinusitis phenotypes and predicts surgical outcome. *J Allergy Clin Immunol* 136, 2: 334–42 e1.

Rosenthal, M., D. Goldberg, A. Aiello, E. Larson, and B. Foxman. 2011. Skin microbiota: microbial community structure and its potential association with health and disease. *Infect Genet Evol* 11, 5: 839–48.

Rup, L. 2012. The human microbiome project. *Ind J Microbiol* 52, 3: 315.

Salzano, F. A., L. d'Angelo, S. Motta, A. del Prete, M. Gentile, and G. Motta, Jr. 1992. Allergic rhinoconjunctivitis: diagnostic and clinical assessment. *Rhinology* 30, 4: 265–75.

Saunders, C. W., A. Scheynius, and J. Heitman. 2012. Malassezia fungi are specialized to live on skin and associated with dandruff, eczema, and other skin diseases. *PLoS Pathog* 8, 6: e1002701.

Simon-Soro, A., I. Tomas, R. Cabrera-Rubio, M. D. Catalan, B. Nyvad, and A. Mira. 2013. Microbial geography of the oral cavity. *J Dent Res* 92, 7: 616–21.

Singh, S., S. Kaye, M. E. Gore, M. O. McClure, and C. B. Bunker. 2009. The role of human endogenous retroviruses in melanoma. *Br J Dermatol* 161, 6: 1225–31.

Srinivas, G., S. Moller, J. Wang, S. Kunzel, D. Zillikens, J. F. Baines, and S. M. Ibrahim. 2013. Genome-wide mapping of gene-microbiota interactions in susceptibility to autoimmune skin blistering. *Nat Commun* 4: 2462.

Srinivasan, S., C. Liu, C. M. Mitchell, T. L. Fiedler, K. K. Thomas, K. J. Agnew, J. M. Marrazzo, et al. 2010. Temporal variability of human vaginal bacteria and relationship with bacterial vaginosis. *PLoS One* 5, 4: e10197.

Stearns, J. C., M. D. Lynch, D. B. Senadheera, H. C. Tenenbaum, M. B. Goldberg, D. G. Cvitkovitch, K. Croitoru, et al. 2011. Bacterial biogeography of the human digestive tract. *Sci Rep* 1: 170.

Tarabichi, Y., K. Li, S. Hu, C. Nguyen, X. Wang, D. Elashoff, K. Saira, et al. 2015. The administration of intranasal live attenuated influenza vaccine induces changes in the nasal microbiota and nasal epithelium gene expression profiles. *Microbiome* 3: 74.

Tariq, R., D. S. Pardi, P. K. Tosh, R. C. Walker, R. R. Razonable, and S. Khanna. 2017. Fecal microbiota transplantation for recurrent clostridium difficile infection reduces recurrent urinary tract infection frequency. *Clin Infect Dis* 65, 10: 1745–7.

The Human Microbiome Project Consortium. 2012. Structure, function and diversity of the healthy human microbiome. *Nature* 486, 7402: 207–14.

Tlaskalova-Hogenova, H., R. Stepankova, H. Kozakova, T. Hudcovic, L. Vannucci, L. Tuckova, P. Rossmann, et al. 2011. The role of gut microbiota (commensal bacteria) and the mucosal barrier in the pathogenesis of inflammatory and autoimmune diseases and cancer: contribution of germ-free and gnotobiotic animal models of human diseases. *Cell Mol Immunol* 8, 2: 110–20.

Turnbaugh, P. J., V. K. Ridaura, J. J. Faith, F. E. Rey, R. Knight, and J. I. Gordon. 2009. The effect of diet on the human gut microbiome: a metagenomic analysis in humanized gnotobiotic mice. *Sci Transl Med* 1, 6: 6ra14.

van de Wijgert, J. H., H. Borgdorff, R. Verhelst, T. Crucitti, S. Francis, H. Verstraelen, and V. Jespers. 2014. The vaginal microbiota: what have we learned after a decade of molecular characterization? *PLoS One* 9, 8: e105998.

Vayssier-Taussat, M., E. Albina, C. Citti, J. F. Cosson, M. A. Jacques, M. H. Lebrun, Y. Le Loir, et al. 2014. Shifting the paradigm from pathogens to pathobiome: new concepts in the light of meta-omics. *Front Cell Infect Microbiol* 4: 29.

Wacklin, P., K. Kaukinen, E. Tuovinen, P. Collin, K. Lindfors, J. Partanen, M. Maki, et al. 2013. The duodenal microbiota composition of adult celiac disease patients is associated with the clinical manifestation of the disease. *Inflamm Bowel Dis* 19, 5: 934–41.

Wade, W. G. 2013. The oral microbiome in health and disease. *Pharmacol Res* 69, 1: 137–43.

Wagner Mackenzie, B., D. W. Waite, M. Hoggard, R. G. Douglas, M. W. Taylor, and K. Biswas. 2017. Bacterial community collapse: a meta-analysis of the sinonasal microbiota in chronic rhinosinusitis. *Environ Microbiol* 19, 1: 381–92.

Wang, M., S. Ahrne, B. Jeppsson, and G. Molin. 2005. Comparison of bacterial diversity along the human intestinal tract by direct cloning and sequencing of 16s rRNA genes. *FEMS Microbiol Ecol* 54, 2: 219–31.

Wang, Z. K., and Y. S. Yang. 2013. Upper gastrointestinal microbiota and digestive diseases. *World J Gastroenterol* 19, 10: 1541–50.

Wein, A. J. 2015. Re: the microbiome of the urinary tract–a role beyond infection. *J Urol* 194, 6: 1643–4.

Wickstrom, C., M. C. Herzberg, D. Beighton, and G. Svensater. 2009. Proteolytic degradation of human salivary muc5b by dental biofilms. *Microbiology* 155, 9: 2866–72.

Williams, H. C., R. P. Dellavalle, and S. Garner. 2012. Acne vulgaris. *Lancet* 379, 9813: 361–72.

Wolfe, A. J., E. Toh, N. Shibata, R. Rong, K. Kenton, M. Fitzgerald, E. R. Mueller, et al. 2012. Evidence of uncultivated bacteria in the adult female bladder. *J Clin Microbiol* 50, 4: 1376–83.

Yan, M., S. J. Pamp, J. Fukuyama, P. H. Hwang, D. Y. Cho, S. Holmes, and D. A. Relman. 2013. Nasal microenvironments and interspecific interactions influence nasal microbiota complexity and *S. aureus* carriage. *Cell Host Microbe* 14, 6: 631–40.

Zamakhchari, M., G. Wei, F. Dewhirst, J. Lee, D. Schuppan, F. G. Oppenheim, and E. J. Helmerhorst. 2011. Identification of rothia bacteria as gluten-degrading natural colonizers of the upper gastro-intestinal tract. *PLoS One* 6, 9: e24455.

Zoetendal, E. G., J. Raes, B. van den Bogert, M. Arumugam, C. C. Booijink, F. J. Troost, P. Bork, et al. 2012. The human small intestinal microbiota is driven by rapid uptake and conversion of simple carbohydrates. *ISME J* 6, 7: 1415–26.

Zoetendal, E. G., M. Rajilic-Stojanovic, and W. M. de Vos. 2008. High-throughput diversity and functionality analysis of the gastrointestinal tract microbiota. *Gut* 57, 11: 1605–15.

7

Structure and Function of Healthy Human Microbiome: Role in Health and Disease

Sunil Banskar
University of Arizona

Shrikant Bhute
The University of Nevada

CONTENTS

7.1 Introduction

7.1.1 Structure and Functions of the Healthy Human Microbiome

The concept of a healthy microbiome in humans has been continuously changing. Earlier when the microbiome or microbiological studies were mostly limited to aerobic cultures, the *Escherichia coli* was considered as the most dominant bacteria in the human intestinal tract (Lozupone et al., 2012). In fact, the presence of *E. coli* in the water is still used as an indicator of fecal contamination (Odonkor, 2013). Later, the advent of anaerobic culture techniques led to the understanding of anaerobic bacteria and its presence in the human gut, which led to the understanding that the human digestive tract is far more diverse than anticipated. Afterward, the development of sequencing technologies has revolutionized our understanding of microbial communities in all known environments including the human-associated microbes (Lloyd-Price et al., 2016).

In general terms, a healthy human microbiome is the microbiome present in a healthy human being. So many previous studies has shown that in the healthy human microbiome, all the microbial communities of all the individuals are not the same; instead, every healthy person's microbiome is unique in composition and proportion of microbes. Then, how can one define a healthy human microbiome? For quite some times, scientific communities around the globe have been discussing the core microbiome, i.e. the group of communities comprising a few taxa present in all the healthy (who doesn't have any noticeable disease) individual's microbiome (Llyode-Price et al., 2016) and the absence of which indicated the dysbiosis (Llyode-Price et al., 2016). But plenty of studies have shown that even in a healthy individual, so-called "core microbiota" is not present and the individual is still healthy without any phenotype of the diagnosable disease. Further studies led to an alternative assumption or concept, i.e. functional core (Shafquat et al., 2014; Llyod-Price et al., 2016). Shafquat et al. (2014) reviewed a new definition mentioning the "functional core" which is defined by the metabolic potential of the consortia present in a specific niche within the body site which is dependent on the functional rather than the taxonomic property of a microbial organism. This concept provides the flexibility of explaining the presence of different taxa within a specific body site among healthy individuals. This alternative hypothesis also provides an opportunity to explain the metabolic or functional properties not necessarily present in a specific taxonomic group but acquired by the horizontal gene transfer (Shafquat et al., 2014). There are certain properties that must be present in a core microbiome irrespective of their classification as "taxonomic core" or the "functional core". They must complement the host in terms of functions or the metabolic potential. For example, the microbes might impart important functions by providing the enzymes necessary for the digestion process or by inhibiting the growth of some fungal or bacterial pathogens by producing some active ingredients (e.g. lactic acid produced by *Lactobacillus* in the vagina prevents the growth of potential pathogenic bacteria; Gliniewicz et al., 2019). The relationship should be more symbiotic where both the host and the microbe benefit from each other. It should have resilience, i.e. it should be able to revert to its original composition after perturbations caused by the changes in diet, drugs, or pharmaceutical usage (Shafquat et al., 2014; Llyod-Price et al., 2016).

7.2 Nasal Microbiome

Nostrils (or nares) are anterior to the nasal cavity and doorway opening of the whole respiratory system which extends through nasal cavity, pharynx, and larynx (upper respiratory tract) followed by the lower respiratory tract which continues through the trachea, bronchi, and finally to the lungs. It directly connects the whole respiratory system to the outer environment. The nasal cavity serves as a first line of defense against the pathogens which attempt to make their way through the nasal cavity. Nasal hairs play a critical role in filtering pollens and dirt particles present in the inhaled air. The microbial colonization in nasal cavity starts from as early as the childbirth. The variations to the colonizing communities are contributed by a number of various factors which include mode of birth (cesarean or vaginal), early exposure to the traditional farm, rural environment or the polluted or industrialized environment, the siblings of the individual and their numbers, feeding habits, personal hygiene practices, smoking or nonsmoking habits, geographical location, antibiotic treatment, and lifestyle. Figure 7.1 explains the most factors that affect the nasal microbiome.

The bacterial communities of the nasal cavity are different from those of the other body parts. For example, it has a microbial composition that is very different from the oral or tongue bacterial community. The nasal cavity microbiome is dominated by phyla *Actinobacteria*, *Firmicutes*, and *Proteobacteria*. On the contrary, at lower taxonomic levels the members of family *Comamonadaceae* and members of order *Burkholderiales* are significantly higher in the nasal microbiome of healthy participants compared to their tongue or oral microbiome. Also, the operational taxonomic units (OTUs) belonging to family *Staphylococcaceae*, although in varying proportions, are present in the nasal cavity of all healthy individuals (Bassis et al., 2014).

In an unhealthy state, nasal microbiome changes drastically. A recently published study using 16S rRNA gene sequencing shows that the healthy control nasal microbiome has significantly less *Bacteroidetes* and *Proteobacteria* compared to exacerbated and non-exacerbated asthmatic nasal microbiome. At lower taxonomic levels, healthy nasal microbiome has significantly less *Prevotella buccalis* and *Gardnerella vaginalis* (Fazlollahi et al., 2018). A study conducted on healthy and asthmatic children has shown that the nasal microbiome and bronchial microbiome of healthy children and severe persistent asthmatic children are different. Severe persistent asthmatic children have a higher abundance of nasal *Streptococcus*, and it is positively associated with the ciliary functions of the nasal epithelia, suggesting the increased inflammatory role of the *Streptococcus* in asthmatic children's nasal microbiome. In a group of healthy children, significant increase in the abundance of *Corynebacterium* was observed, which is negatively associated with the inflammatory genes of the host; this indicates that higher abundance of *Corynebacterium* potentially plays a protective role in healthy children (Chun et al., 2020). The functional profile of the bacteria residing the nasal cavity is also different. The metagenomic projection indicates that healthy individuals (controls) have bacteria that have a significantly higher number of genes for glycerolipid metabolism compared to exacerbated and non-exacerbated asthmatic nasal microbiome (Fazlollahi et al., 2018). Further, another recent study suggests that the microbial composition of the nasal cavity affects the olfactory functions; particularly, butyrate-producing bacteria of the nasal cavity have been indicated to be associated with the impaired olfactory functions (Koskinen et al., 2018). Another interesting study performed in more than 800 children of age less than 1 year shown that early-childhood microbiota is associated with acute respiratory infections in the children. Interestingly, *Moraxella*-dominant nasal profiles of the children had a significantly lower bacterial richness and lower alpha diversity. Also, these children with dominant *Moraxella* had a higher rate of acute respiratory infections within 2 years, whereas *Corynebacteria*-dominated children's nasal microbiome were least associated with such acute respiratory infections (Toivonen et al., 2019). These studies indicate the strong relationship between *Moraxella* dominance in the nasal microbiome and respiratory problems such as acute respiratory infections in early ages and severe asthma in late childhood. Similarly, higher dominance of *Streptococcus* seems to be associated with higher incidences of respiratory problems, and higher dominance of *Corynebacteria* seems to be more protective (Toivonen et al., 2019).

The individual's nasal microbiome is hugely affected by the surrounding environment and geographical location where a person lives. For example, a study compared the nasal microbiota of the individuals living in rural vs industrialized settings and found that they have significantly different microbiota. A person living in the rural environment has a distinct nasal microbial community and has a significantly higher microbial diversity. A study performed on the individuals living in the rural village environment vs urban industrialized environment in the Egypt revealed that in the rural environment, persons have significantly increased *Actinobacteria* and *Bacteroidetes*, whereas in the individuals living in industrialized, polluted environment have significantly increased *Proteobacteria* in the nasal cavity. A genus-level analysis of nasal microbial community revealed that a higher abundance of *Staphylococcus*, *Alloicoccus*, *Corynebacterium*, and *Peptoniphilus* were present in the persons living in the village environment, whereas *Moraxella* and *Sphingomonas* were enriched among the persons living in the industrialized environment (Ahmed et al., 2019). It is interesting to note that the dominance of *Moraxella* in the nasal cavity has been associated with the respiratory infections (Toivonen et al., 2019), and it has been reported to be higher in individuals living in the industrialized settings, whereas *Corynebacterium* associated with potentially protective roles is higher in individual living in the rural environment, which indicates that polluted industrialized environment negatively impacts the microbial communities of the nasal cavity.

7.3 Lung Microbiome

Lung microbiota or pulmonary microbiota are the microbial communities present in the lower respiratory tract and lobes of the lungs of an individual. Earlier, the lungs were considered as the sterile organ, primarily based on the classical culture-based techniques. Even the initial studies on the lung microbiome were seen with skepticism because it was considered unusual to have microbes in inner organ tissues (Moffatt and

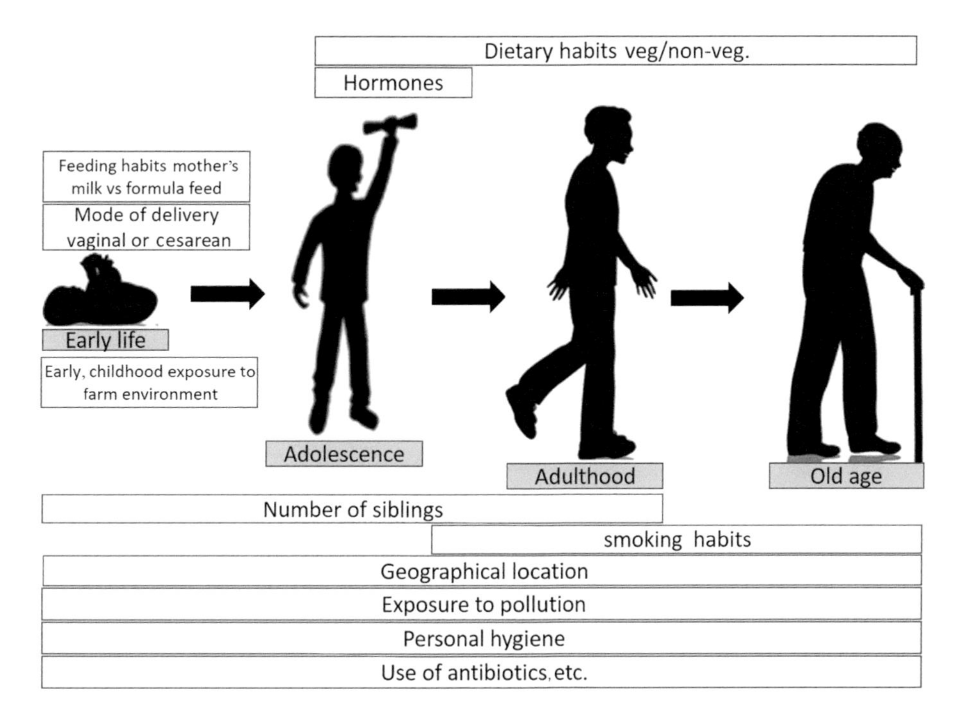

FIGURE 7.1 Factors affecting the microbiome at different stages of life.

Cookson, 2017). However, the advent of the new sequencing technologies has revealed that the lungs are not sterile and contain a variety of bacteria, but their numbers are relatively scarce (Dickson and Huffnagle, 2015).

The air we breathe is not sterile and it contains numerous bacteria, fungi, and their spores. The estimate of bacteria in air is approximately 10^4–10^6 bacteria per mm^2, and the lung mucosa is in continuous contact with this air through the process of inhaling and exhaling, i.e. breathing. This procedure of breathing brings different kinds of bacteria, and some of them are trapped in the lung and airways, while most of them are expelled by the ciliary movement of the upper respiratory tract's ciliated epithelia and with the mucosa through coughing, which is quite frequent even in healthy individuals. Thus, this movement is bidirectional, and it makes the lungs' microbial communities more dynamic than the intestinal microbial communities where the movement is unidirectional, i.e. from the mouth to the anus (Dickson and Huffnagle, 2015). Further, the microbiota of the lungs is more closely related to the oropharyngeal microbiome rather than nasopharyngeal microbiome, even though the nasal cavity is the forepart of the respiratory tract. This is primarily through micro-aspiration (inhalation of microdroplets to the trachea and lungs) of the oral saliva and other liquids, which is common even in healthy individuals (Gleeson et al., 1997; Huxley et al., 1978), thus making translocation of the oral microbial community to the lungs relatively easier. Although both the intestine and lungs have epithelial linings which have the same embryonic and developmental origins, the diversity and number (load) of microbiota they harbor are hugely different. This points out that the other factors such as pH, nutrient availability, oxygen tension, competition between the bacterial communities, and interaction with the immune cells are some of the major factors. In the lungs, there is a high oxygen tension compared to the intestine. Also, the pH of the lungs is about the same as the physiologic pH, i.e. 7.4 (pH: 7.38–7.45), whereas the pH of the intestinal tract varies greatly from highly acidic (pH 2.0) of the stomach to basic (pH ~8.0) of the small intestine. Additionally, surfactant proteins secreted into the lung's linings provide a defense against the potentially pathogenic bacteria by bacteriostasis (Wright, 1997), and the secretory IgA (although in much lesser amounts than in the intestine) also serves to limit the growth and colonization of bacteria in the lungs. Nutritionally, the healthy lungs are not as nutrient-rich as the intestine and therefore can harbor relatively fewer bacteria. Also, the temperature in the intestine is more consistent, i.e. 37°C, whereas the temperature of respiratory tract varies from ambient outside temperature to the normal body temperature of 37°C, thus making a gradient of the temperature (Ingenito et al., 1987; Dickson and Huffnagle, 2015). Thus, there are various factors that affect the microbial composition of the lungs.

7.3.1 The Healthy Human Lung Bacteria

The lungs have relatively fewer bacteria in healthy individuals. Various scientific studies on lung microbiome exploration have provided new information about the lung microbial communities. The major bacterial phyla are *Bacteroides*, *Firmicutes*, and fewer *Proteobacteria* and *Fusobacteria* which contribute to the lung microbiome. The prominent bacterial genera in the lungs of healthy individuals include *Prevotella* (family *Prevotellaceae*), *Veillonella* (*Veillonellaceae*), *Streptococcus* (*Streptococcaceae*), *Pseudomonas* (*Pseudomonadaceae*), and *Methylobacterium* (*Methylobacteriaceae*) (Dickson and Huffnagle, 2015). One recent and interesting study has shown that the microbial communities in the lungs are not homogenous, i.e. all the communities are not present throughout the lungs, but the different lung sites, i.e. upper lung, middle lung, and lower lung, have different bacterial community compositions (Figure 7.2, sites studied are marked with numbers). In fact, the microbial communities of the upper right lobe of the lungs are similar to the upper respiratory tract microbial communities compared to the microbial communities of the distal part of the lungs. As anticipated by the authors, the bacterial community richness decreased with the increase in distance from the upper respiratory tract (Dickson et al., 2015). The micro-aspiration and dispersion through the mucosal lining in the lungs are the important source of microbes in the lungs. Further, anatomically human lungs have relatively straight main stem and relatively short bronchus length (2–3 cm) in the right-side lobes of the lung compared to the left side (Figure 7.2), which makes lung's upper lobe microbial communities more similar to the glottis which is a primary source of microbes (Dickson et al., 2015).

7.4 The Oral Microbiome

The oral cavity, foremost part of the digestive tract, is followed by the other parts of gastrointestinal tract, i.e. esophagus, stomach, small intestine, large intestine, and finally the anus. Unlike the other parts of the gastrointestinal tract, the oral cavity is unique encompassing features such as a muscular tongue, hard tissues and teeth, and different mucosal surfaces within the cavity. These distinct structures along with the site-specific varying conditions enable the different bacterial communities to grow within the oral cavity (Yamashita and Takeshita, 2017). The oral microbiome includes various microbial communities that inhabit different niches within the oral cavity, e.g. which remains in the proximity of teeth, tonsils, soft and hard palates, tongue, and inner cheeks. The ease of access and sampling from the oral cavity has led to numerous articles on the cultivable and uncultivable studies of microbial communities from oral site, but the advanced high-throughput sequencing technologies have enhanced our ability to explore the microbial communities to a much higher depth leading to a better understanding of dynamics, characteristics, and interactions of these communities with the host and other microbial communities. The oral microbiome constitutes more than 700 bacterial species, and some of these are invariably present in all human beings (Zaura et al., 2009).

The initial colonization of an infant's oral cavity begins with the birth canal, mother's milk, and skin microbiome. Therefore, the method of childbirth/delivery has an impact on the microbiome of the newborn. The infant's oral communities are very similar to the mother's oral microbiota and mammary areola (Drell et al., 2017). The initial bacterial colonizers include *Lactococcus*/*Lactobacillus*, *Streptococcus*,

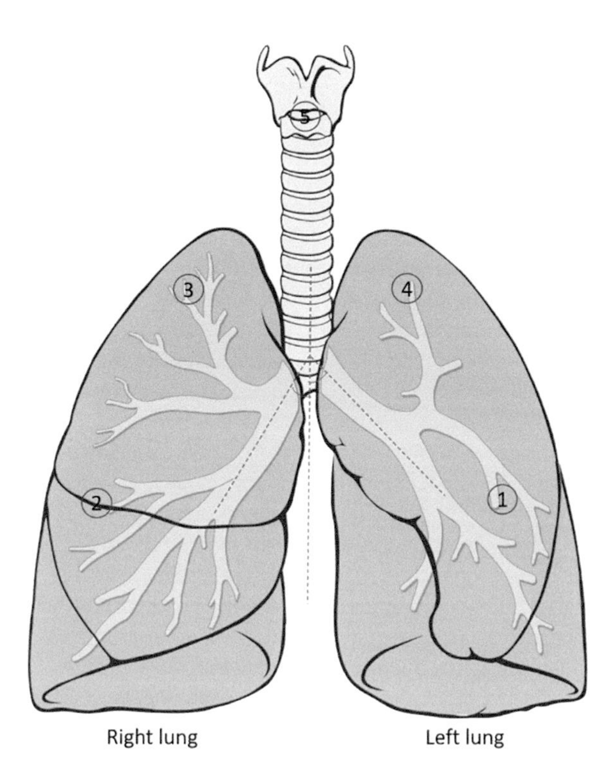

FIGURE 7.2 The human lung showing a relatively straight main stem in the right lung. The marked numbers show the studied sites which showed different microbial compositions.

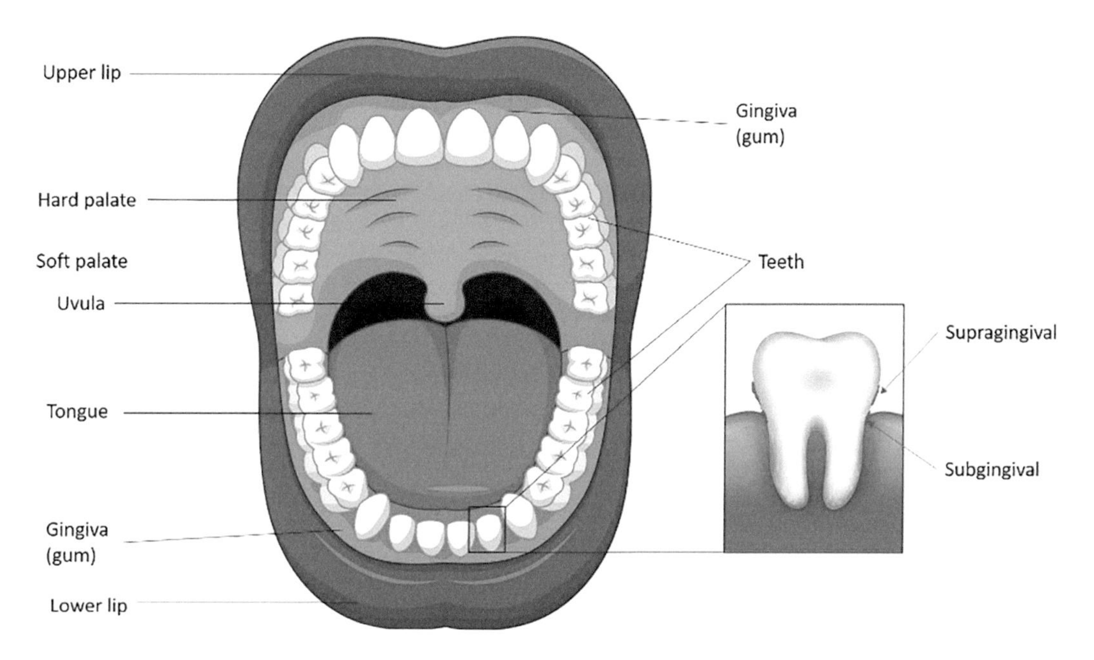

FIGURE 7.3 The human oral cavity with the different bacterial colonization sites.

and *Bifidobacterium*, the growth of which is supported by the mother's milk components such as milk oligosaccharides (Sela et al., 2011). Therefore, there are differences in the microbial communities observed in the mother's milk and formula-fed infants. The mother's milk-fed babies have abundant *Lactobacillus* which inhibits the growth of common pathogens (Reis et al., 2016).

Factors affecting the microbial growth in the oral cavity: The oral cavity is constantly moistened with salivary secretion which contains various biologically active components. The mechanical mastication and movement of saliva helps elimination of colonizing bacteria from the cavity surfaces. However, the movement of saliva also enables the dispersion of the nutrients and the bacteria within the cavity, and thus, helps sustenance of oral communities. The salivary movement helps reduced enamel exposure to the acids and promotes the pH normalization within the cavity (Strużycka, 2014). Any change in the oral environment can cause changes to the oral microbial communities. Such factors that affect the oral microbial communities in short term include type of diet, and often the long-term effect is caused by the long-term usage of antibiotics (Sheiham, 2001). Consumption of alcohol and smoking badly affect the oral communities (Thomas et al., 2014). The oral bacterial communities of the smokers and drinkers or smokers alone in comparison to the control subjects indicate a significant decrease in the oral communities' diversity and richness. Further, the oral communities of smokers and smokers and alcoholics are much more homogenous indicating the progression towards the consequential oral disease, the probability of which is far less in healthy nonsmoker and nonalcoholic individuals (Thomas et al., 2014).

7.4.1 Dental-Associated Microbes

The teeth are hard tissue structures deeply embedded in the gingivae or the gum. It provides a unique opportunity for microbial growth within the oral cavity. The microbial activity in the oral cavity is affected by the various factors (e.g. pH, cariogenic diet). The acidogenic diet has a negative impact on the oral and dental health. It promotes the acidogenic bacteria which produce various types of organic acids from carbohydrate metabolism (Strużycka, 2014), continuous exposure to which leads to dental caries and tooth decay.

The tooth-associated microbial communities form two types of biofilms or plaque, i.e. supragingival plaque (on exposed enamel of the tooth) and the subgingival plaque (below the gum line) (Figure 7.3). These exposed and hidden sites of tooth present the unique conditions for bacterial growth; therefore, both these types of plaques have a very different bacterial composition. The subgingival plaques have anaerobic environment created due to relatively closed conditions and rapid usage of available oxygen by the supragingival communities. Therefore, subgingival plaques are primarily dominated by the Gram-negative anaerobes, for instance, *Porphyromonas gingivalis, Fusobacterium nucleatum, Campylobacter* species, and *Actinobacillus* species (Marsh et al., 2011), whereas the supragingival plaques are dominated by Gram-positive bacteria mostly *Streptococcus* species, e.g. *Streptococcus salivarius, Streptococcus mitis,* and *Streptococcus mutans* and *Lactobacillus* species (Strużycka, 2014; Marsh, 2012). A study published by Mason et al. in 2015 demonstrated that the smoking increases the cariogenic bacterial species in the healthy smoker subjects and their microbiome more closely aligns with the disease-associated community compared to nonsmoker subjects. Probably routine smoking provides plentiful smoke ingredients and carbon dioxide in the oral cavity, which create relatively acidic conditions favorable for cariogenic bacterial growth and hence the increased disease-associated microbial community.

7.4.2 Tongue- and Saliva-Associated Microbes

The tongue provides a large surface area within the oral cavity for microbial colonization which, like other sites of the oral cavity, occurs in the form of microbial biofilm. Although the daily food/water consumption and movement of saliva keep the tongue microbial communities on the check, during the nighttime when such activities are stopped, their population increases drastically. A study by Wilbert et al. (2020) using an advanced multi-spectrum fluorescent imaging attempted to decipher the spatial ecology and pattern of the microbial communities in the tongue dorsum. The results clearly indicated that there is a clear organization of bacterial communities in the tongue dorsum epithelia and tongue itself (Wilbert et al., 2020). The major bacterial communities present in healthy human tongue's surface include *Streptococcus, Actinobacteria, Rothia,* and *Neisseria.* The saliva-associated microbes are the bacteria that have been loosened/detached from different oral cavity sites, though its composition has more resemblance with the tongue coating (Zaura et al., 2009; Segata et al., 2012; Yamashita and Takeshita, 2017). A higher richness of saliva microbial communities has been associated with poor oral health and oral hygiene. Certain bacteria, e.g. the higher abundance of genus *Neisseria,* indicate the healthy periodontal condition, whereas the higher abundance of genera *Prevotella* and *Veillonella* are associated with poor periodontal condition (Yamashita and Takeshita, 2017). In elderly/aging individuals, a higher alpha diversity of oral microbial communities indicates a healthy aging (Singh et al., 2019).

7.4.3 The Human Oral Microbiome Database (HOMD)

The human oral microbiome database (HOMD) is a curated and validated database of microbial communities identified from the oral cavity and its different sites within the cavity including the tongue, teeth, cheeks, etc. The initial version of HOMD encompasses a huge community of bacteria which includes about 619 different identified bacterial species belonging to 13 different bacterial phyla. The example of major phyla includes *Actinobacteria, Bacteroidetes, Chlalmydiae, Firmicutes,* and *Proteobacteria* (Dewhirst et al., 2010). Over time, different updated versions have been released for public and research uses. A publication by Escapa et al. (2018) mentioned the official expansion of the HOMD, and now it is renamed as eHOMD which stands for 'expanded Human Oral Microbiome Database', and it now also includes curated bacterial information from the upper respiratory tract (nostrils) along with the other sites of the oral cavity. The newer and updated released version includes 771 bacteria species from oral cavity microbiome and nostril microbiome (collectively called the aerodigestive tract). The official website of the database mentions that it includes 70% species which can be cultivated (version 15.2). This database provides a great resource for the microbial analysis for microbiome-related studies.

7.5 Gut Microbiome

The human gut microbiome is among the most and vastly studied yet it is not fully understood. Due to relative ease of representative sample collection non-invasively (by fecal collection), the culture-based and non-culture sequencing-based studies have generated an enormous amount of data resulting in a substantial addition to the knowledge of gut microbiome. Further, so many variations of the high-throughput next-generation sequencing (NGS) have been cherry on the cake.

7.5.1 In Utero Colonization

The colonization of bacteria in a human baby starts from the very beginning at the time of delivery, though there are a increasing number of reports mentioning that the colonization might be starting, even earlier than birth, in the uterus (*in utero*) (Lewis et al., 1976; Dong et al., 1987; Jiménez et al., 2005) and probably through umbilical cord and placenta (Jiménez et al., 2005). The experiments performed using the labeled bacteria, i.e. *Enterococcus faecium*, showed that the bacteria can translocate to the uterus and then to the developing child via blood circulation (Jiménez et al., 2008). These observations are important to consider because the meconium or the first stool of the newborn contains the microorganisms that are significantly different from the stool samples collected at later time points (Moles et al., 2013). The combined results of culture and HITChip for meconium microbiome showed the presence of *Staphylococcus*, *Lactobacillus planterum*, and *Streptococcus mitis*, whereas *Enterococcus* and other members of *Enterobacteriaceae* family such as *E. coli*, *E. fergusonii*, and *K. pneumoniae* were more dominant in the fecal samples of the later time points (Moles et al., 2013).

7.5.2 Factors Affecting Colonization – Mode of Delivery

In general, newborns' microbiome mostly resembles the mother's skin, mouth, gut, and vaginal microbiome (Dominguez-Bello et al., 2010). This early colonization affects the later microbiome development in children. The mode of delivery, i.e. natural or vaginal birth and cesarean (or C-section), plays a very important role in deciding what types of bacterial communities will be colonized first (Neu and Rushing, 2011). The babies that are delivered through the natural vaginal route have a higher abundance of vaginal microbiome predominantly represented by *Lactobacillus* and *Bifidobacterium* spp., compared to the babies who have been delivered through C-section or cesarean mode which have higher abundance mother's skin microbiome primarily *Staphylococcus* (Dominguez-Bello et al., 2010), although there are other studies that show contrasting results (Gosalbes et al., 2013). A study performed on 98 Swedish full-term born babies and their mothers for 12 months follow-up has revealed that the vaginally born neonate got 71% of their gut bacterial species matched with the mother's stool samples, whereas babies born by C-section had only about 41% of species common with that of their mother's fecal samples, which shows that the vertical transfer of these bacterial species is far less in the case of cesarean born babies compared to the vaginal born babies. The important *Bacteroidetes* members such as *B. ovatus/B. xylanisolvens*, *B. thetaiotaomicron*, *B. uniformis*, *B. vulgatus*, and *B. dorei* are found to be rare or absent in C-section-delivered babies compared to the vaginally born babies. Such difference continues to exist for a few months after the birth (Bäckhed et al., 2015). There are also reports that suggest that with the age, these differences in microbiome deepen further to reflect the mode of delivery (Chu et al., 2016).

7.5.3 Breastfeeding

Mother's milk is a crucial source of nutrition to a newborn. It not only provides the necessary nutrition to the child but also provides the antibodies that protects the child from infections (Van de Perre, 2003). The mother's milk is also an important prebiotic, which provides the milk oligosaccharides which are digested by the intestinal bacteria and hence provide an important selective pressure for specific bacterial growth, e.g. *Bifidobacterium*, in the baby's intestine (Coppa et al., 2004; Rudloff and Kunz, 2012). The mother's milk also serves as a source of microbial seeding in the infants. Therefore, feeding is an important factor that affects the early colonization in infants, i.e. whether the newborn is either fed with the mother's milk or the commercially available formula food. The infant who consumes mother's milk has significantly different microbiota. The mother's milk feeding causes the change in infants' fecal microbiome in a dose-dependent manner, and nearly 28% of the infants' fecal bacteria are derived from the mother's milk and about 10% from the areola skin (Pannaraj et al., 2017). For clarity, please replace the highlighted sentence with the following one: A study performed on the preterm babies fed either with the mothers milk as a diet or donor mothers milk (i.e. milk donated by other human mothers) or the formula feed and tracked on first 30 days of the birth showed that the fecal microbiome of babies fed with at least 70% of mothers milk had a significant high predominance of members of orders *Clostridiales*, *Lactobacillales*, and *Bacillales* compared to infants groups which are fed with formula diet or other diet combinations feeding groups, which primarily has *Enterobacteriales* (Cong et al., 2017). Further studies performed on the full-term born babies fed with only mother's milk or the formula feed had different gut microbiota. It was observed that at 4 months of age, gut microbiota of the breastfed babies had significantly increased levels of the bacterial taxa which are used as probiotics. The examples include *L. johnsonii/L. gasseri*, *L. paracasei/L. casei*, and *B. longum*. By contrast, the formula fed babies had an increased level of *Clostridium difficile*, *Granulicatella adiacens*, *Citrobacter* spp., *Enterobacter cloacae*, and *Bilophila wadsworthia*. The study also suggests that the formula-fed infants have microbiome that resembles adults' at earlier time points compared to the breastfed infants (Bäckhed et al., 2015). This study also suggests that the cessation of breast milk at the age of 12 months causes a change in microbial communities in the gut of the babies that starts to have an adult-like composition, i.e. enrichment of bacterial genera like *Bacteroides*, *Bilophila*, *Roseburia*, *Clostridium*, and *Anaerostipes* (Bäckhed et al., 2015).

7.5.4 Effect of Environment on the Early Colonization of Microbiome

The environment to which a newborn is exposed is very crucial in deciding which kind of microbiota will colonize the infant. Such early exposures have a lasting effect as these early days of newborns are crucial periods for immune maturation and types of microbial populations exposed during this crucial window of time decide how the immune system develops (Mezouar et al., 2018). This phenomenon of host and environment interactions can be well explained by the phenomenon of "Farm effect", i.e. the children that are grown in or whose early life is exposed to traditional farm environment and are exposed to diverse microbiota such as from farm animals remain protected from the allergic diseases such as asthma (von Mutius and Vercelli, 2010).

Recent studies have demonstrated that the environment contributes about 22%–36% of the microbiome variations among individuals, whereas genetics contributes only about 2.0%–9% (Rothschild et al., 2018) to the variations. It shows the importance of environment in developing an individual's microbiome. Further, studies have revealed that the neonates that are kept at neonatal intensive care units (NICU) have low microbial diversity and relatively fewer anaerobes, which leads to the increased possibility of inflammations and infections in these children (Groer et al., 2014, Pilar Francino, 2014). Additionally, there are several other factors such as use of antibiotics, use of probiotics or prebiotics, the polluted industrial environment, or rural farm environment, exposure to which has an effect that supports or suppresses the growth of specific bacterial communities which may make a host child more susceptible to infections, allergies, and/or other problems, e.g. exposure to antibiotics in early childhood can lead to selective overgrowth of the bacterial communities having antibiotic-resistance genes (Francino, 2016).

7.5.5 Effect of Pregnancy on Microbiota

There are important studies that suggest that during pregnancy, important changes take place in microbial communities of a pregnant mother. The Finnish cohort study performed on more than 90 healthy pregnant women suggested that significant changes happen in the fecal microbiota during pregnancy. These changes are independent of the pre-pregnancy conditions such as age and bodyweight, which shows that these changes are essentially associated with the fetus development. During pregnancy from first to third trimester, members of genus *Faecalibacterium* and *Prausnitzii* were increased, whereas the members of family *Enterobacteriaceae* and members of genus *Streptococcus* were decreased. The bacteria that increase during this time were high energy harvesting bacteria and the ones that protect mothers' fecal and vaginal microbiome from potentially pathogenic bacteria, which is crucial for the healthy development of the fetus and mothers' health (Koren et al., 2012).

7.5.6 Longitudinal Distribution of Healthy Gut Bacteria

The distribution and abundance of the bacteria along the digestive tract are dependent on several factors such as pH of the tract and availability of the air (e.g. the upper digestive tract has higher air tension compared to the lower digestive tract). The oxygen availability decreases from upper digestive tract to colon leading to gradation of more aerobic microbial communities in the upper tract to more anaerobic types of microbes in the colon (Ishikawa et al., 2011). The following section covers the microbiome associated with different parts of the digestive tract.

7.5.6.1 Esophageal Microbiome

The esophagus is a muscular tube and the only connection between the oral cavity and the stomach for the transition of ingested food materials, and therefore very crucial for the survival of an individual.

The normal pH of the lower esophagus in an adult is between 4 and 7, and the lower pH in for longer periods at lower esophagus has been associated with acid reflux in humans (Herbella and Patti, 2010). There are relatively fewer studies that have studied the microbiome of the esophagus in healthy individuals. Most of the available studies indicate that the esophagus has microbiota similar to the oral cavity probably because of the continuous ingestion of the oral saliva (Pei et al., 2004) even in the nighttime when sleeping (Geddes et al., 1974). Though there are studies on bacteria cultivation from the esophagus (Mannell et al., 1983), the bacterial communities present in the esophagus are not easily available for cultivation by luminal washes because the bacteria seem to be closely associated with the mucosa. The 16S rRNA gene cloning and library preparation study from the distal esophageal samples have revealed the presence of six bacterial phyla, namely, *Firmicutes*, *Bacteroides*, *Actinobacteria*, *Proteobacteria*, *Fusobacteria*, and *TM7*. *Firmicutes* were the most dominant phyla, and *TM7* were least represented (Pei et al., 2004). Further, a study showed the presence of members of genera *Streptococcus*, *Prevotella*, *Veillonella*, *Rothia*, *Megasphaera*, etc., and among these, members of the genus *Streptococcus* were most dominant (Pei et al., 2004). A study conducted to assess the microbiome of the esophagus by collecting the brushing samples showed the presence of three clusters each dominated by either *Streptococcus* (*S. mitis*, *S. oralis*, and *S. pneumoniae*) or *Prevotella* (*P. melaninogenica* and *P. pallens*) or *Veillonella*. Additionally, the archaea and micro-eucaryotes were also found; overall the age is the major factor that drives the community composition (Deshpande et al., 2018).

7.5.6.2 Stomach Microbiome

After passing through the esophagus, consumed material reaches the stomach, the first digestive organ of the gastrointestinal tract where the food is mixed mechanically with the gastric acids for further digestions. The gastric secretions make the pH of the stomach very acidic (pH 2–3), which not only serves to digest the consumed food materials, but also serves as a very important filter for the microbes coming along with the food solids and liquids. Studies have shown that the animals that consume on the other animal's remains and/or feed on other phylogenetically close relatives have a lower gastric pH (high acidity) compared to the animals that feed on the herbs or the phylogenetically distant organisms. It proves that the gastric acidity and the low pH serve as the filter, rendering the potentially pathogenic microbes dead coming from relatively close phylogenetic relatives or the decaying flash (Beasley et al., 2015). The study also proves that the gastric pH is a strong modulator of the stomach microbial communities. Therefore, there are fewer bacteria reported from the stomach, i.e. in the range of 10^2–10^3 CFUs (Sheh and Fox, 2013). The high-throughput sequencing-based studies have revealed that stomach microbial communities are much more diverse than expected. The major phyla that are represented in the healthy stomach is *Proteobacteria*, *Firmicutes*, *Actinobacteria*, *Bacteroidetes*, and *Fusobacteria* (Bik et al., 2006). Several studies have indicated the presence of bacterial members of genus *Streptococcus* and *Lactobacillus*, and often an increased abundance of *Helicobacter pylori* has been associated with gastric problems such as peptic ulcers (Sheh and Fox, 2013).

7.5.6.3 Microbiota of the Small Intestine

After the stomach, food material reaches the small intestine which is the primary site of digestion and absorption of nutrients extracted from the digested food materials. The small intestine is divided into three main division, i.e. duodenum, the first part of the small intestine; jejunum, the second or middle part of the small intestine; and ileum, the last part of the small intestine. Duodenum receives the food material (now chyme) propelled from the stomach. The pH is neutralized to near neutral by the pancreatic bi-carbonate and the bile salts secretions into the duodenum. This near-neutral pH allows the growth of the diverse bacterial communities compared to the acidic stomach, but still, bacterial growth needs to be under check by the host for optimal absorption and avoid competition for the nutrients (Walter and Ley, 2011). Therefore, multiple strategies are used such as relatively faster absorption and transit of digested and partially digested contents. Other microbial control strategies include the addition of the bile salts from gall bladder which is strongly bactericidal (Kurdi et al., 2006), production of antimicrobial peptides by enterocytes and intestinal Paneth cells, production of secretory IgA for the obstructions, and control of bacterial overgrowth (Muniz et al., 2012). Additionally, the continuous surveillance by the immune cells primarily present at Peyer's patches keeps the immune check on bacterial growth specifically the pathogens (Walter and Ley, 2011).

Due to the relatively hard to get the samples directly from the parts of the small intestine, there are relatively fewer studies that have studied microbiome directly from the small intestine, though the available studies have shed light onto the typical microbiome of the small intestine. A study performed on the mucosal biopsies collected along the intestinal tract (from jejunum, ileum, colon, and the rectum) reported that among the four sites studied, the jejunum mucosal samples had the significantly least diversity mostly dominated by *Streptococcus* compared to others sites and the diversity increased towards the colon which mostly dominated by *Bacteroidetes* and *Clostridium* clusters XIVa (Wang et al., 2005). The study performed on the samples collected from the ileum has reported a predominance of members of genera *Streptococcus* and *Veillonella*. The members of these genera are reported to have higher genetic machinery to harvest energy from the carbohydrates. The other bacteria isolates include members from *Bacteroides*, *Enterococcus*, and *Lactobacillus* (van den Bogert et al., 2013). Study performed on the aspirates and feces collected from duodenum, jejunum, and from the farthest distance (FD) of intestine (probably representing ileum or samples closest to ileum showed stool samples from FD had high bacterial CFU counts on bacterial cultivation compared at the duodenum and jejunum, although this difference was not visible on bacterial 16S rRNA gene sequencing analysis. There was a similar alpha and beta diversity in these small intestine samples, and in all these sections of the small intestine, the major bacterial phyla found were *Firmicutes*, *Proteobacteria*, *Actinobacteria*, and *Fusobacteria*, which were significantly different from the stool samples (representative of the large intestine microbiome) (Villanueva-Millan, 2019). These results also points out an important question that if a stool microbiome is a true representative of the gut microbiome? It seems that there are multiple facets of gut microbiome all of which need to be studied to understand its relations with the host.

7.5.6.4 Microbiota of the Large Intestine

The nutritionally drained and digested contents with undigestible fibers and escaped nutrients are propelled towards cecum and reached to the large intestine. The large intestine is divided into three parts, i.e. cecum, colon, and rectum. The colon is further recognized as the ascending colon, transverse colon, and descending colon. Colon has the most densely populated microbial communities, and it is the most diverse organ of not just the gastrointestinal tract but also the whole human body. Although the contents here are nutritionally deprived (relatively), their slower transit makes them suitable for microbial growth, and it achieves the bacterial count in the range of 10^{11}–10^{13} CFUs per gram of the contents. Also, the absence of Peyer's patches, unlike the small intestine, limits bacterial surveillance by immune cells; therefore, the high bacterial density can be tolerated (Walter and Ley, 2011). The primary activity the colon performs is the absorption of salts and water content, which concentrates the stool and thus also adds to the higher density of the colonic microbial communities. The fermentative activity of the colonic bacteria produces plenty of short-chain fatty acids (SCFAs) which serve as an important energy source to the enterocytes of the large intestine (Koh et al., 2016; Alexander et al., 2019). SCFAs also have several beneficial effects on host physiology and metabolism (Alexander et al., 2019).

Most studies that are performed on the stool sample mainly represent the colonic microbiota. The bacterial community present in the large intestine is predominated by phyla *Firmicutes* and *Bacteroidetes*. The other phyla present includes *Proteobacteria*, *Fusobacteria*, and *Actinobacteria* (Eckburg et al., 2005; Costello et al., 2009), although the abundance and proportion of *Proteobacteria* and *Actinobacteria* are decreased in the proximal large intestine compared to the small intestine (Friedman, 2018). Studies on the tissue biopsies of the small and large intestines have shown that the bacteria such as the members of *Bifidobacteria* have significant predominance in the large intestine compared to the small intestine, and the members of *Lactobacilli* are more abundant in the distal part of the large intestine. Further, *Eubacterium rectale* and *Faecalibacterium prausnitzii* are also predominant in the large intestine (Ahmed et al., 2007).

7.5.7 Axial Distribution of Gut Bacteria

The whole gastrointestinal tract is like a very large tube having a varying inner-diameter of nearly 2–3 cm in the small intestine and 4–5 cm in the large intestine, the periphery of which is composed of mucosal epithelial cells, goblet cell, enterocytes, etc. (Helander and Fändriks, 2014). The peripheral tissues continue to secrete the mucus which serves as a cushion between the luminal content, microbes, etc. This continuous mucus secretion also makes sure that the luminal microbes are not in direct contact with the intestinal cells though, immune cells, e.g. dendritic cells can samples the microbes (Hapfelmeier et al., 2008). Additionally, local plasma cells maintain continuous production and release of secretory IgA, which is mostly concentrated along the mucus layer of the intestine, but their relative non-specificity renders as many as 74% of luminal bacteria coated with IgA (Van der Waaij et al., 1996; Rogier et al., 2014). Therefore, they maintain a delicate balance between commensal and potentially pathogenic gut bacteria by avoiding direct contact with the healthy intestinal cells.

The oxygen concentrations along the length of the intestine play a significant role in the shaping microbial community. Similarly, the radial distribution of the oxygen also shapes the radial distribution of the bacterial communities. Studies have proved that there is a diffusion of oxygen from the intestinal mucosal surfaces, which makes the site near these surfaces more favorable for the aerotolerant bacteria (Albenberg et al., 2014), whereas the higher consumption of oxygen in luminal content of the intestine by bacteria makes it more anaerobic (Friedman, 2018) favorable for anaerobic bacterial growth. A study performed on the rectal biopsies has shown that there are increased relative proportions of the aerotolerant members of *Proteobacteria* and *Actinobacteria* in the rectal swab microbiota (representatives of mucosa-associated bacteria) compared to the fecal microbiota (Albenberg et al., 2014). Also, there are different bacterial communities on the outer part of the feces compared to the inner-content of the fecal pellet (Swidsinski et al., 2008).

There are bacteria that use host-secreted mucus as their energy source, and such members also increase close to the mucus layers in the intestine (Li et al., 2015). Asaccharolytic bacteria were also reported in the vicinity, which was further supported by the study showing a decreased abundance of carbohydrate-metabolizing genes in the mucus-associated bacteria (Albenberg et al., 2014). This study also found that the mucus-associated bacteria that use the proteinaceous substances belong to the members of genera of phyla *Firmicutes*, *Bacteroidetes*, and *Proteobacteria*. The study also reported that low-abundant bacteria are associated with the outer mucus layers of the intestine, which are mostly absent in the fecal contents (Li et al., 2015). The important example includes some members of cluster XIII of *Clostridium* (in the *Firmicutes* phylum), *Porphyromonas* (*Bacteroidetes*), members of genus *Campylobacter* and family *Enterobacteriaceae* (*Proteobacteria*), and *Corynebacterium* (*Actinobacteria*) (Albenberg et al., 2014).

7.6 The Urogenital Microbiome

The urogenital system includes both the urine excretory and reproductive system together. Both are often referred together because of their use of common ducts, proximity, common embryological, and developmental origin (Levin et al., 2007). The presence of the urogenital system in proximity to the anal opening makes it susceptible to colonization from the fecal bacteria; however, this possibility is manyfold higher in the females compared to the males due to physiological length of the urethra. Also, there are various factors that affect the type and proportion of bacteria that colonize the individual's urogenital tract, which includes gender, hormones levels (estrogen levels in females and testosterone levels in males), personal hygiene, sexual habits, age and medical condition, race or ethnicity, and dietary habits. Also, the microbiome and health state of the sex partner greatly affect the microbiome and its composition in a person (Govender et al., 2019).

7.6.1 Vaginal Microbiome

Vagina, the outermost opening of the female urogenital system, and healthy inner mucosal lining are crucial for healthy reproduction. It undergoes various changes throughout life because of the hormonal, emotional, and sexual statuses of an individual. The important factors that shape and affect the healthy vaginal microbiome include pH of the vagina, genetics and ethnicity of the individual, personal hygiene, the reproductive status of the individual, i.e. pre- or post-menopause, adolescence, and sexual activity. The pH appears the major driving factor in the shaping of the vaginal community which is usually <4.5 in most healthy females (Linhares et al., 2019). There are five major bacterial communities that predominate the vaginal microbiome consortia which includes various facultative anaerobes. Members of family *Lactobacillus* and other lactic acid-producing bacteria are important. The common *Lactobacillus* species include *L. iners*, *L. crispatus*, *L. jensenii*, and *L. gasseri*. *Lactobacillus* plays an important role in maintaining the vaginal environment. It produces lactic acid which helps to decrease the pH and reduce the possibility of colonization by any potential pathogenic bacteria; therefore, it reduces the probability of infections (Linhares et al., 2019). Additionally, it also activates components of the immune system for protection (Fichorova et al., 2011; Kim et al., 2006). Personal hygiene is also an important factor that affects the composition of the microbiome; unhygienic conditions/practices may lead to vaginal infections such as vaginosis (Ness et al., 2002); and sexual habits also shape the composition of the microbes. Ethnicity is an important factor in shaping the vaginal microbiome. For example, Ravel et al. (2011) showed that healthy white and Asian women are dominated by 89.7% and 80.2% of *Lactobacillus*, whereas black women and Hispanic have 61.9% and 59.6% of *Lactobacillus*, respectively.

Menopause is an important part of the female's reproductive cycle. It is strongly driven by the hormonal state, i.e. decreased levels of circulating estrogen in the female body which affects the physiology and metabolism of the female body. The decrease in estrogen levels in the serum leads to decreased glycogen in the vaginal epithelia which resident microbes use as energy source; therefore, this reduced availability of glycogen affects the microbiome composition of the vaginal epithelia (Mitchell et al., 2017). The bacterial abundance of the vaginal microbiome in the post-menopausal women is nearly tenfold lower than the premenopausal women (Gliniewicz et al., 2019). It is also associated with the overall change in vaginal microbiome composition, mainly reduced proportions of *Lactobacillus*, which leads to decreased lactic acid production and therefore increased vaginal pH. This increased pH also makes it susceptible to infections by pathogenic bacteria causing bacterial vaginosis (BV). It often also leads to other problems such as dryness, itching, redness, etc., which are common symptoms of BV (Bautista et al., 2016).

7.6.2 Male Genital Microbiome

There are a handful of studies on the human male genital microbiome. The anatomy of male genitalia provides a unique niche especially in the urethral opening and CS (coronal sulcus) under the prepuce (Mändar et al., 2013). A study performed on the Caucasian males with non-gonococcal urethritis and control revealed that the males in the control group have significantly more aerobic bacteria compared to males with urethritis. These include *Lactobacilli*, alpha-hemolytic *Streptococcus*, *Haemophilus vaginalis* (now renamed as *Gardnerella vaginalis*), and the anaerobic bacteria predominated by *Bacteroides*. In general *Staphylococcus epidermidis*, *Lactococcus*, *Corynebacterium* sp. anaerobic cocci, and *Bacteroides* spp. were more frequently found in healthy individuals (Bowie et al., 1977). Sexual activity has an important impact on male genital microbiome. Interestingly, a study performed on adolescent male subjects showed that *Mycoplasma* sp. and *Ureaplasma urealyticum* were specifically isolated from the urethra of sexually active subjects (Chambers et al., 1988).

The composition of the male urethra and coronal sulcus is also affected by the microbial composition of the sexual partner. A study published to assess the correlation between the urogenital microbiome of male penile and urethra and that of female partners with and without bacterial vaginosis revealed that there is a relatively low correlation between the healthy male penile skin and urethra microbiome and females' vaginal microbiome. A plausible explanation mentioned that the vaginal microbiome is usually predominated by a very few *Lactobacillus* species, notably *L. crispatus* and *L. iners*, probably decreases the discriminatory power of the analysis (Spearman correlation) utilized in the study (Zozaya et al., 2016). Another possible explanation could be the resilience of the healthy urogenital microbiota, i.e. microbiome of healthy sex partners appears to be much resilient; therefore, it reverts to the healthy state quicker. This study, however, most importantly, showed the strong correlation between the male partner's penile and urethral microbiome and BV vaginal microbiome of female partners, which strongly suggested that the BV-associated bacteria can be sexually transmitted (Zozaya et al., 2016). This fact was also supported by a previous study that showed the sexual partners share similar BV-associated bacteria in their penile microbiome (Liu et al., 2015).

The male genital microbiome differs massively between non-circumcised and circumcised males. A study performed on Ugandan males before and after the circumcision has revealed that the males before circumcision has a significantly higher abundance of the anaerobic bacteria of families *Clostridiales* Family XI and *Prevotellaceae*, which decreased drastically post circumcision. On the contrary, a certain family of bacteria, specifically *Pseudomonadaceae* and *Oxalobacteriaceae*, were among the predominant family of bacteria irrespective of the circumcised or non-circumcised status (Price et al., 2010). Further, it was confirmed that circumcision leads to a significant reduction of the overall bacterial diversity and bacterial load, especially the anaerobic bacteria in the coronal sulcus. The 12 anaerobic bacteria decreased significantly in post circumcision includes *Revotella* spp., *Porphyromonas* spp., *Finegoldia* spp., and *Peptostreptococcus* spp. (Liu et al., 2013). There were certain bacteria that increased post circumcision, e.g. *Corynebacterium* spp. and *Staphylococcus* spp. (Liu et al., 2013). The decrease of bacterial load, especially the anaerobic bacteria, has been correlated with the decrease in the possibility of HIV infection/transmission by possibly reduced recruitment of immune cells to the penis (Prodger and Kaul, 2017). Previous studies reviewed by Mändar et al. (2013) suggest that coryneform bacteria appear to be a major component of the male genital microbiome.

7.6.3 Urinary Bladder and Urinary Tract Microbiome

Conventionally, the urinary bladder is considered sterile in healthy individuals, and any presence of microbe (as detected by culture-based methods) was considered as the urinary tract infection. Earlier, traditional methods of urine bacterial detection were aerobic culture-based techniques. Such methods were not good enough to detect anaerobic and uncultivable bacteria. The recent unprecedented advancements in the sequencing technologies have enabled the detection of microbes that are uncultivable or undetectable by any known culture-based technique (primarily because of the lack of knowledge of their physiology). These studies have changed the paradigm of the sterile urinary bladder and sterile urine in healthy humans. The various methods (sample collection to microbiome sequencing and analysis) employed affect the outcome and introduces the variations in the results. However, despite the variations and differences, all studies have mentioned *Lactobacillus* and *Streptococcus* consistently in the studies. Above two species are associated with different body tissues and produce lactic acid which help to minimize the possibility of infections by pathogens. (Aragon et al., 2018). The other important and common bacteria reported include *Corynebacteria* and *Prevotella*. The bacterial genus *Gardnerella* appears to be common in females even in a healthy individual. Previously,

Gardnerella vaginalis has been found in the urine and bladder aspirations of healthy females. However, the presence of *G. vaginalis* in >10^3 CFU count in urine has been used as an indication for its possible presence in bladder aspirates (Lam et al., 1988). Probably, *Gardnerella* could be the part of normal flora in many female individuals, which remain in check because of other bacteria such as *Lactobacillus* sp. which produce lactic acid. Further, its presence has been associated with the levels of estrogen in females as *G. vaginalis* is present in relatively higher abundance in healthy pregnant females compared to non-pregnant females (Lewis et al., 1971). In general, the presence of *Gardnerella vaginalis* is considered unlikely in healthy male individuals, and its presence indicates a possible transfer from its partner.

7.7 The Skin Microbiome

The understanding of skin anatomy, physiology, and microbiome is important for the proper treatment of skin-related problems. Like the other sites of the human body, the high-throughput sequencing technologies have enhanced our understanding of the skin microbiome. Skin is the largest organ of the body covering nearly 1.8 m^2 of area. It covers the entire body parts through the various folds and invaginations. The skins of different body parts have different moisture contents. In some parts, it contains high moisture, and at most other parts, mostly it is dry or oily. Anatomically skin composed of three layers, i.e. epidermis (the outermost layer made up of the dead skin cells, i.e. keratinocytes and skin keratin proteins, which are constantly in the process of drying and shading), dermis (middle layer), and hypodermis (i.e. innermost layer of skin). Being the physical interface of the internal body parts and the outside environment, the skin serves as the primary physical line of defense against the invasion from the invading bacteria, viruses, and fungi. For microbial colonization, the skin is a very harsh and formidable habitat that continuously undergoes drying and shading of dead skin cells. It has varying moisture contents, as the sebaceous and sweat glands secrete oil, water, and salts making it saline and maintaining its pH nearly 5.0 which is hostile for most of the pathogenic microbes which usually grow at normal pH (i.e. ~7.0) and normal saline conditions (Grice and Segre, 2011). Despite the harsh conditions, various microbes are selected or adapted to the extremities present in the skin colonization. Different skin sites, such as invaginations (e.g. sweat glands, hair follicles, and sebaceous glands) and appendages, with different moisture contents may serve as the sites of microbial activity and support their colonization.

Sebaceous glands are the exocrine gland present in the skin of the face, back, shoulders, and upper chest which secrete sebum, an oily (lipid) secretion. Sebaceous glands are associated with hair follicles, and its sebum secretions provide shine and hydrophobicity to the skin and the hairs and protect from the microbes. The oily nature of these secretions makes these glands hypoxic, making its local environment more suitable for the growth of the facultative anaerobic microbial communities (Roth and James, 1988). The very common commensal bacterium of the skin is *Propionibacterium acne*, which is also a common cause of skin acne. It degrades the lipid content of the sebum and the release of free fatty acids which helps them colonize. These released free fatty acids also help in decreasing the skin pH to ~5.0 inhibiting the growth of many pathogenic bacteria, although this reduced pH gives an opportunity to other bacteria to grow which can survive better in low pH (Grace et al., 2008; Grice and Segre, 2011).

7.7.1 Sweat Glands

Another important gland of the human skin is sweat glands. These are an essential part of the skin and secrete sweat which is crucial for thermo-regulation. Salt and water secretion serves to maintain salt balance and maintain the skin pH for protection from the microbial activity. Sweat glands have two major types: eccrine sweat glands and apocrine sweat glands. Eccrine sweat glands are present throughout the body, but the apocrine sweat glands are located primarily in the armpits, mammary areola, and mons pubis (Grice and Segre, 2011). Apocrine glands are about ten times larger on average than eccrine glands and have a common opening with hair pore and sebaceous glands (Hu et al., 2018). The secretions of the eccrine glands mostly contain the salt and water, whereas the apocrine sweat glands secretions are viscous and contain lipids. Although the secretions of the apocrine sweat glands are odorless, the content of these glands provides an opportunity for the bacterial activity to produce different characteristic sweaty smells (Grice and Segre, 2011). The presence of these glands in higher numbers in the parts of the body increases moisture and therefore the abundance of bacteria. The common bacteria of such regions include *Corynebacterium* sp. (Roth and James, 1988).

7.7.2 The Skin Topology

The skin topology is an important factor that determines the type and load of inhabiting bacteria. The parts of the skin that are exposed are usually drier than those which are relatively covered or occluded, e.g. armpits and groins. The drier parts of the body, e.g. legs and forearms, are more prone to temperature fluctuations, and therefore, they harbor relatively fewer bacteria mostly dominated by the gram-negative rods and coryneform bacteria. Further, the drier parts of the body such as buttock, volar forearm (forearm), and hypothenar palm are primarily dominated by *Proteobacteria* followed by *Bacteroides*. At lower taxa levels, the drier parts of the skin are predominated by the *Staphylococcus*, *Micrococcus*, *Propionibacterium*, and the members of *Streptococcus* and members of *Proteobacteriaceae* and *Bacteroides* (Zeeuwen et al., 2012). The moist and intertriginous parts of the body harbor more and diverse bacteria, e.g. *Corynebacterium* and *Staphylococcus* (Callewaert et al., 2013; Roth and James, 1988). Specifically, the axillary vault, antecubital fossa, and interdigital web spaces are predominated by *Proteobacteria* followed by *Bacteroides* and *Firmicutes* (*Staphylococcaceae*), whereas the other moist places, e.g. inguinal crease and gluteal crease, are dominated by the members of *Corynebacteriaceae* family. Interestingly, umbilicus is almost exclusively predominated by *Corynebacteriaceae* (*Actinobacteria*) followed by

Staphylococcaceae (*Firmicutes*). Some other parts remain moist but have quite different compositions. It includes popliteal fossa; plantar heel is dominated by *Staphylococcaceae* (*Firmicutes*) members followed by *Proteobacteria*, and toe web space is predominated by the members of *Actinobacteria* (*Corynebacteriaceae* and *Micrococcaceae*) and *Staphylococcaceae* (*Firmicutes*) family. Further, the presence of sweat glands and sebaceous glands further increases the moisture and lipid contents through secretions; therefore, the presence of sebaceous glands affects the colonization of skin microbiota. These skin sites such as the face, back, and armpits are predominated by lipid-degrading bacteria from the *Propionibacteriaceae* family. The microbial communities of moist parts such as the areas of skin creases and folds are mostly variable. These sebaceous parts of the body are mostly dominated by *Actinobacteria*, i.e. *Corynebacteriaceae* and *Propionibacteriaceae* and *Micrococcaceae* family members. The examples of such sites includes manubrium, alar crease (nose side crease), and retro auricular crease (ear crease). Interestingly, the back which has a high number of sebaceous glands is mostly predominated by *Propionibacteriaceae* (*Actinobacteria*) with few *Bacteroides* and *Proteobacteria* (Grice and Segre, 2011).

At the age of puberty, host's hormonal activities activate the apocrine sweat glands (Sato et al., 1987); therefore, the bacterial activity is affected, and bacteria such as *Propionibacterium acne* become common inhabitants of the skin. With the age, the eccrine sweat glands start to decline especially after the age of 70, resulting in decreased sweat production and therefore decrease in moisture contents and lower bacterial growth in the skin (Wilke et al., 2007; Hu et al., 2018).

The sex of the individual has an important effect on the skin microbiota. For example, the skin of the male carries a relatively higher bacterial load than the female of the same age. This difference potentially could be because of the higher production of sweat in the male (Roth and James, 1988). Although not much data is available, the use of different soaps and detergents has an effect on the skin; therefore, it affects the growth and colonization of bacteria. The detergents affect the skin surface pH and thus affect the microbial activity (Yu et al., 2018). The skin microbiome is a crucial part that affects the skin barrier functions, protects the body from potential invading pathogens, and therefore needs proper attention so that a more holistic approach can be developed to better understand and maintain this human–microbiome system for the better human health.

7.7.3 Role of Microbiome in Diseases

The healthy microbiome has a crucial role in the maintenance of good health in human subjects. It provides various metabolites and genetic capabilities that help a host to maintain healthy conditions. But different known or unknown interventions such as infections, antibiotic consumption, or sudden dietary shift can create a condition that may lead to a shift in healthy community structure, which in long term may lead to a diseased state. In the following section, we'll try to understand the cause and/or effect of this microbial dysbiosis state in health and disease conditions.

7.7.4 Diabetes and Gut Microbiota

With over more than 90% of cases, type 2 diabetes is the most common type of diabetes of the three types of diabetes (type 1 diabetes, type 2 diabetes, and gestational diabetes) (The International Diabetes Federation, 2019). Type 2 diabetes, a group of chronic diseases, occurs when an individual's body does not respond to insulin (a hormone produced by the pancreas that regulates blood glucose level) or does not produce enough insulin to regulate the blood glucose levels (Polonsky, 2012). This leads to a condition called hyperglycemia, where the blood glucose levels remain very high. Increased blood glucose level, in turn, stimulates the pancreas to make more insulin, which over time becomes exhausted and cannot produce enough insulin. Thus, insulin resistance (inability to respond normally to the insulin levels) and pancreatic β-cell dysfunction (inability to produce enough insulin) are considered to be primary pathophysiologic factors driving the development of type 2 diabetes. Studies on various populations of the world suggest that the development of diabetes is linked with specific anthropometric and metabolic characteristics such as obesity, hyperglycemia, dyslipidemia, hypertension, and glucose intolerance (Unnikrishnan et al., 2017). Hence, an individual with an initial normal glucose tolerance may progressively develop diabetes over a period of time, if that individual gains weight, shows signs of increased blood glucose levels, and have an abnormal lipid profile (Ferrannini et al., 2004). However, in the real world, these characteristics are often neglected leading to the delayed diagnosis of diabetes (Dunkley et al., 2014). Prolonged hyperglycemia can directly or indirectly affect the vascular system, leading to macrovascular (coronary artery disease, heart attack, stroke, etc.) and microvascular (retinopathy, nephropathy, neuropathy, etc.) complications (Fowler, 2008). These complications expand the effect of hyperglycemia on other bodily organs and contribute to a major proportion of diabetes-related morbidities and mortalities.

Although it is known that interplay of host genetics, behavior, and environmental factors are responsible for the development of diabetes, they do not fully explain the heightened development of diabetes in Southeast Asian countries and other countries from Europe and North America suggesting the role of additional contributing factors in the development of diabetes (Murea et al., 2012). Microbiome studies in humans and animal models have shown disease-associated patterns across a wide spectrum of disorders including metabolic disorders such as diabetes (Shreiner et al., 2015). Studies have indicated a moderate degree of gut microbial dysbiosis in diabetic individuals, and many studies are being conducted to characterize the gut microbiome of diabetic patients to understand whether dysbiosis is a cause or consequence of diabetes. Below, we provide evidences of how microbes are associated with the development of diabetes. Moreover, we also provide details on how the gut microbiome and its members can be used as a treatment for diabetes.

Microbiome characterization of diabetic individuals from the various populations of the world using the 16S rRNA gene sequencing-based approach and metagenomics sequencing has indicated a correlation between the gut microbiome and their product with diabetes development (Larsen et al., 2010;

Karlsson et al., 2013; Zhang et al., 2013; Bhute et al., 2017). These studies indicate an increased abundance of *Firmicutes* and enrichment of *Lactobacillus* in diabetic patients. Further, these specific lactobacilli were found to be correlated with fasting glucose and glycated hemoglobin. However, the association of *Lactobacillus* with diabetes should be interpreted cautiously; for example, many lactobacilli are routinely consumed as probiotics; hence, it is difficult to point out whether the blown-up proportion of *Lactobacillus* in diabetic patients is due to diabetic condition or it is just a representation of probiotic strains of lactobacilli. In fact, some of the commonly used probiotic strains of lactobacilli are found to have antidiabetic properties that regulate the blood glucose level and body weight (Andreasen et al., 2010; Chen et al., 2014). Further, it has been noted that genus *Lactobacillus* is extremely diverse (comprising 261 species); using whole-genome analysis, clade-specific signature genes, and physiological criteria, genus *Lactobacillus* has been recently reclassified into 25 distinct genera (Zheng et al., 2020), indicating the fact that although many lactobacilli are considered as probiotic species, some residing in the gut may be involved in the disease development.

Often, the microbiome of diabetic individuals has reduced levels of SCFA-producing bacterial families like *Ruminococcaceae* and *Lachnospiraceae* (such as *Roseburia intestinalis* and *Faecalibacterium prausnitzii*), despite the increased abundance of *Firmicutes* to which these bacterial families belong to (Lau and Vaziri, 2019). Further, this effect is profoundly observed in newly diagnosed diabetic patients that are not receiving any treatment compared to those on antidiabetic treatment such as metformin and that of prediabetic individuals (Allin et al., 2018; Gaike et al., 2020). This indicates that treatment-naïve diabetic patients are likely to show a higher level of dysbiosis compared to known diabetic patients receiving antidiabetic treatment. Interestingly, a study showed that the therapeutic effect of metformin may operate through the enhanced levels of SCFA production likely by the gut microbes (Forslund et al., 2015). Similarly, a gram-negative mucin-degrading bacterium, *Akkermansia muciniphilla*, is found to be reduced in diabetic individuals in many populations of the world (Gaike et al., 2020; Li et al., 2020). A study on Indian diabetic individuals has also reported an increased abundance of opportunistic fungi in newly diagnosed diabetic patients; these include fungi such as *Aspergillus*, *Candida*, and *Saccharomyces* (Bhute et al., 2017). The same study also indicated an increased abundance of *Methanobrevibacter*, a dominant methanogenic archaeon, in diabetic patients. This is an interesting observation as *Methanobrevibacter* is known to direct gut bacteria to profoundly utilize polysaccharide, leading to the accumulation of high levels of SCFAs which it uses for methanogenesis with a consequent increase in host adiposity (Samuel and Gordon, 2006).

Although it is not fully understood whether the gut microbial dysbiosis is a cause or consequence of diabetes, the gut microbiota is being considered as a novel therapeutic target for managing diabetes. It has been shown that gut microbiota composition can be modulated using the diets, prebiotics, probiotics, and fecal microbiota transplantation, and nearly all of these approaches have been explored to mitigate the diabetes-associated dysbiosis. For example, due to its probiotic properties, *Akkermansia muciniphilla* is gaining attention and is being popularized as a new probiotic for the treatment of diabetes and obesity (Zhang et al., 2019). A study found a purified membrane protein of *Akkermansia muciniphilla* given either in live or pasteurized form to improve metabolic profile (Plovier et al., 2016). Various prebiotic treatments are also known to improve the abundance of *Akkermansia muciniphilla* in the gut, and this includes agents such as fructooligosaccharides and other fermentable oligo-, di-, and monosaccharides and polyols (FODMAP); thus, these indirect treatments that improve beneficial microbes appear to be promising in managing the diabetic conditions (Zhou, 2017). In addition to targeting the specific organism for diabetes treatment, restructuring the entire microbiome through fecal microbiome transplantation has also been considered. This study in an animal model of diabetes shows promising improvement in insulin resistance and also found to stop beta cell destruction (Wang et al., 2020). However, considering the fair amount of risk associated with the fecal microbiota transplantation, a targeted approach seems viable for diabetes treatment and management (Giles et al., 2019).

7.7.5 Obesity and Gut Microbiota

Obesity is a multifactorial disease in which excessive accumulation of adipose tissue leads to disproportionate body weight for the height that is often accompanied by low-grade systemic inflammation (González-Muniesa et al., 2017). Body mass index (BMI) is commonly used to define overweight and obesity, and it is calculated by taking the ratio of body weight (expressed in kilograms) to height (expressed in meters and squared). BMI between 30 and 40 kg m^{-2} is defined as obesity, and BMI over 40 kg m^{-2} is considered as extreme obesity. Worldwide growing incidences of obesity (~39% of the world's population is considered as obese) is often attributed to the rapid urbanization and economic growth that in turn affecting the diet and lifestyle (Rosin, 2008). Obesity alone leads to a huge economic impact (estimated 2.8% of the world's GDP) in developed and developing countries of the world (Tremmel et al., 2017). Thus, obesity is considered as one of the fastest-growing health challenges of the 21st century. For more information on obesity-related trends and risk factors, readers are encouraged to refer to a review by González-Muniesa (González-Muniesa et al., 2017).

Although host genetics, increased caloric intake and decreased physical activity, and other environmental factors have been directly linked with the development of obesity (Heymsfield and Wadden, 2017), it doesn't fully explain the large explosion of obesity worldwide; clearly, other factors also seem to contribute to the obesity development. Over the past couple of decades, a compelling set of links between altered gut microbiota and the pathophysiology of obesity has emerged (Cornejo-Pareja et al., 2019). Evidence suggests that gut microbiota may be contributing to the development of obesity through two independent mechanisms. First, it may efficiently extract energy from the dietary source of carbohydrates and lipids that is eventually stored as adipose tissues in the host, thus causing the overweight. This is especially true for the complex plant-based polysaccharides which host can't digest and reaches the large intestine where gut microbiota digest it readily converting them to a pool of SCFAs – butyrate, propionate, and acetate which can then be stored in the adipocytes (Samuel et al., 2008). The second mechanism is related to the increased gut permeability resulting in a phenomenon called leaky gut, in which microbial products such as lipopolysaccharides (LPS) cross the gut barrier and can cause the low-grade systemic inflammation, a hallmark of the obesity (Monteiro and Azevedo, 2010).

Early studies linking the gut microbiota with obesity suggested compositional changes at higher taxonomic levels; a reduction of *Bacteroidetes/Firmicutes* ratio (Ley et al., 2006) in obese patients and reduced diversity with increased *Actinobacteria* in obese twins (Turnbaugh et al., 2009). The increased abundance of the members of phylum *Firmicutes* in the gut increases energy-harvesting capacity from the diet, thus contributing to obesity (Turnbaugh et al., 2006). At the lower taxonomic levels, the members of family *Erysipelotrichaceae* were found to be enriched in the mice fed with high-fat Western diet (Turnbaugh et al., 2008); surprisingly, this member of *Firmicutes* diminished with the subsequent dietary manipulation. Similarly, the levels of LPS and LPS-producing bacteria in the gut (members of family *Enterobacteriaceae*) are found to be increased in Sprague Dawley rats fed with high-fat diet (De La Serre et al., 2010). In a recent study, inoculation of LPS-producing bacterium (*Enterobacter cloacae* B29) isolated from a morbidly obese human found to induce obesity and insulin resistance in mice fed high-fat diet (Fei and Zhao, 2013). These observations indicate that LPS-producing gut microbes may be contributing to inflammation-dependent obesity. A recent study has indicated the importance of regulation of expression of miR-181 (a family of microRNA) in the white adipose tissues and its role in the development of obesity (Virtue et al., 2019). Interestingly, microbiota-derived indoles (which they produce using tryptophan) and its derivatives were found to be negatively regulating the expression of miR-181 in adipocytes, thereby mitigating the obesity.

In addition to the increased abundance of few microbial taxa in obese humans and animals, some taxa also decrease in abundance. Decreased levels of *Akkermansia muciniphilla* were observed in obese and type-2-diabetic mice; further, restoring the normal levels of this organism was found to improve diet-induced metabolic disorder (Everard et al., 2013). Similarly, a common bacterial family, *Christensenellaceae*, often associated with a healthy gut is found to be reduced in obese individuals (Peters et al., 2018). In a study on Japanese obese individuals, it has been indicated that gut microbes such as *Bacteroides faecichinchillae*, *Bacteroides thetaiotaomicron*, *Blautia wexlerae*, *Clostridium bolteae*, and *Flavonifractor plautii* decrease in abundance in obese subjects (Kasai et al., 2015).

Obesity-associated gut microbiota can be reversed, and this can be achieved through various mechanisms such as lifestyle modifications, use of pre- and probiotics, and microbiota transplants. A 10-week intervention of energy-restricted diet has shown improved levels of *Bacteroides fragilis* and *Lactobacillus*, but decreased levels of *Clostridium coccoides* and *Bifidobacterium* in obese individuals which are accompanied by the weight loss (Santacruz et al., 2009). Evidences are supporting the anti-obesity properties of classical probiotic organisms. An 8-week supplementation of *L. rhamnosus* PL60 found to reduce body weight and amount of white adipose tissue (Lee et al., 2006). Fecal microbiota transplant has been attempted in obese human subjects. Few studies have indicated short-term (6 weeks post transplant) benefits of fecal transplant in the form of a significant improvement in the gut microbial diversity along with the butyrate-producing microbes in the gut, with no or little metabolic changes after 18 weeks post fecal transplant (Vrieze et al., 2012; Kootte et al., 2017). As stated earlier, although fecal microbiota transplants have potential associated risks, their cost and acceptability need to be considered.

7.7.6 Inflammatory Bowel Disease (IBD) and Gut Microbiota

Inflammatory bowel disease (IBD) is a term used to describe a complex, relapsing, and immune-mediated chronic inflammatory condition of the gastrointestinal tract. IBD is divided into two pathophysiologically distinct conditions that may require long-term treatment, frequent hospitalization, and possibly colectomy (Abraham and Cho, 2009). In its first form, Crohn's disease (CD), IBD can affect any part of the gastrointestinal tract, from the mouth to the anus, but generally small and large intestines are mostly affected. In the CD, the inflammatory ulcers appear in patches along with the healthy gut tissues (Roda et al., 2020). In its second form, ulcerative colitis (UC), inflammation is largely restricted to the large intestine with continuous damaged areas (Ungaro et al., 2017). There are genetic, immunological, environmental, and microbiota-associated factors involved in both forms of IBD (Abraham and Cho, 2009). Although until recently IBD was profoundly common in the Western countries, its increased incidences in African, Asian, and South American countries are an indication that it has become a global disease (Ng et al., 2017).

Over the past decade, the gut microbiome of CD and UC patients has been extensively investigated. A decrease in microbial diversity and a concomitant decrease in several butyrate-producing microbes including *Faecalibacterium prausnitzii* and *Bifidobacterium longum* and increased relative abundance of *Bacteroides fragilis*, *Bacteroides vulgatus*, and inflammation-promoting *Escherichia/Shigella* species have been observed in CD patients (Vich Vila et al., 2018). Metagenome-based functional analysis indicated the enrichment of virulence factors (enterobactin pathways) and antibiotics-resistant genes in patients with CD. Similarly, a major shift in oxidative stress pathways and decreased carbohydrate metabolism and amino acid biosynthesis has been observed in CD patients (Morgan et al., 2012). In terms of host genetics, out of over 140 nonoverlapping CD-associated gene loci, NOD2 variant has been strongly associated with decreased levels of *Faecalibacterium* and increased levels of *Escherichia* in CD patients (Frank et al., 2011). In pediatric CD cohort, an increased abundance of *Enterobacteriaceae*, *Bacteroidales*, *Clostridiales*, *Pasteurellaceae* (*Haemophilus* sp.), *Veillonellaceae*, *Neisseriaceae*, and *Fusobacteriaceae* has been indicated as the biomarker of classification treatment-naïve CD patient (Gevers et al., 2014).

The various treatment options that target gut microbiota in the management of CD are underway. Surprisingly, the use of probiotics seems to confer little benefits (except *Saccharomyces boulardii*) in CD patients (Lichtenstein et al., 2016). Studies using *Lactobacillus rhamnosus* strain GG both in adult and in pediatric CD patients for 1–2 years found ineffective in preventing recurrence of CD (Prantera et al., 2002; Bousvaros et al., 2005). However, this may be due to the improper selection and/or dose of the probiotic strain or perhaps the wrong timing of the administration of the probiotics in the course of CD. It is evident from a study that the nontraditional probiotic *Faecalibacterium prausnitzii* has anti-inflammatory properties against 2,4,6-trinitrobenzenesulphonic acid-induced colitis (Sokol et al., 2008). The FMT (Fecal Microbiota Transplant) has been shown to have some degree of benefits in remission of CD. Few studies have shown that multiple fresh microbiota transplants are needed for clinical remission (He et al., 2017; Sokol et al., 2020), while another study has shown an adverse effect on FMT in a small number of participants (Gutin et al., 2019).

Characteristic microbiota dysbiosis has also been noted in UC patients. In general, a peculiar feature of the UC patient's gut microbiota is the presence of increased pathogenic and pro-inflammatory bacteria and decreased anti-inflammatory bacteria. In pediatric severe UC patient, reduction in microbial diversity and associated decrease in clostridia, and an increase in Gamma-*Proteobacteria* are observed (Michail et al., 2012). Notable reduction of *Faecalibacterium prausnitzii, Clostridium coccoides*, and *Clostridium leptum* is a characteristic feature of UC patient's gut microbiota (Zhang et al., 2017). Similarly, increased *Proteobacteria*, especially *E. coli* at the site of inflammation, is associated with the perpetuation of inflammation in patients with UC (Pilarczyk-Zurek et al., 2013). Increased abundance of sulfate-reducing bacteria (such as *Desulfovibrio indonesiensis*) has been implicated in the pathogenesis of UC possibly through the production of hydrogen sulfide (Coutinho et al., 2017).

Probiotics use seems to be an effective way of controlling UC; according to a review, *E. coli Nissle* 1917 (a non-pathogenic strain of *E. coli*) is effective (and equivalent to a control mesalazine) in preventing relapse of diseases (Losurdo et al., 2015). A probiotic supplementation called VSL3 (containing four *Lactobacillus* species, three *Bifidobacteria* species, and a *Streptococcus thermophiles*) was found to induce remission and reduce the UC disease activity index in 12 weeks in mild to moderately active UC patients (Sood et al., 2009). Fecal microbiota transplant in the treatment of UC appears to have a variable outcome: some studies indicate the complete reversal of UC symptoms in 4 months of FMT (Borody et al., 2003), while others reported only a short-term improvement (Kump et al., 2013).

7.8 Conclusions

Human microbiome is a rapidly evolving field of research. The definition of the healthy microbiome is still evolving just like understanding the significance of microbiome in health and disease. Irrespective of its dynamism, evidences clearly indicate that the host-associated microbiome plays crucial roles in the host well-being and maintenance of good health. Microbial communities of every organ serve to protect the host by multiple mechanisms including but not limited to inhibiting pathogens by providing competition for the available resources like space, stimulating the components of immune systems, and producing various metabolites, thus ensuring a healthy state to the host. Despite all these invaluable contributions by the microbes for healthy state, conditions of dysbiosis may arise due to various factors such as host genetics, diet, environment, antibiotics, and pathogen entry into the body affecting physiological processes. Such dysbiosis may ultimately lead to the unhealthy state of host. To treat such dysbiotic conditions, microbial probiotics and/or prebiotics are being used with positive results, which gives hope for the treatment of microbial dysbiosis-associated health conditions with microbial interference. We hope the content of this chapter would help the readers to understand the human microbiome in the perspective of health and disease.

REFERENCES

Abraham, C. and Cho, J. H. (2009) Inflammatory bowel disease, *New England Journal of Medicine*. Massachussetts Medical Society, 361(21), pp. 2066–2078. doi: 10.1056/NEJMra0804647.

Ahmed, N., Mahmoud, N. F., Solyman, S. and Hanora, A. (2019) Human nasal microbiome as characterized by metagenomics differs markedly between rural and industrial communities in Egypt, *OMICS: A Journal of Integrative Biology*, 23(11), pp. 573–582.

Ahmed, S., Macfarlane, G. T., Fite, A., McBain, A. J., Gilbert, P. and Macfarlane, S. (2007) Mucosa-associated bacterial diversity in relation to human terminal ileum and colonic biopsy samples, *Applied and Environmental Microbiology*, 73(22), 7435–7442. doi: 10.1128/AEM.01143-07.

Albenberg, L., et al. (2014) Correlation between intraluminal oxygen gradient and radial partitioning of intestinal microbiota, *Gastroenterology*, 147, pp. 1055–1063.e8.

Alexander, C., Swanson, K. S., Fahey Jr, G. C. and Garleb, K. A. (2019) Perspective: physiologic importance of short-chain fatty acids from nondigestible carbohydrate fermentation, *Advances in Nutrition*, 10(4), pp. 576–589. doi: 10.1093/advances/nmz004.

Allin, K.H., et al. (2018) Aberrant intestinal microbiota in individuals with prediabetes, *Diabetologia*. Springer Verlag, 61(4), pp. 810–820. doi: 10.1007/s00125-018-4550-1.

Andreasen, A. S., et al. (2010) Effects of *Lactobacillus acidophilus* NCFM on insulin sensitivity and the systemic inflammatory response in human subjects, *British Journal of Nutrition*, 104(12), pp. 1831–1838. doi: 10.1017/S0007114510002874.

Aragón et al., 2018, The urinary tract microbiome in health and disease. *Europian Urology Focus*. 4(1). pp 128–138. doi: 10.1016/j.euf.2016.11.001.

Bäckhed, et al., (2015) Dynamics and stabilization of the human gut microbiome during the first year of life, *Cell Host & Microbe*, 17, pp. 690–703.

Bassis, C. M., Tang, A. L., Young, V. B. and Pynnonen, M. A. (2014) The nasal cavity microbiota of healthy adults, *Microbiome*, 2, p. 27. http://www.microbiomejournal.com/content/2/1/2.

Bautista, C. T., Wurapa, E., Sateren, W. B., Morris, S., Hollingsworth, B. and Sanchez, J. L. (2016) Bacterial vaginosis: a synthesis of the literature on etiology, prevalence, risk factors, and relationship with chlamydia and gonorrhea infections, *Military Medical Research*, 3, p. 4. doi: 10.1186/s40779-016-0074-5.

Beasley, D. E., Koltz, A. M., Lambert, J. E., Fierer, N. and Dunn, R. R. (2015) The evolution of stomach acidity and its relevance to the human microbiome, *PLoS One*, 10(7, p. e0134116. doi: 10.1371/journal.pone.0134116.

Bhute, S. S., et al. (2017) Gut microbial diversity assessment of Indian type-2-diabetics reveals alterations in eubacteria, archaea, and eukaryotes, *Frontiers in Microbiology*, 8. doi: 10.3389/fmicb.2017.00214.

Bik, E. M., et al. (2006) Molecular analysis of the bacterial microbiota in the human stomach, *Proceedings of the National Academy of Sciences of the U S A*, 103(3), pp. 732–737. doi: 10.1073/pnas.0506655103.

Borody, T. J., et al. (2003) Treatment of ulcerative colitis using fecal bacteriotherapy, *Journal of Clinical Gastroenterology*, 37(1), pp. 42–47. doi: 10.1097/00004836-200307000-00012.

Bousvaros, A., et al. (2005) A randomized, double-blind trial of lactobacillus GG versus placebo in addition to standard maintenance therapy for children with Crohn's disease, *Inflammatory Bowel Diseases*, 11(9), pp. 833–839. doi: 10.1097/01.MIB.0000175905.00212.2c.

Bowie, et al. (1977) Bacteriology of the urethra in normal men and men with nongonococcal urethritis, *Journal of Clinical Microbiology*, 6(5), pp. 482–488.

Callewaert, C., et al. (2013) Characterization of *Staphylococcus* and *Corynebacterium* clusters in the human axillary region, *PLoS One*, 8(8), p. e70538.

Chambers, et al. (1988) Microflora of the urethra in adolescent boys: relationships to sexual activity and nongonococcal urethritis. *The Journal of Pediatrics*, 110(2), 314–321.

Chen, P., et al. (2014) Antidiabetic effect of *Lactobacillus casei* CCFM0412 on mice with type 2 diabetes induced by a high-fat diet and streptozotocin, *Nutrition*. Elsevier Inc., 30(9), pp. 1061–1068. doi: 10.1016/j.nut.2014.03.022.

Chun, et al. (2020) Integrative study of the upper and lower airway microbiome and transcriptome in asthma, *JCI Insight*, 5(5), p. e133707.

Chu, D.M., Antony, K.M., Ma, J. et al. The early infant gut microbiome varies in association with a maternal high-fat diet. *Genome Med* 8, 77 (2016). https://doi.org/10.1186/s13073-016-0330-z

Coppa, G. V., et al. (2004) The first prebiotics in humans: human milk oligosaccharides, *Journal of Clinical Gastroenterology*, 38(Suppl 6), pp. S80–S83.

Cornejo-Pareja, I., et al. (2019) Importance of gut microbiota in obesity, *European Journal of Clinical Nutrition*. Springer Nature, pp. 26–37. doi: 10.1038/s41430-018-0306-8.

Costello, E. K., Lauber, C. L., Hamady, M., Fierer, N., Gordon, J. I. and Knight, R. (2009) Bacterial community variation in human body habitats across space and time, *Science*, 326, pp. 1694–1697.

Coutinho, C. M. L. M., et al. (2017) Sulphate-reducing bacteria from ulcerative colitis patients induce apoptosis of gastrointestinal epithelial cells, *Microbial Pathogenesis*. Academic Press, 112, pp. 126–134. doi: 10.1016/j.micpath.2017.09.054.

Deshpande, N. P., et al. (2018) Signatures within the esophageal microbiome are associated with host genetics, age, and disease, *Microbiome*, 6(1), p. 227. doi: 10.1186/s40168-018-0611-4.

Dewhirst, F. E., et al. (2010) The human oral microbiome, *Journal of Bacteriology*, 192, pp. 5002–5017.

Dickson, R. P. and Huffnagle, G. B. (2015) The lung microbiome: new principles for respiratory bacteriology in health and disease, *PLoS Pathog*, 11(7), p. e1004923. doi: 10.1371/journal.ppat.1004923.

Dominguez-Bello, M. G., et al. (2010). Delivery mode shapes the acquisition and structure of the initial microbiota across multiple body habitats in newborns. *Proceedings of the National Academy of Sciences of U.S.A*, 107, pp. 11971–11975. doi: 10.1073/pnas.1002601107.

Dong, Y., St Clair, P. J., Ramzy, I., Kagan-Hallet, K. S. and Gibbs, R. S. (1987) A microbiologic and clinical study of placental inflammation at term, *Obstetrics & Gynecology*, 70, pp. 175–182.

Drell, T., et al. (2017) The influence of different maternal microbial communities on the development of infant gut and oral microbiota, *Scientific Reports*, **7**, p. 9940. doi: 10.1038/s41598-017-09278-y.

Dunkley, A. J., et al. (2014) Diabetes prevention in the real world: effectiveness of pragmatic lifestyle interventions for the prevention of type 2 diabetes and of the impact of adherence to guideline recommendations – a systematic review and meta-analysis, *Diabetes Care*. American Diabetes Association Inc., pp. 922–933. doi: 10.2337/dc13-2195.

Eckburg, P. B., Bik, E. M., Bernstein, C. N., Purdom, E., Dethlefsen, L. and Sargent, M. (2005) Diversity of the human intestinal microbial flora, *Science*, 308, pp. 1635–1638.

Escapa, I. F., Chen, T., Huang, Y., Gajare, P., Dewhirst, F. E. and Lemon, K. P. (2018) New insights into human nostril microbiome from the expanded human oral microbiome database (eHOMD): a resource for the microbiome of the human aerodigestive tract, *mSystems*, 3(6), pp. e00187–18. doi: 10.1128/mSystems.00187-18.

Everard, A., et al. (2013) Cross-talk between Akkermansia muciniphila and intestinal epithelium controls diet-induced obesity, *Proceedings of the National Academy of Sciences of the United States of America*. National Academy of Sciences, 110(22), pp. 9066–9071. doi: 10.1073/pnas.1219451110.

Fazlollahi, M., et al. (2018) The nasal microbiome in asthma, *The Journal of Allergy and Clinical Immunology*, 142(3), pp. 834–843.e2. doi: 10.1016/j.jaci.2018.02.020.

Fei, N. and Zhao, L. (2013) An opportunistic pathogen isolated from the gut of an obese human causes obesity in germfree mice, *ISME Journal*. Nature Publishing Group, 7(4), pp. 880–884. doi: 10.1038/ismej.2012.153.

Ferrannini, E., et al. (2004) Mode of onset of type 2 diabetes from normal or impaired glucose tolerance, *Diabetes*. American Diabetes Association, 53(1), pp. 160–165. doi: 10.2337/diabetes.53.1.160.

Fichorova, et al. (2011) Novel vaginal microflora colonization model providing new insight into microbicide mechanism of action, *mBio*, 2(6), pp. e00168–11.

Forslund, K., et al. (2015) Disentangling the effects of type 2 diabetes and metformin on the human gut microbiota, *Nature*, 528(7581), pp. 262–266. doi: 10.1038/nature15766. Disentangling.

Fowler, M. J. (2008) Microvascular and macrovascular complications of diabetes, *Clinical Diabetes*. American Diabetes Association, pp. 77–82. doi: 10.2337/diaclin.26.2.77.

Francino, M. P. (2016) Antibiotics and the human gut microbiome: dysbioses and accumulation of resistances, *Frontiers in Microbiology*, 6, p. 1543. doi: 10.3389/fmicb.2015.01543.

Frank, D. N., et al. (2011) Disease phenotype and genotype are associated with shifts in intestinal-associated microbiota in inflammatory bowel diseases, *Inflammatory Bowel Diseases*. Oxford Academic, 17(1), pp. 179–184. doi: 10.1002/ibd.21339.

Friedman, E. S. (2018) Microbes vs. chemistry in the origin of the anaerobic gut lumen, *Proceedings of the National Academy of Sciences*, 115(16), 4170–4175. doi: 10.1073/pnas.1718635115.

Gaike, A. H., et al. (2020) The gut microbial diversity of newly diagnosed diabetics but not of prediabetics is significantly different from that of healthy nondiabetics, *mSystems*. American Society for Microbiology, 5(2). doi: 10.1128/msystems.00578-19.

Geddes, D. A. M. and Jenkins, G. N. (1974) Intrinsic and extrinsic factors influencing the flora of the mouth. In: Skinner, F. A. and Carr, J. G. (eds.), *The Normal Microbial Flora of Man*. London, New York: Academic Press, pp. 85–100.

Gevers, D., et al. (2014) The treatment-naive microbiome in new-onset Crohn's disease, *Cell Host and Microbe*. Cell Press, 15(3), pp. 382–392. doi: 10.1016/j.chom.2014.02.005.

Giles, E. M., D'Adamo, G. L. and Forster, S. C. (2019) The future of faecal transplants, *Nature Reviews Microbiology*. Nature Publishing Group, p. 719. doi: 10.1038/s41579-019-0271-9.

Gleeson, K., Eggli, D. F. and Maxwell, S. L. (1997) Quantitative aspiration during sleep in normal subjects, *Chest*, 111, pp. 1266–1272. PMID: 9149581.

Gliniewicz, et al. (2019) Comparison of the vaginal microbiomes of premenopausal and postmenopausal women, *Frontiers in microbiology*. doi: 10.3389/fmicb.2019.00193.

González-Muniesa, P., et al. (2017) Obesity, *Nature Reviews Disease Primers*. Nature Publishing Group, 3(1), pp. 1–18. doi: 10.1038/nrdp.2017.34.

Gosalbes MJ, Llop S, Valles Y, Moya A, Ballester F, et al. (2013) Meconium microbiota types dominated by lactic acid or enteric bacteria are differentially associated with maternal eczema and respiratory problems in infants. Clin Exp Allergy 43: 198–211. DOI: 10.1111/cea.12063.

Govender, Y., Gabriel, I., Minassian, V. and Fichorova, R. (2019) The current evidence on the association between the urinary microbiome and urinary incontinence in women, *Frontiers in Cellular and Infection Microbiology*, 9, p. 133. doi: 10.3389/fcimb.2019.00133.

Grice EA, Kong HH, Renaud G, Young AC; NISC Comparative Sequencing Program, Bouffard GG, Blakesley RW, Wolfsberg TG, Turner ML, Segre JA. A diversity profile of the human skin microbiota. Genome Res. 2008 Jul;18(7):1043-50. doi: 10.1101/gr.075549.107. Epub 2008 May 23. PMID: 18502944; PMCID: PMC2493393.

Grice, E. A. and Segre, J. A. (2011) The skin microbiome [published correction appears in Nat Rev Microbiol. Aug;9(8):626], *Nature Reviews Microbiology*, 9(4), pp. 244–253. doi: 10.1038/nrmicro2537.

Groer, M. W., Luciano, A. A., Dishaw, L. J., Ashmeade, T. L., Miller, E. and Gilbert, J. A. (2014) Development of the preterm infant gut microbiome: a research priority, *Microbiome*, 2, p. 38.

Gutin, L., et al. (2019) Fecal microbiota transplant for Crohn disease: a study evaluating safety, efficacy, and microbiome profile, *United European Gastroenterology Journal*. SAGE Publications Ltd, 7(6), pp. 807–814. doi: 10.1177/2050640619845986.

Hapfelmeier, S., et al. (2008) Microbe sampling by mucosal dendritic cells is a discrete, MyD88-independent stepin *ΔinvG S.* Typhimurium colitis. *Journal of Experimental Medicine*, 205(2), pp. 437–450. doi: 10.1084/jem.20070633.

He, Z., et al. (2017) Multiple fresh fecal microbiota transplants induces and maintains clinical remission in Crohn's disease complicated with inflammatory mass, *Scientific Reports*. Nature Publishing Group, 7(1), pp. 1–10. doi: 10.1038/s41598-017-04984-z.

Helander, H. F. and Fändriks, L. (2014) Surface area of the digestive tract – revisited, *Scandinavian Journal of Gastroenterology*, 49(6), pp. 681–689. doi: 10.3109/00365521.2014.898326.

Herbella, F. A. and Patti, M. G. (2010) Gastroesophageal reflux disease: from pathophysiology to treatment, *World Journal of Gastroenterology*, 16(30), pp. 3745–3749. doi: 10.3748/wjg.v16.i30.3745.

Heymsfield, S. B. and Wadden, T. A. (2017) Mechanisms, pathophysiology, and management of obesity, *New England Journal of Medicine*. Edited by D. L. Longo. Massachussetts Medical Society, 376(3), pp. 254–266. doi: 10.1056/NEJMra1514009.

Hu, Y., Converse, C., Lyons, M. and Hsu, W. (2018) Neural control of sweat secretion: a review, *British Journal of Dermatology*, 178, pp. 1246–1256. doi: 10.1111/bjd.15808.

Huxley, E. J., Viroslav, J., Gray, W. R. and Pierce, A. K. (1978) Pharyngeal aspiration in normal adults and patients with depressed consciousness, *American Journal of Medicine*, 64, pp. 564–568. PMID: 645722.

Ingenito, E. P., et al. (1987) Indirect assessment of mucosal surface temperatures in the airways: theory and tests, *Journal of Applied Physiology*, 63, pp. 2075–2083. PMID: 3693240.

Ishikawa, et al. (2011) Transport phenomena of microbial flora in the small intestine, *Journal of Theoretical Biology*, 279, pp. 63–73.

Jiménez, E., et al. (2005) Isolation of commensal bacteria from umbilical cord blood of healthy neonates born by cesarean section, *Current Microbiology*, 51, pp. 270–274 doi: 10.1007/s00284-005-0020-3.

Jiménez, E., et al. (2008) Is meconium from healthy newborns actually sterile? *Research in Microbiology*, 159, pp. 187–193.

Karlsson, F. H., et al. (2013) Gut metagenome in European women with normal, impaired and diabetic glucose control, *Nature*, 498(7452), pp. 99–103.

Kasai, C., et al. (2015) Comparison of the gut microbiota composition between obese and non-obese individuals in a Japanese population, as analyzed by terminal restriction fragment length polymorphism and next-generation sequencing, *BMC Gastroenterology*. BioMed Central Ltd., 15(1), p. 100. doi: 10.1186/s12876-015-0330-2.

Kim, et al. (2006) Probiotic *Lactobacillus casei* activates innate immunity via NF-kappaB and p38 MAP kinase signaling pathways, *Microbes and Infection*, 8, pp. 994–1005. doi: 10.1016/j.micinf.2005.10.019.

Koh, A., De Vadder, F., Kovatcheva-Datchary, P. and Bäckhed, F. (2016) From dietary fiber to host physiology: short-chain fatty acids as key bacterial metabolites, *Cell*, 165, pp. 1332–1345.

Kootte, R. S., et al. (2017) Improvement of insulin sensitivity after lean donor feces in metabolic syndrome is driven by baseline intestinal microbiota composition, *Cell Metabolism*. Cell Press, 26(4), pp. 611–619.e6. doi: 10.1016/j.cmet.2017.09.008.

Koren, O., et al. (2012) Host remodeling of the gut microbiome and metabolic changes during pregnancy, *Cell*, 150, pp. 470–480.

Koskinen, K., et al. (2018) The nasal microbiome mirrors and potentially shapes olfactory function, *Scientific Reports*, 8, p. 1296. doi: 10.1038/s41598-018-19438-3.

Kump, P. K., et al. (2013) Alteration of intestinal dysbiosis by fecal microbiota transplantation does not induce remission in patients with chronic active ulcerative colitis, *Inflammatory Bowel Diseases*, 19(10), pp. 2155–2165. doi: 10.1097/MIB.0b013e31829ea325.

Kurdi, P., Kawanishi, K., Mizutani, K. and Yokota, A. (2006) Mechanism of growth inhibition by free bile acids in lactobacilli and bifidobacterial, *Journal of Bacteriology*, 188, pp. 1979–1986. doi: 10.1128/JB.188.5.1979-1986.2006.

De La Serre, C. B., et al. (2010) Propensity to high-fat diet-induced obesity in rats is associated with changes in the gut microbiota and gut inflammation, *American Journal of Physiology – Gastrointestinal and Liver Physiology*. American Physiological Society, 299(2), p. G440. doi: 10.1152/ajpgi.00098.2010.

Lam, et al. (1988) Prevalence of gardnerella vaginalis in the urinary tract, *Journal of Clinical Microbiology*, 26(6), pp. 1130–1133.

Larsen, N., et al. (2010) Gut microbiota in human adults with type 2 diabetes differs from non-diabetic adults, *PLoS One*, 5(2), p. e9085.

Lau, W. L. and Vaziri, N. D. (2019) Gut microbial short-chain fatty acids and the risk of diabetes, *Nature Reviews Nephrology*. Nature Publishing Group, 15(7), pp. 389–390. doi: 10.1038/s41581-019-0142-7.

Lee, H. Y., et al. (2006) Human originated bacteria, *Lactobacillus rhamnosus* PL60, produce conjugated linoleic acid and show anti-obesity effects in diet-induced obese mice, *Biochimica et Biophysica Acta – Molecular and Cell Biology of Lipids*. Elsevier, 1761(7), pp. 736–744. doi: 10.1016/j.bbalip.2006.05.007.

Levin, T. L., Han, B. and Little, B. P. (2007) Congenital anomalies of the male urethra, *Pediatric Radiology*, 37(9), pp. 851–945. doi: 10.1007/s00247-007-0495-0.

Lewis, J. F., Johnson, P. and Miller, P. (1976) Evaluation of amniotic fluid for aerobic and anaerobic bacteria, *American Journal of Clinical Pathology*, 65, pp. 58–63.

Lewis, J. F., O'Brien, S. M., Ural, U. M. and Burke, T. (1971) Corynebacterium vaginal e vaginitis in pregnant women, *American Journal of Clinical Pathology*, 56, pp. 580–583.

Ley, R. E., et al. (2006) Microbial ecology: human gut microbes associated with obesity, *Nature*. Nature Publishing Group, 444(7122), pp. 1022–1023. doi: 10.1038/4441022a.

Li, H., et al. (2015) The outer mucus layer hosts a distinct intestinal microbial niche, *Nature Communications*, 6, p. 8292. doi: 10.1038/ncomms9292.

Li, Q., et al. (2020) Implication of the gut microbiome composition of type 2 diabetic patients from northern China, *Scientific Reports*. Nature Research, 10(1), pp. 1–8. doi: 10.1038/s41598-020-62224-3.

Lichtenstein, L., Avni-Biron, I. and Ben-Bassat, O. (2016) Probiotics and prebiotics in Crohn's disease therapies, *Best Practice and Research: Clinical Gastroenterology*. Bailliere Tindall Ltd, pp. 81–88. doi: 10.1016/j.bpg.2016.02.002.

Linhares, I. M, Minis, E., Robial, R. and Witkin, S. S. (2019) The human vaginal microbiome. In: Joel, F. and Salamao, F. (eds.), *Microbiome and Metabolome in Diagnosis, Therapy, and Other Strategic Applications*. Web.

Liu, et al. (2013) Male circumcision significantly reduces prevalence and load of genital anaerobic bacteria, *mBio*. doi: 10.1128/mBio.00076-13.

Liu, C. M., et al. (2015) Penile microbiota and female partner bacterial vaginosis in Rakai, Uganda, *mBio*, 6(3), pp. e00589–15. doi: 10.1128/mBio.00589-15.

Lloyd-Price, J., Abu-Ali, G. and Huttenhower, C. (2016) The healthy human microbiome, *Genome Med*, 8(51). doi: 10.1186/s13073-016-0307-y.

Losurdo, G., et al. (2015) *Escherichia coli* Nissle 1917 in ulcerative colitis treatment: systematic review and meta-analysis, *Journal of Gastrointestinal and Liver Diseases*, 24(4), pp. 499–505. doi: 10.15403/jgld.2014.1121.244.ecn.

Lozupone, C. A., Stombaugh, J. I., Gordon, J. I., Jansson, J. K. and Knight, R. (2012) Diversity, stability and resilience of the human gut microbiota, *Nature*, 489, pp. 220–230.

Mändar R. Microbiota of male genital tract: Impact on the health of man and his partner. Pharmacological Research 2013 69(1): 32 – 41. Doi: https://doi.org/10.1016/j.phrs.2012.10.019

Mannell, et al. (1983) Microflora of the esophagus. Aylwyn Mannell, Mary Plant, Julius Frolich, *Annals of the Royal College of Surgeons of England*, 65.

Marsh, P. D. (2012) Contemporary perspective on plaque control, *British Dental Journal*, 12, pp. 601–606.

Marsh, P.D., Moter, A. and Devine, D.A. (2011), Dental plaque biofilms: communities, conflict and control. Periodontology 2000, 55: 16–35. https://doi.org/10.1111/j.1600-0757.2009.00339.x

Mason, M. R., Preshaw, P. M., Nagaraja, H. N., Dabdoub, S. M., Rahman, A. and Kumar, P. S. (2015) The subgingival microbiome of clinically healthy current and never smokers, *The ISME Journal*, 9, pp. 268–272.

Mezouar, S., et al. (2018) Microbiome and the immune system: from a healthy steady-state to allergy associated disruption, *Human Microbiome Journal*, 10, pp. 11–20.

Michail, S., et al. (2012) Alterations in the gut microbiome of children with severe ulcerative colitis, *Inflammatory Bowel Diseases*. Oxford Academic, 18(10), pp. 1799–1808. doi: 10.1002/ibd.22860.

Mitchell, C. M., et al. (2017) Vaginal microbiota and genitourinary menopausal symptoms: a cross-sectional analysis, *Menopause*, 24(10), pp. 1160–1166. doi: 10.1097/GME.0000000000000904.

Moffatt, M. F. and Cookson, W. O. C. M. (2017) The lung microbiome in health and disease, *Clinical Medicine*, 17(6), pp. 525–529. doi: 10.7861/clinmedicine.17-6-525.

Moles, L., et al. (2013) Bacterial diversity in meconium of preterm neonates and evolution of their fecal microbiota during the first months of life, *PLoS One*, 8, pp. 1–13.

Monteiro, R. and Azevedo, I. (2010) Chronic inflammation in obesity and the metabolic syndrome, *Mediators of Inflammation*, 2010. doi: 10.1155/2010/289645.

Morgan, X. C., et al. (2012) Dysfunction of the intestinal microbiome in inflammatory bowel disease and treatment, *Genome Biology*. BioMed Central, 13(9), p. R79. doi: 10.1186/gb-2012-13-9-r79.

Muniz, L. R., Knosp, C. and Yeretssian, G. (2012) Intestinal antimicrobial peptides during homeostasis, infection, and disease, *Frontiers in Immunology*, 3, p. 310. doi: 10.3389/fimmu.2012.00310.

Murea, M., Ma, L. and Freedman, B. I. (2012) Genetic and environmental factors associated with type 2 diabetes and diabetic vascular complications, *Review of Diabetic Studies*. Society for Biomedical Diabetes Research, pp. 6–22. doi: 10.1900/RDS.2012.9.6.

Ness, et al. (2002) Douching in relation to bacterial vaginosis, lactobacilli, and facultative bacteria in the vagina, *Obstetrics & Gynecology*, 100, p. 765. doi: 10.1016/S0029-7844(02)02184-1.

Neu, J. and Rushing, J. (2011) Cesarean versus vaginal delivery: long-term infant outcomes and the hygiene hypothesis, *Clinics in Perinatology*, 38(2), pp. 321–331. doi: 10.1016/j.clp.2011.03.008.

Ng, S. C., et al. (2017) Worldwide incidence and prevalence of inflammatory bowel disease in the 21st century: a systematic review of population-based studies, *The Lancet*. Lancet Publishing Group, 390(10114), pp. 2769–2778. doi: 10.1016/S0140-6736(17)32448-0.

Odonkor, S. (2013) *E coli* as an indicator of bacteriological quality of water: an overview, *Microbiology Research*, 4. doi: 10.4081/mr.2013.e2.

Pannaraj, P. S., et al. (2017) Association between breast milk bacterial communities and establishment and development of the infant gut microbiome, *JAMA Pediatrics*, 171, pp. 647–654.

Pei, Z., Bini, E. J., Yang, L., Zhou, M., Francois, F. and Blaser, M. J. (2004) Bacterial biota in the human distal esophagus, *Proceedings of the National Academy of Sciences USA*, 101(12), pp. 4250–4255.

Peters, B. A., et al. (2018) A taxonomic signature of obesity in a large study of American adults, *Scientific Reports*. Nature Publishing Group, 8(1), pp. 1–13. doi: 10.1038/s41598-018-28126-1.

Pilarczyk-Zurek, M., et al. (2013) Possible role of *Escherichia coli* in propagation and perpetuation of chronic inflammation in ulcerative colitis, *BMC Gastroenterology*. BioMed Central, 13(1), p. 61. doi: 10.1186/1471-230X-13-61.

Pilar Francino, M. (2014) Early development of the gut microbiota and immune health, *Pathogens*, 4, pp. 769–790. doi: 10.3390/pathogens3030769.

Plovier, H., et al. (2016) A purified membrane protein from *Akkermansia muciniphila* or the pasteurized bacterium improves metabolism in obese and diabetic mice, *Nature Publishing Group*. Nature Publishing Group, (November). doi: 10.1038/nm.4236.

Polonsky, K. S. (2012) The past 200 years in diabetes, *New England Journal of Medicine*, 367(14), pp. 1332–1340. doi: 10.1056/NEJMra1110560.

Prantera, C., et al. (2002) Ineffectiveness of probiotics in preventing recurrence after curative resection for Crohn's disease: a randomised controlled trial with *Lactobacillus* GG, *Gut*, 51(3), pp. 405–409. doi: 10.1136/gut.51.3.405.

Price, et al. (2010) The effects of circumcision on the penis microbiome, *PLoS One*, 5(1), p. e8422. doi: 10.1371/journal.pone.0008422.

Prodger, J. L. and Kaul, R. (2017) The biology of how circumcision reduces HIV susceptibility: broader implications for the prevention field, *AIDS Research and Therapy*, 14, p. 49. doi: 10.1186/s12981-017-0167-6.

Reis, N., Saraiva, M., Duarte, E., de Carvalho, E., Vieira, B. and Evangelista-Barreto, N. (2016) Probiotic properties of lactic acid bacteria isolated from human milk, *Journal of Applied Microbiology*, 121, pp. 811–820. doi: 10.1111/jam.13173.

Roda, G., et al. (2020) Crohn's disease, *Nature Reviews Disease Primers*. Nature Research, 6(1), pp. 1–19. doi: 10.1038/s41572-020-0156-2.

Rogier, E. W., Frantz, A. L., Bruno, M. E. C. and Kaetzel, C. S. (2014) Secretory IgA is concentrated in the outer layer of colonic mucus along with gut bacteria, *Pathogens*, 3, pp. 390–403. doi: 10.3390/pathogens3020390.

Rosin, O. (2008) The economic causes of obesity: a survey, *Journal of Economic Surveys*. John Wiley & Sons, Ltd, 22(4), pp. 617–647. doi: 10.1111/j.1467-6419.2007.00544.x.

Roth, R. R. and James, W. D. (1988) Microbial ecology of the skin, *Annual Review of Microbiology*, 42, pp. 441–464.

Rothschild, D., et al. (2018) Environment dominates over host genetics in shaping human gut microbiota, *Nature*, 555, pp. 210–215.

Rudloff, S. and Kunz, C. (2012) Milk oligosaccharides and metabolism in infants, *Advances in Nutrition*, 3, pp. 398S–405S.

Samuel, B. S., et al. (2008) Effects of the gut microbiota on host adiposity are modulated by the short-chain fatty-acid binding G protein-coupled receptor, Gpr41, *Proceedings of the National Academy of Sciences of the United States of America*. doi: 10.1073/pnas.0808567105.

Samuel, B. S. and Gordon, J. I. (2006) A humanized gnotobiotic mouse model of host-archaeal-bacterial mutualism, *Proceedings of the National Academy of Sciences of the United States of America*. National Academy of Sciences, 103(26), pp. 10011–10016. doi: 10.1073/pnas.0602187103.

Santacruz, A., et al. (2009) Interplay between weight loss and gut microbiota composition in overweight adolescents, *Obesity*. John Wiley & Sons, Ltd, 17(10), pp. 1906–1915. doi: 10.1038/oby.2009.112.

Segata, N., et al. (2012) Composition of the adult digestive tract bacterial microbiome based on seven mouth surfaces, tonsils, throat and stool samples, *Genome Biology*, 13, p. R42.

Sela, D. A., et al. (2011) An infant-associated bacterial commensal utilizes breast milk sialyloligosaccharides, *Journal of Biological Chemistry*, 286(14), pp. 11909–11918.

Shafquat, A., Joice, R., Simmons, S. L. and Huttenhower, C. (2014) Functional and phylogenetic assembly of microbial communities in the human microbiome, *Trends in Microbiology*, 22(5), pp. 261–266. doi: 10.1016/j.tim.2014.01.011.

Sheh, A. and Fox, J. G. (2013) The role of the gastrointestinal microbiome in *Helicobacter pylori* pathogenesis, *Gut Microbes*, 4(6), pp. 505–531. doi: 10.4161/gmic.26205.

Sheiham, A. (2001) Dietary effects on dental diseases, *Public Health Nutrition*, 4(2B), pp. 569–591. doi: 10.1079/phn2001142.

Shreiner, A. B., Kao, J. Y. and Young, V. B. (2015) The gut microbiome in health and in disease, *Current Opinion in Gastroenterology*. Lippincott Williams and Wilkins, pp. 69–75. doi: 10.1097/MOG.0000000000000139.

Singh, H., Torralba, M.G., Moncera, K.J. *et al.* Gastro-intestinal and oral microbiome signatures associated with healthy aging. *GeroScience* **41**, 907–921 (2019). https://doi.org/10.1007/s11357-019-00098-8

Sokol, H., et al. (2008) Faecalibacterium prausnitzii is an anti-inflammatory commensal bacterium identified by gut microbiota analysis of Crohn disease patients, *Proceedings of the National Academy of Sciences of the United States of America*. National Academy of Sciences, 105(43), pp. 16731–16736. doi: 10.1073/pnas.0804812105.

Sokol, H., et al. (2020) Fecal microbiota transplantation to maintain remission in Crohn's disease: a pilot randomized controlled study, *Microbiome*. NLM (Medline), 8(1), p. 12. doi: 10.1186/s40168-020-0792-5.

Sood, A., et al. (2009) The probiotic preparation, VSL#3 induces remission in patients with mild-to-moderately active ulcerative colitis, *Clinical Gastroenterology and Hepatology*. W.B. Saunders, 7(11). doi: 10.1016/j.cgh.2009.07.016.

Strużycka, I. (2014) The oral microbiome in dental caries, *Polish Journal of Microbiology*, 63(2), pp. 127–135.

Swidsinski, A., Loening-Baucke, V., Verstraelen, H., Osowska, S. and Doerffel, Y. (2008) Biostructure of fecal microbiota in healthy subjects and patients with chronic idiopathic diarrhea, *Gastroenterology*, 135, pp. 568–579. doi: 10.1053/j.gastro.2008.04.017.

The International Diabetes Federation. (2019) *IDF Diabetes Atlas*. https://www.idf.org/aboutdiabetes/what-is-diabetes/facts-figures.html

Thomas, A. M., et al. (2014) Alcohol and tobacco consumption affects bacterial richness in oral cavity mucosa biofilms, *BMC Microbiology*, 14, p. 250. doi: 10.1186/s12866-014-0250-2.

Toivonen, L., et al. (2019) Early nasal microbiota and acute respiratory infections during the first years of life, *Thorax*, 74, pp. 592–599. doi: 10.1136/thoraxjnl-2018-212629.

Tremmel, M., et al. (2017) Economic burden of obesity: a systematic literature review, *International Journal of Environmental Research and Public Health*. MDPI AG. doi: 10.3390/ijerph14040435.

Turnbaugh, P. J., et al. (2006) An obesity-associated gut microbiome with increased capacity for energy harvest, *Nature*, 444(7122), pp. 1027–1031.

Turnbaugh, P. J., et al. (2008) Diet-induced obesity is linked to marked but reversible alterations in the mouse distal gut microbiome, *Cell Host and Microbe*. Cell Press, 3(4), pp. 213–223. doi: 10.1016/j.chom.2008.02.015.

Turnbaugh, P. J., et al. (2009) A core gut microbiome in obese and lean twins, *Nature*. Nature Publishing Group, 457(7228), pp. 480–484. doi: 10.1038/nature07540.

Ungaro, R., et al. (2017) Ulcerative colitis, *The Lancet*. Lancet Publishing Group, pp. 1756–1770. doi: 10.1016/S0140-6736(16)32126-2.

Unnikrishnan, R., et al. (2017) Type 2 diabetes: demystifying the global epidemic, *Diabetes*. American Diabetes Association Inc., pp. 1432–1442. doi: 10.2337/db16-0766.

Van de Perre, P. (2003) Transfer of antibody via mother's milk, *Vaccine*, 21(24), pp. 3374–3376.

van den Bogert, B., et al. (2013) Diversity of human small intestinal *Streptococcus* and *Veillonella* populations, *FEMS Microbiology Ecology*, 85, pp. 376–388

Van der Waaij, L. A., Limburg, P. C., Mesander, G. and van der Waaij, D. (1996) In vivo IgA coating of anaerobic bacteria in human faeces, *Gut*, 38, pp. 348–354.

Vich Vila, A., et al. (2018) Gut microbiota composition and functional changes in inflammatory bowel disease and irritable bowel syndrome, *Science Translational Medicine*. American Association for the Advancement of Science (AAAS), 10(472), p. eaap8914. doi: 10.1126/scitranslmed.aap8914.

Villanueva-Millan, M. J. et al (2019) Sa1909 – Deep Sequencing of the Entire Human Small Intestine: A Comparison to Stool Microbiome in the Reimagine Study, Gastroenterology, 156(6), pp. S-448–S-449.

Virtue, A. T., et al. (2019) The gut microbiota regulates white adipose tissue inflammation and obesity via a family of microRNAs, *Science Translational Medicine*. American Association for the Advancement of Science, 11(496). doi: 10.1126/scitranslmed.aav1892.

von Mutius, E., Vercelli, D. (2010) Farm living: effects on childhood asthma and allergy, *Nature Reviews Immunology*, 10, pp. 861–868. doi: 10.1038/nri2871.

Vrieze, A., et al. (2012) Transfer of intestinal microbiota from lean donors increases insulin sensitivity in individuals with metabolic syndrome, *Gastroenterology*. W.B. Saunders, 143(4), pp. 913–916.e7. doi: 10.1053/j.gastro.2012.06.031.

Walter, J. and Ley, R. (2011) The human gut microbiome: ecology and recent evolutionary changes, *Annual Review of Microbiology*, 65, pp. 411–429.

Wang, H., et al. (2020) Promising treatment for type 2 diabetes: fecal microbiota transplantation reverses insulin resistance and impaired islets, *Frontiers in Cellular and Infection Microbiology*. Frontiers Media S.A., 9, p. 455. doi: 10.3389/fcimb.2019.00455.

Wang, M., Ahrne, S., Jeppsson, B. and Molin, G. (2005) Comparison of bacterial diversity along the human intestinal tract by direct cloning and sequencing of 16S rRNA genes, *FEMS Microbiology Ecology*, 54, pp. 219–231.

Wilbert, et al. (2020) Spatial ecology of the human tongue dorsum microbiome, *Cell Reports*, 30, pp. 4003–4015.

Wilke, K., Martin, A., Terstegen, L. and Biel, S.S. (2007), A short history of sweat gland biology. International Journal of Cosmetic Science, 29: 169–179. https://doi.org/10.1111/j.1467-2494.2007.00387.x

Wright, J. R. (1997) Immunomodulatory functions of surfactant, *Physiological Reviews*, 77, pp. 931–962.

Yamashita, Y. and Takeshita, T. (2017) The oral microbiome and human health, *Journal of Oral Science*, 59(2), 201–206. doi: 10.2334/josnusd.16-0856.

Yu, J. J., Manus, M. B., Mueller, O., Windsor, S. C., Horvath, J. E. and Nunn, C. L. (2018) Antibacterial soap use impacts skin microbial communities in rural Madagascar, *PLoS One*, 13(8, p. e0199899. doi: 10.1371/journal.pone.0199899.

Zaura, E., Keijser, B. J., Huse, S. M. and Crielaard, W. (2009) Defining the healthy "core microbiome" of oral microbial communities, *BMC Microbiology*, 9, p. 259.

Zeeuwen, P. L., et al. (2012) Microbiome dynamics of human epidermis following skin barrier disruption, *Genome Biology*, 13, p. R101.

Zhang, T., et al. (2019) Akkermansia muciniphila is a promising probiotic, *Microbial Biotechnology*. John Wiley and Sons Ltd, 12(6), pp. 1109–1125. doi: 10.1111/1751-7915.13410.

Zhang, X., et al. (2013) Human gut microbiota changes reveal the progression of glucose intolerance, *PLoS One*, 8(8), p. e71108.

Zhang, S. L., Wang, S. N. and Miao, C. Y. (2017) Influence of microbiota on intestinal immune system in Ulcerative colitis and its intervention, *Frontiers in Immunology*. Frontiers Media S.A., p. 1674. doi: 10.3389/fimmu.2017.01674.

Zheng, J., et al. (2020) A taxonomic note on the genus *Lactobacillus*: description of 23 novel genera, emended description of the genus *Lactobacillus Beijerinck* 1901, and union of *Lactobacillaceae* and *Leuconostocaceae*, *International Journal of Systematic and Evolutionary Microbiology*. Microbiology Society, 70(4), pp. 2782–2858. doi: 10.1099/ijsem.0.004107.

Zhou, K. (2017) Strategies to promote abundance of *Akkermansia muciniphila*, an emerging probiotics in the gut, evidence from dietary intervention studies, *Journal of Functional Foods*. Elsevier Ltd, pp. 194–201. doi: 10.1016/j.jff.2017.03.045.

Zozaya, et al. (2016) Bacterial communities in penile skin, male urethra, and vaginas of heterosexual couples with and without bacterial vaginosis, *Microbiome*, 4, p. 16. doi: 10.1186/s40168-016-0161-6.

8

Human Microbiome's Role in Disease

Sahabram Dewala and Yogesh S. Shouche
National Centre for Cell Science

CONTENTS

8.1 Introduction

The human intestinal tract harbors a wide range of microorganisms whose collective genome is called 'microbiome' which contains at least 100 times as many genes as human genome (Tremaroli and Bäckhed 2012). These microbes coevolve with their host and act as key determinants of host health by influencing nutrient absorption, barrier function, and immune development (Hall, Tolonen, and Xavier 2017). The composition of the microbiota depends on the diet, host genetic background, and geographical location. Gut is the first place of interaction between microorganisms and host immune system, and thus it plays an important role in overall host health. Luminal surface of the small intestine is covered by a thick mucus layer which contains binding sites for commensal microbes (Gilbert et al. 2018; Flint et al. 2012). Most of the information that is available regarding the composition of gut microbiota derives from fecal sample. A broad consensus on microbial diversity has become possible by sequencing of 16S rRNA gene from fecal sample; therefore, the human gut is dominated by three bacterial phyla in the healthy state, namely, Firmicutes, Bacteroidetes and Actinobacteria; in addition, Verrucomicrobia and Proteobacteria are also present in lower numbers (Grice and Segre 2012). Interest in the role of the microbiome in human health has flourished over the past decade with the advent of high-throughput technology for interrogating complex microbial communities. When human microbiome project was initiated, two big questions "who they are?" and "what they are doing?" were the challenges to researchers. Current understanding of microbiome is centered on how microbes influence human health. A catalog of healthy microbiome from metagenomic sequencing was established, but the disease-associated microbiome was not very clear (Qin et al. 2010). The microbiota is involved in energy harvest and storage, as well as in a variety of metabolic functions such as fermenting and absorbing undigested carbohydrates (Lee et al. 2018). Perhaps even more importantly, the gut microbiota interacts with the immune system, providing signals to promote the maturation of immune cells and the normal development of immune functions. Some members of the gut microbiota such as B. fragilis, Eubacterium rectale, and Faecalibacterium prausnitzii have the ability to ferment undigested plant polysaccharides resulting in the production of short-chain fatty acids (SCFAs), including acetate, propionate, and butyrate. SCFAs have shown to induce IL-18 production from epithelial cell and promote dendritic cells to produce IL-10 and retinoic acid (Verdu, Galipeau, and Jabri 2015). SCFAs-producing bacteria were reported to decrease in gut inflammation. In general, host diet, host genetics, antibiotic consumption, and phylogeny contribute to manipulating the composition of gut microbial community (Hasan and Yang 2019). Genome-scale metabolic modeling or genome reconstruction shows that variations in the diet of the host significantly modifies the composition of the three representative human gut bacteria (B. thetaiotaomicron, E. rectale and M. smithii). In return, different compositions of the representative human gut bacteria influence host metabolism and cause related diseases (Baart and Martens 2012; Gu et al. 2019). This chapter will brief current understanding of the microbiota on human health and metabolic disorders. Perturbations can be described in a number of chronic inflammatory diseases such as obesity, diabetes, inflammatory bowel disease (IBD), celiac disease, and inflammatory bowel syndrome (IBS). The overall contribution of dysbiosis from disruption of homeostasis to disease is not well known (Figure 8.1).

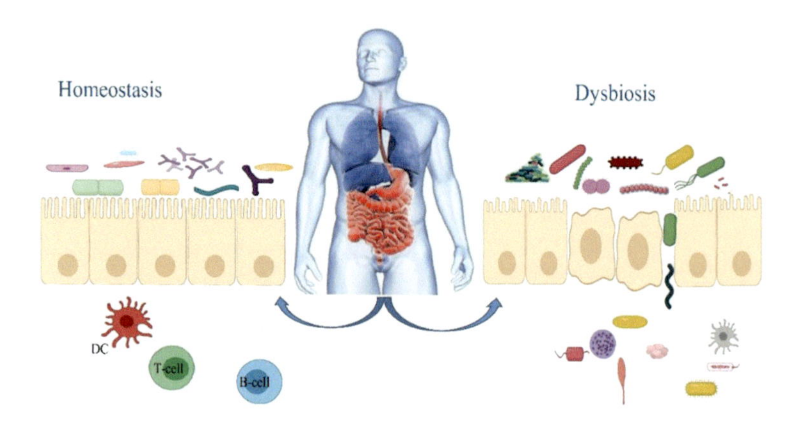

FIGURE 8.1 Consequences of alteration of healthy gut microbiome. A homeostasis state represents balance between human gut commensal and pathogenic microbiome, with maintained mucosal integrity, stimulus of immune maturation, and colonization resistance against invading pathogens. Dominant gut commensals microflora enhance the production of SCFAs, vitamin, and neurotransmitters. Change in diet, consumption of antibiotics, and genetic factors cause microbial dysbiosis and pathogen domination, which eventually turns to be gastrointestinal inflammation and tumor diseases. Abbreviation: SCFAs, short-chain fatty acids.

8.2 Human Gut Microbiome and Diabetes

Type 2 diabetes mellitus (T2DM) has become a major public health issue throughout the world (Wang et al. 2012). It is a metabolic disorder characterized by insulin resistance, an imbalance in blood glucose level, and high blood pressure (Sharma and Tripathi 2019). Genetic constituents, high-fat and high-energy dietary habits, and a sedentary lifestyle are the three major factors that contribute to high risk of T2DM. Several studies have reported gut microbiome dysbiosis as a factor in rapid progression of insulin resistance in T2DM that accounts for about 90% of all diabetes cases worldwide (Zhang and Zhang 2013a). The global diabetes prevalence was estimated in 2019 to be 9.3% (463 million people), rising to 10.2% (578 million) by 2030 and 10.9% (700 million) by 2045. The prevalence is higher in urban areas (estimated 10.8%) than in rural areas (7.2%), and in high-income (10.4%) than low-income countries (4.0%). One in two (50.1%) people living with diabetes do not know that they have diabetes. The global prevalence of impaired glucose tolerance was estimated to be 7.5% (374 million) in 2019 (Saeedi et al. 2019). It has been shown that obesity contributes to T2DM by decreasing insulin sensitivity in adipose tissues, liver, and skeletal muscle, and subsequently leads to impaired beta-cell function, and more than 90% of people with T2DM are overweight or obese (Okazaki et al. 2019). However, the underlying mechanism is not well understood, but it has been known that surgical removal of adipose tissues does not improve insulin sensitivity in obese subjects despite weight loss induction (Fabbrini et al. n.d.). The intestinal microbiota associated with chronic inflammation has been shown to contribute to the onset of T2DM (Saeedi et al. 2019). Moreover, the microbiota is altered in the development of T2DM and its complications including diabetic retinopathy, kidney toxicity, atherosclerosis, hypertension, diabetic foot ulcers, cystic fibrosis, and Alzheimer's disease (Zhang and Zhang 2013a). Rapid increase of type 2 diabetes cases made it a widespread metabolic disease in the past few decades (Gareau, Sherman, and Walker 2010). In recent years, the increase in our understanding of how human gut microbiome is associated with T2DM provides a new therapeutic option for T2DM. Studies on germ-free mice have shown that gut microbiome is essential for early immune development (Urination and Urination 2014). As one of the most concerned obesity-related disorders, T2DM is associated with abnormal energy metabolism and low-level chronic inflammation in fat tissues (McArdle et al. 2013). Growing clinical evidence suggests that obese people with insulin resistance are characterized by an altered composition of gut microbiota, particularly an increased *Firmicutes/Bacteroidetes* ratio compared with healthy people (Zhang and Zhang 2013b). Furthermore, transplantation of the obese gut microbiota in animals greatly affected the energy harvest of hosts (Tremaroli and Bäckhed 2012). Consequently, it is proposed that altered microbiota in obesity modulates intestinal cell permeability and increases metabolic endotoxin secretion that leads to low-grade inflammation, the pathogenesis of insulin resistance, and onset of T2DM (Diamant, Blaak, and de Vos 2011). Recently, commensal bacterial species such as *Bacteroidetes thetaiotaomicron*, *Akkermansia muciniphila*, and *Escherichia coli* were showed to have different influences on the intestinal mucus and glycocalyx layer, which may affect intestinal permeability (Devaraj, Hemarajata, and Versalovic 2013). Besides, microbiota-dependent changes in gut tight-junction proteins, endocannabinoid system, and intestinal alkaline phosphatase activity may also be involved in altered intestinal permeability and the pathogenesis of insulin resistance (Cani et al. 2009). Firstly, alteration in the *Bacteroidetes/Firmicutes* ratio may modify the host energy metabolism by a specific polysaccharide utilization loci mechanism. Moreover, bacterial cell components (e.g. LPS, peptidoglycans, and flagellin) in the intestine are thought to accelerate the inflammation in T2DM (Kootte et al. 2017). Besides, gastric bypass surgery, an effective way to restore the blood glucose level to treat T2DM, could reduce body weight due to the alteration of the microbiome at the distal gut (Liou et al. 2013a). It has been confirmed that some human intestine commensal strains are able to modulate blood glucose homeostasis, and hence improve against chronic inflammation induced by altered microbiota (Liou et al. 2013b). It is well established that obesity-induced chronic low-level inflammation is a leading cause of the development of T2DM (Donath 2011). Human commensal bacteria have been confirmed to prevent the onset of diabetes through down-regulating proinflammatory cytokines IFN- γ and IL-2 or IL-1 or enhancing anti-inflammatory cytokines IL-10 production in diabetic mice (Matsuzaki et al. 1997). A recent study showed that *Lactobacillus reuteri* GMNL-263 helps to normalize the serum glucose, insulin, leptin, C-peptide, glycated hemoglobin, GLP-1 level, inflammatory IL-6 and TNF-α in adipose tissues, and PPAR-γ and GLUT4 gene expression in high-fructose-fed rats (Hsieh et al. 2013). In addition, some commensal strains act as antioxidants that are apparently able to alleviate pancreatic oxidative stress which can lead to chronic inflammation and apoptosis of pancreatic β-cells (Ejtahed et al. 2012). Moreover, the supplementation of certain probiotic strains can potentially modulate the lipid metabolism and result in the reduction of the serum total cholesterol level and LDL cholesterol, which will reduce the risk of T2DM (Ooi and Liong 2010). *Bifidobacterium* appears to be the most consistently supported genus in the literature, playing a protective role against T2DM by improving glucose tolerance (Kim Chung et al. 2015). Some studies also found a negative association between specific species such as *B. adolescentis*, *B. bifidum*, *B. pseudocatenulatum*, *B. longum*, and *B. dentium* and diseases in patients treated with metformin or after undergoing gastric bypass surgery (Moya-Pérez, Neef, and Sanz 2015). Despite multiple studies supporting the importance of gut microbiota in pathophysiology of T2DM, the field is still in its early state.

8.3 Microbiome and Obesity

Obesity is a physiological state that has become a major health concern in populations that have adopted Western diet and is closely associated to the microbiota (Ley et al. 2016). Obesity is defined as an accumulation of excessive fat that spoils health status (Hou et al. 2017). The obesity is dramatically increased over the past decades, and it has been estimated that about 1.9 billion adults are overweight (Ghoorah et al. 2016). Dysbiosis in the gut microflora is strongly associated with the pathogenesis of obesity and T2DM (Ahmad et al. 2019). Studies of germ-free mice colonized with a defined microbial community have provided substantial evidence that the diversity, as well as presence and relative proportion, of different microbes in the intestine plays active roles in energy homeostasis (Bäckhed et al. 2004). The importance of the microbiome is highlighted by the consequences of its absence and repletion. Therefore, germ-free mice are less efficient in processing foods, and they gain weight when colonized with almost any gut microbes. Transplantation of germ-free mice with 'conventional' microbes results in weight gain and shows similar levels of fatness to the donor mice (Bäckhed et al. 2004). The amount of weight gained by gnotobiotic rodents differs depending on which microbes are inoculated (Turnbaugh et al. n.d.). Obesity is associated with phylum-level changes in the microbiota, reduced bacterial diversity, and altered representation of bacterial genes and metabolic pathways (Turnbaugh et al. n.d.). Recent studies show that microbiota may regulate weight gain and adiposity through many ways, i.e. energy harvest and production of microbial metabolites (Duranti et al. n.d.). One of the most important metabolic activities of intestine microbiome is production of non-gaseous SCFAs (acetate, propionate, and butyrate), through fermentation of undigested plant polysaccharides (Wong et al. 2006). The dominant commensal bacteria that produce SCFAs are represented by *Faecalibacterium prausnitzii*, *Akkermansia muciniphilia*, *Prevotella* spp., *Ruminococus* spp., *Coprococcus* spp., *Eubacterium rectale*, and *Roseburia* spp. (Duranti et al. n.d.). SCFAs can have multiple effects on host and are important modulators of gut health and immune response, intestinal hormone production, and lipogenesis. Manipulation in host genetics and environmental factors has been used to induce obesity in animal models and thus provoke a change in microbiota composition (Zhang and Zhang 2013a). Obesity results in low-grade chronic inflammation which is very distinct from classical inflammation (Gregor and Hotamisligil 2011). This pattern includes moderate induction of inflammatory cytokines such as TNF-α, IL-1b, and CCL2, as well as an increase in the number of mast cells, T cells, and macrophages (Gregor and Hotamisligil 2011). An increase in *Bifidobacterium* spp. has also been shown to modulate inflammation in obese mice by increasing the production of glucagon-like peptide-2 (GLP-2), which reduces intestinal permeability and thus reduces the translocation of lipopolysaccharides (Cani et al. 2009). Genetically obese (*ob*/*ob*) mice have decreased *Bacteroidetes/Firmicutes* ratios compared with their lean (*ob*/+ and +/+ wild-type) siblings (Zhang and Zhang 2013a). Transplantation of gut microbiota from the obese (*ob*/*ob*) to germ-free mice conferred an obese phenotype, demonstrating transmissibility of metabolic phenotypes; the transferred microbiomes had increased capacities for energy harvest. In humans, the relative proportions of members of the *Bacteroidetes* phylum increase with weight loss (Murphy et al. 2010). In studies of monozygotic and dizygotic twins, obesity was associated with smaller populations of *Bacteroidetes*, diminished bacterial diversity, and enrichment of genes related to lipid and carbohydrate metabolisms (Turnbaugh, Ridaura, et al. 2009). The same trend is observed within individual humans on weight loss diets. A study with human twins showed that in obese individuals, the decrease in *Bacteroidetes* was accounted by an increase in *Actinobacteria* rather than in *Firmicutes* (Murphy et al. 2010). Despite substantial taxonomic variations, functional metagenomic differences were minor. Modern lifestyles that change the selection pressures on microbiomes could alter exposures to bacteria during the early lives of hosts and thus may contribute to the development of obesity (Turnbaugh, Hamady, et al. 2009). Antibiotic use in human infancy was significantly associated with obesity development. Studies have proven that early childhood exposure to antibiotics has been associated with a significantly increased risk of metabolic disorders. By contrast, perinatal administration of a *Lactobacillus rhamnosus* probiotic decreased excessive weight gain during childhood. These early studies provide support for the concept that dysbiosis in microbiota could lead to childhood-onset obesity. The interaction between the microbiota and the immune system in obesity was demonstrated in mice which were genetically modified lacking TLR5, which recognizes flagellin is one of the major microbial receptors of the innate immune system (Fernandez-Feo et al. 2013). These mice develop characteristics of metabolic syndrome along with significant changes in their gut microbiota, and showed that alterations in the gut flora induce a low-grade inflammatory signaling that eventually results in the development of metabolic syndrome. Moreover, this obesity phenotype is transmissible to wild-type mouse simply by transferring the intestine microbiota (Vijay-Kumar et al. 2010). Hence, the exact mechanisms responsible for obesity are still an open challenge, and several mice studies demonstrate that an imbalanced gut microbiota associate with diseased states and suggest hypotheses to be tested in future research.

8.4 Role of Human Gut Microbiome in IBD

IBDs are proposed to result from an inappropriate immune response to the gut microbes in a genetically susceptible host. It is a chronic inflammatory disorder of the intestinal tract of an unknown cause. The incidence of IBD has increased in the Western world since the midst of the 20th century (Molodecky et al. 2012; Rocchi et al. 2012; Hammer et al. 2016). At the turn of the 21st century, it plateaued in some developed nations with a prevalence of up to 0.5% of the general population, while it is continuing to rise in developing nations (Benchimol et al. 2017). Etiological studies on IBD are centered on several factors, including host genetics, immune responses, the gut microbiota dysbiosis, and the environmental stimuli in disease. Gut dysbiosis has been consistently shown to be associated with IBD. Due to the expansion in application of high-throughput deep sequencing technology in the past decade, we are able to gradually decoding the role of the microbiome in development of IBD, but little is known about the individual nature of microbiome dynamics in IBD. These findings have improved our knowledge on the functional mechanisms of the microbiome in the pathogenesis of IBD (Zuo and Ng 2018). Cross-sectional study of IBD revealed microbial signature for different IBD subtypes, including ulcerative colitis (UC), colonic Crohn's disease, and ileal Crohn's disease (Halfvarson et al. 2017). Crohn's disease (CD) can affect any portion of the GI tract from the mouth to the anus (Durack and Lynch 2019). It shows deep ulceration and might extend to all layers of bowel wall. Granuloma formation is the best isolating symptom of CD from UC. The most important differentiating feature of CD from UC is involvement of all layers of bowel not just mucosa or submucosa. On the other hand, UC is characterized by the thinning and continuous inflammation of colon and wall without intermittent healthy tissue. It shows ulcerated mucus lining of the large intestine and never progresses beyond the inner lining, restricted only to the large intestine. Earlier, incidents of UC were higher than CD, but recent reports found more cases affected with CD than with UC. An estimate of 1–2 million people of the United States are suffering from UC or CD with an incident rate of 1/1,000 individuals. Intestinal microorganisms provoke the immune reaction through releasing some chemicals that induce uncontrolled inflammation by releasing proinflammatory cytokines INF-γ, TNF-α, and other mediators. Early-childhood exposure to antibiotics has been associated with a significantly increased risk for CD, suggesting that gut microbiome perturbations are important for disease risk. Microbial diversity is significantly reduced in CD, suggesting that a perturb gut microbiome population could affect immune interactions. A broad microbial alteration pattern was identified including reduction in diversity, decreased abundances of bacterial taxa within the phyla *Firmicutes* and *Bacteroides*, and increases in the Gamma *Proteobacteria* (Frank et al. 2007; Putignani et al. 2015; Casén et al. 2015). In IBD, it has been consistently shown that there is a decrease in biodiversity, knowingly alpha diversity, and in species richness. Patients with CD displayed a reduced alpha diversity in the fecal microbiome compared with healthy controls (Manichanh et al. 2006). This decreased diversity was partly linked to the temporal instability of the dominant taxa in IBD (Martinez et al. 2008). There is also reduced diversity in inflamed vs. non-inflamed tissues even within the same patient, and a lower bacterial load was observed at the inflamed regions in CD patients (Sepehri et al. 2007). A multicenter study analyzed >1,000 treatment naïve pediatric CD samples collected from multiple concurrent gastrointestinal locations (Gevers et al. 2014). It was found that changes in bacteria, including increased *Veillonellaceae*, *Pasteurellacaea*, *Enterobacteriaceae*, and *Fusobacteriaceae*, and decreased *Bacteroidales*, *Erysipelotrichales*, and *Clostridiales*, strongly correlated with disease status. This study also showed that rectal mucosa-associated microbiome profiling offered a feasible biomarker for the diagnosis of CD (Gevers et al. 2014). 16S rRNA and shotgun metagenome study from fecal samples have shown that specific bacterial taxa were altered in IBD. *Enterobacteriaceae* bacteria were found dominant both in patients with IBD and in mice. *Escherichia coli*, particularly adherent-invasive strains, was isolated from ileal CD biopsies and was also found in UC patients (Darfeuille-Michaud et al. 2004; Sokol et al. 2006). *Fusobacteria* is another clade of adherent and invasive bacteria. Bacteria of the genus *Fusobacterium* principally colonize both the oral cavity and the gut but present at a higher abundance in the colonic mucosa of patients with UC compared to healthy controls (Ohkusa et al. n.d.; 2002). It indicates that the inflammatory environment in IBD may favor the growth of this bacterial clade. Administering *Fusobacterium varium* by rectal enema in mice caused colonic mucosal inflammation. The invasive ability of human *Fusobacterium* bacteria positively correlates with host IBD severity (Lee et al. 2016). The study has found that *Fusobacterium* species were abundantly present in tumor tissues than in adjacent normal tissues in colorectal cancer. Moreover, *Fusobacterium* isolated from humans were reported to have a tumorigenesis role in mice (Kostic et al. 2013). A recent study found that TLR-4 dependent inflammatory glucorhamnan polysaccharide made by *Ruminococcus gnavus*, gut microbes associated with CD, and other diseases. The study has further identified that this molecule enhances the production of inflammatory cytokines like TNF-α by dendritic cells and may contribute to the association between *R. gnavus* and CD. This work establishes a plausible molecular mechanism that may explain the association between a member of the gut microbiome and an inflammatory disease (Henke et al. 2019). There are specific groups of human gut commensal bacteria that may play a protective role against IBD. A range of bacterial species, most notably the genera *Lactobacillus*, *Bifidobacterium*, and *Faecalibacterium*, have been reported to play a protective role in the host from mucosal inflammation via several mechanisms, including the stimulation of the anti-inflammatory cytokine, including IL-10 and down-regulation of inflammatory cytokines. *Faecalibacterium prausnitzii* has been shown to have anti-inflammatory properties in IBD. The abundance of *F. prausnitzii* is significantly decreased, while the abundance of *E. coli* is increased in ileal biopsies of CD specimens (Willing et al. 2009). CD patients with low abundances of *F. prausnitzii* in the mucosa are more likely to have relapse after surgery (Harry Sokol et al. 2008). By contrast, restoration of *F. prausnitzii* after recurrence is associated with maintenance of clinical remission of UC (Varela et al. 2013). In mice with chemically induced colitis, the colitis phenotype is more severe in germ-free mice than in conventionally reared mice (Kitajima et al. 2001). The commensal microbes protect the host via colonization resistance, where commensals occupy niches within the host and prevent the invasion of pathogenic bacteria (Callaway et al. 2008). In addition, the gut microbiota can modulate the host mucosal immune response. *Clostridium* and *Bacteroides* spp. could induce the expansion of regulatory T cells (Treg) and mitigate intestinal inflammation. Other gut bacterial members can alleviate mucosal inflammation by regulating NF-kB activation (Kelly et al. 2004). The gut microbiota coevolves with the polysaccharide-rich diet, is able to efficiently extract energy from dietary fiber, and protects the host from inflammation. Lack of dietary fiber intake has been associated with the development of IBD (Kitajima et al. 2001).

8.5 Therapeutic Implications of the Human Microbiome

The intestinal microbiota has emerged as an attractive therapeutic target in intestinal metabolic disorder or microbiome dysbiosis. The majority of microbiota-based therapies aim at engineering the intestinal ecosystem by means of probiotics or prebiotics and fecal microbiome transplantation (FMT).

8.6 Probiotics

One possible approach to achieving a healthy microbiota is to directly administer beneficial bacteria, i.e. probiotics, which is defined by the International Scientific Association for Probiotics and Prebiotics as "live microorganisms which when administered in adequate amounts confer a health benefit on the host" (Hill et al. 2014). One common approach to developing probiotics to benefit particular disorder is to administer bacterial taxa whose reduced abundance is associated with disease. For example, IBD, probiotic approaches have generally sought to administer bacteria whose reduced abundance is associated with disease such as *F. prausnitzii*, which is depleted in this disorder (Adeshirlarijaney and Gewirtz 2020). Hence, this approach has generally focused on administering bacteria with anti-inflammatory properties or bacteria with beneficial metabolic properties such as propensity to produce SCFAs. Various clinical trials support the hypothesis that restoring the intestinal microbiota in this manner could be effective in diabetes, IBD, and other inflammatory disorders. Most widely studied probiotics with respect to diabetes is the *Bifidobacterium* and *Lactobacillus* phyla. Specific strains of *L. casei*, *L. acidophilus*, *L. gasseri*, *L. rhamnosus*, and *L. plantarum* have been demonstrated to exert antidiabetic effects (Andreasen et al. 2010). Hence, several strains of species *L. plantarum* have been shown to improve the glycemic control in obese and diabetic patients, via their cluster of complex carbohydrate-utilizing genes. Moreover, administration of *Bifidobacterium animalis*, *B. breve*, *B. longum*, and *P. copri* led to amelioration of glucose intolerance (Razmpoosh et al. 2016). The underlying mechanisms by which probiotics modulate the host metabolism have not been well understood, but they may have the potential to improve intestinal barrier function, inhibit α-glucosidase activity, regulate bile acid metabolism, and produce anti-microbial activity, immune modulation, and SCFAs (Han et al. 2019). *F. prausnitzii* and *A. muciniphila* are other probiotics, which are negatively associated with hyperglycemia and overweight, that may be potential candidates for next-generation probiotics. More recently, genetically engineered bacteria have been studied which secrete immunosuppressive substances such as interleukin-10 (IL-10). Similarly, *Bifidobacterium*-fermented milk decreased the fecal amount of organisms associated with UC. A recent placebo-controlled trial showed that the direct administration of *A. muciniphilia* improved glycemic control in persons with metabolic syndrome, although, unexpectedly, the impact of heat-killed *Akkermansia* appeared more significant than that of the live organism, highlighting that further development is needed before deploying this strategy on a large scale (Dao et al. n.d.) (Figure 8.2).

8.7 Prebiotics

Prebiotics are dietary substances, usually nondigested polysaccharides, that stimulate the growth of protective commensal enteric bacteria. Fructo-oligosaccharides, lactosucrose, inulin, bran, psyllium, and germinated barley extracts boost the growth of lactic acid bacilli, *Lactobacillus* and *Bifidobacterium* spp., and stimulate the production of SCFAs, especially butyrate. Oligosaccharides are of great interest as prebiotics capable of

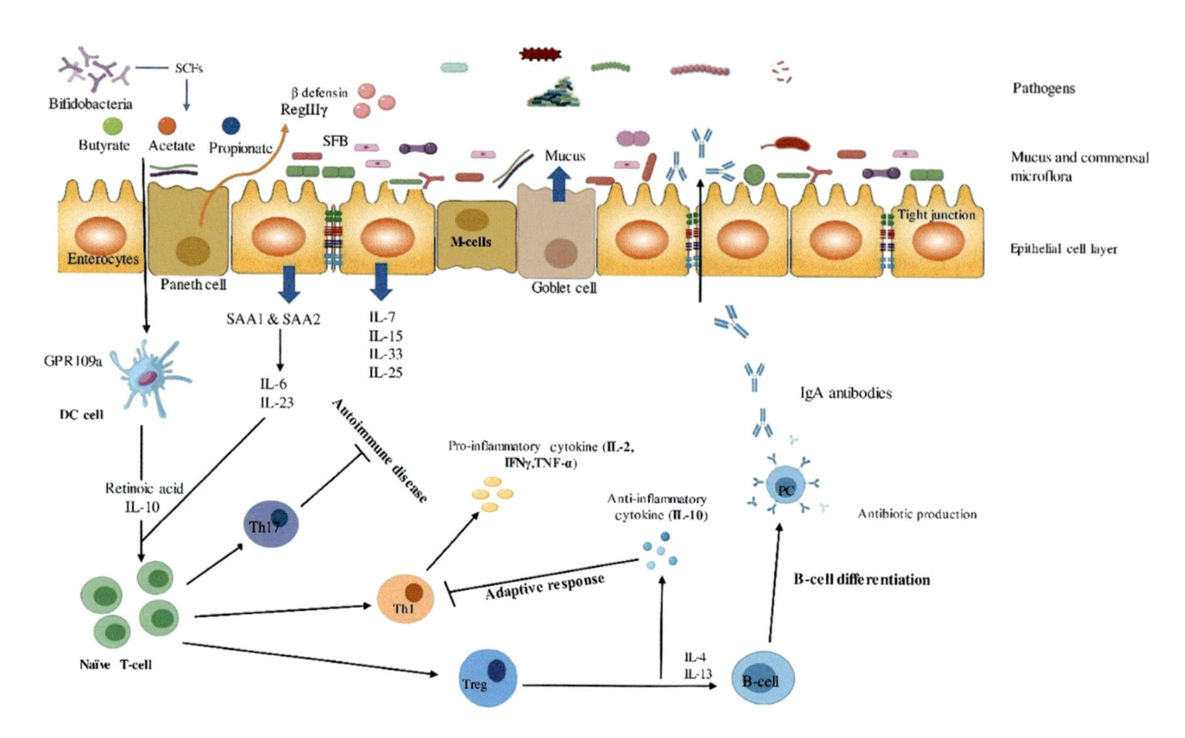

FIGURE 8.2 Probiotic mechanism of shaping host immune system and maintaining mucosal homeostasis. Interaction between the immune system and commensal microflora minimizes bacterial epithelial invasion. These include the secretion of mucus by goblet cells, antibacterial proteins (RegIIIγ and β defensing) by Paneth cells, and IgA secreted by lamina propria plasma cells. Notably, epithelial-derived cytokines IL-7, IL-15, IL-33, and IL-25 have various key functional properties in the regulation and expansion of innate and adaptive immune cell populations. SFB promote intestinal epithelial cells (IECs) production of SAA1 and SAA2 through IL-6 and -23 that foster the differentiation of naïve T-cells into Th17 cells and production of IgA. SCFAs, especially propionate and butyrate, are essential for Treg accumulation and differentiation. It also regulates reinforcing the IEC barrier and exerts anti-inflammatory effect. Abbreviations: IL, interleukin; SFB, segmented filamentous bacteria; IgA, immunoglobulin A; SCFAs, short-chain fatty acids.

modulating the colonic microbiota in humans and animals. Thus, these prebiotic food additives potentially restore the commensal bacterial in the colon of IBD patients by several mechanisms. These substances stimulate the growth of lactic acid bacilli which suppress detrimental species by decreasing the luminal pH and inducing colonization resistance, blocking epithelial attachment and invasion, and secreting bactericidal substances. In addition, increased intake of fiber-rich substances stimulates the growth of genera *Lactobacilli* and *Bifidobacteria* which increase bacterial fermentation with resultant SCFAs production that improves epithelial barrier function. The net result of prebiotic administration is functionally equivalent to administering probiotic bacteria. Although this is an attractive concept for treating intestinal inflammation, experimental support for these prebiotics is even less extensive than that for probiotics.

8.8 Fecal Microbiota Transplantation

Fecal microbiota transplantation (FMT) has become an effective therapy for recurrent *Clostridium difficile* infection (CDI) and has also gained substantial interest as a novel treatment for other disorders (Drekonja et al. 2015). The success of FMT in treating CDI is described to restore the healthy microbiome in patients with dysbiosis. This approach was later extended to study the treatment of other diseases such as IBD and metabolic syndrome (Khoruts and Sadowsky 2016; Vrieze et al. 2012a). The evidence that links gut microbial dysbiosis with IBD has led to the exploration of FMT as therapy for the disease (Damman et al. 2012). A systematic review and meta-analysis of 18 studies that recruited 122 patients with IBD found that around 36%–45% of patients accomplished clinical remission during follow-up. Subgroup analyses demonstrated that a pooled estimate of clinical remission rate in UC patients was 22%, while younger patients (aged 7–20 years) had a rate of 64.1%, and patients with CD 61%. The results suggest that FMT may be more effective for CD in younger patients than for UC infection; however, it is difficult to draw absolute conclusions due to the small sample sizes, short follow-up times, and heterogeneous results (Kelly et al. 2015). Two placebo-controlled studies of FMT on UC patients have shown contradicting results with one study demonstrating the importance of the donor effect. It was documented that some patients with UC experienced fevers and increased levels of creative protein post-FMT. A recent double-blind placebo-randomized controlled trial showed that intensive dosing with multidonor FMT induces clinical remission and endoscopic improvement in active CD and is associated with distinct microbial changes related to the outcome (Angelberger et al. 2013). Importantly, the microbial diversity increased and persisted after FMT (Angelberger et al. 2013). These data are consistent with a more recent study showing that the low-intensity pooled donor FMT is also effective in active UC remission (Costello et al. 2019). It has been observed that 30%–50% of the donor's microbiota persists in the recipient after FMT. Two studies have shown bacteriophage transfer from donor to recipient, and in a pilot study, donor virome richness is associated with the success of FMT (Chehoud et al. 2016; Zuo et al. n.d.). When donor virome richness especially with Caudovirales was higher than that of the recipient, the recipient was more likely to be cured after FMT treatment (Zuo et al. n.d.). Patients with IBD have been observed to harbor significantly higher virome richness than healthy controls which may account for the higher failure rate of FMT in treating IBD than in treating CDI (Abstract 2015; Colman and Rubin 2014). These data show the importance of donor selection, where inclusion of a donor with high virome richness or pooled multiple donors is preferred. A FMT-TRIM double-blind placebo-controlled pilot trial of weekly oral FMT from healthy lean donors to obese recipient mice showed no clinically significant metabolic effects (Yu et al. 2020). In another study, mice receiving FMT from normal fat diet (NFD)-exercised donors showed remarkably reduced food efficacy, and also mitigated metabolic profiles ($p<0.05$). 16S rRNA amplicon sequencing showed that beneficial effects of FMT were associated with the increased number of genera *Helicobacter*, *Odoribacter* and AF12 (Lai et al. 2018). Another placebo-controlled pilot study found that FMT capsules (derived from a lean donor) were safe to be given to obese metabolically uncompromised patients but did not reduce BMI. Hence, the FMT capsules led to change in the intestinal microbiome and bile acid profiles that were similar to that of the lean donor. FMT is associated with improvements in RD and HbA1c in 6 weeks. The beneficial effects of FMT on RD and HbA1c were needed to maintain long-term assessment (Kootte et al. 2017). Transplantation of the entire fecal microbiota from the healthy donor to the obese could influence host metabolism by modulating microbial composition and/or functions (Koren et al. 2019). Short-term improvement in insulin sensitivity increases in relative abundance of *Ruminococcus bromii* and *Roseburia intestinalis*, species which are well known as dietary fiber degraders and butyrate producers, respectively and may play a role in improving insulin sensitivity through regulation of GLP-1 and intestinal gluconeogenesis (Vrieze et al. 2012b). A recent study showed that FMT could alleviate the symptoms associated with type 2 diabetes. Humanized mouse models were consumed of a high-fat diet combined with streptozotocin (100 mg kg^{-1}), and their fecal was transferred to diabetic mice with monitoring of some blood parameters. FMT improved the insulin resistance and reshaped normal gut microbiome (Marotz and Zarrinpar 2016). Another study reported a 46-year-old female with diabetic neuropathy (DN) with 8 years of diabetic history achieved remission by FMT treatment. It proposed that FMT could be a promising treatment in patients with diabetes or diabetes-related complications like DN (Cai et al. 2018). A study conducted on Chinese Kazak ethnic groups demonstrates that FMT from individuals with normal glucose tolerance (NGT) significantly improves the intestinal microbiome composition in *db/db* mice and reduces plasma glycolipid levels. They also found that *Desulfovibrio* and *Clostridium coccoides* levels were significantly reduced and the level of *A. muciniphila* was increased, which correlated with HDAC3 expression (Zhang et al. 2020). They suggest that fecal microbiome from Kazak ethic groups with NGT could be used as a potential source for the diagnosis of T2DM. Indeed, FMT from obese or lean twins into germ-free mice induced differential metabolic and body weight improvement according to the diet administrated to mouse receivers. Interestingly, however, response to donor FMT showed major inter-individual variability among receivers in both studies, with some patients displaying major improvements, while others remaining stable (Fang, Fu, and Wang 2018). Lean donor FMT induced differential microbiota modifications in good and poor responders. For example, after FMT, good responders displayed an increase in *Akkermansia muciniphila*, which has been previously associated with improved health metabolism and increased insulin sensitivity not only in mice but also in humans. The beneficial effects may be due to a microbes-induced increase in the intestinal levels of endocannabinoids and epithelial toll-like receptor 2, which regulate epithelial barrier function and inflammation (Dao et al. n.d.). However, these shifts in bacterial species following transplantation were not consistent across all studies. FMT also appeared to be definitely safer and more tolerable than the pharmacologic treatment in patients with DN. In addition, FMT from lean donors can increase intestinal microbial diversity and improve insulin sensitivity in patients suffering from metabolic syndrome (Fang, Fu, and Wang 2018). The long-term consequence of FMT in treating diseases remains unclear. In the future, FMT will be likely substituted by the use of defined microbial consortia. It was proven in animals that such an approach was feasible and effective for the treatment of IBD.

8.9 Conclusion

Interest in the role of the microbiome in human health and disease has been gained huge interest over the past decade owing to the advent of high-throughput sequencing and humanized gnotobiotic model for interrogating complex microbial communities. Recent research suggests that human gut microbiome is a key player in human health and disease. Microbiota is a crucial regulator in maintaining the intestinal barrier integrity, sustaining a normal metabolic homeostasis, protecting the host from infection by pathogens, and enhancing host defense system. Alterations to human intestinal microbiome composition lead to the onset of metabolic disorders such as IBD, obesity, and diabetes. However, health human gut microbiome is highly personalized, but individual nature of disease-associated microbiota is not very clear. Probiotics and prebiotics have become a common way to achieve a healthy microbiota by directly administering beneficial bacteria or nondigested polysaccharides that stimulate the growth of protective commensal enteric bacteria. Fecal microbiota transplantation is another potential approach that could improve insulin sensitivity, enhance anti-inflammatory cytokines production, and intestinal gluconeogenesis. Furthermore, while FMT has demonstrated its cost-effectiveness in different severe diseases including CD infection, obesity, and IBD, it remains to be proven whether this approach can be extended to T2D and particularly the most severe patients for whom glucose control remains above target, despite intensive medical therapy. However, the potential mechanisms linking the microbiota to metabolic disorders have not been fully elucidated, and therefore, continuing research efforts are needed. Hence, FMT probiotics and prebiotics are emerging as a therapeutic option to reshape the healthy gut microbiome.

REFERENCES

Adeshirlarijaney, A., & Gewirtz, A. T. (2020). Considering gut microbiota in treatment of type 2 diabetes mellitus. *Gut Microbes*, 11(3), 1–12.

Ahirwar, R., & Mondal, P. R. (2019). Prevalence of obesity in India: a systematic review. *Diabetes & Metabolic Syndrome: Clinical Research & Reviews, 13*(1), 318–321.

Ahmad, A., Yang, W., Chen, G., Shafiq, M., Javed, S., Ali Zaidi, S. S., & Bokhari, H. (2019). Analysis of gut microbiota of obese individuals with type 2 diabetes and healthy individuals. *PLoS One, 14*(12), e0226372.

Andreasen, A. S., Larsen, N., Pedersen-Skovsgaard, T., Berg, R. M., Møller, K., Svendsen, K. D., & Pedersen, B. K. (2010). Effects of *Lactobacillus acidophilus* NCFM on insulin sensitivity and the systemic inflammatory response in human subjects. *British Journal of Nutrition, 104*(12), 1831–1838.

Angelberger, S., Reinisch, W., Makristathis, A., Lichtenberger, C., Dejaco, C., Papay, P., & Berry, D. (2013). Temporal bacterial community dynamics vary among ulcerative colitis patients after fecal microbiota transplantation. *American Journal of Gastroenterology, 108*(10), 1620–1630.

Aron-Wisnewsky, J., Clement, K., & Nieuwdorp, M. (2019). Fecal microbiota transplantation: a future therapeutic option for obesity/diabetes?. *Current Diabetes Reports*, 19(8), 51.

Baart, G. J., & Martens, D. E. (2012). Genome-scale metabolic models: reconstruction and analysis. In Christodoulides, M., editor. *Neisseria meningitides* (pp. 107–126). London: Humana Press.

Backhed, F., Ding, H., Wang, T., Hooper, L. V., Koh, G. Y., Nagy, A., & Gordon, J. I. (2004). The gut microbiota as an environmental factor that regulates fat storage. *Proceedings of the National Academy of Sciences*, 101(44), 15718–15723.

Belkaid, Y., & Hand, T. W. (2014). Role of the microbiota in immunity and inflammation. *Cell, 157*(1), 121–141.

Benchimol, E. I., Bernstein, C. N., Bitton, A., Carroll, M. W., Singh, H., Otley, A. R., & Mack, D. R. (2017). Trends in epidemiology of pediatric inflammatory bowel disease in Canada: distributed network analysis of multiple population-based provincial health administrative databases. *The American Journal of Gastroenterology, 112*(7), 1120.

Bi, Y., Wang, T., Xu, M., Xu, Y., Li, M., Lu, J., & Ning, G. (2012). Advanced research on risk factors of type 2 diabetes. *Diabetes/Metabolism Research and Reviews, 28*, 32–39.

Cai, T. T., Ye, X. L., Yong, H. J., Song, B., Zheng, X. L., Cui, B. T., … & Ding, D. F. (2018). Fecal microbiota transplantation relieve painful diabetic neuropathy: a case report. *Medicine, 97*(50), e13543.

Callaway, T. R., Edrington, T. S., Anderson, R. C., Harvey, R. B., Genovese, K. J., Kennedy, C. N., & Nisbet, D. J. (2008). Probiotics, prebiotics and competitive exclusion for prophylaxis against bacterial disease. *Animal Health Research Reviews, 9*(2), 217.

Cani, P. D., Possemiers, S., Van de Wiele, T., Guiot, Y., Everard, A., Rottier, O., & Muccioli, G. G. (2009). Changes in gut microbiota control inflammation in obese mice through a mechanism involving GLP-2-driven improvement of gut permeability. *Gut, 58*(8), 1091–1103.

Casen, C., Vebø, H. C., Sekelja, M., Hegge, F. T., Karlsson, M. K., Ciemniejewska, E., & Munkholm, P. (2015). Deviations in human gut microbiota: a novel diagnostic test for determining dysbiosis in patients with IBS or IBD. *Alimentary Pharmacology & Therapeutics, 42*(1), 71–83.

Chehoud, C., Dryga, A., Hwang, Y., Nagy-Szakal, D., Hollister, E. B., Luna, R. A., & Bushman, F. D. (2016). Transfer of viral communities between human individuals during fecal microbiota transplantation. *MBio, 7*(2), e003224-16.

Colman, R. J., & Rubin, D. T. (2014). Fecal microbiota transplantation as therapy for inflammatory bowel disease: a systematic review and meta-analysis. *Journal of Crohn's and Colitis, 8*(12), 1569–1581.

Costello, S. P., Hughes, P. A., Waters, O., Bryant, R. V., Vincent, A. D., Blatchford, P., & Rosewarne, C. P. (2019). Effect of fecal microbiota transplantation on 8-week remission in patients with ulcerative colitis: a randomized clinical trial. *Jama, 321*(2), 156–164.

Crommen, S., & Simon, M. C. (2018). Microbial regulation of glucose metabolism and insulin resistance. *Genes, 9*(1), 10.

Damman, C. J., Miller, S. I., Surawicz, C. M., & Zisman, T. L. (2012). The microbiome and inflammatory bowel disease: is there a therapeutic role for fecal microbiota transplantation?. *American Journal of Gastroenterology, 107*(10), 1452–1459.

Dao, M. C., Everard, A., Aron-Wisnewsky, J., Sokolovska, N., Prifti, E., Verger, E. O., & Dumas, M. E. (2016). *Akkermansia muciniphila* and improved metabolic health during a dietary intervention in obesity: relationship with gut microbiome richness and ecology. *Gut, 65*(3), 426–436.

Darfeuille-Michaud, A., Boudeau, J., Bulois, P., Neut, C., Glasser, A. L., Barnich, N., & Colombel, J. F. (2004). High prevalence of adherent-invasive Escherichia coli associated with ileal mucosa in Crohn's disease. *Gastroenterology, 127*(2), 412–421.

Devaraj, S., Hemarajata, P., & Versalovic, J. (2013). The human gut microbiome and body metabolism: implications for obesity and diabetes. *Clinical Chemistry, 59*(4), 617–628.

Diamant, M., Blaak, E. E., & De Vos, W. M. (2011). Do nutrient-gut-microbiota interactions play a role in human obesity, insulin resistance and type 2 diabetes?. *Obesity Reviews*, 12(4), 272–281.

Donath, M. Y., & Shoelson, S. E. (2011). Type 2 diabetes as an inflammatory disease. Nature *Reviews Immunology*, 11(2), 98–107.

Drekonja, D., Reich, J., Gezahegn, S., Greer, N., Shaukat, A., MacDonald, R., & Wilt, T. J. (2015). Fecal microbiota transplantation for Clostridium difficile infection: a systematic review. *Annals of Internal Medicine, 162*(9), 630–638.

Durack, J., & Lynch, S. V. (2019). The gut microbiome: relationships with disease and opportunities for therapy. *Journal of Experimental Medicine, 216*(1), 20–40.

Duranti, S., Ferrario, C., van Sinderen, D., Ventura, M., & Turroni, F. (2017). Obesity and microbiota: an example of an intricate relationship. *Genes & Nutrition, 12*(1), 1–15.

Ejtahed, H. S., Mohtadi-Nia, J., Homayouni-Rad, A., Niafar, M., Asghari-Jafarabadi, M., & Mofid, V. (2012). Probiotic yogurt improves antioxidant status in type 2 diabetic patients. *Nutrition, 28*(5), 539–543.

Fabbrini, E., Tamboli, R. A., Magkos, F., Marks-Shulman, P. A., Eckhauser, A. W., Richards, W. O., & Abumrad, N. N. (2010). Surgical removal of omental fat does not improve insulin sensitivity and cardiovascular risk factors in obese adults. *Gastroenterology, 139*(2), 448–455.

Flint, H. J., Scott, K. P., Louis, P., & Duncan, S. H. (2012). The role of the gut microbiota in nutrition and health. *Nature Reviews Gastroenterology & Hepatology, 9*(10), 577.

Frank, D. N., Amand, A. L. S., Feldman, R. A., Boedeker, E. C., Harpaz, N., & Pace, N. R. (2007). Molecular-phylogenetic characterization of microbial community imbalances in human inflammatory bowel diseases. *Proceedings of the National Academy of Sciences, 104*(34), 13780–13785.

Gareau, M. G., Sherman, P. M., & Walker, W. A. (2010). Probiotics and the gut microbiota in intestinal health and disease. *Nature Reviews Gastroenterology & Hepatology, 7*(9), 503.

Gevers, D., Kugathasan, S., Denson, L. A., Vázquez-Baeza, Y., Van Treuren, W., Ren, B., & Morgan, X. C. (2014). The treatment-naive microbiome in new-onset Crohn's disease. *Cell Host & Microbe, 15*(3), 382–392.

Gilbert, J. A., Blaser, M. J., Caporaso, J. G., Jansson, J. K., Lynch, S. V., & Knight, R. (2018). Current understanding of the human microbiome. *Nature Medicine, 24*(4), 392–400.

Gregor, M. F., & Hotamisligil, G. S. (2011). Inflammatory mechanisms in obesity. *Annual Review of Immunology, 29*, 415–445.

Grice, E. A., & Segre, J. A. (2012). The human microbiome: our second genome. *Annual Review of Genomics and Human Genetics, 13*, 151–170.

Gu, C., Kim, G. B., Kim, W. J., Kim, H. U., & Lee, S. Y. (2019). Current status and applications of genome-scale metabolic models. *Genome Biology, 20*(1), 121.

Halfvarson, J., Brislawn, C. J., Lamendella, R., Vázquez-Baeza, Y., Walters, W. A., Bramer, L. M., & McClure, E. E. (2017). Dynamics of the human gut microbiome in inflammatory bowel disease. *Nature Microbiology, 2*(5), 1–7.

Hall, A. B., Tolonen, A. C., & Xavier, R. J. (2017). Human genetic variation and the gut microbiome in disease. *Nature Reviews Genetics, 18*(11), 690.

Hasan, N., & Yang, H. (2019). Factors affecting the composition of the gut microbiota, and its modulation. *PeerJ, 7*, e7502.

Henke, M. T., & Clardy, J. (2019). Molecular messages in human microbiota. *Science, 366*(6471), 1309–1310.

Hill, C., Guarner, F., Reid, G., Gibson, G. R., Merenstein, D. J., Pot, B., & Calder, P. C. (2014). Expert consensus document: The International Scientific Association for Probiotics and Prebiotics consensus statement on the scope and appropriate use of the term probiotic. *Nature Reviews Gastroenterology & Hepatology, 11*(8), 506.

Hsieh, F. C., Lee, C. L., Chai, C. Y., Chen, W. T., Lu, Y. C., & Wu, C. S. (2013). Oral administration of Lactobacillus reuteri GMNL-263 improves insulin resistance and ameliorates hepatic steatosis in high fructose-fed rats. *Nutrition & Metabolism, 10*(1), 35.

Kelly, D., Campbell, J. I., King, T. P., Grant, G., Jansson, E. A., Coutts, A. G., … & Conway, S. (2004). Commensal anaerobic gut bacteria attenuate inflammation by regulating nuclear-cytoplasmic shuttling of PPAR-γ and RelA. *Nature Immunology, 5*(1), 104–112.

Khoruts, A., & Sadowsky, M. J. (2016). Understanding the mechanisms of faecal microbiota transplantation. *Nature Reviews Gastroenterology & Hepatology, 13*(9), 508–516.

Kim, S. H., Huh, C. S., Choi, I. D., Jeong, J. W., Ku, H. K., Ra, J. H., & Ahn, Y. T. (2014). The anti-diabetic activity of *Bifidobacterium lactis* HY 8101 in vitro and in vivo. *Journal of Applied Microbiology, 117*(3), 834–845.

Kitajima, S., Morimoto, M., Sagara, E., Shimizu, C., & Ikeda, Y. (2001). Dextran sodium sulfate-induced colitis in germ-free IQI/Jic mice. *Experimental Animals, 50*(5), 387–395.

Kloting, N., & Bluher, M. (2014). Adipocyte dysfunction, inflammation and metabolic syndrome. *Reviews in Endocrine and Metabolic Disorders, 15*(4), 277–287.

Kootte, R. S., Levin, E., Salojärvi, J., Smits, L. P., Hartstra, A. V., Udayappan, S. D., & Knop, F. K. (2017). Improvement of insulin sensitivity after lean donor feces in metabolic syndrome is driven by baseline intestinal microbiota composition. *Cell Metabolism, 26*(4), 611–619.

Kostic, A. D., Chun, E., Robertson, L., Glickman, J. N., Gallini, C. A., Michaud, M., & El-Omar, E. M. (2013). *Fusobacterium nucleatum* potentiates intestinal tumorigenesis and modulates the tumor-immune microenvironment. *Cell Host & Microbe, 14*(2), 207–215.

Lai, Z. L., Tseng, C. H., Ho, H. J., Cheung, C. K., Lin, J. Y., Chen, Y. J., & Wu, C. Y. (2018). Fecal microbiota transplantation confers beneficial metabolic effects of diet and exercise on diet-induced obese mice. *Scientific Reports, 8*(1), 1–11.

Lee, J. J., Wedow, R., Okbay, A., Kong, E., Maghzian, O., Zacher, M., & Fontana, M. A. (2018). Gene discovery and polygenic prediction from a genome-wide association study of educational attainment in 1.1 million individuals. *Nature Genetics, 50*(8), 1112–1121.

Ley, S. H., Pan, A., Li, Y., Manson, J. E., Willett, W. C., Sun, Q., & Hu, F. B. (2016). Changes in overall diet quality and subsequent type 2 diabetes risk: three US prospective cohorts. *Diabetes Care, 39*(11), 2011–2018.

Liou, A. P., Paziuk, M., Luevano, J. M., Machineni, S., Turnbaugh, P. J., & Kaplan, L. M. (2013). Conserved shifts in the gut microbiota due to gastric bypass reduce host weight and adiposity. *Science Translational Medicine, 5*(178), 178ra41.

Manichanh, C., Rigottier-Gois, L., Bonnaud, E., Gloux, K., Pelletier, E., Frangeul, L., & Roca, J. (2006). Reduced diversity of faecal microbiota in Crohn's disease revealed by a metagenomic approach. *Gut, 55*(2), 205–211.

Marotz, C. A., & Zarrinpar, A. (2016). Focus: microbiome: treating obesity and metabolic syndrome with fecal microbiota transplantation. *The Yale Journal of Biology and Medicine, 89*(3), 383.

Martinez, C., Antolin, M., Santos, J., Torrejon, A., Casellas, F., Borruel, N., & Malagelada, J. R. (2008). Unstable composition of the fecal microbiota in ulcerative colitis during clinical remission. *American Journal of Gastroenterology, 103*(3), 643–648.

McArdle, M. A., Finucane, O. M., Connaughton, R. M., McMorrow, A. M., & Roche, H. M. (2013). Mechanisms of obesity-induced inflammation and insulin resistance: insights into the emerging role of nutritional strategies. *Frontiers in Endocrinology, 4*, 52.

Molodecky, N. A., Soon, S., Rabi, D. M., Ghali, W. A., Ferris, M., Chernoff, G., & Kaplan, G. G. (2012). Increasing incidence and prevalence of the inflammatory bowel diseases with time, based on systematic review. *Gastroenterology, 142*(1), 46–54.

Moya-Pérez, A., Neef, A., & Sanz, Y. (2015). *Bifidobacterium pseudocatenulatum CECT* 7765 reduces obesity-associated inflammation by restoring the lymphocyte-macrophage balance and gut microbiota structure in high-fat diet-fed mice. *PLoS One, 10*(7), e0126976.

Murphy, E. F., Clarke, S. F., Marques, T. M., Hill, C., Stanton, C., Ross, R. P., & Cotter, P. D. (2013). Antimicrobials: strategies for targeting obesity and metabolic health?. *Gut Microbes*, 4(1), 48–53.

Ng, S. C., Shi, H. Y., Hamidi, N., Underwood, F. E., Tang, W., Benchimol, E. I., & Sung, J. J. (2017). Worldwide incidence and prevalence of inflammatory bowel disease in the 21st century: a systematic review of population-based studies. *The Lancet, 390*(10114), 2769–2778.

Ohkusa, T., Sato, N., Ogihara, T., Morita, K., Ogawa, M., & Okayasu, I. (2002). Fusobacterium varium localized in the colonic mucosa of patients with ulcerative colitis stimulates species-specific antibody. *Journal of Gastroenterology and Hepatology, 17*(8), 849–853.

Okazaki, S., Otsuka, I., Numata, S., Horai, T., Mouri, K., Boku, S., & Hishimoto, A. (2019). Epigenetic clock analysis of blood samples from Japanese schizophrenia patients. *NPJ Schizophrenia, 5*(1), 1–7.

Ooi, L. G., & Liong, M. T. (2010). Cholesterol-lowering effects of probiotics and prebiotics: a review of in vivo and in vitro findings. *International Journal of Molecular Sciences*, 11(6), 2499–2522.

Qin, J., Li, R., Raes, J., Arumugam, M., Burgdorf, K. S., Manichanh, C., & Mende, D. R. (2010). A human gut microbial gene catalogue established by metagenomic sequencing. *Nature*, 464(7285), 59–65.

Rocchi, A., Benchimol, E. I., Bernstein, C. N., Bitton, A., Feagan, B., Panaccione, R., & Ghosh, S. (2012). Inflammatory bowel disease: a Canadian burden of illness review. *Canadian Journal of Gastroenterology, 26*, 811–817.

Round, J. L., & Mazmanian, S. K. (2009). The gut microbiota shapes intestinal immune responses during health and disease. *Nature Reviews Immunology, 9*(5), 313–323.

Saeedi, P., Petersohn, I., Salpea, P., Malanda, B., Karuranga, S., Unwin, N., & Shaw, J. E. (2019). Global and regional diabetes prevalence estimates for 2019 and projections for 2030 and 2045: results from the International *Diabetes Federation Diabetes Atlas. Diabetes Research and Clinical Practice, 157*, 107843.

Sepehri, S., Kotlowski, R., Bernstein, C. N., & Krause, D. O. (2007). Microbial diversity of inflamed and noninflamed gut biopsy tissues in inflammatory bowel disease. *Inflammatory Bowel Diseases, 13*(6), 675–683.

Sharma, S., & Tripathi, P. (2019). Gut microbiome and type 2 diabetes: where we are and where to go?. *The Journal of Nutritional Biochemistry, 63*, 101–108.

Siljander, H., Honkanen, J., & Knip, M. (2019). Microbiome and type 1 diabetes. *EBioMedicine, 46*, 512–521.

Sokol, H., Pigneur, B., Watterlot, L., Lakhdari, O., Bermúdez-Humarán, L. G., Gratadoux, J. J., & Grangette, C. (2008). *Faecalibacterium prausnitzii* is an anti-inflammatory commensal bacterium identified by gut microbiota analysis of Crohn disease patients. *Proceedings of the National Academy of Sciences, 105*(43), 16731–16736.

Strauss, J., Kaplan, G. G., Beck, P. L., Rioux, K., Panaccione, R., DeVinney, R., & Allen-Vercoe, E. (2011). Invasive potential of gut mucosa-derived Fusobacterium nucleatum positively correlates with IBD status of the host. *Inflammatory Bowel Diseases, 17*(9), 1971–1978.

Tremaroli, V., & Bäckhed, F. (2012). Functional interactions between the gut microbiota and host metabolism. *Nature, 489*(7415), 242–249.

Turnbaugh, P. J., Ley, R. E., Mahowald, M. A., Magrini, V., Mardis, E. R., & Gordon, J. I. (2006). An obesity-associated gut microbiome with increased capacity for energy harvest. *Nature, 444*(7122), 1027.

Turnbaugh, P. J., Ridaura, V. K., Faith, J. J., Rey, F. E., Knight, R., & Gordon, J. I. (2009). The effect of diet on the human gut microbiome: a metagenomic analysis in humanized gnotobiotic mice. *Science Translational Medicine, 1*(6), 6ra14.

Varela, E., Manichanh, C., Gallart, M., Torrejón, A., Borruel, N., Casellas, F., & Antolin, M. (2013). Colonization by *Faecalibacterium prausnitzii* and maintenance of clinical remission in patients with ulcerative colitis. *Alimentary Pharmacology & Therapeutics, 38*(2), 151–161.

Verdu, E. F., Galipeau, H. J., & Jabri, B. (2015). Novel players in coeliac disease pathogenesis: role of the gut microbiota. *Nature Reviews Gastroenterology & Hepatology, 12*(9), 497.

Vijay-Kumar, M., Aitken, J. D., Carvalho, F. A., Cullender, T. C., Mwangi, S., Srinivasan, S., & Gewirtz, A. T. (2010). Metabolic syndrome and altered gut microbiota in mice lacking toll-like receptor 5. *Science, 328*(5975), 228–231.

Vrieze, A., Van Nood, E., Holleman, F., Salojärvi, J., Kootte, R. S., Bartelsman, J. F., & Derrien, M. (2012). Transfer of intestinal microbiota from lean donors increases insulin sensitivity in individuals with metabolic syndrome. *Gastroenterology, 143*(4), 913–916.

Walker, W. A. (2008). Mechanisms of action of probiotics. *Clinical Infectious Diseases, 46*(Supplement_2), S87–S91.

Wang, X. Q., Zhang, A. H., Miao, J. H., Sun, H., Yan, G. L., Wu, F. F., & Wang, X. J. (2018). Gut microbiota as important modulator of metabolism in health and disease. *RSC Advances, 8*(74), 42380–42389.

Willing, B., Halfvarson, J., Dicksved, J., Rosenquist, M., Järnerot, G., Engstrand, L., & Jansson, J. K. (2009). Twin studies reveal specific imbalances in the mucosaassociated microbiota of patients with ileal Crohn's disease. *Inflammatory Bowel Diseases, 15*(5), 653–660.

Wullaert, A., Bonnet, M. C., & Pasparakis, M. (2011). NF-κB in the regulation of epithelial homeostasis and inflammation. Cell Research, 21(1), 146–158.

Yu, E. W., Gao, L., Stastka, P., Cheney, M. C., Mahabamunuge, J., Torres Soto, M., & Hohmann, E. L. (2020). Fecal microbiota transplantation for the improvement of metabolism in obesity: the FMT-TRIM double-blind placebo-controlled pilot trial. *PLoS Medicine*, 17(3), e1003051.

Zhang, P. P., Li, L. L., Han, X., Li, Q. W., Zhang, X. H., Liu, J. J., & Wang, Y. (2020). Fecal microbiota transplantation improves metabolism and gut microbiome composition in db/db mice. *Acta Pharmacologica Sinica, 41*(5), 678–685.

Zhang, Y., & Zhang, H. (2013). Microbiota associated with type 2 diabetes and its related complications. *Food Science and Human Wellness, 2*(3–4), 167–172.

Zhang, Z., Mocanu, V., Cai, C., Dang, J., Slater, L., Deehan, E. C., & Madsen, K. L. (2019). Impact of fecal microbiota transplantation on obesity and metabolic syndrome–a systematic review. *Nutrients, 11*(10), 2291.

Zuo, T., & Ng, S. C. (2018). The gut microbiota in the pathogenesis and therapeutics of inflammatory bowel disease. *Frontiers in Microbiology, 9*, 2247.

9

Role of Human Gut Microbiome in Health and Disease

Nazar Reehana
Jamal Mohamed College (Autonomous)

Mohamed Yousuff Mohamed Imran
Srimad Andavan Arts and Science College (Autonomous)

Nooruddin Thajuddin and D. Dhanasekaran
Bharathidasan University

CONTENTS

9.1 Introduction

Microbes that reside in the human gut are key contributors to host metabolism and are considered potential sources of novel therapeutics. Undeniably, it is due to the advent of genetic tools and the metagenomic revolution of the last 15 years that are now able to characterize the composition and function of microbiomes from different parts of the body and link them to potential diseases, risks, or even to the clear onset of clinical symptoms. In recent decades, microbes have mostly been used to develop disease-specific diagnostics. Currently, the mechanisms of interactions or of defense against potential pathogens are often described at the molecular level.

Gut microbiota of healthy adults is commonly dominated by two bacterial phyla, *Firmicutes* and *Bacteroidetes*, with inter-individual variability in their proportions. *Proteobacteria*, *Actinobacteria*, and *Verrucomicrobia* are present at lower levels (Lozupone et al. 2012). In spite of the great inter- and intraindividual variability, in the past years, classification of subjects based on the most abundant genera in their gut microbiome was proposed. Some years ago, it was observed that all the subjects may be classified into three discrete clusters, named "enterotypes," based on the prevalence of *Prevotella*, *Bacteroides*, or *Ruminococcus* in their gut microbiome (Arumungam et al. 2011). However, the "enterotype" concept was later criticized, since a rigorous categorization may lead to an oversimplified vision of the gut microbiome (Knights et al. 2014). On the contrary, although this classification may be attractive for understanding microbial variation in health and disease, the existence of a smooth gradient of the dominant taxa is more plausible, where the abundance of dominant genera varies continuously in the human population going from an enterotype to another. The important role of gut microbiota in influencing human well-being is widely recognized. It plays a primary function in host health by shaping the development of the immune system, metabolizing dietary nutrients and drugs, and synthesizing vitamins, bioactive molecules, and other beneficial or detrimental metabolites.

In the past decades, gut dysbiosis has been linked to the development of several kinds of diseases, including obesity (Le Chatelier et al. 2013),diabetes (Qin et al. 2012),inflammatory bowel disease (IBD) (Marchesi et al. 2016),and cardiovascular diseases (Koeth et al. 2013). Moreover, the gut microbiome may influence human behavior by the bidirectional communication path between the gastrointestinal (GI) tract and the central nervous system, namely, the gut–brain axis. This happens by a microbiome-mediated production of molecules that have neuroactive effects, such as serotonin and g-aminobutyric acid (GABA). Nutrients and microbial metabolites interact with the enteroendocrine cells (EECs) located along the GI tract and containing most of the nutrient receptors. Interactions with EEC receptors are crucial in mechanisms such as the regulation of appetite and insulin secretion (Furness et al. 2013). Indeed, recent studies suggest that bacterial proteins may influence the appetite-controlling pathways, acting locally in the gut with a short-term effect on satiation. Moreover, plasmatic levels of specific bacterial proteins may activate host anorexigenic circuitries with a long-term regulation of the feeding pattern (Breton et al. 2016).

Moreover, the current understanding is that some gut bacteria may also achieve this goal by communicating with human cells and mostly by promoting immune interactions. A large number of recent papers and reviews have covered different aspects of the microbiome and its potential role in human health, including the early life but also specific diseases, such as cardiometabolic disorders, IBDs, neuropsychiatric diseases, and cancer. The gut microbiota is now considered an important partner of human cells, interacting with virtually all human cells. Therefore, this simple finding highlights the fact that this field of research is not only blossoming but also strongly suggests the necessity for advancement.

9.2 Distribution of Microbial Communities in the Human Gastrointestinal Tract

Environmental conditions in the human GI tract are not uniform but differ considerably between the stomach and the colon. The human stomach is a J-shaped structure with a volume of approximately 1.5 L. The main functions of the human stomach are temporary food storage, a mixture of food and gastric juice to chyme, pre-digestion of proteins by acidic pH and pepsin, and disinfection of the ingested food. The environmental conditions in the stomach are eutrophic – due to ingested food, mucus, desquamated epithelial cells, and dead microbes – aerobic, and acidic, with a more or less constant temperature of 37°C, i.e. the body temperature of the host. Pronounced daily fluctuations in temperature, pH (from pH 1 to 5), and available nutrients are common and linked to ingestions of food and beverages. Bacterial viable counts are strongly dependent on the actual gastric pH and range from 10^3 to 10^6/mL (Wilson2008). It's therefore not surprising that the microbial communities reside in the various sections of the digestive tract differ in several aspects including cell density, composition, and metabolic activity.

9.2.1 The Microbiota of the Stomach

Data on the human stomach microbiome are usually collected by investigating biopsies, taken endoscopically after several hours of fasting. Despite the harsh and antimicrobial environment, recent molecular diversity studies have shown that the human stomach contains a diverse, unevenly distributed microbial community dominated by *Proteobacteria*, *Firmicutes*, *Bacteroidetes*, and *Actinobacteria* (Bik et al. 2006). In the thick mucous layer overlying the gastric epithelium, non-acidophilic bacteria can also be found, in particular *Helicobacter pylori*. So far, *H. pylori* is the only bacterium of the human stomach that can be considered unambiguously as a true resident and is considered to contribute to the development of gastritis, peptic ulcers, and even gastric cancer (Dorer et al. 2009).

Approximately 10^{10} microorganisms enter the human stomach every day. As a consequence, a clear differentiation of true residents from just transient (swallowed) microbial species is difficult. Indeed, the majority of the 33 phylotypes identified in the stomach of all three patients investigated by Andersson et al. (2008) were affiliated with the genera *Streptococcus*, *Actinomyces*, *Prevotella*, and *Gemella*, which were also abundant in the throat community. Acid tolerance is clearly a prerequisite for (even just transient) microbial survival in the stomach lumen, and this is why particularly acid-tolerant *Streptococcus*, *Lactobacillus*, *Staphylococcus*, and *Neisseria* spp. have frequently been found in the stomach lumen. It was suggested that some of these bacteria be investigated in more detail for potentially beneficial (probiotic) properties (Ryan et al. 2008).

9.2.2 The Microbiota of the Small Intestine

Due to its restricted accessibility, the microbiota of the human stomach, and particularly of the small intestine, has been investigated much less intensively than that of the mouth and large intestine or feces. In particular, data on the small intestinal microbiota of healthy individuals are scarce. Until a few years ago, it was common knowledge that the lumen and mucosa of duodenum and jejunum were colonized at low density by only a few microorganisms, including acid-tolerant streptococci and lactobacilli. Toward the end of the ileum, the lumen was described as being dominated by streptococci, enterococci, and coliforms, while in the mucosa, obligate anaerobes

(*Bacteroides* spp., *Clostridium* spp., *Bifidobacterium* spp.) could also be found (Wilson2008). This knowledge has been broadened during the past few years. The small intestinal microbiota is characterized by molecular methods of the ileal effluent of patients called Brooke ileostomies. They showed that the small intestine was characterized by a less diverse and temporarily more fluctuating microbial community than the large intestine (Booijink et al. 2010).

9.2.3 The Microbiota of the Large Intestine

Of all microorganisms inhabiting the large intestine, bacteria are by far the dominating ones, although an archaeal, viral, and eukaryotic community is also present (Minot et al. 2011). The results obtained from culture-dependent and -independent approaches show that the microbiota of the colon is very complex. Culture-based approaches show that many species are present in very small numbers, and it has been estimated that only some 40 species, belonging to a handful of genera, make up almost 90% of the colonic microbiota. Recent results from 16S rRNA gene-sequencing studies suggested even higher numbers, with up to 1,800 genera and 15,000 species-level phylotypes, for the collective human GI tract microbiota (Peterson et al. 2008).

Regarding the proportions of the different organisms present, the microbiota of the colon is dominated by obligate anaerobes. Over the last years, it became evident that two taxa, representing more than 80% of all phylotypes, dominated the colonic microbiota. These taxa are *Firmicutes* and *Bacteroidetes*. The majority of colonic bacteria affiliated with *Bacteroidetes* belong to the genus *Bacteroides*, while the majority of *Firmicutes* detectable in the human GI tract fall mainly into two groups: the *Clostridium coccoides* group (also referred to as *Clostridium* cluster XIVa) and the *Clostridium leptum* group (also referred to as *Clostridium* cluster IV). However, members of *Proteobacteria*, *Actinobacteria*, *Verrucomicrobia*, and *Fusobacteria*are also present, albeit in lower numbers (Nam et al. 2011).

9.3 The Function of the Human Gut Microbiota

In the mammalian gut, considerable numbers and species of microorganisms make up a bacterial community referred to as the gut microbiota. The number of microbes present in the gut exceeds the number of human cells and represents a weight of 2 kg, which is more than the weight of the human brain. Recent studies have demonstrated that the relationship between gut microbiota and humans is not merely commensal; rather, it is a symbiotic relationship. Germ-free (GF) animals have been instrumental in elucidating the contribution of gut microbiota to host health. Comparisons between GF and conventional (CV) animals have suggested that the bacterial community plays a role in fermenting unused energy substrates, synthesizing and converting bioactive compounds, altering intestinal morphology and motility, inducing maturation of the immune system, preventing pathogenic infection, and influencing host behavior, among other effects. Thus, the importance of the intestinal microbiota in human health is increasingly acknowledged, and there has been a surge in the number of research papers on these subjects in recent years. In this chapter, the functions of the gut microbiota will be reviewed in six categories, as summarized below.

9.3.1 Fermentation of Undigested Polysaccharides, Synthesis, and Conversion of Bioactive Compounds

The microbiota can be viewed as a metabolically active 'organ' that provides beneficial products for its host. The production of short-chain fatty acids (SCFAs) by the microbiota represents one major example that has been the subject of much research over the decades (Cummings 1981). The gut microbiota also produces essential nutrients such as vitamins and is involved in the biotransformation of drugs, conjugated bile acids, and other xenobiotics (Wilson and Nicholson 2009). The production of SCFAs, conversion of equol, biosynthesis of vitamin K, drug metabolism, and other metabolic activities of the gut microbiota will be described below.

9.3.1.1 Production of SCFAs

The intestinal microbiota produces acetate, propionate, and butyrate as its main fermentation end products from undigested dietary carbohydrates and endogenous substrates such as mucin. The SCFAs thus produced are then assimilated by the mammalian host, and it has been estimated that they can account for approximately 10% of the total caloric requirements of humans (McNeil 1984). Although the concentration and composition of SCFAs varies between individuals, the concentration of SCFAs in the lumen generally ranges between 70 and 130 mmol L^{-1}, with molar ratios of acetate:propionate:butyrate varying from approximately 75:15:10 to 40:40:20 (Bergman 1990). Although acetate, propionate, and butyrate are all taken up by the colonic mucosa, butyrate is the preferred energy source for colonic epithelial cells. In addition to its role as a fuel, butyrate is notable for its ability to inhibit histone deacetylases, which leads to the hyperacetylation of chromatin and influences gene expression. Indeed, butyrate exerts potent effects on a variety of colonic mucosal functions, such as inhibiting inflammation and carcinogenesis, reinforcing various components of the colonic defense barrier, and decreasing oxidative stress (Hamer et al. 2008). In contrast, the majority of propionate is delivered to hepatocytes and consumed through gluconeogenesis, while acetate serves as a substrate for de novo lipogenesis in both hepatocytes and adipocytes.

9.3.1.2 Conversion of Daidzein to Bioactive Equol

Equol (4′,7-isoflavandiol) is a bioactive isoflavonoid produced from daidzein (one of the major soy isoflavones) by the bacterial microbiota in the gut (Figure 9.1). Equol is known to possess a stronger estrogenic activity than other isoflavone derivatives and has been reported to prevent hormone-dependent and age-related diseases, including prostate cancer,

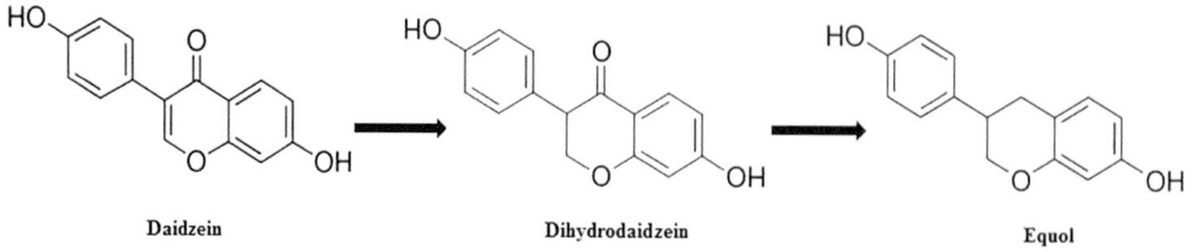

HO O OH O Daidzein → Dihydrodaidzein → Equol

FIGURE 9.1 Equol production by the intestinal microbiota (Biotrans-conversion from daidzein to equol).

breast cancer, menopausal symptoms, osteoporosis, and cardiovascular diseases. It should be noted, however, that not all people are hosts to intestinal bacteria that produce equol. For instance, the percentage of healthy equol-producing Japanese and Korean individuals has been reported to be 46%–59%. In contrast, only 29%–30% of prostate cancer patients are colonized by bacteria that produce equol (Akaza et al. 2004). Therefore, controlling the blood and intestinal equol concentration is a promising strategy for the primary prevention of hormone-dependent and age-related diseases, although until recently, little has been known about which intestinal bacteria are involved in the conversion of daidzein to equol or the distribution of equol-producing bacteria in the human intestinal tract.

9.3.1.3 *Vitamin K Production*

Vitamin K (VK) is a group of structurally similar, fat-soluble vitamins that are required for the modification and activation of a number of VK-dependent proteins involved in coagulation and absorption of insoluble calcium salts. VK deficiency is rare in the adult human population, although hypoprothrombinemia has been reported after the administration of a large number of different antibiotics (Savage and Lindenbaum1983). The onset of hemorrhagic disease due to VK deficiency has also been observed in infants at 3–8 weeks of age (Matsuda et al.1991). Furthermore, early animal studies have shown that GF animals experience VK deficiency and, in consequence, have an increased dietary requirement for this vitamin. Thus, it has been recognized that the microbial synthesis of menaquinones in the large intestine is an important source of the vitamin. The distribution of the different menaquinones synthesized by various microorganisms have been investigated are used as a taxonomic tool, which are synthesized and utilized in the large intestine is poorly understood. It will therefore be interesting to investigate the relationship between gut microbiota composition, the profile of the menaquinones, and their utilization by the host.

9.3.1.4 *Drug Metabolism and Other Metabolic Phenotypes of the Gut Microbiota*

An important example of drugs affecting gut microbiota is digoxin, which is widely used for the treatment of various heart diseases. Due to the presence of a certain bacteria in the gut, digoxin was found to undergo a significant metabolic conversion to cardio-inactive metabolites in approximately 10% of patients (Lindenbaum et al. 1981). It has also been shown that serum digoxin concentrations rose as much as twofold after antibiotics treatment. Subsequent studies identified that *Eggerthella lenta*, a common anaerobe of the human colonic microbiota, was responsible for the conversion of digoxin in the presence of arginine (Saha et al. 1983). In addition to digoxin, the gut microbiota has been implicated in the conversion of other compounds in the intestine, although these functions have largely been underexplored (Wikoff et al. 2009). The gut microbiota is also known to be capable of degrading proteins, which releases ammonia, sulfur-containing compounds, amines, indoles, p-cresol, and phenol. The production of these compounds (putrefaction) is considered less favorable for the host because these toxic metabolites have the potential to lead to diseases such as cancers and chronic kidney diseases (Evenepoel et al. 2009).

9.3.2 Alterations of Intestinal Morphology and Angiogenesis

Early studies demonstrated that there were a number of morphological and histological differences between the intestinal tracts of GF and CV animals (Smith et al. 2007). For example, differences in the regulation of the crypt-villus structure of the small intestine (Khoury et al. 1969) and angiogenesis (Stappenbeck et al. 2002) have been identified.

9.3.3 Maturation of the Immune System

The intestinal epithelia are constantly exposed to a high load of commensal bacteria and play a role in physically defining the barrier between the host and the external environment. Collectively, the intestinal tract is known to constitute a highly developed immunological self-defense system. The mucosal immune system includes both innate and adaptive immune systems, and microorganisms are constantly sampled from the intestinal lumen and taken into the intestinal immune system from inductive sites (mainly M cells in Peyer's patches and lymphoid follicles). Following uptake, commensal microorganisms are involved in the maturation of the gut and its immune system, as well as the generation of a state of 'physiological inflammation' (Sansonetti2004). Indeed, a comparison between GF and CV animals has revealed that the immune system is not fully developed in GF animals, whereby the content of intestinal IgA-secreting plasma cells is reduced and Peyer's patches are reduced in size and contain fewer lymphoid follicles. The T-cell content of the mucosal immune system, particularly the CD4+ cells of the lamina propria, is also reduced in these animals, while their spleen and lymph nodes are relatively structureless, with poorly formed B- and T-cell zones (Macpherson and Harris2004).

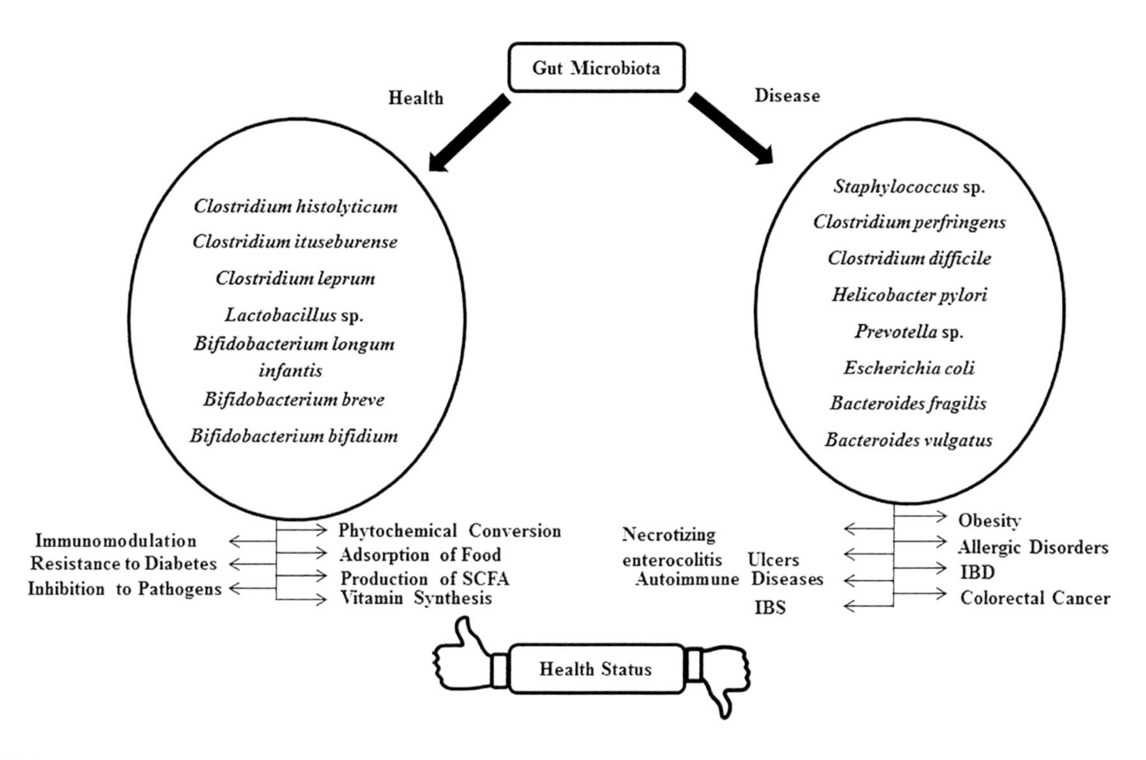

FIGURE 9.2 Schematic representation of the key microbial players in promoting health/diseases in humans.

9.3.4 Prevention of Pathogenic Infection

It is well known that the intestinal microbiota provides protection against colonization by many pathogenic infectious agents. The term 'colonization resistance' was first used by van der Waaij and co-workers in 1971 to denote the concept that the indigenous anaerobic microbiota limited the concentration of potentially pathogenic bacteria in the digestive tract (van der Waaij et al. 1971). Commensal bacteria occupy important ecological niches and compete with ingested pathogens for nutrients; some commensal bacteria, such as *Lactobacillus*, secrete molecules, such as lactic acid, that inhibit the growth of competing microorganisms (Asahara et al. 2004).

To date, many attempts have been made to identify the microbiota that confers resistance or susceptibility to infection and inflammation. Recent studies have demonstrated that treatment with antibiotics alters the intestinal microbiota composition, leading to increased susceptibility to *Salmonella* infection and intestinal pathology (Ferreira et al. 2011). The acetate produced by protective bifidobacteria has also been found to improve intestinal defenses mediated by epithelial cells, thereby protecting the host against lethal infection by *Escherichia coli* O157 (Fukuda et al. 2011). The gut microbiota plays a critical role in both the fermentation of indigestible complex plant polysaccharides and host-produced glycans (e.g. mucin), as well as in the protection against pathogenic bacteria (Figure 9.2).

9.3.5 The Gut Microbiota Affects Energy Homeostasis, Adiposity, and Obesity

Studies have indicated that CV animals have an increased total body fat when compared to their GF counterparts (Backhed et al. 2004). So far, this observation has been explained by the presence of SCFAs produced by the microbiota and their utilization by the host. However, recent studies have suggested that several other mechanisms also exist that could potentially allow gut microbes to interact with host tissues and regulate energy metabolism. The composition of the gut microbiota has also been shown to differ between lean and obese types in mice and humans (Ley et al. 2006), and the traits of obese-type gut microbiota that allows efficient energy extraction have been found to be transmissible (Turnbaugh et al. 2006). Furthermore, lipopolysaccharides (LPS) derived from intestinal Gram-negative bacteria are able to act as a triggering factor that induces inflammation and systemic insulin resistance (Cani et al. 2007). These findings suggest that the gut microbiota could be a new therapeutic target for the prevention of obesity and its associated metabolic syndrome.

9.3.6 Effects of Gut Microbiota on Host Behavior

Epidemiological studies have indicated that there is an association between neurodevelopmental disorders, such as autism and schizophrenia, and microbial pathogen infections during the perinatal period (Mittal et al. 2008). Recently, several mechanistic investigations have examined the relationship between the gut microbiota and behavior. Heijtz and colleagues have more recently evaluated the behavioral differences in these mice by measuring the motor activity and anxiety-like behavior (Heijtz et al. 2011). When compared with conventionalized mice, GF mice were found to display increased motor activity and reduced anxiety. This phenotype was found to be associated with the altered expression of genes involved in second messenger pathways and synaptic long-term potentiation in brain regions implicated in motor control and anxiety-like behavior.

These findings suggest that the gut microbiota could potentially be a new therapeutic target for psychiatric disorders. Indeed, Bravo and colleagues have recently demonstrated that probiotic bacteria have the potential to alter brain neurochemistry and treat anxiety and depression-related disorders (Bravo et al. 2011). They reported that mice fed with a certain *Lactobacillus* strain showed significantly fewer stress, anxiety, and depression-related behaviors than mice fed with no *Lactobacillus* strain. Moreover, the ingestion of lactobacilli resulted in significantly lower levels of the stress-induced hormone, corticosterone, being detected in the plasma.

While the mechanisms underlying these phenotypes have yet to be defined, the results presented emphasize the importance of studying how gut microbiota impacts brain development and subsequent behavior. It will be an exciting challenge to identify neurologically active compounds associated with the microbiota and to develop unique microbiota-based strategies for the treatment of stress-related psychiatric disorders such as anxiety and depression.

9.4 Gut Microbiome and Diseases

External factors (such as antibiotic consumption, a dietary component, and psychological and physical stress) and host factors can induce dysbiosis in the gut microbiome. Dysbiosis is likely to impair the normal functioning of gut microbiota in maintaining host wellness, and potentially induce selective enumeration of certain microbiota member including pathobionts, leading to dysregulated production of microbial-derived products or metabolites which might be harmful to the host, causing a diverse range of diseases on local, systemic, or remote organ (Table 9.1), with some of the notable diseases, which are discussed below along with their respective microbiome-based therapy.

9.4.1 Inflammatory Bowel Diseases

IBDs, such as Crohn's disease and ulcerative colitis, are chronically relapsing, immune-mediated disorders of the GI tract which have been steadily increasing over the past decades. Additionally, epidemiologic studies pointed toward the Western lifestyle and habits as part of central environmental factors contributing to both development and maintenance of intestinal inflammation. In this regard, the gut microbiota is thought to play a decisive role in disease progression. The intestinal microbiota was unequivocally shown to be indispensable in orchestrating the development and the functionality of the immune system further having a critical impact on both intestinal homeostasis and inflammation in preclinical models. Even though profound changes in the composition of the intestinal microbiota have been frequently observed in human IBD, unraveling cause and consequences of intestinal dysbiosis need further understanding of the interaction between host genetics, microbial ecosystems, and environmental triggers.

Although our understanding of the mechanism of pathogenesis for this disease is still lacking, crosstalk between gut microbiota and host factors shows great potential in contributing to the disease development, as demonstrated in Figure 9.3. The inappropriate host immune response against GI microbiota in a genetically predisposed individual is speculated to be the main culprit in causing severe inflammation.

One of the possible etiologies for IBD is due to the hyperresponsiveness of T-lymphocyte toward non-pathogenic antigens presented on gut microbiota. Several studies have observed the developments of antibodies against commensal microbial antigens and autoantigens such as anti-*Saccharomyces cerevisiae*, anti-OmpC, perinuclear anti-neutrophil cytoplasmic antibody, and anti-*Pseudomonas fluorescens*-associated sequence 12 (Landers et al. 2002). Also, it has been reported that each

TABLE 9.1

Gut Microbiome-Associated Human Diseases and Their Respective Dysbiotic Features

Disease Categories	Specific Diseases	Associated Dysbiotic Features
Immune-mediated/ autoimmune diseases	Inflammatory bowel disease (IBD)	i. Increase in virulent gut microbes (Enterobacteriaceae, *Bacteroides fragilis*) and mucolytic *Ruminococcus* sp. ii. Decrease in butyrate-producing *Firmicutes* (such as *Faecalibacterium prausnitzii* and *Roseburia hominis*)
	Irritable bowel syndrome (IBS)	i. Increase in *Escherichia coli* ii. Decrease in *C. leptum* group of bacteria and *Bifidobacterium*
	Celiac disease	i. Increase in *Bacteroides–Prevotella* group ii. Decrease in *Bifidobacterium*
	Systemic lupus erythematosus (SLE)	i. Increase in *Blautia* sp. and *Proteobacteria* ii. Decrease in gut microbiota diversity, *Odoribacter* sp., and *Alistipes* sp.
	Type1 diabetes	i. Increase in *Bacteroidetes* ii. Decrease in *Actinobacteria*, *Firmicutes*, and *Firmicutes/Bacteroidetes* ratio
	Rheumatoid arthritis (RA)	i. Increase in *Prevotella copri* and decrease in *Bacteroides* sp. In new-onset RA ii. Increase in microbiota diversity of *Lactobacillus* genus in early RA
Metabolic disorders/ cardiovascular disorders	Obesity	i. Increase in *Firmicutes* and *Actinobacteria* ii. Varying observation (decrease, no change, increase) in *Bacteroidetes*
	Type-2 diabetes	i. Decrease in *C. coccoides*, *Atopobium cluster*, and *Prevotella* ii. Decrease in butyrate biosynthesis
	Hypertension	i. Increase in the *Firmicutes/Bacteroidetes* ratio, lactate producer ii. Decrease in microbiota diversity, acetate and butyrate producers
	Atherosclerosis	i. Increase in metabolites TMAO, endotoxin level (risk factor for early atherosclerosis)

(*Continued*)

TABLE 9.1 (*Continued*)

Gut Microbiome-Associated Human Diseases and Their Respective Dysbiotic Features

Disease Categories	Specific Diseases	Associated Dysbiotic Features
Cancer	Colorectal cancer (CRC)	i. Increase in enterotoxigenic *Bacteroides fragilis*, and pathobionts *Fusobacterium* and *Campylobacter* sp. ii. Decrease in butyrate producer (*Faecalibacterium* and *Roseburia*)
Neuropsychiatric	Autism spectrum disorder (ASD)	i. Increase in *Clostridium* sp., *Bacteroidetes*, *Lactobacillus*, *Desulfovibrio* ii. Decrease in *Bifidobacteria*
	Alzheimer's disease	i. Possible connection between gut microbiota-synthesized amyloids, LPS, g-aminobutyric acid and the increased permeability of gut barrier and blood-brain barrier with age
	Depression	i. Increase in genera *Eggerthella*, *Holdemania*, *Gelria*, *Turicibacter*, *Paraprevotella*, and *Anaerofilum* ii. Decrease in gut microbiota diversity, *Prevotella* and *Dialiste*
	Parkinson's disease	i. Increase in anti-inflammatory butyrate producers from genus *Blautia*, *Coprococcus*, and *Roseburia* in patient fecal sample, pro-inflammatory *Proteobacteria* in patient mucosa
Infectious disease	*Clostridium difficile* infection (CDI)	i. Increase in *Clostridium difficile* ii. Decrease in gut microbiota diversity and secondary bile acids-producing *Clostridium scindens*
Uremic disease	Chronic kidney disease	i. Increase in *Firmicutes*, *Proteobacteria*, and *Actinobacteria* ii. Decrease in *Lactobacillus*

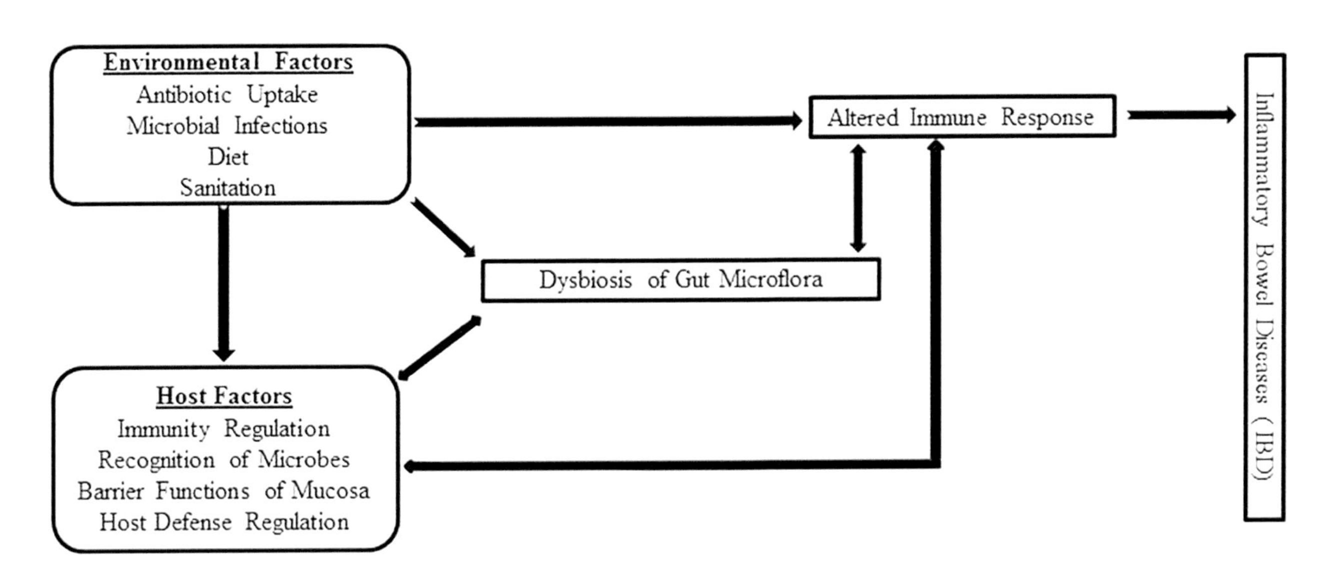

FIGURE 9.3 Proposed complex interactive relationships among gut microbiome, host factors, and environmental factors in IBD pathogenesis.

of these antibodies' response patterns highly correlate with distinctive clinical characteristics, disease onset and severity, suggesting that loss of different microbiota species that would affect the gut barrier function and gut immunity results in different degrees of gut inflammation (Vasiliauskas et al. 2000).

Conversely, Nagao-Kitamoto et al. (2016) suggested that gut dysbiosis potentially contributes to IBD pathogenesis. In this study, elevated pro-inflammatory gene expression was observed in GF mice colonized by gut microbiota isolated from IBD patients. Colitis-prone genetically predisposed GF mice colonized by IBD-associated microbiota developed severe colitis compared to those that were colonized by healthy human microbiota. Together these findings strongly indicate a bidirectional relationship between such disease and gut dysbiosis, in which dysbiosis potentially contributes to the onset of IBD and also serves as a secondary consequence of gut inflammation. An example of the dysbiotic features that are commonly observed in IBD patients is the reduction in gut *Firmicutes* such as *Faecalibacterium prausnitzii* and *Roseburia* sp. (Machiels et al. 2014). These bacteria play an important anti-inflammatory role in reducing pro-inflammatory cytokines (IL-12, IFN-g) and increasing anti-inflammatory IL-10 (Sokol et al. 2008). Besides, *Firmicutes* is an important producer of butyrate, the primary energy substrate for colonocytes. Therefore, reduction in *Firmicutes* could elicit or heighten local inflammation by decreasing anti-inflammatory cytokine, an important regulator of mucosal immunity, and/or by SCFA-deficiency-induced impairment in colonic barrier function (Machiels et al. 2014). As such, it may be interesting to explore the therapeutic use of probiotic *F. prausnitzii* in managing IBD. Another dysbiotic feature observed in IBD patients is the elevation of virulent gut microbes such as Enterobacteriaceae and *Bacteroides fragilis* with both having high endotoxic LPS in their outer membranes (Darfeuille-Michaud et al. 1998).

Another suggestion regarding the IBD–gut microbiome link is the initial impairment in gut mucus barrier function (due to either dysbiosis or other factors), resulting in elevation of mucus-eating (mucolytic) gut microbial species, which, in turn, aggravates the barrier function and stimulates severe inflammatory response. GI mucus layer and antimicrobial peptides (such as human defensins) secreted from epithelium work together as a barrier in preventing direct contact of luminal gut microbiota with GI epithelial cells and inhibit aberrant inflammation (Salzman et al. 2010).

9.4.2 Obesity

In recent years, the newly identified factor – gut microbiome which is largely involved in host metabolism regulation –has been integrated into crosstalk studies between genetic factors, behavior, and environmental factors as a possible contributor to obesity. Obesity is a global health hazard affecting more than 600 million people worldwide in 2014. It is associated with elevated energy intake and decreased energy expenditure, causing excessive fat accumulation with raised body mass index (BMI $\geq$ 30kgm^{-2}),and is linked to metabolic syndrome, posing obese individuals to have a higher risk of developing obesity-associated disorders (for example cardiovascular disease, type-II diabetes, and liver abnormalities), low-grade inflammation, and premature mortality (Cani et al. 2007).

Metagenomics studies had discovered a significant increase in butyrate-producing *Firmicutes* and generally, a decrease in *Bacteroidetes* was observed in distal colonic microbiome of obese patients and genetically obese mice, compared to their normal lean counterparts. These obesity-associated dysbiosis features were also accompanied by an elevation in starch-degrading glycoside hydrolase and SCFAs (butyrate and acetate),and increased energy harvest capability, as evidenced by markedly decreased fecal energy in obese mice (Turnbaugh et al. 2006),speculating significant elevation in the host metabolism-related microbial communities would abnormally increase energy harvest, thus increasing the risk of developing obesity.

Moreover, several studies suggest that increase in endotoxic LPS of Gram-negative gut bacteria contributes to obesity-associated metabolic syndrome. For instance, it was reported that elevated LPS led to obesity-associated insulin resistance and low-grade inflammation as demonstrated by Cani et al. (2007). Similarly,plasma LPS (metabolic endotoxemia) showed a marked increase in high-fat-fed mice, along with the reduction of *Bifidobacteria*, a potential down-regulator of intestinal endotoxin (Griffiths et al. 2004). In the investigation of the association between metabolic endotoxemia and obesity-associated metabolic disorders, continuously LPS-infused mice were shown to develop fasting glycemia, insulinemia, hepatic insulin resistance, and increased hepatic and whole-body fat gains, similar to high-fat-fed mice phenotypes. In addition, the absence of LPS receptor resisted these adverse features regardless of LPS-infusion or high-fat-diet treatments as shown in CD14 (LPS-receptor) knockout mice, indicating that the LPS/CD14 system mediates insulin insensitivity, thus inducing the onset of obesity and obesity-related metabolic disorders (Cani et al. 2007).

9.4.3 Colorectal Cancer

Cancer is a multifactorial disease caused by genetic predispositions and environmental factors resulting in accumulation of genetic and epigenetic mutations leading to uncontrolled cell proliferation. According to the World Health Organization (WHO), Cancer-related death (CRC) was the third most common cause of death worldwide in 2015. CRC is known to result from a well-established sequential cascade of mutations leading to loss of function of tumor suppressor genes or activation of oncogenes, where key genetic changes are associated with various stages of cancer progression (Simon Fearon and Vogelstein 1990). In addition, common mutations include the epidermal growth factor (EGF) pathway, p53 mutations, and mutations in the transforming growth factor-β (TGF-β) signaling pathway. Genetic lesions can be hereditary or acquired. Hereditary mutation in the adenomatous polyposis coli (APC) gene leads to a polyposis syndrome that is associated with the development of CRC in all affected individuals. Thus, CRC is a genetic disease, but the vast majority of cases occur sporadically without a known genetic predisposition. Induction, accumulation, and persistence of mutations in the tissue are influenced by a variety of environmental factors such as diet, lifestyle, comorbidities, and especially the history of IBD.

The human colon is a densely populated microbial ecosystem (Simon and Gorbach 1984). The GI microbiota plays a crucial role in health and disease. The microbiota can be beneficial to the host by providing protection against the pathogen, helping with digestive processes, contributing to metabolic pathways, and shaping the GI immune system. The absence of a healthy microbiota can be a result of active disease or caused due to dietary habits. CRC is more prevalent in developed countries and believed to be associated with lower nutritional diversity in the food intake. An infection or nutritional imbalance leads to reduced colonization by beneficial microorganisms and enrichment of pro-carcinogenic bacterial groups. Dysbiosis disturbs the immune system homeostasis causing inflammation, disrupting mucosal barrier, increasing epithelial permeability, and creating a microenvironment that further perpetuates pro-carcinogenic dysbiosis.

9.4.4 Autism Spectrum Disorders (ASD)

This neuropsychiatric disorder manifests primarily during early childhood, and in most cases, symptoms persist throughout adulthood. The gut microbiome of children having autism spectrum disorders (ASD) has been analyzed in cross-sectional studies with multiple different cohorts (Kelly et al. 2017). A recent meta-analysis of 15 microbiome studies found differences in the abundance of bacteria in the phyla *Firmicutes*, *Bacteroidetes*, and *Proteobacteria* in ASD patients vs. controls (Cao et al. 2013).

Another line of evidence for the impact of the gut microbiota on ASD comes from a small, interventional clinical study where children with ASD received oral, non-absorbed vancomycin for 12weeks (Sandler et al. 2000). During the treatment, eight out of ten children showed significant improvements in behavioral symptoms, but these gains largely waned after discontinuation of the treatment. Another study investigated the effects of probiotics in ASD patients using a formulation of *Lactobacillus*, *Bifidobacterium*, and *Streptococcus* but observed only effects on the patient's microbiome and fecal cytokine levels and no significant change in behavior (Tomova et al. 2015).

A recent study conducted in patients with first-episode psychosis revealed an elevated abundance of bacteria from the *Lactobacillus* family vs. controls that also correlated with the severity of symptoms (Schwarz et al. 2018). A subgroup of these patients that displayed the strongest microbiota differences during the initial assessment also showed a poorer response to treatment after 12months of antipsychotic administration. Interestingly, *Lactobacillus* and *Bifidobacteria* were also found in another study to be over-represented in oropharyngeal samples from schizophrenic patients vs. controls (Castro-Nallar et al. 2015). Another clinical study analyzed a blood-specific microbiome in schizophrenic patients and observed an increased alpha and beta diversity compared to controls and other patients with neuropsychiatric disorders (Olde Loohuis 2018).

Given the potential of probiotics to restore a disturbed gut microbiota, a study was carried out in a controlled 14-week probiotic intervention study in schizophrenic patients using a combination of *Lactobacillus rhamnosus* and *Bifidobacterium animalis* strains (Dickerson et al. 2014). This intervention only alleviated bowel movement difficulties associated with schizophrenia and/or medication in this cohort and did not improve psychiatric symptoms. In these studies, it is noteworthy that the vast majority of patients in these studies received antipsychotic medication, which can impact gut microbiota composition and, therefore, can profoundly bias disease–microbiome associations or microbiota intervention trials (Bahr et al. 2015a, b).

9.4.5 *Clostridium difficile* Infection (CDI)

Clostridium difficile is a Gram-positive toxin and spore-producing anaerobe. It is also one of the *Firmicutes* members in normal gut microbiota. The catalytic activity of *C. difficile* toxins A (TcdA) and B (TcdB) damages cytoskeleton and colonic epithelial barrier integrity, thereby inducing aberrant inflammatory response and cell death (Pruitt et al. 2012). *C. difficile* infection (CDI)-associated symptoms include diarrhea, pseudo-membranous colitis, sepsis, and death in severe cases (Bartlett 1982). Intriguingly, antibiotic administration was found to be the major risk factor for CDI (Pear et al. 1994). Around 5%–35% of antibiotic-treated individuals developed diarrhea as a common side effect. Association of *C. difficile* with antibiotic-associated diarrhea is most frequent, which accounts for 10%–20% of total incidences, compared to other pathogens such as *Staphylococcus aureus* that greater incidence of diarrhea was correlated with uptake of antibiotics with broad-spectrum antibacterial effect (Bartlett2002). Also, *C. difficile* acquisition of resistant genes toward broad-spectrum clindamycin, erythromycin, chloramphenicol, and linezolid mediated by multiple horizontal gene transfer modes (potentially via mobilizable transposon, phage transduction, and conjugative plasmids) within *C. difficile* strains and possibly among commensal microbes had been reported (Johanesen et al. 2015). A cohort study by Pépin et al. (2005) discovered that broad-spectrum fluoroquinolone appeared to be the most potent risk contributor to *C. difficile*-associated diarrhea compared to other antibiotics. The precise mechanism of this antibiotic-associated diarrhea remains unknown; however, its noteworthy correlation with CDI inspires research on the relationship between the gut microbiome and its pathogenic member *C. difficile* in healthy non-disease state.

Currently, it is postulated that dominant gut microbiota species confer protection to the host by employing colonization resistance mechanisms against overgrowth of *C. difficile* in normal microbiome. One of the proposed mechanisms is via the bio-conversion of primary bile acid to secondary bile acids. Primary bile acids (cholate derivatives) serve as germinant for *C. difficile* spores, whereas secondary bile acid (deoxycholate) inhibits vegetative growth of *C. difficile* (Sorg and Sonenshein2008). Antibiotics administration perturbs the gut microbial communities and reduces their diversity, especially secondary bile acids-synthesizing dominant microbes such as *Clostridium scindens* (Antonopoulos et al. 2009). As a result, there is a significant reduction in microbial bioconversion of primary bile acids into a *C. difficile* vegetative growth, allowing *C. difficile* outgrowth and colonization of the empty niches, leading to higher susceptibility of the host toward CDI (Theriot et al. 2014). The increased toxin secretion by a greater amount of vegetative *C. difficile* exerts greater damage on the intestinal barrier, stimulating severe inflammatory response and causing impairment in intestinal ion absorption that leads to diarrhea.

9.5 Conclusion

Genetic and environmental factors are known to shape the composition of the gut microbiota, which plays a key role in modulating the immune response at both intestinal and extra-intestinal sites, as well as in controlling the development of certain types of autoimmune and allergic diseases and particular types of cancers. The relationship between the gut microbiota, immunity, and disease is highly complex, since the same commensal bacteria can induce either a protective response or a pathogenic response, depending on the susceptibility of the individual. So far, the specific microorganisms contributing to either the etiology of or the protection from different types of diseases, i.e. microbial biomarkers, have not yet been fully characterized. Future investigations aimed at the discovery of microbial biomarkers as well as effective prebiotic compounds will be crucial in order to establish biotherapeutical protocols for the prevention and cure of many diseases. Furthermore, the establishment of profiles of the gut microbiota in humans based on bacterial composition or enterotypes (Arumungam et al. 2011) will allow the development of a new kind of 'biological fingerprint' similar to blood or tissue typing that may, in the future, be used to predict the response to drugs or a specific diet, and that might lead ultimately to the development of personalized therapies.

REFERENCES

Akaza, H., Miyanaga, N., Takashima, N., et al. 2004. Comparisons of percent equol producers between prostate cancer patients and controls: case-controlled studies of isoflavones in Japanese, Korean and American residents. *Japanese Journal of Clinical Oncology*, 34: 86–89.

Andersson, A. F., Lindberg, M., Jakobsson, H., Bäckhed, F., Nyrén, P., and Engstrand, L. 2008. Comparative analysis of human gut microbiotaby barcoded pyrosequencing. *PLoS One*, 3: e2836.

Antonopoulos, D. A., Huse, S. M., Morrison, H. G., Schmidt, T. M., Sogin, M. L., and V. Young, B. 2009. Reproducible community dynamics of the gastrointestinal microbiota following antibiotic perturbation. *Infection and Immunity*, 77: 2367–2375.

Arumungam, M., Raes, J., Pelletier, E., et al. 2011. Enterotypes of the human gut microbiome. *Nature*, 473(7346): 174–180.

Asahara, T., Shimizu, K., Nomoto, K., Hamabata, T., Ozawa, A., and Takeda, Y. 2004. Probiotic *Bifidobacteria* protect mice from lethal infection with Shiga toxin-producing *Escherichia coli* O157:H7. *Infection and Immunity*, 72: 2240–2247.

Backhed, F., Ding, H., Wang, T., et al. 2004. The gut microbiota as an environmental factor that regulates fat storage. *Proceedings of the National Academy of Sciences of the United States of America*, 101: 15718–15723.

Bahr, S. M., Tyler, B. C., Wooldridge, N., et al. 2015a. Use of the second-generation antipsychotic, risperidone and secondary weight gain are associated with an altered gut microbiota in children. *Translational Psychiatry*, 5: e652.

Bahr, S. M., Weidemann, B. J., Castro, A. N., et al. 2015b. Risperidone-induced weight gain is mediated through shifts in the gut microbiome and suppression of energy expenditure. *eBioMedicine*, 2: 1725–1734.

Bartlett, J. G. 1982. Antimicrobial agents implicated in *Clostridium difficile* toxin-associated diarrhea or colitis. *Obstetrical & Gynecological Survey*, 37: 46.

Bartlett, J. G. 2002. Clinical practice. Antibiotic-associated diarrhea. *The New England Journal of Medicine*, 346: 334–339.

Bergman, E. N. 1990. Energy contributions of volatile fatty acids from the gastrointestinal tract in various species. *Physiological Reviews*, 70: 567–590.

Bik, E. M., Eckburg, P. B., Gill, S. R., et al. 2006. Molecular analysis of the bacterial microbiota in the human stomach. *Proceedings of the National Academy of Sciences of the United States of America*, 103: 732–737.

Booijink, C.C.G.M., Boekhorst, J., Zoetendal, E.G., Smidt, H., Kleerebezem, M., and De Vos, W. M. 2010. Metatranscriptome analysis of the human fecal microbiota reveals subject-specific expression profiles, with genes encoding proteins involved in carbohydrate metabolism being dominantly expressed. *Applied and Environmental Microbiology*, 76: 5533–5540.

Bravo, J. A., Forsythe, P., Chew, M. V., et al. 2011. Ingestion of *Lactobacillus* strain regulates emotional behavior and central GABA receptor expression in a mouse via the vagus nerve. *Proceedings of the National Academy of Sciences of the United States of America*, 108: 16050–16055.

Breton, J., Tennoune, N., Lucas, N., et al. 2016. Gut commensal *E.* coli proteins activate host satiety pathways following nutrient-induced bacterial growth. *Cell Metabolism*, 23(2): 324–334.

Cani, P.D., Amar, J., Iglesias, M.A., et al. 2007. Metabolic endotoxemia initiates obesity and insulin resistance. *Diabetes*, 56: 1761–1772.

Cao, X., Lin, P., Jiang, P., and Li, C. 2013. Characteristics of the gastrointestinal microbiome in children with autism spectrum disorder: a systematic review. *Shanghai Archives of Psychiatry*, 25:342–353.

Castro-Nallar, E., Bendall, M. L., Pérez-Losada, M., et al. 2015. Composition, taxonomy and functional diversity of the oropharynx microbiome in in dividuals with schizophrenia and controls. *Peer J*, 3: e1140.

Cummings, J.H. 1981. Short chain fatty acids in the human colon. *Gut*, 22: 763–779.

Darfeuille-Michaud, A., Neut, C., Barnich, N., et al. 1998. Presence of adherent *Escherichia coli* strains in ileal mucosa of patients with Crohns disease. *Gastroenterology*, 115: 1405–1413.

Dickerson, F.B., Stallings, C., Origoni, A., et al. 2014. Effect of probiotic supplementation on schizophrenia symptoms and association with gastrointestinal functioning: A randomized, placebo-controlled trial. *The Primary Care Companion for CNS Disorders*, 16.

Dorer, M.S., Talarico, S., and Salama, N. R. 2009. *Helicobacter pylori*'s unconventional role in health and disease. *PLoS Pathogens*, 5:e1000544.

Evenepoel, P., Meijers, B.K., Bammens, B.R., and Verbeke, K. 2009. Uremic toxins originating from colonic microbial metabolism. *Kidney International Supplements*, 114: S12–S19.

Ferreira, R.B., Gill, N., Willing, B.P., et al. 2011. The intestinal microbiota plays a role in *Salmonella* induced colitis independent of pathogen colonization. *PLoS One*, 6: e20338.

Fukuda, S., Toh, H., Hase, K., et al. 2011. Bifidobacteria can protect from enteropathogenic infection through production of acetate. *Nature*, 469: 543–547.

Furness, J. B., Rivera, L. R., Cho, H. J., Bravo, D. M., and Callaghan, B. 2013. The gut as a sensory organ. *Nature Reviews Gastroenterology & Hepatology*, 10: 729–740.

Griffiths, E. A., Duffy, L. C., Schanbacher, F. L., et al. 2004. *In vivo* effects of *Bifidobacteria* and lactoferrin on gut endotoxin concentration and mucosal immunity in balb/c mice. *Digestive Diseases and Sciences*, 49: 579–589.

Hamer, H.M., Jonkers, D., Venema, K., Vanhoutvin, S., Troost, F.J., and Brummer, R. J. 2008. Review article: the role of butyrate on colonic function. *Alimentary Pharmacology and Therapeutics*, 27: 104–119.

Heijtz, R. D., Wang, S., Anuar, F., et al. 2011. Normal gut microbiota modulates brain development and behavior. *Proceedings of the National Academy of Sciences of the United States of America*, 108: 3047–3052.

Johanesen, P.A., Mackin, K.E., Hutton, M.L., et al. 2015. Disruption of the gut microbiome: *Clostridium difficile* infection and the threat of antibiotic resistance. *Genes*, 6: 1347–1360.

Kelly, J. R., Minuto, C., Cryan, J. F., Clarke, G., and Dinan, T. G. 2017. Cross talk: the microbiota and neuro developmental disorders. *Frontiers in Neuroscience*, 11: 490.

Khoury, K.A., Floch, M.H., and Hersh, T. 1969. Small intestinal mucosal cell proliferation and bacterial microbiota in the conventionalization of the germfree mouse. *Journal of Experimental Medicine*, 130: 659–670.

Knights, D., Ward, T. L., McKinlay, C. E., et al. 2014. Rethinking "enterotypes". *Cell Host & Microbe*, 16(4): 433–437.

Koeth, R. A., Wang, Z., Levison, B. S., et al. 2013. Intestinal microbiota metabolism of L-carnitine, anutrient in red meat, promotes atherosclerosis. *Nature Medicine*, 19(5): 576–585.

Landers, C. J., Cohavy, O., Misra, R., et al. 2002. Selected loss of tolerance evidenced by Crohns disease–associated immune responses to auto- and microbial antigens. *Gastroenterology*, 123: 689–699.

Le Chatelier, E., Nielsen, T., Qin, J., et al. 2013. Richness of human gut microbiome correlates with metabolic markers. *Nature*, 500(7464): 541–546.

Ley, R.E., Turnbaugh, P.J., Klein, S., and Gordon, J. I. 2006. Microbial ecology: human gut microbes associated with obesity. *Nature*, 444: 1022–1023.

Lindenbaum, J., Rund, D.G., Butler, V.P. Jr., Tse-Eng, D., and Saha, J. R. 1981. Inactivation of digoxin by the gut microbiota: reversal byantibiotic therapy. *New England Journal of Medicine*, 305: 789–794.

Lozupone, C. A., Stombaugh, J. I., and Gordon, J. I., et al. 2012. Diversity, stability and resilience of the humangut microbiota. *Nature*, 489(7415): 220–230.

Machiels, K., Joossens, M., Sabino, J., et al. 2014. A decrease of the butyrate-producing species *Roseburia hominis* and *Faecalibacterium prausnitzii* defines dysbiosis in patients with ulcerative colitis. *Gut*, 63: 1275–1283.

Macpherson, A.J., and Harris, N.L. 2004. Interactions between commensal intestinal bacteria and the immune system. *Nature Reviews Immunology*, 4: 478–485.

Marchesi, J. R., Adams, D. H., Fava, F., et al. 2016. The gut microbiota and host health: a new clinical frontier. *Gut*, 65: 330–339.

Matsuda, I., Endo, F., and Motohara, K. 1991. Vitamin K deficiency in infancy. *World Review of Nutrition and Dietetics*, 64: 85–108.

McNeil, N.I. 1984. The contribution of the large intestine to energy supplies in man. *The American Journal of Clinical Nutrition*, 39: 338–342.

Minot, S., Sinha, R., Chen, J., et al. 2011. The human gut virome: inter-individual variation and dynamic response to diet. *Genome Research*, 21: 1616–1625.

Mittal, V.A., Ellman, L.M., and Cannon, T. D. 2008. Gene–environment interaction and covariation in schizophrenia: the role of obstetric complications. *Schizophrenia Bulletin*, 34: 1083–1094.

Nagao-Kitamoto, H., Shreiner, A. B., Gilliland, M. G. I. I. I., et al. 2016. Functional characterization of inflammatory bowel disease–associated gut dysbiosis in gnotobiotic mice. *Cellular and Molecular Gastroenterology and Hepatolog*, 2: 468–481.

Nam, Y. D., Jung, M. J., Roh, S.W., Kim, M. S., and Bae, J. W. 2011. Comparative analysis of Korean human gut microbiota by barcoded pyrosequencing. *PLoS One*, 6: e22109.

Olde Loohuis, L. M. 2018. Transcriptome analysis in whole blood reveals increased microbial diversity in schizophrenia. *Translational Psychiatry*, 8: 96.

Pear, S. M., Williamson, T. H., Bettin, K. M., Gerding, D. N., and Galgiani, J. N. 1994. Decrease in nosocomial *Clostridium difficile*–associated diarrhea by restricting clindamycin use. *Annals of Internal Medicine*, 120: 272–277.

Pépin, J., Saheb, N., Coulombe, M.A., et al. 2005. Emergence of fluoroquinolones as the predominant risk factor for *Clostridium difficile* associated diarrhea: a cohort study during an epidemic in Quebec. *Clinical Infectious Diseases*, 41: 1254–1260.

Peterson, D.A., Frank, D.N., Pace, N.R., and Gordon, J. I. 2008. Metagenomic approaches for defining the pathogenesis of inflammatory bowel diseases. *Cell Host and Microbe*, 3: 417–427.

Pruitt, R. N., Chumbler, N. M., Rutherford, S. A., et al. 2012. Structural determinants of *Clostridium difficile* toxina glucosyl transferase activity. *Journal of Biological Chemistry*, 287: 8013–8020.

Qin, J., Li, Y., Cai, Z., et al. 2012. A metagenome-wide association study of gut microbiota in type 2 diabetes. *Nature*, 490(7418): 55–60.

Ryan, K.A., Jayaraman, T., Daly, P., et al. 2008. Isolation of *Lactobacilli* with probiotic properties from the human stomach. *Letters in Applied Microbiology*, 47: 269–274.

Saha, J.R., Butler, V.P.Jr., Neu, H.C., and Lindenbaum, J. 1983. Digoxin-inactivating bacteria: identification in human gut microbiota. *Science*, 220: 325–327.

Salzman, N. H., Hung, K., Haribhai, D., et al. 2010. Enteric defensins are essential regulators of intestinal microbial ecology. *Nature Immunology*, 11: 76–83.

Sandler, R. H., Finegold, S. M., Bolte, E. R., Buchanan, C. P., Maxwell, A. P., and Väisänen, M. L. 2000. Short-term benefit from oral vancomycin treatment of regressive-onset autism. *Journal of Child Neurology*, 15: 429–435.

Sansonetti, P.J. 2004. War and peace at mucosal surfaces. *Nature Reviews Immunology*, 4: 953–964.

Savage, D., and Lindenbaum, J. 1983. Clinical and experimental human vitamin K deficiency. In: Lindenbaum, J. (ed.) *Nutrition in Hematology*, 271–320. Churchill Livingstone, New York.

Schwarz, E., Maukonen, J., Hyytiäinen, T., et al. 2018. Analysis of microbiota in first episode psychosis identifies preliminary associations with symptom severity and treatment response. *Schizophrenia Research*, 192: 398–403.

Simon, G. L., and Gorbach, S. L. 1984. Intestinal flora in health and disease. *Gastroenterology*, 86: 174–193.

Simon Fearon, E. R., and Vogelstein, B. 1990. A genetic model for colorectal tumorigenesis. *Cell*, 61: 759–767.

Smith, K., McCoy, K. D., and Macpherson, A. J. 2007. Use of axenic animals in studying the adaptation of mammals to their commensal intestinal microbiota. *Seminars in Immunology*, 19: 59–69.

Sokol, H., Pigneur, B., Watterlot, L., Lakhdari, O., et al. 2008. *Faecalibacterium prausnitzii* is an anti-inflammatory commensal bacterium identified by gut microbiota analysis of Crohn disease patients. *Proceedings of the National Academy of Sciences*, 105:16731–16736.

Sorg, J. A., and Sonenshein, A. L. 2008. Bile salts and glycine as cogerminants for *Clostridium difficile* spores. *Journal of Bacteriology*, 190: 2505–2512.

Stappenbeck, T. S., Hooper, L. V., and Gordon, J. I. 2002. Developmental regulation of intestinal angiogenesis by indigenous microbes via Paneth cells. *Proceedings of the National Academy of Sciences of the United States of America*, 99: 15451–15455.

Theriot, C. M., Koenigsknecht, M. J., Carlson, P. E., et al. 2014. Antibiotic-induced shifts in the mouse gut microbiome and metabolome increase susceptibility to *Clostridium difficile* infection. *Nature Communications*, 5: 3114.

Tomova, A., Husarova, V., Lakatosova, S., et al. 2015. Gastrointestinalmicrobiota in children with autism in Slovakia. *Physiology and Behavior*, 138: 179–187.

Turnbaugh, P. J., Ley, R. E., Mahowald, M. A., Magrini, V., Mardis, E. R., and Gordon, J. I. 2006. An obesity-associated gut microbiome with increasedcapacity for energy harvest. *Nature*, 444: 1027–1031.

van der Waaij, D., Berghuis-de Vries, J.M., and Lekkerkerk, L. V. 1971. Colonization resistance of the digestive tract in conventional and antibiotic treated mice. *Journal of Hygiene*, 69: 405–411.

Vasiliauskas, E. A., Kam, L. Y., Karp, L. C., Gaiennie, J., Yang, H., and Targan, S. R. 2000. Marker antibody expression stratifies Crohns disease into immunologically homogeneous subgroups with distinct clinical characteristics. *Gut*, 47: 487–496.

Wikoff, W.R., Anfora, A.T., Liu, J., et al. 2009. Metabolomics analysis reveals large effects ofgut microbiota on mammalian blood metabolites. *Proceedings of the National Academy of Sciences of the United States of America*, 106: 3698–3703.

Wilson, I.D., and Nicholson, J. K. 2009. The role of gut microbiota in drug response. *Current Pharmaceutical Design*, 15: 1519–1523.

Wilson, M. 2008. The indigenous microbiota of the gastrointestinal tract. In: Wilson, M. (ed.) *Bacteriology of Humans: An Ecological Perspective*, 266–326. Blackwell Publishing, Oxford, UK.

10

The Role of Probiotics and Prebiotics in the Composition of the Gut Microbiota and Their Influence on Inflammatory Bowel Disease, Obesity, and Diabetes

Rafael Resende Maldonado
Technical College of Campinas (COTUCA – UNICAMP)

Ana Lúcia Alves Caram
UNIMOGI

Daniela Soares de Oliveira
Municipal College Professor Franco Montoro (FMPFM)

Eliana Setsuko Kamimura and Mônica Roberta Mazalli
University of São Paulo (USP)

Elizama Aguiar-Oliveira
State Univeristy of Santa Cruz (UESC)

CONTENTS

10.1 Gut Microbiota, Probiotics, and Their Impacts on the Human Health

In recent decades, the number of studies on the composition and role of the intestinal environment on human health has increased incredibly fast. It is estimated that there are thousands of species of microorganisms in the gastrointestinal tract, including bacteria, fungi, viruses, and protozoa. Other than those, inactive cells, genetic material, microbial and human metabolites, nutrients from the diet, and other substances that add enormous complexity to this environment are also present. Much has been said about gut microbiota and microbiome. These terms are often used without a clear distinction though. However, some authors employ the term "gut microbiota" referring to the microorganisms that inhabit the intestine, while the term "microbiome" is applied in a broader concept, which also includes the genetic material related to the gut microbiota. The participation in mechanisms that favor health or the development of diseases has often been attributed to the gut microbiota. It interacts with the host symbiotically, modulating inflammation; the immune system; the absorption of micronutrients; the synthesis of vitamins, enzymes, and proteins; the fermentation of energetic substrates; the resistance to pathogens; etc. (Wagner et al., 2018; Moraes et al., 2014).

Despite the numerous functions attributed to the gut microbiota, its composition is not thoroughly known and varies from an individual to another. Part of its composition is influenced by genetic factors. However, the factors, both intrinsic and extrinsic, that define its composition and performance are numerous. The literature indicates that eating habits, stress, age, nutritional status (malnutrition, eutrophy, or obesity), the use of antibiotics and other medications, consumption of alcohol, exposure to radiation, hormonal changes, and many other factors are of great relevance for altering the composition of the gut microbiota (Moraes et al., 2014).

Alterations in the composition or function of the microbiota, generically called dysbiosis, can have an impact on immune responses, intestinal metabolism and permeability, and digestive motility, and they can also promote bacterial overgrowth, produce toxins, and compromise the absorption and synthesis of numerous nutrients, among other consequences. Such alterations are indicated as the causes of inflammation, insulin resistance, increased cardiometabolic risk, and induction of abnormal immune responses, among other problems. All of these are factors related to the appearance of several diseases such as diabetes; obesity; digestive, neurological, autoimmune, and neoplastic diseases; etc. (Sircana et al., 2018; Passos and Moraes-Filho, 2017; Schmidt et al., 2017; Andrade et al., 2015; Mafra et al., 2014; Moraes et al., 2014; Everard and Cani, 2013; Cani and Delzenne, 2011).

Among the numerous factors that can positively or negatively affect the gut microbiota, the consumption of probiotics is one that has attracted great interest in the last decades. The current definition of the term, made by the International Scientific Association for Probiotics and Prebiotics (ISAPP), says that probiotics are live microorganisms that, when administered in adequate quantities, confer benefits to the health of the consumer. This update of the concept means that traditional fermented foods containing live microorganisms should not be called probiotics due to the difficulties in defining the composition, stability, and action of such microorganisms in this food matrix. Despite the new definition, probiotics are fairly popular and well-known for their health effects and constitute a rapidly expanding market, whether in the form of food or supplements, for being used by humans and animals and with a market projection above 70 billion dollars in 2024 (Zucko et al., 2020).

Lactobacillus and *Bifidobacterium* are the most studied probiotics with positive effects on health, being the live bacteria administered orally in sufficient quantity to allow the colonization of the colon (Tilg et al., 2020). There is a vast literature on the subject, which will be briefly discussed below. It shows a much larger number of studies on the effects of probiotics in animal models than in humans, which indicates that many studies still need to be conducted to validate the effectiveness and understand the mechanisms of action of these microorganisms on human health.

Studies on animals indicate that the utilization of probiotics (especially *Lactobacillus* and *Bifidobacterium*), associated with prebiotics or not, was able to reduce the severity of alcoholic steatohepatitis, oxidative stress, intestinal and liver inflammation, and damage caused by endotoxins (Forsyth et al., 2009; Mutlu et al., 2009; Marotta et al., 2005). Although older studies point to the benefits of probiotics on liver disorders, a recent review conducted by Zhou et al. (2020) indicated that the direct effect of gut microbiota on the carcinogenesis of hepatocellular carcinoma (HCC) has not yet been elucidated, in spite of the fact that there are many indications that changes in the microbiota and intestine and liver inflammation may contribute to the evolution of the disease. These authors list the therapeutic application of probiotics as a potential alternative for the treatment or prevention of HCC, but reinforce the need for further studies on this subject.

Probiotics also demonstrate clinical efficiency for treating or preventing various intestinal diseases such as traveler's diarrhea, antibiotic-associated diarrhea, and necrotizing enterocolitis, among others. However, there are studies that indicate the inefficiency of these microorganisms on inflammatory bowel disease (IBD) and even adverse effects in patients with Crohn's disease (CD), so that the administration of probiotics for gastrointestinal disorders is not always recommended (Lee and Bak, 2011).

The use of probiotics to prevent and reduce the severity of type 2 diabetes (DM2) has presented better results in studies with animals than in humans. The number of randomized clinical studies that evaluate the relationship between the use of probiotics and factors related to DM2 such as markers of oxidative stress, inflammation, and incretins is still relatively small (Tonucci et al., 2017).

Cavalcanti Neto et al. (2018) and Thushara et al. (2016) report the existence of several studies in the literature on the ability of probiotics to improve risk factors for cardiovascular disease. Among these are the modulation of the intestinal metabolism of the host; reduction of hypercholesterolemia and hypertension; control of blood glucose, obesity, and the effects of DM2; anti-inflammatory and antioxidant action; etc. However, there are still ambiguities related to the choice of strains and dosages to be administered and the impact of the immunity and genetics of the individual on the efficacy of probiotics and their mechanisms of action on cardiovascular diseases.

Falcinelli et al. (2017) reported that the addition of *Lactobacillus rhamnosus* to a high-fat diet administered to zebrafish had a positive effect on several factors connected to obesity such as transcriptional reduction of orexigenic genes and genes for cholesterol and triglycerides metabolism, positive regulation of anorexigenic genes, attenuation of weight gain, etc. These suggest a potential mitigating effect of probiotics on metabolic disorders related to the high-fat diet.

Autoimmune diseases can also be alleviated by the administration of probiotics. Chae et al. (2012) verified the prophylactic action of a mixture of five probiotics in the evolution of EAMG (experimental autoimmune myasthenia gravis) in mice. Abhari et al. (2016), in turn, observed an improvement in the clinical picture of rheumatoid arthritis and reduction of biochemical markers related to damage to the joints and inflammation in rats which received supplementation with *Bacillus coagulans*. Rinaldi et al. (2018) found divergent results in studies on the administration of some probiotics to patients with rheumatoid arthritis. However, they identified *Lactobacillus casei* as a possible adjuvant in the treatment of the disease.

The administration of different probiotic strains can also have a positive effect on HIV patients whose disease is controlled by antiretroviral therapy (ART). Even people with HIV on ART present a very different microbiota from uninfected individuals, and the administration of probiotics can improve antioxidant defenses and help restore the immune function, reducing side effects of treatment (such as diarrhea) and helping to reduce the chances of the patient developing AIDS (D'Angelo et al., 2017).

Studies show that the health of women can also be improved with the administration of probiotics. According to López-Moreno and Aguilera (2020), different studies point to a positive clinical impact of the use of probiotics in endocrine disorders in women, such as mastitis, vaginal dysbiosis, and polycystic ovary syndrome. The oral administration of *Lactobacillus* has proved to be efficient in the clinical modulation of those disorders in different studies. However, there are still numerous questions yet to be elucidated regarding the type of strain and the doses to be administered, the impact on the individual autochthonous microbiota, hormonal values, and endocrine regulation. The authors emphasize the need to focus studies not only on the final result but on personalizing the administration of probiotics as well. Regarding the action of probiotics during pregnancy, the data are conflicting. Some authors have reported benefits of high doses of probiotics during pregnancy (10^9–10^{10} CFU $g^{-1}day^{-1}$), with effects on weight control, levels of triglycerides, and cholesterolaemia during pregnancy. Other authors, however, have not been able to prove such effects (López-Moreno and Aguilera, 2020; Badehnoosh et al., 2018; Wickens et al., 2017; Jamilian et al., 2016; Luoto et al., 2010).

Probiotics can also have varied effects according to life stages. Alcon-Giner et al. (2019) obtained positive results with the oral supplementation of *Bifidobacterium* and *Lactobacillus* to pre-term babies. An increase in *Bifidobacterium*, acetate, and lactate (short-chain fatty acids); a reduction in pathogenic bacteria in the feces; improvement in the capacity of metabolizing oligosaccharide from breast milk; and reduced intestinal pH in pre-term were observed in babies who received supplementation with probiotics in comparison to the control group. The authors stated that probiotic supplementation was able to modify the gut microbiome of pre-term babies, thus making it more similar to that of full-term babies.

Tran et al. (2019) achieved panic anxiety reduction results, neurophysiological anxiety, negative affect, preoccupation, and increase in regulation of negative mood in university students of average age 20.6 years and predominantly female, treated with probiotic supplementation for 28 days. The authors verified that the effects depended more on the total bacterial count than on the variety of species administered, and on the participants who had a higher degree of initial suffering, more significant effects were obtained.

Gao et al. (2019) conducted a long-term study on the administration of probiotics in the elderly and observed that the ingestion of high doses decreased the microbial richness, but had a positive effect on the control of bowel inflammation. For them, the increase in the composition of beneficial microorganisms contributed to the maintenance of health and the control of homeostasis of the gut microenvironment.

Yang et al. (2020) found that the oral administration of a commercial probiotic supplement, containing different species of *Bifidobacterium* and *Lactobacillus*, resulted in numerous beneficial effects in elderly mice such as the reduction of memory deficits, neuronal and synaptic injuries; glial activation; improvement of the composition of the microbiota in the feces and the brain; improvement in intestinal and blood-brain barriers; decrease in tumor necrosis factor-α; and decreased plasma and cerebral lipopolysaccharide concentration, among others. The results demonstrated the therapeutic potential of probiotics on deficits in the microbiota–gut–brain axis, cognitive function in ageing, and inflammatory responses.

Bonfili et al. (2020), in turn, elucidated a series of mechanisms according to which the supplementation with probiotics contributes to neutralizing the progression of Alzheimer's disease (AD) in mice. According to the authors, probiotics were able to neutralize the time-dependent increase of glycated hemoglobin and the increase in final products of advanced glycation, which had a positive impact on the memory and contributed to delaying the AD.

The data in the literature show the important role of the microbiota in human health and the prevention of diseases. They also demonstrate the potential of probiotics for therapeutic usage. However, there are numerous questions yet to be answered, such as the choice of strains, definition of dosages, elucidation of mechanisms, and greater evidence of clinical effects on humans.

10.2 Gut Microbiota, Prebiotics, and Their Impacts on the Human Health

Prebiotics were defined more precisely in 1995 by the emblematic study of Gibson and Roberfroid as compounds able to selectively stimulate the probiotics located in the small intestine promoting colonic fermentation, which results in the beneficial effects on the health of the host associated with the ideal composition of probiotics in the gut microbiota. However, with the accumulation of research in the area over the years, it is possible to observe a variety of terms such as dietary fiber, prebiotics, or FODMAPs (fermentable oligosaccharides, disaccharides, monosaccharides, and polyols) in the literature (Yan et al., 2018).

Based on the Codex Alimentarium definition, dietary fibers are a heterogeneous group of natural or synthetic carbohydrate polymers with a degree of polymerization higher than 3 which are not metabolized by the body throughout the gastrointestinal tract. They are able to stimulate the intestinal transit by increasing the fecal volume (such as insoluble vegetable fibers) or to stimulate and modulate the growth of the probiotic microbiota present in the small intestine (such as soluble fibers or prebiotic). In turn, the definition of FODMAPs encompasses different compounds such as disaccharides (degree of polymerization less than 3). The reduction in the intake of these compounds has been extensively investigated due to the adverse effects they caused to human health, such as intestinal intolerance or sensitivity and high caloric intake (Eswaran et al., 2017).

Compounds with prebiotic activity can be extracted by different techniques from different sources, such as fruits, barks and roots (Gullón et al., 2013), seaweeds (Praveen et al., 2019), mushrooms (Cheung, 2013), lignocellulosic compounds (Samanta et al., 2015), the cellular structure of microorganisms (Freimund et al., 2003) and others. Some examples of the most investigated and applied prebiotics are the fructooligosaccharides (FOS), xylooligosaccharides (XOS), galactooligosaccharides (GOS), mannanooligosaccharides (MOS), pectic oligosaccharides (POS), beta-glucan, glycomannans, inulin, apple pectin, polydextrose, resistant starch, etc. Many prebiotics can also be produced by enzymatic action (Botvynko et al., 2019; de Oliveira et al., 2020; Tiangui et al., 2018).

Prebiotics may be ingested isolatedly or made available in formulation with probiotics, thus being called symbiotics (Mohanty et al., 2018), which enables the stimulation of probiotics from the moment of ingestion and, depending on their concentration, they can still stimulate gut microbiota. Symbiotics have been developed in different areas such as food and medicines (Segura-Badilla et al., 2020). The inclusion of compounds with prebiotic activity (and other compounds) in formulations with probiotics may also serve to maintain the viability of the cells during the storage time or to enhance the quality of the fermented product obtained (Oliveira et al., 2009).

The effects reported arising from the frequent consumption of prebiotics or symbiotics are as many as those reported for probiotics. For example, the reduction of symptoms of inflammatory conditions, constipation, control of cholesterol and insulin (Bengmark, 2012) among others. In vitro and in vivo tests (animals and humans) have been made in different studies and the results confirm the ability to promote health and well-being arising from the intake of prebiotics (Paesani et al., 2020).

In a study by Kondo et al. (2020), it was observed that the intake of a diet containing 10% (w/w) of FOS for 8 weeks could be related to the control of sensorineural hearing loss (SNHL) in mice and broiler chickens. The administration of probiotics and FOS or MOS prebiotics also resulted in the good development of the animals, without any sensory changes in the meat cuts (Al-Khalaifa et al., 2019). The administration of probiotics with inulin in older women fed a high-protein diet is an example of tests performed on humans. The symbiotic supplement helped to maintain balance in the gut microflora, which tends to be impaired by excessive protein intake (Ford et al., 2020).

Antibiotic treatments also cause damage to the ideal composition of the gut microbiota and the ingestion of prebiotics and/or symbiotics has been highly recommended to promote the recovery of the intestinal colonization. The associated intake of prebiotics and probiotics is also reported as having an important role in combating infectious diseases as an alternative to the treatment with antibiotics only followed by the recomposition of the gut microbiota. However, each organism reacts differently to prebiotics, and the mechanism of action of these compounds against infectious diseases is still little understood (Yang et al., 2019).

It is also important to emphasize that the excessive intake of prebiotics can induce intestinal discomfort (such as bloating, colic, gas, mild diarrhea, etc.) due to the overgrowth of probiotics (Gibson, 2004). Additionally, it has also been suggested to reduce the consumption of prebiotics by those with conditions such as irritable bowel syndrome (IBS), although some studies indicate that treatments with probiotics and prebiotics are effective in relieving the symptoms of this type of condition (Ooi et al., 2019).

Recent studies have evaluated the administration of prebiotics for treating different types of diseases with varying effects. Wang et al. (2020) evaluated the effect of inulin in a colonic model with faeces from omnivorous and vegetarian donors added with casein as a protein source. The addition of inulin was effective in reducing the production of two undesirable metabolites of protein degradation (branched-chain fatty acids and ammonia) in the groups evaluated. Furthermore, inulin also reduced the production of p-cresol in stool samples from vegetarians.

Amiriani et al. (2020) evaluated the use of a symbiotic supplement for complementary therapeutic action in patients undergoing treatment for ulcerative colitis. In the group that received supplementation, there was an improvement in the intestinal condition in 64.3% of patients versus 47% of the patients in the control group after 8 weeks of treatment, thus demonstrating the efficiency of the action of the symbiotic product.

Miyoshi et al. (2020) evaluated prebiotic supplementation with hydrolysed guar gum in a group of 15 patients on hemodialysis. The degree of constipation of the patients was evaluated, and it was found that the addition of the prebiotic improved the individual shape of the stool, reduced the constipation index by 40%, caused an increase in the beneficial microbiota (*Bifidobacterium* and *Bacteroides*) up to 3.6 times and the concentration of short-chain fatty acids by 1.58 times.

Kang et al. (2020) evaluated the effect of a natural prebiotic on the symptoms and biochemical markers of atopic dermatitis in mice. The authors verified effects such as the decrease in the thickness of the ears and the dermal and epidermal layers, of pro-inflammatory cytokines and other biochemical markers related to the disease and the increase in the population of *Bifidobacterium* in the microbiota of the animals. Despite the results presented, as well as in the case of probiotics, further research is needed for prebiotics to evaluate the dosage and efficacy of the proposed therapeutic applications for different types of diseases and individuals.

10.3 Paraprobiotics, Postbiotics, Psychobiotics and More to Come

Starting from the classic concepts of probiotics, prebiotics and symbiotics and, with the advancement and specificity of research over the years, new concepts as paraprobiotics, postbiotics and psychobiotics have emerged (Barros et al., 2019) and are expanding the areas of research and application. As mentioned earlier, by definition, probiotics are necessarily live/active microorganisms capable of promoting health/well-being when ingested regularly. However, it has been proven that these benefits are not restricted to viable cells. It is known that both the ingestion of inactive cells – non-viable (de Almada et al., 2016), and the ingestion of certain bioproducts from the cellular metabolism of probiotics (Aguilar-Toalá et al., 2018) may result in health benefits. Thus, the nomenclatures: paraprobiotics (inactive probiotic cells) and postbiotics (bioproducts of probiotic metabolism or parts of inactivated cells) have already been defined and are widely known. Additionally, to cover a specific area of promoted beneficial effects, the third nomenclature mentioned, psychobiotics, emerged. It defines those probiotics capable of presenting psychotropic properties and thus promoting mental health (Dinan et al., 2013).

The distinction between paraprobiotics and posbiotics is confusing at times since both the inactivated cell (or parts of it) and compounds secreted by these cells are capable of resulting in different health benefits (Collado et al., 2019). However, the differentiation between these two classes tends to be increasingly recognised. The great advantage regarding the possibility of applying inactivated probiotics is the fact that its administration to individuals with a compromised immune system can be performed without major risks (Chuang et al., 2007; Villena et al., 2009). Furthermore, as many researchers emphasise, its maintenance and commercialization are simpler. There would be no concern with ingesting or producing any compounds by cellular metabolism with the cell inactivated as well (Sawada et al., 2016).

Paraprobiotics were initially classified as "inactivated probiotics" or "ghost probiotics" (Taverniti and Guglielmetti, 2011). For the classification to be appropriate, the probiotic cell must be exposed to an external factor that causes irreversible damage to its cellular structure or its genetic material, resulting in loss of biological activity. In spite of this fact, paraprobiotics still maintain some residual metabolic activity which is responsible for the health benefits for the host, depending on the integrity of the cells, although the mechanism involved in these situations remains unknown. This principle is very similar to that of "bacterial ghosts" applied in the elaboration of non-living vaccines (Tian et al., 2019). Any strain that has a proven probiotic action can be applied as a paraprobiotic. Lactic acid bacteria are employed very frequently (Xavier-Santos et al., 2020). However, non-saccharomyces yeasts have also been suggested (Saadat et al., 2020).

The application of appropriate time/temperature binomials that result in the controlled precipitation of important proteins/enzymes (Ou et al., 2011); UV radiation (van Hoffen et al., 2010) which damages the genetic material of the cells; and high-intensity ultrasound (Guimarães et al., 2019) that can cause damage to the cell membrane exposing the cytoplasmic content are examples of probiotic inactivation treatments to obtain paraprobiotics. The extent of the damage caused to the biological activity of the cells must be sufficient only for their inactivation, maintaining the properties of the paraprobiotic. The alterations caused by the inactivation treatments may be accompanied, for example, by the use of specific dyes which allow the analysis of the integrity of the cytoplasmic membrane, breathing capacity and other characteristics as mentioned by de Almada et al. (2016).

Progress in research enlightened that certain metabolites of interest (postbiotics) may be found in cell-free media and are capable of acting specifically, presenting health benefits. Postbiotics may then be understood as soluble bioactive compounds originated from the cellular metabolism of probiotics or the breaking of the cellular structure during the production of paraprobiotics (Barros et al., 2019). Some examples of cell metabolites classified as posbiotics are: gamma-aminobutyric acid or GABA, which are important, for example, in the prevention of type 1 diabetes and neurological diseases (Diez-Gutiérrez et al., 2020); pyrrolo [1,2-a] pyrazine-1,4-dione which presents antimicrobial and antioxidant action (Moradi et al., 2019); and indole-3-aldehyde which helps to reduce oxidative stress (Puccetti et al., 2018). Exopolysaccharides, lipoteichoic acids, polar lipids, peptidoglycans and proteins are other examples of postbiotics obtained by breaking the cellular structure and are able to contribute to a better modulation of the immune system (Pyclik et al., 2020).

As mentioned earlier, the inactivation of probiotic cells may produce paraprobiotics, and the removal of these cells, viable or not, can be used to obtain supernatants containing the postbiotic compounds. Different strains have been used to obtain supernatants rich in cell metabolites produced by probiotics, such as the postbiotic of *Lactobacillus salivarius* which, according to a study conducted by Moradi et al. (2019), presented antibacterial and antibiofilm action against *Listeria monocytogenes*. In the study by Kareem et al. (2017) the *L. plantarum* RG14 postbiotic used in chickens, in combination with the prebiotic inulin, was able to assist in the expression of cytokines, fundamental substances to inflammatory responses. In addition to these, many other health benefits have been reported in animal and human models (Aguilar-Toalá et al., 2018; Barros et al., 2019). Some researchers argue that postbiotics should be classified according to their benefits. This could, however, result in an excessive number of nomenclatures given the wide variety of benefits reported.

Among all these benefits related to the frequent consumption of probiotics (and other derivatives), there is an improvement in humour (Tran et al., 2019). However, one of the first mentions that certain probiotics are able to influence the central nervous system and thus assist, for example, in the treatment of the major depressive disorder (MDD), was made by Logan and Katzman (2005). Therefore, the increasingly explored definition of an axis of brain-intestine relationship has been observed (Rogers et al., 2016). Although psychobiotics can assist the treatment of diseases such as MDD, Alzheimer's, autism, schizophrenia, postpartum depression, cognitive dysfunction, Parkinson's, anxiety and others based on studies conducted until the present day (Cheng et al., 2019; Kim and Shin, 2019; Sarkar et al., 2016), they should not be seen as the only form of treatment. Sarkar et al. (2016) emphasise that the effects of psychobiotics also extend to healthy individuals, and research should not be restricted to specific clinical groups. In addition, these researchers alert to the fact that any substance that stimulates the gut microbiota and consequently results in psychophysiological benefits can potentially be classified as a psychobiotic.

Different strains of *Lactobacillus* and *Bifidobacterium* have been reported as psychobiotics. This can be seen in the studies by O'Hagan et al. (2017), Slykerman et al. (2018) and Tian et al. (2019). One of the ways in which a psychobiotic improves mental health is associated with the ability of these strains to produce certain neurotransmitters, whose deficiency in the body can result, among other effects, in depression and anxiety. Examples of these compounds include gamma-aminobutyric acid (GABA) (Yunes et al., 2016), serotonin (Xie et al., 2020) and the endocannabinoids (Cani et al., 2014). Another important form of action of psychobiotics refers to the reduction of oxidative stress and levels of pro-inflammatory cytokines, both with a strongly related to the development of depression and anxiety symptoms which can be caused, for example, by the inflammation of the gastrointestinal tract or imbalances in the hypothalamic-pituitary-adrenal axis (Kim and Shin, 2019; Tran et al., 2019).

Regardless of the mechanism of action of the bacteria classified as psychobiotics, they are gaining more and more the interest of Science, and eventually, of the general public. Therefore, it is possible to expect that in the years yet to come, new information and new nomenclatures will be defined. However, it is necessary to reinforce the need for a greater volume of studies in humans, as these are still fewer than those conducted in animal models. As an example of the frequent expansion of knowledge in this area, we cite the study by Nishida et al. (2017) in which the ability to reduce the symptoms of chronic stress in students (25 years of age on average) was demonstrated thanks to the frequent consumption of a paraprobiotic (heat-inactivated *Lactobacillus gasseri* CP2305), proving that paraprobiotics can also result in effects similar to postbiotics.

Much more is yet to come as an expansion of the initial knowledge of probiotics and prebiotics. This area of knowledge is rich in information and only contributes to the well-being of humans and animals. Therefore, research conducted with ethics and seriousness must be stimulated.

10.4 Influence of Probiotics, Prebiotics and Postbiotics in Chronic Non-Communicable Diseases

10.4.1 Inflammatory Bowel Diseases

Recent technological advances have made it possible to identify that alterations in the composition and function of the microbiota affect the intestinal health directly, favoring the appearance of diseases. IBDs, including ulcerative colitis (UC) and CD, are chronic and recurrent inflammatory bowel processes whose incidence is increasing in the world population. The etiology is uncertain, however, the hypotheses suggest complex interactions between genetic, and environmental factors and the immune system of the host, which lead to aberrant immune responses and chronic intestinal inflammation (Nishida et al., 2018).

CD is an inflammatory bowel of unknown origin, characterised by a segmental, asymmetrical and transmural striking of any portion of the digestive tract, from the mouth to the anus, and can appear in three main forms: inflammatory, fistulous and fibrostenotic. It can also present extraintestinal manifestations (ophthalmological, dermatological and rheumatological), in which the natural evolution of the disease is marked by activations and remissions (Brasil, 2017). Silva et al. (2011) consider the excessive production of pro-inflammatory cytokines and the imbalance of the gut microbiota as the probable causes of the disease.

Nutritional care is essential in IBD, both in the prevention and treatment of malnutrition and/or specific nutrient deficiencies resulting from the development of the disease. Several patients in remission may present good nutritional state, however, some may be overweight and have an abnormal body composition. It is important to emphasise that adipose tissue influences the regulation of immunity and inflammation; is a source of cytokines and produces about 30% of circulating interleukin-6 (IL-6). Surgical CD patients present a mesentery that is often thickened and hardened and with high accumulation of fat. In addition, the intra-abdominal fat observed in these patients may be related to the development and progression of this disease (Silva et al., 2010).

The expenditure of energy at rest may vary depending on the inflammatory activity, the extension of the CD or the nutritional state. The caloric intake should vary from 25 to 30 kcal. kg body-1. day-1. Measuring vitamins and minerals is also essential, especially in the acute phase of the CD or after extended surgery that reduces the size of the intestine. Nutritional changes depend on the extent and severity of the disease and they can worsen the prognosis of patients undergoing clinical treatment or those undergoing surgery besides compromising immune function (Silva et al., 2010).

According to Moraes et al. (2014), the intake of probiotics and prebiotics in the diet can be a preventive or therapeutic measure for IBD, as it can improve the composition and functionality of the microbiota with impacts on the sensitivity of the intestinal epithelium and the central nervous system, influencing the central regulation of appetite and satiety. However, the effects of the administration of probiotics on IBD are controversial, as there are cases in which they have no effect or cause negative effects on health.

Shadnoush et al. (2015) conducted a study with 305 patients with IBD to evaluate the administration of probiotic yoghurt. After 8 weeks, the group that ate the probiotic yoghurt showed a significant increase in the count of *Lactobacillus*, *Bifidobacterium* and *Bacteroides* in the faeces, indicating an improvement in the intestinal microbiota. However, there was no significant variation in the average weight in the body mass index when compared with the control group (without IBD and yoghurt) and with the placebo group (with IBD and without yoghurt).

Ganji-Arjenaki and Rafieian-Kopaei (2018) evaluated 27 studies of the effect of probiotics in patients with IBD and concluded that some probiotic strains (with or without association with prebiotics) presented a positive effect for UC, but there were no significant effects for patients with CD, considering a 95% confidence level. Iannitti and Palmieri (2010), observed in another review that, in most of the studies evaluated, there was a little or no effect of the therapeutic administration of probiotics in patients with CD.

Derwa et al. (2017) evaluated different studies with a commercial probiotic formulation and concluded that the use of probiotics was effective to induce remission in some patients with active UC and the presence of quiescent UC. However, the efficacy in patients with CD proved uncertain.

Zhang et al. (2019) studied the influence of oral administration of *Lactobacillus plantarum* in mice with colitis triggered by dextran sodium sulfate (DSS). The authors found that the animals treated with the probiotic had less severe symptoms of colitis, diversification of microbial species in the colon and restriction of the activity of pathogenic bacteria in the intestine, which improved the stability of the gastrointestinal tract.

Ballini et al. (2019) evaluated the therapeutic action of probiotics in 20 patients with IBD for 90 days. The authors observed that there were significant improvements in the global oxidative capacity and in the antioxidant response, taking oxidative stress to non-pathological values in the group treated with probiotic in comparison to the control group.

Alard et al. (2018) evaluated different probiotic strains in rats with induced colitis. The results indicated that *Bifidobacterium bifidum* PI22 was more protective against chronic colitis, *Bifidobacterium lactis* LA804 was more efficient against acute colitis. *Lactobacillus helveticus* PI5 did not have an anti-inflammatory effect in vitro, but it strengthened the epithelial barrier and reduced the effects of acute colitis in vivo. Finally, the *Lactobacillus salivarius* strain LA307 protected the animals against both types of colitis.

Silva et al. (2018) evaluated, by means of a questionnaire, the knowledge, use and effect of probiotics, fibres and prebiotics in patients with IBD (57% with UC, 39% with CD and 4% with indeterminate colitis). The results showed that 87% of the participants knew probiotics, 74%, fibres and only 43% prebiotics. Of those surveyed, 65% said they had used any of these products in the past, but only 31% reported improving their quality of life after the consumption. Despite the high utilisation of these products, it was not possible to prove their effectiveness in improving intestinal health.

From a theoretical point of view, the modification of the microbiota in IBD by the use of probiotics and prebiotics makes sense. However, clinical studies have not demonstrated this conclusively yet. There are conflicting results concerning UC, slightly more convincing evidence about pouchitis, and there is no substantial evidence for use in DC. So much so that the clinical protocol and therapeutic guidelines of the CD disagree with the use of probiotics, precisely because there is no scientific evidence of its effectiveness on the disease. There are also conflicting data in the literature on the role of postbiotics on IBD. Limitations such as the small size of the populations studied, heterogeneity of the strains, dosages and formulations, besides the lack of explanation regarding the interference of drugs and diet, hinder the interpretation of most studies (Russo et al., 2019; Brasil, 2017; Abraham and Quigley, 2016).

10.4.2 Diabetes Mellitus

Diabetes mellitus is a group of metabolic diseases characterised by chronic hyperglycemia resulting from abnormalities in the secretion and/or action of insulin which affects directly the metabolism of carbohydrates, lipids and proteins. Type 1 diabetes mellitus (DM1) is an autoimmune disease associated with a complete loss of the ability of the pancreatic beta cells to secrete insulin, resulting in the need for its external administration for survival. Regarding type 2 diabetes mellitus (DM2), although there is some production of pancreatic insulin, the resistance to it is what plays the most important role in the pathogenesis (Zheng et al., 2018; Farsani et al., 2017; Kharroubi and Darwish, 2015).

Despite the genetic predisposition for the development of DM2, an unhealthy diet and a sedentary lifestyle are important factors in the current global epidemic of the disease, in which the number of people with suffering from it has quadrupled in the last three decades. Continued weight gain and body fat accumulation have effects on blood glucose levels, and these are independent of changes in sensitivity to insulin or the function of pancreatic beta-cells. Many cases of DM2 could be prevented with changes in the lifestyle, including bodyweight control, a healthy diet and moderate physical activity. Most DM2 patients have at least one other comorbidity besides diabetes, especially cardiovascular diseases, which are the main cause of morbidity and mortality in these patients (Gong et al., 2019; Zheng et al., 2018; Low Wang et al., 2016).

The gut microbiota has been considered the new organ system in the body, and the occurrence of its dysbiosis has shown an important role in the development of diabetes. In this context, approaches that suggest a healthy host-microbiota relationship may be relevant to reducing the problems caused by diabetes (Adeshirlarijaney and Gewirtz, 2020; Liu and Lou, 2020; Anwar et al., 2019; Eid et al., 2017).

A study conducted in germ-free mice (GF) fed a high-fat diet showed that this group had less weight gain and did not show resistance to insulin and intolerance to glucose, in comparison to the control group. This result was attributed to the activity of protein kinase activated by 5′adenosine monophosphate (AMPK), which plays a central role in energy homeostasis. This enzyme contributes to greater oxidation of fatty acids in the muscles and liver, and greater energy expenditure, hence the result with less body weight gain in the GF group. High-fat diets usually worsen the health of the gut microbiome, leading to dysbiosis, metabolic dysregulation, increased resistance to insulin and inflammation, all key factors in the development of DM2 (Sikalidis and Maykish, 2020; Saad et al., 2016; Bäckhed et al., 2004).

It is important to note that gut bacteria can influence the colonic mucus layer. This covers the interior of the colon, acting as a physical barrier to trillions of gut bacteria, separating them from the host by size exclusion. In this sense, the intestinal barrier includes intercellular junctions that directly prevent the penetration of bacterial products into the bloodstream. However, it also hosts systems of mucus implantation and innate immunity, which maintain bacteria at a safe distance from the epithelium, thus conserving the composition of the microbiota stable. The presence of distinct bacteria is crucial for the preservation of the mucus and the proper functioning of the intestine. Another factor in maintaining this balance is the intestinal pH, whose role in the composition of bacteria in the gut microbiota is determinant (Liu and Lou, 2020; Kamphuis et al., 2017).

Another aspect related to the gut microbiota is the production of short-chain fatty acids, especially butyrate (the ionised form of butyric acid). This substance is produced by fermentation in the large intestine by the gut microbiota (bacteria *Clostridiales* spp., *Eubacterium recatle*, *Faecalibacterium prausnitzii*, *Roseburia intestinalis* and *Roseburia inulinivorans*). This acid has a protective effect against the resistance to insulin and the fatty liver; increases the function of the intestinal barrier and facilitates the assembly of narrow junctions of the epithelium, which causes the activation of AMPK. Low production of butyrate has an effect on DM1-associated autoimmunity and may decrease the gut microbiota of DM2 patients in comparison to healthy individuals. Therefore, the administration of bacteria with anti-inflammatory properties and producing short-chain fatty acids are promising in the treatment and prevention of diabetes (Gonzalez et al., 2019; Tanca et al., 2018; Zheng et al., 2018; Qin et al., 2012).

Nonetheless, the presence of other microorganisms in the intestine represents a risk factor for DM2, as they cause the sensitivity to insulin to decrease. The *Bacteroides enterotype* and proteobacteria can cause low-grade inflammation in diabetics by releasing flagella and/or other surface components. This process triggers an inflammatory response and contributes to the development of diabetes. Firmicutes and increased intestinal populations of Bacteroides are also commonly found in the gut microbiota and associated with compromising the health of DM1 patients (Adeshirlarijaney and Gewirtz, 2020; Han et al., 2018).

Approaches to enhance the microbiota and permeability of the intestinal barrier include the administration of probiotic bacteria supplementation, and prebiotic compounds or treatments with postbiotics such as bile salts, which improve the gut microbiome. Although much is said about the potential benefits of therapies with microorganisms, substances and metabolites for human health, it is important to note that most studies in the literature have been done in animals. The evidence in humans is still insufficient.

A study conducted with gliclazide, an antidiabetic drug of oral administration, in patients with DM1, presented a synergistic effect of the drug with probiotics and bile acids in the reduction of blood glucose levels and in improving diabetic complications (Mikov et al., 2018).

Concerning postbiotics, an alternative approach has been the direct administration of bile acids, which demonstrates a beneficial effect on the host. These acids interact locally or systemically with cellular receptors, such as FXR (nuclear farnesoid X receptor) and TGR5 (Takeda G protein-coupled receptor 5), influencing the systemic metabolism of lipids and cholesterol, metabolism of energy, immune homeostasis and intestinal electrolyte balance. Through specific enzymatic activities, the gut microbiota can modify signaling properties of bile acid and have a positive impact on the health of the host (Fiorucci et al., 2018; Joyce and Gahan, 2016).

Prebiotic substances such as resistant starch, resistant dextrin and inulin enriched with oligofructose have proven effective in improving DM2-related metabolic and inflammatory biomarkers in women with at least 18 years of age (Colantonio et al., 2019).

Non-digestible polysaccharides from acorn, quinoa, sunflower, pumpkin and sago seeds caused an increase in the production of short-chain fatty acids and the diversity of the gut microbiota when offered to mice submitted to a hyperlipidic diet. In this study, the authors observed that feeding with acorn and sago prebiotics reduced the intolerance to glucose and resistance to insulin without affecting adiposity. Furthermore, the beneficial prebiotic effects were superior to those of inulin, which indicates that the administration of freshly isolated prebiotics reduces damage to the metabolism of glucose caused by the high-fat diet, through the modulation of the microbiome-gut-brain axis. These prebiotics were considered to be beneficial for preventing and treating diet-induced obesity and diabetes (Ahmadi et al., 2018).

The alternative prebiotic therapy with the administration of a polysaccharide obtained from the sea cucumber (*Holothuria leucospilota*) has shown promise in the treatment of diabetes. The study was conducted with diabetic rats that consumed a dietary supplementation in instant tea. The polysaccharide utilised improved the gut microbiota, reduced the intolerance to glucose, regulated blood lipids and hormones, thus relieving the symptoms of DM2. In addition, deficiencies in the pancreas and colon were repaired, the population of bacteria producing short-chain fatty acids was increased, and there was a decrease in opportunistic bacterial pathogens in the faeces of the rats (Zhao et al., 2020).

Research has recognised that the gut microbiota has a direct effect on the brain and that the brain also influences the microbiota. Comorbid brain disorders associated with diabetes, oxidative stress and hyperglycemia are responsible for depression, anxiety and memory impairment in diabetics. The gut-brain axis includes the central, autonomic and enteric, neuroendocrine, neuroimmune and gut microbiota nervous systems. Probiotics and prebiotics have demonstrated a beneficial effect on the brain through the modulation of neurotransmitters and glucose homeostasis. This indicates a bidirectional link between the gut microbiota and the gut-brain axis, which can be very positive in the treatment and prevention of chronic diseases such as diabetes (Thakur et al., 2014, 2019).

10.4.3 Obesity

Obesity is a chronic disease characterized by excessive accumulation of body fat, with pathological consequences in the medium and long term. It is one of the most relevant public health problems due to its high global prevalence, and its contribution to the high rates of morbidity and mortality related to approximately two-thirds of deaths worldwide. The aetiology of the disease is multifactorial. However, the interaction between genetics, environmental factors, diet (high energy intake and excessive accumulation of body fat) and the low level of physical activity are considered the main contributors to the development of obesity (Schmidt et al., 2017; Andrade et al., 2015; Moreira et al., 2012).

Obesity is also associated with other diseases such as metabolic syndrome, type 2 diabetes mellitus, development of resistance to insulin, dyslipidemia and systemic arterial hypertension. Other factors that influence negatively the quality of life of the individual are psychosocial disorders, depression, anxiety disorders and alterations in body image (Andrade et al., 2015; Nadal et al., 2009).

The gut microbiota is considered an important endogenous factor that influences the epidemiology of obesity. It exerts an important function in converting food into nutrients and energy, breaking undigested food molecules into metabolites such as short-chain fatty acids, as well as synthesising some vitamins. Therefore, imbalances in the gut microbiota (called intestinal dysbiosis, characterised by the dominance of pathogenic bacteria over beneficial bacteria) may have harmful effects on human health, starting with damage to intestinal integrity (Schmidt et al., 2017; Harakeh et al., 2016; dos Santos and Ricci, 2016; Andrade et al., 2015; da Silva et al., 2013; Moreira et al., 2012; Almeida et al., 2009).

The development of obesity may be related to the composition of the gut microbiota, which is different between eutrophic and obese humans. It is also influenced by the relative proportions of two main phyla of bacteria of the gut microbiota, the Bacteroidetes and the Firmicutes. In most individuals, approximately 90% of the phyla are Firmicutes and Bacteroidetes, the others being Actinobacteria (family Bifidobacteriaceae) and Proteobacteria (family Enterobacteriaceae). Subsequently, in order of frequency, the phyla Synergistetes, Verrucomicrobia, Fusobacteria and Euryarchaeota appear representing a small percentage of our microbiota. The highest proportion of Firmicutes in relation to Bacteroidetes is related to obesity and metabolic disorders. Currently, the concept of "obesogenic" microbiota is widely reported in the literature, in which the two main phyla of the gut microbiota, Bacteroidetes and Firmicutes, facilitate the removal and storage of the calories ingested, extracting the energy from the diet with a higher frequency than the microbiota of the eutrophic individual (Schmidt et al., 2017; dos Santos and Ricci, 2016; Correria and Percegoni, 2014; Moraes et al., 2014).

Diet is a determining factor in the intestinal colonization, having relevance in the metabolic modulation and the regulation of body adiposity. Therefore, qualitative and quantitative changes in the ingestion of specific food components (fatty acids, carbohydrates, micronutrients, prebiotics and probiotics) can alter the composition of the gut microbiota. Intestinal bacterial colonization begins when the newborn is exposed to different species of microorganisms present in the mother and becomes stable (adult microbiota) by the age of two, depending on the balance between beneficial and pathogenic bacteria and the diet acquired since early life. In addition, studies demonstrate that obese and slim people have distinct microbiotas, which may contribute to the development of obesity (Stanislawski et al., 2019; Schmidt et al., 2017; dos Santos and Ricci, 2016; Andrade et al., 2015; Angelakis et al., 2012; Delzenne et al., 2011).

Studies demonstrated that diets high in fat and low in fibre alter the gut microbiota negatively, reducing the beneficial bacteria (Bacteroidetes – mainly *Bifidobacterium*) and increasing the pathogenic ones (Firmicutes), which stimulates the secretion of pro-inflammatory cytokines. Therefore, reducing the ratio of Firmicutes to Bacteroidetes in obese individuals benefits the treatment of obesity. Diets rich in fibre promote greater loss of energy through faeces. This can decrease the food intake; synthesis of lipids and adipogenesis, and increase lipolysis leading to a reduction in body fat. Furthermore, obese individuals present a small proportion of Bacteroidetes and a greater proportion of Firmicutes when compared to slim individuals. The increased amount of Bacteroidetes appears to be the most important factor in the percentage of body weight loss than the caloric amount of the diet in obese individuals (Chakraborti, 2015; Perpétuo et al., 2015; Kotzampassi et al., 2014; Ridaura et al., 2013; da Silva et al., 2013; Jumpertz et al., 2011; Vrieze et al., 2010; Brinkworth et al., 2009; Cani and Delzenne, 2009; Turnbaugh et al., 2009; Bäckhed et al., 2004).

In the study by Duncan et al. (2008), there was no difference between Bacteroidetes in the faecal samples of obese and non-obese individuals. Nonetheless, there were reductions in the Firmicutes group in obese subjects submitted to low-calorie diets. These studies suggest that the manipulation of the composition of the gut microbiota may prevent weight gain or facilitate weight loss in humans (Angelakis et al., 2012).

The abundance of *Bifidobacterium* has been indicated as a critical factor for improving the gut microbiota with an impact on obesity. Diets containing probiotics may contribute to the increase of this population, while the intake of prebiotic fibre can potentiate the development of this group of beneficial bacteria. Fibre is an important fermentative fuel for the growth of the gut microbiota, and for the production of short-chain fatty acids that impact positively on the reduction of obesity. There are a large number of studies related to probiotics and prebiotics and obesity. However, further studies are still necessary to provide support for their application in terms of public health (Klancic and Reimer, 2020; Delgado and Tamashiro, 2018; Kobyliak et al., 2016).

A study with 50 individuals whose body mass index was above 25 kg m^{-2} treated with probiotic supplementation for 12 weeks presented a significant reduction in obesity markers (waist circumference, total fat area, visceral fat and ratio of visceral fat area to subcutaneous), and improvement in the composition of the gut microbiota, while the placebo group worsened in terms of body fat, blood glucose, inulin and the presence of pathogenic bacteria in the microbiota (Song et al., 2020). In another study, obese pregnant women until the 36th week of gestation received probiotic supplementation, and there was no improvement in the indicators of mental health of the analyzed cohort (anxiety, depression, health and functional well-being) (Dawe et al., 2020).

Another study evaluated the influence of probiotics, prebiotics and symbiotics in Wistar rats (n=48) fed a high-lipid diet for 12 weeks. The rats developed obese-insulin resistance, dyslipidemia and decreased sensitivity to insulin due to the increased body weight. In addition, there were alterations in the bone marrow and bone metabolism. The animals treated with probiotics, prebiotics or symbiotics presented better parameters than those that only had a high fat intake. Nevertheless, the bone alterations were similar in all groups evaluated, with no improvement by supplementation (Eaimworawuthikul et al., 2017). An *in vitro* study of faecal cultures of normal-weight and morbidly obese adults evaluated the supplementation with oligosaccharides and inulin. The result of the supplementation was different in the two groups. The obese individuals had an increase in Bacteroides and *Faecalibacterium*, whereas eutrophic individuals presented an increase in *Bifidobacterium* and *Faecalibacterium*. 1-kestose was the prebiotic evaluated that impacted the alterations observed in the faecal cultures the most. The results of this study indicate the necessity to develop specific prebiotic products for supplementation in different populations (Nogacka et al., 2019). The application of probiotics, prebiotics and other forms of supplementation seems to have great therapeutic potential on obesity. However, there is much yet to be studied for elucidating mechanisms of action and defining more specific and effective therapies.

10.5 Final Considerations

The last few decades have resulted in significant advances in the understanding of how gut microbiota influences the proper functioning of the entire organism of humans and different animals. However, there is much yet to be investigated. It is undeniable that certain microorganisms and their metabolites play an important modulating role in different bodily functions. There are also numerous substances obtained from food which can contribute or harm the gut microbiota. Probiotics, prebiotics and the other categories derived from these two initial concepts may be important supporting factors in the prevention and/or treatment of various diseases. Nonetheless, the mechanisms involved in the action of these microorganisms and substances are not completely understood, and there is much to be studied.

The benefits reported in the scientific literature are numerous and go far beyond intestinal health. They may have a systemic impact on the entire organism of humans or other animals. For being this an area still in expansion, frequent analyses of terms, definitions, results and applications are necessary in order to always try to converge the information obtained so that this expansion occurs cohesively. From the current state of the literature, it should be noted that there are still relatively few studies on humans with sufficiently robust data to validate all the properties attributed to probiotics, prebiotics and other groups. In addition, the literature lacks studies that specify the types of microorganisms and substances to be used, dosages and influence of genetic and environmental factors. Finally, there is much yet to be developed in terms of products for therapeutic application, including targeting specific groups, that is, products for use by healthy individuals or for those who have specific morbidities or diseases.

REFERENCES

Abhari, K., Shekarforoush, S. S., Hosseinzadeh, S., Nazifi, S., Sajedianfard, J., and Eskandari, M. H. 2016. The effects of orally administered *Bacillus coagulans* and inulin on prevention and progression of rheumatoid arthritis in rats. *Food & Nutrition Research*, 60(1): 30876.

Abraham, B. P., and Quigley, E. M. 2016. Prebiotics and probiotics in inflammatory bowel disease (IBD). In *Nutritional Management of Inflammatory Bowel Diseases*, ed. A. Ananthakrishnan, 131–147. Springer: Cham.

Adeshirlarijaney, A., and Gewirtz, A. T. 2020. Considering gut microbiota in treatment of type 2 diabetes mellitus. *Gut Microbes*: 1–12.

Aguilar-Toalá, J. E., Garcia-Varela, R., Garcia, H. S., Mata-Haro, V., González-Córdova, A. F., Vallejo-Cordoba, B., and Hernández-Mendoza, A. 2018. Postbiotics: an evolving term within the functional foods field. *Trends in Food Science & Technology*, 75: 105–114.

Ahmadi, S., Nagpal, R. K., Wang, S., and Yadav, H. 2018. New prebiotics to ameliorate high-fat diet-induced obesity and diabetes via modulation of microbiome-gut-brain axis. *Diabetes*, 67(S1): doi: 10.2337/db18-264-LB.

Alard, J., Peucelle, V., Boutillier, D., Breton, J., Kuylle, S., Pot, B., Holowacz, S., and Grangette, C. 2018. New probiotic strains for inflammatory bowel disease management identified by combining in vitro and in vivo approaches. *Beneficial Microbes*, 9(2): 317–331.

Alcon-Giner, C., Dalby, M. J., Caim, S., Ketskemety, J., Shaw, A., Sim, K., Lawson, M., Kiu, R., Leclaire, C., Chalklen, L., Kujawska, M., Mirtra, S., Kroll, F.S., Clarke, P, Hall, L. J. 2019. Microbiota supplementation with Bifidobacterium and Lactobacillus modifies the preterm infant gut microbiota and metabolome. *bioRxiv*, 698092.

Al-Khalaifa, H., Al-Nasser, A., Al-Surayee, T., Al-Kandari, S., Al-Enzi, N., Al-Sharrah, T., Ragheb, G., Al-Qalaf, S., and Mohammed, A. 2019. Effect of dietary probiotics and prebiotics on the performance of broiler chickens. *Poultry Science*, 98(10): 4465–4479.

de Almada, C. N., Almada, C. N., Martinez, R. C., and Sant'Ana, A. S. 2016. Paraprobiotics: evidences on their ability to modify biological responses, inactivation methods and perspectives on their application in foods. *Trends in Food Science & Technology*, 58: 96–114.

Almeida, L. B., Marinho, C. B., Souza, C. D. S., and Cheib, V. B. P. 2009. Disbiose intestinal. *Revista Brasileira de Nutrição Clínica*, 24(1): 58–65.

Amiriani, T., Rajabli, N., Faghani, M., Besharat, S., Roshandel, G., Tabib, A. A., and Joshaghani, H. 2020. Effect of Lactocare® synbiotic on disease severity in ulcerative colitis: a randomized placebo-controlled double-blind clinical trial. *Middle East Journal of Digestive Diseases*, 12(1): 27.

Andrade, V. L. A., Regazzoni, L. A. D. A., Moura, M. T. R. S., Anjos, E. M. S. D., Oliveira, K. A. D., Pereira, M., de Amorim, N. R., and Iskandar, S. M. 2015. Obesidade e microbiota intestinal. *Revista Média de Minas Gerais*, 25(4): 583–589.

Angelakis, E., Armougom, F., Million, M., and Raoult, D. 2012. The relationship between gut microbiota and weight gain in humans. *Future Microbiology*, 7(1): 91–109.

Anwar, H., Irfan, S., Hussain, G., Faisal, M. N., Muzaffar, H., Mustafa, I., Mukhtar, I., Malik, S., and Ullah, M. I. 2019. Gut microbiome: a new organ system in body. *Eukaryotic Microbiology*. IntechOpen. doi: 10.5772/intechopen.89634.

Bäckhed, F., Ding, H., Wang, T., Hooper, L. V., Koh, G. Y., Nagy, A., Semenkovich, C. F., and Gordon, J. I. 2004. The gut microbiota as an environmental factor that regulates fat storage. *Proceedings of the National Academy of Sciences*, 101(44): 15718–15723.

Badehnoosh, B., Karamali, M., Zarrati, M., Jamilian, M., Bahmani, F., Tajabadi-Ebrahimi, M., Jafari, P., Rahmani, E., and Asemi, Z. 2018. The effects of probiotic supplementation on biomarkers of inflammation, oxidative stress and pregnancy outcomes in gestational diabetes. *The Journal of Maternal-Fetal & Neonatal Medicine*, 31(9): 1128–1136.

Ballini, A., Santacroce, L., Cantore, S., Bottalico, L., Dipalma, G., Topi, S., Saini, R., de Vito, D., and Inchingolo, F. 2019. Probiotics efficacy on oxidative stress values in inflammatory bowel disease: a randomized double-blinded placebo-controlled pilot study. *Endocrine, Metabolic & Immune Disorders-Drug Targets (Formerly Current Drug Targets-Immune, Endocrine & Metabolic Disorders)*, 19(3): 373–381.

Barros, C. P., Guimarães, J. T., Esmerino, E. A., Duarte, M. C. K., Silva, M. C., Silva, R., Ferreira, B. M., Sant'Ana, A. S., de Freitas, M. Q., and da Cruz, A. G. 2019. Paraprobiotics, postbiotics and psychobiotics: concepts and potential applications in dairy products. *Current Opinion in Food Science*, in press. doi: 10.1016/j.cofs.2019.12.003.

Bengmark, S. 2012. Integrative medicine and human health-the role of pre-, pro-and synbiotics. *Clinical and Translational Medicine*, 1(1): 1–13.

Bonfili, L., Cecarini, V., Gogoi, O., Berardi, S., Scarpona, S., Angeletti, M., Rossi, G., and Eleuteri, A. M. 2020. Gut microbiota manipulation through probiotics oral administration restores glucose homeostasis in a mouse model of Alzheimer's disease. *Neurobiology of Aging*, 87: 35–43.

Botvynko, A., Bednářová, A., Henke, S., Shakhno, N., and Čurda, L. 2019. Production of galactooligosaccharides using various combinations of the commercial β-galactosidases. *Biochemical and Biophysical Research Communications*, 517(4): 762–766.

Brasil. 2017. Ministério da Saúde. Protocolo Clínico e Diretrizes Terapêuticas da Doença de Crohn. https://www.saude.gov.br/images/pdf/2017/dezembro/08/420112-17-61-MINUTA-de-Portaria-Conjunta-PCDT-Doenca-de-Crohn-27-11-2017 (Accessed April 06, 2020).

Brinkworth, G. D., Noakes, M., Buckley, J. D., Keogh, J. B., and Clifton, P. M. 2009. Long-term effects of a very-low-carbohydrate weight loss diet compared with an isocaloric low-fat diet after 12 mo. *The American Journal of Clinical Nutrition*, 90(1): 23–32.

Cani, P. D., and Delzenne, N. M. 2009. The role of the gut microbiota in energy metabolism and metabolic disease. *Current Pharmaceutical Design*, 15(13): 1546–1558.

Cani, P. D., and Delzenne, N. M. 2011. The gut microbiome as therapeutic target. *Pharmacology & Therapeutics*, 130(2): 202–212.

Cani, P. D., Geurts, L., Matamoros, S., Plovier, H., and Duparc, T. 2014. Glucose metabolism: focus on gut microbiota, the endocannabinoid system and beyond. *Diabetes & Metabolism*, 40(4): 246–257.

Cavalcanti Neto, M. P., de Souza Aquino, J., da Silva, L. D. F. R., de Oliveira Silva, R., de Lima Guimaraes, K. S., de Oliveira, Y., de Souza, E. L., Magnani, M., Vidal, H., and de Brito Alves, J. L. 2018. Gut microbiota and probiotics intervention: a potential therapeutic target for management of cardiometabolic disorders and chronic kidney disease? *Pharmacological Research*, 130: 152–163.

Chae, C. S., Kwon, H. K., Hwang, J. S., Kim, J. E., and Im, S. H. 2012. Prophylactic effect of probiotics on the development of experimental autoimmune myasthenia gravis. *PLoS One*, 7(12): e52119.

Chakraborti, C. K. 2015. New-found link between microbiota and obesity. *World Journal of Gastrointestinal Pathophysiology*, 6(4): 110.

Cheng, L. H., Liu, Y. W., Wu, C. C., Wang, S., and Tsai, Y. C. 2019. Psychobiotics in mental health, neurodegenerative and neurodevelopmental disorders. *Journal of Food and Drug Analysis*, 27: 632–648.

Cheung, P. C. 2013. Mini-review on edible mushrooms as source of dietary fiber: preparation and health benefits. *Food Science and Human Wellness*, 2(3–4): 162–166.

Chuang, L., Wu, K. G., Pai, C., Hsieh, P. S., Tsai, J. J., Yen, J. H., and Lin, M. Y. 2007. Heat-killed cells of lactobacilli skew the immune response toward T helper 1 polarization in mouse splenocytes and dendritic cell-treated T cells. *Journal of Agricultural and Food Chemistry*, 55(26): 11080–11086.

Colantonio, A. G., Werner, S. L., and Brown, M. 2019. The effects of prebiotics and substances with prebiotic properties on metabolic and inflammatory biomarkers in individuals with type 2 diabetes mellitus: a systematic review. *Journal of the Academy of Nutrition and Dietetics*, 20(4): 587–607 e2.

Collado, M. C., Vinderola, G., and Salminen, S. 2019. Postbiotics: facts and open questions. A position paper on the need for a consensus definition. *Beneficial Microbes*, 10(7): 711–719.

Correria, S. S., and Percegoni, N. 2014. Microbiota intestinal e Ganho de Peso Corporal-Uma Revisão. *Universidade Federal de Juiz de Fora*. http://www. ufjf. br/gradnutricao/files/2015/03/MICROBIOTA-INTESTINAL-E-GANHODE-PESO-CORPORAL-UMA-REVIS% C3% 83O. pdf (Accessed April 19, 2020).

D'Angelo, C., Reale, M., and Costantini, E. 2017. Microbiota and probiotics in health and HIV infection. *Nutrients*, 9(6): 615.

Dawe, J. P., McCowan, L. M., Wilson, J., Okesene-Gafa, K. A., and Serlachius, A. S. 2020. Probiotics and maternal mental health: a randomised controlled trial among pregnant women with obesity. *Scientific Reports*, 10(1): 1–11.

Delgado, G. T. C., and Tamashiro, W. M. D. S. C. 2018. Role of prebiotics in regulation of microbiota and prevention of obesity. *Food Research International*, 113: 183–188.

Delzenne, N. M., Neyrinck, A. M., and Cani, P. D. 2011. Modulation of the gut microbiota by nutrients with prebiotic properties: consequences for host health in the context of obesity and metabolic syndrome. *Microbial Cell Factories*, 10(S1): S10.

Derwa, Y., Gracie, D. J., Hamlin, P. J., and Ford, A. C. 2017. Systematic review with meta-analysis: the efficacy of probiotics in inflammatory bowel disease. *Alimentary Pharmacology & Therapeutics*, 46(4): 389–400.

Diez-Gutiérrez, L., San Vicente, L., Barrón, L. J. R., del Carmen Villarán, M., and Chávarri, M. 2020. Gamma-aminobutyric acid and probiotics: multiple health benefits and their future in the global functional food and nutraceuticals market. *Journal of Functional Foods*, 64: 103669.

Dinan, T. G., Stanton, C., and Cryan, J. F. 2013. Psychobiotics: a novel class of psychotropic. *Biological Psychiatry*, 74(10): 720–726.

Duncan, S. H., Lobley, G. E., Holtrop, G., Ince, J., Johnstone, A. M., Louis, P., and Flint, H. J. 2008. Human colonic microbiota associated with diet, obesity and weight loss. *International Journal of Obesity*, 32(11): 1720–1724.

Eaimworawuthikul, S., Thiennimitr, P., Chattipakorn, N., and Chattipakorn, S. C. 2017. Diet-induced obesity, gut microbiota and bone, including alveolar bone loss. *Archives of Oral Biology*, 78: 65–81.

Eid, H. M., Wright, M. L., Anil Kumar, N. V., Qawasmeh, A., Hassan, S. T., Mocan, A., Nabavi, S. M., Rastrelli, L., Atanasov, A. G., and Haddad, P. S. 2017. Significance of microbiota in obesity and metabolic diseases and the modulatory potential by medicinal plant and food ingredients. *Frontiers in Pharmacology*, 8: 387.

Eswaran, S., Farida, J. P., Green, J., Miller, J. D., and Chey, W. D. 2017. Nutrition in the management of gastrointestinal diseases and disorders: the evidence for the low FODMAP diet. *Current Opinion in Pharmacology*, 37: 151–157.

Everard, A., and Cani, P. D. 2013. Diabetes, obesity and gut microbiota. *Best Practice & Research Clinical Gastroenterology*, 27(1): 73–83.

Falcinelli, S., Rodiles, A., Hatef, A., Picchietti, S., Cossignani, L., Merrifield, D. L., Unniappan, S., and Carnevali, O. 2017. Dietary lipid content reorganizes gut microbiota and probiotic *L.* rhamnosus attenuates obesity and enhances catabolic hormonal milieu in zebrafish. *Scientific Reports*, 7(1): 1–15.

Farsani, S. F., Brodovicz, K., Soleymanlou, N., Marquard, J., Wissinger, E., and Maiese, B. A. 2017. Incidence and prevalence of diabetic ketoacidosis (DKA) among adults with type 1 diabetes mellitus (T1D): a systematic literature review. *BMJ Open*, 7(7): e016587.

Fiorucci, S., Biagioli, M., Zampella, A., and Distrutti, E. 2018. Bile acids activated receptors regulate innate immunity. *Frontiers in Immunology*, 9: 1853.

Ford, A. L., Nagulesapillai, V., Piano, A., Auger, J., Girard, S. A., Christman, M., Tompkins, T. A., and Dahl, W. J. 2020. Microbiota stability and gastrointestinal tolerance in response to a high-protein diet with and without a prebiotic, probiotic, and synbiotic: a randomized, double-blind, placebo-controlled trial in older women. *Journal of the Academy of Nutrition and Dietetics*, 120(4): 500–516.

Forsyth, C. B., Farhadi, A., Jakate, S. M., Tang, Y., Shaikh, M., and Keshavarzian, A. 2009. *Lactobacillus* GG treatment ameliorates alcohol-induced intestinal oxidative stress, gut leakiness, and liver injury in a rat model of alcoholic steatohepatitis. *Alcohol*, 43(2): 163–172.

Freimund, S., Sauter, M., Käppeli, O., and Dutler, H. 2003. A new non-degrading isolation process for 1, 3-β-D-glucan of high purity from baker's yeast Saccharomyces cerevisiae. *Carbohydrate Polymers*, 54(2): 159–171.

Ganji-Arjenaki, M., and Rafieian-Kopaei, M. 2018. Probiotics are a good choice in remission of inflammatory bowel diseases: a meta analysis and systematic review. *Journal of Cellular Physiology*, 233(3): 2091–2103.

Gao, R., Zhang, X., Huang, L., Shen, R., and Qin, H. 2019. Gut microbiota alteration after long-term consumption of probiotics in the elderly. *Probiotics and Antimicrobial Proteins*, 11(2): 655–666.

Gibson, G. R. 2004. Fibre and effects on probiotics (the prebiotic concept). *Clinical Nutrition Supplements*, 1(2): 25–31.

Gibson, G. R., and Roberfroid, M. B. 1995. Dietary modulation of the human colonic microbiota: introducing the concept of prebiotics. *The Journal of Nutrition*, 125(6): 1401–1412.

Gong, Q., Zhang, P., Wang, J., Ma, J., An, Y., Chen, Y., Zhang, B., Feng, X., Li, H., Chen, X., Cheng, Y. J., Gregg, E. W., Hu, Y., Bennett, P. H., and Li, G. 2019. Morbidity and mortality after lifestyle intervention for people with impaired glucose tolerance: 30-year results of the Da Qing diabetes prevention outcome study. *The Lancet Diabetes & Endocrinology*, 7(6): 452–461.

Gonzalez, A., Krieg, R., Massey, H. D., Carl, D., Ghosh, S., Gehr, T. W., and Ghosh, S. S. 2019. Sodium butyrate ameliorates insulin resistance and renal failure in CKD rats by modulating intestinal permeability and mucin expression. *Nephrology Dialysis Transplantation*, 34(5): 783–794.

Guimarães, J. T., Balthazar, C. F., Scudino, H., Pimentel, T. C., Esmerino, E. A., Ashokkumar, M., Freitas, M. Q., and Cruz, A. G. 2019. High-intensity ultrasound: a novel technology for the development of probiotic and prebiotic dairy products. *Ultrasonics Sonochemistry*, 57: 12–21.

Gullón, B., Gómez, B., Martínez-Sabajanes, M., Yáñez, R., Parajó, J. C., and Alonso, J. L. 2013. Pectic oligosaccharides: manufacture and functional properties. *Trends in Food Science & Technology*, 30(2): 153–161.

Han, H., Li, Y., Fang, J., Liu, G., Yin, J., Li, T., and Yin, Y. 2018. Gut microbiota and type 1 diabetes. *International Journal of Molecular Sciences*, 19(4): 995.

Harakeh, S. M., Khan, I., Kumosani, T., Barbour, E., Almasaudi, S. B., Bahijri, S. M., Alfadul, S. M., Ajabnoor, G. M. A., and Azhar, E. I. 2016. Gut microbiota: a contributing factor to obesity. *Frontiers in Cellular and Infection Microbiology*, 6: 95.

van Hoffen, E., Korthagen, N. M., de Kivit, S., Schouten, B., Bardoel, B., Duivelshof, A., Knol, J., Garssen, J., and Willemsen, L. E. M. 2010. Exposure of intestinal epithelial cells to UV-killed *Lactobacillus* GG but not *Bifidobacterium* breve enhances the effector immune response in vitro. *International Archives of Allergy and Immunology*, 152(2): 159–168.

Iannitti, T., and Palmieri, B. 2010. Therapeutical use of probiotic formulations in clinical practice. *Clinical Nutrition*, 29(6): 701–725.

Jamilian, M., Bahmani, F., Vahedpoor, Z., Salmani, A., Tajabadi-Ebrahimi, M., Jafari, P., Dizaji, S. H., and Asemi, Z. 2016. Effects of probiotic supplementation on metabolic status in pregnant women: a randomized, double-blind, placebo-controlled trial. *Archives of Iranian Medicine*, 19(10): 0–0.

Joyce, S. A., and Gahan, C. G. 2016. Bile acid modifications at the microbe-host interface: potential for nutraceutical and pharmaceutical interventions in host health. *Annual Review of Food Science and Technology*, 7: 313–333.

Jumpertz, R., Le, D. S., Turnbaugh, P. J., Trinidad, C., Bogardus, C., Gordon, J. I., and Krakoff, J. 2011. Energy-balance studies reveal associations between gut microbes, caloric load, and nutrient absorption in humans. *The American Journal of Clinical Nutrition*, 94(1): 58–65.

Kamphuis, J. B., Mercier-Bonin, M., Eutamene, H., and Theodorou, V. 2017. Mucus organisation is shaped by colonic content; a new view. *Scientific Reports*, 7(1): 1–13.

Kang, L. J., Oh, E., Cho, C., Kwon, H., Lee, C. G., Jeon, J., Lee, H., Choi, S., Jae Han, S., Nam, J., Song, C-U., Jung, H., Kim, H. Y., Park, E-J., Choi, E-J., Kim, J., Eyun, S., and Song, C. U. 2020. 3′-Sialyllactose prebiotics prevents skin inflammation via regulatory T cell differentiation in atopic dermatitis mouse models. *Scientific Reports*, 10(1): 1–13.

Kareem, K. Y., Loh, T. C., Foo, H. L., Asmara, S. A., and Akit, H. 2017. Influence of postbiotic RG14 and inulin combination on cecal microbiota, organic acid concentration, and cytokine expression in broiler chickens. *Poultry Science*, 96(4): 966–975.

Kharroubi, A. T., and Darwish, H. M. 2015. Diabetes mellitus: the epidemic of the century. *World Journal of Diabetes*, 6(6): 850.

Kim, C. S., and Shin, D. M. 2019. Probiotic food consumption is associated with lower severity and prevalence of depression: a nationwide cross-sectional study. *Nutrition*, 63: 169–174.

Klancic, T., and Reimer, R. A. 2020. Gut microbiota and obesity: impact of antibiotics and prebiotics and potential for musculoskeletal health. *Journal of Sport and Health Science*, 9(2): 110–118.

Kobyliak, N., Conte, C., Cammarota, G., Haley, A. P., Styriak, I., Gaspar, L., Fusek, J., Rodrigo, L, and Kruzliak, P. 2016. Probiotics in prevention and treatment of obesity: a critical view. *Nutrition & Metabolism*, 13(1): 14.

Kondo, T., Saigo, S., Ugawa, S., Kato, M., Yoshikawa, Y., Miyoshi, N., and Tanabe, K. 2020. Prebiotic effect of fructo-oligosaccharides on the inner ear of DBA/2J mice with early-onset progressive hearing loss. *The Journal of Nutritional Biochemistry*, 75: 108247.

Kotzampassi, K., Giamarellos-Bourboulis, E. J., and Stavrou, G. 2014. Obesity as a consequence of gut bacteria and diet interactions. *ISRN Obesity*, 2014: ID651895.

Lee, B. J., and Bak, Y. T. 2011. Irritable bowel syndrome, gut microbiota and probiotics. *Journal of Neurogastroenterology and Motility*, 17(3): 252–266.

Liu, Y., and Lou, X. 2020. Type 2 diabetes mellitus-related environmental factors and the gut microbiota: emerging evidence and challenges. *Clinics*, 75: 1–7.

Logan, A. C., and Katzman, M. 2005. Major depressive disorder: probiotics may be an adjuvant therapy. *Medical Hypotheses*, 64(3): 533–538.

López-Moreno, A., and Aguilera, M. 2020. Probiotics dietary supplementation for modulating endocrine and fertility microbiota dysbiosis. *Nutrients*, 12(3): 757.

Low Wang, C. C., Hess, C. N., Hiatt, W. R., and Goldfine, A. B. 2016. Clinical update: cardiovascular disease in diabetes mellitus: atherosclerotic cardiovascular disease and heart failure in type 2 diabetes mellitus–mechanisms, management, and clinical considerations. *Circulation*, 133(24): 2459–2502.

Luoto, R., Kalliomäki, M., Laitinen, K., and Isolauri, E. 2010. The impact of perinatal probiotic intervention on the development of overweight and obesity: follow-up study from birth to 10 years. *International Journal of Obesity*, 34(10): 1531–1537.

Mafra, D., Lobo, J. C., Barros, A. F., Koppe, L., Vaziri, N. D., and Fouque, D. 2014. Role of altered intestinal microbiota in systemic inflammation and cardiovascular disease in chronic kidney disease. *Future Microbiology*, 9(3): 399–410.

Marotta, F., Barreto, R., Wu, C. C., Naito, Y., Gelosa, F., Lorenzetti, A., Yoshioka, M., and Fesce, E. 2005. Experimental acute alcohol pancreatitis-related liver damage and endotoxemia: synbiotics but not metronidazole have a protective effect. *Chinese Journal of Digestive Diseases*, 6(4): 193–197.

Mikov, M., Đanić, M., Pavlović, N., Stanimirov, B., Goločorbin-Kon, S., Stankov, K., and Al-Salami, H. 2018. Potential applications of gliclazide in treating type 1 diabetes mellitus: formulation with bile acids and probiotics. *European Journal of Drug Metabolism and Pharmacokinetics*, 43(3): 269–280.

Miyoshi, M., Shiroto, A., Kadoguchi, H., Usami, M., and Hori, Y. 2020. Prebiotics improved the defecation status via changes in the microbiota and short-chain fatty acids in hemodialysis patients. *Kobe Journal of Medicine Science*, 66(1): E12–E21.

Mohanty, D., Misra, S., Mohapatra, S., and Sahu, P. S. 2018. Prebiotics and synbiotics: recent concepts in nutrition. *Food Bioscience*, 26: 152–160.

Moradi, M., Mardani, K., and Tajik, H. 2019. Characterization and application of postbiotics of *Lactobacillus* spp. on *Listeria monocytogenes* in vitro and in food models. *LWT*, 111: 457–464.

Moraes, A. C. F. D., Silva, I. T. D., Almeida-Pititto, B. D., and Ferreira, S. R. G. 2014. Microbiota intestinal e risco cardiometabólico: mecanismos e modulação dietética. *Arquivos Brasileiros de Endocrinologia & Metabologia*, 58(4): 317–327.

Moreira, A. B., Teixeira, T. F. S., and Alfenas, R. D. C. G. 2012. Gut microbiota and the development of obesity. *Nutrición Hospitalaria*, 27(5): 1408–1414.]

Mutlu, E., Keshavarzian, A., Engen, P., Forsyth, C. B., Sikaroodi, M., and Gillevet, P. 2009. Intestinal dysbiosis: a possible mechanism of alcohol-induced endotoxemia and alcoholic steatohepatitis in rats. *Alcoholism: Clinical and Experimental Research*, 33(10): 1836–1846.

Nadal, A., Alonso-Magdalena, P., Soriano, S., Ropero, A. B., and Quesada, I. 2009. The role of oestrogens in the adaptation of islets to insulin resistance. *The Journal of Physiology*, 587(21): 5031–5037.

Nishida, A., Inoue, R., Inatomi, O., Bamba, S., Naito, Y., and Andoh, A. 2018. Gut microbiota in the pathogenesis of inflammatory bowel disease. *Clinical Journal of Gastroenterology*, 11(1): 1–10.

Nishida, K., Sawada, D., Kuwano, Y., Tanaka, H., Sugawara, T., Aoki, Y., Fujiwara, S., and Rokutan, K. 2017. Daily administration of paraprobiotic *Lactobacillus gasseri* CP2305 ameliorates chronic stress-associated symptoms in Japanese medical students. *Journal of Functional Foods*, 36: 112–121.

Nogacka, A., Salazar, N., Endo, A., Suárez, A., Martinez-Faedo, C., González de los Reyes-Gavilán, C., and Gueimonde Fernández, M. 2019. A fecal-culture model, monitoring gas production, for assessing prebiotics' fermentability in normal-weight and obese subjects. http://hdl.handle.net/10261/202056 (Accessed April 29, 2020).

O'Hagan, C., Li, J. V., Marchesi, J. R., Plummer, S., Garaiova, I., and Good, M. A. 2017. Long-term multi-species *Lactobacillus* and *Bifidobacterium* dietary supplement enhances memory and changes regional brain metabolites in middle-aged rats. *Neurobiology of Learning and Memory*, 144: 36–47.

Oliveira, R. P., Florence, A. C., Silva, R. C., Perego, P., Converti, A., Gioielli, L. A., and Oliveira, M. N. 2009. Effect of different prebiotics on the fermentation kinetics, probiotic survival and fatty acids profiles in nonfat symbiotic fermented milk. *International Journal of Food Microbiology*, 128(3): 467–472.

de Oliveira, R. L., da Silva, M. F., da Silva, S. P., de Araújo, A. C. V., Cavalcanti, J. V. F. L., Converti, A., and Porto, T. S. 2020. Fructo-oligosaccharides production by an Aspergillus aculeatus commercial enzyme preparation with fructosyltransferase activity covalently immobilized on Fe_3O_4–chitosan-magnetic nanoparticles. *International Journal of Biological Macromolecules*, 150: 922–929.

Ooi, S. L., Correa, D., and Pak, S. C. 2019. Probiotics, prebiotics, and low FODMAP diet for irritable bowel syndrome–what is the current evidence? *Complementary Therapies in Medicine*, 43: 73–80.

Ou, C. C., Lin, S. L., Tsai, J. J., and Lin, M. Y. 2011. Heat-killed lactic acid bacteria enhance immunomodulatory potential by skewing the immune response toward Th1 polarization. *Journal of Food Science*, 76(5): M260–M267.

Paesani, C., Degano, A. L., Salvucci, E., Zalosnik, M. I., Fabi, J. P., Sciarini, L., and Perez, G. T. 2020. Soluble arabinoxylans extracted from soft and hard wheat show a differential prebiotic effect in vitro and in vivo. *Journal of Cereal Science*, 93: 102956.

Passos, M. D. C. F., and Moraes-Filho, J. P. 2017. Intestinal microbiota in digestive diseases. *Arquivos de Gastroenterologia*, 54(3): 255–262.

Perpétuo, J. P., de Albuquerque Wilasco, M. I., and Schneider, A. C. R. 2015. The role of intestinal microbiota in energetic metabolism: new perspectives in combating obesity. *Clinical & Biomedical Research*, 35(4): 196–199.

Praveen, M. A., Parvathy, K. K., Balasubramanian, P., and Jayabalan, R. 2019. An overview of extraction and purification techniques of seaweed dietary fibers for immunomodulation on gut microbiota. *Trends in Food Science & Technology*, 92: 46–64.

Puccetti, M., Giovagnoli, S., Zelante, T., Romani, L., and Ricci, M. 2018. Development of novel indole-3-aldehyde–loaded gastro-resistant spray-dried microparticles for postbiotic small intestine local delivery. *Journal of Pharmaceutical Sciences*, 107(9): 2341–2353.

Pyclik, M., Srutkova, D., Schwarzer, M., and Górska, S. 2020. Bifidobacteria cell wall-derived exo-polysaccharides, lipoteichoic acids, peptidoglycans, polar lipids and proteins–their chemical structure and biological attributes. *International Journal of Biological Macromolecules*, 147: 333–349.

Qin, J., Li, Y., Cai, Z., Li, S., Zhu, J., Zhang, F., Liang, S., Zhang, W., Guan, Y., Shen, D., Peng, Y., Zhang, D., Jie, Z., Wu, W., Qin, Y., Xue, W., Li, J., Han, L., Lu, D., Wu, P., Dai, Y., Sun, X., Lu, d., Wu, P., Dai, Y., Sun, X., Li, Z., Tang, A., Zhong, S., Li, X., Chen, W., Xu, R., Wang, M., Feng, Q., Gong, M., Yu, J., Zhang, Y., Zhang, M., Hansen, T., Sanchez, G., Raes, J., Falony, G., Okuda, S., Almeida, M., LeChatelier, E., Renault, P., Pons, N., Batto, J-M., Zhang, Z., Chen, H., Yang, R., Zheng, W., Li, S., Yang, H., Wang, J., Ehrilich, S. D., Nielsen, R., Pedersen, O., Kristiansen, K., and Wang, J. 2012. A metagenome-wide association study of gut microbiota in type 2 diabetes. *Nature*, 490(7418): 55–60.

Ridaura, V. K., Faith, J. J., Rey, F. E., Cheng, J., Duncan, A. E., Kau, A. L., Griffin, N. W., Lombard, V., Henrissat, B., Bain, J. R., Muehlbauer, M. J., Ilkayeva, O., Semenkovich, C. F., Funai, K., Hayashi, D. K., Lyle, B. J., Martini, M. C., Ursell, L. K., Clemente, J. C., van Treuren, W., Walters, W. A., Knight, R., Newgard, C. B., Heath, A. C., and Gordon, J. I. 2013. Gut microbiota from twins discordant for obesity modulate metabolism in mice. *Science*, 341(6150): 1241214.

Rinaldi, E., Consonni, A., Guidesi, E., Elli, M., Mantegazza, R., and Baggi, F. 2018. Gut microbiota and probiotics: novel immune system modulators in myasthenia gravis? *Annals of the New York Academy of Sciences*, 1413(1): 49–58.

Rogers, G. B., Keating, D. J., Young, R. L., Wong, M. L., Licinio, J., and Wesselingh, S. 2016. From gut dysbiosis to altered brain function and mental illness: mechanisms and pathways. *Molecular Psychiatry*, 21(6): 738–748.

Russo, E., Giudici, F., Fiorindi, C., Ficari, F., Scaringi, S., and Amedei, A. 2019. Immunomodulating activity and therapeutic effects of short chain fatty acids and tryptophan post-biotics in inflammatory bowel disease. *Frontiers in Immunology*, 10: 2754.

Saad, M. J. A., Santos, A., and Prada, P. O. 2016. Linking gut microbiota and inflammation to obesity and insulin resistance. *Physiology*, 31(4): 283–293.

Saadat, Y. R., Khosroushahi, A. Y., Movassaghpour, A. A., Talebi, M., and Gargari, B. P. 2020. Modulatory role of exopolysaccharides of Kluyveromyces marxianus and Pichia kudriavzevii as probiotic yeasts from dairy products in human colon cancer cells. *Journal of Functional Foods*, 64: 103675.

Samanta, A. K., Jayapal, N., Jayaram, C., Roy, S., Kolte, A. P., Senani, S., and Sridhar, M. 2015. Xylooligosaccharides as prebiotics from agricultural by-products: production and applications. *Bioactive Carbohydrates and Dietary Fibre*, 5(1): 62–71.

dos Santos, K. E. R., and Ricci, G. C. L. 2016. Microbiota intestinal e a obesidade. *Revista Uningá Review*, 26(1): 74–82.

Sarkar, A., Lehto, S. M., Harty, S., Dinan, T. G., Cryan, J. F., and Burnet, P. W. 2016. Psychobiotics and the manipulation of bacteria–gut–brain signals. *Trends in Neurosciences*, 39(11): 763–781.

Sawada, D., Sugawara, T., Ishida, Y., Aihara, K., Aoki, Y., Takehara, I., Takano, K., and Fujiwara, S. 2016. Effect of continuous ingestion of a beverage prepared with *Lactobacillus gasseri* CP2305 inactivated by heat treatment on the regulation of intestinal function. *Food Research International*, 79: 33–39.

Schmidt, L., Soder, T. F., Deon, R. G., and Benetti, F. 2017. Obesidade e sua relação com a microbiota intestinal. *Revista Interdisciplinar de Estudos em Saúde*, 6(2): 29–43.

Segura-Badilla, O., Lazcano-Hernández, M., Kammar-García, A., Vera-López, O., Aguilar-Alonso, P., Ramírez-Calixto, J., and Navarro-Cruz, A. R. 2020. Use of coconut water (Cocus nucifera L) for the development of a symbiotic functional drink. *Heliyon*, 6(3): e03653.

Shadnoush, M., Hosseini, R. S., Khalilnezhad, A., Navai, L., Goudarzi, H., and Vaezjalali, M. 2015. Effects of probiotics on gut microbiota in patients with inflammatory bowel disease: a double-blind, placebo-controlled clinical trial. *The Korean Journal of Gastroenterology*, 65(4): 215–221.

Sikalidis, A. K., and Maykish, A. 2020. The gut microbiome and type 2 diabetes mellitus: discussing a complex relationship. *Biomedicines*, 8(1): 8.

Silva, M., Chibbar, R., Wine, E., Walter, J., Goodman, K., Keshteli, A. H., Valcheva, R. S., and Dieleman, L. A. 2018. A110 use of probiotics, prebiotics and dietary fibre supplements in patients with inflammatory bowel disease. *Journal of the Canadian Association of Gastroenterology*, 1(suppl_2): 167–168.

Silva, M. L. T., Dias, M. C. G., Vasconcelos, M. I. L., Sapucahy, M. V., Catalani, L. A., Miguel, B. Z. B., and Buzzini, R. 2011. Terapia Nutricional na Doença de Chron. Sociedade Brasileira de Nutrição Parenteral e Enteral Associação Brasileira de Nutrologia. https://diretrizes.amb.org.br/_BibliotecaAntiga/terapia_nutricional_na_doenca_de_crohn.pdf (Accessed March 28, 2020).

da Silva, S. T., dos Santos, C. A., and Bressan, J. 2013. Intestinal microbiota; relevance to obesity and modulation by prebiotics and probiotics. *Nutricion Hospitalaria*, 28(4): 1039–1048.

Silva, A. F. D., Schieferdecker, M. E. M., Rocco, C. S., and Amarante, H. M. B. D. S. 2010. Relação entre estado nutricional e atividade inflamatória em pacientes com doença inflamatória intestinal. *ABCD. Arquivos Brasileiros de Cirurgia Digestiva (São Paulo)*, 23(3): 154–158.

Sircana, A., Framarin, L., Leone, N., Berrutti, M., Castellino, F., Parente, R., De Michieli, F., Pachetta, E., and Musso, G. 2018. Altered gut microbiota in type 2 diabetes: just a coincidence? *Current Diabetes Reports*, 18(10): 98.

Slykerman, R. F., Hood, F., Wickens, K., Thompson, J. M. D., Barthow, C., Murphy, R., Kang, J., Rowden, J., Stanislawski, M. A., Dabelea, D., Wagner, B. D., Iszatt, N., Dahl, C., Sontag, M. K., Lozupone, C. A., and Eggesbø, M. 2018. Gut microbiota in the first 2 years of life and the association with body mass index at age 12 in a Norwegian birth cohort. *MBio*, 9(5): e01751–18.

Song, E. J., Han, K., Lim, T. J., Lim, S., Chung, M. J., Nam, M. H., Kim, H., and Nam, Y. D. 2020. Effect of probiotics on obesity-related markers per enterotype: a double-blind, placebo-controlled, randomized clinical trial. *EPMA Journal*, 11(1): 31–51.

Stanislawski, M. A., Dabelea, D., Wagner, B. D., Iszatt, N., Dahl, C., Sontag, M. K., Knight, R., Lozupone, C.A., and Eggesbø, M. 2019. Reply to Moossavi and Azad, "Quantifying and Interpreting the Association between Early-Life Gut Microbiota Composition and Childhood Obesity". *Mbio*, 10(1).

Tanca, A., Palomba, A., Fraumene, C., Manghina, V., Silverman, M., and Uzzau, S. 2018. Clostridial butyrate biosynthesis enzymes are significantly depleted in the gut microbiota of nonobese diabetic mice. *mSphere*, 3(5): e00492–18.

Taverniti, V., and Guglielmetti, S. 2011. The immunomodulatory properties of probiotic microorganisms beyond their viability (ghost probiotics: proposal of paraprobiotic concept). *Genes & Nutrition*, 6(3): 261.

Thakur, A. K., Shakya, A., Husain, G. M., Emerald, M., and Kumar, V. 2014. Gut-microbiota and mental health: current and future perspectives. *Journal of Pharmacol Clinical Toxicology*, 2(1): 1016.

Thakur, A. K., Tyagi, S., and Shekhar, N. 2019. Comorbid brain disorders associated with diabetes: therapeutic potentials of prebiotics, probiotics and herbal drugs. *Translational Medicine Communications*, 4(1): 12.

Thushara, R. M., Gangadaran, S., Solati, Z., and Moghadasian, M. H. 2016. Cardiovascular benefits of probiotics: a review of experimental and clinical studies. *Food & Function*, 7(2): 632–642.

Tian, P., Wang, G., Zhao, J., Zhang, H., and Chen, W. 2019. *Bifidobacterium* with the role of 5-hydroxytryptophan synthesis regulation alleviates the symptom of depression and related microbiota dysbiosis. *The Journal of Nutritional Biochemistry*, 66: 43–51.

Tiangui, W. A. N. G., Chuang, L. I., Rongrong, F. A. N., and Mengmeng, S. O. N. G. 2018. Xylo-oligosaccharide preparation through enzyme hydrolysis of hemicelluloses isolated from press lye. *Grain & Oil Science and Technology*, 1(4): 171–176.

Tilg, H., Zmora, N., Adolph, T. E., and Elinav, E. 2020. The intestinal microbiota fuelling metabolic inflammation. *Nature Reviews Immunology*, 20: 40–54.

Tonucci, L. B., dos Santos, K. M. O., de Luces Fortes Ferreira, C. L., Ribeiro, S. M. R., De Oliveira, L. L., and Martino, H. S. D. 2017. Gut microbiota and probiotics: focus on diabetes mellitus. *Critical Reviews in Food Science and Nutrition*, 57(11): 2296–2309.

Tran, N., Zhebrak, M., Yacoub, C., Pelletier, J., and Hawley, D. 2019. The gut-brain relationship: investigating the effect of multispecies probiotics on anxiety in a randomized placebo-controlled trial of healthy young adults. *Journal of Affective Disorders*, 252: 271–277.

Turnbaugh, P. J., Hamady, M., Yatsunenko, T., Cantarel, B. L., Duncan, A., Ley, R. E., Sogin, M. L., Jones, W. J., Roe, B. A., Afourtit, J. P., Egholm, M., Henrissat, B., Heath, A. C., Knight, R., and Gordon, J. I. 2009. A core gut microbiome in obese and lean twins. *Nature*, 457(7228): 480–484.

Villena, J., Barbieri, N., Salva, S., Herrera, M., and Alvarez, S. 2009. Enhanced immune response to pneumococcal infection in malnourished mice nasally treated with heat-killed *Lactobacillus casei*. *Microbiology and Immunology*, 53(11): 636–646.

Vrieze, A., Holleman, F., Zoetendal, E. G., De Vos, W. M., Hoekstra, J. B. L., and Nieuwdorp, M. 2010. The environment within: how gut microbiota may influence metabolism and body composition. *Diabetologia*, 53(4): 606–613.

Wagner, N. R. F., Zaparolli, M. R., Cruz, M. R. R., Shcieferdecker, M. E. M., and Campos, A. C. L. 2018. Postoperative changes in intestinal microbiota and use of probiotics in Roux-en-y Gastric bypass and sleeve vertical gastrectomy: an integrative review. *Arquivos Brasileiros de Cirurgia Digestiva (São Paulo)*, 31(4): 1–5. http://www.scielo.br/scielo.php?script=sci_arttext&pid=S0102-67202018000400500&lng=pt

Wang, X., Gibson, G. R., Sailer, M., Theis, S., and Rastall, R. A. 2020. Prebiotics inhibit proteolysis by gut bacteria in a host diet-dependent manner: a three-stage continuous in vitro gut model experiment. *Applied and Environmental Microbiology*, 86(10): e02730–19.

Wickens, K. L., Barthow, C. A., Murphy, R., Abels, P. R., Maude, R. M., Stone, P. R., Mitchell, E. A., Stanley, T. V., Purdie, G. L., Kang, J. M., Hood, F. E., Rowden, J. L., Barnes, P. K., Fitzharris, P. F., and Crane, J. 2017. Early pregnancy probiotic supplementation with *Lactobacillus rhamnosus* HN001 may reduce the prevalence of gestational diabetes mellitus: a randomised controlled trial. *British Journal of Nutrition*, 117(6): 804–813.

Xavier-Santos, D., Bedani, R., Lima, E. D., and Saad, S. M. I. 2020. Impact of probiotics and prebiotics targeting metabolic syndrome. *Journal of Functional Foods*, 64: 103666.

Xie, R., Jiang, P., Lin, L., Jiang, J., Yu, B., Rao, J., Liu, H., Wei, W., and Qiao, Y. 2020. Oral treatment with Lactobacillus reuteri attenuates depressive-like behaviors and serotonin metabolism alterations induced by chronic social defeat stress. *Journal of Psychiatric Research*, 122: 70–78.

Yan, Y. L., Hu, Y., and Gänzle, M. G. 2018. Prebiotics, FODMAPs and dietary fiber-conflicting concepts in development of functional food products? *Current Opinion in Food Science*, 20: 30–37.

Yang, H., Sun, Y., Cai, R., Chen, Y., and Gu, B. 2019. The impact of dietary fiber and probiotics in infectious diseases. *Microbial Pathogenesis*, 140: 103931.

Yang, X., Yu, D., Xue, L., Li, H., and Du, J. 2020. Probiotics modulate the microbiota–gut–brain axis and improve memory deficits in aged SAMP8 mice. *Acta Pharmaceutica Sinica B*, 10(3): 475–487.

Yunes, R. A., Poluektova, E. U., Dyachkova, M. S., Klimina, K. M., Kovtun, A. S., Averina, O. V., Orlova, V. S., and Danilenko, V. N. 2016. GABA production and structure of gadB/gadC genes in *Lactobacillus* and *Bifidobacterium* strains from human microbiota. *Anaerobe*, 42: 197–204.

Zhang, J., Chen, X., Song, J. L., Qian, Y., Yi, R., Mu, J., Zhao, X., and Yang, Z. 2019. Preventive effects of *Lactobacillus plantarum* CQPC07 on colitis induced by dextran sodium sulfate in mice. *Food Science and Technology Research*, 25(3): 413–423.

Zhao, F., Liu, Q., Cao, J., Xu, Y., Pei, Z., Fan, H., Yuan, Y., Shen, X., and Li, C. 2020. A sea cucumber (Holothuria leucospilota) polysaccharide improves the gut microbiome to alleviate the symptoms of type 2 diabetes mellitus in Goto-Kakizaki rats. *Food and Chemical Toxicology*, 135: 110886.

Zheng, Y., Ley, S. H., and Hu, F. B. 2018. Global aetiology and epidemiology of type 2 diabetes mellitus and its complications. *Nature Reviews Endocrinology*, 14(2): 88.

Zhou, A., Tang, L., Zeng, S., Lei, Y., Yang, S., and Tang, B. 2020. Gut microbiota: a new piece in understanding hepatocarcinogenesis. *Cancer Letters*, 474: 15–22.

Zucko, J., Starcevic, A., Diminic, J., Oros, D., Mortazavian, A. M., and Putnik, P. 2020. Probiotic–friend or foe? *Current Opinion in Food Science*, 32: 45–49.

11

Skin Microbiome, Its Impact on Dermatological Diseases, and Intervention of Probiotics

Mitesh Dwivedi, Firdosh Shah, and Prashant S. Giri
Uka Tarsadia University

CONTENTS

11.1 Introduction

The skin is a largest ecosystem of the body that consists of specific niches of a diverse range of microorganisms including bacteria, fungi, and viruses. Many of the skin habitants are harmless microorganisms (Grice & Segre, 2011). The surface areas of skin, oral mucosa, and the gastrointestinal (GI) tract are the major residing hub for the gut microbiota which has been reported to play an important role in either maintaining health or, in some cases, contributing to different diseases (The Human Microbiome Project Consortium, 2012). The human skin microbes colonize the stratum corneum of the epidermis and skin appendages such as sweat glands and hair follicles.

The skin's symbiotic microorganisms protect the skin against invasion by pathogenic microbes. However, the disruption of the homeostasis between harmless and pathogenic microorganisms can result in skin diseases or infections. Nevertheless, the exact mechanism of maintaining this balance is yet to be known. The composition of skin microbiota varies from individual to individual but remains stable over time (The Human Microbiome Project Consortium, 2012; Costello et al., 2009). It has been suggested that the competition within and between microbial species is important for the development and maintenance of a healthy microbiome (Schommer & Gallo, 2013).

In addition, altered host–microbe relationship can be harmful for the human health which can be contributed by host susceptibility factors and genetic variation which further contributes to a specific microbial composition. The other characteristics such as physical and chemical natures of the skin also result in a unique microbial community. Therefore, complete characterization of the skin microbiota and understanding of how it interacts with the host is necessary for maintaining skin health and preventing many skin diseases and infections. The gut microbiome has been investigated extensively (Doré & Blottière, 2015); however, investigations of the skin microbiome has started recently (Oh et al., 2016; Flores et al., 2014; Perez et al., 2016; Troccaz et al., 2015). In recent years, the development of next-generation sequencing (NGS) technologies has allowed for the study of these communities with an unprecedented depth and resolution.

This chapter provides an update on the skin microbiome, its impact on different skin diseases, and the mechanisms of biological functions involved of the skin microbiome and gut microbiome, since there is a link between the gut microbiota and the skin health. Therefore, the chapter also discusses the role of skin–gut axis in the pathogenesis as well as improvement of several skin-related diseases. In addition to the current clinical understanding of host–microbe interaction on the skin, the chapter also discusses therapeutic roles of probiotics and prebiotics in different disease conditions such as dermatitis, skin infections, psoriasis, and vitiligo.

11.2 The Skin Microbiome

There is the highest degree of variation from one individual to another in terms of human skin microbiome. Even within the study subjects, one could find differences while comparing symmetrical sites, i.e. left and right antecubital creases and site-specific differences (The Human Microbiome Project Consortium, 2012; Costello et al., 2009; Grice et al., 2009). The human skin contains several niches wherein various microbial populations inhabit. Few of the microbial flora are the part of environment which are temporarily present on the skin, whereas there are also other sections of microbes that maintain homeostasis within the host, which are considered as normal or commensal organisms of the skin (Chiller et al., 2001). The ecological factors such as humidity, temperature, and the composition of antimicrobial peptides and lipids might also be responsible for influencing these microorganisms. Moreover, there are also some parts of the skin such as hair follicles, sebaceous, eccrine, and apocrine glands that also have an enormous amount of specific microbiota (Grice et al., 2009). For instance, the moist areas such as inner elbow, armpit, and the portion behind the knee consist of *Staphylococcus* spp. and *Corynebacterium* spp. in abundance, whereas the areas with large sebaceous secretion such as the back, forehead, and behind the ear has a larger population of *Propionibacterium* spp. Similarly, the areas such as the forearm and other parts of the hands have dry skin possessing the highest diversity of gut microbiota but very few proportion of numbers (Costello et al., 2009; Grice et al., 2009). The innate immunity functions are also accountable for individual specific levels of the host microbiota present on normal or wounded skin. A study was conducted in order to analyze the dynamics of recolonization of skin microbiota through tape stripping–mediated skin barrier disruption (Zeeuwen et al., 2012). This study showed that the microbes are not distributed uniformly throughout the stratum corneum. The stratum granulosum contains very less density of bacteria, whereas the surface layers of the skin consist of the highest bacterial density. These reports also suggest that microbiome can extend their spread to the dermal compartments as well (Nakatsuji et al., 2013).

The human skin microbiota is crucial for improving and maintaining optimal skin immunity (Naik et al., 2012). It modulates cutaneous immune system (Lai et al., 2009; Wanke et al., 2011; Li et al., 2013) and also controls colonization of potent pathogenic microorganisms (Iwase et al., 2010; Shu et al., 2013). Overall, these findings suggest that skin microbiota plays beneficial roles in maintaining our health (Blaser, 2011). Reports on culture studies revealed that cutaneous microbiota has diversified organisms (Grice & Segre, 2011). The below section describes the bacterial, fungal, and other microbiome composition of the skin.

11.2.1 Bacterial Skin Microbiome

A report suggests detection of *Proteobacteria* (16.5%), *Firmicutes* (24.4%), *Actinobacteria* (51.8%), and *Bacteroidetes* (6.3%) among the 19 phyla of the skin microbiome through 16S rRNA gene phylotyping. The major genera identified were *Corynebacterium*, *Propionibacterium*, and *Staphylococcus* (Grice et al., 2009). Sebaceous glands support the growth of facultative anaerobes such as *Propionibacterium acnes* due to its anoxic nature (Leeming et al., 1984). The *P. acnes* genome sequencing reported multiple genes for lipases that degrade sebum lipids, and the bacteria colonize on the sebum gland due to the free fatty acids released (Brüggemann et al., 2004; Gribbon et al., 1993). Free fatty acids are also responsible for the acidic nature of skin surface (pH 5.0) (Roth & James, 1988). This acidic environment prevents the growth of *Staphylococcus aureus* and *Streptococcus pyogenes*; however, few bacteria such as *Staphylococci* and *Corynebacteria* are still able to withstand this acidic environment (Korting et al., 1990). Further, due to skin occlusion, the elevated pH supports the growth of *S. aureus* and *S. pyogenes* (Aly et al., 1978). The species of *Corynebacterium* and *Staphylococcus* mostly dominate the moist areas of the body. However, in dry sites of the body, mixed populations of *β-Proteobacteria* and *Flavobacteriales* are found (Grice et al., 2009). A 16S rRNA metagenomic study on the skin showed less microbial

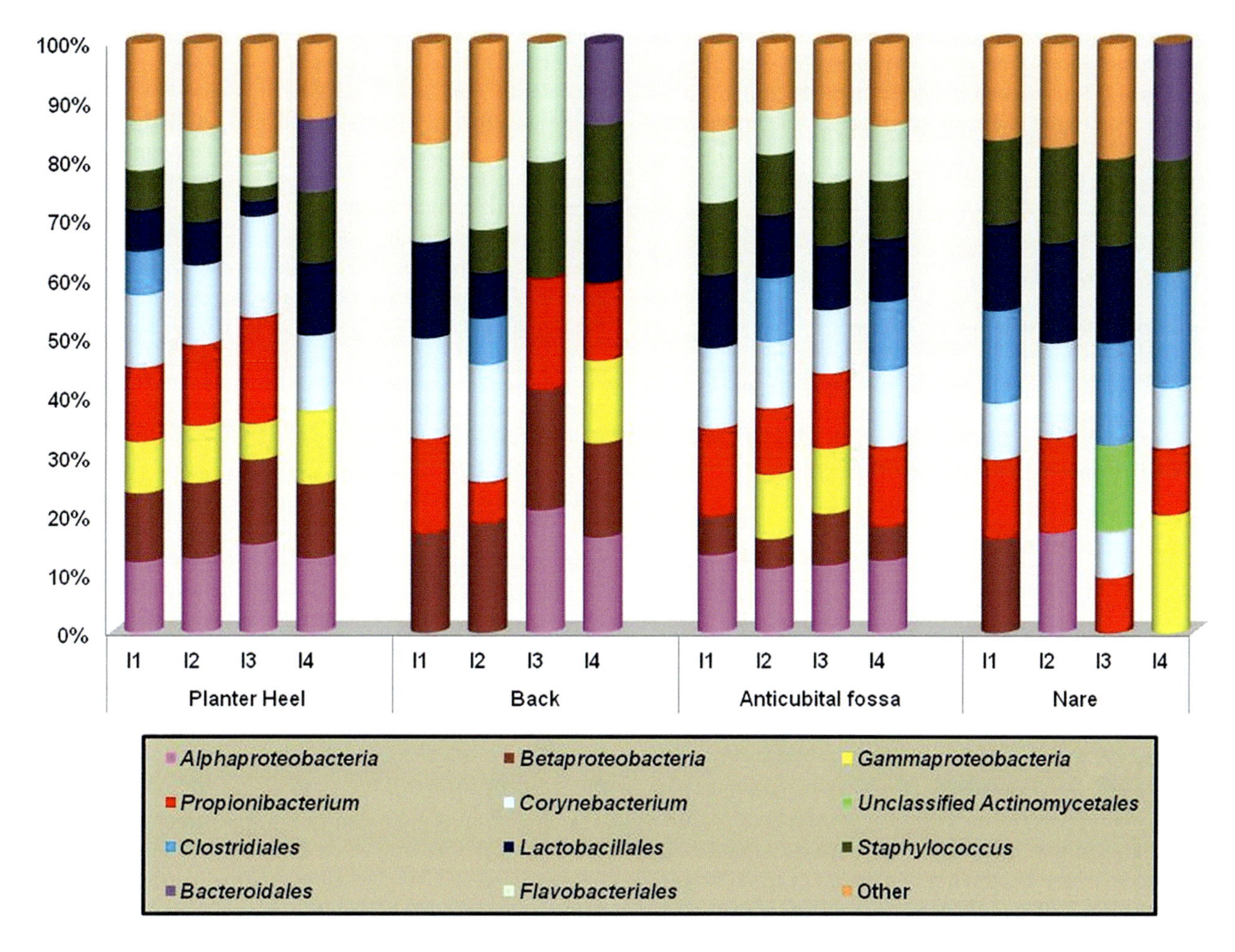

FIGURE 11.1 The skin microbiome's interpersonal variation. The distribution of microbiota at four sites on healthy individuals (I1, I2, I3, and I4) is shown at the plantar heel, back, antecubital fossa, and nare. The variation in skin microbiota is more dependent on the site than on the individual. The relative abundance of bacterial taxa was determined by 16S ribosomal RNA sequencing and represented by bars (Grice et al., 2009).

interpersonal variation between the symmetric skin sites (Costello et al., 2009; Grice et al., 2009). The individual variation in bacterial skin microbiome is shown in Figure 11.1 (Grice et al., 2009). The skin surface regions of plantar heel, back, nare, and antecubital fossa are dissimilar to any other site on the same individual, whereas it is more similar to the same site on another individual (Figure 11.1). In this context, it has been reported that the individual genetic differences among healthy volunteers could not be considered to play a greater role as an influencing factor for the determination of microbial composition within the skin than the ecological body site (Grice et al., 2009; Grice & Serge, 2011).

11.2.2 Fungal Skin Microbiome

The other important microbiome of the human skin is fungal microbiome. The certain species of *Malassezia* fungi such as *Malassezia restricta*, *Malassezia globosa*, and *Malassezia sympodialis* were considered reportedly dominating after analyzing through 18S rRNA phylotyping (Gioti et al., 2013; Findley et al., 2013; Paulino et al., 2008). These *Malassezia* species are frequently found in sebum-rich areas of the skin (Xu et al., 2007). *M. restricta* is prevalent on the surface area of the skin like scalp, external auditory canal, retroauricular crease, and glabella, whereas *M. globosa* generally exists on the occiput, back, and in the inguinal crease (Clavaud et al., 2013; Findley et al., 2013). *Aspergillus*, *Rhodotorula*, *Cryptococcus*, and *Epicoccum* colonize the skin areas of the foot region (Findley et al., 2013).

11.2.3 Other Microbiomes

In addition to bacterial and fungal microbiome, the human skin is also a habitat for viruses and arthropods. The eukaryotes that colonize the human skin belong to the phylum Arthropoda. The lipid component of the sebum helps in the survival of *Demodex* mites (Lacey et al., 2011). Generally, *Demodex folliculorum* is found in hair follicles in clusters with other mites of the same species. At the rim of the eyelid, there are sebaceous or meibomial glands in which smaller mites *Demodex brevis* solely reside (Lacey et al., 2009).

There is less data available regarding the human skin virome due to difficulties in establishment of virus cell culture, limited antigenic/serological cross-reactivity, and the lack of nucleic acid hybridization to known viral sequences. The metagenomic sequencing of total DNA is usually carried out for detecting virus genome. With the help of advanced high-throughput metagenomic sequencing, a wide range of diversified DNA viruses were determined on the human skin surface (Foulongne et al., 2012). Moreover, the human papillomavirus have also been suggested as normal part of the skin microbiome (Foulongne et al., 2012; Antonsson et al., 2003; Delwart, 2007).

11.2.4 Factors Affecting Skin Microbiome

Apart from the distinctive body characteristics of an individual such as salinity, moisture, pH, and sebum content, the intrinsic factors (such as age, sex, and genotype) and extrinsic factors (such as lifestyle, cosmetics, geographical location, and occupation) are also partly responsible in giving specific skin microbial composition of an individual (Fierer et al., 2010).

11.2.4.1 Host-Related Factors

The variability of skin microbiota may be due to host-related factors such as age, location, genotype, and sex. The microenvironment of the skin can be affected majorly by age and thus may affect the colonizing microbiota (Leyden et al., 1975). The colonization generally occurs only after birth, either during vaginal delivery or by caesarian section (Dominguez-Bello et al., 2010). Apart from this, the levels of lipophilic bacteria on the skin were also observed to be affected during the puberty because of the changes caused within the sebum production (Somerville, 1969). The gender-based differences in microbiota have also been documented due to physiological and anatomical differences between male and female cutaneous environments, for example, production of sweat, sebum, and hormone, surface pH, skin thickness, and hair growth (Giacomoni et al., 2009; Marples, 1982). A study conducted on human epidermis (after the skin barrier disruption) significantly showed higher microbial diversity within hands of females than males. The reason for this outcome was associated with the use of make-up and lower pH of the female skin (Giacomoni et al., 2009).

Several studies reported the role of skin physiology in deciding the colonization of bacteria, since specific bacteria are associated with a particular skin microenvironment like moist, dry, and sebaceous area of the skin (Costello et al., 2009; Grice et al., 2009). The selection of organisms is on the basis of their tolerating capacity under different conditions of these areas. Therefore, the sebaceous sites are reported to harbor bacteria with the lowest diversity.

11.2.4.2 Environmental Factors

The variability of skin microbiota may include the individual's environmental factors such as occupation, clothing choice, lifestyle, and cosmetics. The antibiotic treatment generally affects the gut microbiota (Dethlefsen & Relman, 2011); however, no study has reported antibiotic effects of skin microbiota. The skin microbiota can vary due to use of different kinds of cosmetics, soaps, hygienic products, and moisturizers. Although these products are considered to bring alteration within the condition of skin barrier, their effects on skin microbiota remain veiled. The comparison of bacterial proportion on the axillary vaults, feet, and back during high-temperature and low-humidity conditions showed increase in the quantity of bacterial population on these surfaces as compared to the high-temperature and high-humidity conditions (McBride et al., 1977). A large amount of Gram-negative bacteria were found on the surfaces of back and feet during the low-temperature and high-humidity conditions (McBride et al., 1977). Moreover, the longitudinal and/or latitudinal variation in UV exposure can also lead to geographical variability in the skin (Faergemann & Larkö, 1987).

An individual's lifestyle also affects the skin microbiota composition. According to a comparative study on cutaneous bacterial communities, a significant difference within the forearm skin bacterial communities of Amerindian samples was observed as compared to the US volunteers (Blaser et al., 2013). There were two clusters detected among Amerindian skin microbiota upon performing 16S rRNA gene pyrosequencing which was completely lacking in the samples of US volunteers. The alpha diversity showed relative equivalence in species richness within the US volunteers, whereas in the case of Amerindian samples, there was no recognizable dominant bacterial taxon found due to cluster B which possesses larger species richness (Blaser et al., 2013).

11.2.5 Skin Microbiome and Host Interaction

The interaction of host with skin microbiome is an essential relationship that leads to either a healthy skin or a disease condition of a skin. This interactive relationship of microorganisms inside or on a host can be under three major categories. The impact of this relationship can be negative, positive, or no impact at all on one of the species involved (Faust & Raes, 2012). One of such interactions is called 'commensalism', in which only one species benefits from the relationship and the other one is unaffected. The other interaction where both the partners are benefitted by each other is termed as mutualism. On the other hand, in the case of parasitism, only one of the organisms is benefitted and the other one is harmed. Figure 11.2 describes a crosstalk between skin microbiome and host.

The skin microbiota play an indispensable role in maintaining homeostasis of cutaneous immunity and maturation (Figure 11.2). It has been reported that gut flora regulates skin immunity independently in an autonomous manner (Naik et al., 2012). The interaction of bacteria with skin results in the development of innate immune system in epithelium through toll-like receptors (TLRs) which are involved in detection of microorganisms. The stimulation of TLRs leads to various kinds of gene expression leading to activation of a variety of immune responses including pro-inflammatory response that provides defense against infectious microorganisms. For example, the expressions of various innate factors such as antimicrobial peptides (AMPs), IL-α, and the components of complement proteins are modulated by the skin flora such as *Cutibacterium thiopeptides* and *Staphylococcus epidermidis* (Naik et al., 2012; Chehoud et al., 2013) (Figure 11.2). Despite the fact that *S. epidermidis* could be considered as an opportunistic pathogen in several immunodeprived conditions, it functions mostly as a mutualist. *S. epidermidis* that produces

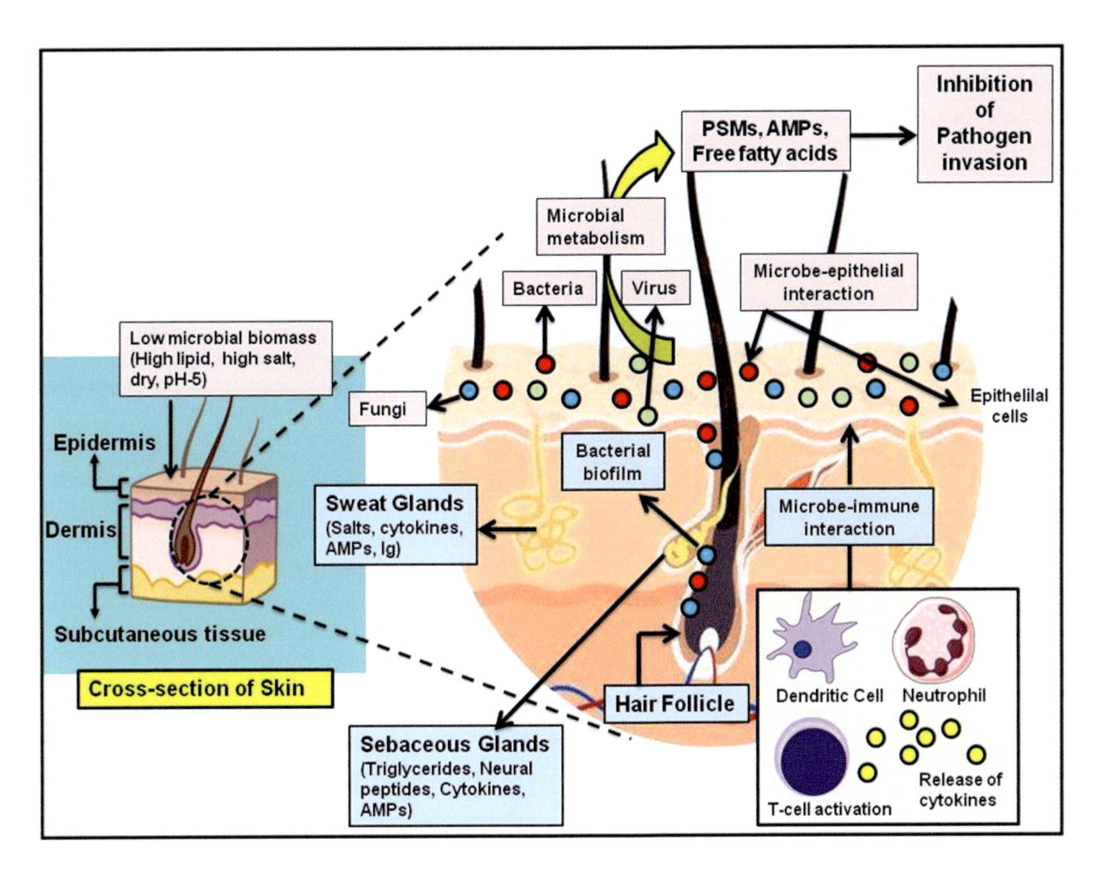

FIGURE 11.2 Skin microbiome and host interaction. The cross section of skin depicts that it is divided into three major parts, i.e. epidermis, dermis, and subcutaneous tissues. Skin covers structures such as hair follicles, sweat, and sebaceous glands which are responsible for forming biofilms at some sites. The skin consists of low microbial biomass due to its acidic nature (pH 5) and dry environment which is also rich in lipid and salt concentrations. The skin serves as a resident surface for various microbes such as viruses, fungi, and bacteria. The microbes are considered to produce free fatty acids, phenol-soluble modulins (PSMs), and antimicrobial peptides (AMPs) which are released upon metabolizing host bioactive compounds, lipids, and proteins. These microbial products further inhibit the pathogen invasion through the skin. The host microbial–immune interaction stimulates keratinocyte-derived immune mediators like complement proteins and pro-inflammatory cytokines (IL-1). It also modulates the expression of different innate factors such as neutrophils, dendritic, and T-helper cells which influence immune cell activity of microbes present within the skin.

AMPs which prevent cutaneous infection caused by *S. aureus* through inducing activation of IL-17$^+$CD8$^+$ T cells within the keratinocytes (Naik et al., 2015). Furthermore, the adaptive responses to the members of the skin community develop in the absence of inflammation (Naik et al., 2015), and this process is termed as 'homeostatic immunity', which can be induced by the endogenous network of skin-resident antigen-presenting cells (Naik et al., 2015). In addition, the role of commensals in epithelial integrity during tissue repair has also been reported. The binding of cell wall components of *S. epidermidis* to TLR2 initiates inflammation which leads to the promotion of wound healing and further limits damages caused within tissue (Lai et al., 2009).

During the state of inflammation, there is a remarkable change within the composition of skin microbiome (Kong et al., 2012). However, the exact mechanism through which pathogens interact with commensal skin microbial populations remains unexplored. The microorganisms belonging to *Clostridium* clusters IV and XI, *Faecalibacterium prausnitzii*, and *Bacteroides fragilis* induce anti-inflammatory responses via accumulation of regulatory T cells (Tregs) (Forbes et al., 2016). Few microbes such as segmented filamentous bacteria promote accumulation of pro-inflammatory Th1 and Th17 cells.

One of the most dominating genera of skin-resident bacteria is *Corynebacterium*. It is generally found at all the body sites, especially on moist surfaces. Among the various species of *Corynebacterium*, only two of them cause skin diseases superficially, namely, *Corynebacterium tenuis* (trichomycosis axillaris) and *Corynebacterium minutissimum* (erythrasma). The *Corynebacterium* cell wall contains lipomannan and lipoarabinomannan ligands that bind to the host TLRs and C-type lectin receptors resulting in pro- or anti-inflammatory responses (Fukuda et al., 2013). One more skin-resident bacteria, namely *Corynebacterium accolens*, is considered to inhibit the growth of *Streptococcus pneumonia* by secreting lipase that hydrolyses triolein to release oleic acid (Bomar et al., 2016). This study suggests that microbe–microbe interactions also have impacts on human health in addition to the host–skin microbe interaction. However, it still remains unexplored whether immune responses against the skin microbiota also influence microbiota composition or function. According to a study, *S. epidermidis* colonizes the skin of mouse during the early life and induces tolerance against the same microbe once it attains adulthood; however, accumulation of *S. epidermidis*-specific regulatory T cells in neonatal skin was also reported

(Scharschmidt et al., 2015, 2017). The skin homeostasis is the process that involves secretion of complex lipids which helps in signaling and solves the purpose of skin barrier (Feingold, 2009), and also maintains the tight junctions (Brandner, 2016; Natsuga, 2014) and produces lipid protein that prevents transepidermal water loss (Madison, 2003). The disruption caused within any of these processes results in the phenotypes such as blistering, progerias, ichthyoses, and diffuse fibrosis (McLean, 2016; Has & Bruckner-Tuderman, 2014; Capell et al., 2009) (Figure 11.3). SCFAs such as butyrate, acetate, and propionate play an indispensable role in understanding the predominance of microbial profiles of the skin which ultimately influences cutaneous immune defense mechanism of the skin. Propionic acid, one of the SCFAs produced by *Propionibacterium*, has shown to exhibit a strong antimicrobial effect against most commonly acquired methicillin-resistant *S. aureus* (USA300) (Shu et al., 2013; Samuelson et al., 2015; Schwarz et al., 2017). Among several cutaneous microorganisms, *S. epidermidis* and *P. acnes* are known to possess higher tolerance of SCFA shifts than other commensals. Overall, all these studies provide shreds of evidences supporting functional interactive mechanism between the skin and its microbiota for maintaining skin homeostasis (Figure 11.3). The effects of skin microbiome on development of different skin disease conditions are discussed in the below sections.

11.3 Skin Microbiome of Different Dermatological Conditions

There are various factors such as diet, humidity, pH, temperature, and medications which fluctuate at daily basis and influence the skin microbiome (Spadoni et al., 2015; Foster et al., 2017; Sonnenburg & Backhed, 2016). Skin microbiota is responsible for maintaining homeostasis by influencing the inflammation and immune cell pathways (Lai et al., 2009; Naik et al., 2015). There are various species of bacteria that predominate on the various regions of the skin such as *S. aureus*, *Propionibacterium*, and *Corynebacterium* (Hillion et al., 2013).

Certain pathogenic bacteria such as *S. aureus* and *Pseudomonas aeruginosa* may initiate and propagate an opportunistic infection which further leads to a skin disease (Malic et al., 2011; Gan et al., 2002). Therefore, it is not surprising that such bacterial imbalances may result in various skin diseases such as psoriasis, acne, atopic dermatitis (AD), seborrheic dermatitis (SD), vitiligo, and Rosacea. The below section describes the skin microbiome involved in such dermatological conditions.

11.3.1 Psoriasis

Psoriasis is a chronic inflammatory skin disease that causes erythematous or extracutaneous lesions on the external surface

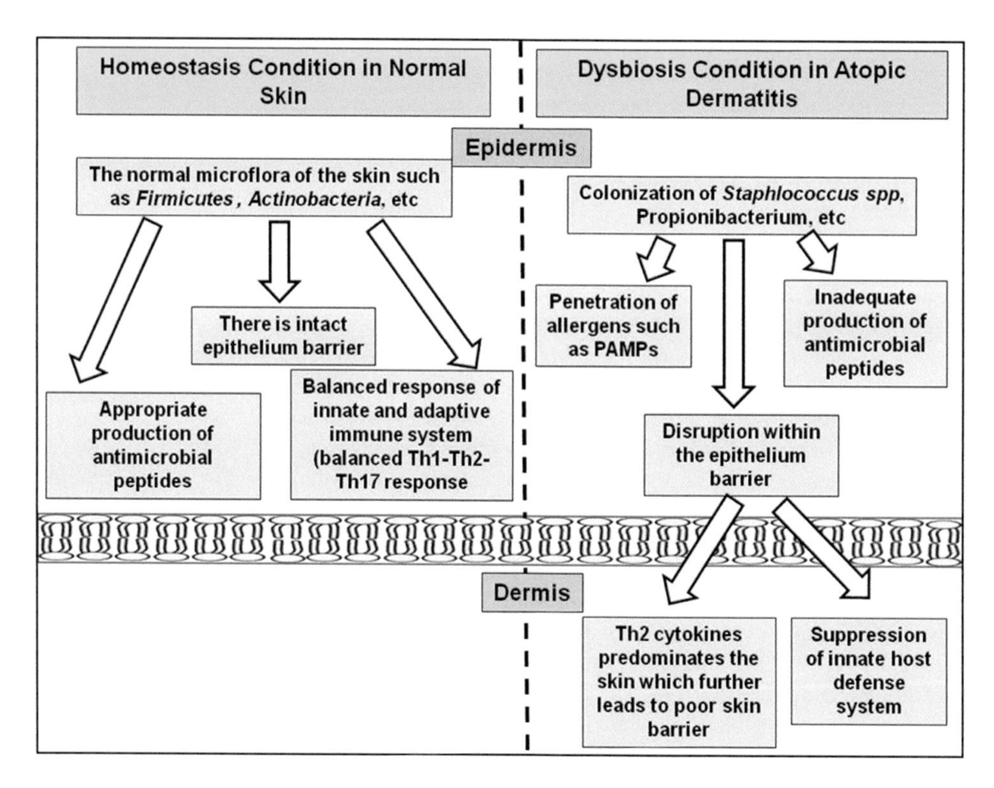

FIGURE 11.3 Role of microorganisms in homeostasis and dysbiosis within the skin. In the normal healthy skin, the gut microbiota composition consists of bacteria such as *Firmicutes* and *Actinobacteria*. Generally, normal skin consists of intact epithelial barrier and the production of antimicrobial peptides along with balanced innate and adaptive immune responses. However, the colonization of *Staphylococcus* spp., *Propionibacterium*, etc. results in the bacterial dysbiosis in the dermatological disease. In the case of atopic dermatitis, the colonization of gut microbiota results in the production of antimicrobial peptides. The penetration of allergens and pathogen-associated molecular patterns results in the impairment of gut epithelium barrier. This disruption within the gut epithelium results in the production of Th2 cytokines and also suppresses the innate host defense mechanism.

of the skin in the age group between 15 and 25. It is estimated to cause disease in 2%–3% of the world population (Pariser et al., 2007). Although the exact mechanism behind the cause of psoriasis is unclear, it is said to be multifactorial, which involves impairment in immune system, environmental, and genetic factors (Xu & Zhang, 2017).

Several studies have been attempted to investigate and characterize the bacterial composition of microbiota in skin lesions in order to understand their role in causing dysbiosis in psoriasis. According to a study by Gao et al. (2008), *Firmicutes* were significantly higher in psoriatic lesions compared to healthy individuals' skin, whereas *Actinobacteria* were found in significantly reduced numbers. According to another study, there was decrease in the population of *Firmicutes* and *Staphylococcus* in the psoriatic skin as compared to non-affected skin (Assarsson et al., 2018). Similarly, other studies also identified many such microorganisms such as *Staphylococcus*, *Corynebacterium*, *Propionibacterium*, and *Streptococcus* as major bacterial species on the lesional and non-lesional regions of the psoriatic patients, among which *Firmicutes* and *Actinobacteria* were present in abundance (Yan et al., 2017). One study showed gradual decrease in the bacterial population diversity within lesional and non-lesional skins of the psoriatic patient with respect to healthy individuals (Alekseyenko et al., 2013).

Although it is rarely seen that *S. aureus* can cause any skin infection, few reports showed the colonization of *S. aureus* more abundantly in the skin of psoriasis patients compared to healthy controls (Ng et al., 2017; Chang et al., 2018). A study by Chang et al. (2018) reported Th17 polarization in the mice colonized with *S. aureus* compared to the control. This result shows that there is release of pro-inflammatory cytokines in the psoriasis patients due to upregulation of Th17 through *S. aureus*, indicating that psoriasis might have role in promoting dysregulation within the composition of cutaneous microbiota. Furthermore, studies have also tried to find the effect of treatment of psoriasis on the skin microbiota. According to a study by Assarsson et al. (2018), there is a significant decrease in *Firmicutes*, *Anaerococcus*, *Gardnerella*, *Prevotella*, *Staphylococcus*, *Pseudomonas*, *Clostridium*, and *Finegoldia* after treating with narrowband Ultraviolet -B (nbUV-B).

Psoriasis involves activation of various immune pathways which leads to elevation of pro-inflammatory cytokines. Even the gut microbiota is responsible for the induction of pro-inflammatory Th17 cells, which results in the development of obesity and inflammatory bowel disease (Tan et al., 2018). According to Tan et al. (2018), patients with psoriasis had a decreased number of *Akkermansia muciniphila*, which are responsible for preventing inflammatory diseases such as obesity, atherosclerosis, and inflammatory bowel disease and maintaining the gut epithelium integrity (Li et al., 2016; Reunanen et al., 2015). In addition, a decrease in the microbial diversity was observed within the gut of patients with skin-limited psoriasis and psoriatic arthritis (Scher et al., 2015). There was increase in *Firmicutes* to *Bacteroidetes* proportion in psoriatic patients in one group, whereas the other group showed decrease in *Actinobacteria* compared to healthy controls due to increase in inflammation (Scher et al., 2015; Benhadou et al., 2018). However, a study by Codoner et al. (2018) reported contrary findings from the above-mentioned studies. The study reported an elevation in microbial variation among *Faecalibacterium*, *Akkermansia*, and *Ruminococcus* and a decline in *Bacteroides* within the psoriatic patients.

A study by Zakostelska et al. (2016) reported an increase in lactic acid-producing bacteria *Lactobacilalles* with anti-inflammatory effects in healthy gut, whereas the germ-free mice treated with antibiotics showed a reduction of Th17 cells. Although the treatments used presently for psoriasis do not involve gut microbiota, the studies suggest that gut microbiota can be considered as a potential treatment in the upcoming years. Therefore, these above-mentioned studies propose the development of gut microbiota–based treatment for psoriasis possibly by targeting Th17 cells and pro-inflammatory pathways.

11.3.2 Seborrheic Dermatitis

Seborrheic dermatitis (SD) is the common cause for rashes in the skin, especially in the regions with higher density of sebaceous glands such as scalp and face (An et al., 2017). SD is often seen to be caused through specific triggers such as depression, emotional stress, or weather change (Berg, 1989). SD is observed at three different ages, i.e. infant, teenage, and adults over 50 years, and the possible reason behind this could be the changes in the production of sebum due to fluctuations in hormonal function. Although SD is commonly observed in healthy individuals, it is also seen in the patients of human immunodeficiency virus (HIV) and Parkinson's disease (Gary, 2013).

Moreover, SD is also associated with ubiquitous fungi which are normally a part of human skin microbiome, namely *Malassezia*. However, the exact function of *Malassezia* in the development or cause of SD is yet to be known (Paulino, 2017). According to one study, it was reported that lesions of SD-affected skin are dominated by *Acinetobacter*, *Staphylococcus*, and *Streptococcus* (Tanaka et al., 2016). Another study by An et al. (2017) on patients with SD showed an increased colonization of *S. epidermidis* compared with the controls. In order to understand the association of skin microbiota with SD, more clinical studies are needed for demonstrating microbial compositions before and after the treatment.

11.3.3 Atopic Dermatitis

Atopic dermatitis (AD) is a chronic allergic dermatological disease which is also called as eczema. The common symptoms of AD are pruritic rash and erythematous (Bieber, 2008). This disease is mostly prevalent in infants with 20% and 3% in adults across the globe, and it is also associated with other diseases such as allergic rhinitis and asthma (Nutten, 2015; Kapoor et al., 2008). Similar to the case of psoriasis, the cause of AD is also multifactorial due to contribution of both epigenetic and genetic factors (Bieber, 2008). Interestingly, AD is seen in the countries with higher industrialization, suggesting that excessive hygiene could be one of the reasons inhibiting beneficial immune responses against pathogens (Braback et al., 2004).

Several studies investigated the role of the skin microbiota and its effect on the pathogenesis of AD. The genetic defects cause both physical and Th2-mediated immunological disruptions in the skin barrier of AD patients, which further increases the patients' susceptibility to allergens and infection (Ong et al., 2002). *S. aureus* are often seen to colonize the AD skin with higher numbers in both lesional and non-lesional atopic skins (Breuer et al., 2002; Gong et al., 2006; Masenga et al., 1990; Goh et al., 1997). The higher density of *S. aureus* is also related to increased disease severity and inflammation (Goh et al., 1997; Ogawa et al., 1994; Tauber et al., 2016). According to a report, there is overall decrease in the diversity of the microorganisms in the lesional AD skin (Kong et al., 2012). In another study, infants of 2 months old showed reduction in the risk of developing eczema among the infants colonized with *S. epidermidis* and *Staphylococcus cohnii* (Kennedy et al., 2017).

Although AD is said to be a dermatological disease, there are evidential studies depicting the role of gut microbiota in disease pathogenesis. Many researches are based on finding relationship between the age group and the demographic distribution of the disease (Bieber, 2008). There are several reports that suggest lower levels of microbial diversity among adults with AD who had showed colonic microbiota during their infant stage (Forno et al., 2008; Abrahamsson et al., 2012; Ismail et al., 2012; Wang et al., 2008).

Studies have shown the prevalence of several types of gut microbiota in patients ailing with or without eczema. Studies reported higher density of *S. aureus* in fecal samples of AD patients (Bjorksten et al., 1999; Watanabe et al., 2003). Apart from *S. aureus*, *Clostridium difficile* were also found in more density in children and adults suffering from AD (Bjorksten et al., 1999; Watanabe et al., 2003; Mah et al., 2006). According to reports, higher risk of AD is associated with *C. difficile* to the age of seven, and at age two, there is colonization of *Escherichia coli* (Penders et al., 2007; van Nimwegen et al., 2011). *Enterobacteriaceae* in AD are observed in higher prevalence in both adults and infants (Matsumoto et al., 2004; Yap et al., 2014). Along with *Enterobacteriaceae*, a significant increase in *F. prausnitzii* was found in AD patients (Song et al., 2016; Zheng et al., 2016); however, *Bifidobacteria* were found to be less colonized in AD population (Yap et al., 2014; Bjorksten et al., 1999). In order to clearly elucidate and understand the interaction of skin microbiota with AD, there is a need of large studies with long-term follow-up.

11.3.4 Acne

In the past decades, there are many studies that are reported on acne microbiome (Marples, 1974; Kishishita et al., 1979; Puhvel, 1968; Takizawa, 1977; Whiteside & Voss, 1973; Dagnelie et al., 2018). *Cutibacterium acnes* are considered as a causative agent for acne vulgaris, though there are very few reports supporting this fact. There are three sub-groups of *C. acnes* (I, II, and III) which have been reported to contribute to acne (Dagnelie et al., 2018; Fitz-Gibbon et al., 2013; Higaki et al., 2000; Lomholt & Kilian, 2010; McDowell et al., 2005). Studies have reported that *C. acnes* bacteriophages are more prevalent with higher abundance in subjects (Tomida et al., 2013; Barnard et al., 2016).

Although the role of *S. epidermis* in pathology has not been explored, there are several studies that mark its abundance on acne (Bek-Thomsen et al., 2008; Dreno et al., 2017; Nishijima et al., 2000). *Malassezia* species are also responsible for causing folliculitis in the acne pathology (Akaza et al., 2016; Cheikhrouhou et al., 2017; Lévy et al., 2007; Omran & Mansori, 2018; Prohic et al., 2016; Song et al., 2014).

11.3.5 Rosacea

Rosacea is a chronic inflammatory condition in which facial skin of an individual is majorly affected. The symptoms of rosacea are facial erythema, telangiectasia, and/or inflammatory papules and pustules (Wilkin et al., 2002). The exact mechanism behind the clinical manifestation of rosacea is so far not understood. Researchers consider multifactorial reasons behind rosacea such as abnormal neurovascular activation and release of inflammatory molecules (Two et al., 2015).

Scientific interest has been increased in finding the difference between cutaneous microbiome and rosacea. A mite called *D. folliculorum* resides within sebaceous glands of healthy skin. There are several studies on *D. folliculorum* which implicate its role in the pathogenesis of rosacea (Jarmuda et al., 2012; Casas et al., 2012; Chang & Huang, 2017). The higher *Demodex*-containing skin biopsies showed abundance in inflammatory cell populations around hair follicles, which also encodes higher expression of genes responsible for inflammatory peptides, and the mite exoskeleton is postulated to stimulate synthesis of pathogenic mediators (Casas et al., 2012; Roihu & Kariniemi, 1998; Koller et al., 2011).

In addition, a decrease in *Demodex* was observed upon treating rosacea with topical anti-*Demodex* cream, suggesting the role of bacterial pathogens in contribution of rosacea disease (Kocak et al., 2002). Researchers address *Demodex* as a suspect carrier of *Bacillus oleronius*, which are susceptible towards various antibiotics such as doxycycline used in the treatment of rosacea condition (Lacey et al., 2007; Szkaradkiewicz et al., 2012; O'Reilly et al., 2012). A healthy skin commensal, *S. epidermidis*, was also isolated from pustules of rosacea patients. Unlike non-hemolytic *S. epidermidis* found on healthy individuals, this *S. epidermidis* depicted β-hemolytic variant with increased virulence (Whitfeld et al., 2011; Dahl et al., 2004).

One of the most studied bacteria with respect to association of rosacea is *Helicobacter pylori*, which mostly resides within the stomach (Holmes, 2013). It is somewhat difficult to assess *H. pylori* contribution in rosacea, since the IgG seropositivity is largely related to the rosacea (Lazaridou et al., 2010; Bonamigo et al., 2000). Though several studies showed that eradication of *H. pylori* decreases the severity of rosacea (Utaş et al., 1999; Szlachcic, 2002; De Miquel et al., 2006), still the exact pathway between *H. pylori* and rosacea is yet to be explored. Reports suggest that there might be pro-inflammatory virulence peptides which may be responsible for symptoms of rosacea (Argenziano et al., 2003; Szlachcic et al., 1999). But the relationship of *H. pylori* with rosacea remains a debatable topic whether dysbiosis causes rosacea or dysbiosis occurs due to rosacea (Sharma et al., 1998; Son et al., 1999; Jorgensen et al., 2017). Therefore, it is advisable to conduct more studies on the relationship between the microbiome and rosacea in order to come up on any conclusion.

11.3.6 Vitiligo

Vitiligo is an immune-mediated skin disease that causes skin depigmentation (Dwivedi et al., 2015a). The disease occurs due to destruction of melanocytes by T cells, pro-inflammatory cytokines, or autoantibodies (Laddha et al., 2013). The role of microorganisms in vitiligo cannot be denied as they are involved in induction of inflammatory cytokines. However, the study of skin microbiome in vitiligo is lacking. A recent study was conducted by Ganju et al. (2016) in which the cutaneous microbiota of the vitiligo patients was analyzed. According to this report, there was a decline observed in the microbial variation in lesional sites compared with non-lesional sites. *Firmicutes* and *Proteobacteria* were found in abundance in the lesional sites, whereas *Actinobacteria* are found in higher density at non-lesional skin (Ganju et al., 2016). Many such studies with skin microbiome in vitiligo are required to confirm the microbiota link to the vitiligo pathogenesis.

11.4 Relationship between Skin and Gut Microbiome

The gut microbiota can be either transient or resident on the skin. The disruption caused within the gut flora may contribute to the pathogenesis of autoimmune and inflammatory diseases in skin despite the fact that skin is an organ that is distant from the gut (Kosiewicz et al., 2011). There are several shreds of evidences that report intestinal dysbiosis in some common inflammatory skin diseases such as atopic dermatitis (AD) (Williams & Flohr, 2006; Rather et al., 2016), psoriasis (Hidalgo-Cantabrana et al., 2019; Tan et al., 2018), acne vulgaris (Bowe & Logan, 2011), and rosacea (Nam et al., 2018). These studies gave birth to a notion that is termed as gut–skin axis (Lee et al., 2018; O'Neill et al., 2016). In 1930, two dermatologists proposed a concept regarding an inter-relationship between gut microbiota and skin inflammation, which eventually was stated as a model for explaining gut–brain–skin axis models (Arck et al., 2010; Shanahan, 1999; Stokes & Pillsbury, 1930). Despite the several studies predicting the association of gut microbiota with human health and diseases, the exact mechanism behind the cause of such conditions remains veiled.

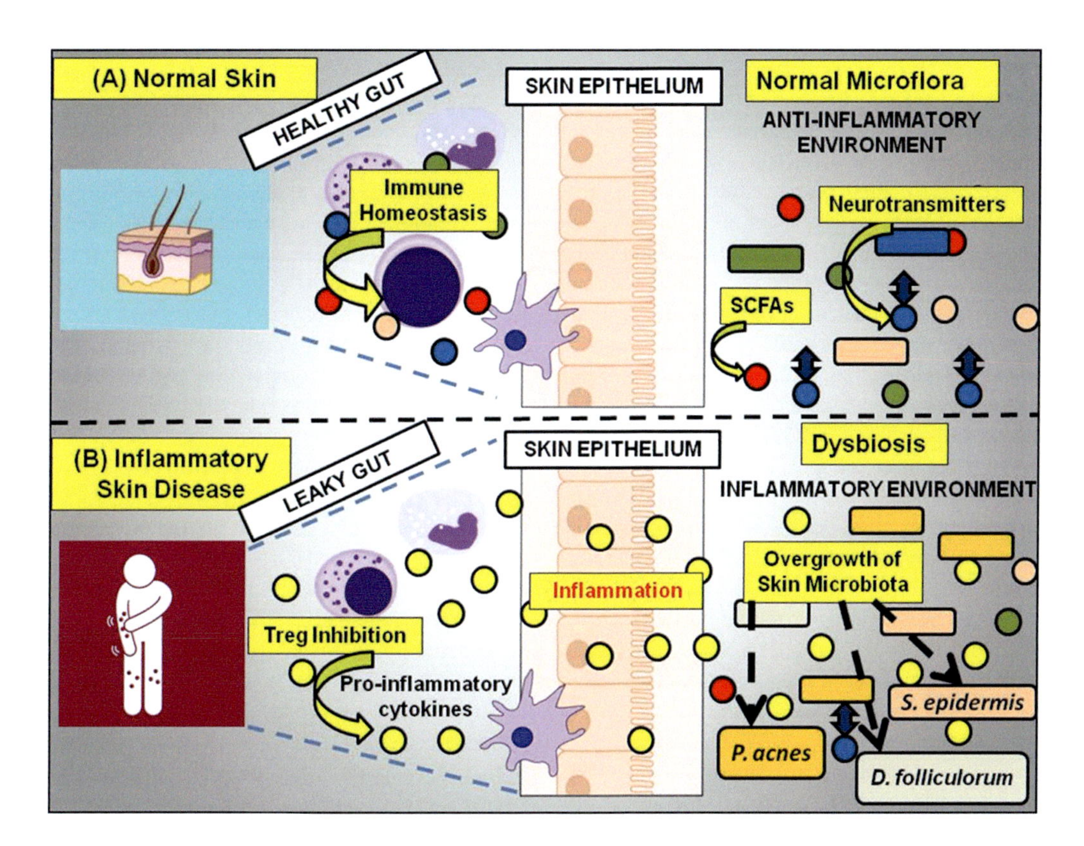

FIGURE 11.4 The relationship between skin and gut microbiota. (a) In a normal (healthy) skin, the immune cell contributes to maintaining the homeostasis condition within the system. The neurotransmitters cross the gut epithelium, enter the bloodstream, and induce systemic effects. Intestinal microorganisms release neurotransmitters in response to stress, which can modulate skin functions through production of short-chain fatty acids (SCFAs) resulting in the beneficial effect on skin functions. (b) In the inflammatory skin diseases, the disturbance caused within the immune cell regulation such as inhibition of regulatory T cell further leads to the impairment of the enteric system due to production of pro-inflammatory cytokines and excessive microbial metabolites. Inflammation within the skin may also release long-chain fatty acids resulting in the excessive stimulation of regulatory binding proteins such as phenol p-cresol, which may increase the synthesis of fatty acids promoting overgrowth of the microorganisms responsible for causing dermatological diseases such as psoriasis, acne, and rosacea.

Studies have suggested that changes in the human microbiome are associated with our modern lifestyle. There is an increase of the allergic and inflammatory diseases due to the widespread acceptance of western lifestyle (Ehlers & Kaufmann, 2010). In 1989, the "hygiene hypothesis" for the first time reported that improvement in household facilities, upliftment in the standards of personal hygiene, and decline in the family size have resulted in the spread of clinical atopic diseases to a certain extent (Strachan, 1989). According to von Hertzen et al.'s "biodiversity hypothesis", the environmental biodiversity is linked to microbial diversity and human health because there is an aberrant development of immune system in the case of limited exposure to several environmental microbial populations (von Hertzen et al., 2011). The studies on atopic individuals showed lower skin microbiota diversity of *Gammaproteobacteria* due to lower environmental biodiversity in the vicinity of their homes (Hanski et al., 2012). In few of the inflammatory skin diseases such as AD and psoriasis, the disturbance caused within the homeostatic relation across the host–skin–gut microbiota axis could also be a possible contributing factor responsible for such inflammatory skin conditions (Kuo et al., 2013). The relationship between skin and gut microbiota is shown by representing the role of gut microbiota in healthy and diseased skin conditions (Figure 11.4).

Studies strongly recommend that penetration of microbes or pathogen-associated molecular patterns (PAMPs) plays an indispensable role in causing variation within the function of cutaneous sensitization and skin barrier (De Benedetto et al., 2012). Few groups of skin microbiome were investigated from diseased and injured skin in order to understand the role of microbes in inflammatory and allergic skin diseases. Surprisingly, the role of *S. aureus* was pointed out in the pathogenesis of AD as the proportion of *S. aureus* was found to be 90% on the skin of AD patients (Boguniewicz & Leung, 2011). Apart from this, there is one more report that suggests that upon receiving certain treatment, there is reduction in the colonization of *S. aureus* which eventually leads to decline in the severity of AD (Huang et al., 2009). However, it is also true that *S. aureus* is not only a single species that causes AD. A study analyzed the microbial composition among lesional AD patients ranging from moderate to severe pediatric and the period during the onset of disease and after the treatment (Kong et al., 2012). The study reveals that there was a temporal shift in skin microbial composition during both the periods. In particular, there was a rapid increase in the clinical severity of the disease along with the inclination of skin commensal organisms such as *S. epidermidis* and *S. aureus*. On the other hand, species belonging to genera *Streptococcus*, *Propionibacterium*, and *Corynebacterium* seem to be in relatively higher number after patients going into therapy (Kong et al., 2012). Therefore, it could be clearly speculated that in lesional AD, there is increase in *S. epidermidis* which selectively inhibits the growth of *S. aureus* (Iwase et al., 2010). There is less microbiota diversity in the areas of skin of the inner elbow and the back of the knee of the AD patients as compared to the skin area of healthy controls (Kong et al., 2012). Moreover, just like in the case of AD, alteration in skin microbiota also contributes to the pathogenesis of psoriasis (Fry & Baker, 2007). A pioneer study on effects of microbiota within patients with psoriasis revealed that the microbial composition of the lesional skin was comparatively much higher in the skin of healthy individuals or non-lesional skin obtained from psoriatic individuals (Gao et al., 2008). On the contrary, *Firmicutes* were found to be more in the psoriatic lesions compared to the *Propionibacterium* as they were significantly less in the non-lesional skin samples of psoriatic patients. However, this study was contradicted upon determining skin microbiota of whole skin biopsies (Fry et al., 2013).

Few of the studies report that bacteria are not only the part of the superficial layer of the total host indigenous skin microbiota but also of the deeper parts of the skin (Zeeuwen et al., 2012; Nakatsuji et al., 2013). But still the role of the microbiota composition remains unclear whether disease development has a casual role in abnormal skin biology.

11.5 Role of Prebiotics in the Improvement of Skin Health

Prebiotics are considered to be an effective therapeutic component in modulating the gut microbiome. Prebiotics are the dietary compounds such as fructooligosaccharides (FOS), galactooligosaccharides (GOS), inulin, xylitol or sorbitol, polydextrose, and lactulose which support the growth of microbes (Collins & Reid, 2016). The conventional cosmetic strategy effectively reduces *P. acnes* by not only using antibacterial agents but also by promoting the growth of beneficial commensal bacteria such as *S. epidermis* which prevents human skin infections. With respect to this, a prebiotic-based treatment approach could be used in rebalancing skin's microflora by preventing the growth of *P. acnes* and further promoting the growth of beneficial bacteria. In a study, a successful approach was developed through prebiotic cosmetic approach in order to attain a balanced composition of the cutaneous microflora (Bockmuhl et al., 2006). These methods are not only precise as it gives direct observation of bacteria on the skin, but they also avoid limitations of cultural methods (Harmesen, 2000). It was seen that the application of cosmetic product composed of selected plant extracts such as ginger or pine on human skin twice a day for 3 weeks reduces the proliferation of *P. acnes* on the skin surfaces (Bockmuhl et al., 2006). Therefore, these studies clearly depict that prebiotic-based cosmetic approach is far better than antibacterial cosmetic products which inhibit the growth of bacteria without any means of antimicrobial agents (Holland & Bojar, 2002). But still these studies are needed to be confirmed with a more advanced approach. There are several reports on *S. epidermis* as a beneficial bacteria in improving skin barrier function and development of innate immune responses in the skin. Therefore, such a study on the relationship between the skin and its microbiota can serve as a foundation for building prebiotic strategies that could serve as a treatment of skin diseases such as atopic eczema (de Jongh et al., 2005). There are several ongoing and clinical trials that also depict the potential role of prebiotics in addressing the issues related to dermatological diseases (Table 11.1).

TABLE 11.1

Clinical Trials on Prebiotics in the Treatment of Dermatological Diseases

Prebiotic	Title of Study	Skin Condition	ClinicalTrials.Gov Indentifier (NCT Number)	Recruitment Status[a]	Key Findings
Mixture of galacto-oligo-saccharide/inulin	A multicenter clinical trial to assess the efficacy of antenatal maternal supplementation with GOS/inulin prebiotics on atopic dermatitis prevalence in high-risk 1-year-old children	Atopic dermatitis	NCT03183440	Recruiting	No results reported so far
Addition of food-grade GRAS (generally regarded as safe) galactooligosaccharides	Double-blind, randomized study on the effect of prebiotics on incidence of atopic manifestations in infants	Atopic dermatitis	NCT02077088	Completed	• The primary outcome of the study was a difference in the severity of atopic dermatitis measured using SCORAD criteria. • Secondary outcomes were anthropometry (length, weight, and head circumference), together with the tolerance and incidence of infections. • Both groups showed a decrease of average SCORAD values, but no statistically significant difference between the evaluated groups was observed.
50:50 mixture of GOS/PDX formula	Effects of GOS/PDX-supplemented formula in preventing and modifying the history of allergy and acute infections in a population of infants at risk of atopy	Atopic dermatitis	NCT02116452	Completed	The early administration of supplementation formula protects against respiratory infections and mediates a species-specific modulation of the intestinal microbiota.

[a] Status is based on the review of https://www.clinicaltrials.gov as on May 10, 2020.

11.6 *In vivo* Animal and Human Clinical Studies on Probiotics for Skin Health

Probiotics are defined as "live microorganisms, which when consumed in adequate amounts, confer a health effect on the host" (FAO/WHO, 2001). The bacterial strains such as *Lactobacillus*, *Streptococcus*, *Saccharomyces*, *Bifidobacterium*, *Enterococcus*, *Pediococcus*, *Leuconostoc*, *E. coli*, and *Bacillus* are the probiotics which are reported to exert beneficial health effects in humans (Fijan, 2014). As far as skin health is concerned, probiotics such as *Bifidobacterium* and *Lactobacillus* are responsible to exert a crucial role in maintaining skin health (Roudsari et al., 2015). According to a study by Levkovich et al. (2013), administration of yogurt containing *Lactobacillus reuteri* to mice induces an increase in folliculogenesis and sebocytogenesis, which leads to development of thick and radiant fur. Probiotics contribute to the modulation of the immune system in order to improve immune tolerance and epithelial integrity by secreting health-stimulating hormones and increasing the production of anti-inflammatory cytokine IL-10 which further induces regulatory T cells (Tregs) (Levkovich et al., 2013). Unfortunately, this study does not seep into detail mechanism or cause of modulation within gut microbiota upon administration of probiotics. But one thing could be seen clearly that supplementation of probiotics can bring desirable improvement in human skin health and can be applied as a therapy in different inflammatory skin conditions.

Currently, oral probiotics are not casually used in the dermatological practices, but considering gut dysbiosis as a common symptom in skin-related diseases, probiotics can be considered as a therapeutic strategy by targeting pathways of skin pathologies. According to our understanding, probiotics can act on gut epithelial innate immune response by influencing intestinal inflammation through local stimulation (Pagnini et al., 2010). Activation of nuclear factor (NF)-kappaB pathway and increase in the production of tumor necrosis factor (TNF)-alpha through epithelial cells are considered to play important roles in improving epithelial barrier function via probiotics (Pagnini et al., 2010).

There is a necessity of bringing evolving strategies such as orally administered probiotics, prebiotics, and fecal microbiota transplantation (FMT) while conducting studies related to relationship between gut microbiota and dermatological diseases in order to explore causative mechanisms holding behind such skin diseases as well to bring novel treatment approaches in dealing with such dermatological conditions (Miko et al., 2016). So far, antibiotics have been used in the

TABLE 11.2

Animal Model Studies Demonstrating the Relationship between the Probiotics and Dermatological Conditions

Animal Models	Probiotic Strain	Type or Dose of Treatments	Key Findings	References
BALB/c imiquimod-induced psoriasis-like mice	*L. pentosus* GMNL-77	5×10^7 CFU/0.2 mL d^{-1} or 5×10^8 CFU/0.2 mL $d^{-1}\times7$ days	• Decrease in erythema & scaling • Decrease in expression of pro-inflammatory cytokines (TNF-α, IL-6, and IL-23/IL-17A) • Effect on differentiation or proliferation of T cell	Chen et al. (2017)
SKH-1 hairless mice (AD mouse model)	*L. rhamnosus* (Lcr35)	1×10^9 CFU d^{-1}	• Reduces TEWL, erythema and inflammation after exposure to topical allergen ovalbumin • Decrease in IL-4 and TSLP via increasing mechanism involved in $CD4^+CD25^+Foxp3^+$ regulatory T cells	Kim et al. (2012)
Hairless Skh:hr1 mice	*L. johnsonii* (La1)	1×10^8 CFU $d^{-1}\times10$ days	• Protected against UVR-induced contact hypersensitivity • Decrease in epidermal LCs density and increase in IL-10 plasma levels	Guéniche et al. (2009)
NC/Nga mice (AD mouse model)	*L. rhamnosus* IDCC 3201	1×10^8, 1×10^9, or 1×10^{10} cells $d^{-1}\times8$ weeks	• Decrease in dermatitis scores, frequency of scratching, and epidermal thickness • Suppression of mast cell–mediated inflammation	Lee et al. (2016)
Hairless mice	*L. plantarum* HY7714	100 µl PBS* d^{-1} with 1×10^9 CFU, 1 hour prior to UVB irradiation	• Decrease in development of wrinkles following UVB radiation; decrease in UVB-induced epidermal thickness • Inhibition of MMP-13 expression, MMP-2 activity, and MMP-9 activity in dermal tissue	Kim et al. (2014)
C57BL/6 wild type and IL-10-deficient mice	*L. reuteri* ATCC 6475	(3.5×10^5 organisms $d^{-1}\times20$–24 weeks)	• Thicker, shinier fur. • Increase in dermal thickness, folliculogenesis, and sebocyte production. • IL-10-dependent anti-inflammatory pathway	Levkovich et al. (2013)
Hos:HR-1 hairless mice	*L. helveticus*	Fermented milk whey (in distilled water ad libitum$\times5$ weeks)	• Decrease in TEWL, severity of sodium dodecyl sulfate-induced dermatitis • Decrease in Keratinocyte differentiation and expression of pro-filaggrin (pro-FLG)	Baba et al. (2010)
Wistar rats and hairless Wistar Yagi (HWY) rats	*L. brevis* SBC8803	0.1 mg mL^{-1} in drinking water	• Decrease in cutaneous arterial sympathetic nerve activity and TEWL • Increase in cutaneous blood flow • Activation of 5-HT3 receptors	Horii et al. (2014)
C57BL/6 female mice, MHC class II-deficient (Aβ %) mice	*L. casei* DN-114 001	200 µl fermented milk d^{-1} with 2×10^8 CFU* $d^{-1}\times26$ days	• 50% inhibition of contact hypersensitivity response to 2,4-DNFB* • Decrease in hapten-specific $CD8^+$ T cell proliferation	Chapat et al. (2004)
NC/Nga mice (AD mouse model)	*L. plantarum* CJLP55, CJLP133 and CJLP136	1×10^{10} CFU $d^{-1}\times55$ days	• Suppression of house-dust mite-induced dermatitis • Decrease in epidermal thickening • Increase in IL-10 production and alteration of the Th1/Th2 balance	Won et al. (2011)
C57BL/6 wild type, oxytocin-deficient WT and KO *B6; 129S-Oxttm1Wsy/J mice	*L. reuteri* ATCC-PTA-6475	3.5×10^5 organisms d^{-1} in drinking water$\times2$–3 weeks	• Accelerated wound healing • Increase in $Foxp3^+$ regulatory T cells	Poutahidis et al. (2013)

* CFU, colony-forming units; DNFB, dinitrofluorobenzene; LCs, Langerhans cells; TEWL, transepidermal water loss; TSLP, thymic stromal 7lymphopoietin; PBS, phosphate-buffered saline; APCs, antigen-presenting cells.

management of cutaneous inflammation, but there is a major setback that limits its use due to the risk of developing resistance. In recent years, FMT has emerged as a budding therapeutic approach against the infection caused by *C. difficile* species (Smits et al., 2013). There are several studies conducted on animals and humans that depict the beneficial roles of probiotics in the improvement of dermatological diseases (Tables 11.2 and 11.3).

11.6.1 Animal Studies

Several studies have suggested the ameliorating effects of probiotics therapy on the rodents. Few of these studies are mentioned in Table 11.2. A study on the evaluation of *Lactobacillus pentosus* GMNL-77 showed significant reduction in erythema, thickening of epidermal cells, and scaling within the probiotics-treated psoriasis mouse model (Chen et al., 2017). This study proposed that these effects are mostly seen due to modulation of Treg cells via suppression of TNF-α, interleukin (IL)-6, 23 and17, and CD 103^{+} dendritic cells (DCs) within the GI tract.

According to few reports, after exposed to ultraviolet (UV) radiation, the gut microbiota has shown to restore skin homeostasis in the mouse model which received probiotics therapy. In a study conducted on hairless mice, the mice became resistant to UV hypersensitivity along with increase in systemic IL-10 levels upon receiving oral supplementation of *Lactobacillus johnsonii* (Guéniche et al., 2006).

TABLE 11.3

Human Studies Demonstrating Relationship between the Probiotics and Dermatological Conditions

Study Population	Study Design	Method of Therapy or Analysis	Key Findings	References
126 subjects with elevated TEWL	Randomized, double-blind, placebo-controlled	*L. brevis* SBC8803 (25 or 50 mg $d^{-1} \times 12$ weeks)	• Decrease in TEWL & increase in corneal hydration • Stimulation of serotonin release from intestinal enterochromaffin cells → ↑ vagal nerve activity	Ogawa et al. (2016)
64 females with sensitive skin	Randomized, double-blind, placebo-controlled	*L. paracasei* NCC2461 (ST11) ($1 \times 1{,}010$ CFU $d^{-1} \times 2$ months)	• Decrease in skin sensitivity, TEWL • Decrease in skin sensitivity and neurogenic inflammation • Positive effect on skin barrier function via increase in circulating TGF-β	Gueniche et al. (2014)
54 healthy subjects	Randomized, double-blind, placebo-controlled	*L. johnsonii* (La1) for 6 weeks	• Increase in skin immune homeostasis following UV-induced immunosuppression	Peguet-Navarro et al. (2008)
129 females with dry skin and wrinkles	Randomized, double-blind, placebo-controlled	*L. plantarum* HY7714 ($1 \times 1{,}010$ CFU $d^{-1} \times 12$ weeks)	• Increase in skin hydration, skin elasticity & decrease in TEWL wrinkle depth • Molecular control of signaling pathways and gene expression in skin cells	Lee et al. (2015)
26 subjects with plaque psoriasis	Randomized, double-blind, placebo-controlled	*B. infantis* *35624 ($1 \times 1{,}010$ CFU* $d^{-1} \times 8$ weeks)	• Decrease in systemic inflammation (decrease in CRP & TNF-α) • Induction of mucosal immunoregulatory responses that can exert systemic effects	Groeger et al. (2013)
300 subjects with acne	Intervention group only	*L. acidophilus* and *L. bulgaricus** (probiotic × 8 days, 2-weeks washout, then re-introduction × 8 days)	• clinical improvement in 80% of patients, particularly those with inflammatory acne • Mechanism not established	Siver (1961)
47-year-old female with severe pustular psoriasis	Case report	*L. sporogenes* (supplementation 3× d^{-1})	• Clinical improvement at 15 days, almost complete clearance at 4 weeks • Mechanism not established	Vijayashankar and Raghunath (2012)
Healthy population	Randomized control trial	*S. thermophilus* Unknown dose, 0.5 g, twice day^{-1} for 7 days	• Probiotics increased skin ceramides	Di Marzio et al. (1999)
AD	Randomized control trail	*S. thermophilus*	• 1.7 g/5 mL in 20 mL lotion, twice day^{-1}	Di Marzio et al. (2003)
Elderly healthy	Pilot study	*S. thermophilus*, 1.7 g/5 mL in 20 mL lotion, twice day^{-1}	• Increase in ceramides, hydration	Di Marzio et al. (2008)
AD	Randomized control trail	*S hominis* or *S epidermidis*	• 1 × 105 CFU cm^{-2} one dose	Nakatsuji et al. (2017)

* CFU, colony-forming units; CRP, C-reactive protein; TNF-α, tumor necrosis factor alpha; TEWL, transepidermal water loss; TGF-β, transforming growth factor beta; AD, atopic dermatitis.

If we gaze upon the probiotics tested for the treatment of AD, then *Lactobacillus* and *Bifidobacterium* would be the most commonly studied species among them (Enomoto et al., 2014). There was an observation of downregulation in IL-4 and thymic stromal lymphopoietin along with the upregulation of the Treg cells ($CD4^{+}CD25^{+}Foxp3^{+}$) in the AD mouse model upon oral consumption of *Lactobacillus rhamnosus* Lcr35 (Kim et al., 2012). In another study, AD mouse resulted in an increase in the production of IL-10 upon supplementation of *Lactobacillus plantarum* CJLP55, CJLP133, and CJLP136 which not only altered the balance of Th1 and Th2 but also inhibited induced dermatitis condition within the mouse (Won et al., 2011). A study by Lee et al. (2016) also reported suppression of inflammation mediated by mast cells in the same mouse model after supplementation of *L. rhamnosus* IDCC 3201. In another study, *L. plantarum* HY7714 inhibited expression of MMP-1 in the dermal fibroblasts which depicted prevention of mice from UV-induced photoaging (Kim H.J. et al., 2014).

A study conducted on *L. reuteri* supplementation on mice showed improvement in dermal thickness and fur. It also enhanced folliculogenesis and sebocyte production which gave more shiner fur to the mice (Levkovich et al., 2013). A study by Horii et al. (2014) reported significant declination in the trans epidermal water loss (TEWL), which is an essential marker of skin barrier function upon oral administration of *Lactobacillus brevis* SBC8803 in rats. This study also showed increase in the cutaneous blood flow. The possible reason stated for this flow is the release of serotonin from enterochromaffin cells which later activates parasympathetic pathways (Horii et al., 2014).

There are also few rodent studies that debate about the beneficial impact of gut microbiota in improving the function of disturbed skin barrier. The study conducted by Baba et al. (2010) falls under the same category which demonstrates decrease in severity of sodium dodecyl sulfate-induced dermatitis and TEWL upon supplementation of *Lactobacillus helveticus*. Apart from this, there are also studies that supports the fact that gut microbiota has the potential to influence T cell differentiation and hence promote skin allostasis. A supporting study to this is the oral administration of *Lactobacillus casei* DN-114 001, which has shown to decrease recruitment of the skin and cause impairment within $CD8^{+}$ T cells when exposed to 2–4-dinitrofluorobenzene (DNFB) (Chapat et al., 2004). These microorganisms are also considered important in improving skin homeostasis by modulating immune mechanisms through increasing $FoxP3^{+}$ Treg cells within the skin (Hacini-Rachinel et al., 2009).

Another study on *L. reuteri* depicted enhancement in the wound healing. The histomorphological examination of wound healing in the probiotics-treated mice showed maximum population of $Foxp3^{+}$ Treg cells on the wound sites, whereas neutrophils were completely absent from this site. There was also decrease in the healing time of wounds in treated group due to induction of oxytocin-mediated Treg cells via *L. reuteri* (Poutahidis et al., 2013). All these animal studies indicate the beneficial roles of probiotics in different dermatological conditions.

11.6.2 Human Studies

There are several investigations carried out on humans for finding the role of probiotics in the treatment of skin-related diseases. Some of these human studies are mentioned in Table 11.3. The *L. brevis* SBC8803 was orally administered in human subjects for the time span of 12 weeks. This resulted in significant increase in corneal hydration and decrease in TEWL (Ogawa et al., 2016). Besides this study, a placebo-controlled human study was conducted on *Lactobacillus paracasei* NCC2461 for 2 months in order to elucidate its positive effect on skin barrier function. This study depicted increase in circulating TGF-β and decrease in skin sensitivity (Guéniche et al., 2014).

Furthermore, in a placebo-controlled study with 54 healthy volunteers, cutaneous immune homeostasis was observed in the individuals receiving *L. johnsonii* La1. The reason behind this observation was stated by authors as the normalization of epidermal expression of CD1a (Peguet-Navarro et al., 2008). In another study, 110 middle-aged subjects consumed *L. plantarum* HY7714 for 12 weeks orally, and the study reported improved cutaneous elasticity and also marked better skin hydration, along with anti-aging effect of this probiotics strain on the skin (Lee et al., 2015). The study by You et al. (2013) also suggests the similar anti-aging effect of *Lactobacillus sakei* LTA which modulates effect on monocytes and further reverses UV-induced skin aging.

A study conducted on psoriatic patients showed significant decrease in the plasma levels of TNF-α in the groups that were treated with the probiotics *Bifidobacterium infantis* 35624 (Groeger et al., 2013). Another study on severe pustular psoriatic patient showed clinical improvement after giving *Lactobacillus sporogenes* thrice a day for 2 weeks (Vijayashankar & Raghunath, 2012). A study on 300 acne patients showed improvement in 80% of subjects, especially in the patients with inflammatory lesions after probiotics supplementation of *Lactobacillus acidophilus* and *Lactobacillus bulgaricus* tablets (Siver, 1961; Bowe & Logan, 2011; Bowe et al., 2014). Probiotics such as *Streptococcus salivarius* and *Lactococcus* HY449 can suppress *Propionibacterium acnes* by producing a bacteriocin-like inhibitory substance (Bowe & Logan, 2011; Bowe et al., 2014; Kober & Bowe, 2015). *Lactobacillus* and *Bifidobacterium* species are also reported to reduce acne lesion count in the study subjects compared with oral antibiotics (Volkova et al., 2001; Jung et al., 2013).

In the case of AD and SD, the probiotics have shown to improve erythema and scaling, increase skin ceramides, and also lower the concentration of *S. aureus*. Few of the other probiotics reported for the clinical improvements include *Streptococcus thermophilus*, *Vitreoscilla filiformis*, *Staphylococcus hominis*, *S.* epidermidis, and *L. johnsonii* (Di Marzio et al., 2003; Cinque et al., 2006; Guéniche et al., 2008; Blanchet-Réthoré et al., 2017).

11.7 Beneficial Effects of Probiotics in Amelioration of Dermatological Diseases

So far, several studies have suggested the role of probiotics in various skin conditions. Bacterial strains such as *V. filiformis*, *S. hominis*, *L. johnsonii*, *Enterococcus faecelis*, *Bifidobacteria longum*, *S. thermophilus*, and *L. plantarum* were investigated in order to treat dermatological conditions such as AD, acne, and skin infections (Figure 11.5). Apart from this, there are also various ongoing clinical trials that investigate the role of probiotics in the treatment of skin-related diseases (Table 11.4).

11.7.1 Dermatitis

11.7.1.1 Atopic Dermatitis

One of the most common symptoms of AD is eczema, an itchiness which also results in reduction of TEWL and barrier function leading to dry skin (McPherson, 2016). The balanced microbial composition of the mucosa results in the production of immunoglubulin A (IgA) which helps in maintaining membrane of the skin barrier further enhancing the expression of TGF (Czarnecka-Operacz & Sadowska-Przytocka, 2017). In infants, the higher risk of developing AD was observed due to increase in the number of *Clostridium* species compared to the healthy controls (Kalliomaki et al., 2001). Probiotic strains specifically prevent the pathogenesis and the symptoms of AD

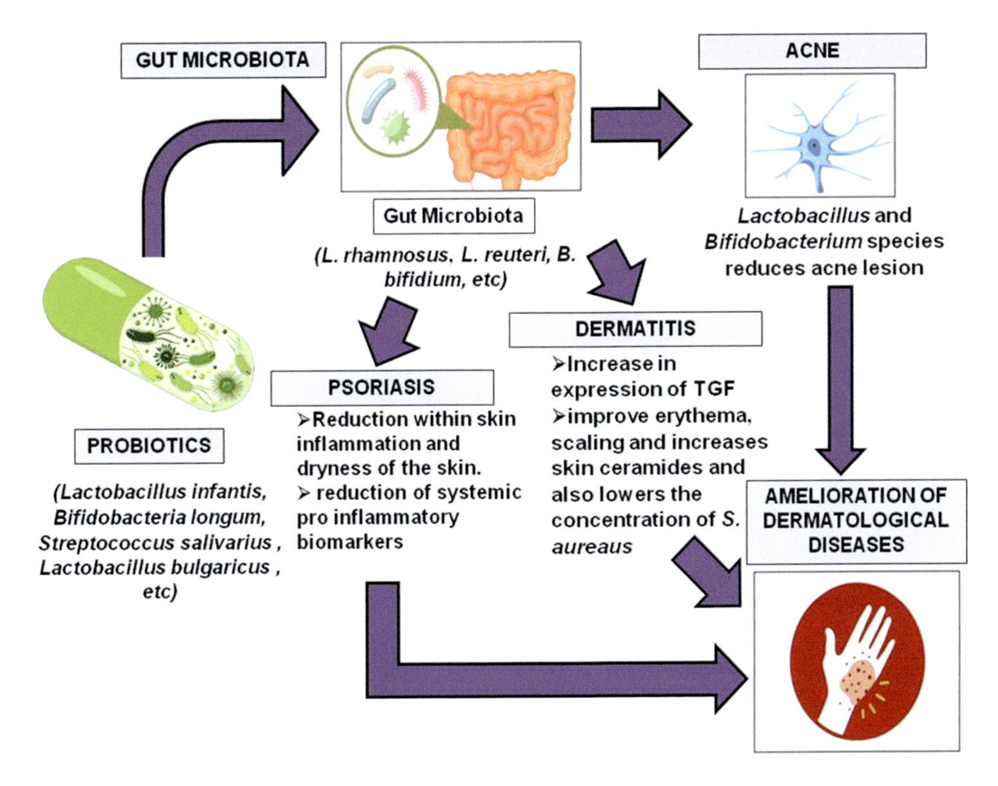

FIGURE 11.5 Proposed mechanism of modulation caused in skin-gut axis through probiotics. Ingestion of selective probiotics such as *Lactobacillus infantis*, *B. longum*, *S. salivarius*, and *L. bulgaricus* exhibits beneficial effects on human skin health. These probiotics along with the beneficial gut flora produce several microbial peptides as short-chain fatty acids (SCFAs) which can modulate the function of the signaling pathways and enteric systems. Probiotics are considered to reduce inflammation and dryness in the skin of psoriatic patient along with a reduction in systematic pro-inflammatory biomarkers. It also decreases the concentration of *S. aureus*, which further improves erythema condition. The microorganisms, such as *Lactobacillus* and *Bifidobacterium* species, reduce lesions caused in the acne and thus further leading to the amelioration of skin disease.

TABLE 11.4

Clinical Studies on Probiotics in the Treatment of Dermatological Diseases

Probiotics	Title of Study	Skin Condition	Clinical Trials Gov Identifiers	Recruitment Status[a]	Key Findings
Probiotic capsules	A double blind, randomized, placebo-controlled intervention study to evaluate the effectiveness and safety of a treatment with a probiotic in adult patients diagnosed with mild to moderate plaque psoriasis	Psoriasis	NCT02576197	Completed	Number of patients with a PASI score reduction higher than 75% from the basal value
Probiotic Bths-003	Randomized, double-blind and placebo-controlled pilot study to evaluate the effect of a probiotic mixture in acne vulgaris	Acne	NCT03878238	Completed	No results reported so far

(*Continued*)

TABLE 11.4 (*Continued*)

Clinical Studies on Probiotics in the Treatment of Dermatological Diseases

Probiotics	Title of Study	Skin Condition	Clinical Trials Gov Identifiers	Recruitment Status[a]	Key Findings
Probiotic® pur (four types of probiotic bacteria: *Bifidobacterium bifidity*, *L. acidophilus*, *Lactobacillus casei* and *Lactobacillus salivarium*)	Effect of probiotics in the treatment of children with atopic dermatitis	Atopic dermatitis	NCT01224132	Completed	No results reported so far
Probiotic mixture with three bacterial strains with maltodextrin as a carrier	Randomized, double-blind, placebo-controlled study to evaluate efficiency and safety on the use of a probiotic in SCORAD reduction in 4–17 years old patients with atopic dermatitis	Atopic dermatitis	NCT03822624	Recruiting	No results reported so far
Lactobacillus paracasei Lpc-37, *L. acidophilus* 74-2, *Bifidobacterium animalis* subsp. lactis DGCC 420	Intervention study on the effects of a probiotic yoghurt drink on the immune system and further physiological parameters in patients with atopic dermatitis and healthy persons	Atopic dermatitis	NCT00550472	Completed	SCORAD change in patients with AD taken at baseline and after 4 and 8 weeks, respectively – change in phagocytic and oxidative burst activity of monocytes and granulocytes at baseline and after 8 weeks
Probiatop (probiotic comprising the mixture of strains: *Lactobacillus rhamnosus*, *L. acidophilus*, *Lactobacillus paracasei*, and *B. lactis*)	Randomized, double-blind, controlled trial on effectiveness combined probiotics in the treatment of atopic dermatitis in children up to 6 months	Atopic dermatitis	NCT02519556	Completed	No results reported so far
Topical ointment containing live *L. reuteri* DSM17938	Clinical study for the evaluation and comparison of cutaneous acceptability and the efficacy of two cosmetic products, under normal conditions of use, in adult participants with atopic dermatitis	Atopic dermatitis	NCT03632174	Completed	A combination of *L. rhamnosus* 19070-2 and *L. reuteri* DSM 122460 was beneficial in the management of AD. The effect was more pronounced in patients with a positive skin prick test response and increased IgE levels.

[a] Status is based on the review of https://www.clinicaltrials.gov as on May 10, 2020.

by influencing several biological processes in AD (Figure 11.5). Apart from AD, probiotics also influence other dermatological diseases such as psoriasis, acne, and wounds in the similar manner (Table 11.4). In a study, the severity of eczema was reduced by 56% in children with AD when supplemented with the combination of *L. rhamnosus* and *L. reuteri* (Rosenfeldt et al., 2004). In another study, *L. rhamnosus* reduced the risk of causing AD in children during the first 7 years of age upon administrating *L. rhamnosus* orally to women 4 weeks before parturition and after 6 months of delivery (Kalliomaki et al., 2001). Infants with a higher risk of developing AD showed reduction of immunoglobulin E (IgE) by 40% upon administration of mixed probiotic strains such as *Bifidobacterium lactis*, *Bifidobacterium bifidium*, *L. acidophilus*, etc. which further improved AD condition as IgE is associated with the cause of AD (Pite, 2010).

11.7.1.2 Allergic Dermatitis

Allergic contact dermatitis (ACD) is also referred as eczema which causes an allergic reaction when the skin comes in the direct contact with an allergic substance. ACD causes blisters, itchiness, skin inflammation, etc. These allergic reactions are regulated by $CD4^+$ T cells where peptides derived from allergens induce Th2-type cytokines release such as IL-4, IL-5, and IL-13 (Woodfolk, 2007). Probiotics and prebiotics have been shown to have the potential to prevent pathogenesis and symptoms even in the case of ACD. *L. casei* is responsible for inhibiting INF-γ by involving regulatory $CD4^+$ T cells, thus reducing skin inflammation (Chapat et al., 2004; Hacini-Rachinel et al., 2009). The bacteria also activate $CD4^+CD25^+$ Tregs which increase the production of IL-10 that acts against skin inflammation (Hacini-Rachinel et al., 2009). A probiotics strain *E. coli Nissle* 1917 (EcN) induces the expression of regulatory cytokines such as TGF-β, INF-γ, and IL-10 that further exert immunomodulatory activity against allergen-induced dermatitis, and it also prevents ACD by increasing the number of FOXP3$^+$ cells (Weise et al., 2011). A study also showed induction of CD4$^+$CD25$^+$FOXP3$^+$ Tregs in prevention of ACD after administration of *L. acidophilus* strain L-92 (Shah et al., 2012). Moreover, a study on prebiotics suggests that the administration of FOS reduces contact hypersensitivity within the intestinal tract of the mice by increasing the proliferation of *Bifidobacterium pseudolongum* and further suppressing skin inflammation (Watanabe et al., 2008). In addition, the impact of *L. rhamnosus* on the rapid increase of the clinical signs within AD to allergic asthma has been shown by suppressing Th2 and Th17 responses via CD4$^+$CD25$^+$FoxP3$^+$ Tregs (Kim Y.G. et al., 2014).

11.7.2 Skin Infections

11.7.2.1 Acne

There are not enough evidences on the effect of prebiotics and probiotics in acne. But few studies have been reported to show ameliorating effects of probiotics on the symptoms of acne. A study on mixture of probiotics containing *Lactobacillus acidophilus*, *Lactobacillus delbrueckii*, and *Bifidobacterium bifidium* showed reduction in anti-inflammatory effect. Therefore, this study suggests that supplementation of probiotics serves as a potent alternative treatment option against *Acne vulgaris* (Jung et al., 2013).

S. epidermis showed an antagonizing activity on *P. acnes*, and hence, it plays a potential therapeutic role against acne (Wang et al., 2014). A study conducted by Kang et al. (2009) showed the therapeutic role of *E. faecalis* SL-5 in reducing inflammation, thereby limiting the use of antibiotics. Additionally, Konjac glucomannan hydrolysate (GMH) has also been reported to prevent the growth of *A. vulgaris* by stimulating the growth of probiotic microorganisms such as *Lactobacilli* (Al-Ghazzewi & Tester, 2010; Bateni et al., 2013).

11.7.2.2 Chronic Wounds

Most of the skin diseases are caused through the infection of pathogens. When the skin is disrupted, it results into torn skin which is termed as wound. The torn tissues undergo four stages of healing process, namely, maintaining homeostasis, initiating an inflammatory response, production of growth factors, and proliferation of fibroblasts and extracellular matrix proteins like collagen and hyaluronan (Flanagan, 2000). The healing process also involves generation of oxidative stress (Rieger et al., 2015). According to reports, the absence of skin microbiota has shown decrease in the wound-healing time (Canesso et al., 2014). Researchers have developed interest in finding the effect of specific probiotics in improving the speed of healing and improvement of wound inflammation. In a study, *Saccharomyces cerevisiae* was observed to improve the healing process of burn wounds (Oryan et al., 2018). Methicillin-resistant *S. aureus* (MRSA) is considered among the most frequently infecting pathogens which causes infection of wounds (Sikorska & Smoragiewicz, 2013). Probiotics such as *L. casei* and *L. acidophilus* have been shown to have antibacterial activities against MRSA (Karska-Wysocki et al., 2010). A study by Prince et al. (2012) reported that three probiotics *L. reuteri*, *L. rhamnosus*, and *Lactobacillus salivarius* reduced the ability of pathogens to induce keratinocyte cell death, indicating that the integrin-binding sites of the skin cells competitively remove the adherence of *S. aureus* on the keratinocyte cells situated upon epidermal tissues.

Kefir is a fermented milk product, and it has been reported to possess antimicrobial and healing properties (Codex, 2011; Farnworth, 2006). It is composed of acetic acid, hydrogen peroxide, lactic acid, and bacteriocins. These components are responsible for showing the antimicrobial activity of the kefir (Satir & Guzel-Seydim, 2016). There are other numerous reports that also suggest the role of kefir in the healing process of wounds (Huseini et al., 2012; Rahimzadeh et al., 2015; Tsiouris et al., 2017). *P. aeruginosa* is one of the pathogens causing the highest risk of serious illness such as sepsis syndromes (Bassetti et al., 2018). These are antibiotic-resistant pathogens and have the potential to cause infection in burn wounds (Defez et al., 2004; Livermore, 2002). Apart from this, the pathogenic bacterial strains such as *S. salivarius*, *S. pyogenes*, *Salmonella typhimurium*, *Listeria monocytogenes*, *S. aureus*, *P. aeruginosa*, *Candida albicans*, and *E. coli* were found to be reduced in the cases of burn wounds, when the subjects were treated with kefir (Rodrigues et al., 2005).

11.7.3 Psoriasis

There are numerous studies that report the effect of probiotics and prebiotics in prevention of the psoriasis disease. Studies suggested that microbial load of skin with *S. aureus*, *S. epidermis*, and *P. acnes* is associated with skin diseases such as psoriasis (Pariser et al., 2007). These studies also reported beneficial immunomodulatory roles of the yeast species where treating psoriasis patients with dimethylfumurate (DMF) resulted in restoration of *S. cerevisiae* (Pariser et al., 2007). The study indicated an important link between the T cell activation and the proliferation of psoriasis disease. Most importantly, CD4$^+$ T cells were reported to possess an association with the development of psoriatic arthritis. Moreover, reduction of skin inflammation and dryness of the skin upon administration of probiotics has also been shown (Schon & Boehncke, 2005). Probiotics such as *L. sporogenes* were reported to treat pustular psoriasis through improving lesions (Penders et al., 2007). A study on the immunoregulatory effect of *B. infantis* in the patients with ulcerative colitis and psoriasis reported decreased plasma levels of C-reactive protein (CRP) and TNF-α in psoriatic patients resulting in the reduction of systemic pro-inflammatory biomarkers indicating that *B. infantis* could also act as a therapeutic approach in treating psoriatic disease (Groeger et al., 2013).

Furthermore, the enhancement in keratinocyte differentiation has been shown upon administration of combined therapy with sodium butyrate and PD153035 (Szkaradkiewicz et al., 2012). Overall, this study suggests that sodium butyrate can be utilized in managing skin diseases such as psoriasis by modulating apoptosis, proliferation, and differentiation. SCFAs such as butyrate have also shown impact on anti-inflammatory response within psoriatic patients which therefore depicts the ameliorating effect of SCFAs (Gao et al., 2008; Seite et al., 2014). The *F. prausnitzii*, which is also an important source of butyrate, has been reported to reduce inflammatory and oxidative stress responses caused within psoriatic patients (van Nimwegen et al., 2011; Matsumoto et al., 2004).

11.7.4 Vitiligo

There are no evidences on the effect of prebiotics and probiotics in vitiligo. However, Dwivedi et al. (2016) suggested that inducing Tregs in vitiligo patients could reduce the autoimmune loss of melanocytes. The Tregs in vitiligo patients have shown reduced expression of FOXP3, IL-10, and TGF-β and reduction in the number of Tregs (Dwivedi et al., 2013, 2015b; Giri et al., 2020). The probiotics have been suggested to induce the expression of proteins such as Forkhead box P3 (FoxP3), IL-10, and TGF-β in Tregs. The oral consumption of *B. infantis* 35624 has been reported to enhance secretion of IL-10 and expression of FOXP3 in the human peripheral blood (Konieczna et al., 2012). The *B. infantis* administration in murine resulted in increased $CD4^{+}CD25^{+}FoxP3^{+}$ lymphocytes (O'Mahony et al., 2008). Moreover, it has been seen that *in vitro* human dendritic cells stimulated with *B. infantis* selectively promoted the upregulation of FoxP3 expression in naïve lymphocytes (Konieczna et al., 2012). The probiotics produce SCFAs which have been reported to show beneficial effects on immune regulation and prevent autoimmune responses by inducing the production of Tregs through the expression of G protein-coupled receptors (GPR) – 43, 109A, prostaglandin E2 (PGE2) – and inhibition of histone deacetylases (HDACs) (Dwivedi et al., 2016). Furthermore, a probiotic mixture (including *Lactobacillus casei*, *L. reuteri*, *L. acidophilus*, *Bifidobacterium*, and *S. thermophilus*) has also been reported to induce the production of $CD4^{+}FOXP3^{+}$ Tregs from the $CD4^{+}CD25^{-}$ population and to elevate the suppression effect of naturally occurring $CD4^{+}CD25^{+}$ Tregs in mice (Kwon et al., 2010). These data suggest that probiotics can be used for therapeutic purpose in inflammatory and other autoimmune diseases including vitiligo through induction of Tregs. However, further animal and human studies are needed to provide evidence for the possible use of probiotics in the therapy of skin depigmentation.

11.8 Future Perspectives

Research in skin microbiome is a new emerging area, and the studies carried out till date are not sufficient. Hence, further studies are needed with larger data sets for skin microbiome in order to make better comparisons within the same group with different analysis criteria. Additionally, future studies are required in order to reveal how the microbial strains of the skin and other sites are established during the initial years of life, as a new-born baby explores its environment, matures its immune system, and further gets stabilized with it. In addition, research studies with animal models and human clinical trials are required for the use of probiotics and prebiotics in different dermatological conditions. It is also necessary to consider patients receiving probiotics treatment separately in a different group from the ones who are not receiving. In addition, each patient should be monitored and repeatedly analyzed over time in order to elucidate the metabolic parameters associated with disease symptoms.

Several reports have suggested the clinical importance of microbial population in different diseases. As far as dermatological diseases are concerned, it has been evident that intestinal microbiota controls the health and well-being of individuals. So far, it is not clear whether dysbiosis is the primary reason for impacting skin function and development or it is the secondary aberration that is responsible for causing impairment in gut function through alteration in skin regulation. We also suggest that the studies in the future should also focus on the confounding factors of the study group such as high-fat diet, SCFAs which may be helpful in finding the relationship between gut and skin microbiota, and different dermatological conditions. Among all the therapeutic options available for dermatological conditions, different probiotics can be investigated as a potential treatment in reducing skin-related diseases.

11.9 Conclusions

Researchers have gain interest in the past decade in determining the effects of skin microbiome in different dermatological conditions. The specific dermatological condition represents a unique microbiome that interacts with host skin as well as the other skin microbiota. The homeostasis between the beneficial and pathogenic skin microorganisms decides the health of skin. There is emerging potential seen in the probiotics and prebiotics for the prevention and therapy of skin diseases. There are numerous studies that either consider the role of microorganisms in specific processes involved within pathophysiology of skin diseases or mixture of naturally obtained probiotics to have a beneficial role in the improvement of skin diseases. However, the exact mechanism behind ameliorating effects of such beneficial bacteria remains to be elucidated. Although clinical trials on patients with several diseases have shown promising results, it is still considered to have some limitations in safety, tolerability evaluations, and in the method of analysis of microbiota. Therefore, there is a requirement of well-designed, randomized, placebo-controlled clinical trials in order to identify accurate dose and time of treatment along with a more appropriate strain of microbe. The emerging growth in gaining insights into microbial pathophysiology and therapeutic aspects of dermatological diseases may bring novel approaches in managing the clinical severity of the disease and may help in reducing the burden of these skin diseases.

Acknowledgments

We are grateful to Uka Tarsadia University, Maliba Campus, Tarsadi, Gujarat, India for providing the facilities needed for the preparation of this chapter. We are thankful to Dr. Elizabeth A. Grice, Associate Professor of Dermatology and Microbiology, Perelman School of Medicine, University of Pennsylvania, Philadelphia for giving us kind permission to use a figure of one of her articles on skin microbiome.

Abbreviations

ACD, allergic contact dermatitis
AD, atopic dermatitis
CRP, C-reactive protein
CTLA-4, cytotoxic T lymphocyte antigen-4
DCs, dendritic cells
DMF, dimethylfumurate
DNFB, dinitrofluorobenzene
ET-1, endothelin-1
FLG, filaggrin
FMT, fecal microbiota transplantation
FOS, fructooligosaccharides
FoxP3, forkhead box P3
GI, gastrointestinal
GOS, galactooligosaccharides
GPRs, G protein-coupled receptors
HDACs, histone deacetylase
HIV, human immunodeficiency virus
IgA, immunoglubulin A
IL, interleukin
NF, nuclear factor
PAMPs, pathogen-associated molecular patterns
PGE2, prostaglandin E2
SCF, stem cell factor
SCFAs, short-chain fatty acids
SD, seborrheic dermatitis
TEWL, transepidermal water loss
TGF, transforming growth factor
TGF-α, transforming growth factor alpha
TGF-β, transforming growth factor beta
TNF, tumor necrosis factor
Tregs, regulatory T cells
UV, ultraviolet

REFERENCES

Abrahamsson, T.R., Jakobsson, H.E., Andersson, A.F., Björkstén, B., Engstrand, L., Jenmalm, M.C. 2012. Low diversity of the gut microbiota in infants with atopic eczema. *Journal of Allergy and Clinical Immunology* 129(2):434–440.

Akaza, N., Akamatsu, H., Numata, S., Yamada, S., Yagami, A., Nakata, S., et al. 2016. Microorganisms inhabiting follicular contents of facial acne are not only *Propionibacterium* but also *Malassezia* spp. *The Journal of Dermatology* 43(8):906–911.

Alekseyenko, A.V., Perez-Perez, G.I., De Souza, A., Strober, B., Gao, Z., Bihan, M., et al. 2013. Community differentiation of the cutaneous microbiota in psoriasis. *Microbiome* 1(1):31.

Al-Ghazzewi, F.H., Tester, R.F. 2010. Effect of konjac glucomannan hydrolysates and probiotics on the growth of the skin bacterium *Propionibacterium* acnes in vitro. *International Journal of Cosmetic Science* 32:139–142.

Aly, R., Shirley, C., Cunico, B., Maibach, H.I. 1978. Effect of prolonged occlusion on the microbial flora, pH, carbon dioxide and transepidermal water loss on human skin. *Journal of Investigative Dermatology* 71(6):378–381.

An, Q., Sun, M., Qi, R.Q., Zhang, L., Zhai, J.L., Hong, Y.X., et al. 2017. High staphylococcus epidermidis colonization and impaired permeability barrier in facial seborrheic dermatitis. *Chinese Medical Journal* 130(14):1662.

Antonsson, A., Erfurt, C., Hazard, K., Holmgren, V., Simon, M., Kataoka, A., et al. 2003. Prevalence and type spectrum of human papillomaviruses in healthy skin samples collected in three continents. *Journal of General Virology* 84(7):1881–1886.

Arck, P., Handjiski, B., Hagen, E., Pincus, M., Bruenahl, C., Bienenstock, J., et al. 2010. Is there a 'gut-brain-skin axis'? *Experimental Dermatology* 19(5):401–405.

Argenziano, G., Donnarumma, G., Maria Rosaria Iovene, B.D., Arnese, P., Assunta Baldassarre, M., Baroni, A. 2003. Incidence of anti-*Helicobacter pylori* and anti-CagA antibodies in rosacea patients. *International Journal of Dermatology* 42(8): 601–604.

Assarsson, M., Duvetorp, A., Dienus, O., Söderman, J., Seifert, O. 2018. Significant changes in the skin microbiome in patients with chronic plaque psoriasis after treatment with narrowband ultraviolet B. *Acta Dermato-Venereologica* 98(3–4):428–436.

Baba, H., Masuyama, A., Yoshimura, C., Aoyama, Y., Takano, T., Ohki, K. 2010. Oral intake of *Lactobacillus helveticus*-fermented milk whey decreased transepidermal water loss and prevented the onset of sodium dodecylsulfate-induced dermatitis in mice. *Bioscience, Biotechnology, and Biochemistry* 74(1):18–23.

Barnard, E., Shi, B., Kang, D., Craft, N., Li, H. 2016. The balance of metagenomic elements shapes the skin microbiome in acne and health. *Scientific Reports* 6:39491.

Bassetti, M., Vena, A., Croxatto, A., Righi, E., Guery, B. 2018. How to manage *Pseudomonas aeruginosa* infections. *Drugs Context* 7:212527.

Bateni, E., Tester, R., Al-Ghazzewi, F., Bateni, S., Alvani, K., Piggott, J. 2013. The use of konjac glucomannan hydrolysates (GMH) to improve the health of the skin and reduce acne vulgaris. *American Journal of Dermatology and Venereology* 2:10–14.

Bek-Thomsen, M., Lomholt, H.B., Kilian, M. 2008. Acne is not associated with yet-uncultured bacteria. *Journal of Clinical Microbiology* 46(10):3355–3360.

Benhadou, F., Mintoff, D., Schnebert, B., Thio, H.B. 2018. Psoriasis and microbiota: a systematic review. *Diseases* 6(2):47.

Berg, M. 1989. Epidemiological studies of the influence of sunlight on the skin. *Photo-Dermatology* 6(2):80–84.

Bieber, T. 2008. Atopic dermatitis. *The New England Journal of Medicine* 358(14):1483–1494.

Bjorksten, B., Naaber, P., Sepp, E., Mikelsaar, M. 1999. The intestinal microflora in allergic Estonian and Swedish 2-year-old children. *Clinical and Experimental Allergy* 29(3):342–346.

Blanchet-Réthoré, S., Bourdès, V., Mercenier, A., Haddar, C.H., Verhoeven, P.O., Andres, P. 2017. Effect of a lotion containing the heat-treated probiotic strain *Lactobacillus johnsonii* NCC 533 on *Staphylococcus aureus* colonization in atopic dermatitis. *Clinical, Cosmetic and Investigational Dermatology* 10:249.

Blaser, M. 2011. Stop the killing of beneficial bacteria. *Nature* 476(7361):393–394.

Blaser, M.J., Dominguez-Bello, M.G., Contreras, M., Magris, M., Hidalgo, G., Estrada, I., Gao, Z., Clemente, J.C., Costello, E.K., Knight, R. 2013. Distinct cutaneous bacterial assemblages in a sampling of South American Amerindians and US residents. *The ISME Journal* 7(1):85–95.

Bockmuhl, D., Jasoy, C., Nieveler, S., Scholtyssek, R., Wadle, A., Waldmann-Laue, M. 2006. Prebiotic cosmetics: an alternative to antibacterial products. *IFSCC Magazine* 9:1–5.

Boguniewicz, M., Leung, D.Y. 2011. Atopic dermatitis: a disease of altered skin barrier and immune dysregulation. *Immunological Reviews* 242(1):233–246.

Bomar, L., Brugger, S.D., Yost, B.H., Davies, S.S., Lemon, K.P. 2016. Corynebacterium accolens releases antipneumococcal free fatty acids from human nostril and skin surface triacylglycerols. *MBio* 7:e01725–e15.

Bonamigo, R.R., Leite, C.S., Wagner, M., Bakos, L. 2000. Rosacea and *Helicobacter pylori*: interference of systemic antibiotic in the study of possible association. *Journal of the European Academy of Dermatology and Venereology* 14(5):424–425.

Bowe, W.P., Logan, A.C. 2011. Acne vulgaris, probiotics and the gut-brain-skin axis-back to the future? *Gut Pathogens* 3(1):1.

Bowe, W., Patel, N.B., Logan, A.C. 2014. Acne vulgaris, probiotics and the gut-brain-skin axis: from anecdote to translational medicine. *Beneficial Microbes* 5(2):185–199.

Braback, L., Hjern, A., Rasmussen, F. 2004. Trends in asthma, allergic rhinitis and eczema among Swedish conscripts from farming and non-farming environments. A nationwide study over three decades. *Clinical & Experimental Allergy* 34(1):38–43.

Brandner, J.M. 2016. Importance of tight junctions in relation to skin barrier function. *Current Problems in Dermatology* 49:27–37.

Breuer, K., Häussler, S., Kapp, A., Werfel, T. 2002. *Staphylococcus aureus*: colonizing features and influence of an antibacterial treatment in adults with atopic dermatitis. *British Journal of Dermatology* 147(1):55–61.

Brüggemann, H., Henne, A., Hoster, F., Liesegang, H., Wiezer, A., Strittmatter, A., et al. 2004. The complete genome sequence of *Propionibacterium* acnes, a commensal of human skin. *Science* 305(5684):671–673.

Canesso, M.C., Vieira, A.T., Castro, T.B., Schirmer, B.G., Cisalpino, D., Martins, F.S. 2014. Skin wound healing is accelerated and scarless in the absence of commensal microbiota. *The Journal of Immunology* 193:5171–5180.

Capell, B.C., Tlougan, B.E., Orlow, S.J. 2009. From the rarest to the most common: insights from progeroid syndromes into skin cancer and aging. *Journal of Investigative Dermatology* 129(10):2340–2350.

Casas, C., Paul, C., Lahfa, M., Livideanu, B., Lejeune, O., Alvarez-Georges, S., et al. 2012. Quantification of demodex folliculorum by PCR in rosacea and its relationship to skin innate immune activation. *Experimental Dermatology* 21(12):906–910.

Chang, Y.S., Huang, Y.C. 2017. Role of Demodex mite infestation in rosacea: a systematic review and meta-analysis. *Journal of the American Academy of Dermatology* 77(3):441–447.

Chang, H.W., Yan, D., Singh, R., Liu, J., Lu, X., Ucmak, D., et al. 2018. Alteration of the cutaneous microbiome in psoriasis and potential role in Th17 polarization. *Microbiome* 6(1):154.

Chapat, L., Chemin, K., Dubois, B., Bourdet-Sicard, R., Kaiserlian, D. 2004. *Lactobacillus casei* reduces $CD8^+$ T cell-mediated skin inflammation. *European Journal of Immunology* 34(9):2520–2528.

Chehoud, C., Rafail, S., Tyldsley, A.S., Seykora, J.T., Lambris, J.D., Grice, E.A. 2013. Complement modulates the cutaneous microbiome and inflammatory milieu. *Proceedings of the National Academy of Sciences* 110(37): 15061–15066.

Cheikhrouhou, F., Guidara, R., Masmoudi, A., Trabelsi, H., Neji, S., Sellami, H., et al. 2017. Molecular identification of *Malassezia* species in patients with *Malassezia folliculitis* in Sfax, Tunisia. *Mycopathologia* 182(5–6):583–589.

Chen, Y.H., Wu, C.S., Chao, Y.H., Lin, C.C., Tsai, H.Y., Li, Y.R., et al. 2017. *Lactobacillus pentosus* GMNL-77 inhibits skin lesions in imiquimod-induced psoriasis-like mice. *Journal of Food and Drug Analysis* 25(3):559–566.

Chiller, K., Selkin, B.A., Murakawa, G.J. 2001. Skin microflora and bacterial infections of the skin. *Journal of Investigative Dermatology Symposium Proceedings* 6(3):170–174.

Cinque, B., Di Marzio, L., Della Riccia, D.N., Bizzini, F., Giuliani, M., Fanini, D., et al. 2006. Effect of *Bifidobacterium* infantis on interferon-gamma-induced keratinocyte apoptosis: a potential therapeutic approach to skin immune abnormalities. *International Journal of Immunopathology and Pharmacology* 19:775–786.

Clavaud, C., Jourdain, R., Bar-Hen, A., Tichit, M., Bouchier, C., Pouradier, F., et al. 2013. Dandruff is associated with disequilibrium in the proportion of the major bacterial and fungal populations colonizing the scalp. *PLoS One* 8(3):e58203.

Codex Alimentarius. 2011. Codex standards for fermented milks. In *Milk and Milk Products*, 2nd ed.; The European Community and its Member States (ECMS): Queenstown, New Zealand:6–16.

Codoner, F.M., Ramírez-Bosca, A., Climent, E., Carrión-Gutierrez, M., Guerrero, M., Pérez-Orquín, J.M., et al. 2018. Gut microbial composition in patients with psoriasis. *Scientific Reports* 8(1):1–7.

Collins, S., Reid, G. 2016. Distant site effects of ingested prebiotics. *Nutrients* 8(9):523.

Costello, E.K., Lauber, C.L., Hamady, M., Fierer, N., Gordon, J.I., Knight, R. 2009. Bacterial community variation in human body habitats across space and time. *Science* 326(5960):1694–1697.

Czarnecka-Operacz, M., Sadowska-Przytocka, A. 2017. Probiotics for the prevention of atopic dermatitis and other allergic diseases: what are the real facts? *Alergologia Polska - Polish Journal of Allergology* 4:89–92.

Dagnelie, M.A., Khammari, A., Dréno, B., Corvec, S. 2018. Cutibacterium acnes molecular typing: time to standardize the method. *Clinical Microbiology and Infection* 24(11):1149–1155.

Dahl, M.V., Ross, A.J., Schlievert, P.M. 2004. Temperature regulates bacterial protein production: possible role in rosacea. *Journal of the American Academy of Dermatology* 50(2):266–272.

De Benedetto, A., Kubo, A., Beck, L.A. 2012. Skin barrier disruption: a requirement for allergen sensitization? *The Journal of Investigative Dermatology* 132:949–963.

de Jongh, G.J., Zeeuwen, P.L., Kucharekova, M., Pfundt, R., van der Valk, P.G., Blokx, W., et al. 2005. High expression levels of keratinocyte antimicrobial proteins in psoriasis compared with atopic dermatitis. *Journal of Investigative Dermatology* 125(6):1163–1173.

De Miquel, D.B., Romero, M.V., Sequeiros, E.V., Olcina, J.F., de Miquel, P.B., Roman, A.L.S., Villanueva, S.A., de Argila de Prados, C.M. 2006. Effect of *Helicobacter pylori* eradication therapy in rosacea patients. *Revista Espanola de Enfermedades Digestivas* 98(7):501.

Defez, C., Fabbro-Peray, P., Bouziges, N., Gouby, A., Mahamat, A., Daurès, J.P., et al. 2004. Risk factors for multidrug-resistant *Pseudomonas aeruginosa* nosocomial infection. *Journal of Hospital Infection* 57:209–216.

Delwart, E.L. 2007. Viral metagenomics. *Reviews in Medical Virology* 17(2):115–131.

Dethlefsen, L., Relman, D.A. 2011. Incomplete recovery and individualized responses of the human distal gut microbiota to repeated antibiotic perturbation. *Proceedings of the National Academy of Sciences* 108(Supplement 1):4554–4561.

Di Marzio, L., Centi, C., Cinque, B., Masci, S., Giuliani, M., Arcieri, A., et al. 2003. Effect of the lactic acid bacterium *Streptococcus thermophilus* on stratum corneum ceramide levels and signs and symptoms of atopic dermatitis patients. *Experimental Dermatology* 12(5):615–620.

Di Marzio, L., Cinquel, B., Cupelli, F., De Simone, C., Cifone, M.G., Giuliani, M. 2008. Increase of skin-ceramide levels in aged subjects following a short-term topical application of bacterial sphingomyelinase from *Streptococcus thermophilus*. *International Journal of Immunopathology and Pharmacology* 21(1):137–143.

Di Marzio, L., Cinque, B., De Simone, C., Cifone, M.G. 1999. Effect of the lactic acid bacterium *Streptococcus thermophilus* on ceramide levels in human keratinocytes in vitro and stratum corneum in vivo. *Journal of Investigative Dermatology* 113(1):98–106.

Dominguez-Bello, M.G., Costello, E.K., Contreras, M., Magris, M., Hidalgo, G., Fierer, N., Knight, R. 2010. Delivery mode shapes the acquisition and structure of the initial microbiota across multiple body habitats in newborns. *Proceedings of the National Academy of Sciences* 107(26):11971–11975.

Doré, J., Blottière, H. 2015. The influence of diet on the gut microbiota and its consequences for health. *Current Opinion in Biotechnology* 32:195–199.

Dreno, B., Martin, R., Moyal, D., Henley, J.B., Khammari, A., Seité, S. 2017. Skin microbiome and acne vulgaris: *Staphylococcus*, a new actor in acne. *Experimental Dermatology* 26(9):798–803.

Dwivedi, M., Kemp, E.H., Laddha, N.C., Mansuri, M.S., Weetman, A.P., Begum, R. 2015b. Regulatory T cells in vitiligo: implications for pathogenesis and therapeutics. *Autoimmunity Reviews* 14(1):49–56.

Dwivedi, M., Kumar, P., Laddha, N.C., Kemp, E.H. 2016. Induction of regulatory T cells: a role for probiotics and prebiotics to suppress autoimmunity. *Autoimmunity Reviews* 15(4):379–392.

Dwivedi, M., Laddha, N.C., Arora, P., Marfatia, Y.S., Begum, R. 2013. Decreased regulatory T-cells and CD 4^+/CD 8^+ ratio correlate with disease onset and progression in patients with generalized vitiligo. *Pigment Cell & Melanoma Research* 26(4):586–591.

Dwivedi, M., Laddha, N.C., Weetman, A.P., Begum, R., Kemp, H. 2015a. Vitiligo–a complex autoimmune skin depigmenting disease. In *Autoimmunity: Pathogenesis, Clinical Aspects and Therapy of Specific Autoimmune Diseases*. K. Chatzidionysiou (Ed.); InTech:153–173. doi: 10.5772/59762.

Ehlers, S., Kaufmann, S.H. 2010. Infection, inflammation, and chronic diseases: consequences of a modern lifestyle. *Trends in Immunology* 31(5):184–190.

Enomoto, T., Sowa, M., Nishimori, K., Shimazu, S., Yoshida, A., Yamada, K., et al. 2014. Effects of bifidobacterial supplementation to pregnant women and infants in the prevention of allergy development in infants and on fecal microbiota. *Allergology International* 63(4):575–585.

Faergemann, J., Larkö, O. 1987. The effect of UV-light on human skin microorganisms. *Acta Dermato-Venereologica* 67(1):69–72.

Farnworth, E.R. 2006. Kefir a complex probiotic. *Food Science & Technology Bulletin Functional Foods* 2:1–17.

Faust, K., Raes, J. 2012. Microbial interactions: from networks to models. *Nature Reviews Microbiology* 10(8):538–550.

Feingold, K.R. 2009. The outer frontier: the importance of lipid metabolism in the skin. *Journal of Lipid Research* 50(Supplement):S417–S422.

Fierer, N., Lauber, C.L., Zhou, N., McDonald, D., Costello, E.K., Knight, R. 2010. Forensic identification using skin bacterial communities. *Proceedings of the National Academy of Sciences of the United States of America* 107(14):6477–6481.

Fijan, S. 2014. Microorganisms with claimed probiotic properties: an overview of recent literature. *International Journal of Environmental Research and Public Health* 11(5):4745–4767.

Findley, K., Oh, J., Yang, J., Conlan, S., Deming, C., Meyer, J.A., et al. 2013. Topographic diversity of fungal and bacterial communities in human skin. *Nature* 498(7454):367–370.

Fitz-Gibbon, S., Tomida, S., Chiu, B.H., Nguyen, L., Du, C., Liu, M., et al. 2013. *Propionibacterium* acnes strain populations in the human skin microbiome associated with acne. *Journal of Investigative Dermatology* 133(9):2152–2160.

Flanagan, M. 2000. The physiology of wound healing. *Journal of Wound Care* 9:299–300.

Flores, G.E., Caporaso, J.G., Henley, J.B., Rideout, J.R., Domogala, D., Chase, J., et al. 2014. Temporal variability is a personalized feature of the human microbiome. *Genome Biology* 15(12):531.

Food and Agriculture Organization of the United Nations/World Health Organization (FAO/WHO). 2001. Evaluation of health and nutritional properties of powder milk and live lactic acid bacteria. Food and Agriculture Organization of the United Nations (FAO)/World Health Organization (WHO). http://www.fao.org/3/a-a0512e.pdf

Forbes, J.D., Van Domselaar, G., Bernstein, C.N. 2016. The gut microbiota in immune-mediated inflammatory diseases. *Frontiers in Microbiology* 7:1081.

Forno, E., Onderdonk, A.B., McCracken, J., Litonjua, A.A., Laskey, D., Delaney, M.L., et al. 2008. Diversity of the gut microbiota and eczema in early life. *Clinical and Molecular Allergy* 6(1):11.

Foster, J.A., Rinaman, L., Cryan, J.F. 2017. Stress & the gut-brain axis: regulation by the microbiome. *Neurobiology of Stress* 7:124–136.

Foulongne, V., Sauvage, V., Hebert, C., Dereure, O., Cheval, J., Gouilh, M.A., et al. 2012. Human skin microbiota: high diversity of DNA viruses identified on the human skin by high throughput sequencing. *PLoS One* 7(6):e38499.

Fry, L., Baker, B.S. 2007. Triggering psoriasis: the role of infections and medications. *Clinics in Dermatology* 25(6):606–615.

Fry, L., Baker, B., Powles, A., Fahlen, A., Engstrand, L. 2013. Is chronic plaque psoriasis triggered by microbiota in the skin? *British Journal of Dermatology* 169:47–52.

Fukuda, T., Matsumura, T., Ato, M., Hamasaki, M., Nishiuchi, Y., Murakami, Y., et al. 2013. Critical roles for lipomannan and lipoarabinomannan in cell wall integrity of mycobacteria and pathogenesis of tuberculosis. *MBio* 4(1):e00472–12.

Gan, B.S., Kim, J., Reid, G., Cadieux, P., Howard, J.C. 2002. *Lactobacillus fermentum* RC-14 inhibits *Staphylococcus aureus* infection of surgical implants in rats. *The Journal of Infectious Diseases* 185(9):1369–1372.

Ganju, P., Nagpal, S., Mohammed, M.H., Kumar, P.N., Pandey, R., Natarajan, V.T., et al. 2016. Microbial community profiling shows dysbiosis in the lesional skin of Vitiligo subjects. *Scientific Reports* 13(6):18761.

Gao, Z., Tseng, C.H., Strober, B.E., Pei, Z., Blaser, M.J. 2008. Substantial alterations of the cutaneous bacterial biota in psoriatic lesions. *PLoS One* 3(7):e2719.

Gary, G. 2013. Optimizing treatment approaches in seborrheic dermatitis. *The Journal of Clinical and Aesthetic Dermatology* 6(2):44.

Giacomoni, P.U., Mammone, T., Teri, M. 2009. Gender-linked differences in human skin. *Journal of Dermatological Science* 55(3):144–149.

Gioti, A., Nystedt, B., Li, W., Xu, J., Andersson, A., Averette, A.F., et al. 2013. Genomic insights into the atopic eczema-associated skin commensal yeast *Malassezia sympodialis*. *MBio* 4(1):e00572–12.

Giri, P.S., Dwivedi, M., Laddha, N.C., Begum, R., Bharti, A.H. 2020. Altered expression of nuclear factor of activated T cells, Forkhead Box P3 and immune suppressive genes in regulatory T cells of generalized vitiligo patients. *Pigment Cell & Melanoma Research* 33(4):566–578. doi: 10.1111/pcmr.12862.

Goh, C.L., Wong, J.S., Giam, Y.C. 1997. Skin colonization of *Staphylococcus aureus* in atopic dermatitis patients seen at the National Skin Centre, Singapore. *International Journal of Dermatology* 36(9):653–657.

Gong, J.Q., Lin, L., Lin, T., Hao, F., Zeng, F.Q., Bi, Z.G., et al. 2006. Skin colonization by *Staphylococcus aureus* in patients with eczema and atopic dermatitis and relevant combined topical therapy: a double-blind multicentre randomized controlled trial. *British Journal of Dermatology* 155(4):680–687.

Gribbon, E.M., Cunliffe, W.J., Holland, K.T. 1993. Interaction of *Propionibacterium* acnes with skin lipids in vitro. *Microbiology* 139(8):1745–1751.

Grice, E.A., Kong, H.H., Conlan, S., Deming, C.B., Davis, J., Young, A.C., et al. 2009. Topographical and temporal diversity of the human skin microbiome. *Science* 324(5931):1190–1192.

Grice, E.A., Segre, J.A. 2011. The skin microbiome. *Nature Reviews Microbiology* 9(4):244–253.

Groeger, D., O'Mahony, L., Murphy, E.F., Bourke, J.F., Dinan, T.G., Kiely, B., et al. 2013. Bifidobacterium infantis 35624 modulates host inflammatory processes beyond the gut. *Gut Microbes* 4(4):325–339.

Guéniche, A., Benyacoub, J., Buetler, T.M., Smola, H., Blum, S. 2006. Supplementation with oral probiotic bacteria maintains cutaneous immune homeostasis after UV exposure. *European Journal of Dermatology* 16(5):511–517.

Guéniche, A., Cathelineau, A.C., Bastien, P., Esdaile, J., Martin, R., Queille Roussel, C., et al. 2008. Vitreoscilla filiformis biomass improves seborrheic dermatitis. *Journal of the European Academy of Dermatology and Venereology* 22(8):1014–1015.

Guéniche, A., Philippe, D., Bastien, P., Blum, S., Buyukpamukcu, E., Castiel-Higounenc, I. 2009. Probiotics for photoprotection. *Dermato-Endocrinology* 1(5):275–279.

Gueniche, A., Philippe, D., Bastien, P., Reuteler, G., Blum, S., Castiel-Higounenc, I., et al. 2014. Randomised double-blind placebo-controlled study of the effect of *Lactobacillus paracasei* NCC 2461 on skin reactivity. *Beneficial Microbes* 5(2):137–145.

Hacini-Rachinel, F., Gheit, H., Le Luduec, J.B., Dif, F., Nancey, S., Kaiserlian, D. 2009. Oral probiotic control skin inflammation by acting on both effector and regulatory T cells. *PLoS One* 4(3):e4903.

Hanski, I., von Hertzen, L., Fyhrquist, N., Koskinen, K., Torppa, K., Laatikainen, T., et al. 2012. Environmental biodiversity, human microbiota, and allergy are interrelated. *Proceedings of the National Academy of Sciences* 109(21):8334–8339.

Harmesen, H.J. 2000. Comparison of viable cell counts and fluorescence in situ hybridization using specific rRNA-based probes for the quantification of human feel bacteria. *FEMS Microbiology Letters* 183:125–129.

Has, C., Bruckner-Tuderman, L. 2014. The genetics of skin fragility. *Annual Review of Genomics and Human Genetics* 15:245–268.

Hidalgo-Cantabrana, C., Gomez, J., Delgado, S., Requena-López, S., Queiro-Silva, R., Margolles, A., et al. 2019. Gut microbiota dysbiosis in a cohort of patients with psoriasis. *British Journal of Dermatology* 181(6):1287–1295.

Higaki, S., Kitagawa, T., Kagoura, M., Morohashi, M., Yamagishi, T. 2000. Correlation between *Propionibacterium* acnes biotypes, lipase activity and rash degree in acne patients. *The Journal of Dermatology* 27(8):519–522.

Hillion, M., Mijouin, L., Jaouen, T., Barreau, M., Meunier, P., Lefeuvre, L., et al. 2013. Comparative study of normal and sensitive skin aerobic bacterial populations. *Microbiologyopen* 2(6):953–961.

Holland, K.T., Bojar, R.A. 2002. Cosmetics. *American Journal of Clinical Dermatology* 3(7):445–449.

Holmes, A.D. 2013. Potential role of microorganisms in the pathogenesis of rosacea. *Journal of the American Academy of Dermatology* 69(6):1025–1032.

Horii, Y., Kaneda, H., Fujisaki, Y., Fuyuki, R., Nakakita, Y., Shigyo, T., et al. 2014. Effect of heat-killed *Lactobacillus brevis* SBC 8803 on cutaneous arterial sympathetic nerve activity, cutaneous blood flow and transepidermal water loss in rats. *Journal of Applied Microbiology* 116(5): 1274–1281.

Huang, J.T., Abrams, M., Tlougan, B., Rademaker, A., Paller, A.S. 2009. Treatment of *Staphylococcus aureus* colonization in atopic dermatitis decreases disease severity. *Pediatrics* 123(5):e808–14.

Huseini, H.F., Rahimzadeh, G., Fazeli, M.R., Mehrazma, M., Salehi, M. 2012. Evaluation of wound healing activities of kefir products. *Burns* 38:719–723.

Ismail, I.H., Oppedisano, F., Joseph, S.J., Boyle, R.J., Licciardi, P.V., Robins-Browne, R.M., et al. 2012. Reduced gut microbial diversity in early life is associated with later development of eczema but not atopy in high-risk infants. *Pediatric Allergy and Immunology* 23(7):674–681.

Iwase, T., Uehara, Y., Shinji, H., Tajima, A., Seo, H., Takada, K., et al. 2010. *Staphylococcus* epidermidis Esp inhibits *Staphylococcus aureus* biofilm formation and nasal colonization. *Nature* 465(7296):346–349.

Jarmuda, S., O'Reilly, N., Żaba, R., Jakubowicz, O., Szkaradkiewicz, A., Kavanagh, K. 2012. Potential role of Demodex mites and bacteria in the induction of rosacea. *Journal of Medical Microbiology* 61(11):1504–1510.

Jorgensen, A.H., Egeberg, A., Gideonsson, R., Weinstock, L.B., Thyssen, E.P., Thyssen, J.P. 2017. Rosacea is associated with *Helicobacter pylori*: a systematic review and meta-analysis. *Journal of the European Academy of Dermatology and Venereology* 31(12):2010–2015.

Jung, G.W., Tse, J.E., Guiha, I., Rao, J. 2013. Prospective, randomized, open-label trial comparing the safety, efficacy, and tolerability of an acne treatment regimen with and without a probiotic supplement and minocycline in subjects with mild to moderate acne. *Journal of Cutaneous Medicine and Surgery* 17(2):114–122.

Kalliomaki, M., Kirjavainen, P., Eerola, E., Kero, P., Salminen, S., Isolauri, E. 2001. Distinct patterns of neonatal gut microflora in infants in whom atopy was and was not developing. *The Journal of Allergy and Clinical Immunology* 107:129–134.

Kang, B.S., Seo, J.G., Lee, G.S., Kim, J.H., Kim, S.Y., Han, Y.W., et al. 2009. Antimicrobial activity of enterocins from *Enterococcus faecalis* SL-5 against *Propionibacterium* acnes, the causative agent in acne vulgaris, and its therapeutic effect. *The Journal of Microbiology* 47(1):101–109.

Kapoor, R., Menon, C., Hoffstad, O., Bilker, W., Leclerc, P., Margolis, D.J. 2008. The prevalence of atopic triad in children with physician-confirmed atopic dermatitis. *Journal of the American Academy of Dermatology* 58(1):68–73.

Karska-Wysocki, B., Bazo, M., Smoragiewicz, W. 2010. Antibacterial activity of Lactobacillus acidophilus and *Lactobacillus casei* against methicillin-resistant *Staphylococcus aureus* (MRSA). *Microbiological Research* 165:674–686.

Kennedy, E.A., Connolly, J., Hourihane, J.O., Fallon, P.G., McLean, W.I., Murray, D., et al. 2017. Skin microbiome before development of atopic dermatitis: early colonization with commensal staphylococci at 2 months is associated with a lower risk of atopic dermatitis at 1 year. *Journal of Allergy and Clinical Immunology* 139(1):166–172.

Kim, H.J., Kim, Y.J., Kang, M.J., Seo, J.H., Kim, H.Y., Jeong, S.K., et al. 2012. A novel mouse model of atopic dermatitis with epicutaneous allergen sensitization and the effect of *Lactobacillus rhamnosus. Experimental Dermatology* (9):672–675.

Kim, H.J., Kim, Y.J., Lee, S.H., Yu, J., Jeong, S.K., and Hong, S.J. 2014. Effects of *Lactobacillus rhamnosus* on allergic march model by suppressing Th2, Th17, and TSLP responses via $CD4^+$ $CD25^+$ $Foxp3^+$ Tregs. *Clinical Immunology* 153(1):178–186.

Kim, Y.G., Udayanga, K.G., Totsuka, N., Weinberg, J.B., Núñez, G., Shibuya, A. 2014. Gut dysbiosis promotes M2 macrophage polarization and allergic airway inflammation via fungi-induced PGE2. *Cell Host & Microbe* 15(1):95–102.

Kishishita, M.A., Ushijima, T., Ozaki, Y., Ito, Y. 1979. Biotyping of *Propionibacterium* acnes isolated from normal human facial skin. *Applied and Environmental Microbiology* 38(4):585–589.

Kober, M.M., Bowe, W.P. 2015. The effect of probiotics on immune regulation, acne, and photoaging. *International Journal of Women's Dermatology* 1(2):85–89.

Kocak, M., Yagli, S., Vahapoğlu, G., Ekşioğlu, M. 2002. Permethrin 5% cream versus metronidazole 0.75% gel for the treatment of papulopustular rosacea. *Dermatology* 205(3):265–270.

Koller, B., Müller-Wiefel, A.S., Rupec, R., Korting, H.C., Ruzicka, T. 2011. Chitin modulates innate immune responses of keratinocytes. *PLoS One* 6(2):e16594.

Kong, H.H., Oh, J., Deming, C., Conlan, S., Grice, E.A., Beatson, M.A., et al. 2012. Temporal shifts in the skin microbiome associated with disease flares and treatment in children with atopic dermatitis. *Genome Research* 22(5):850–859.

Konieczna, P., Groeger, D., Ziegler, M., Frei, R., Ferstl, R., Shanahan, F., et al. 2012. Bifidobacterium infantis 35624 administration induces Foxp3 T regulatory cells in human peripheral blood: potential role for myeloid and plasmacytoid dendritic cells. *Gut* 61(3):354–366.

Korting, H.C., Hübner, K., Greiner, K., Hamm, G., Braun-Falco, O. 1990. Differences in the skin surface pH and bacterial microflora due to the long-term application of synthetic detergent preparations of pH 5.5 and pH 7.0. Results of a crossover trial in healthy volunteers. *Acta Dermato-Venereologica* 70(5):429–431.

Kosiewicz, M.M., Zirnheld, A.L., Alard, P. 2011. Gut microbiota, immunity, and disease: a complex relationship. *Frontiers in Microbiology* 2:180.

Kuo, I.H., Yoshida, T., De Benedetto, A., Beck, L.A. 2013. The cutaneous innate immune response in patients with atopic dermatitis. *Journal of Allergy and Clinical Immunology* 131(2):266–278.

Kwon, H.K., Lee, C.G., So, J.S., Chae, C.S., Hwang, J.S., Sahoo, A., et al. 2010. Generation of regulatory dendritic cells and $CD4^+$ $Foxp3^+$ T cells by probiotics administration suppresses immune disorders. *Proceedings of the National Academy of Sciences* 107(5):2159–2164.

Lacey, N., Delaney, S., Kavanagh, K., Powell, F.C. 2007. Mite-related bacterial antigens stimulate inflammatory cells in rosacea. *British Journal of Dermatology* 157(3):474–481.

Lacey, N., Kavanagh, K., Tseng, S.C. 2009. Under the lash: Demodex mites in human diseases. *The Biochemist* 31(4):20–24.

Lacey, N., Raghallaigh, S.N., Powell, F.C. 2011. Demodex mites–commensals, parasites or mutualistic organisms. *Dermatology* 222(2):128–130.

Laddha, N.C., Dwivedi, M., Mansuri, M.S., Gani, A.R., Ansarullah, Md., Ramachandran, A.V., Dalai, S., Begum, R. 2013. Vitiligo: interplay between oxidative stress and immune system. *Experimental Dermatology* 22(4):245–250.

Lai, Y., Di Nardo, A., Nakatsuji, T., Leichtle, A., Yang, Y., Cogen, A.L., et al. 2009. Commensal bacteria regulate Toll-like receptor 3-dependent inflammation after skin injury. *Nature Medicine* 15(12):1377–1382.

Lazaridou, E., Apalla, Z., Sotiraki, S., Ziakas, N.G., Fotiadou, C., Ioannides, D. 2010. Clinical and laboratory study of rosacea in northern Greece. *Journal of the European Academy of Dermatology and Venereology* 24(4):410–414.

Lee, D.E., Huh, C.S., Ra, J., Choi, I.D., Jeong, J.W., Kim, S.H., et al. 2015. Clinical evidence of effects of *Lactobacillus plantarum* HY7714 on skin aging: a randomized, double blind, placebo-controlled study. *Journal of Microbiology and Biotechnology* 25(12):2160–2168.

Lee, S.Y., Lee, E., Park, Y.M., Hong, S.J. 2018. Microbiome in the gut-skin axis in atopic dermatitis. *Allergy Asthma & Immunol Research* 10(4):354–362.

Lee, S.H., Yoon, J.M., Kim, Y.H., Jeong, D.G., Park, S., Kang, D.J. 2016. Therapeutic effect of tyndallized *Lactobacillus rhamnosus* IDCC 3201 on atopic dermatitis mediated by down-regulation of immunoglobulin E in NC/Nga mice. *Microbiology and Immunology* 60(7):468–476.

Leeming, J.P., Holland, K.T., Cunliffe, W.J. 1984. The microbial ecology of pilosebaceous units isolated from human skin. *Microbiology* 130(4):803–807.

Levkovich, T., Poutahidis, T., Smillie, C., Varian, B.J., Ibrahim, Y.M., Lakritz, J.R., et al. 2013. Probiotic bacteria induce a 'glow of health'. *PLoS One* 8(1):e53867.

Lévy, A., de Chauvin Feuilhade, M., Dubertret, L., Morel, P., Flageul, B. 2007. *Malassezia folliculitis*: characteristics and therapeutic response in 26 patients. *Annales de Dermatologie et de Vénéréologie* 134(11):823–828.

Leyden, J.J., McGiley, K.J., Mills, O.H., Kligman, A.M. 1975. Age-related changes in the resident bacterial flora of the human face. *Journal of Investigative Dermatology* 65(4):379–381.

Li, D., Lei, H., Li, Z., Li, H., Wang, Y., Lai, Y. 2013. A novel lipopeptide from skin commensal activates TLR2/CD36-p38 MAPK signaling to increase antibacterial defense against bacterial infection. *PLoS One* 8(3):e58288.

Li, J., Lin, S., Vanhoutte, P.M., Woo, C.W., Xu, A. 2016. Akkermansia muciniphila protects against atherosclerosis by preventing metabolic endotoxemia-induced inflammation in Apoe−/− mice. *Circulation* 133(24):2434–2446.

Livermore, D.M. 2002. Multiple mechanisms of antimicrobial resistance in *Pseudomonas aeruginosa*: our worst nightmare? *Clinical Infectious Diseases* 34:634–640.

Lomholt, H.B., Kilian, M. 2010. Population genetic analysis of *Propionibacterium* acnes identifies a subpopulation and epidemic clones associated with acne. *PLoS One* 5(8):e12277.

Madison, K.C. 2003. Barrier function of the skin: "la raison d'être" of the epidermis. *Journal of Investigative Dermatology* 121(2):231–241.

Mah, K.W., Björkstén, B., Lee, B.W., Van Bever, H.P., Shek, L.P., Tan, T.N., et al. 2006. Distinct pattern of commensal gut microbiota in toddlers with eczema. *International Archives of Allergy and Immunology* 140(2):157–163.

Malic, S., Hill, K.E., Playle, R., Thomas, D.W., Williams, D.W. 2011. In vitro interaction of chronic wound bacteria in biofilms. *Journal of Wound Care* 20(12):569–570, 572, 574–577.

Marples, R.R. 1974. The microflora of the face and acne lesions. *Journal of Investigative Dermatology* 62(3):326–331.

Marples, R.R. 1982. Sex, constancy, and skin bacteria. *Archives of Dermatological Research* 272(3–4):317–320.

Masenga, J., Garbe, C., Wagner, J., Orfanos, C.E. 1990. *Staphylococcus aureus* in atopic dermatitis and in non-atopic dermatitis. *International Journal of Dermatology* 29(8):579–582.

Matsumoto, M., Ohishi, H., Kakizoe, K., Benno, Y. 2004. Faecal microbiota and secretory immunogloblin a levels in adult patients with atopic dermatitis. *Microbial Ecology in Health and Disease* 16(1):13–17.

McBride, M.E., Duncan, W.C., Knox, J.M. 1977. The environment and the microbial ecology of human skin. *Applied and Environmental Microbiology* 33(3):603–608.

McDowell, A., Valanne, S., Ramage, G., Tunney, M.M., Glenn, J.V., McLorinan, G.C., et al. 2005. *Propionibacterium* acnes types I and II represent phylogenetically distinct groups. *Journal of Clinical Microbiology* 43(1):326–334.

McLean, W.H. 2016. Filaggrin failure – from ichthyosis vulgaris to atopic eczema and beyond. *The British Journal of Dermatology* 175(Suppl 2):4–7.

McPherson, T. 2016. Current understanding in pathogenesis of atopic dermatitis. *Indian Journal of Dermatology* 61:649–655.

Miko, E., Vida, A., Bai, P. 2016. Translational aspects of the microbiome-to be exploited. *Cell Biology and Toxicology* 32(3):153–156.

Naik, S., Bouladoux, N., Linehan, J.L., Han, S.J., Harrison, O.J., Wilhelm, C., et al. 2015. Commensal-dendritic-cell interaction specifies a unique protective skin immune signature. *Nature* 520(7545):104–108.

Naik, S., Bouladoux, N., Wilhelm, C., Molloy, M.J., Salcedo, R., Kastenmuller, W., et al. 2012. Compartmentalized control of skin immunity by resident commensals. *Science* 337(6098):1115–1119.

Nakatsuji, T., Chen, T.H., Narala, S., Chun, K.A., Two, A.M., Yun, T., et al. 2017. Antimicrobials from human skin commensal bacteria protect against *Staphylococcus aureus* and are deficient in atopic dermatitis. *Science Translational Medicine* 9(378):eaah4680.

Nakatsuji, T., Chiang, H.I., Jiang, S.B., Nagarajan, H., Zengler, K., Gallo, R.L. 2013. The microbiome extends to subepidermal compartments of normal skin. *Nature Communications* 4(1):1–8.

Nam, J.H., Yun, Y., Kim, H.S., Kim, H.N., Jung, H.J., Chang. Y., et al. 2018. Rosacea and its association with enteral microbiota in Korean females. *Experimental Dermatology* 27(1):37–42.

Natsuga, K. 2014. Epidermal barriers. *Cold Spring Harbor Perspectives in Medicine* 4(4):a018218.

Ng, C.Y., Huang, Y.H., Chu, C.F., Wu, T.C., Liu, S.H. 2017. Risks for *Staphylococcus aureus* colonization in patients with psoriasis: a systematic review and meta-analysis. *British Journal of Dermatology* 177(4):967–977.

Nishijima, S., Kurokawa, I., Katoh, N., Watanabe, K. 2000. The bacteriology of acne vulgaris and antimicrobial susceptibility of *Propionibacterium* acnes and *Staphylococcus* epidermidis isolated from acne lesions. *The Journal of Dermatology* 27(5):318–323.

Nutten, S. 2015. Atopic dermatitis: global epidemiology and risk factors. *Annals of Nutrition and Metabolism* 66(Suppl. 1):8–16.

Ogawa, T., Katsuoka, K., Kawano, K., Nishiyama, S. 1994. Comparative study of staphylococcal flora on the skin surface of atopic dermatitis patients and healthy subjects. *The Journal of Dermatology* 21(7):453–460.

Ogawa, M., Saiki, A., Matsui, Y., Tsuchimoto, N., Nakakita, Y., Takata, Y., et al. 2016. Effects of oral intake of heat-killed *Lactobacillus brevis* SBC8803 (SBL88™) on dry skin conditions: a randomized, double-blind, placebo-controlled study. *Experimental and Therapeutic Medicine* 12(6):3863–3872.

Oh, J., Byrd, A.L., Park, M., Kong, H.H., Segre, J.A., NISC Comparative Sequencing Program. 2016. Temporal stability of the human skin microbiome. *Cell* 165(4):854–866.

O'Mahony, C., Scully, P., O'Mahony, D., Murphy, S., O'Brien, F., Lyons, A., et al. 2008. Commensal-induced regulatory T cells mediate protection against pathogen-stimulated NF-κB activation. *PLoS Pathogens* 4(8):e1000112.

Omran, A.N., Mansori, A.G. 2018. Pathogenic yeasts recovered from acne vulgaris: molecular characterization and antifungal susceptibility pattern. *Indian Journal of Dermatology* 63(5):386.

O'Neill, C.A., Monteleone, G., McLaughlin, J.T., Paus, R. 2016. The gut-skin axis in health and disease: a paradigm with therapeutic implications. *BioEssays* 38(11):1167–1176.

Ong, P.Y., Ohtake, T., Brandt, C., Strickland, I., Boguniewicz, M., Ganz, T., et al. 2002. Endogenous antimicrobial peptides and skin infections in atopic dermatitis. *New England Journal of Medicine* 347(15):1151–1160.

O'Reilly, N., Menezes, N., Kavanagh, K. 2012. Positive correlation between serum immunoreactivity to Demodex-associated *Bacillus* proteins and erythematotelangiectatic rosacea. *British Journal of Dermatology* 167(5): 1032–1036.

Oryan, A., Jalili, M., Kamali, A., Nikahval, B. 2018. The concurrent use of probiotic microorganism and collagen hydrogel/scaffold enhances burn wound healing: an in vivo evaluation. *Burns* 44:1775–1786.

Pagnini, C., Saeed, R., Bamias, G., Arseneau, K.O., Pizarro, T.T., Cominelli, F. 2010. Probiotics promote gut health through stimulation of epithelial innate immunity. *Proceedings of the National Academy of Sciences of the United States of America* 107(1):454–459.

Pariser, D.M., Bagel, J., Gelfand, J.M., Korman, N.J., Ritchlin, C.T., Strober, B.E. 2007. National Psoriasis Foundation clinical consensus on disease severity. *Archives of Dermatology* 143(2):239–242.

Paulino, L.C. 2017. New perspectives on dandruff and seborrheic dermatitis: lessons we learned from bacterial and fungal skin microbiota. *European Journal of Dermatology* 27(1):4–7.

Paulino, L.C., Tseng, C.H., Blaser, M.J. 2008. Analysis of Malassezia microbiota in healthy superficial human skin and in psoriatic lesions by multiplex real-time PCR. *FEMS Yeast Research* 8(3):460–471.

Peguet-Navarro, J., Dezutter-Dambuyant, C., Buetler, T., Leclaire, J., Smola, H., Blum, S., et al. 2008. Supplementation with oral probiotic bacteria protects human cutaneous immune homeostasis after UV exposure-double blind, randomized, placebo controlled clinical trial. *European Journal of Dermatology* 18(5):504–511.

Penders, J., Thijs, C., van den Brandt, P.A., Kummeling, I., Snijders, B., Stelma, F., et al. 2007. Gut microbiota composition and development of atopic manifestations in infancy: the KOALA birth cohort study. *Gut* 56:661.

Perez, G.I.P., Gao, Z., Jourdain, R., Ramirez, J., Gany, F., Clavaud, C., et al. 2016. Body site is a more determinant factor than human population diversity in the healthy skin microbiome. *PLoS One* 11(4):e0151990.

Pite, H. 2010. Effect of probiotic mix (*Bifidobacterium bifidum, Bifidobacterium Lactis, Lactobacillus acidophilus*) in the primary prevention of eczema: a double-blind, randomized, placebo -controlled trial. *Revista Portuguesa de Imunoalergologia* 18:385–386.

Poutahidis, T., Kearney, S.M., Levkovich, T., Qi, P., Varian, B.J., Lakritz, J.R., et al. 2013. Microbial symbionts accelerate wound healing via the neuropeptide hormone oxytocin. *PLoS One* 8(10):e78898.

Prince, T., Mcbain, A.J., O'Neill, C.A. 2012. *Lactobacillus reuteri* protects epidermal keratinocytes from *Staphylococcus aureus*-induced cell death by competitive exclusion. *Applied and Environmental Microbiology* 78:5119–5126.

Prohic, A., Jovovic Sadikovic, T., Krupalija-Fazlic, M., Kuskunovic-Vlahovljak, S. 2016. Malassezia species in healthy skin and in dermatological conditions. *International Journal of Dermatology* 55(5):494–504.

Puhvel, S.M. 1968. Characterization of Corynebacterium acnes. *Microbiology* 50(2):313–320.

Rahimzadeh, G., Fazeli, M.R., Mozafari, A.N., Mesbahi, M. 2015. Evaluation of anti-microbial activity and wound healing of kefir. *International Journal of Pharmaceutical Sciences and Research* 6:286–293.

Rather, I.A., Bajpai, V.K., Kumar, S., Lim, J., Paek, W.K., Park, Y.H. 2016. Probiotics and atopic dermatitis: an overview. *Frontiers in Microbiology* 12(7):507.

Reunanen, J., Kainulainen, V., Huuskonen, L., Ottman, N., Belzer, C., Huhtinen, H., et al. 2015. Akkermansia muciniphila adheres to enterocytes and strengthens the integrity of the epithelial cell layer. *Applied and Environmental Microbiology* 81(11):3655–3662.

Rieger, S., Zhao, H., Martin, P., Abe, K., Lisse, T.S. 2015. The role of nuclear hormone receptors in cutaneous wound repair. *Cell Biochemistry and Function* 33:1–13.

Rodrigues, K.L., Gaudino Caputo, L.R., Tavares Carvalho, J.C., Evangelista, J., Schneedorf, J.M. 2005. Antimicrobial and healing activity of kefir and kefiran extract. *The International Journal of Antimicrobial Agents* 25:404–408.

Roihu, T., Kariniemi, A.L. 1998. Demodex mites in acne rosacea. *Journal of Cutaneous Pathology* 25(10):550–552.

Rosenfeldt, V., Benfeldt, E., Valerius, N.H., Pærregaard, A., Michaelsen, K.F. 2004. Effect of probiotics on gastrointestinal symptoms and small intestinal permeability in children with atopic dermatitis. *The Journal of Pediatrics* 145:612–616.

Roth, R.R., James, W.D. 1988. Microbial ecology of the skin. *Annual Reviews in Microbiology* 42(1):441–464.

Roudsari, M.R., Karimi, R., Sohrabvandi, S., Mortazavian, A.M. 2015. Health effects of probiotics on the skin. *Critical Reviews in Food Science and Nutrition* 55(9):1219–1240.

Samuelson, D.R., Welsh, D.A., Shellito, J.E. 2015. Regulation of lung immunity and host defense by the intestinal microbiota. *Frontiers in Microbiology* 6:1085.

Satir, G., Guzel-Seydim, Z.B. 2016. How kefir fermentation can affect product composition? *Small Ruminant Research* 134:1–7.

Scharschmidt, T.C., Vasquez, K.S., Pauli, M.L., Leitner, E.G., Chu, K., Truong, H.A., et al. 2017. Commensal microbes and hair follicle morphogenesis coordinately drive Treg migration into neonatal skin. *Cell Host & Microbe* 21(4):467–477.

Scharschmidt, T.C., Vasquez, K.S., Truong, H.A., Gearty, S.V., Pauli, M.L., Nosbaum, A., et al. 2015. A wave of regulatory T cells into neonatal skin mediates tolerance to commensal microbes. *Immunity* 43(5):1011–1021.

Scher, J.U., Ubeda, C., Artacho, A., Attur, M., Isaac, S., Reddy, S.M., et al. 2015. Decreased bacterial diversity characterizes the altered gut microbiota in patients with psoriatic arthritis, resembling dysbiosis in inflammatory bowel disease. *Arthritis & Rheumatology* 67(1):128–139.

Schommer, N.N., Gallo, R.L. 2013. Structure and function of the human skin microbiome. *Trends in Microbiology* 21(12):660–668.

Schon, M.P., Boehncke, W.H. 2005. Psoriasis. *The New England Journal of Medicine* 352:1899–1912.

Schwarz, A., Bruhs, A., Schwarz, T. 2017. The short-chain fatty acid sodium butyrate functions as a regulator of the skin immune system. *Journal of Investigative Dermatology* 137(4):855–864.

Seite, S., Flores, G.E., Henley, J.B., Martin, R., Zelenkova, H., Aguilar, L., et al. 2014. Microbiome of affected and unaffected skin of patients with atopic dermatitis before and after emollient treatment. *The Journal of Drugs in Dermatology* 13:1365–1372.

Shah, M.M., Saio, M., Yamashita, H., Tanaka, H. 2012. *Lactobacillus acidophilus* Strain L-92 induces CD4 CD25 Foxp3 regulatory T cells and suppresses allergic contact dermatitis. *Biological and Pharmaceutical Bulletin* 35:612–616.

Shanahan, F. 1999. Brain-gut axis and mucosal immunity: a perspective on mucosal psychoneuroimmunology. *Seminars in Gastrointestinal Disease* 10(1):8–13.

Sharma, V.K., Lynn, A., Kaminski, M., Vasudeva, R., Howden, C.W. 1998. A study of the prevalence of *Helicobacter pylori* infection and other markers of upper gastrointestinal tract disease in patients with rosacea. *The American Journal of Gastroenterology* 93(2):220–222.

Shu, M., Wang, Y., Yu, J., Kuo, S., Coda, A., Jiang, Y., et al. 2013. Fermentation of *Propionibacterium* acnes, a commensal bacterium in the human skin microbiome, as skin probiotics against methicillin-resistant *Staphylococcus aureus*. *PLoS One* 8(2):e55380.

Sikorska, H., Smoragiewicz, W. 2013. Role of probiotics in the prevention and treatment of meticillin-resistant *Staphylococcus aureus* infections. *The International Journal of Antimicrobial Agents* 42:475–481.

Siver, R.H. 1961. Lactobacillus for the control of acne. *The Journal of the Medical Society of New Jersey* 59:52–53.

Smits, L.P., Bouter, K.E., de Vos, W.M., Borody, T.J., Nieuwdorp, M. 2013. Therapeutic potential of fecal microbiota transplantation. *Gastroenterology* 145(5):946–953.

Somerville, D.A. 1969. The normal flora of the skin in different age groups. *British Journal of Dermatology* 81(4):248–258.

Son, S.W., Kim, I.H., Oh, C.H., Kim, J.G. 1999. The response of rosacea to eradication of *Helicobacter pylori*. *British Journal of Dermatology* 140(5):984–985.

Song, H.S., Kim, S.K., Kim, Y.C. 2014. Comparison between *Malassezia folliculitis* and non-*Malassezia folliculitis*. *Annals of Dermatology* 26(5):598–602.

Song, H., Yoo, Y., Hwang, J., Na, Y.C., Kim, H.S. 2016. *Faecalibacterium prausnitzii* subspecies–level dysbiosis in the human gut microbiome underlying atopic dermatitis. *Journal of Allergy and Clinical Immunology* 137(3):852–860.

Sonnenburg, J.L., Backhed, F. 2016. Diet-microbiota interactions as moderators of human metabolism. *Nature* 535(7610):56–64.

Spadoni, I., Zagato, E., Bertocchi, A., Paolinelli, R., Hot, E., Di Sabatino, A., et al. 2015. A gut-vascular barrier controls the systemic dissemination of bacteria. *Science* 350(6262):830–834.

Stokes, J.H., Pillsbury, D.H. 1930. The effect on the skin of emotional and nervous states: theoretical and practical consideration of a gastrointestinal mechanism. *Archives of Dermatology* 22:962–993.

Strachan, D.P. 1989. Hay fever, hygiene, and household size. *British Medical Journal* 299(6710):1259.

Szkaradkiewicz, A., Chudzicka-Strugala, I., Karpinski, T.M., Goslinska-Pawlowska, O., Tulecka, T., Chudzicki, W., et al. 2012. *Bacillus oleronius* and Demodex mite infestation in patients with chronic blepharitis. *Clinical Microbiology and Infection* 18:1020–1025.

Szlachcic, A. 2002. The link between *Helicobacter pylori* infection and rosacea. *Journal of the European Academy of Dermatology and Venereology* 16(4):328–333.

Szlachcic, A., Sliwowski, Z., Karczewska, E., Bielański, W., Pytko-Polonczyk, J., Konturek, S.J. 1999. *Helicobacter pylori* and its eradication in rosacea. *Journal of Physiology and Pharmacology: An Official Journal of the Polish Physiological Society* 50(5):777–786.

Takizawa, K. 1977. A study on the characterization of Corynebacterium acnes. *The Journal of Dermatology* 4(5): 193–202.

Tan, L., Zhao, S., Zhu, W., Wu, L., Li, J., Shen, M. et al. 2018. The Akkermansia muciniphila is a gut microbiota signature in psoriasis. *Experimental Dermatology* 27(2):144–149.

Tanaka, A., Cho, O., Saito, C., Saito, M., Tsuboi, R., Sugita, T. 2016. Comprehensive pyrosequencing analysis of the bacterial microbiota of the skin of patients with seborrheic dermatitis. *Microbiology and Immunology* 60(8):521–526.

Tauber, M., Balica, S., Hsu, C.Y., Jean-Decoster, C., Lauze, C., Redoules, D., et al. 2016. *Staphylococcus aureus* density on lesional and nonlesional skin is strongly associated with disease severity in atopic dermatitis. *Journal of Allergy and Clinical Immunology* 137(4):1272–1274.

The Human Microbiome Project Consortium. 2012. Structure, function and diversity of the healthy human microbiome. *Nature* 486:207–214.

Tomida, S., Nguyen, L., Chiu, B.H., Liu, J., Sodergren, E., Weinstock, G.M., et al. 2013. Pan-genome and comparative genome analyses of *Propionibacterium* acnes reveal its genomic diversity in the healthy and diseased human skin microbiome. *MBio* 4(3):e00003–13.

Troccaz, M., Gaïa, N., Beccucci, S., Schrenzel, J., Cayeux, I., Starkenmann, C., Lazarevic, V. 2015. Mapping axillary microbiota responsible for body odours using a culture-independent approach. *Microbiome* 3(1):3.

Tsiouris, C.G., Kelesi, M., Vasilopoulos, G., Kalemikerakis, I., Papageorgiou, E.G. 2017. The efficacy of probiotics as pharmacological treatment of cutaneous wounds: meta-analysis of animal studies. *European Journal of Pharmaceutical Sciences* 104:230–239.

Two, A.M., Wu, W., Gallo, R.L., Hata, T.R. 2015. Rosacea: part I. Introduction, categorization, histology, pathogenesis, and risk factors. *Journal of the American Academy of Dermatology* 72(5):749–758.

Utaş, S., Özbakir, Ö., Turasan, A., Utaş, C. 1999. *Helicobacter pylori* eradication treatment reduces the severity of rosacea. *Journal of the American Academy of Dermatology* 40(3):433–435.

Van Nimwegen, F.A., Penders, J., Stobberingh, E.E., Postma, D.S., Koppelman, G.H., Kerkhof, M., et al. 2011. Mode and place of delivery, gastrointestinal microbiota, and their influence on asthma and atopy. *Journal of Allergy and Clinical Immunology* 128:948–955.

Vijayashankar, M., Raghunath, N. 2012. Pustular psoriasis responding to probiotics – a new insight. *Our Dermatology Online* 3:326–329.

Volkova, L.A., Khalif, I.L., Kabanova, I.N. 2001. Impact of the impaired intestinal microflora on the course of acne vulgaris. *Klinicheskaia meditsina* 79(6):39–41.

Von Hertzen, L., Hanski, I., Haahtela, T. 2011. Biodiversity loss and inflammatory diseases are two global megatrends that might be related. Natural immunity. *EMBO Reports* 12(11):1089–1093.

Wang, M., Karlsson, C., Olsson, C., Adlerberth, I., Wold, A.E., Strachan, D.P., et al. 2008. Reduced diversity in the early fecal microbiota of infants with atopic eczema. *Journal of Allergy and Clinical Immunology* 121(1):129–134.

Wang, Y., Kuo, S., Shu, M., Yu, J., Huang, S., Dai, A., et al. 2014. Staphylococcus epidermidis in the human skin microbiome mediates fermentation to inhibit the growth of *Propionibacterium* acnes: implications of probiotics in acne vulgaris. *Applied Microbiology and Biotechnology* 98:411–424.

Wanke, I., Steffen, H., Christ, C., Krismer, B., Götz, F., Peschel, A., et al. 2011. Skin commensals amplify the innate immune response to pathogens by activation of distinct signaling pathways. *Journal of Investigative Dermatology* 131(2):382–390.

Watanabe, S., Narisawa, Y., Arase, S., Okamatsu, H., Ikenaga, T., Tajiri, Y., et al. 2003. Differences in fecal microflora between patients with atopic dermatitis and healthy control subjects. *Journal of Allergy and Clinical Immunology* 111(3):587–591.

Watanabe, J., Sasajima, N., Aramaki, A., Sonoyama, K. 2008. Consumption of fructo-oligosaccharide reduces 2,4-dinitrofluorobenzene-induced contact hypersensitivity in mice. *British Journal of Nutrition* 100:339–346.

Weise, C., Zhu, Y., Ernst, D., Kühl, A.A., Worm, M. 2011. Oral administration of *Escherichia coli* Nissle 1917 prevents allergen-induced dermatitis in mice. *Experimental Dermatology* 20:805–809.

Whiteside, J.A., Voss, J.G. 1973. Incidence and lipolytic activity of *Propionibacterium* acnes (Corynebacterium acnes group I) and *P. granulosum* (C. acnes group II) in acne and in normal skin. *Journal of Investigative Dermatology* 60(2):94–97.

Whitfeld, M., Gunasingam, N., Leow, L.J., Shirato, K., Preda, V. 2011. Staphylococcus epidermidis: a possible role in the pustules of rosacea. *Journal of the American Academy of Dermatology* 64(1):49–52.

Wilkin, J., Dahl, M., Detmar, M., Drake, L., Feinstein, A., Odom, R., et al. 2002. Standard classification of rosacea: report of the National Rosacea Society Expert Committee on the classification and staging of Rosacea. *Journal of the American Academy of Dermatology* 46(4): 584–587.

Williams, H., Flohr, C. 2006. How epidemiology has challenged 3 prevailing concepts about atopic dermatitis. *Journal of Allergy and Clinical Immunology* 118(1): 209–213.

Won, T.J., Kim, B., Lim, Y.T., Song, D.S., Park, S.Y., Park, E.S., et al. 2011. Oral administration of *Lactobacillus* strains from Kimchi inhibits atopic dermatitis in NC/Nga mice. *Journal of Applied Microbiology* 110(5):1195–1202.

Woodfolk, J.A. 2007. T-cell responses to allergens. *The Journal of Allergy and Clinical Immunology* 119:280–294.

Xu, J., Saunders, C.W., Hu, P., Grant, R.A., Boekhout, T., Kuramae, E.E., et al. 2007. Dandruff-associated Malassezia genomes reveal convergent and divergent virulence traits shared with plant and human fungal pathogens. *Proceedings of the National Academy of Sciences* 104(47):18730–18735.

Xu, X., Zhang, H.Y. 2017. The immunogenetics of psoriasis and implications for drug repositioning. *International Journal of Molecular Sciences* 18(12):2650.

Yan, D., Issa, N., Afifi, L., Jeon, C., Chang, H.W., Liao, W. 2017. The role of the skin and gut microbiome in psoriatic disease. *Current Dermatology Reports* 6(2):94–103.

Yap, G.C., Loo, E.X., Aw, M., Lu, Q., Shek, L.P., Lee, B.W. 2014. Molecular analysis of infant fecal microbiota in an Asian at-risk cohort–correlates with infant and childhood eczema. *BMC Research Notes* 7(1):166.

You, G.E., Jung, B.J., Kim, H.R., Kim, H.G., Kim, T.R., Chung, D-K. 2013. *Lactobacillus sakei* lipoteichoic acid inhibits MMP-1 induced by UVA in normal dermal fibroblasts of human. *Journal of Microbiology and Biotechnology* 23(10):1357–1364.

Zakostelska, Z., Malkova, J., Klimešová, K., Rossmann, P., Hornová, M., Novosádová, I., et al. 2016. Intestinal microbiota promotes psoriasis-like skin inflammation by enhancing Th17 response. *PLoS One* 11(7):e0159539.

Zeeuwen, P.L., Boekhorst, J., van den Bogaard, E.H., de Koning, H.D., van de Kerkhof, P.M., Saulnier, et al. 2012. Microbiome dynamics of human epidermis following skin barrier disruption. *Genome Biology* 13(11):R101.

Zheng, H., Liang, H., Wang, Y., Miao, M., Shi, T., Yang, F., et al. 2016. Altered gut microbiota composition associated with eczema in infants. *PLoS One* 11(11):e0166026.

12

Role of Dysregulation of the Human Oral and Gastrointestinal Microbiome in Chronic Inflammatory Disease

Diana R. Cundell and Manuela Tripepi
Jefferson University

CONTENTS

12.1 Oral Flora during Health

The relationship between the oral microbiome and human during health is a symbiotic one (Kumar and Mason, 2015). Indigenous populations of oral commensal bacteria primarily maintain health by preventing the adhesion of pathogenic bacteria to the surface of the mouth—a phenomenon termed "colonization resistance" (Vollaard and Clasener, 1994). If the equilibrium between the host and the commensal microbiota is disturbed, e.g., by the use of antimicrobials, bacteria can breach the symbiotic relationship and become pathogenic—a situation termed "dysbiosis" (Wade, 2013; Deo and Deshmukh, 2019). Of the oral bacteria, *Streptococcus salivarius* has emerged as one of the primary players involved in maintaining a healthy equilibrium in the oral microbial community (Masdea et al., 2012). Studies have reported that the *S. salivarius* K12 strain produces a bacteriocin that inhibits the growth of Gram-negative species, often associated with halitosis and periodontitis (Masdea et al., 2012). Oral commensal bacteria can also help maintain an effective cardiovascular health through the local conversion of ingested nitrates to nitrites (Kapil et al., 2010; Wade, 2013). Nitrites enter the circulation and break down to result in the antihypertensive, anti-inflammatory agent nitric oxide (NO) (Kapil et al., 2010).

12.1.1 Composition of the Oral Microbiome

Changes in the host's diet, lifestyle, and pH all influence oral microbial composition and interactions between the microbiome and the host (Könönen, 2000; McLean, 2014). The oral microbiome is divided into core microbiome—which is common to all individuals and therefore the less variable portion—and the variable microbiome, which differs between individuals and can be influenced by environmental factors (Deo and Deshmukh, 2019; Palmer, 2014). The human core and variable oral microbiome consists of as many as several thousand species of bacteria, fungi, viruses, archaea, and protozoa, making the oral cavity the second most populated habitat after the gastrointestinal tract (Arweiler and Netuschil, 2016; Verma et al., 2018; Zhang et al., 2018). Many studies on the core microbiome have identified six major phyla, present in healthy individuals: Firmicutes, Proteobacteria, Actinobacteria, Bacteroidetes, Spirochetes, and Fusobacteria (Rosier et al., 2017). Bacteria can colonize the hard surface of teeth and the soft parts of the oral mucosa (Jenkinson and Lamont, 2005). Gingival crevices and the folds present on the tongue are habitats that allow for anaerobic growth (Jenkinson and Lamont, 2005). Both gram-positive and gram-negative bacterial species inhabit the healthy oral cavity of adults (Kilian et al., 2016; Deo and Deshmukh, 2019). Within the gram-positive group alone, just under 30 genera constitute the healthy oral microbiome, including potentially pathogenic *Streptococcus, Veillonella, Fusobacterium*, and *Prevotella* species (Kilian et al., 2016; Deo and Deshmukh, 2019).

12.1.2 Development of the Oral Microbiome

While *in utero*, the fetus is in a sterile environment and only encounters bacteria during birth (Dominguez-Bello et al., 2010; Sampaio-Maia and Monteiro-Silva, 2014). Studies report that different birth delivery methods influence the species that will constitute the microbiome of an individual (Dominguez-Bello et al., 2010; Sampaio-Maia and Monteiro-Silva, 2014). A few hours after birth, the first colonization wave of the mouth begins with pioneer gram-positive aerobic species of *Streptococcus* and *Staphylococcus* (Sampaio-Maia and Monteiro-Silva, 2014; Krishnan et al., 2017). These organisms facilitate the subsequential colonization of the second wave of colonizers by modifying the oral environment and creating better conditions for adherence and biofilm formation (Marsh, 2000). During the first year of life, oral colonization continues, mostly by species belonging to *Streptococcus, Lactobacillus, Actinomyces, Neisseria*, and *Veillonella* (Deo and Deshmukh, 2019). When teeth began to erupt, they provide new surfaces for colonization, the teeth, non-shedding part of the mouth, and the gingival crevices (Cephas et al., 2011). The availability of different surfaces creates different niches ideal for colonization by diverse communities and a gateway for pathogenic species (Patil et al., 2013). At puberty, a new change in the microbial flora results from hormonal and dietary changes (Jenkinson and Lamont, 2015). During this time, a higher rate of colonization by anaerobic species of gram-negative and spirochete occurs (Jenkinson and Lamont, 2015).

12.1.3 Methods of Studying the Oral Microbiome

Prior to 1987, the only microbes studied in the microbiome were those able to be cultured (Woese, 1987). With the development of the 16S rRNA gene-based cloning method, microbiologists were able to isolate nonculturable species and explore their phylogenetic relationship to one another (Woese, 1987). Since then, the development of DNA-based techniques such as polymerase chain reaction (PCR) amplification techniques, random amplicon cloning, PCR-restriction fragment length polymorphism (RFLP), T-RFLP, denaturing gradient gel electrophoresis (DGGE), and DNA microarray and with the advance of next-generation sequencing (NGS) information on the human microbiomes has increased rapidly (Sharma et al., 2018; Verma et al., 2018). The creation of the Human Oral Microbiome Database (HOMD), renamed the expanded Human Oral Microbiome Database (eHOMD) in 2018, now contains over 600 species identified by these methods and cataloged with a provisional naming system (Chen et al., 2010). The eHOMD lists microbes living in the aerodigestive tract sites, including the nasal passages, sinuses, throat, esophagus, and mouth (Escapa et al., 2018).

12.2 Dysbiosis of Oral Flora: Dental Caries to Periodontal Disease

Oral dysbiosis is very common with dental caries and periodontal disease affecting at least one in three (Yadav and Prakash, 2016; Peres et al., 2019) members of the global population as a whole. Individuals can present with dental caries without periodontal illness, but it is now understood that periodontal disease (PD) only appears after dental caries and gingivitis (Lu et al., 2019). Several bacterial species are associated with the pathogenesis of each species, and their interaction is described in Figure 12.1.

Dental caries appears in many forms, each associated with different oral flora patterns, but begins with the formation of plaque (bacterial biofilm) on the teeth and gingiva of the mouth (Lu et al., 2019). The process of caries production often begins in childhood with the initial colonization of the tooth surface by *Streptococcus mutans* (Lu et al., 2019). *S. mutans* is an effective "recruiter" of other oral species into the biofilm, and these multiple species then act synergistically to promote their growth and resistance to removal (Lu et al., 2019). For example, *Streptococci* are at least ten times more prevalent in the tooth plaque of dental caries patients, when compared with surfaces in the healthy mouth (Peterson et al., 2013). Dental caries results in inflammatory changes to the gingiva (gingivitis), which are associated with a change in the microflora species and their level of activation (Nowicki et al., 2018).

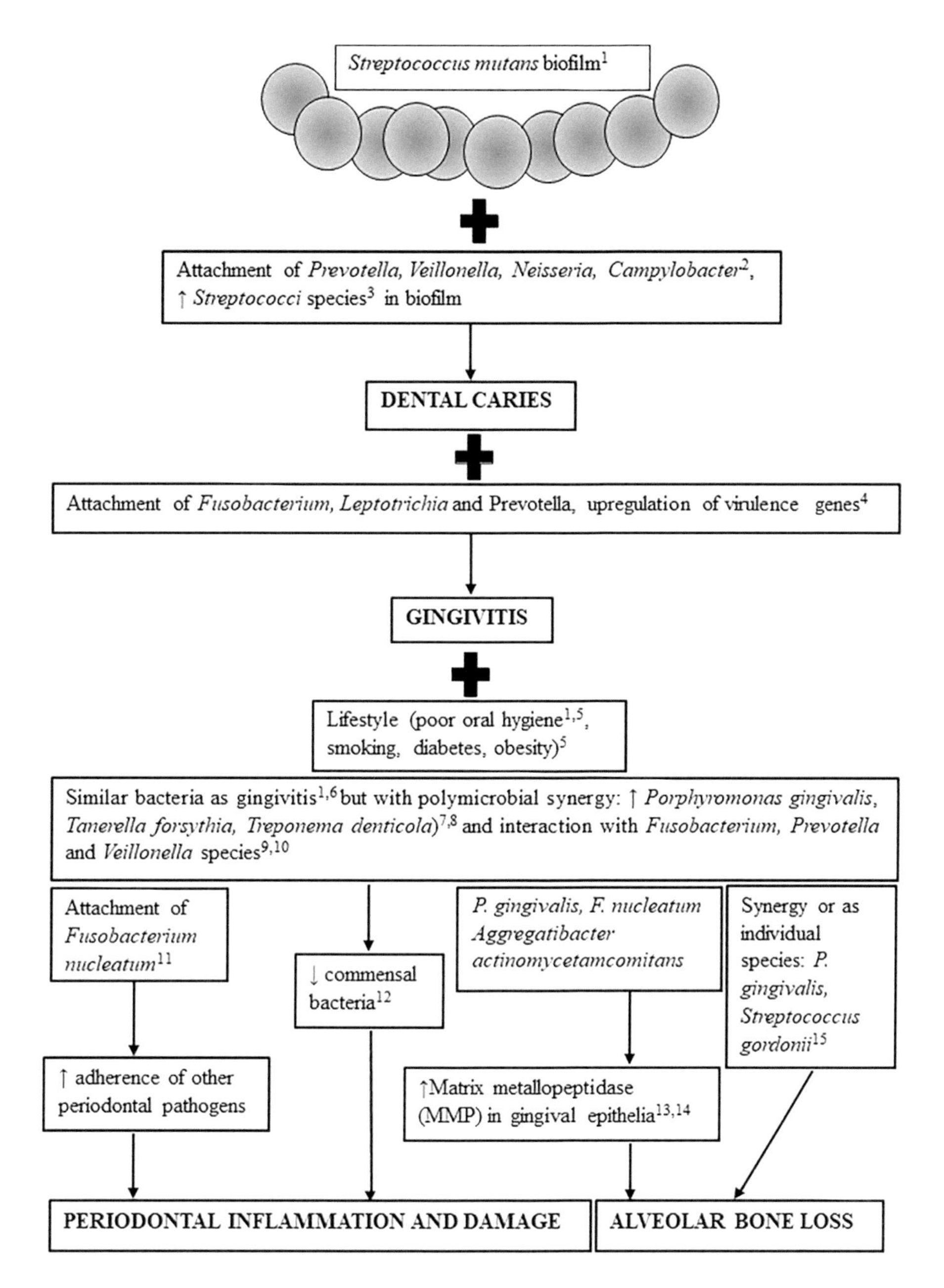

FIGURE 12.1 How dental caries becomes gingivitis and ultimately periodontal disease.

References: [1]Lu et al., 2019; [2]Tanner et al., 2019; [3]Peterson et al., 2013; [4]Nowicki et al., 2018; [5]Könönen et al., 2019; [6]Malhotra et al., 2010; [7]Lasserre et al., 2018; [8]Hajishengallis and Lamont, 2012; [9]Kumar et al., 2005; [10]Ramsey et al., 2011; [11]Bourgeois et al., 2019; [12]Scher et al., 2012; [13]Hou et al., 2013; [14]Franco et al., 2017; [15]Daep et al., 2011.

PD flora are similar to those of gingivitis, but demonstrate increased levels of interaction and virulence (Lu et al., 2019; Malhotra et al., 2010). As these species increase, normal commensals decrease (Scher et al., 2012). Several species play direct roles in PD pathology (Daep et al., 2011; Hou et al., 2013; Franco et al., 2017). *Porphyromonas gingivalis*, *Fusobacterium nucleatum*, and *Aggregatibacter actinomycetemcomitans* increase the expression of matrix metallopeptidase (MMP) enzymes in gingival epithelia (Hou et al., 2013). When upregulated, MMP enzymes can degrade type I collagen, resulting in the destruction of the periodontal ligament, cement, and underlying bone (Franco et al., 2017). Both *P. gingivalis* and *Streptococcus gordonii* can directly produce alveolar bone loss, and together produce significantly greater ($p<0.001$) level of PD and alveolar bone loss than each independently (Daep et al., 2011).

12.2.1 Leaving the Mouth: Blood and Gastrointestinal Colonization

Once PD has developed, treatment involves painful scaling and root planing (Sweeting et al., 2008). Pockets of PD are very close to the oral mucosal circulation, and as a result, bacteria and their contents can escape easily into the circulation during any event that induces bleeding (Bourgeois et al., 2019). Several of the bacteria involved in PD appear to possess the propensity to regularly "escape" into the bloodstream and may additionally be swallowed, thereby entering the gastrointestinal tract microbiome (Chukkapalli et al., 2015; Lu et al., 2019). Good evidence exists for these bacteria then colonizing the large intestine and initiating chronic inflammation of the mucosa and microbiome dysregulation (Hatton et al., 2018; du Teil et al., 2019).

Release of oropathogens into the circulation commonly occurs following periodontal treatment (40% of patients) and dental extractions (35%) (Bale et al., 2016). Many of these bacteria associate with the blood vessel endothelium, especially when it is damaged by atherosclerotic lesions (Bale et al., 2016). Several species have been identified with these capabilities, including the "red complex," *A. actinomycetemcomitans*, and *F. nucleatum* (Bale et al., 2016). Some periodontal biofilm formers may also enter synovial joints via the bloodstream and be responsible for prosthetic loosening in osteoarthritis (OA) patients (Ehrlich et al., 2014).

P. gingivalis gingipains may even cross the blood–brain barrier and be involved in the pathophysiology of Alzheimer's disease (AD) according to a recent literature study (Dominy et al., 2019). Gingipains were found in the middle temporal gyri of patients with AD, and their level was found to be correlated with the degree of tau fragmentation and ubiquitin deposition in these patients (Dominy et al., 2019). The authors also reported that a novel small-molecule inhibitor of *P. gingivalis* gingipains decreased the production of amyloid plaques in a murine model of AD (Dominy et al., 2019).

12.2.2 Pathogenic Oral Bacteria and Chronic Inflammatory Disease

Periodontal oral pathogens require virulence mechanisms to produce disease, with *F. nucleatum, P. gingivalis*, and *A. actinomycetemcomitans* able to colonize a diversity of cell types, including epithelium and endothelium (Tribble and Lamont, 2010). These three bacterial species are also the most likely to survive leaving the oral cavity and are the most frequently isolated from the distal sites (Han, 2015; Konkel et al., 2019; Belibasakis et al., 2019).

12.2.2.1 Fusobacterium Nucleatum

F. nucleatum is a normal commensal in the oral cavity (Han, 2015). Levels of the bacterium rise as oral disease progresses from gingivitis to PD when it can be isolated from both saliva and periodontal pockets (Han, 2015). *F. nucleatum* adherence to and invasion of oral epithelium is facilitated by the highly conserved adhesin FadA (Xu et al., 2007; Han, 2015). FadA induces cell–cell separation by interacting with epithelial and endothelial cell cadherins, thereby increasing cell permeability (Fardini et al., 2011; Rubinstein et al., 2013). In patients with PD and other comorbidities, *F. nucleatum* is also the most commonly isolated bacterium from atherosclerotic plaques, colorectal tumors, inflammatory bowel disease (IBD) mucosa, and the synovia of rheumatoid arthritis (RA) patients (Han and Wang, 2013). Interestingly, levels of the FadA gene directly correlate with the progression from normal tissue to adenomas in colonic epithelia (Rubinstein et al., 2013).

12.2.2.2 Porphyromonas Gingivalis and Aggregatibacter actinomycetemcomitans

Much of the understanding of how oropathogens elicit periodontal and chronic inflammatory disease has come from studies of *P. gingivalis* (Nakhjiri et al., 2001; Gyorgy et al., 2006; Foulquier et al., 2007; Lundberg et al., 2008; Pischon et al., 2009; Wegner et al., 2010; Nesse et al., 2012; Harvey et al., 2013). Frequently isolated from periodontal biofilms, *P. gingivalis* can prevent gingival cell death by decreasing epithelial apoptosis (Nakhjiri et al., 2001). *P. gingivalis* can rapidly adhere to, enter, and increase the apoptosis of human chondrocytes *in vitro* (Pischon et al., 2009). The ability of *P. gingivalis* to produce and maintain biofilms is directly connected to its secretion of gingipains, which agglutinate and lyse red blood cells, suppress immune system removal of *P. gingivalis*, and digest the periodontal matrix and bone (Li and Collyer, 2011).

P. gingivalis can directly subvert the immune response through the production of citrullinated human fibrinogen and α-enolase proteins (Wegner et al., 2010). *A. actinomycetemcomitans* also increases citrullinated protein production (Konig et al., 2016; McHugh, 2017). Antibodies against these proteins recognize both human and bacterial proteins and have long been considered the initiating factors for RA (Nesse et al., 2012). Interestingly, elevated levels of citrullinated proteins are present in patients with PD, with or without RA, suggesting their involvement in the pathophysiology of both conditions (Foulquier et al., 2007; Nesse et al., 2012; Harvey et al., 2013).

12.2.3 Oral Bacteria and Autoimmune Disease

Recently reviewed by Nikitakis and colleagues (2016), oral pathogens are believed to be involved in the initiation of autoimmune disease through different mechanisms, including pathogen persistence, molecular mimicry, epitope spreading, and Toll-like receptor (TLR) activation of immune system cells. They can also alter gene function through DNA methylation, posttranslational histone modification, and microRNA alteration without changes in DNA sequence (Nikitakis et al., 2016; Takahashi, 2014). Oropathogen persistence may also worsen the existing autoimmune conditions (van der Meulen et al., 2016; Nikitakis et al., 2016).

12.2.3.1 Rheumatoid Arthritis

RA is a chronic relapsing chronic inflammatory systemic condition that primarily damages the synovial joints and results in a significant morbidity as the disease progresses (Guo et al., 2018. Successful intervention involves the reduction of the major disease markers of rheumatoid factor (RF) as well as anti-citrullinated protein antibodies (ACPA) (Malmstrom et al., 2017). RF is an immunoglobulin M (IgM) antibody that is raised against the circulating immunoglobulin IgG (Aletaha and Smollen, 2018). ACPA are unique to RA (van de Stadt et al., 2011), and for the two-thirds of all patients that are ACPA-positive, they signal a more aggressive disease course (Malmstrom et al., 2017).

Several studies have connected the appearance of RF and/or ACPA with the presence of the primary periodontal pathogens

P. gingivalis, *A. actinomycetemcomitans*, and *F. nucleatum* (Thé and Ebersole, 1991; Hara et al., 1996; Wegner et al., 2010; Konig et al., 2016). Studies of human (Thé and Ebersole, 1991) and murine-induced PD (Hara et al., 1996) confirmed that RF antibodies were primarily induced against *F. nucleatum*. Patients with RA produce anti-*P. gingivalis* antibodies, and their titer corresponds to both the classic ACPA and RA autoantibodies (Mikuls et al., 2012). The *P. gingivalis* α-enolase shares over 80% homology with the human variant (Lundberg et al., 2008) and also appears to be a major factor in the initiation of RA (Lundberg et al., 2008).

Oral dysbiosis has been connected with RA pathology by at least six separate human studies involving over 300 patients with active disease (Moen et al., 2006; Scher et al., 2012; Zhang et al., 2015; Chen et al., 2018; Eriksson et al., 2019; Corrêa et al., 2019). Elevated levels of *Prevotella* species were observed in all but one of the studies (Zhang et al., 2015), and all studies reported changes in levels of at least two separate oropathogens.

Scher and colleagues (2012) used high-throughput bacterial DNA sequencing to relate RA oropathogens to both disease duration and PD level. Oral flora from patients with early RA disease were more likely to contain the red complex bacteria than from those with chronic disease (Scher *et al*, 2012). Chen and colleagues (2018) found that the RA oral microbiome had higher microbial diversity compared with healthy subjects. Thirteen oropathogens were more common in the RA patients, and four of them were the PD bacteria *Porphyromonas, P. melaninogenica, Actinomyces*, and *Streptococcus* (Chen et al., 2012). Small molecules, including lipopolysaccharide (LPS) and heat shock proteins (hsp), released from these oropathogens can easily escape the mouth and enter the circulation, then becoming a driver for RA (Yoshida et al., 2001; Lyu et al., 2018).

More clues on the relationship between PD and RA have come from a murine model of combined disease (de Aquino et al., 2014). Using a collagen-induced arthritis murine model of RA, de Aquino and colleagues (2014) reported that PD only developed following oral, co-inoculation of *P. gingivalis* and *Prevotella nigrescens*. Most importantly, de Aquino and colleagues (2014) observed that the addition of these PD pathogens resulted in a more rapid, aggressive joint erosion in the animals than that seen in the RA model alone.

Realistically if these two conditions are connected, then PD treatment should improve RA symptoms, and *vice versa*. A meta-analysis of five separate human clinical trials by Kaur and colleagues (2014) reported that nonsurgical PD treatment was associated with a significant improvement ($p<0.01$) in RA disease markers. Interestingly, although disease markers did show an improvement in patients, in these studies patients with PD did not report feeling less RA symptoms or change the amount of medication being taken (Okada et al., 2013; Kaur et al., 2014).

In contrast, the effects of patient RA treatment on PD symptoms have yielded the mixed results (Pers et al., 2008; Mayer et al., 2009; Kobayashi et al., 2014a, b). Improvements of PD symptoms were seen following treatment with monoclonal antibodies against tumor necrosis factor (TNF; adalimumab) and the anti-interleukin 6 receptor (IL-6R; tocilizumab) (Kobayashi et al., 2014a, b). Two studies with the McAb infliximab (nonhumanized anti-TNF antibody) produced conflicting data regarding PD symptoms in RA patients after the long-term administration (3 months) (Pers et al., 2008; Mayer et al., 2009). Most importantly, none of the trials have been the gold standard of double-blind, randomized, and placebo-controlled so it remains to be seen whether classic RA therapies can affect the underlying PD in these patients.

12.2.3.2 Systemic Lupus Erythematosus (SLE) and Sjögren's Syndrome (SS)

SLE patients produce a diversity of autoantibodies, which attack and damage oral, renal, dermal, and orthopedic tissues, among others (Brennan et al., 2005; Zucchi et al., 2019). Although SLE is predominantly associated with gastrointestinal dysbiosis (see Section 4.1 below), the recent literature reports have suggested an oral dysbiosis in patients with SLE disease (Corrêa et al., 2017; Pessoa et al., 2019). Corrêa and colleagues (2017) reported that SLE patients displayed a higher prevalence of PD, which was more severe and began at a younger age, when compared to the healthy controls. Pessoa and colleagues (2019) reported elevated levels of the pathogens *Treponema denticola* and *T. forsythia* in the subgingival PD plaque of patients with active SLE versus those in remission or age-matched subjects without SLE. Interestingly, PD patients with SLE also exhibited greater carriage of several other oropathogens, most notably *P. nigrescens*, the cariogenic "yellow complex," and the "orange complex" (*F. nucleatum* and *F. polymorphum*) (Pessoa et al., 2019).

Patients with SS exhibit poor salivary secretions and saliva flow, and frequently suffer from increased oral infections and tooth decay as a result (Pasoto et al., 2019). Plaque samples from these patients have elevated levels of the cariogenic bacteria *S. mutans* as well as *Lactobacillus* species (de Paiva et al., 2016). Patients with primary SS (no SLE) have also been reported to show increased saliva levels of *Capnocytophaga, Dialister, Fusobacterium, Helicobacter, Streptococcus*, and *Veillonella* species, while the classic PD pathogens *P. gingivalis* and *A. actinomycetemcomitans* were not detected in any patient (Sandhya et al., 2005). To complicate matters further, one in seven patients with SLE also exhibits symptoms of SS (Pasoto et al., 2019). Recent studies by van der Meulen and colleagues (2019) reported a difference between the two conditions with increased numbers of salivary *Actinomyces* and *Lactobacillus* species in SLE patients when compared with primary SS patients. Clearly, future studies are needed to separate SLE patients with poor salivary flow (i.e., a potential SS comorbidity) from those with normal saliva when determining the effects of oral dysbiosis on the pathophysiology of this autoimmune condition.

12.2.4 Oral Bacteria and Cardiovascular Disease

Cardiovascular disease (CVD) is understood to begin with the deposition of low-density lipoprotein (LDL) cholesterol on the endothelial lining of blood vessels as "fatty streaks" (Ambrose and Singh, 2015). Hyperlipidemia and cholesterol deposition is then followed by a cascade of endothelial damage, resulting in the formation of atherosclerotic plaques (Ambrose and Singh, 2015). Studies of atherosclerotic plaques from patients with PD have reported that at least two of the oropathogenic species *P. gingivalis* and *A. actinomycetemcomitans* could be detected in at least two-thirds of all lesions (Gaetti-Jardim et al., 2009; Figuero et al., 2011). Figuero and colleagues (2011) also reported that at least half of all atherosclerotic plaques from these patients contained *T. forsythia* and *F. nucleatum*.

Another sequela of atherosclerotic plaque formation is that the uneven surface supports platelet activation and clot formation (Ambrose and Singh, 2015). Oral PD pathogens, including *P. gingivalis*, can directly interact with platelets and induce local thrombus formation (Barrington and Lusis, 2017). At least seven large studies of patients with extant CVD have reported an increasing risk for myocardial infarction with more aggressive PD and resultant tooth loss (DeStefano et al., 1993; Joshipura et al., 1996; Loesche et al., 1998; Bahekar et al., 2007; Humphrey et al., 2008; Mucci et al., 2009; Wilson et al., 2018). Interestingly, one study of twins suggested that shared genetic factors may exist between CVD, tooth loss, and PD (Mucci et al., 2009). Demmitt and colleagues (2017) recently reported that loci on chromosomes 7 and 12 were associated with inherited oral microbiome constituents (Demmitt et al., 2017). Although this study did not delineate which alterations at these loci predisposed towards oral health or dysbiosis, the study by Demmitt and colleagues (2017) does add an additional facet of controlling the oral microbiome.

12.2.5 Oral Bacteria and Tumors

While commensal, probiotic bacteria prevent the development of tumors, dysbiosis and the emergence of oropathogens are hypothesized to indirectly induce tumorigenesis by generating a chronic inflammatory cascade and the production of carcinogens (Figure 12.2).

Dysbiosis of the oral microbiome and the emergence of oropathogens have been associated with at least three types of tumors (Table 12.1).

Although oropharyngeal OSCC (oral squamous cell carcinomas) tumors are primarily associated with tobacco and alcohol use (Karpiński, 2019), cariogenic *Streptococci* and *Prevotella* species, opportunistic *C. gingivalis*, and the primary PD pathogen *P. gingivalis* have all been isolated from their neoplastic lesions (Table 12.1). Precancerous squamous lesions yielded the PD-promoter bacterium *F. nucleatum* together with other normal oral commensals (Nagy et al., 1998). Oral cariogenic

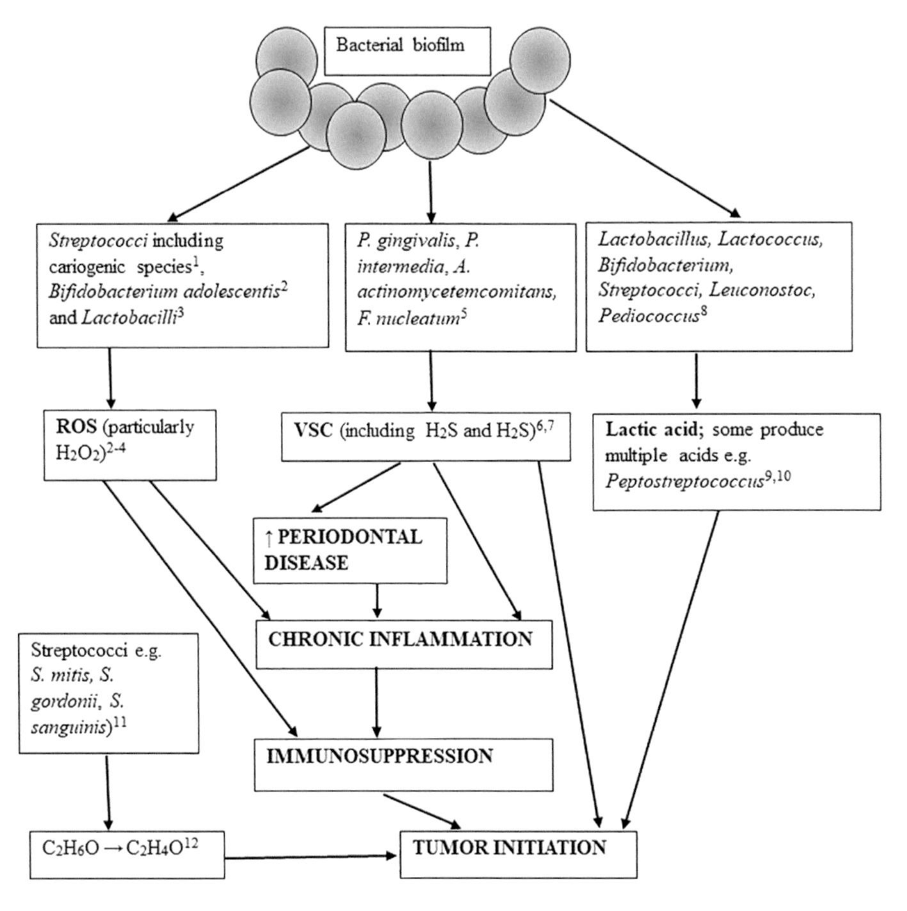

FIGURE 12.2 Mechanisms involved in oral microbiome prevention and induction of tumors.

Abbreviations: C_2H_4O, acetaldehyde; C_2H_6O, ethanol; H_2S, hydrogen sulfide; CH_3SH, methyl mercaptan; ROS, reactive oxygen species; VSC, volatile sulfur compounds.

References: [1]Cundell, 2020; [2]Abranches et al., 2018; [3]Brauncajs et al., 2001; [4]Hussain et al., 2003; [5]Karpiński, 2019; [6]Milella, 2015; [7]Hellmich and Szabo, 2015; [8]Karpiński and Szkaradkiewicz, 2013; [9]Downes and Wade, 2006; [10]Lunt et al., 2009; [11]Pavlova et al., 2013; [12]Meurman and Uittano, 2008.

TABLE 12.1

Oral Microbiome Dysbiosis and Tumors: Evidence from Multiple Reports

Fusobacterium	*Prevotella*	*Streptococci*	Other Species
Oral Squamous Cell Carcinomas (OSCC)			
↑ *[1]	↑*[1], PM[3,4]	↑*[1], SA[2], SG, SP, SS[5] Saliva ↑ SM[3,4]	↑*Clostridia*, *Veillonella*, *Porphyromonas**[1] saliva ↑ CG[3,4], ↑ PG[4] Tumor ↑ GH, GM, PS, JI[5], *Bacillus*, *Enterococcus*, *Parvimonas*, *Peptostreptococcus*, *Slackia*[6]
Pancreatic Tumors			
↑ *[7–12]↑ FN[13]			↓ *Clostridia* sp.[7,11]↑ *Porphyromonas* sp.[7] ↑ *Lactococcus* sp., ↓ *Pseudomonas*, *Shigella*, *Escherichia* sp.[12]
Colorectal Tumors			
↑ *[16, 17]		↑ SM[14]	↑ *Neisseria elongata*[14], ↑ AA[16] ↑ Antibodies to PG = ↑ cancer risk[15]

Abbreviations: *, no individual species determined; AA, *Aggregatibacter actinomycetemcomitans*; CG, *Capnocytophaga gingivalis*; GH, *Gemella haemolysans*; FN, *Fusobacterium nucleatum*; GM, *G. morbillorum*; JI, *Jonsonella ignava*; PG, *Porphyromonas gingivalis*; PM, *Prevotella melaninogenica*; PS, *Peptostreptococcus* stomatis, *Streptococcal* species; SA, *Streptococcus anginis*; SC, *S. constellatus*; SG, *S. gordonii*; SI, *S. intermedius*; SM, *S. mitis*; SO, *S. oralis*; SP, *S. parasanguinis*; SS, *S. sanguis*.

References: [1]Nagy et al., 1998; [2]Sasaki et al., 2005; [3]Magee et al., 2005; [4]Galvão-Moreira and da Cruz, 2016; [5]Pushalkar et al., 2012; [6]Lee et al., 2017; [7]Castellarin et al., 2012; [8]Kostic et al., 2012; [9]Kostic et al., 2013; [10]Tahara et al., 2014; [11]Ahn et al., 2013; [12]Gao et al., 2015; [13]Flanagan et al., 2014; [14]Farrell et al., 2012; [15]Michaud et al., 2013; [16]Mitsuhashi et al., 2015; [17]Fan et al., 2018.

bacteria were also isolated from the lymph node metastases of patients with OSCC (Sakamoto et al., 1999).

Seven studies of colorectal cancer lesions all reported isolating either *Fusobacterium* species or *F. nucleatum* with two associating these changes with a fall in beneficial *Clostridia* species (Table 12.1). Studies by Castellarin and colleagues (2012) found a correlation between elevated levels of *Fusobacterium* species or *F. nucleatum* and the development of lymph node metastases. Mima and colleagues (2016) also suggested that elevated levels of *F. nucleatum* were more common in cecal tumors (11% of those sampled) than at other large intestinal sites. At least four studies have tied pancreatic tumor development with the presence of oral bacteria in the lesions (Table 12.1). *Fusarium* species were isolated in two of the studies and *A. actinomycetemcomitans* in a third (Table 12.1), but it appeared that *P. gingivalis* contributed to the most risk of disease (Michaud et al., 2013). Indeed, antibody titers to *P. gingivalis* were found to be directly correlated with pancreatic cancer risk (Michaud et al., 2013). Oral bacteria are hypothesized to indirectly induce tumorigenesis by generating a chronic inflammatory cascade and the production of carcinogens as outlined in Figure 12.2.

12.3 Gastrointestinal Flora during Health

Over the last several years, studies focused on the composition and functions of the human gut microbiota have increased exponentially, especially aided by the improvement in the technology of DNA sequencing (Marchesi et al., 2016). The bacteria that inhabit the gut interact with virtually all human cells and perform many beneficial functions for the host (Cani, 2018). Structurally, healthy flora preserves the architecture of the epithelium (Natividad and Verdu, 2013), regulates the development and function of the immune system (Gensollen et al., 2016; Wu and Wu, 2012), and metabolizes nutrients and metabolites (Rowland et al., 2018). Furthermore, the gut microbiome plays a crucial role in preventing the growth of pathogenic bacteria by competing for resources, by preventing biofilm formation, and in some cases by producing bacteriocins (Brestoff and Artis, 2013; Garcia-Gutierrez et al., 2019).

12.3.1 Composition of the Gastrointestinal Microbiome

The human gut microbiome is composed of many species of microorganisms, including bacteria, archaea, yeasts, and viruses (Shreiner et al., 2015; Thursby and Juge, 2017). Interestingly, the colon is one of the most densely populated microbial habitats of our planet, with a bacterial cell density of around 10^{11}–10^{12} per milliliter (Ley et al., 2006). In a healthy individual, the most prevalent phyla are Firmicutes and Bacteroidetes, which comprise almost 90% of organisms, with lesser numbers belonging to the phyla Actinobacteria, Fusobacteria, Proteobacteria, and Verrucomicrobia (Arumugam et al., 2011). The ratio of Firmicutes to Bacteroidetes may vary from person to person without affecting the overall functioning of the gut microbiome (Marchesi et al., 2016). The gastrointestinal tract is composed of several anatomical regions, which harbor microbial communities with a diversity of dominant species (Jandhyala et al., 2015; Rinninella et al., 2019). *Streptococcus* species are the primary colonists of the distal esophagus, duodenum, and jejunum, due to their acidic pH (Pei et al., 2004). If present, *Helicobacter* are the predominant stomach genus (Blaser, 1999). As a healthy commensal, *Helicobacter pylori* supports the gut flora species diversity, but as a pathogen, the number of genera diminishes (Blaser, 1999; Rinninella et al., 2019).

12.3.2 Development of the Gastrointestinal Microbiome

Until birth, the intestine is sterile (Dominguez-Bello et al., 2010; Salminen et al., 2004). Infants delivered via natural birth methods acquire bacteria from the mother's vagina, such as *Lactobacillus*, *Prevotella*, and *Sneathia* (Mueller et al., 2015).

After a cesarean section, the gut microbiota contains species present in hospital settings or on the mother's skin such as *Staphylococcus, Corynebacterium*, and *Propionibacterium* sp. (Mueller et al., 2015). Throughout childhood, different factors influence the composition of the gut microbiome and its impact in later life (Penders et al., 2006; Tanaka and Nakayama, 2017). Diet is always an essential factor to consider both in early life (such as methods of milk feeding, introduction of solid food) and in adulthood (vegan or meat based) (De Filippo et al., 2010; David et al., 2014).

12.3.3 Methods of Studying the Gastrointestinal Microbiome

Current methodologies are essentially the same as for the oral microbiome (see Section 1.3 above). Researchers now understand that detailing the composition of the healthy gut microbiota represents a baseline to use in studies of dysbiosis (King et al., 2019). In order to better understand the microbes involved, a database of healthy human microbiome species and abundance profiles termed the "GutFeeling KnowledgeBase" (GutFeelingKB) has therefore been created (King et al., 2019).

12.4 Dysbiosis of Gastrointestinal Flora and Disease

Many environmental factors can affect the composition of the gastrointestinal microbiome, including diet, medications such as antibiotics, and acquisition of pathogens. As with the oral microbiome, much of the information regarding species variation in health versus disease has now been expanded due to the 16S rRNA sequencing technologies that have permitted the identification of both culturable and nonculturable organisms. Dysregulation of the gut biome typically involves a decrease in probiotic bacteria such as *Clostridia* (groups IV and IXV) and an increase in bacteria such as *Bacteroides, Streptococci*, and enterobacteria as well as a decrease in the overall diversity of species (Costello et al., 2013). A shift away from probiotic bacteria also decreases the production of their anti-inflammatory short-chain fatty acid (SCFA) metabolites. SCFA metabolites provide two-thirds of the energy needs for the colonic epithelial cells (colonocytes) (Roediger, 1982) and about one-tenth of human daily caloric needs (Bergman, 1990). More importantly, they are also involved in systemic lipid and glucose regulation (den Besten et al., 2013).

Food consumption can also affect the gut microflora (Cardona et al., 2013; Koeth et al., 2013; Queipo-Ortuño et al., 2012; Bu and Wang, 2018). Plant polyphenols are metabolized by beneficial bacteria and promote their growth (Cardona et al., 2013). Commensals and beneficial species, including *Enterococci, Bacteroides,* and *Blautia coccoides–Eubacterium rectale*, are increased significantly following red wine polyphenol consumption (Queipo-Ortuño et al., 2012). Ingestion of diets high in complex carbohydrates also promotes the growth of probiotic species, including *Bifidobacteria* (Shattat, 2014). In contrast, diets high in refined sugars and fats result in the overgrowth of opportunists such as *Clostridium difficile* and *C. perfringens,* and promote dysbiosis (Brown et al., 2012; Bu and Wang, 2018).

Omnivorous diets are more likely to reduce gastrointestinal microbiome diversity, which in turn increases the secretion of the microbial phosphatidylcholine metabolite trimethylamine or trimethylamine *N*-oxide (TMAO) (Brown et al., 2012; Bu and Wang, 2018). TMAO upregulates inflammatory receptors on endothelial cells, and elevated TMAO levels are associated with a more "proinflammatory" gut biome (Brown et al., 2012; Bu and Wang, 2018). Interestingly, red meat contains *L*-carnitine, a precursor of trimethylamine molecule (TMA), that is rapidly metabolized to TMAO (Koeth et al., 2013). Murine studies have demonstrated that foods rich in TMA molecules like *L*-carnitine can accelerate the progression of atherosclerotic disease, thus providing another link between gut dysbiosis and systemic illness (Koeth et al., 2013).

12.4.1 Gastrointestinal Dysbiosis and Autoimmune Diseases

Various bacteria that are either normal gastrointestinal commensals or pathogens possess antigenic structures that under permissive circumstances can initiate autoimmune diseases (Campbell, 2014). A classic example of this is the highly conserved Ro60 RNA binding protein that is involved in allowing cellular adaptation to stress (Greiling et al., 2018). *Corynebacteria, Propionibacteria*, and *Bacteroides* species possess Ro60 that are virtually homologous with their human hosts (Greiling et al., 2018). Bacterial Ro60 may stimulate the initiation of antibodies against them in susceptible individuals (Greiling et al., 2018). Anti-Ro antibodies are found in around half of all patients with SLE and 80% of all patients with SS (Greiling et al., 2018).

12.4.1.1 Ankylosing Spondylitis

Ankylosing spondylitis (AS) is an inherited abnormality associated with chronic inflammatory damage to the sacroiliac joints and the axial skeleton (Shamji et al., 2008). Patients display damage and swelling at the sites where the tendons enter bone (enthesitis), and ultimately, their joints become stiffened or fused (ankylosed) (Shamji et al., 2008). Most AS patients inherit a misfolded self-antigen (HLA-B27), which then elicits an autoimmune assault on these tissues following exposure to particular bacterial gastrointestinal pathogens that include *Campylobacter, Shigella, Salmonella*, and *Klebsiella* (Ebringer and Wilson, 2000; Shamji et al., 2008). AS patients often demonstrate gastrointestinal inflammation, and there are considerable numbers of patients with IBD (Costello et al., 2013).

Particular interest has been paid to *Klebsiella pneumoniae*, which appears to contribute several mechanisms to AS induction and may even be the primary pathogenic cause of active flares of this disease (Husby et al., 1989; Fielder et al., 1995; Rashid and Ebringer, 2007) as well as complications such as eye inflammation (uveitis) (Ebringer et al., 1979). There is a significant structural homology between the autoantigen HLA-B27 and the *K. pneumoniae* enzymes nitrogenase reductase (Husby et al., 1989) and pullulanase (Fielder et al., 1995). Part of the *K. pneumoniae* pullulanase enzyme also shares homology with human collagens (Fielder et al., 1995). Antibodies against these components are known to cross-react with synovial membrane tissues in the joints of HLA-B27-expressing AS patients and are elevated during active disease (Rashid and Ebringer, 2007). Interestingly, as many as one in ten AS patients goes on to develop IBDs, including Crohn's disease (CD), and nearly three-quarters have subclinical gastrointestinal inflammation or "leaky gut"—data further linking the dysbiotic gut biome to this condition (Costello et al., 2013).

12.4.1.2 Crohn's Disease

CD belongs to a group of IBD characterized by flares involving white cell damage of the colon or ileum, and symptoms of abdominal pain, diarrhea, and weight loss (Ha and Khalil, 2015). Much of the research into the pathophysiology of CD suggests it is an inappropriately aggressive response to bacterial components of the gastrointestinal biome (Yu and Huang, 2013). Unlike other dysbiosis-driven conditions discussed in this chapter, the current evidence has failed to link an individual species to CD, instead pointing to reduced species diversity, especially Firmicute numbers, and coincidental increase of Proteobacteria as contributory factors (Frank et al., 2007; Yu and Huang, 2013; Matsuoka and Kanai, 2015).

Studies by Sokol and colleagues (2009), using 16S rRNA gene analysis, reported that numbers of one particular bacterium, *Faecalibacterium prausnitzii*, were significantly reduced ($p<0.004$) in patients with CD flares. *F. prausnitzii* has been shown to ameliorate CD inflammation in a rodent model of the disease (Sokol et al., 2008) and can create an anti-inflammatory environment by upregulating T-regulator (Treg) and T-suppressor (Foxp3) cells (Qiu et al., 2013). Interestingly, *F. prausnitzii* deficit has also been associated with a second IBD, ulcerative colitis (Machiels et al., 2011)—results suggesting the bacterium may have therapeutic applications in this group of conditions.

12.4.1.3 Other Autoimmune Diseases

At least five studies have demonstrated an increase in *Prevotella copri* bacterial numbers in fecal samples from patients with RA (Scher et al., 2012, 2013; Zhang et al., 2015; Pianta et al., 2017; Alpizar-Rodriguez et al., 2019). *P. copri* was demonstrated in at least two studies to be elevated in more than 75% of patients with early-stage RA disease (Scher et al., 2012; Alpizar-Rodriguez et al., 2019). Interestingly, increases in the diversity of *Lactobacilli* have also been reported in early-stage RA patients (Liu et al., 2013). Other possible gastrointestinal bacterial changes reported in patients with RA are either a combined decrease of *Haemophilus* species and increase in *Lactobacillus salivarius* (Scher et al., 2012) or a decrease in *Bacteroides* species (Zhang et al., 2015).

At least four studies have reported a significant decrease in Firmicutes and an increase in Bacteroidetes in the gastrointestinal microbiome of patients with SLE when compared with healthy controls (Hevia et al., 2014; He et al., 2016; Greiling et al., 2018; van der Meulen et al., 2019). The largest and most detailed study compared the fecal microbiome of patients with SLE with that of those with SS (van der Meulen et al., 2019). Unlike the oral microbiome of these patients that demonstrated clear differences between patients with these two autoimmune conditions (see Section 2.5 for details), van der Meulen and colleagues found the SLE and SS gastrointestinal flora to be very similar. In addition to the altered Firmicutes/Bacteroidetes ratio, van der Meulen and colleagues (2019) reported an increase in a specific Bacteroidetes genus *Alistipes* in these patients. Several *Bacteroides* species were also more prominent in both SLE and SS fecal flora, including *B. ovatus*, *B. uniformis*, and *B. vulgatus* (van der Meulen et al., 2019). *Bacteroides* species are known to secrete the tumor-preventing SCFAs that both prevent tumors (Jan et al., 2002; Wei et al., 2016) and are known to bind to anti-Ro antibodies (van der Meulen et al., 2019). Since the van der Meulen study is the first differential analysis of its kind, it is difficult to know how the shifts in bacteria relate to patient symptoms but gastrointestinal dysbiosis is clearly present in both SLE and SS.

12.4.2 Gastrointestinal Dysbiosis and Osteoarthritis

At least two studies have connected OA with an increased passage of gastrointestinal bacteria or their products, e.g., LPS into the circulation (Metcalfe et al., 2012; Collins et al., 2015). Collins and colleagues (2015) reported that increasing prevalence of *Lactobacillus* and *Methanobrevibacter* species has shown a strong predictive relationship with the severity of OA disease (Mankin score). Obesity is a major risk factor for OA development (Mooney et al., 2011) and may in turn dysregulate the gastrointestinal microflora (Schott et al., 2018). Studies of a murine model of knee OA by Schott and colleagues (2018) reported a fall in beneficial *Bifidobacteria* in these patients accompanied by an increase in Firmicutes, particularly pathogenic *Clostridia* species. Schott and colleagues (2018) found that this shift in gastrointestinal pathogens was also associated with chronic systemic inflammation in which macrophages were recruited to the mouse knee synovium.

12.4.3 Gastrointestinal Dysbiosis and Tumors

Just as beneficial probiotic species, including *Propionibacteria* and *Lactobacilli* species, can prevent tumor initiation, the emergence of several different species of gastrointestinal pathogens as dominant components of the gut biome is associated with the development of stomach, breast, and colon cancer (Figure 12.3). Gastric cancer results from prior *H. pylori* infection and the ability of this bacterium to secrete cytotoxin-associated gene A (CagA) (Lara-Tejero and Galán, 2000). CagA has a diversity of effects, including an impairment of apoptosis of damaged cells, and separation of cells by disrupting E-cadherin (Figure 12.3). These effects then activate the host intracellular mitogen-associated protein kinase pathways (MAPK) enhancing the proliferation of transformed tumor cells (Bronte-Tinkew et al., 2009).

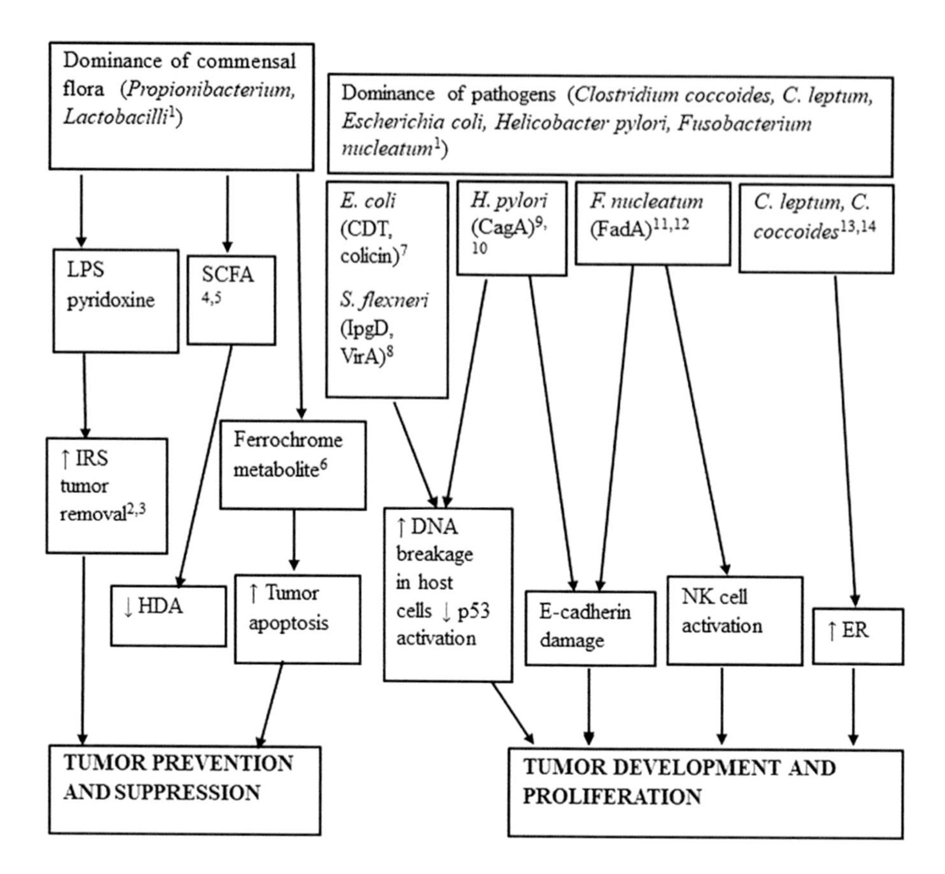

FIGURE 12.3 Mechanisms of gastrointestinal biome prevention and induction of tumors.

Abbreviations: AvrA, avirulence protein A; CagA, cytotoxin-associated gene A; CDT, cytolethal distending toxin; ER, estrogen receptors; FadA, *Fusobacterium nucleatum* effector adhesin A; HDA, histone deacetylases; IpgD, inositol phosphate phosphatase D; LPS, lipopolysaccharide; NK, natural killer; SCFA, short-chain fatty acids; VirA, cysteine protease-like virulence gene A.

References: [1]Vivarelli et al., 2019; [2]Paulos et al., 2007; [3]Aranda et al., 2015; [4]Wei et al., 2016; [5]Jan et al., 2002; [6]Konishi et al., 2016; [7]Lara-Tejero and Galán, 2000; [8]Bergounioux et al., 2012; [9]Buti et al., 2011; [10]Murata-Kamiya et al., 2007; [11]Rubinstein et al., 2013; [11]Lu et al., 2014; [12]Gur et al., 2015; [13]Doisneau-Sixou et al., 2003; [14]Plottel and Blaser, 2011.

Colorectal cancer risk is associated with increased levels of the periodontal pathogen *F. nucleatum* (see Table 12.1 and Section 2.7 for details). The effector adhesin A protein (FadA) from *F. nucleatum* disrupts E-cadherin and impairs early removal of tumors by preventing natural killer cell proliferation (Figure 12.3). Most uniquely certain *Clostridia* species possess an enzyme (β-glucuronidase) able to liberate free estrogen from endogenous and plant-derived sources (Plottel and Blaser, 2011). Increased estrogen levels upregulate estrogen receptor expression on breast tissue—an event directly correlated with breast cancer (Fernández et al., 2018).

12.5 Future Directions and Final Thoughts

This chapter has presented evidence linking chronic inflammatory disease with the overgrowth of six bacterial pathogens, namely, *S. mutans*, *A. actinomycetemcomitans*, *F. nucleatum*, *K. pneumoniae*, *P. gingivalis*, and *P. copri* (Gyorgy et al., 2006; Shamji et al., 2008; Figuero et al., 2011; Scher et al., 2012; Karpiński, 2019). Oral and/or gut dysbiosis shows strong connections to the development of at least five autoimmune diseases, namely, RA (Konig et al., 2016; Chen et al., 2018; Alpizar-Rodriguez et al., 2019), CD (Yu and Huang, 2013), SLE (Greiling et al., 2018), SS (van der Meulen et al., 2019), and AS (Shamji et al., 2008). The common chronic inflammatory conditions of CVD (Figuero et al., 2011), OA (Metcalfe et al., 2012; Collins et al., 2015) and tumors are also associated with microbiome dysbiosis (Karpiński, 2019; Figures 12.2 and 12.3). So apart from recommending normal good oral hygiene, regular dental cleanings (Lu et al., 2019), a balanced diet (Bu and Wang, 2018), and consumption of probiotics (Korotkyi et al., 2019), what other strategies might remove or decrease these "trigger" pathogens in the at-risk patient?

12.5.1 Removal of Oropathogens by Vaccines

Since pathogens are still in the oral cavity and can recolonize teeth and gums once plaque is removed, this is not a permanent solution (Lu et al., 2019). As a result, several groups have been trying to produce a vaccine to eliminate the primary cariogenic oropathogens: *S. mutans* (Shanmugam et al., 2013). Early trials of active, mucosal immunization with the glucosyl transferase (CA-gtf) from *S. mutans* (Haas et al., 1997) or passively with monoclonal antibodies against *S. mutans* (Childers et al., 1994; Smith et al., 2001) have at least delayed the appearance of the bacterium in the mouth. Nasal and tonsillar administration was found to be the most effective but the efficacy was very short-lived (less than 2 months) (Shanmugam et al., 2013). Newer recombinant and DNA vaccines against *S. mutans* have produced successful prevention of animal caries in preclinical studies (Patel, 2020).

Given their importance in the initiation and perpetuation of PD, vaccine development has been attempted against the bacterial pathogen *P. gingivalis* (O'Brien-Simpson et al., 2016; Huang et al., 2018). O'Brien-Simpson and colleagues (2016) produced a successful gingipain vaccine that elicited *P. gingivalis* neutralizing antibodies in an experimental PD model. Later studies by Huang and colleagues (2018) employed a recombinant vaccine comprising *P. gingivalis* hemagglutinin gingipains (HA1 and HA2) and major fimbriae protein.

The creation of a third, this time divalent, vaccine against both *P. gingivalis* and *F. nucleatum* was recently described by Puth and colleagues (2019). Administered intranasally, this adjuvant-based vaccine contains the FadA adhesin from *F. nucleatum* as well as the major *P gingivalis* gingipain Hgp44 (Puth et al., 2019). Healthy mice produced effective mucosal antibody responses when immunized with the vaccine (Puth et al., 2019). Sera from the immunized mice prevented *P. gingivalis* and *F. nucleatum* forming *in vitro* biofilms and co-aggregating the PD pathogen *T. denticola* (Puth et al., 2019). Most importantly, the divalent vaccine decreased alveolar bone loss in a murine PD model (Puth et al., 2019). Attempts to make an individual vaccine against *F. nucleatum* have been directed primarily towards the prevention of colon cancer (Lipkin and Monroe, 2016). Funded through the National Institutes of Health (NIH), the vaccine is being created from *F. nucleatum* outer membrane vesicles (Lipkin and Monroe, 2016). The current status of this project is unknown. Finally, a targeted approach to the respiratory and gastrointestinal pathogen *K. pneumoniae* was recently described by Feldman and colleagues (2019). A bioconjugate vaccine produced using the capsule of *K. pneumoniae* was able to protect mice from pulmonary infection with the bacteria (Feldman et al., 2019). None of these vaccines have yet been tested in human subjects.

12.5.2 Removal and Reduction of Oropathogens by Antimicrobial Agents

Antibiotic therapy, even when only applied locally to diseased gum pockets, is problematic as around 20% of PD bacterial species are resistant to the most commercially available antibiotics (Gerits et al., 2017). Other avenues are now being explored, including the use of natural plant-based derivatives, probiotic lozenges, and quorum-sensing inhibitors (Chaturvedi et al., 2016; Gerits et al., 2017; Basu et al., 2018).

Plant products—including polyphenols from cranberries and green tea, carvacrol from thyme, and capsaicin from chili peppers—have also been found to inhibit *P. gingivalis* biofilms and gingipain secretion (Gerits et al., 2017). Freshly squeezed pomegranate juice, which is high in polyphenols, has been reported as a highly effective mouthwash and reduced numbers of *Lactobacillus* and *Streptococci* in healthy volunteers (Basu et al., 2018). Short-term use of chewable tablets containing *Lactobacillus* probiotics has lowered some bacterial pathogens for up to 2 months (Jayaram et al., 2016). A small study by Chaturvedi and colleagues (2016) reported that orthodontic patients who ate lozenges containing the probiotic *Lactococcus brevis* daily were able to eliminate *S. mutans* after 30 days. The last decade has also seen a growth in the development of small molecules that are able to prevent the development of *P. gingivalis* alone as well as mixed biofilms containing *F. nucleatum* or *S. mutans* (Gerits et al., 2017). No quorum-sensing agents have yet moved past animal phase trials.

Antimicrobial photodynamic therapy (APDT) kills bacteria by generating reactive oxygen species and is in current use for caries, gingival inflammation, and endodontic instrument sterilization (Al-Shammery et al., 2019). The cold atmospheric plasma (CAP) technique, previously used in managing multiresistant skin flora in wound infections, has been recently adapted for dental restorations, including root canals and implants (Liu et al., 2016). Short-term (less than 10 minute) exposure to CAP therapy has been shown to eliminate apical periodontal pathogens and prevent their recolonization of root canal surgeries (Wang et al., 2011). Finally, research into dental implants and adhesives that contain "contact killing" nanoparticulate silver (NPAg), quaternary ammonium compounds (QAC), or antimicrobial peptides (AMPs) has shown promise as they are able to prevent plaque formation (Park, 2014; Padovani et al., 2015; Jiao et al., 2019). Unfortunately, the agents created to date have limitations of either short duration of efficacy, discoloration of the dental product, or possible oral toxicity (Jiao et al., 2019).

12.5.3 Re-Establishing Gastrointestinal Health

Several animal studies have suggested that the administration of the probiotic bacteria *L. casei* and *L. acidophilus* can both help re-establish normal gastrointestinal flora and reduce exacerbations of OA (So et al., 2011; Amdekar et al., 2013; Korotkyi et al., 2019). The most recent studies, using a rodent model of knee OA, demonstrated that 1 month of a probiotic diet plus chondroitin sulfate administration reduced the expression of the markers of inflammation and collagen degradation (Korotkyi et al., 2019). General dysbiosis, which is the hallmark of IBDs like CD, may also be addressed by the use of a fecal transplant (Gutin et al., 2019). As fecal transplant was previously only used in cases of repeated *C. difficile* growth, ten CD patients were given a fecal transplant by endoscopy in a small pilot study (Gutin et al., 2019). Patients were evaluated after 1 month, and only three of the ten responded well, demonstrating a significant lessening of abdominal symptoms ($p<0.05$) as well as colonization of the gastrointestinal microbiome with beneficial bacteria (Gutin et al., 2019). The effects of fecal transplants on abdominal symptoms in AS patients are currently being explored in a Swedish 20- patient randomized, double-blind, placebo-controlled clinical trial (Hitunen, 2018). The results of this study are expected in 2021. More research is needed to identify which patients with gastrointestinal dysbiosis would benefit from either or both of these strategies.

12.5.4 Final Thoughts

Although the targeted solutions that are being developed do hold promise, it is important to remember when and how these species identified as linked with chronic inflammation enter the irrespective biomes (Caufield and Walker, 1989; DiRienzo, 1991; van Steenbergen et al., 1993; Broadley, 2017; Tett et al., 2019). Cariogenic *S. mutans* is probably passed from mother to infant (Caufield and Walker, 1989) and *A. actinomycetemcomitans* in early childhood between family members (DiRienzo, 1991). *P. gingivalis* and *F. nucleatum* are transmissible through saliva contact, throughout life, and between genetically unrelated people (van Steenbergen et al., 1993). Both *F. nucleatum* (Broadley, 2017) and *K. pneumoniae* (Podschun and Ullman, 1998) can also survive in soil and water. *K. pneumoniae* and *P. copri* are present in the healthy gastrointestinal biome (Lau et al., 2008; Tett et al., 2019). *K. pneumoniae* is also a normal oral flora commensal (Lau et al., 2008).

Repeated exposure to the oral bacteria *S. mutans, P. gingivalis*, and *F. nucleatum* is why strategies like probiotics, lozenges, antibiotics, and dental cleanings can only provide a temporary alleviation of oral dysbiosis (Lu et al., 2019). In contrast, vaccines given at a point before colonization has occurred hold the most long-term promise for protection against *S. mutans, F. nucleatum, A. actinomycetemcomitans*, and *P. gingivalis* (Lu et al., 2019). Any vaccine against the cariogenic *S. mutans* would have to precede the teeth development (Shanmugam et al., 2013). In contrast, *P. gingivalis* is absent from the oral cavity until after childhood (van Steenbergen et al., 1993). Therefore, administration of the *F. nucleatum* and *P. gingivalis* divalent vaccine in childhood or even infancy would prevent both these bacteria and the associated pathogen *A. actinomycetemcomitans* from colonizing the oral cavity (Puth et al., 2019).

For gastrointestinal dysbiosis, diet and probiotics would appear to have more potential to maintain a healthy biome and reduce obesity-induced OA and CVD, for example (Schott et al., 2018; Bu and Wang, 2018). *F. nucleatum*-mediated colon cancer and *K. pneumoniae* infections (Feldman et al., 2019) may soon be vaccine-preventable (Lipkin and Monroe, 2016; Puth et al., 2019). This leaves *P. copri*, which has been avoided until now as it has always been considered a beneficial microbe (Alpizar-Rodriguez et al., 2019). However, recent studies by Tett and colleagues (2019) have suggested that *P. copri* rather than being one species has at least four separate clades. At least two of these are disease-associated (A and C) with a third (B) maintaining gut health (Tett et al., 2019). This latter clade is also underrepresented in individuals eating "Western" diets (Tett et al., 2019). Transplantation of group B *P. copri* and potentially more specific mixtures of beneficial species such as *Lactobacilli* may therefore provide future alternative strategies that replace fecal transplants in optimizing the long-term gastrointestinal health for patients with general dysbiotic diseases such as AS and CD.

REFERENCES

Abranches, J., L. Zeng, J.K. Kajfasz, S.R. Palmer, B. Chakraborty, Z.T. Wen, V.P. Richards, L.J. Brady, and J.A. Lemos. 2018. Biology of oral streptococci. *Microbiology Spectrum* 6. doi: 10.1128/microbiolspec.GPP3-0042-2018.

Ahn, J., R. Sinha, Z. Pei, C. Dominianni, J. Wu, J. Shi, J.J. Goedert, R.B. Hayes, and L. Yang. 2013. Human gut microbiome and risk for colorectal cancer. *JNCI: Journal of the National Cancer Institute* 105(24): 1907–1911. doi: 10.1093/jnci/djt300.

Aletaha, D., and J.S. Smolen. 2018. Diagnosis and management of rheumatoid arthritis: a review. *JAMA* 320(13): 1360–1372. doi: 10.1001/jama.2018.13103.

Al-Shammery, D., D. Michelogiannakis, Z.U. Ahmed, H.B. Ahmed, P.E. Rossouw, G.E. Romanos, and F. Javed. 2019. Scope of antimicrobial photodynamic therapy in orthodontics and related research: a review. *Photodiagnosis and Photodynamic Therapy* 25: 456–459. doi: 10.1016/j.pdpdt.2019.02.011.

Alpizar-Rodriguez D., T.R. Lesker, A. Gronow, B. Gilbert, E. Raemy, C. Lamacchia, C. Gabay A. Finckh, T. Strowig. 2019. *Prevotella copri* in individuals at risk for rheumatoid arthritis. *Annals of the Rheumatic Diseases* 78(5): 590–593. doi: 10.1136/annrheumdis-2018-214514.

Ambrose, J.A., and M. Singh. 2015. Pathophysiology of coronary artery disease leading to acute coronary syndromes. *F1000prime Reports* 7: 8. doi: 10.12703/P7-08.

Amdekar, S., V. Singh, A. Kumar, P. Sharma, and R. Singh. 2013. *Lactobacillus casei* and *Lactobacillus acidophilus* regulate inflammatory pathway and improve antioxidant status in collagen-induced arthritic rats. *Journal of Interferon & Cytokine Research* 33: 1–8. doi: 10.1089/jir.2012.0034.

Aranda, F., N. Bloy, J. Pesquet, B. Petit, K. Chaba, A. Sauvat, O. Kepp, N. Khadra, D. Enot, C. Pfirschke, M. Pittet, L. Zitfogel, G. Croemer, and L. Senovilla. 2015. Immune-dependent antineoplastic effects of cisplatin plus pyridoxine in non-small-cell lung cancer. *Oncogene* 34: 3053–3062. doi: 10.1038/onc.2014.234.

Arumugam, M., J. Raes, E. Pelletier, D. Le Paslier, T. Yamada, D.R. Mende, G.R. Fernandes, J. Tap, T. Bruls, J.M. Batto, M. Bertalan, N. Borruel, F. Casellas, L. Fernandez, L. Gautier, T. Hansen, M. Hattori, T. Hayashi, M. Kleerebezem, K. Kurokawa, M. Leclerc, F. Levenez, C. Manichanh, H.B. Nielsen, T. Nielsen, N. Pons, J. Poulain, J. Qin, T. Sicheritz-Ponten, S. Tims, D. Torrents, E. Ugarte, E.G. Zoetendal, J. Wang, F. Guarner, O. Pedersen, W.M. de Vos, S. Brunak, J. Doré, MetaHIT Consortium, M. Antolín, F. Artiguenave, H.M. Blottiere, M. Almeida, C. Brechot, C. Cara, C. Chervaux, A. Cultrone, C. Delorme, G. Denariaz, R. Dervyn, K.U. Foerstner, C. Friss, M. van de Guchte, E. Guedon, F. Haimet, W. Huber, J. van Hylckama-Vlieg, A. Jamet, C. Juste, G. Kaci, J. Knol, O. Lakhdari, S. Layec, K. Le Roux, E. Maguin, A. Mérieux, R. Melo Minardi, C. M'rini, J. Muller, R. Oozeer, J. Parkhill, P. Renault, M. Rescigno, N. Sanchez, S. Sunagawa, A. Torrejon, K. Turner, G. Vandemeulebrouck, E. Varela, Y. Winogradsky, G. Zeller, J. Weissenbach, S.D. Ehrlich, and P. Bork. 2011. Enterotypes of the human gut microbiome. *Nature* 473(7346): 174–180. doi: 10.1038/nature09944.

Arweiler, N.B., and L. Netuschil. 2016. The oral microbiota. In: Schwiertz, A. (ed) *Microbiota of the Human Body. Advances in Experimental Medicine and Biology*, 902. Springer, Cham.

Bahekar, A.A., S. Singh, S. Saha, J. Molnar, and R. Arora. 2007. The prevalence and incidence of coronary heart disease is significantly increased in periodontitis: a meta-analysis. *American Heart Journal* 154(5): 830–837. doi: 10.1016/j.ahj.2007.06.037.

Bale, B.F., A.L. Doneen, and D.J. Vigerust. 2016. High risk periodontal pathogens contribute to the pathogenesis of atherosclerosis. *Postgraduate Medical Journal* 0: 1–6. https://pmj.bmj.com/content/postgradmedj/early/2016/11/29/postgradmedj-2016-134279.full.pdf

Barrington, W.T., and A.J. Lusis. 2017. Atherosclerosis: association between the gut microbiome and atherosclerosis. *Nature Reviews Cardiology* 14(12): 699–700.

Basu, A., E. Masek, and J.E. Ebersole. 2018. Dietary polyphenols and periodontitis – a mini review of the literature. *Molecules* 23: 1786. doi: 10.3390/molecules23071786.

Belibasakis, G.N., T. Maula, K. Bao, M. Lindholm, N. Bostanci, J. Oscarsson, R. Ihalin, and A. Johansson. 2019. Virulence and pathogenicity properties of *Aggregatibacter actinomycetemcomitans*. *Pathogens* 8(4): 222. doi: 10.3390/pathogens8040222.

Bergman, E.N. 1990. Energy contributions of volatile fatty-acids from the gastrointestinal-tract in various species. *Physiological Reviews* 70: 567–590.

Bergounioux, J., R. Elisee, A.L. Prunier, F. Donnadieu, B. Sperandio, P. Sansonetti, and L. Arbibe. 2012. Calpain activation by the *Shigella flexneri* effector VirA regulates key steps in the formation and life of the bacterium's epithelial niche. *Cell Host Microbe* 11: 240–252. doi: 10.1016/j.chom.2012.01.013.

Blaser, M.J. 1999. Hypothesis: the changing relationships of *Helicobacter pylori* and humans: implications for health and disease. *The Journal of Infectious Diseases* 179: 1523–1530.

Bourgeois, D., C. Inquimbert, L. Ottolenghi, and F. Carrouel. 2019. Periodontal pathogens as risk factors of cardiovascular diseases, diabetes, rheumatoid arthritis, cancer, and chronic obstructive pulmonary disease—is there cause for consideration? *Microorganisms* 7: 424. doi: 10.3390/microorganisms7100424.

Brauncajs, M., D. Sakowska, and Z. Krzemiński. 2001. Production of hydrogen peroxide by *lactobacilli* colonising the human oral cavity. *Med Dośw Mikrobiol* 53: 331–336.

Brennan, M.T., M.A. Valerin, J.J. Napeñas, and P.B. Lockhart. 2005. Oral manifestations of patients with lupus erythematosus. *Dental Clinics of North America* 49(1): 127–141. doi: 10.1016/j.cden.2004.07.006.

Brestoff, J.R., and D. Artis. 2013. Commensal bacteria at the interface of host metabolism and the immune system. *Nature Immunology* 14: 676–684.

Broadley, M. 2017. Get the facts about Fusobacterium. *Nursing 2019* 47(5): 64–65. doi: 10.1097/01.NURSE.0000515524.23032.d5.

Bronte-Tinkew, D. M., M. Terebiznik, A. Franco, M. Ang, D. Ahn, H. Mimuro, C. Sasakawa, M.J. Ropeleski, R.M. Peek, and N.L. Jones. 2009. *Helicobacter pylori* cytotoxin-associated gene A activates the signal transducer and activator of transcription 3 pathway in vitro and in vivo. *Cancer Research* 69(2): 632–639. doi: 10.1158/0008-5472.CAN-08-1191.

Brown, K., D. DeCoffe, E. Moican, and D.L. Gibson. 2012. Diet-induced dysbiosis of the intestinal microbiota and the effects on immunity and disease. *Nutrients* 4: 1095–1119.

Bu, J., and Z. Wang. 2018. Cross-talk between gut microbiota and heart via the routes of metabolite and immunity. *Gastroenterology Research and Practice*. Article ID 6458094. doi: 10.1155/2018/6458094.

Buti, L., E. Spooner, A.G. Van der Veen, R. Rappuoli, A. Covacci, and H.L. Ploegh. 2011. *Helicobacter pylori* cytotoxin-associated gene A (CagA) subverts the apoptosis-stimulating protein of p53 (ASPP2) tumor suppressor pathway of the host. *Proceedings of the National Academy of Sciences of USA* 108: 9238–9243. doi: 10.1073/pnas.1106200108.

Campbell, A.W. 2014. Autoimmunity and the gut. *Autoimmune Diseases* 2014: 152428. doi: 10.1155/2014/152428.

Cani, P.D. 2018. Human gut microbiome: hopes, threats and promises. *Gut* 67(9): 1716–1725.

Cardona, F., C. Andrés-Lacueva, S. Tulipani, F.J. Tinahones, and M.I. Queipo-Ortuño. 2013. Benefits of polyphenols on gut microbiota: role in human health. *Journal of Nutritional Biochemistry* 24: 1415–1422.

Castellarin, M., R.L. Warren, J.D. Freeman, L. Dreolini, M. Krzywinski, J. Strauss, R. Barnes, P. Watson, E. Allan-Varcoe, R.A. Moore, and R.A. Holt. 2012. *Fusobacterium nucleatum* infection is prevalent in human colorectal carcinoma. *Genome Res* 22: 299–306. doi: 10.1016/j.oraloncology.2015.11.013.

Caufield, W.P., and T.M. Walker. 1989. Genetic diversity within *Streptococcus* mutans evident from chromosomal DNA restriction fragment polymorphisms. *Journal of Clinical Microbiology* 27: 274–278.

Cephas, K.D., J. Kim, R.A. Mathai, K.A. Barry, S.E. Dowd, B.S. Meline, and K.S. Swanson. 2011. Comparative analysis of salivary bacterial microbiome diversity in edentulous infants and their mothers or primary care givers using pyrosequencing. *PLoS One* 6(8): e23503. doi: 10.1371/journal.pone.0023503.

Chaturvedi, S., U. Jain, A. Prakash, A. Sharma, C. Shukla, and R. Chhajed. 2016. Efficacy of probiotic lozenges to reduce *Streptococcus* mutans in plaque around orthodontic brackets. *Journal of Indian Orthodontic Society* 50(4): 222–227. doi: 10.4103/0301-5742.192620.

Chen, B., Y. Zhao, S. Li, L. Yang, H. Wang, T. Wang, B. Shi, Z. Gai, X. Heng, C. Zhang, J. Yang, and L. Zhang. 2018. Variations in oral microbiome profiles in rheumatoid arthritis and osteoarthritis with potential biomarkers for arthritis screening. *Scientific Reports* 8: 17126. doi: 10.1038/s41598-018-35473-6.

Chen, T., W-H. Yu, J. Izard, O.V. Baranova, A. Lakshmanan, and F.E. Dewhirst. 2010. The human oral microbiome database: a web accessible resource for investigating oral microbe taxonomic and genomic information. *Database* 2010: baq013. doi: 10.1093/database/baq013.

Childers, N.K., S.S. Zhang, and S.M. Michalek. 1994. Oral immunization of humans with dehydrated liposomes containing *Streptococcus mutans* glucosyltransferase induces salivary immunoglobulin A2 antibody responses. *Oral Microbiology and Immunology* 9: 146–153.

Chukkapalli, S.S., M.F. Rivera-Kweh, I.M. Velsko, H. Chen, D. Zheng, I. Bhattacharyya, P.R. Gangula, A.R. Lucas, and L. Kesavalu. 2015. Chronic oral infection with major periodontal bacteria *Tannerella forsythia* modulates systemic atherosclerosis risk factors and inflammatory markers. *Pathogens and Disease* 73(3): ftv009. doi: 10.1093/femspd/ftv009.

Collins, K.H., H.A. Paul, R.A. Reimer, R.A. Seerattan, D.A. Hart, and W. Herzog. 2015. Relationship between inflammation, the gut microbiota, and metabolic osteoarthritis development: studies in a rat model. *Osteoarthr. Cartil.* 23: 1989–1998. doi: 10.1016/j.joca.2015.03.014.

Corrêa, J.D., D.C. Calderaro, G.A. Ferreira, S.M.S. Mendonca, G.R. Fernandes, E. Xiao, E.L. Teixeira, E.J. Leyes, D.T. Graves, and T.A. Silva. 2017. Subgingival microbiota dysbiosis in systemic lupus erythematosus: association with periodontal status. *Microbiome* 5: 34. doi: 10.1186/s40168-017-0252-z.

Corrêa, J.D., G.R. Fernandez, D.C. Calderaro, S.M.S. Mendonça, J.M. Silva, M.L. Albiero, F.Q. Cunha, E. Xiao, G.A. Ferreira, A.L. Teixeira, C. Mukherjee, E.J. Leys, T.J. Silva, and D.T. Graves. 2019. Oral dysbiosis linked to worsened periodontal condition in rheumatoid arthritis patients. *Scientific Reports* 9: 8379. doi: 10.1038/s41598-019-44674-6.

Costello, M., D. Elewaut, T.J. Kenna, and M.A. Brown. 2013. Microbes, the gut and ankylosing spondylitis. *Arthritis Research & Therapy* 15: 214. doi: 10.1186/ar4228.

Cundell, D.R. 2020. Tumoricidal activities of allicin. In: Cundell D.R. (ed) *Allicin the Natural Sulfur Compound from Garlic with Many Uses*, pp. 301–322. Nova Science Publishers, Hauppage, NY.

Daep, C.A, E.A. Novak, R.J. Lamont, and D.R. Demuth. 2011. Structural dissection and *in vivo* effectiveness of a peptide inhibitor of *Porphyromonas gingivalis* adherence to *Streptococcus gordonii*. *Infection and Immunity* 79: 67–74.

David, L.A., C.F. Maurice, R.N. Carmody, D.B. Gootenberg, J.E. Button, B.E. Wolfe, A.V. Ling, A.S. Devlin, Y. Varma, M.A. Fischbach, M.A.S.B. Biddinger, R.J. Dutton, and P.J. Turnbaugh. 2014. Diet rapidly and reproducibly alters the human gut microbiome. *Nature* 505: 559–563.

de Aquino, S.G., S.A-R.M.I. Koenders, F.A.J. van de Loo, G.J.M. Pruijn, R.J. Marijnissen, B. Walgreen, M.M. Helsen, L.A. van den Bersselaar, R.S. de Molon, M.J. Avila Campos, F.Q. Cunha, J.A. Cirelli, and W.B. van den Berg. 2014. Periodontal pathogens directly promote autoimmune experimental arthritis by inducing a TLR2- and IL-1-driven Th17 response. *The Journal of Immunology* 192(9): 4103–4111. doi: 10.4049/jimmunol.1301970.

De Filippo, C., D. Cavalieri, M. Di Paola, M. Ramazzotti, J.B. Poullet, S. Massart, S. Collini, G. Pieraccini, and P. Lionetti. 2010. Impact of diet in shaping gut microbiota revealed by a comparative study in children from Europe and rural Africa. *Proceedings of the National Academy of Sciences of USA* 107: 14691–14696.

de Paiva, C., D. Jones, M. Stern, F. Bian, Q.L. Moore, S. Corbiere, C.F. Streckfus, D.S. Hutchinson, N.J. Ajami, J.F. Petrosino, and S.C. Pflugfelder. 2016. Altered mucosal microbiome diversity and disease severity in Sjögren syndrome. *Scientific Reports* 6: 23561. doi: 10.1038/srep23561.

Demmitt, B.A., R.P. Corley, B.M. Huibregtse, M.C. Keller, J.K. Hewitt, M.B. McQueen, R. Knight, I. McDermott, and K.S. Krauter. 2017. Genetic influences on the human oral microbiome. *BMC Genomics* 18(1): 659. doi: 10.1186/s12864-017-4008-8.

den Besten, G., K. van Eunen, A.K. Grown, K. Venema, D.J. Reijngoud, and B.M. Bakker. 2013. The role of shirt-chain fatty acids in the interplay between diet, gut microbiota, and host energy metabolism. *Journal of Lipid Research* 54: 2325–2340.

Deo, P.N., and R. Deshmukh. 2019. Oral microbiome: unveiling the fundamentals. *Journal of Oral and Maxillofacial Pathology JOMFP* 23(1): 122–128. doi: 10.4103/jomfp.JOMFP_304_18.

DeStefano, F., R.F. Anda, H.S. Kahn, D.F. Williamson, and C.M. Russell. 1993. Dental disease and risk of coronary heart disease and mortality. *BMJ* 306(6879): 688–691. doi: 10.1136/bmj.306.6879.688.

DiRienzo, J.M. 1991. Probe-specific DNA fingerprinting applied to the epidemiology of periodontal bacteria and disease activity of periodontitis. In: Hamada, S., Holt, S.C., and McGhee, J.R. (eds) *Periodontal Disease: Pathogens and Host Immune Responses*, pp. 269–278. Quintessence Publishing Co, Tokyo.

Doisneau-Sixou, S.F., C.M. Sergio, J.S. Carroll, R. Hui, E.A. Musgrove, and R.L. Sutherland. 2003. Estrogen and antiestrogen regulation of cell cycle progression in breast cancer cells. *Endocrine-Related Cancer* 10: 179–186. doi: 10.1677/erc.0.0100179.

Dominguez-Bello, M.G., E.K. Costello, M. Contreras, M. Magris, G. Hidalgo, N. Fierer, and R. Knight. 2010. Delivery mode shapes the acquisition and structure of the initial microbiota across multiple body habitats in newborns. *Proceedings of the National Academy of Sciences of USA* 107(26): 11971–11975. doi: 10.1073/pnas.1002601107.

Dominy, S.S., C. Lynch, F. Ermini, M. Bendyk, A. Marczyk, A. Konradi, M. Nguyen, U. Haditsch, D. Raha, C. Griffin, L.J. Holsinger, S. Arastu-Kapur, S. Kaba, A. Lee, M.I. Ryder, B. Potempa, P. Mydel, A. Hellvard, K. Adamowicz, H. Hasturk, G.D. Waler, E.C. Reynolds, R.L.M. Faull, M.A. Curtis, M. Dragunow, and J. Potempa. 2019. *Porphyromonas gingivalis* in Alzheimer's disease brains: evidence for disease causation and treatment with small molecule inhibitors. *Science Advances* 5(1): eaau3333. doi: 10.1126/sciadv.aau3333.

Downes, J., and W.G. Wade. 2006. *Peptostreptococcus stomatis* sp. nov., isolated from the human oral cavity. *International Journal of Systematic and Evolutionary Microbiology* 56: 751–754. doi: 10.1099/ijs.0.64041-0.

du Teil, E.M., G. Gabarrini, H.J.M. Harmsen, J. Westra, A.J. van Winkelhoff, and J. Maarten van Dijl. 2019. Talk to your gut: the oral-gut microbiome axis and its immunomodulatory role in the etiology of rheumatoid arthritis. *FEMS Microbiology Reviews* 43(1): 1–18. doi: 10.1093/femsre/fuy035.

Ebringer, A., and C. Wilson. 2000. HLA molecules, bacteria and autoimmunity. *Journal of Medical Microbiology* 49: 305–311.

Ebringer, R., D. Cawdell, and A. Ebringer. 1979. Klebsiella pneumoniae and acute anterior uveitis in ankylosing spondylitis. *British Medical Journal* 1: 383.

Ehrlich, G.D., F.Z. Hu, N. Sotereanos, J. Sewicke, J. Parvizi, and P.L. Nara. 2014. What role do periodontal pathogens play in osteoarthritis and periprosthetic joint infections of the knee? *Journal of Applied Biomaterials & Functional Materials* 12(1): 13–20. doi: 10.5301/jabfm.5000203.

Eriksson, K., G. Fei, A. Lundmark, D. Benchimol, L. Lee, Y. Hu, A. Kats, S. Saevarsdottir, A.I. Catrina, B. Klinge, A.F. Andersson, L. Klareskog, K. Lundberg, L. Jannson, and T. Yucel-Lindberg. 2019. Periodontal health and oral microbiota in patients with rheumatoid arthritis. *Journal of Clinical Medicine* 8: 630.

Escapa, I.F., T. Chen, Y. Huang, P. Gajare, F.E. Dewhirst, and K.P. Lemon. 2018. New insights into human nostril microbiome from the expanded human oral microbiome database (eHOMD): a resource for the microbiome of the human aerodigestive tract. *mSystems* 3: e00187–18. doi: 10.1128/mSystems.00187-18.

Fan, X., A.V. Alekseyenko, J. Wu, A.B. Peters E.J. Jacobs, S.M. Gapstur, M.P. Purdue, C.C. Abnet, R. Stolzenberg-Solomon, G. Miller, J. Ravel, R.B. Hayes, and J. Ahn. 2018. Human oral microbiome and prospective risk for pancreatic cancer: a population-based nested case-control study. *Gut* 67: 120–127. doi: 10.1136/gutjnl-2016-312580.

Fardini, Y., X. Wang, S. Temoin, S. Nithianantham, D. Lee, M. Shoham, and Y.W. Han. 2011. Fusobacterium nucleatum adhesin FadA binds vascular endothelial cadherin and alters endothelial integrity. *Molecular Microbiology* 82: 1468–1480.

Farrell, J.J., L. Zhang, H. Zhou, D. Chia, D. Elashoff, D. Akin, B.J. Paster, K. Joshipura, and D.T. Wong. 2012. Variations of oral microbiota are associated with pancreatic diseases including pancreatic cancer. *Gut* 61: 582–588. doi: 10.1136/gutjnl-2011-300784.

Feldman, M.F., A.E.M. Bridwell, N.E. Scott, E. Vinogradov, S.R. McKee, S.M. Chavez, J. Twentyman, C.L. Stallings, D.A. Rosen, and C.M. Harding. 2019. A promising bioconjugate vaccine against hypervirulent *Klebsiella pneumoniae*. *Proceedings of the National Academy of Sciences of USA* 116(37): 18655–18663. doi: 10.1073/pnas.1907833116.

Fernández, M.F., I. Reina-Pérez, J.M. Astorga, A. Rodríguez-Carrillo, J. Plaza-Díaz, and L. Fontana. 2018. Breast cancer and its relationship with the microbiota. *International Journal of Environmental Research and Public Health* 15: 1747. doi: 10.3390/ijerph15081747.

Fielder, M., S.J. Pirt, I. Tarpey, C. Wilson, P. Cunningham, C. Ettelaie, A. Binder, S. Bansal, and A. Ebringer. 1995. Molecular mimicry and ankylosing spondylitis: possible role of a novel sequence in pullulanase of *Klebsiella pneumoniae*. *FEBS Letters*: 369. doi: 10.1016/0014-5793(95)00760-7.

Figuero, E., M. Sánchez-Beltrán, S. Cuesta-Frechoso, J.M. Tejerina, J.A. del Castro, J.M. Gutiérrez, D. Herrera, and M. Sanz. 2011. Detection of periodontal bacteria in atheromatous plaque by nested polymerase chain reaction. *Journal of Periodontology* 82: 1469–1477. doi: 10.1902/jop.2011.100719.

Flanagan, L., J. Schmid, M. Ebert, P. Soucek, T. Kunicka, V. Liska, J. Bruha, P. Neary, N. Dezeeuw, M. Tommasino, M. Jenab, J.H.M. Prehn, and D.J. Hughes. 2014. *Fusobacterium nucleatum* associates with stages of colorectal neoplasia development, colorectal cancer and disease outcome. *European Journal of Clinical Microbiology & Infectious Disease* 33: 1381. doi: 10.1007/s10096-014-2081-3.

Foulquier, C., M. Sebbag, C. Clavel, S. Chapuy-Regaud, R. Al Badine, M.C. Mechin, C. Vincent, R. Nachat, M. Yamada, H. Takahara, M. Simon, M. Guerrin, and G. Serre. 2007. Peptidyl arginine deiminase type 2 (PAD-2) and PAD-4 but not PAD-1, PAD-3, and PAD-6 are expressed in rheumatoid arthritis synovium in close association with tissue inflammation. *Arthritis & Rheumatology* 56: 3541–3553. doi: 10.1002/art.22983.

Franco, C., P. Hernández-Rios, S. Timo, C. Biguetti, and M.H. Hernández. 2017. Matrix metalloproteinases as regulators of periodontal inflammation. *International Journal of Molecular Sciences* 18(2): 440. doi: 10.3390/ijms18020440.

Frank, D.N., A.L. St. Amand, R.A. Feldman, E.C. Boedeker, N. Harpaz, and N.R. Pace. 2007. Molecular-phylogenetic characterization of microbial community imbalances in human inflammatory bowel diseases. *Proceedings of the National Academy of Sciences of USA* 104(34): 13780–13785. doi: 10.1073/pnas.0706625104.

Gaetti-Jardim, E., S.L. Marcelino, A.C.R. Feitosa, G.A. Romito, and M.J. Avila-Campos. 2009. Quantitative detection of periodontopathic bacteria in atherosclerotic plaques from coronary arteries. *Journal of Clinical Microbiology* 58(12). https://www.microbiologyresearch.org/content/journal/jmm/10.1099/jmm.0.013383-0.

Galvão-Moreira, L.V., and M.C. da Cruz. 2016. Oral microbiome, periodontitis and risk of head and neck cancer. *Oral Oncology* 53: 17–19. doi: 10.1016/j.oraloncology.2015.11.013.

Gao, Z., B. Guo, R. Gao, Q. Zhu, and H. Qin. 2015. Microbial disbiosis is associated with colorectal cancer. *Frontiers in Microbiology* 6: 20. doi: 10.3389/fmicb.2015.00020.

Garcia-Gutierrez, E., M.J. Mayer, P.D. Cotter, and A. Narbad. 2019. Gut microbiota as a source of novel antimicrobials. *Gut Microbes* 10(1): 1–21. doi: 10.1080/19490976.2018.1455790.

Gensollen, T., S.S. Iyer, D.L. Kasper, and R.S. Blumberg. 2016. How colonization by microbiota in early life shapes the immune system. *Science* 352: 539–544. doi: 10.1126/science.aad9378.

Gerits, E., N. Verstraeten, and J. Michaels. 2017. New approaches to combat *Porphyromonas gingivalis* biofilms. *Journal of Oral Microbiology* 9(1). doi: 10.1080/200002297.2017.1300366.

Greiling, T.M., C. Dehnger, X. Chen, K. Hughes, A.J. Iñiquez, M. Boccitto, D. Z. Ruiz, S.C. Renfroe, S.M. Vierra, W.E. Ruff, S. Sim, C. Kreigel, J. Glanternik, X. Chen, M. Girardi, P. Degnan, K.H. Costenbader, A.L. Goodman, S.L. Wolin, and M.A. Kriegel. 2018. Commensal orthologs of the human autoantigen Ro60 as triggers of autoimmunity in lupus. *Science Translational Medicine* 10(434): eaan2306. doi: 10.1126/scitranslmed.aan2306.

Guo, Q., Y. Wang, D. Xu, J. Nossent, N.J. Pavlos, and J. Xu. 2018. Rheumatoid arthritis: pathological mechanisms and modern pharmacologic therapies. *Bone Research* 6: 15. doi: 10.1038/s41413-018-0016-9.

Gur, C., Y. Ibrahim, B. Isaacson, R. Yamin, J. Abed, M. Gamliel, J. Enk, Y. Bar-On, N. Stanietsky-Kaynan, S. Coppenhagen-Glazer, N. Shussman, G. Almogy, A. Cuoapio, E. Hofer, D. Mevorach, A. Tabib, R. Ortenberg, G. Markel, K. Miclic, S. Jonjic, C.A. Brennan, W.S. Garrrett, G. Bachrach, and O. Mandelboim. 2015. Binding of the Fap2 protein of *Fusobacterium nucleatum* to human inhibitory receptor TIGIT protects tumors from immune cell attack. *Immunity* 42: 344–355. doi: 10.1016/j.immuni.2015.01.010.

Gutin, L., Y. Piceno, D. Fadrosh, K. Lynch, M. Zydek, Z. Kassam, B. LaMere, J. Terdiman, A. Ma, M. Somsouk, S. Lynch, and N. El-Nachef. 2019. Fecal microbiota transplant for Crohn disease: a study evaluating safety, efficacy, and microbiome profile. *United European Gastroenterology Journal* 7(6): 807–814. doi: 10.1177/2050640619845986.

Gyorgy, B., E. Toth, E. Tarcsa, A. Falus, and E.I. Buzas. 2006. Citrullination: a posttranslational modification in health and disease. *The International Journal of Biochemistry & Cell Biology* 38: 1662–1677. https://www.sciencedirect.com/science/article/abs/pii/S1357272506000938.

Ha, F., and H. Khalil. 2015. Crohn's disease: a clinical update. *Therapeutic Advances in Gastroenterology* 8(6): 352–359. doi: 10.1177/1756283X15592585.

Haas, W., J.A. Banas, P. Brandtzaeg, and B. Haneberg. 1997. Role of nasal associated lymphoid tissue in the human mucosal immune system. *Mucosal Immunol Update* 5: 4–8.

Hajishengallis, G., and R.J. Lamont. 2012. Beyond the red complex and into more complexity: the polymicrobial synergy and dysbiosis (PSB) model of periodontal disease etiology. *Molecular Oral Microbiology* 27(6): 409–419. doi: 10.1111/j.2041-1014.2012.00663.x.

Han, Y.W. 2015. *Fusobacterium nucleatum*: a commensal turned pathogen. *Current Opinion in Microbiology* 0: 141–147. https://www.ncbi.nlm.nih.gov/pmc/articles/PMC4323942/.

Han, Y.W., and X. Wang. 2013. Mobile microbiome: oral bacteria in extra-oral infections and inflammation. *J Dent Res* 92(6): 485–491. doi: 10.1177/0022034513487559.

Hara, Y., T. Kaneko, A. Yoshimura, and I. Kato. 1996. Serum rheumatoid factor induced by intraperitoneal administration of periodontopathic bacterial lipopolysaccharide. *Journal of Periodontal Research* 31(7): 502–507.

Harvey, G.P., T.R. Fitzsimmons, A.A.S.S.K. Dhamarpatni, C. Marchant, D.R. Haynes, and P.M. Bartold. 2013. Expression of peptidylarginine deiminase-2 and -4, citrullinated proteins and anti-citrullinated protein antibodies in human gingiva. *Journal of Periodontal Research* 48: 252–261. doi: 10.1111/jre.12002.

Hatton, G.B., C.M. Madla, S.C. Rabbie, and A.W. Basit. 2018. All disease begins in the gut: influence of gastrointestinal disorders and surgery on oral drug performance. *International Journal of Pharmaceutics* 548(1): 408–422. doi: 10.1016/j.ijpharm.2018.06.054.

He, Z., T. Shao, H. Li, Z. Xie, and C. Wen. 2016. Alterations of the gut microbiome in Chinese patients with systemic lupus erythematosus. *Gut Pathogenes* 8: 64. doi: 10.1186/s13099-016-0146-9.

Hellmich, M.R., and C. Szabo. 2015. Hydrogen sulfide and cancer. *Handbook of Experimental Pharmacology* 230: 233–241. https://www.ncbi.nlm.nih.gov/pmc/articles/PMC4665975/?report=reader.

Hevia, A., C. Milani, P. López, A. Cuervo, S. Arboleya, S. Duranti, F. Turroni, S. González, A. Suárez, M. Gueimonde, M. Ventura, B. Sánchez, and A. Margolles. 2014. Intestinal dysbiosis associated with systemic lupus erythematosus. *mBio* 5(5): e01548-14. doi: 10.1128/mBio.01548-14.

Hitunen, J. 2018. The safety of fecal microbiota transplantation in ankylosing spondylitis (AS) patients (ASGUT). ClinicalTrials.gov Identifier: NCT03726645. https://clinicaltrials.gov/ct2/show/NCT03726645.

Hou, G-Q., C. Guo, G-H. Song, N. Fang, W-J. Fan, X-D. Chen, L. Yuan, and Z-Q. Wang. 2013. Lipopolysaccharide (LPS) promotes osteoclast differentiation and activation by enhancing the MAPK pathway and COX-2 expression in RAW264.7 cells. *International Journal of Molecular Medicine* 32: 503–510. doi: 10.3892/ijmm.2013.1406.

Huang, N., E. Shimomura, G. Yin, C. Tran, A. Sato, A. Steiner, T. Heibeck, M. Tam, J. Fairman, and F.C. Gibson. 2018. Immunization with cell-free-generated vaccine protects from *Porphyromonas gingivalis*-induced alveolar bone loss. *Journal of Clinical Periodontology* 46: 197–205. doi: 10.1111/jcpe.13047.

Humphrey, L.L., R. Fu, D.I. Buckley, M. Freeman, and M. Helfand. 2008. Periodontal disease and coronary heart disease incidence: a systematic review and meta-analysis. *Journal of General Internal Medicine* 23(12): 2079–2086. doi: 10.1007/s11606-008-0787-6.

Husby, G., N. Tsuchiya, P.L. Schwimmbeck, A. Keat, J.A. Pahle, M.B.A. Oldstone, and R.C. Williams, Jr. 1989. Cross-reactive epitope with *Klebsiella pneumoniae* nitrogenase in articular tissue of HLA–B27+ patients with ankylosing spondylitis. *Arthritis & Rheumatism* 32: 437–445. doi: 10.1002/anr.1780320413.

Hussain, S.P., L.J. Hofseth, and C.C. Harris. 2003. Radical causes of cancer. *Nature Reviews Cancer* 3: 276–285. doi: 10.1038/nrc1046.

Jan, G., A.S. Belzacq, D. Haouzi, A. Rouault, D. Métivier, G. Kroemer, and C. Brenner. 2002. Propionibacteria induce apoptosis of colorectal carcinoma cells via short-chain fatty acids acting on mitochondria. *Cell Death & Differentiation* 9: 179–188. doi: 10.1038/sj.cdd.4400935.

Jandhyala, S.M., R. Talukdar, C. Subramanyam, H. Vuyyuru, M. Sasikala, and D. Nageshwar Reddy. 2015. Role of the normal gut microbiota. *World Journal of Gastroenterology* 21(29): 8787–8803.

Jayaram, P., A. Chatterjee, and V. Raghunathan. 2016. Probiotics in the treatment of periodontal disease: a systematic review. *Journal of Indian Society of Periodontology* 20(5): 488–495. doi: 10.4103/0972-124X.207053.

Jenkinson, H.F., and R.J. Lamont. 2005. Oral microbial communities in sickness and in health. *Trends in Microbiology* 13: 589–595.

Jiao, Y., F.R. Tay, L. Niu, and J. Chen. 2019. Advancing antimicrobial strategies for managing oral biofilm infections. *International Journal of Oral Science* 11: 28. doi: 10.1038/s41368-019-0062-1.

Joshipura, K.J., E.B. Rimm, C.W. Douglass, D. Trichopoulos, A. Ascherio, and W.C. Willett. 1996. Poor oral health and coronary heart disease. *Journal of Dental Research* 75(9): 1631–1636. doi: 10.1177/00220345960750090301.

Kapil, V., A.B. Milsom, M. Okorie, S. Maleki-Toyserkani, F. Akram, F. Rehman, S. Arghandawi, V. Pearl, N. Benjamin, S. Loukogeorgakis, R. MacAllister, A.J. Hobbs, A.J. Webb, and A. Ahluwalia. 2010. Inorganic nitrate supplementation lowers blood pressure in humans: role for nitrite-derived no. *Hypertension* 56: 274–281.

Karpiński, T.M. 2019. Role of oral microbiota in cancer development. *Microorganisms* 7(1): 20. https://www.ncbi.nlm.nih.gov/pmc/articles/PMC6352272/.

Karpiński, T.M., and A.K. Szkaradkiewicz. 2013. Characteristic of bacteriocines and their application. *Polish Journal of Microbiology* 62: 223–235. http://www.pjm.microbiology.pl/archive/vol6232013223.pdf.

Kaur, S., R. Bright, S.M. Proudman, and P.M. Bartold. 2014. Does periodontal treatment influence clinical and biochemical measures for rheumatoid arthritis? A systematic review and meta-analysis. *Seminars in Arthritis and Rheumatism* 44(2): 113–122. doi: 10.1016/j.semarthrit.2014.04.009.

Kilian, M., I. Chapple, M. Hannig, P.D. Marsh, V. Meuric, A.M.L. Pedersen, M.S. Tonnetti, W.G. Wade, and E. Zaura. 2016. The oral microbiome – an update for oral healthcare professionals. *British Dental Journal* 221: 657–666. doi: 10.1038/sj.bdj.2016.865.

King, C.H., H. Desai, A.C. Sylvetsky, J. LoTempio, S. Ayanyan, J. Carrie, K.A. Crandall, B.C. Fochtman, L. Gasparyan, N. Gulzar, P. Howell, N. Issa, K. Krampis, L. Mishra, H. Morizono, J.R. Pisegna, S. Rao, Y. Ren, V. Simonyan, K. Smith, S. VedBrat, M.D. Yao, and R. Mazumder. 2019. Baseline human gut microbiota profile in healthy people and standard reporting template. *PLoS One* 14(9): e0206484. doi: 10.1371/journal.pone.0206484.

Kobayashi, T.T., M. Okada, S. Ito, D. Kobayashi, K. Ishida, A. Kojima. 2014b. Assessment of interleukin-6 inhibition therapy on periodontal condition in patients with rheumatoid arthritis and chronic periodontitis. *Journal of Periodontology* 85: 57–67. https://aap.onlinelibrary.wiley.com/doi/abs/10.1902/jop.2013.120696.

Kobayashi, T.T., T. Yokoyama, S. Ito, D. Kobayashi, M. Okada, K. Oofusa, I. Narita, A. Murasawa, K. Nakazano, and H. Yoshie. 2014a. Periodontal and serum protein profiles in patients with rheumatoid arthritis treated with tumor necrosis factor inhibitor adalimumab. *Journal of Periodontology* 85: 1480–1499. https://dialnet.unirioja.es/servlet/articulo?codigo=5311068.

Koeth, R.A., Z. Wang, B.S. Levison, J.A. Buffa, E. Org, B.T. Sheehy, E.B. Britt, X. Fu, Y. Wu, L. Li, J.D. Smith, J.A. DiDonato, J. Chen, H. Li, G.D. Wu, J.D. Lewis, M. Warrier, J.M. Brown, R.M. Krauss, W.H. Tang, F.D. Bushman, A.J. Lusis, and S.L. Hazen. 2013. Intestinal microbiota metabolism of L-carnitine, a nutrient in red meat, promotes atherosclerosis. *Nature Medicine* 19(5): 576–585. doi: 10.1038/nm.3145.

Konig, M.F., L. Abusleme, J. Reinholdt, R.J. Palmer, R.P. Teles, K. Sampson, A. Rosen, P.A. Nigrovic, J. Sokolove, N.M. Moutsopoulos, and J.T. Giles. 2016. Aggregatibacter actinomycetemcomitans-induced hypercitrullination links periodontal infection to autoimmunity in rheumatoid arthritis. *Sci. Transl. Med.* 8: 369ra176. https://www.ncbi.nlm.nih.gov/pmc/articles/PMC5384717/.

Konishi, H., M. Fujiya, H. Tanaka, N. Ueno, K. Moriichi, J. Sasajima, K. Ikuta, H. Akutsu, H. Tanabe, and Y. Kohgo. 2016. Probiotic-derived ferrichrome inhibits colon cancer progression via JNK-mediated apoptosis. *Nature Communications* 7: 12365. doi: 10.1038/ncomms12365.

Konkel, J.E., C. O'Boyle, and K. Siddharth. 2019. Distal consequences of oral inflammation. *Frontiers in Immunology* 10: 1403. doi: 10.3389/fimmu.2019.01403.

Könönen, E. 2000. Development of oral bacterial flora in young children. *Annals of Medicine* 32(2): 107–112.

Könönen, E., M. Gursoy, and U.K. Gursoy. 2019. Periodontitis: a multifaceted disease of tooth-supporting tissues. *Journal of Clinical Medicine* 8(8): 1135. doi: 10.3390/jcm8081135.

Korotkyi, O.H., A.A. Vovk, A.S. Dranitsina, T.M. Falalyeyeva, K.O. Dvorshchenko, S. Fagoonee, and L.I. Ostapchenko. 2019. The influence of probiotic diet and chondroitin sulfate administration on Ptgs2, Tgfb1 and Col2a1 expression in rat knee cartilage during monoiodoacetate-induced osteoarthritis. *Minerva Medica* 110: 419–424. doi: 10.23736/S0026-4806.19.06063-4.

Kostic, A.D., D. Gevers, C.S. Pedamallu, M. Michaud, F. Duke, A.M. Earl, A.I. Ojesina, J. Jung, A.J. Bass, J. Taberno, J. Baselga, C. Liu, R.A. Shivdasani, S. Ogino, B.W. Birren, C. Huttenhower, W.S. Garrett, and M. Meyerson. 2012. Genomic analysis identifies association of *Fusobacterium* with colorectal carcinoma. *Genome Research* 22: 292–298. doi: 10.1101/gr.126573.111.

Kostic, A.D., E. Chun, L. Robertson, J.N. Glickman, C.A. Gallini, M. Michaud, T.E. Clancy, D.C. Chung, P. Lockhead, G.L. Hold, E.M. El-Omar, D. Brenner, C.S. Fuchs, M. Meyerson, and W.S. Garrett. 2013. *Fusobacterium nucleatum* potentiates intestinal tumorigenesis and modulates the tumor-immune microenvironment. *Cell Host Microbe* 14: 207–215. doi: 10.1016/j.chom.2013.07.007.

Krishnan, K., T. Chen, and P.J. Paster. 2017. A practical guide to the oral microbiome and its relation to health and disease. *Oral Diseases* 23(3): 276–286. doi: 10.1111/odi.12509.

Kumar, P.S., A.L. Griffen, M.L. Moeschberger, and E.J. Leys. 2005. Identification of candidate periodontal pathogens and beneficial species by quantitative 16S analysis. *Journal of Clinical Microbiology* 43(8): 3944–3955. doi: 10.1128/JCM.43.8.3944-3955.2005.

Kumar, P.S., and M.R. Mason. 2015. Mouthguards: does the indigenous microbiome play a role in maintaining oral health? *Frontiers in Cellular and Infection Microbiology* 5: 35. doi: 10.3389/fcimb.2015.00035.

Lara-Tejero, M., and J.E. Galán. 2000. A bacterial toxin that controls cell cycle progression as a deoxyribonuclease I-like protein. *Science* 290: 354–357. doi: 10.1126/science.290.5490.354.

Lasserre, J.F., M.C. Brecx, and S. Toma. 2018. Oral microbes, biofilms and their role in periodontal and peri-implant disease. *Materials (Basel)* 11(10): 1802. doi: 10.3390/ma11101802.

Lau, H.Y., G.B. Huffnagle, and T.A. Moore. 2008. Host and microbiota factors that control Klebsiella pneumoniae mucosal colonization in mice. *Microbes Infect.* 10(12–13): 1283–1290. doi: 10.1016/j.micinf.2008.07.040.

Lee, W.M., H.M. Chen, S.F. Yang, C. Liang, C.Y. Peng, F.M. Lin, L.L. Tsai, B.C. Wu, C.H. Hsin, C.Y. Huang, T. Yang, T.L. Yang, S-Y. Ho, W-L. Chen, K-C. Ueng, H-D. Huang, C-N. Huang, and Y-J. Jong. 2017. Bacterial alterations in salivary microbiota and their association in oral cancer. *Scientific Reports* 7: 16540. doi: 10.1038/s41598-017-16418-x.

Ley, R.E., P.J. Turnbaugh, S. Klein, and J.I. Gordon. 2006. Microbial ecology: human gut microbes associated with obesity. *Nature* 444: 1022–1023.

Li, N., and C.A. Collyer. 2011. Gingipains from *Porphyromonas gingivalis* – complex domain structures confer diverse functions. *European Journal of Microbiology & Immunology* 1(1): 41–58. doi: 10.1556/EuJMI.1.2011.1.7.

Lipkin, and S. Monroe. 2016. Igf:OT:iGf anti *Fusobacterium nucleatum* vaccine for colorectal cancer immunoprevention. NIH Grant 2612015000391-0-26100004-1. http://grantome.com/grant/NIH/261201500039I-0-26100004-1.

Liu, X., Q. Zou, B. Zeng, Y. Fang, and H. Wei. 2013. Analysis of fecal lactobacillus community structure in patients with early rheumatoid arthritis. *Current Microbiology* 67: 170–176. doi: 10.1007/s00284-013-0338-1.

Liu, Y., Q. Liu, Q.S. Yu, and Y. Wang. 2016. Nonthermal atmospheric plasmas in dental restoration. *Journal of Dental Research* 95: 496–505.

Loesche, W.J., A. Schork, M.S. Terpenning, Y-M. Chen, B.N. Dominguez, and N. Grossman. 1998. Assessing the relationship between dental disease and coronary heart disease in elderly U.S. veterans. *Journal of the American Dental Association* 129(3): 301–311. doi: 10.14219/jada.archive.1998.0204.

Lu, M., S. Xuan, and Z. Wang. 2019. Oral microbiota: a new view of body health. *Food Science and Human Health* 8: 8–15. doi: 10.1016/j.fshw.2018.12.001.

Lu, R., S. Wu, Y.G. Zhang, Y. Xia, X. Liu, Y. Zheng, H. Chen, K.L. Schaefer, Z. Zhou, M. Bissonnette, L. Li, and J. Sun. 2014. Enteric bacterial protein AvrA promotes colonic tumorigenesis and activates colonic beta-catenin signaling pathway. *Oncogenesis* 3: e105. doi: 10.1038/oncsis.2014.20.

Lundberg, K., A. Kinloch, B.A. Fisher, N. Wegner, R. Wait, P. Charles, T.R. Mikuls, and P.J. Venables. 2008. Antibodies to citrullinates alpha-enolase peptide 1 are specific for rheumatoid arthritis and cross-react with bacterial enolase. *Arthritis & Rheumatology* 58: 3009–3019. doi: 10.1002/art.23936.

Lunt, S.J., N. Chaudary, and R.P. Hill. 2009. The tumor microenvironment and metastatic disease. *Clinical and Experimental Metastasis* 26: 19–34. doi: 10.1007/s10585-008-9182-2.

Lyu, M., J. Chen, Y. Jiang, W. Dong, Z. Fang, and S. Li. 2018. KDiamend: a package for detecting key drivers in a molecular ecological network of disease. *BMC Systems Biology* 12: 5. doi: 10.1186/s12918-018-0531-8.

Machiels, K., Joossens, M., De Preter, V., Arijs, I., Ballet, V., Organe, S., Coopmans, T., Van Assche, G., Verhaegen, J., Rutgeerts, P., and Vermeire, S. 2011. Association of *Faecalibacterium prausnitzii* and disease activity in ulcerative colitis. *Gastroenterology* 140(5): 860.

Magee, D., A. Haffajee, P. Devlin, C. Norris, M. Posner and J. Goodson. 2005. The salivary microbiota as a diagnostic indicator of oral cancer. A descriptive, non-randomized study of cancer free and oral squamous cell carcinoma subjects. *Journal of Translational Medicine* 3: 27. doi: 10.1186/1479-5876-3-27.

Malhotra, R., A. Kapoor, V. Grover, and S. Kaushal. 2010. Nicotine and periodontal issues. *Journal of Indian Society of Periodontology* 14(1): 72–79. doi: 10.4103/0972-124X.65442.

Malmstrom, V., A.I. Catrina, and L. Klareskog. 2017. The immunopathogenesis of seropositive rheumatoid arthritis: from triggering to targeting. *Nature Reviews Immunology* 17: 60–75.

Marchesi, J.R., D.H. Adams, F. Fava, G.D. Hermes, G.M. Hirschfield, G. Hold, M.N. Quraishi, J. Kinross, H. Smidt, K.M. Tuohy, L. V. Thomas, E.G. Zoetendal, and A. Hart. 2016. The gut microbiota and host health: a new clinical frontier. *Gut* 65(2): 330–339. doi: 10.1136/gutjnl-2015-309990.

Marsh, P.D. 2000. Role of the oral microflora in health. *Microbial Ecology in Health and Disease* 12(3): 130–137. doi: 10.1080/089106000750051800.

Masdea, L., E.M. Kulik, I. Hauser- Gerspach, A.M. Ramseier, A. Filippi, and T. Waltimo. 2012. Antimicrobial activity of *Streptococcus salivarius* K12 on bacteria involved in oral malodour. *Arch Oral Biol* 57(8): 1041–1047.

Matsuoka, K. and T. Kanai. 2015. The gut microbiota and inflammatory bowel disease. *Seminars in Immunopathology* 37(1): 47–55. doi: 10.1007/s00281-014-0454-4.

Mayer, Y., A. Balbir-Gurman, and E.E. Machtei. 2009. Anti-tumor necrosis factor-alpha therapy and periodontal parameters in patients with rheumatoid arthritis. *Journal of Periodontology* 80: 1414–1420. https://aap.onlinelibrary.wiley.com/doi/abs/10.1902/jop.2009.090015.

McHugh, J. 2017. Rheumatoid arthritis: new model linking periodontitis and RA. *Nature Reviews Rheumatology* 13: 66. https://www.nature.com/articles/nrrheum.2016.221.

McLean, J.S. 2014. Advancements toward a systems level understanding of the human oral microbiome. *Frontiers in Cellular and Infection Microbiolog* 4: 98.

Metcalfe, D., A.L. Harte, M.O. Aletrari, N.M. Al Daghri, D. Al Disi, G. Tripathi, and P.G. McTernan. 2012. Does endotoxaemia contribute to osteoarthritis in obese patients? *Clinical science (London, England* 123: 627–634. doi: 10.1042/CS20120073.

Meurman, J.H., and J. Uittamo. 2008. Oral micro-organisms in the etiology of cancer. *Acta Odontologica Scandinavica* 66: 321–326. doi: 10.1080/00016350802446527.

Michaud, D.S., J. Izard, C.S. Wilhelm-Benarzti, D.H. You, V.A. Grote, A. Tjønnenland, C.C. Dahm, K. Overad, M. Jenab, V. Fedirko, M.C. Boutron-Roualt, F. Clavel-Chapelon, A. Racine, R. Kaaks, H. Boeing, J. Foerster, A. Trichopolou, P. Lagiou, D. Trichopoulos, C. Sacerdote, S. Sieri, D. Palli, R. Tumino, S. Panico, P.D. Siersema, P.H.M. Peeters, E. Lund, A. Barricarte, J-H. Huerta, E. Molina-Montes, M. Dorronsorro, J.R. Quíros, E.J. Duell, W. Ye, M. Sund, B. Lindqvist, B. Johansen, K-T. Khaw, N. Wareham, R.C. Travis, P. B. Vineis, H.B. Buena-da-Mesquita, and E. Riboli. 2013. Plasma antibodies to oral bacteria and risk of pancreatic cancer in a large European cohort study. *Gut* 62: 1764–1770. https://gut.bmj.com/content/62/12/1764.

Mikuls, T.R., G.M. Thiele, K.D. Deane, J.B. Payne, J.R. O'Dell, F. Yu, H. Sayles, M.H. Weisman, P.K. Gregersen, J.H. Buckner, R.M. Keating, L.A. Derber, W.H. Robinson, V. M. Holers, and J. M. Norris. 2012. Porphyromonas gingivalis and disease-related autoantibodies in individuals at increased risk of rheumatoid arthritis. *Arthritis and Rheumatism* 64(11): 3522–3530. doi: 10.1002/art.34595.

Milella, L. 2015. The negative effects of volatile sulphur compounds. *Journal of Veterinary Dentistry* 32: 99–102. doi: 10.1177/089875641503200203.

Mima, K., Y. Cao, A.T. Chan, Z.R. Qian, J.A. Nowak, Y. Masugi, Y. Shi, M. Song, A. da Silva, M. Gu, W. li, T. Hamada, K. Kosumi, A. Hanuyada, L. Liu, A.D. Kostic, M. Giannakis, S. Bullman, C.A. Brennan, D.A. Milner, H. Baba, L.A. Garraway, J.A. Meyerhardt, W.S. Garrett, C. Huttenhower, M. Meyerson, E.L. Giovannucci, C.S. Fuchs, R. Nishihara, and S. Ogino. 2016. *Fusobacterium nucleatum* in colorectal carcinoma tissue according to tumor location. *Clinical and Translational Gastroenterology* 7: e200. doi: 10.1038/ctg.2016.53.

Mitsuhashi, K., K. Nosho, Y. Sukawa, Y. Matsunaga, M. Ito, H. Kurihara, S. Kanno, H. Igarashi, T. Naito, Y. Adachi, M. Tachibana, T. Tanuma, H. Maguchi, T. Shinohara, T. Hasegawa, M. Imamura, Y. Kimura, K. Hirata, R. Maruyama, H. Suzuki, K. Imai, H. Yamamoto, and Y. Shinomura. 2015. Association of *Fusobacterium* species in pancreatic cancer tissues with molecular features and prognosis. *Oncotarget* 6: 7209–7220. http://www.oncotarget.com/index.php?journal=oncotarget&page=article&op=view&path%5B%5D=3109&path%5B%5D=7330.

Moen, K, J.G. Brun, M. Valen, S. Skartveit, E.K. Ribs Eribe, I. Olsen, and R. Jonssen. 2006. Synovial inflammation in active rheumatoid arthritis and psoriatic arthritis facilitates trapping of a variety of oral bacterial DNAs. *Clinical and Experimental Rheumatology* 24(6): 656–663. https://www.clinexprheumatol.org/abstract.asp?a=2980.

Mooney, R.A., E.R. Sampson, J. Lerea, R.N. Rosier, and M.J. Zuscik. 2011. High-fat diet accelerates progression of osteoarthritis after meniscal/ligamentous injury. *Arthritis Research & Therapy* 13: R198. doi: 10.1186/ar3529.

Mucci, L.A., C-c. Hsieh, P.L. Williams, M. Arora, H-O. Adami, U. da Faire, C.W. Douglass, and N.L. Pedersen. 2009. Do genetic factors explain the association between poor oral health and cardiovascular disease? A prospective study among Swedish twins. *American Journal of Epidemiology* 170(5): 615–621. doi: 10.1093/aje/kwp177.

Mueller, N.T., E. Bakacs, J. Combellick, Z. Grigoryan, and M.G. Dominguez-Bello. 2015. The infant microbiome development: mom matters. *Trends in Molecular Medicine* 21: 109–117.

Murata-Kamiya N., Y. Kurashima, Y. Teishikata, Y. Yamahashi, Y. Saito, H. Higashi, H. Aburatani, T. Akiyama, R.M. Peek, T. Azuma, and M. Hatakeyama. 2007. *Helicobacter pylori* CagA interacts with E-cadherin and deregulates the beta-catenin signal that promotes intestinal transdifferentiation in gastric epithelial cells. *Oncogene* 26: 4617–4626. doi: 10.1038/sj.onc.1210251.

Nagy, K.N., I. Sonkodi, I. Szöke, E. Nagy, and H.N. Newman. 1998. The microflora associated with human oral carcinomas. *Oral Oncology* 34: 304–308. doi: 10.1016/S1368-8375(98)80012-2.

Nakhjiri, S.F., Y. Park, O. Yilmaz, W.O. Chung, K. Watanabe, A. El-Sabaeny, K. Park, and R.J. Lamont. 2001. Inhibition of epithelial cell apoptosis by *Porphyromonas gingivalis*. *FEMS Microbiology Letters* 200: 145–149. doi: 10.1111/j.1574-6968.2001.tb10706.x.

Natividad, J.M., and E.F. Verdu. 2013. Modulation of intestinal barrier by intestinal microbiota: pathological and therapeutic implications. *Pharmacological Research* 69(1): 42–51.

Nesse, W., J. Westra, J.E. van der Wal, F. Abbas, A.P. Nicholas, A. Vissink, and E. Brouwer. 2012. The periodontium of periodontitis patients contains citrullinated proteins which may play a role in ACPA (anti-citrullinated protein antibody) formation. *Journal of Clinical Periodontology* 39: 599–607. doi: 10.1111/j.1600-051X.2012.01885.x.

Nikitakis, N.G., W. Papaioannou, L.I. Sakkas, and E. Kousvelari. 2016. The autoimmunity–oral microbiome connection. *Oral Diseases* 23(7): 828–839. doi: 10.1111/odi.12589.

Nowicki, E.M., R. Shroff, J.A. Singleton, D.E. Renaud, D. Wallace, J. Drury, J. Zirnheld, B. Colletti, A.D. Ellington, R.J. Lamont, D.A. Scott, and M. Whitely. 2018. Microbiota and metrascriptome changes accompanying the onset of gingivitis. *mBio* 9(2): e00575–18. https://www.ncbi.nlm.nih.gov/pmc/articles/PMC5904416/.

O'Brien-Simpson, N., J. Holden, L. Lenzo, Y. Tan, G.C. Brammar, K.A. Walsh, W. Singleton, R.K.H. Orth, N. Slakeski, K.J. Cross, I.B. Darby, D. Becher, T. Rowe, A.B. Morelli, A. Hammet, A. Nash, A. Brown, B. Ma, D. Vingadassalom, J. McCluskey, H. Kleanthous, and E.C. Reynolds. 2016. A herapeutic *Porphyromonas gingivalis* gingipain vaccine induces neutralising IgG1 antibodies that protect against experimental periodontitis. *npj Vaccines* 1: 16022. doi: 10.1038/npjvaccines.2016.22.

Okada, M., T. Kobayashi, S. Ito, T. Yokoyama, A. Abe, A. Murasawa, and H. Yoshie. 2013. Periodontal treatment decreases levels of antibodies to *Porphyromonas gingivalis* and citrulline in patients with rheumatoid arthritis. *Journal of Periodontology* 84(12): e74–e84. doi: 10.1902/jop.2013.130079.

Ortiz, P., N.F. Bissada, L. Palomo, Y.W. Han, M.S. Al-Zahrani, A. Panneerselvam, and A. Askari. 2009. Periodontal therapy reduces the severity of active rheumatoid arthritis in patients treated with or without tumor necrosis factor inhibitors. *Journal of Periodontology* 80(4): 535–540. doi: 10.1902/jop.2009.080447.

Padovani, G.C., V.P. Feitosa, S. Sauro, F.R. Tay, G. Durán, A.J. Paula and N. Durán. 2015. Advances in dental materials through nanotechnology: facts, perspectives and toxicological aspects. *Trends in Biotechnology* 33: 621–636.

Palmer Jr, R.J. 2014. Composition and development of oral bacterial communities. *Periodontology 2000* 64(1): 20–39. doi: 10.1111/j.1600-0757.2012.00453.x.

Park, K. 2014. Controlled drug delivery systems: past forward and future back. *Journal of Controlled Release* 190: 3–8.

Pasoto, S.G., V. Adriano de Oliveira Martins, and E. Bonfa. 2019. Sjögren's syndrome and systemic lupus erythematosus: links and risks. *Open Access Rheumatology: Research and Reviews* 11: 33–45. doi: 10.2147/OARRR.S167783.

Patel, M. 2020. Dental caries vaccine: are we there yet? *Letters in Applied Microbiology* 70: 2–12. doi: 10.1111/lam.13218.

Patil, S., R.S. Rao, N. Amrutha, and D.S. Sanketh. 2013. Oral microbial flora in health. *World Journal of Dentistry* 4: 262–266.

Paulos, C.M., C. Wrzesinski, A. Kaiser, C.S. Hinrichs, M. Chieppa, L. Cassard, D.C. Palmer, A. Boni, P. Muranski, Z. Yu, L. Gattinoni, P.A. Antony, S.A. Rosenberg, and N.P. Restifo. 2007. Microbial translocation augments the function of adoptively transferred self/tumor-specific CD8+ T cells via TLR4 signaling. *Journal of Clinical Investigation* 117: 2197–2204. doi: 10.1172/JCI3220.

Pavlova, S.I., L. Jin, S.R. Gasparovich, and L. Tao. 2013. Multiple alcohol dehydrogenases but no functional acetaldehyde dehydrogenase causing excessive acetaldehyde production from ethanol by oral streptococci. *Microbiology* 159(Pt 7): 1437–1446. doi: 10.1099/mic.0.066258-0.

Pei, Z., E.J. Bini, L. Yang, M. Zhou, F. Francois, and Blaser M.J. 2004. Bacterial biota in the human distal esophagus. *Proceedings of the National Academy of Sciences of USA* 101: 4250–4255.

Penders, J., C. Thijs, C. Vink, F.F. Stelma, B. Snijders, I. Kummeling, P.A. van den Brandt, and E.E. Stobberingh. 2006. Factors influencing the composition of the intestinal microbiota in early infancy. *Pediatrics* 118: 511–521.

Peres, M.A., L.M.D. Macpherson, R.J. Weyant, B. Daly, R. Venturelli, M.R. Mathur, S. Listl, R.K. Celeste, C.C. Guarnizo-Herreño, C. Kearns, H. Benzian, P. Allison, and R.G. Watt. 2019. Oral diseases: a global public health challenge. *The Lancet* 394(10194): P249–P260. doi: 10.1016/S0140-6736(19)31146-8.

Pers, J-O., A. Saraux, R. Pierre, and P. Youinou. 2008. Anti–TNF-α immunotherapy is associated with increased gingival inflammation without clinical attachment loss in subjects with rheumatoid arthritis. *Journal of Periodontology* 79: 1645–1651. doi: 10.1902/jop.2008.070616.

Pessoa, L., G. Aleti, S. Choudhury, D. Nguyen, T. Yaskell, Y. Zhang, W. Li, K.E. Nelson, L.L.S. Neto, A.C.P. Sant'Ana, and M. Freire. 2019. Host-microbe interactions in systemic lupus erythematosus and periodontitis. *Frontiers in Immunology*. doi: 10.3389/fimmu.2019.02602.

Peterson, S.N., E. Snesrud, J. Liu, A.C. Ong, M. Kilian, N.S. Schork, and W. Bretz. 2013. The dental plaque microbiome in health and disease. *PLoS One* 8(3): e58487. doi: 10.1371/journal.pone.0058487.

Pianta, A., S. Arvikar, K. Strle, E.E. Drouin, Q. Wang, C.E. Costello, and A.C. Steere. 2017. Evidence of the immune relevance of *Prevotella copri*, a gut microbe, in patients with rheumatoid arthritis. *Arthritis Rheumatology* 69: 964–975.

Pischon, N., E. Röhner, A. Hocke, P. N'Guessan, H.C. Muller, G. Matziolis, V. Kanitz, P. Purucker, B.M. Kleber, J.P. Bernimoulin, G. Burmester, F. Buttgereit, and J. Detert. 2009. Effects of *Porphyromonas gingivalis* on cell cycle progression and apoptosis of primary human chondrocytes. *Annals of the Rheumatic Diseases* 68: 1902–1907.

Plottel, C.S., and M.J. Blaser. 2011. Microbiome and malignancy. *Cell Host Microbe* 10: 324–335. doi: 10.1016/j.chom.2011.10.003.

Podschun, R., and U. Ullmann. 1998. *Klebsiella* spp. as nosocomial pathogens: epidemiology, taxonomy, typing methods, and pathogenicity factors. *Clinical Microbiology Reviews* 11(4): 589–603. doi: 10.1128/CMR.11.4.589.

Pushalkar, S., X. Ji, Y. Li, C. Estilo, R. Yegnanaranya, B. Singh, X. Li, and D. Saxena. 2012. Comparison of oral microbiota in tumor and non-tumor tissues of patients with oral squamous cell carcinoma. *BMC Microbiology* 12: 144. doi: 10.1186/1471-2180-12-144.

Puth, S., S.H. Hong, H.S. Na, H.H. Lee, Y.S. Lee, S.Y. Kim, W. Tan, H.S. Hwang, S. Sivasamy, K. Jeong, J-K. Kook, S-J. Ahn, I-C. Kang, J-H. Ryu, J.T. Koh, J.H. Rhee, and S.E. Lee. 2019. A built-in adjuvant-engineered mucosal vaccine against dysbiotic periodontal diseases. *Mucosal Immunology* 12: 565–579. doi: 10.1038/s41385-018-0104-6.

Qiu, X., M. Zhang, X. Yang, N. Hong, and C. Yu. 2013. *Faecalibacterium prausnitzii* upregulates regulatory T cells and anti-inflammatory cytokines in treating TNBS-induced colitis. *Journal of Crohn's and Colitis* 7(11): e558–e568. doi: 10.1016/j.crohns.2013.04.002.

Queipo-Ortuño, M.I., M. Boto-Ordonez, M. Murri, J.M. Zumaquero, M. Clemente-Postigo, R. Estruch, F.C. Diaz, C. Andrés-Lacueva, and F.J. Tinahones. 2012. Influence of red wine polyphenols and ethanol on the git microbiota ecology and biochemical biomarkers. *American Journal of Clinical Nutrition* 95: 1323–1324.

Ramsey, M.M., K.P. Rumbaugh, and M. Whiteley. 2011. Metabolite cross-feeding enhances virulence in a model polymicrobial infection. *PLoS Pathogenes* 7: e1002012.

Rashid, T., and A. Ebringer. 2007. Ankylosing spondylitis is linked to Klebsiella – the evidence. *Clinical Rheumatology* 26: 858–864.

Rinninella, E., P. Raoul, M. Cintoni, F. Franceschi, G. Miggiano, A. Gasbarrini, and M.C. Mele. 2019. What is the healthy gut microbiota composition? A changing ecosystem across age, environment, diet, and diseases. *Microorganisms* 7(1): 14. doi: 10.3390/microorganisms7010014.

Rosier, B., P. Marsh, and A. Mira. 2017. Resilience of the oral microbiota in health: mechanisms that prevent dysbiosis. *Journal of Dental Research* 97(4): 371–380. doi: 10.1177/0022034517742139.

Rowland, I., G. Gibson, A. Heinken, K. Scott, J. Swann, I. Thiele, and K. Tuohy. 2018. Gut microbiota functions: metabolism of nutrients and other food components. *European Journal of Nutrition* 57(1): 1–24. doi: 10.1007/s00394-017-1445-8.

Rubinstein, M.R., X. Wang, W. Liu, Y. Hao, G. Cai, and Y.W. Han. 2013. *Fusobacterium nucleatum* promotes colorectal carcinogenesis by modulating E-cadherin/β-catenin signaling via its FadA adhesin. *Cell Host Microbe* 14: 195–206.

Sakamoto, H., H. Naito, Y. Ohta, R. Tanakna, N. Maeda, J. Sasaki, and C.E. Nord. 1999. Isolation of bacteria from cervical lymph nodes in patients with oral cancer. *Archives of Oral Biology* 44: 789–793. doi: 10.1016/S0003-9969(99)00079-5.

Salminen, S., G.R. Gibson, A.L. McCartney, and E. Isolauri. 2004. Influence of mode of delivery on gut microbiota composition in seven year old children. *Gut* 53: 1388–1389.

Sampaio-Maia, B., and F. Monteiro-Silva. 2014. Acquisition and maturation of oral microbiome throughout childhood: an update. *Dental Research Journal* 11(3): 291–301.

Sandhya, P., D. Sharma, S. Vellarikkal, A.K. Surin, R. Jayrajan, A. Verma, V. Dixit, S. Sivasubbu, D. Danda, and V. Scaria. 2015. AB0188 systematic analysis of the oral microbiome in primary Sjögren's syndrome suggest enrichment of distinct microbes. *Annals of the Rheumatic Diseases* 74: 953–954. https://ard.bmj.com/content/74/Suppl_2/953.3.

Sasaki, M., C. Yamamura, Y. Ohara-Nemoto, S. Tajika, Y. Kodama, T. Ohya, R. Harada, and S. Yimura. 2005. *Streptococcus anginius* infection in oral cancer and its infection route. *Journal of Oral Diseases* 11: 151–156. doi: 10.1111/j.1601-0825.2005.01051.x.

Scher, J.U., A. Sczesnak, R.S. Longman, N. Segeta, C. Ubeda, C. Bielski, T. Rostron, V. Cerundolo, E.G. Pamer, S.B. Abramson, C. Huttenhower, and D.R. Littman. 2013. Expansion of intestinal *Prevotella copri* correlates with enhanced susceptibility to arthritis. *eLife* 2: 01202. doi: 10.7554/eLife.1202.

Scher, J.U., C. Ubeda, M. Equinda, R. Khanin, Y. Buischi, A. Viale, L. Lipuma, M. Attur, M.H. Pillinger, G. Weissman, D.R. Littman, E.G. Pamer, W.A. Bretz, and S.B. Abramson. 2012. Periodontal disease and the oral microbiota in new-onset rheumatoid arthritis. *Arthritis & Rheumatology* 64: 3083–3094. doi: 10.1002/art.34539.

Schott, E., C.W. Farnsworth, A. Grier, J.A. Lillis, S. Soniwara, G.H. Dadouriyan, R.D. Bell, M.L. Doolittle, D.A. Villani, H. Awad, J.P. Ketz, F. Kamal, C. Ackert-Bickell, J.M. Ashton, S.R. Gill, R.A. Mooney, and M.J. Zusik. 2018. Targeting the gut microbiome to treat the osteoarthritis of obesity. *J.C.I. Insight* 3(8): e95997. https://insight.jci.org/articles/view/95997.

Shamji, M.F., M. Bafaquh, and E. Tsai. 2008. The pathogenesis of ankylosing spondylitis. *Neurosurgical Focus* 24(1): E3. doi: 10.3171/FOC/2008/24/1/E3.

Shanmugam, K.T., K.M.K. Masthan, N. Balachander, S. Jimson, and R. Sarangarajan. 2013. Dental caries vaccine – a possible option? *Journal of Clinical and Diagnostic Research: JCDR* 7(6): 1250–1253. doi: 10.7860/JCDR/2013/5246.3053.

Sharma, N., S. Bhatia, A.S. Sodhi, and N. Batra. 2018. Oral microbiome and health. *AIMS Microbiology* 4(1): 42–66. doi: 10.3934/microbiol.2018.1.42.

Shattat, G.F. 2014. A review article on hyperlipidemia: types, treatments and new drug targets. *Biomedical and Pharmacology Journal* 7(2). doi: 10.13005/bpj/504.

Shreiner, A.B., J.Y. Kao, and V.B. Young. 2015. The gut microbiome in health and in disease. *Current Opinion in Gastroenterology* 31: 69–75.

Smith, D.J., W.F. King, and R. Godiska. 2001. Passive transfer of IgY antibody to *Streptococcus mutans* glucan binding protein-B can be protective for experimental dental caries. *Infection and Immunity* 69: 3135–3142.

So, J-S., M-K. Song, H-K. Kwon, C-G. Lee, C-S. Chae, A. Sahoo, A. Jash, S.H. Lee, Z.Y. Park, and S-H. Im. 2011. *Lactobacillus casei* enhances type II collagen/glucosamine-mediated suppression of inflammatory responses in experimental osteoarthritis. *Life Sciences* 88: 358–366. doi: 10.1016/j.lfs.2010.12.013.

Sokol, H., B. Pigneur, L. Watterlot, O. Lakhdari, L.G. Bermúdez-Humarán, J-J. Gratadoux, S. Blugeon, C. Bridonneau, J-P. Furet, G. Corthier, C. Grangette, N. Vasquez, P. Pochart, G. Trugnan, G. Thomas, H.M.M. Blottière, J. Doré, P. Marteau, P. Seksik, and P. Langella. 2008. *Faecalibacterium prausnitzii* is an anti-inflammatory commensal bacterium identified by gut microbiota analysis of Crohn disease patients. *Proceedings of the National Academy of Sciences of USA* 105(43): 16731–16736. doi: 10.1073/pnas.0804812105.

Sokol, H., P. Seksik, J. Furet, O. Firmesse, I. Nion-Larmurier, L. Beaugerie, J. Cosnes, G. Corthier, P. Marteau, and J. Doré. 2009. Low counts of *Faecalibacterium prausnitzii* in colitis microbiota. *Inflammatory Bowel Disease* 15: 1183–1189. doi: 10.1002/ibd.20903.

Sweeting, L.A., K. Davis, and C.M. Cobb. 2008. Periodontal treatment protocol for the general dental practice. *The Journal of Dental Hygiene* 82(3): 16–26. https://jdh.adha.org/content/jdenthyg/82/suppl_2/16.full.pdf.

Tahara, T., E. Yamamoto, H. Suzuki, R. Maruyama, W. Chung, J. Garriga, J. Jelinek, H-o. Yamano, T. Sugai, B. An, I. Shureiqi, M. Toyota, Y. Kondo, M.R.H. Estécio, and J-P.J. Issa. 2014. *Fusobacterium* in colonic flora and molecular features of colorectal carcinoma. *Cancer Research* 74(5): 1311–1318. doi: 10.1158/0008-5472.CAN-13-1865.

Takahashi, K. 2014. Influence of bacteria on epigenetic gene control. *Cellular and Molecular Life Sciences* 71: 1045–1054. doi: 10.1007/s00018-013-1487-x.

Tanaka, M., and J. Nakayama. 2017. Development of the gut microbiota in infancy and its impact on health in later life. *Allergology International* 66: 515–522.

Tanner, A., C.A. Kressirer, S. Rothmiller, I. Johanssen, and N.I. Chalmers. 2019. The caries microbiome: implications for reversing dysbiosis. *Advances in Dental Research* 29(1): 78–85.

Tett, A., K.D. Huang, F. Asnicar, H. Fehlner-Peach, E. Pasolli, N. Karcher, F. Armanini, P. Manghi, K. Bonham, M. Zolfo, F. De Filippis, C. Magnabosco, R. Bonneau, J. Lusingu, J. Amuasi, K. Reinhard, T. Rattei, F. Boulund, L. Engstrand, A. Zink, M. Carmen Collado, D.R. Littman, D. Eibach, D. Ercolini, O. Rota-Stabelli, C. Huttenhower, F. Maixner, and N. Segata. 2019. The *Prevotella copri* complex comprises four distinct clades underrepresented in westernized populations. *Cell Host & Microbe* 26(5): 666–679.e7. doi: 10.1016/j.chom.2019.08.018.

Thé, J., and J.L. Ebersole. 1991. Rheumatoid factor (RF) distribution in periodontal disease. *Journal of Clinical Immunology* 11: 132–142.

Thursby, E., and N. Juge. 2017. Introduction to the human gut microbiota. *Biochemical Journal* 474: 1823–1836.

Tribble, G.D., and R.J. Lamont. 2010. Bacterial invasion of epithelial cells and spreading in periodontal tissue. *Periodontol 2000* 52: 68–83. doi: 10.1111/j.1600-0757.2009.00323.x.

van de Stadt, L.A., M.H.M.T. de Konig, R.J. van de Stadt, G. Wolbink, B.A.C. Dijkmans, D. Hamann, and D. van Shaadenberg. 2011. Development of the anti-citrullinated protein antibody repertoire prior to the onset of rheumatoid arthritis. *Arthritis & Rheumatology* 63: 3226–3233. doi: 10.1002/art.30537.

van der Meulen, T.A., H.J.M. Harmsen, A.V. Vila, A. Kurilshikov, S.C. Liefers, A. Zhernakova, J. Fu, C. Wijmenga, R.K. Weersma, K. de Leeuw, H. Bootsma, F.K.L. Spijkervet, A. Vissink, and F.G.M. Kroese. 2019. Shared gut, but distinct oral microbiota composition in primary Sjögren's syndrome and systemic lupus erythematosus. *Journal of Autoimmunity* 97: 77–87. doi: 10.1016/j.jaut.2018.10.009.

van der Meulen, T.A., H.J.M. Harmsen, H. Bootsma, F.K.L. Spijkervet, F.G.M. Kroese, and A. Vissink. 2016. The microbiome-systemic diseases connection. *Oral Diseases* 22(8): 719–734. doi: 10.1111/odi.12472.

van Steenbergen, T.J.M., M.D.A. Petit, L.H.M. Scholte, U. van der Velden, and J. de Graaff. 1993. Transmission of *Porphyromonas gingivalis* between spouses. *Journal of Clinical Periodontology* 20: 340–345. doi: 10.1111/j.1600-051X.1993.tb00370.x.

Verma, D., P.K. Garg, and A.K. Dubey. 2018. Insights into the human oral microbiome. *Archives of Microbiology* 200: 525–540. doi: 10.1007/s00203-018-1505-3.

Vivarelli, S., R. Salemi, S. Candido, L. Falzone, M. Santagati, S. Stefani, F. Torino, G.L. Banna, G. Tonini, and M. Libra. 2019. Gut microbiota and cancer: from pathogenesis to therapy. *Cancers* 11(1): 38. doi: 10.3390/cancers11010038.

Vollaard, E.J., and H.A. Clasener. 1994. Colonization resistance. *Antimicrobial Agents and Chemotherapy* 38: 409–414.

Wade, W.G. 2013. The oral microbiome in health and disease. *Pharmacological Research* 69(1): 137–143. doi: 10.1016/j.phrs.2012.11.006.

Wang, R., H. Zhou, P. Sun, H. Wu, J. Pan, W. Zhu, J. Zhang, and J. Fang. 2011. The effect of an atmospheric pressure, dc nonthermal plasma microjet on tooth root canal, dentinal tubules infection and reinfection prevention. *Plasma Medicine* 1: 143–155.

Wegner, N., R. Wait, A. Sroka, S. Eick, K-A. Nguyen, K. Lundberg, A. Kinloch, S. Culshaw, J. Potempa, and P. J. Venables. 2010. Peptidylarginine deiminase from *Porphyromonas gingivalis* citrullinates human fibrinogen and α-enolase: implications for autoimmunity in rheumatoid arthritis. *Arthritis and Rheumatism* 62(9): 2662–2672. doi: 10.1002/art.27552.

Wei, W., W. Sun, S. Yu, Y. Yang, and L. Ai. 2016. Butyrate production from high-fiber diet protects against lymphoma tumor. *Leukemia & Lymphoma* 57: 2401–2408. doi: 10.3109/10428194.2016.1144879.

Wilson, K., Z. Liu, J. Huang, A. Roosaar, T. Axéll, and W. Ye. 2018. Poor oral health and risk of incident myocardial infarction: a prospective cohort study of Swedish adults, 1973–2012. *Scientific Reports* 8(1): 11479. doi: 10.1038/s41598-018-29697-9.

Woese, C.R. 1987. Bacterial evolution. *Microbiological Reviews* 51: 221–271.

Wu, H.J., and E. Wu. 2012. The role of gut microbiota in immune homeostasis and autoimmunity. *Gut Microbes* 3(1): 4–14. doi: 10.4161/gmic.19320.

Xu, M., M. Yamada, M. Li, H. Liu, S.G. Chen, and Y.W. Han. 2007. FadA from *Fusobacterium nucleatum* utilizes both secreted and nonsecreted forms for functional oligomerization for attachment and invasion of host cells. *Journal of Biological Chemistry* 282: 25000–25009.

Yadav, K., and S. Prakash. 2016. Dental caries – a review. *Asian Journal of Biomedical and Pharmaceutical Sciences* 6(53): 01–07. doi: 10.15272/ajps.v6i53.773.

Yoshida, A., Y. Nakano, T. Yamashita, T. Oho, H. Ito, M. Kondo, M. Ohishi, and T. Koga. 2001. Immunodominant region of *Actinobacillus actinomycetemcomitans* 40-kilodalton heat shock protein in patients with rheumatoid arthritis. *Journal of Dental Research* 80: 346–350. https://journals.sagepub.com/doi/abs/10.1177/00220345010800010901.

Yu, C.G., and Q. Huang. 2013. Gut microbiota in pathogenesis of IBD. *Journal of Digestive Diseases* 14: 513–517. doi: 10.1111/1751-2980.12087.

Zhang, X., D. Zhang, H. Jia, Q. Feng, D. Wang, D. Liang, X. Wu, J. Li, L. Tang, Y. Li, Z. Lan, B. Chen, Y. Li, H. Zhong, H. Xie, Z. Jie, W. Chen, S. Tang, X. Xu, X. Wang, X. Cai, S. Liu, Y. Xia, J. Li X. Qiao, J.Y. Al-Aama, H. Chen, L. Wang, Q-j. Wu, F. Zhang, W. Zheng, Y. Li, M. Zhang, G. Luo, W. Xue, Y. Yin, H. Yang, J. Wang, K. Kristiansen, L. Liu, T. Li, Q. Huang, Y. Li, and J. Wang. 2015. The oral and gut microbiomes are perturbed in rheumatoid arthritis and partly normalized after treatment. *Nature Medicine* 21: 895–905. https://www.nature.com/articles/nm.3914.

Zhang, Y., X. Wang, H. Li, C. Ni, Z. Du, and F. Yan. 2018. Human oral microbiota and its modulation for oral health. *Biomedecine and Pharmacotherapie* 99: 883–893.

Zucchi, D., E. Elefante, D. Calabresi, V. Signorinii, A. Bortoluzzi, and C. Tani. 2019. One year in review 2019: systemic lupus erythematosus. *Clinical and Experimental Rheumatology* 37: 715–722. https://www.clinexprheumatol.org/article.asp?a=14436.

13

Microbiome in Women Reproductive Health

C. Anchana Devi
Women's Christian College

T. Ramani Devi
Ramakrishna Medical Center

Pavithra Amritkumar
Women's Christian College

CONTENTS

13.1 Introduction

The term "microbiome" was first coined by the Nobel Laureate Joshua Lederberg in 2001, describing as "the ecological community of commensal, symbiotic, and pathogenic microorganisms that literally share our body space and have been all but ignored as determinants of health and disease" (Prescott, 2017). Often referred to as "second genome," it gained importance post the publication of human genome sequence in 2001. Human Microbiome Project (HMP) was initiated in 2007 involving several elite research groups to understand microbiome's characteristics, evolution, range, and influence (Peterson et al., 2009). Molecular-based techniques used in HMP have provided new information to detect fastidious organisms which are difficult to culture easily. These techniques have improved our knowledge on microbiota to the extent that today we know 3.3 million microbial genes are present in the human gut alone, while the protein-coding genes of human genome range between 20,000 and 50,000 (Kay et al., 2004; Yang et al., 2009). Of the 2,776 bacteria isolated from the HMP, 581 have been found in the vaginal microbiota (Diop et al., 2019). These microbiomes contain a consortium of microorganisms unique to the site they are located and are involved in a variety of microbe–microbe interactions such as physical contact, and chemical and metabolic exchanges (Keller and Surette, 2006). These exchanges in the various consortia help in carrying out complex activities bringing about maintaining the well-being of the environment by resisting invasion by other organisms and controlling various kinds of stresses (Brenner et al., 2008; Burmolle et al., 2006). Thus, these diverse consortia of microorganisms inhabiting the different sites of the human body play an important role in human physiology, immunity, and nutrition (Ley et al., 2006a, b; Ling et al., 2010; Mazmanian et al., 2005; Rosenstein et al., 1996).

There exists a marked difference in the composition of the microbiome in different anatomical sites of our body such as skin, oral cavity, and gastrointestinal (GI), respiratory, and urogenital tracts and also within the various microniche of a given site. They also show a large degree of variation between individuals even in the absence of any disease (Faust et al., 2012). With these variations, it is not possible to define a "healthy core human microbiome" (Qin et al., 2010). Despite these variations, many studies conducted across the world in different populations have concluded that an overall alteration in the microbial profile of an individual can be linked with a number of pathogenic states, including cardiovascular diseases, type 2 diabetes, obesity, and metabolic disorders and infertility (Franasiak and Scott, 2015; Howitt and Garrett, 2012; Jayasinghe et al., 2016; Larsen et al., 2010; Ordovas and Mooser, 2006; Qin et al., 2012; Tang and Hazen, 2014).

13.2 Microbiome in Female Reproductive System

Microbiome in urogenital tract makes up 9% of the total human microbiome (Sirota et al., 2014). Figure 13.1 gives an overview of the female reproductive system. Colonization of microbes in the human body commences after the parturition when newborn comes in physical contact with the maternal womb, vaginal, fecal, and skin microbes (Collado et al., 2012; Dunlop et al., 2015). The main force that confers the postnatal immunity to the newborn is believed to be through this colonization (Gomezde-Aguero et al., 2016). Certain microbial species (*Escherichia coli, Escherichia faecalis*, and *Staphylococcus epidermidis*) have been isolated from the meconium of healthy neonates born to healthy mothers within 2 hours of delivery, indicating that the transfer of bacteria from the maternal body initiates through the amniotic fluid to the fetus by direct contact *in utero* (Jimenez et al., 2008). Figure 13.2 shows an overview of the various microbial species located in different parts of the female reproductive system.

13.2.1 Microbiome in Vagina

Vaginal microenvironment represents one of the most dynamic ecosystems which play an important role in providing antimicrobial defense mechanism and maintaining healthy reproductive environment (Sobel, 1999). It is affected by a number of endogenous (e.g., age, physiology, body size) and exogenous (e.g., sexual behavior, substrate use, and routine association with microbes) factors. Recent numerous studies have demonstrated that menstruation, hormonal fluctuation, sexual behaviors, hygiene practices, new sexual partner, and vaginal microbiota composition contribute to the fluctuation of bacterial communities. About 250 bacterial taxonomic

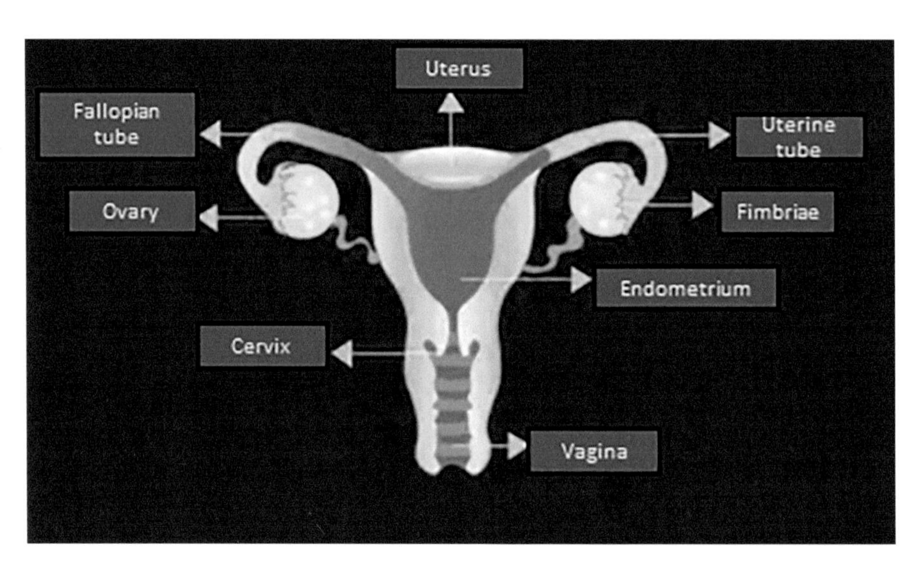

FIGURE 13.1 Structure of female reproductive system.

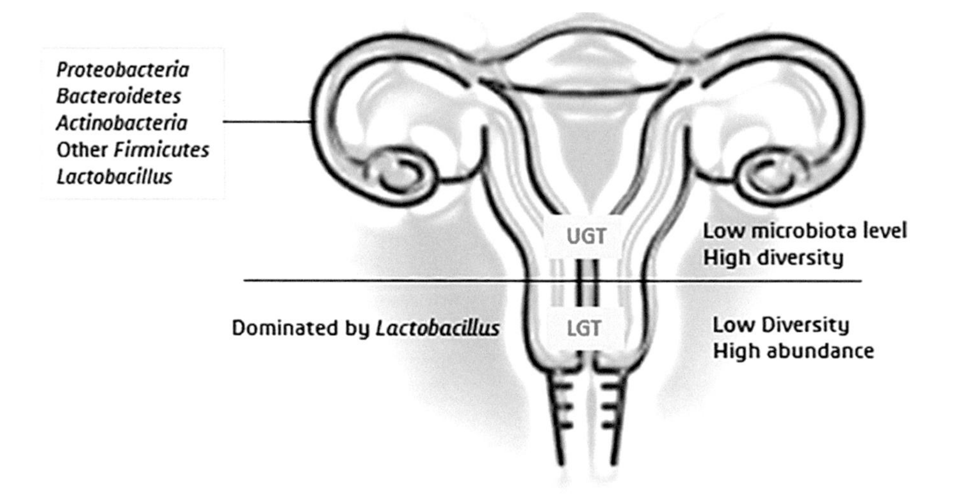

FIGURE 13.2 Microbiota in different parts of female reproductive system.

units have been identified from the vagina of women of various ages, health status, and countries of origin. Recent next-generation sequencing (NGS) and metagenomic analyses have revealed that the vaginal microbiome harbors a high proportion of *Firmicutes* and a low percentage of *Proteobacteria*, *Bacteroidetes*, *Fusobacteria*, and *Actinobacteria*. Vaginal microbiota belong to seven community types (I–VII), of which majority (types I, II, III, and V) are predominated by one or more species of *Lactobacillus*. Frequently detected members include *Lactobacillus crispatus*, *Lactobacillus gasseri*, *Lactobacillus iners*, and *Lactobacillus jensenii* (Haldar et al., 2016). Among the vaginal Lactobacilli, *L. crispatus* and *L. iners* are the main producers of *D*-lactic acid and responsible for maintaining an acidic environment (pH 3.5–5.0). The human vagina and the colonizing microbes interact in a mutualistic relationship. Acidic pH prevents the growth of various pathogens, including human immunodeficiency virus (HIV), yeast, *Neisseria gonorrhoeae*, *Atopobium*, *Megasphaera*, *Mobiluncus*, *Prevotella*, *Sneathia*, and *Gardnerella vaginalis* (*G. vaginalis*)—the latter being the causative agent of bacterial vaginosis (BV) and help in maintaining host fitness for the reproductive ability. *Lactobacillus* sp. also produce a number of antimicrobial compounds, including hydrogen peroxide, lactic acid, acidocin, and gassericin (bacteriocins by *L. gasseri*), which facilitate the adhesion of *Lactobacilli* to the vaginal epithelial cells and competition for the nutrients in the niche discouraging the growth of pathogenic bacteria and preventing pathogen colonization. Any imbalance in the vaginal microbiome composition, also known as "dysbiosis," could play a role in common conditions like BV, sexually transmitted diseases, urinary infections, and preterm birth (Fredricks et al., 2005). It is important to understand the composition of the vaginal microbial ecosystem in order to study its dynamics and comprehensively dissect the etiology of vaginal diseases and their role in women's reproductive health (Ling et al., 2010).

13.2.2 Microbiota in Other Parts of Female Reproductive System

Vulvar microbiota composition is similar to that of vagina. In addition, *Fusobacterium*, *Murdochiella*, and *Segniliparus*, which together account for 1.0%–3.5% of the total microbial population, were also seen in the vulva. Endometrium and upper endocervix have been evaluated for the presence of bacterial species. Bacteroidetes (*Bacteroides xylanisolvens, Bacteroides thetaiotaomicron,* and *Bacteroides fragilis*) was reported to be a ubiquitous component of microbiome in the uterine endometrium. *Pelomonas* of class Betaproteobacteria, *L. iners*, *Prevotella* sp., and *L. crispatus* have been reported recently in few studies.

Previous studies investigating the relationship between bacteria and fallopian tube pathology have largely assumed this site to be sterile. But recent molecular studies on fallopian tubes of asymptomatic women have shown diverse microbial communities, which are affected by hormones and antibiotics, and display biogeographical tropism. Microbial communities were dominated by members of the phyla Firmicutes, such as *Staphylococcus* sp., *Enterococcus* sp., and *Lactobacillus* sp. Other highly abundant and prevalent taxa included Pseudomonads (*Pseudomonas* sp. and *Burkholderia* sp.) and the known genital tract anaerobes such as *Propionibacterium* sp. and *Prevotella* sp.

Placenta appears to harbor a low biomass microbiome composed of nonpathogenic commensal phyla such as *Firmicutes*, *Tenericutes*, *Proteobacteria*, *Bacteroidetes*, and *Fusobacteria* (Haldar et al., 2016).

13.3 Microbial Dysbiosis and Women Health Issues

Microbial dysbiosis is referred to as "an alteration of the normal microbiota (bacterial or fungal species) due to the exposure to disruptive factors such as antibiotics, chronic disease, stress, medical procedures, or medications" (McFarland, 2014).

In this section, we will discuss some of the major health issues related to women in association with microbial dysbiosis.

13.3.1 Bacterial Vaginosis

BV is characterized by vaginal microbial dysbiosis due to the displacement of beneficial *Lactobacilli* by a population of gram-negative anaerobic bacteria. BV has been identified as an independent risk factor for the acquisition of sexually transmitted infections (STIs), HIV, pelvic inflammatory disease (PID), and a wide range of reproductive and obstetric complications, including miscarriage and preterm births.

Conventional cultivation-based methods have identified microbial diversity, including anaerobic bacteria such as *Gardnerella*, *Prevotella*, *Mobiluncus*, *Ureaplasma*, and *Mycoplasma*, to be associated with BV. Recent cloning and sequencing methods have confirmed a strong association of *G. vaginalis* and *Atopobium vaginae* with BV. 16S rRNA sequencing methods have also identified fastidious anaerobic, clostridia like BV-associated bacteria (BVAB) 1, 2, and 3, *Leptotrichia* and *Sneathia genera*, and uncultivated *Megasphaera*-like phylotype. Newer molecular techniques have paved the way for many novel bacterial species previously undetected by conventional cultivation techniques, thus displaying microbial diversity in etiology of BV. Another study (Soksawatmaekhin et al., 2004) outlines the use of several biogenic amines (BAs), which can serve as important biomarker that contributes to the "fishy" odor, a characteristic feature of BV. BA production requires protons, which increases the pH of the habitat (vagina). This in turn contributes to the initial colonization of the BA-producing taxa that nullifies the host immune defenses paving way for the colonization of a wide variety of pathogens (Figure 13.3). Despite efforts to characterize BV using epidemiological, microscopic, microbiological culture, and sequence-based methods, it has not been possible to determine the common etiology for BV. This may be attributed to the complexity of the vaginal microbiome, host immunity, and the variability in individual responses to potentially inflammatory mediators produced by an array of microorganisms. Prospective longitudinal studies with frequently collected samples and a detailed behavioral metadata will help to understand the causes of BV. These methods will help in identifying women at risk of acquiring BV, paving the

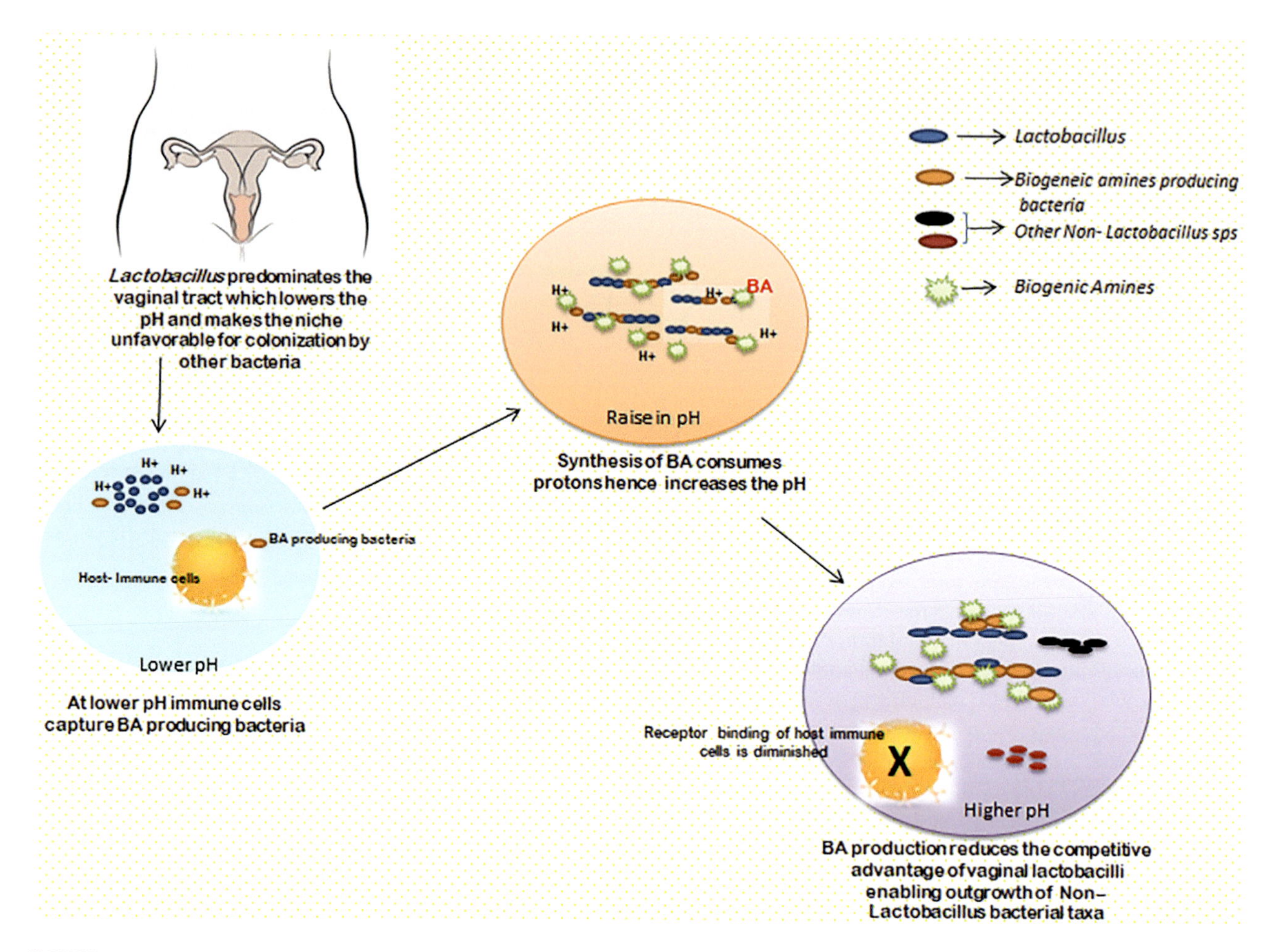

FIGURE 13.3 Role of vaginal biogenic amines (BAs) in bacterial vaginosis.

way for better diagnostic, intervention, and prevention strategies (Onderdonk et al., 2016; Sharma et al., 2014).

13.3.2 Pelvic Inflammatory Disease

PID is inflammation of the female upper reproductive tract, including the endometrium, fallopian tubes, ovaries, and pelvic peritoneum, induced by microbial infection. PID has wide clinical manifestations ranging from tubo-ovarian abscesses (TOAs) leading to fatal consequences such as multiple organ failure and septicemia (Figure 13.4). It can cause the long-term complications such as infertility, ectopic pregnancy (EP), and chronic pelvic pain, resulting in serious damage to the female reproductive system. Numerous studies using Nugent score and cultures have demonstrated that BV-associated vaginal microbiota predisposes to PID and STI. Because PID is usually secondary to STI, there has long been an "ascending infection hypothesis" that pathogenic microbes spread from the cervix or vagina to the upper reproductive tract and cause PID. Disturbance of the genital tract microbiota has been linked to an increased risk of pelvic infections. Under normal circumstances, a healthy microbiota defends against invading pathogens or opportunistic resident microbes. In patients with PID, the microbial communities of the vagina are altered, resulting in overgrowth of harmful microorganisms, which breach the cervical barrier and ascend, causing dysbiosis and infections of the upper reproductive tract. Studies have shown that PID induces a selective loss of ciliated epithelial cells along the fallopian tube epithelium, which impedes ovum transport, resulting in tubal-factor infertility or EP.

The cervix and pelvis represent two biological niches and exert different selective pressures on bacteria, which probably trigger transformation in the microbiota profile during the ascending migration of the bacteria. In addition, the vagina represents a complicated niche that allows for the survival and propagation of both beneficial bacteria (such as *Lactobacillus*) and opportunistic pathogens (such as Group B *Streptococcus, Bacteroides, Enterococcus, E. coli*, and *Prevotella*). An imbalance of vaginal microflora, like an elevated level of species diversity, may result in BV. Through conventional culture-based methods, *Neisseria gonorrhoeae* and *Chlamydia trachomatis* have been commonly reported to be responsible for 16%–75% of PID cases. However, 16S rRNA sequencing method and NGS-based approaches have also isolated *Acinetobacter, Aeromonas, Pseudomonas*, and *Shewanella* from PID patients. Newer approaches and larger PID cohort studies in different populations can bring more clarity to the etiology of PID (Sharma et al., 2014).

13.3.3 Endometriosis

Endometriosis is a condition in which endometrial epithelial and stromal tissues exist outside the uterine cavity. The presence of ectopic endometrial tissue and the resultant

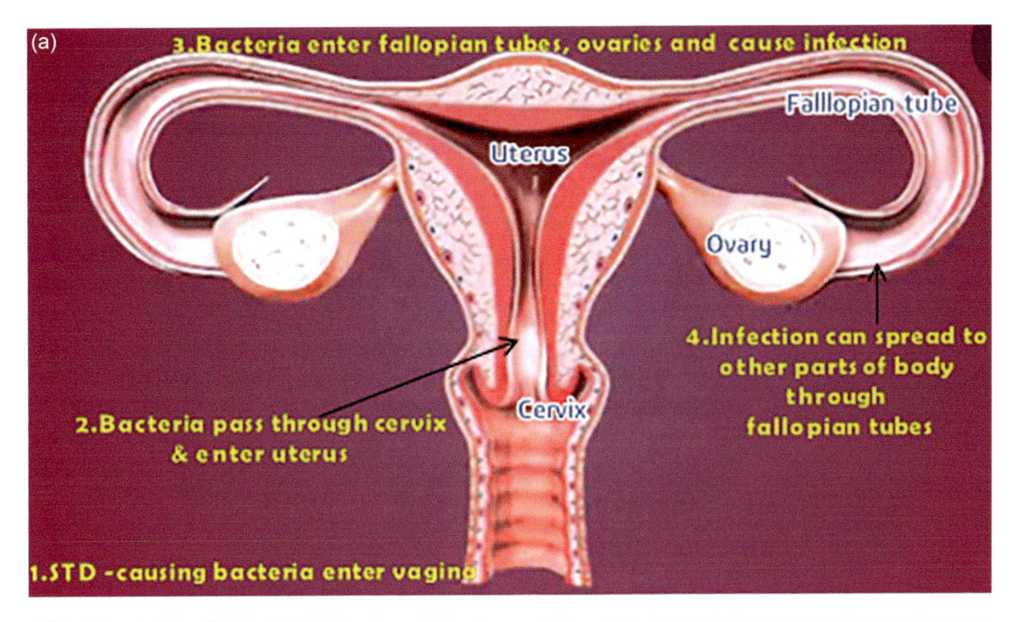

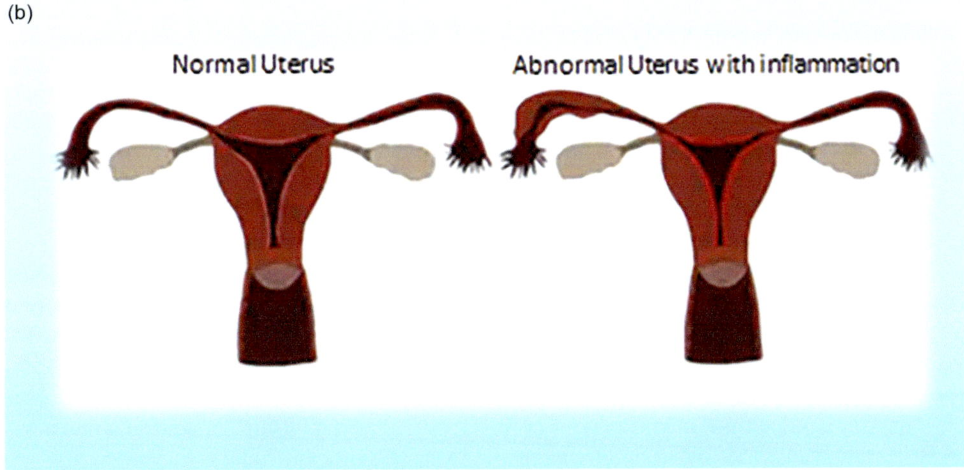

FIGURE 13.4 Mechanism underlying pelvic inflammatory disease (PID).

inflammation cause serious symptoms, including chronic pelvic pain, dysmenorrhea, dyspareunia, and infertility.

The growth and cyclical bleeding of endometrial implants and the resultant inflammation increase the production of proinflammatory cytokines. These processes eventually lead to the development of pelvic adhesions and endometriomas, which distort anatomical structure and interfere with reproductive function. The mechanisms involved in the pathogenesis of endometriosis are still unclear. At present, the "retrograde menstruation theory" wherein the endometrial cells shed during menstruation flow backward through the fallopian tubes and are "regurgitated" into the peritoneal cavity, where they implant and develop into ectopic lesions, is widely accepted. It is important to note that although retrograde menstruation occurs in up to 90% of women, only 6%–10% of all menstruating women develop endometriosis. This suggests that other factors contribute to the establishment and survival of ectopic endometriotic implants. Previous studies have shown that inflammatory cytokines such as interleukin-1 beta, interleukin-6, and tumor necrosis factor promote the implantation of ectopic endometrial tissue. It is possible that systemic or local inflammatory changes related to upper reproductive tract infections, such as increased production of inflammatory mediators because of bacterial colonization, influence the development of endometriosis (Wang et al., 2018).

A genome-wide expression analysis of autologous, paired eutopic and ectopic endometrial tissues from fertile women with various stages of endometriosis revealed a dysfunctional expression of immuno-neuro-endocrine behavior in the endometrium. The pathognomonic characteristics in women with stage IV ovarian endometriosis showed a marked indication of neoplastic potential. The etiology of endometriosis is thus complex. The activation of innate immune system due to microbial infection/disruption of normal flora leading to inflammation is also believed to play a role. Women with stage III–IV endometriosis have been reported to have higher rates of TOA, a polymicrobial PID in which *E. coli*, *N. gonorrhoeae*, *C. trachomatis*, and other obligate anaerobic bacteria are reported to be associated. A significantly higher colonization of *Gardnerella sp.*, *alpha Streptococcus sp.*, *Staphylococcus sp.*, and *Enterococci sp.* with a concomitant decline in *Lactobacillus* sp. was also noticed in the endometrial smears of women diagnosed with endometriosis. Other bacterial species remarkably associated with endometriosis include *Actinomyces sp., Fusobacterium sp., Staphylococcus sp., Propionibacterium sp., Prevotella sp., and Corynebacterium sp.* (Haldar et al., 2016).

Tai et al. (2018) have demonstrated that patients with PID had a threefold increase in the risk of developing endometriosis compared with women without PID. The correlation between the presence of PID and the development of endometriosis might be due to a causal connection between the two conditions (i.e., that PID somehow predisposes an individual to the development of endometriosis). Or, there could be a confounding third variable—a common pathological process or risk factor that influences the development of both conditions. Dysbiosis and inflammation may be the underlying mechanisms linking PID and endometriosis. Khan et al. (2016) have compared the bacteria present in endometrial swab samples and ovarian cystic fluid (from endometriomas or non-endometriotic cystadenomas) obtained from women with or without endometriosis. Using 16S rRNA sequencing techniques, the study found evidence of subclinical infection in the uterine cavity and also in ovarian endometriomas. Significantly higher numbers of bacteria from the Streptococcal and Staphylococcal families were present in the cystic fluid of women with ovarian endometriosis. Certain bacterial markers are thought to be associated with an increased risk of developing endometriosis. Immunologic abnormalities, including chronic inflammation and inadequate immune surveillance in the peritoneum, are also involved in the development of endometriosis. A large body of evidence has associated endometriosis with a higher risk of diseases linked to an altered immune response. These include ovarian and breast cancers, asthma, and other atopic, autoimmune, and cardiovascular diseases. In addition, numerous studies have demonstrated that the production of inflammatory cytokines such as interleukin-1 beta, interleukin-6, and tumor necrosis factor enhances the adhesion of endometrial tissue fragments to the peritoneum. Though it is unclear whether an altered microbiome is the cause or the effect of impaired host immunity, the presence of pathogenic bacteria in the pelvic cavity of patients with PID could promote the development of endometriosis by causing excessive peritoneal inflammation.

These findings suggest that the overgrowth of certain bacterial species in the uterine cavity can disrupt the delicate immunological balance of the endometrial microbiota and contribute to the pathogenesis of endometriosis. Diagnosis and timely management of PID can possibly reduce the occurrence of endometriosis in a significant way (Tai et al., 2018).

13.3.4 Polycystic Ovarian Syndrome

Polycystic ovarian syndrome (PCOS) is a common gynecological endocrine disease, affecting up to one in five women of reproductive age. It has significant and diverse clinical implications, including reproductive (infertility, hyperandrogenism, hirsutism), metabolic (insulin resistance (IR), impaired glucose tolerance, type 2 diabetes mellitus, adverse cardiovascular risk profiles), and psychological features (increased anxiety, depression, and worsened quality of life). The phenotype in PCOS varies widely depending on life stage, genotype, ethnicity, and environmental factors, including lifestyle and bodyweight.

The exact pathophysiology of PCOS is complex and remains largely unclear. The underlying hormonal imbalance created by a combination of increased androgens and/or IR is associated with PCOS. Genetic and environmental factors to hormonal disturbances combined with other factors (including obesity, ovarian dysfunction, and hypothalamic–pituitary abnormalities) contribute to the etiology of PCOS (Teede et al., 2010).

Although there are key diagnostic features—namely, oligomenorrhea/amenorrhea, clinical or biochemical hyperandrogenism, and PCO on ultrasound—there are also many potential phenotypes. This heterogeneity of the condition is further aggravated by degree of obesity, IR, ethnicity, and other factors. Both the heterogeneity of PCOS and the lack of an understanding of its etiology contribute to the evolving diagnostic criteria. Currently, the European Society of Human Reproduction and Embryology (ESHRE)/American Society for Reproductive Medicine (ASRM) or Rotterdam criteria are the agreed international diagnostic criteria for PCOS, although further research is needed (Zhao et al., 2020).

In recent years, a new consensus on the diagnosis and treatment of PCOS has emerged, and the idea that metabolic abnormalities and other complications of PCOS should be included in the diagnosis. Treatment of PCOS has been emphasized by several reproductive endocrine institutions, including the National Institutes of Health (NIH), the American Endocrine Society, and the European Endocrine Society. It is very important to systematically analyze changes to metabolic states when exploring the pathogenesis of PCOS. Modern research shows that gut microbiome (GM) disorders are closely related to the occurrence and development of metabolic diseases. In the past two decades, some studies have reported a relationship between GM and metabolic syndrome (MS) (Zhao et al., 2020).

The microecosystem of the human GM is diverse and dynamic. It consists of about 1,000–1,500 species of bacteria, around 100 trillion bacteria in all, which colonize the human intestinal tract. Colonization by GM occurs at birth, when the newborn acquires complex bacteria as they pass through the birth canal. Therefore, the infant GM composition shares maternal characteristics for a time after birth, gradually changing to become more individualized after about 1 year. A healthy individual can have at least 160 kinds of dominant GM. Although there are individual differences in terms of GM species, the dominant species are basically the same. More than 90% are *Firmicutes* and *Bacteroides*, but typically all humans harbor *Bacteroides*, *Prevotella*, *Porphyromonas*, *Clostridium*, and *Eubacterium*. Other less abundantly present bacteria include actinomycetes, proteobacteria, and methanogenic archaea. GM species are divided into three categories according to their interaction with the host: beneficial bacteria, conditioned pathogens, and harmful bacteria. Beneficial bacteria include *Lactobacilli* and *Bifidobacterium*, which can inhibit harmful microorganisms, enhance immunity, promote absorption, synthesize vitamins, inhibit tumors, reduce infection, and relieve allergic reactions. Harmful bacteria include *Staphylococcus*, *Salmonella*, and *Campylobacter*, which can produce toxins, increase the incidence of cancer, and induce infections. Opportunistic pathogens include *Enterobacter*, *Escherichia*, and *Bacteroides*. Under normal circumstances, the bacteria in the GM maintain a delicate dynamic balance to prevent the development of various diseases. A range of diverse factors can lead to changes in GM, including age, diet, and the use of antibiotics. With increasing age, the beneficial bacteria in the intestinal tract decrease, while the toxins produced by harmful bacteria increase, leading to accelerated aging.

PCOS is a complex endocrine and metabolic disease. Several PCOS-related genes have been found to be associated with carbohydrate metabolism and the steroid synthesis pathway, suggesting an important correlation between metabolic factors and the pathological mechanism of PCOS. GM species are involved in various metabolic activities, and accordingly, GM species are closely associated with the pathogenesis and clinical manifestations of PCOS.

Several studies to this effect have been carried out to understand the correlation between GM and PCOS. Analysis of GM of women with PCOS and normal women showed that PCOS women had fewer types of GM and that this trend was related to an increase in androgens. To study the effect of GM on the host's PCOS phenotype, feces from healthy women and women with PCOS were, respectively, transplanted into two groups of mice by oral lavage. Compared with control mice, the mice transplanted with fecal samples from women with PCOS developed IR, and had an increased number of ovarian cyst-like follicles, decreased corpus luteum, and higher levels of testosterone and luteinizing hormone. Fertility tests were carried out, in which the number of pups in the first litters after mating was counted. Mice given fecal transplants from women with PCOS produced fewer pups than healthy control mice. These studies indicate that GM may be involved in the occurrence and development of PCOS. This could offer a new therapeutic strategy for the clinical diagnosis and treatment of PCOS. Of course, the exact mechanism of the GM in PCOS needs to be supported by further evidence.

There are several pathways proposed for the link between GM and PCOS. GM degrades carbohydrates into simple sugars and then ferments them into hydrogen, CO_2, CH_4, and short-chain fatty acids (SCFAs) to provide energy to the host. Obese individuals take more energy from food, which then aggravates the symptoms of obesity, creating a vicious cycle. This phenomenon is associated with a disordered GM. GM disorders can accelerate the process of PCOS by affecting energy absorption.

Gut microbiota decompose organic materials to produce three major types of SCFAs, namely, acetate, propionate, and butyrate. An abnormal SCFA metabolism caused by an abnormal GM is associated with IR and hyperandrogenemia present in PCOS. High androgen levels in turn can aggravate the abnormal GM, creating a vicious cycle (Figure 13.5).

Lipopolysaccharides (LPS), also known as "bacterial endotoxins," are a unique component in the cell wall of gram-negative bacteria. *Bacteroides* and *Escherichia* in the human

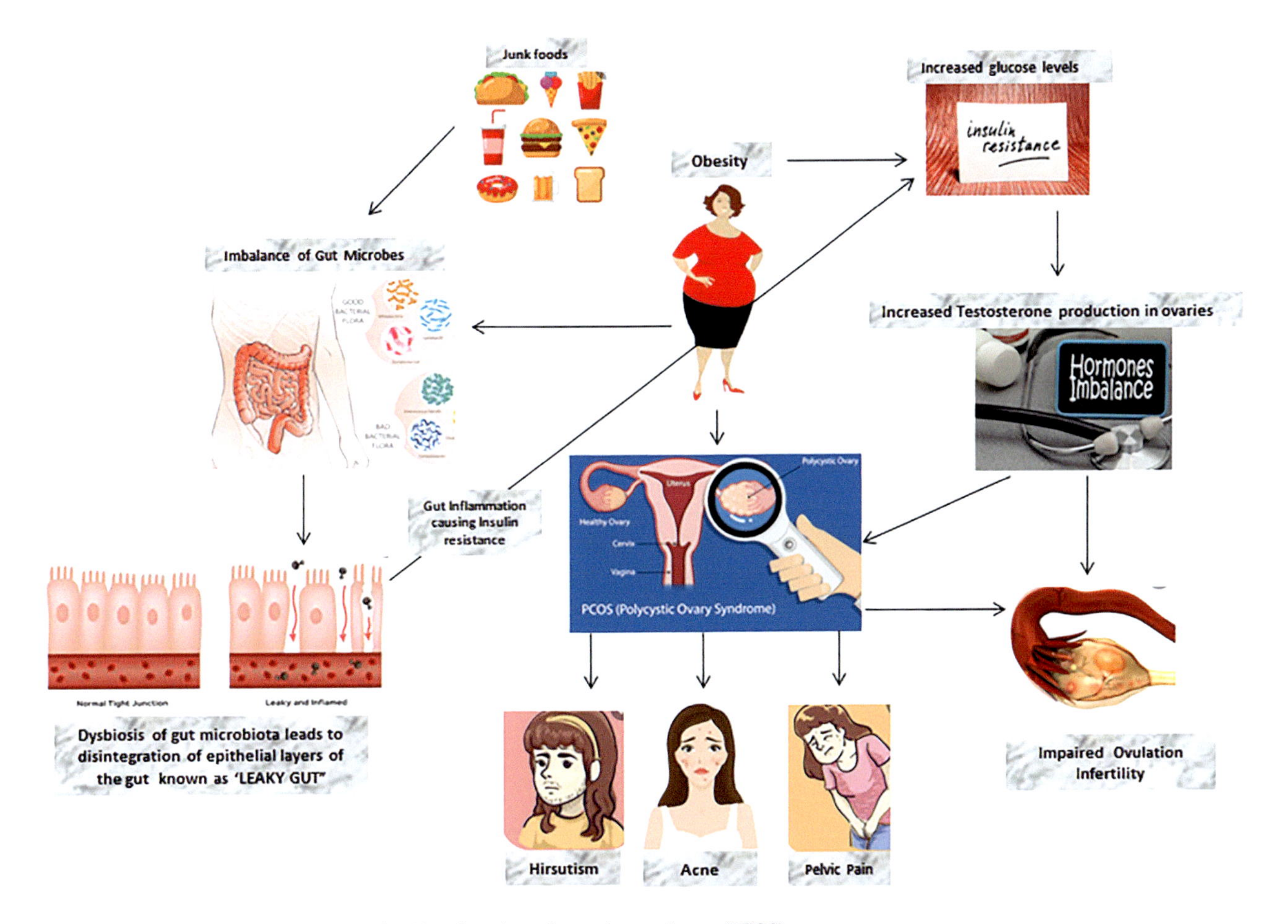

FIGURE 13.5 Role of dysbiosis of gut microbiota in polycystic ovarian syndrome (PCOS).

intestine are gram-negative bacteria. After being absorbed into the blood, LPS can bind to Toll-like receptor 4 (TLR4) on the surface of immune cells through the transmission of LPS-binding protein (LBP), CD14, and bone marrow differentiation factor (MD-2). This can lead to the activation of downstream signaling pathways which promote the expression of TNF-α, IL-6, etc., affecting insulin sensitivity, leading to IR. Studies comparing the levels of serum LBP in PCOS patients with that of healthy women found that LBP levels in PCOS patients were significantly increased, irrespective of body mass, providing strong evidence for the involvement of LPS in the pathogenesis of PCOS. Studies have also found that GM affects glucose and lipid metabolism and chronic inflammatory states in PCOS patients by influencing bile acids, leading to other endocrine disorders.

The brain–gut axis is the information-communication system between the intestine and the brain, a neuroendocrine–immune network formed by the central nervous system, the intestinal nervous system (ENS), the hypothalamic–pituitary–adrenal axis, and the intestinal tract. GM communicate with the host through the gut–brain axis via immune, neuroendocrine, and vagal pathways. Under pathological conditions, GM disorders can result in abnormal secretions of peptides, cytokines, and inflammatory factors in the intestinal canal, which transmit information to the brain. Some studies have speculated that the emotional disorders exhibited by some PCOS patients may be related to abnormal transmission of information along the brain–gut–GM axis. The intestinal microbiome can also interfere with host immune regulation through the brain–gut axis, although research evidence for this hypothesis is inadequate.

All the above hypotheses lead to a link between imbalance in GM and PCOS. This may possibly help in devising better treatment strategies for PCOS, including prebiotic and probiotic therapies leading to stabilizing the GM (Zhao et al., 2020).

13.3.5 Ectopic Pregnancy and Spontaneous Abortion

Ectopic Pregnancy (EP) is a major female reproductive problem in the developing countries resulting in much morbidity and mortality (Figure 13.6). Though the exact etiology is still unclear, a relation of EP with PID is well known. Chlamydia infection in the fallopian tissues is another risk factor for EP. It is postulated that Chlamydia infections may reduce the activity of the cilia in the epithelium of the fallopian tubes and subsequently the contraction of the smooth muscles, resulting in EP. Recently, molecular studies have identified the presence of *Ureaplasma urealyticum* and *Mycoplasma hominis* and *C. trachomatis* from menstrual tissue samples of patients with EP. Antimicrobial prophylaxis prior to gynecologic surgery also alters the fallopian tube microbiota, resulting in undetectable *Atopobium* sp., *Porphyromonas* sp., *Prevotella* sp., and *Clostridium* sp., as well as a reduced detection of *Staphylococci*. These bacterial species are frequently associated with reproductive tract infections. Studies have indicated alterations in the fallopian tube microbiota in pre- and post-menopausal women, and in response to exogenous hormone treatment. These findings provide important insights into the labile nature of the fallopian tube microbiota, which should be considered as a potential source of microbial seeding in post-surgical infection, and a possible cause of reproductive pathology (Pelzer et al., 2018).

The association between Chlamydia with spontaneous abortion is well documented. Release of proteases due to invasion of Chlamydia into the choriodecidual space and subsequent arousal of placental immune response and inflammation (chorioamnionitis) are the major factors responsible for premature rupture of the membranes, activation of arachidonic acid pathway, and uterine contractions in spontaneous abortion and preterm labor. The disruption of *Lactobacillus* sp. during BV and the presence of *U. urealyticum* in the genital tract in combination with other microorganisms have also been found to provoke spontaneous abortion and cervical incompetence. Endometrial dysbiosis is also associated with miscarriage or obstetrical complication due to alterations of the immune and inflammatory response. Alterations of the endometrial microbiome could be related to pregnancy complications involving infection and inflammation of the maternal or fetal membranes such as deciduitis, chorioamnionitis, and other obstetrical conditions. In addition, it has been proposed that an unfavorable bacterial composition may activate antiangiogenic pathways during placentation and embryo–fetal development, resulting in altered trophoblast and endothelial functions, which are characteristic of preeclampsia.

The composition and function of microbiome in the genital tract of women with abnormal pregnancy outcomes and spontaneous abortions are still unclear, indicating the urgent need for more enhanced studies. Molecular studies can be useful in early detection of infections in gynecological problems such as tubal obstruction, PID, EP, spontaneous abortions, and unexplained infertility (Haldar et al., 2016).

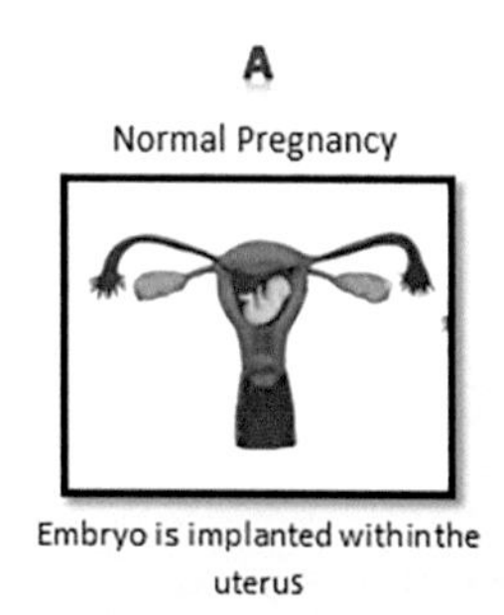

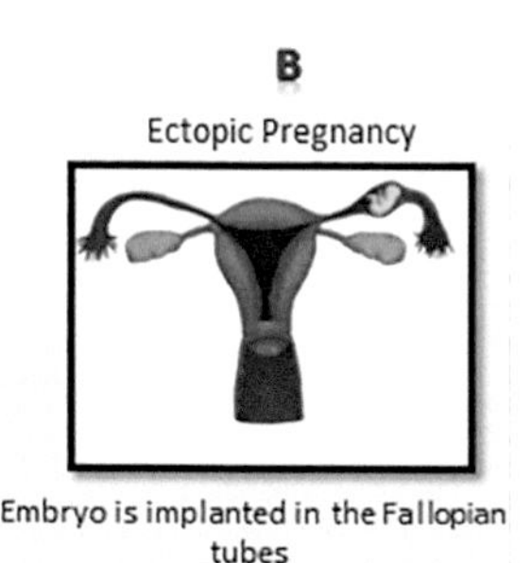

FIGURE 13.6 Normal pregnancy *vs* ectopic pregnancy.

13.3.6 Infertility and Implantation Failures

The absence of a definable cause makes idiopathic infertility extremely frustrating for couples trying to conceive, thus highlighting the need of dedicated studies in this specific field. The role played by genital microbial community on reproductive outcome is not fully understood yet, despite recent evidence suggesting an association between an altered vaginal and seminal microbiome composition and infertility. Evidence from several studies shows a differential vaginal microbiota composition in infertile patients compared with healthy and fertile women. An abnormal vaginal microbiota has been associated with a poor reproductive outcome in patients undergoing *in vitro* fertilization (IVF), as well as with early spontaneous abortion in patients undergoing IVF treatment.

The immune system is intimately involved in all aspects of the reproductive process, particularly around the time of conception, and in the peri-implantation period. Different stages during the initiation of pregnancy hold specific immunologic challenges; to enable a healthy pregnancy, the maternal immune system must support the introduction of semen to allow fertilization, the initial contact of blastocyst and endometrium, and correct formation of the placenta. Bacterial metabolites and/or compounds may cause an inflammatory response in the endometrium that could interfere with embryo implantation after the intrauterine implantation (IUI) procedure during IVF treatment. Vaginal microbiome characterization could be useful for women with idiopathic infertility. The modulation of vaginal microbiota could improve the likelihood of either spontaneous or IUI-induced pregnancy. Probiotics may be an option, but their clinical effectiveness is still debated depending on several issues, including the selected strain, dosing, and delivery route (oral or topical). Instead of administering single probiotic strains, fluids from fertile women with *L crispatus*-dominated microbiota could be transferred to women with vaginal dysbiosis to restore a healthier vaginal flora and a more "pregnancy-friendly" environment. More interventional studies in the pre-IUI modulation of vaginal microbiota are necessary for the success of IVF treatment (Amato et al., 2019).

Recent studies have also associated endometrial dysbiosis as an emerging cause of implantation failure and pregnancy loss. *Actinomyces, Corynebacterium, Enterococcus, E. coli, Fusobacterium, Gardnerella, Prevotella, Propionibacterium, Staphylococcus* and *Streptococcus* to the detriment of *Lactobacillus spp.* in endometrial samples and menstrual blood of patients with endometriosis, and Staphylococcal and Streptococcal species have been identified at the molecular level in ovarian endometrial fluid corroborating endometrial dysbiosis. Chronic endometritis, which is often asymptomatic, is seldom suspected and diagnosed, leading to important inconsistencies in the estimated prevalence reported for this disease. This subclinical disease could affect up to 45% of infertile patients with recurrent implantation failure (RIF) and recurrent pregnancy loss (RPL). Studies have shown that antibiogram-driven treatment of chronic endometritis in RIF and RPL patients has improved their reproductive outcomes. Hence, unequivocal diagnosis of chronic endometritis in infertile patients, using objective and reliable methods, could help to improve the clinical management of asymptomatic patients in whom chronic endometritis is not suspected or diagnosed (Moreno and Simon, 2018).

13.3.7 Gynecological Cancers

Many researches on the GI microbiome have shown that commensal bacteria can promote or inhibit carcinogenesis. Recent evidences now suggest that the state of the vaginal microbiome could influence the development of cancer. Available research on the relation between the vaginal microbiome and gynecological cancers (Figure 13.7) is still in its infancy, although they have suggested a potential link. One such is how the microbiome mediates the risk of human papillomavirus (HPV)-induced cervical cancer. It is well known that certain strains of HPV significantly increase the risk of cervical cancer. Studies conducted on women infected with HPV showed microbiomes with greater bacterial diversity, specifically abundant in *L. gasseri* and *G. vaginalis* and lower levels of other *Lactobacillus*. *Fusobacteria*, including *Sneathia*, are most strongly correlated with HPV infection. The rate of clearance of HPV, and thus the risk of developing HPV-associated malignant transformation, may also be dependent on the composition of the microbiome. Specifically, vaginal microbiomes with greater amounts of *L. gasseri* or *L. iners* were associated with a rapid remission of HPV, whereas microbiomes with low amounts of *Lactobacillus* and high amounts of *Atopobium* were associated with slower HPV clearance. Although *L.*

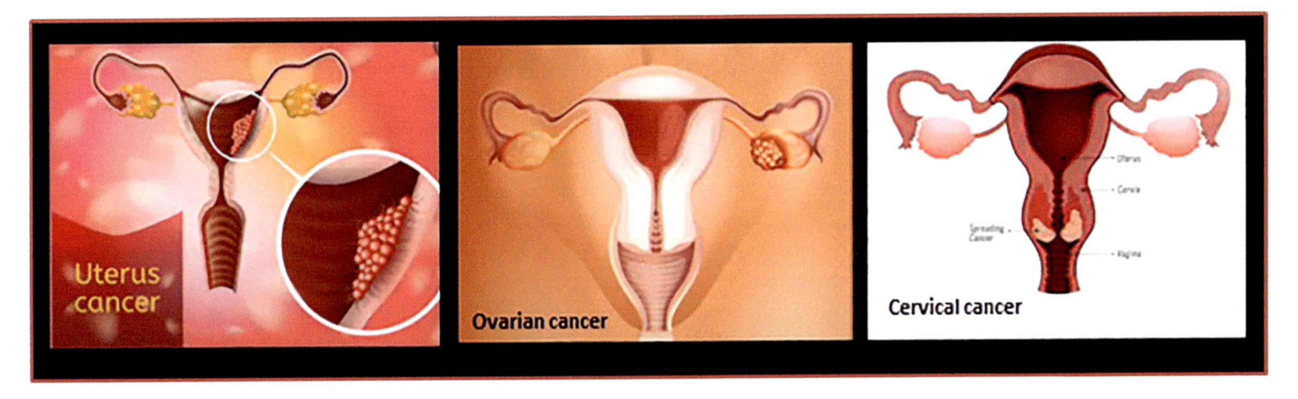

FIGURE 13.7 Different types of gynecological cancers.

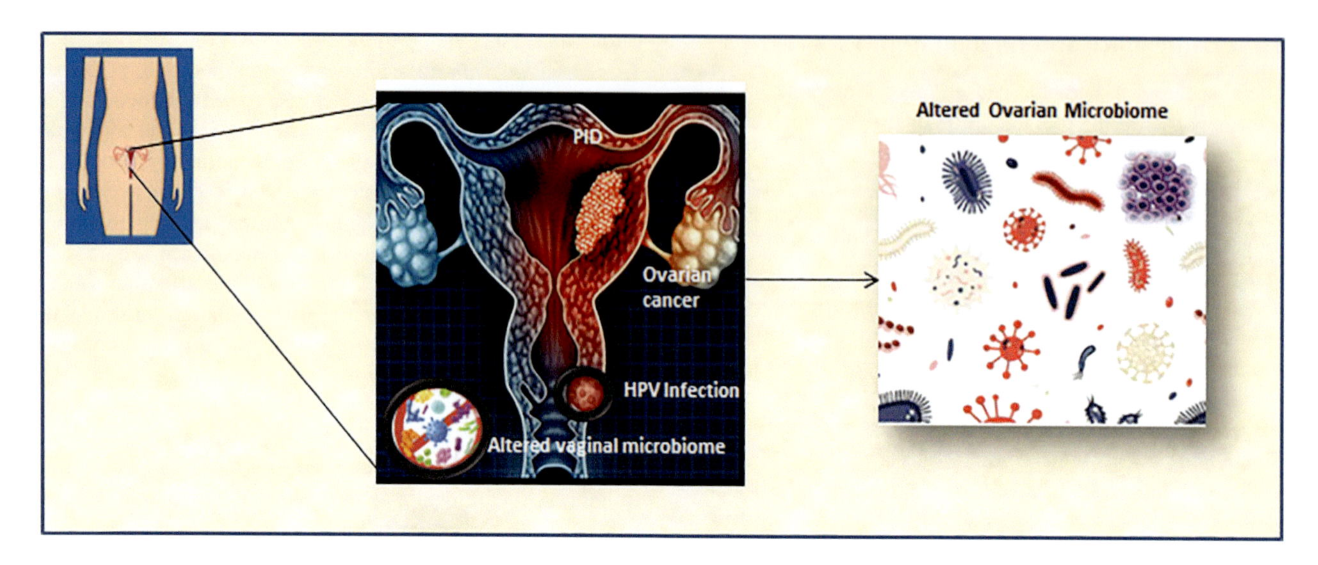

FIGURE 13.8 Potential link between altered microbiome and ovarian cancer onset.

gasseri was positively correlated with a higher level of HPV infection, it was also associated with a higher remission rate. It has been suggested that vaginal infection by *C. trachomatis* may also predispose individuals to HPV infection by altering the microbiome.

Vaginal microbiome can also be directly correlated with gynecological cancers. Studies to date have documented an overall increase in diversity in cervical cancer. Studies have suggested that cervical intraepithelial neoplasia (CIN) progression is correlated with increasing vaginal microbiota diversity. This increased diversity may possibly be because of the epithelial barrier rupture and the host's immune dysregulation. Among the microbiome, *L. iners*, in particular, was associated with cervical cancer alone or in conjunction with HPV infection, as well as higher grades of CIN in HPV-positive patients.

PID is also widely established as a risk factor for epithelial ovarian cancer (Figure 13.8). Among the organisms responsible for PID onset, *Chlamydia* is the most common, followed by *N. gonorrhoeae*. Previous studies have found that antibodies against *Chlamydia* infection are associated with ovarian cancer (Xu et al., 2020).

Researches have also identified potential microbiotas contributing to the genesis of endometrial cancer. A recently identified *A. vaginae* and *Porphyromonas spp.* in the reproductive tract in combination with a high vaginal pH are found to be statistically related to the occurrence of endometrial cancer. It was further demonstrated that *Porphyromonas spp.* combined with high pH in the vagina could be a promising biomarker for endometrial cancer. These findings are significant as they put forth a promising biomarker for early detection and pave the way for possible primary preventive interventions. Molecular mechanisms underlying the interaction between microbiome and pathogenesis of endometrial cancer still need elucidation (Łaniewski et al., 2020).

Emerging evidence shows that genital dysbiosis and/or specific bacteria might have an active role in the development and/or progression and metastasis of gynecological malignancies, such as cervical, endometrial, and ovarian cancers, through direct and indirect mechanisms, including modulation of estrogen metabolism. Cancer therapies might also alter microbiota at sites throughout the body. Reciprocally, microbiota composition can influence the efficacy and toxic effects of cancer therapies, as well as quality of life following cancer treatment. Modulation of the microbiome via probiotics or microbiota transplant might prove useful in improving responsiveness to cancer treatment and quality of life. Elucidating these complex host–microbiome interactions, including the crosstalk between distal and local sites, will translate into interventions for prevention, therapeutic efficacy, and toxic effects to enhance health outcomes for women with gynecological cancers (Łaniewski et al., 2020).

13.4 Manipulation of Microbiome/ Restoration of Microbiome

Manipulation of the vaginal microbiota is essential for women's health. Oral or intravaginal antibiotic is a routine treatment of BV, which often relapses and fails to offer a long-term defensive barrier. Application of "probiotics" and cationic antimicrobial peptides (AMPs) is gaining popularity as an alternative strategy of treatment. Probiotics are defined as live microorganisms which, when administered in adequate amounts, confer a health benefit to the host. Oral or intravaginal administration of different species of *Lactobacillus* has shown to increase the number of vaginal *Lactobacilli* which presumably provides a mechanical barrier against *G. vaginalis* and prevents the adhesion of the pathogens to the vaginal epithelium. *L. rhamnosus* GR-1 is one such probiotic with proven inhibitory activity against the growth of *G. vaginalis*. *L. crispatus* has also demonstrated potentials as a hopeful probiotic in preventing *N. gonorrhoeae* infections through counteracting *N. gonorrhoeae* viability. Vaginal microbiome modulation via probiotics, novel antimicrobials, and/or vaginal microbiota transplantation might be a novel approach to the prevention of gynecological cancers and/or the reduction of vaginal toxicities related to cancer treatment (Haldar et al., 2016; Łaniewski et al., 2020; Xu et al., 2020).

13.5 Conclusion

Multiple socioeconomic, behavioral, environmental, hormonal, and genetic factors can affect the female genital microbiome by disrupting homeostasis and promoting dysbiosis. Female reproductive tract microbiome is intimately interconnected with other mucosal sites. Researches over the years have shown the importance of vaginal microbiota as the first line of defense against infection in females. There are still limitations in knowledge about the functional mechanisms that maintain a robust and stable ecosystem of microbiome in the female reproductive organs. This offers major challenges to develop therapeutics for the reproductive diseases and infertility that are common in many parts of the world due to various socioeconomic and lifestyle-related factors. Since one-solution-fits-all approach cannot be an ideal remedy, there is an increasing demand for personalized treatment taking into account the host genetics, health, immunological, and nutritional status and lifestyle. Future research should hence focus on antibiotic-sparing, individual-centric strategies to combat many of the female reproductive diseases with the goal to restore niche-specific structure of microbiome.

REFERENCES

Amato, V., Papaleo, E., Pasciuta, R., et al. (2019). Differential composition of vaginal microbiome, but not of seminal microbiome, is associated with successful intrauterine insemination in couples with idiopathic infertility: a prospective observational study. *Open Forum Infectious Diseases®*: 1–9.

Brenner, K., You, L. & Arnold, F.H. (2008). Engineering microbial consortia: a new frontier in synthetic biology. *Trends in Biotechnology*, 26(9): 483–489.

Burmolle, M., Webb, J.S., Rao, D., Hansen, L.H., Sorensen, S.J. & Kjelleberg, S. (2006). Enhanced biofilm formation and increased resistance to antimicrobial agents and bacterial invasion are caused by synergistic interactions in multispecies biofilms. *Applied and Environmental Microbiology*, 72(6): 3916–3923.

Collado, M.C., Cernada, M., Bauerl, C., Vento, M. & Perez-Martinez, G. (2012). Microbial ecology and host–microbiota interactions during early life stages. *Gut Microbes*, 3(4): 352–365.

Diop, Dufour, Levasseur et al. (2019) Exhaustive repertoire of human vaginal microbiota. Human microbiome journal 11:100051.

Dunlop, A.L., Mulle, J.G., Ferranti, E.P., Edwards, S., Dunn, A.B. & Corwin, E.J. (2015). Maternal microbiome and pregnancy outcomes that impact infant health: a review. *Advances in Neonatal Care*, 15(6): 377–385.

Faust, K., Sathirapongsasuti, J.F., Izard, J., et al. (2012). Microbial co-occurrence relationships in the human microbiome. *PLoS Computational Biology*, 8(7): e1002606.

Franasiak, J.M. & Scott Jr., R.T. (2015). Introduction: microbiome in human reproduction. *Fertility and Sterility*, 104(6): 1341–1343.

Fredricks, D.N., Fiedler, T.L. & Marrazzo, J.M. (2005). Molecular identification of bacteria associated with bacterial vaginosis. *New England Journal of Medicine*, 353: 1899–1911. doi: 10.1056/NEJMoa043802.

Gomezde-Aguero, M., Ganal-Vonarburg, S.C., Fuhrer, T., et al. (2016). The maternal microbiota drives early postnatal innate immune development. *Science*, 351(6279): 1296–1302.

Haldar, S., Kapil, A., Sood, S. & Sengupta, S. (2016). Female reproductive tract microbiome in gynecological health and problems. *Journal of Reproductive Health and Medicine*, 2, S48–S54. doi: 10.1016/j.jrhm.2016.11.007.

Howitt, M.R. & Garrett, W.S. (2012). A complex microworld in the gut: gut microbiota and cardiovascular disease connectivity. *Nature Medicine*, 18(8): 1188–1189.

Jayasinghe, T.N., Chiavaroli, V., Holland, D.J., Cutfield, W.S. & O'Sullivan, J.M. (2016). The new era of treatment for obesity and metabolic disorders: evidence and expectations for gut microbiome transplantation. *Frontiers in Cellular and Infection Microbiology*, 6: 15.

Jimenez, E., Marin, M.L., Martin, R., et al. (2008). Is meconium from healthy newborns actually sterile? *Research in Microbiology*, 159(3): 187–193.

International Human Genome Sequencing Consortium. (2004). Finishing the euchromatic sequence of the human genome. *Nature*, 431(7011): 931–945. doi: 10.1038/nature03001.

Keller, L. & Surette, M.G. (2006). Communication in bacteria: an ecological and evolutionary perspective. *Nature Reviews Microbiology*, 4(4): 249–258.

Khan, K.N., Fujishita, A., Masumoto, H., Muto, H., Kitajima, M., Masuzaki, H. & Kitawaki, J. (2016). Molecular detection of intrauterine microbial colonization in women with endometriosis. *European Journal of Obstetrics & Gynecology and Reproductive Biology*, 199: 69–75.

Łaniewski, P., Ilhan, Z.E. & Herbst-Kralovetz, M.M. (2020). The microbiome and gynaecological cancer development, prevention and therapy. *Nature Reviews Urology*, 17: 232–250. doi: 10.1038/s41585-020-0286-z.

Larsen, N., Vogensen, F.K., Van den Berg, F.W., et al. (2010). Gut microbiota in human adults with type 2 diabetes differs from non-diabetic adults. *PLoS One*, 5(2): e9085.

Ley, R.E., Peterson, D.A. & Gordon, J.I. (2006a). Ecological and evolutionary forces shaping microbial diversity in the human intestine. *Cell*, 124: 837–848. doi: 10.1016/j.cell.2006.02.017.

Ley, R.E., Turnbaugh, P.J., Klein, S. & Gordon, J.I. (2006b). Microbial ecology: human gut microbes associated with obesity. *Nature*, 444: 1022–1023. doi: 10.1038/4441022a.

Ling, Z., Kong, J., Liu, F., et al. (2010). Molecular analysis of the diversity of vaginal microbiota associated with bacterial vaginosis. *BMC Genomics*, 11: 488. doi: 10.1186/1471-2164-11-488.

Mazmanian, S.K., Liu, C.H., Tzianabos, A.O. & Kasper, D.L. (2005). An immunomodulatory molecule of symbiotic bacteria directs maturation of the host immune system. *Cell*, 122: 107–118. doi: 10.1016/j.cell.2005.05.007.

McFarland, L.V. (2014). Use of probiotics to correct dysbiosis of normal microbiota following disease or disruptive events: a systematic review. *BMJ Open*, 4(8): e005047.

Moreno, I. & Simon, C. (2018). Relevance of assessing the uterine microbiota in infertility. *Fertility and Sterility®*, 110: 337–343.

Onderdonk, A.B., Delaney, M.L. & Fichorova, R.N. (2016). The human microbiome during bacterial vaginosis. *Clinical Microbiology Reviews*, 29: 223–238. doi: 10.1128/CMR.00075-15.

Ordovas, J.M. & Mooser, V. (2006). Metagenomics: the role of the microbiome in cardiovascular diseases. *Current Opinion in Lipidology*, 17(2): 157–161.

Pelzer, E.S., Willner, D., Buttini, M., Hafner, L.M., Theodoropoulos, C. & Huygens, F. (2018). The fallopian tube microbiome: implications for reproductive health. *Oncotarget*, 9(30): 21541–21551. doi: 10.18632/oncotarget.25059.

Peterson, J., Garges, S., Giovanni, M., et al; NIH HMP Working Group. (2009). The NIH Human Microbiome Project. *Genome Research*, 19(12): 2317–2323.

Prescott, S.L. (2017). History of medicine: origin of the term microbiome and why it matters. *Human Microbiome Journal*, 4: 24–25.

Qin, J., Li, R., Raes, J., et al. (2010). A human gut microbial gene catalogue established by metagenomic sequencing. *Nature*, 464(7285): 59–65.

Qin, J., Li, Y., Cai, Z., et al. (2012). A metagenome-wide association study of gut microbiota in type 2 diabetes. *Nature*, 490(7418): 55–60.

Rosenstein, I.J., Morgan, D.J., Sheehan, M., Lamont, R.F. & Taylor-Robinson, D. (1996). Bacterial vaginosis in pregnancy: distribution of bacterial species in different gram-stain categories of the vaginal flora. *Journal of Medical Microbiology*, 45: 120–126. doi: 10.1099/00222615-45-2-120.

Sharma, H., Tal, R., Clark, N.A. & Segars, J.H. (2014). Microbiota and pelvic inflammatory disease. *Seminars in Reproductive Medicine*, 32: 43–49.

Sirota, I., Zarek, S.M. & Segars, J.H. (2014). Potential influence of the microbiome on infertility and assisted reproductive technology. *Seminars in Reproductive Medicine*, 32(1): 35–42.

Sobel, J.D. (1999). Is there a protective role for vaginal flora? *Current Infectious Disease Reports*, 1: 379–383. doi: 10.1007/s11908-999-0045-z.

Soksawatmaekhin, W., Kuraishi, A., Sakata, K., Kashiwagi, K. & Igarashi, K. (2004). Excretion and uptake of cadaverine by CadB and its physiological functions in *Escherichia coli*. *Molecular Microbiology*, 51: 1401–1412. doi: 10.1046/j.1365-2958.2003.03913.x.

Tai, F.W., Yi Chang, C., Chiang, J.H., Lin, W.C. & Wan, L. (2018). Association of pelvic inflammatory disease with risk of endometriosis: a nationwide cohort study involving 141,460 individuals. *Journal of Clinical Medicine*, 7: 379. doi: 10.3390/jcm7110379.

Tang, W.H. & Hazen, S.L. (2014). The contributory role of gut microbiota in cardiovascular disease. *Journal of Clinical Investigation*, 124(10): 4204–4211.

Teede, H., Deeks, A. & Moran, L. (2010). Polycystic ovary syndrome: a complex condition with psychological, reproductive and metabolic manifestations that impacts on health across the lifespan. *BMC Medicine*, 8: 41.

Wang, Y., Zhang, Y., Zhang, Q., Chen, H. & Feng, Y. (2018). Characterization of pelvic and cervical microbiotas from patients with pelvic inflammatory disease. *Journal of Medical Microbiology*, 67(10): 1519–1526. doi: 10.1099/jmm.0.000821.

Xu, J., Peng, J.J., Yang, W., Fu, K. & Zhang, Y. (2020). Vaginal microbiomes and ovarian cancer: a review. *American Journal of Cancer Research*, 10(3): 743–756. www.ajcr.us/ISSN:2156–6976/ajcr0108589.

Yang, X., Xie, L., Li, Y. & Wei, C. (2009). More than 9,000,000 unique genes in human gut bacterial community: estimating gene numbers inside a human body. *PLoS One*, 4(6): e6074.

Zhao, X., Jiang, Y., Xi, H., Chen, L. & Feng, X. (2020). Exploration of the relationship between gut microbiota and polycystic ovary syndrome (PCOS): a review. *Geburtshilfe Frauenheilkd*, 80(2): 161–171.

14

Crosstalk between Bacteria and Host Immune System with Special Emphasis on Foodborne Pathogens

A.A.P. Milton, G. Bhuvana Priya, M. Angappan, S. Ghatak, and Vivek Joshi
ICAR Research Complex for NEH Region
ICAR – National Research Centre on Mithun

CONTENTS

14.1 Introduction

Bacteria are ubiquitously present everywhere in nature. Bacterial diseases are drastically increasing in prevalence due to resistance to all the known antibiotics. Host–pathogen interaction is the interaction between a pathogen and its host. Bacteria have developed interesting strategies by expressing virulence factors that interact with the host, leading to adherence, colonization, invasion, and proliferation; latent stage; and commensalism. Successful infection occurs only if the pathogen is able to survive in the host environment and modulate the host mechanism to prevent it from being killed. For modulating the effects of pathogen, host also makes some efforts to overcome it (Falkow 1997). The pathogens evade the host responses, e.g., by altering their surface proteins (Lange and Ferguson 2009; Sasaki 1994; Haraguchi and Sasaki 1997; Drake et al. 1998). In an environment of complex interaction between the host and the pathogen, the victory goes to the favored interaction whether it is of host or of pathogen (Casadevall and Pirofski 2000). The outcome of the host–pathogen interaction will be either disease or death or clearance of pathogen by the host. Understanding the host–pathogen interaction is not as simple; a molecular-level understanding of each individual component is necessary (Forst 2006). To obtain an effective solution for the prevention and control of infectious diseases, there should be a clear knowledge about the crisscross interaction between host and pathogen at the molecular level.

Identification of the key element of disease pathogenesis, i.e., virulence factor, will lead to combat the disease effectively through the development of novel drug and vaccine. During pre-genomic era, gene expression techniques such as *in vivo* expression technology (IVET) (Mahan et al. 1993) and signature-tagged mutagenesis (STM) (Hensel et al. 1995) helped in the identification of virulence factors under *in vivo* conditions. Genomics, transcriptomics, and proteomics approaches in the post-genomic era resulted in the accelerated discovery of virulence factors (Hsing et al. 2008). Biological data obtained from transcriptomic and proteomic study can be useful for understanding the host–pathogen interactions using appropriate computational tools (Mukherjee et al. 2013). In this chapter, host–bacteria interaction is discussed with special emphasis on foodborne pathogens.

14.2 Virulence Factors

Identification of novel virulence factor is helpful in expanding the horizon of knowledge about disease pathogenesis and providing clear-cut therapeutic target for striking out the disease successfully and effectively. An extensive review on virulence factors of bacterial pathogens is done by Finlay and Falkow (1997). There are several virulence factors like capsule—which has antiphagocytic property—and secretory proteins secreted by the secretion system that transports the protein toxins into the host cell (China and Goffaux 1999). Autotransporters are group of virulence proteins produced by gram-negative bacteria, which are secreted by type 5 SS which translocates these proteins to the cell surface (Dautin and Bernstein 2007). Outer membrane proteins belong to the group of virulence factor determinants, which have a role in adhesion of bacteria to the cell surface, invasion, and colonization. Lipopolysaccharide (LPS), a major outer membrane component in gram-negative bacteria, protects the cell from complement-mediated opsonization (Finlay and Falkow 1997). Siderophores and biofilm-forming proteins are the other virulence factors involved in disease pathogenesis. *Salmonella enterica* serovar Enteritidis, TlpA (TIR-like protein A), a virulence factor, modulates the host defense mechanisms (Newman et al. 2006).

14.3 Capsule

Capsule is one of the most important virulence factors which bacteria use to evade the host immune response. Capsules are produced by bacteria such as *Bacillus anthracis*, *Pseudomonas aeruginosa*, *Pasteurella multocida*, and *Streptococcus*. The capsule protects the bacteria from the host's opsonizing antibodies. Although capsules of different organisms differ in their composition and immune-modulating activities, they increase their antiphagocytic property invariably (Wilson et al. 2002). The polysaccharide capsule of *Campylobacter jejuni* plays vital roles in the chicken gut colonization, epithelial cell invasion, and serum resistance (Kim et al. 2018)

14.4 Cell Wall

Based on the structure of cell wall, bacteria are divided into two groups: gram-positive and gram-negative. The toxic components are embedded within their cell wall, which are released into the extracellular medium only if the bacteria burst or if the infected cell dies off. These components are effective virulence determinants and are important in the pathogenesis of septic shock. Both gram-positive and gram-negative bacteria share a common pathway for septic shock to come into play (Periti 2000; Horn et al. 2000; Verhoef and Mattson 1995). Proteinaceous exotoxins, which are the enzymes secreted by the bacteria, are highly toxic even at lower concentrations. Non-proteinaceous substances such as LPS (endotoxin) (Rietschel et al. 1994) become biologically active only at higher concentrations, and LPS is a significant component of the gram-negative bacterial cell wall and teichoic acid for gram-positive bacteria (Finlay and Falkow 1997). Exotoxins such as cholera toxin and diphtheria toxin play an essential role in the disease pathogenesis. They solely cannot induce a visible potential effect on the host. They rely on other virulence factors. For instance, apart from cholera toxin, adhesion is also required for *Vibrio cholerae* to establish their virulence (Srivastava et al. 1980). Both gram-positive and gram-negative bacteria after infecting the host stimulate the monocytes to release proinflammatory mediators—mainly TNF and IL-1 (Verhoef and Mattson 1995)—and activate the complement cascade reaction.

14.5 Lipopolysaccharides (LPS)

LPS is one of the significant virulence factors that is known to cause septic shock, a severe form of infection. Endotoxins are released during the growth of pathogens in host niche (Nitsche et al. 1994) due to either death of the organism or antibiotic treatment (Shenep and Mogan 1984). The receptor for LPS is present on the surface of macrophages. LPS, a major virulence determinant (Burns and Hull 1998; Pollack 1999), consists of regions of pathogen-associated molecular patterns (PAMPs) on its surface (Pugin et al. 1994; Krug et al. 2001; Seya et al. 2001), which interact with the host innate immune system (Janssens and Beyaert 2003). Moran and Prendergast (2001) reported that *C. jejuni* and Lewis antigens by *Helicobacter pylori* alter their LPS structure, mimic the host ganglioside structure, and escape the defense mechanism. Endotoxins modify their structure under *in vivo* conditions due to the addition of sialic acid to their galactose moieties (Smith et al. 1992; Vanputten and Robertson 1995). Smith et al. (1992) reported that the mucosal pathogens such as *Neisseria* and *Haemophilus* also undergo sialylation. Lipid A of LPS has the toxic property (Morrison and Ulevitch 1978; Raetz et al. 1991), and on its release, there will be uncontrolled release of cytokines, complements, and other mediators leading to septic shock (Shalaby et al. 1986; Whiteley et al. 1990, 1991; Maheswaran et al. 1992). The lipid A portion of the LPS interacts with the Toll-like receptor 4 (TLR4) of the host (Netea et al. 2002). It also interacts with lipopolysaccharide-binding protein (LBP), CD (cluster of differentiation) 14 (a surface molecule of macrophages) (Kitchens 2005), CD18 (Paape et al. 1996), and antimicrobial peptides (Werling et al. 1996; Vreugdenhil et al. 1999; Wu et al. 2004; Hari-Dass et al. 2005). The uncontrolled release of these inflammatory mediators results in an imbalance between the anti-inflammatory and proinflammatory pathways (Goldman et al. 1990; Borovikova et al. 2000). The LPS interacts with CD14, and this interaction is catalyzed after association with LBP (Viriyakosol and Kirkland 1995). Takeda (2005) and Trianta¢lou and Trianta¢lou (2005) described the sequence of interaction between TLR and LPS. Werling and Jungi (2003) reported that CD14 and TLR 4 in association with MD2 form a complex, and this complex binds with LPS, which leads to an intracellular signaling pathway that mediates a series of reactions that include MyD88 (myeloid differentiation primary response 88), IRAK (interleukin-1 receptor-associated kinases), and TRAF6 (tumor necrosis factor receptor (TNFR)-associated factor 6). These mediators activate NF-kB (nuclear factor kappa B) that enables the release of cytokine mediators. CD80 and CD86—the co-stimulatory molecules on the surface of antigen-presenting cells—are also expressed by the induction of TLR along with the processed antigen that has been bounded to MHC (major histocompatibility complex) class II molecules that activate the adaptive immune system to respond to the infection (Medzhitov and Janeway 2000; Werling and Jungi 2003). TLR4 specifically uses MyD88 as signaling adapter while interacting with the LPS (Fitzgerald et al. 2001; O'Neill 2002). However, the complex mechanism of recognition and cellular-level response is not well documented (Raetz and Whitfield 2002; Janssens and Beyaert 2003). Some of the endotoxin-related diseases include salmonellosis in a wide range of animals (Fierer and Guiney 2001), colibacillosis in cattle and sheep (Hodgson 1994), septic shock in humans (Barron 1993), pasteurellosis in sheep (Hodgson 1993), and systemic inflammatory response syndrome (SIRS) (Young 1985; Hamill and Maki 1986; Jorgensen 1986; van Deventer et al. 1988; Brandtzaeg et al. 1989; Vincent 1996). The severity of endotoxin-related diseases is well documented (Hodgson 2006).

14.6 Secretion System

Secretion system is considered to be one of the virulence determinants of the bacteria. To enjoy a healthy and comfortable stay in the adverse host system, the bacteria itself secretes a range of proteins. Gram-negative bacteria have six secretion systems (type I, type II, type III, type IV, type V, and type VI), which secrete the proteins that play a role in the disease pathogenesis (Wooldridge 2009). These proteins altogether are termed as "effectors." Type I, II, and V secretion systems secrete the protein transversely from the envelope of bacteria to the extracellular environment. However, type III and type IV secretion systems directly deliver the proteins across the eukaryotic cell membranes. Type III secretion system is like a molecular syringe, which injects the secretory proteins into the cells. Examples of type III secretion system include *Shigella*, *Salmonella*, and *Yersinia* (Salyers and Whitt 2002). Type VI secretion system apart from pathogenesis also has a role in inter-bacterial interactions (Coulthurst 2013; Schwarz et al. 2010). Type VII secretion system was identified in gram-positive bacteria, which is also known as "ESAT-6/ESX." Simeone et al. (2009) reported that *Mycobacterium tuberculosis* has ESX-I secretory system that secretes two proteins, namely, ESAT-6 and CFP-10, which form a heterodimeric complex. This protein complex is an important virulence factor that plays a role in host–pathogen interaction by eliciting a strong T cell-mediated immune response, which eventually results in cell or membrane lysis. This ESX system is also found in other gram-positive bacteria like *Actinobacteria*. Pukatzki et al. (2007) reported that type VI secretory proteins are injected into the host cell in a similar manner, and they also share the structure of the injection system similar to that of T4 bacteriophage. As the effector proteins reveal virulence properties to the organism, their cellular-level microbiology in understanding the host–pathogen interaction is new area of interest. Ehsani et al. (2009) discussed the regulation, expression, and secretion of proteins in the infected cell using fluorescent tags and light microscopy. *Pseudomonas aeruginosa*, a gram-negative bacteria, encodes all the secretion systems, among which type III is highly critical (Joanne and Priya 2009). *Bordetella* uses type II, type IV, and type III to deliver the effector proteins into the host cell (Shrivastava and Miller 2009). Next comes the enteric bacteria in which type III secretion system (T3SS) plays a key role in the pathogenesis. Enteropathogenic *Escherichia coli* (EPEC) encodes about 21 type III effectors (Dean and Kenny 2009). These effectors escape the phagocytic activity of the M cells and weaken the function of enterocytes (Matsumoto and Young 2006). Matsumoto and Young (2006) also discussed that yop proteins are involved in systemic infection, whereas ysp proteins are involved in intestinal infection. Along with the ysp proteins, the type VI effectors are also found to be involved in pathogenicity of *Yersinia* infection. McGhie et al. (2009) discussed that *Salmonella* encodes two T3SSs, namely, SPI-1 and SPI-2 having about 30 proteins in repeats. These SPI-1 and SPI-2 work cooperatively in all the stages of infection, and they endorse bacterial internalization, and alter the membrane transport pathway and the inflammatory responses. *Shigella flexneri* encodes type 3SS having 25–30 effectors. The primary effectors are fore-synthesized through molecular chaperons and are involved in invading the cell. One among these effectors regulates the expression of second set of effectors, which inhibits the innate immune response of the host. Common themes emerging about the effector proteins are that their objective is to target GTPases and modify the ubiquitin and the transcriptional response of host. Their ultimate aim is to enter the host cell, mute the immune response, and uphold dissemination (Coburn et al. 2007). Understanding the biochemical functions of the effector proteins helps in the development of new therapy and measures to prevent the disease.

14.7 Outer Membrane Vesicles (OMVs)

Mostly, the virulence factors of gram-negative bacteria are secretory proteins. OMVs are produced by both pathogenic and nonpathogenic gram-negative bacteria (Mayrand and Grenier 1989; Kadurugamuwa and Beveridge 1996; Li et al. 1998; Beveridge 1999) that include *E. coli* (Hoekstra et al. 1976; Gankema et al. 1980), *Brucella melitensis* (Gamazo and Moriyon 1987), *C. jejuni* (Logan and Trust 1982; Blaser et al. 1983), *H. pylori* (Fiocca et al. 1999), *Borrelia burgdorferi* (Shoberg and Thomas 1993), *Salmonella* spp. (Vesy et al. 2000; Wai et al. 2003), and *P. multocida* (Hatfaludi et al. 2010). OMVs are distinct, whereas closed outer membrane blebs are produced by the growing bacteria (Mug-Opstelten and Witholt 1978; Zhou et al. 1998; Yaganza et al. 2004; McBroom and Kuehn 2005). The OMV of the pathogenic bacteria contains toxins, adhesin, host immune modulator LPS, and other innate immune-activating ligands. OMV is composed of proteins and lipids of the outer membrane of the gram-negative bacteria (McBroom and Kuehn 2005). The OMVs have been identified in the body fluids during infection, suggesting their ability of dissemination. The contents of the OMVs are delivered to the target site through endocytosis (Furuta et al. 2009; Kesty and Kuehn 2004). *Salmonella* Typhimurium, an intracellular pathogen, secretes OMVs inside the host cell (Garcia-del Portillo and Finlay 1995; Vesy et al. 2000; Bergman et al. 2005). During infection, gram-negative bacteria express surface factors such as adhesions that mediate the adhesion of OMV to the host cell followed by internalization. On interaction, the vesicles are internalized into the host cell by a nondegradative cholesterol-dependent pathway. OMVs induce immune response by stimulating the TLRs and enhance leukocyte migration (Galdiero et al. 1999; Akira et al. 2001). Alaniz et al. (2007) demonstrated that the OMVs trigger the inflammatory response in their study with *Salmonella enterica* serovar Typhimurium. *Salmonella* OMVs induced the production of proinflammatory cytokines (such as TNF-α, IL-12) and also activated dendritic cells and macrophages to express more MHC-II molecules on their surface. Ismail et al. (2003) studied that *H. pylori* vesicles elicited IL-8 response in the host and constitutively shed OMVs that play a role in stimulating the low-grade gastritis related to *H. pylori* infection. Pettit and Judd (1992) proposed that the *Neisseria* vesicles bind and remove the complement factors in the serum. LPS-O antigen alters its expression to evade the host defense mechanism (Pier 2000; Lerouge and Vanderleyden 2002) and enhance their survival in nutrient-scarce host niche.

14.8 Adhesins

The bacteria have adherence factors called "adhesins," which are made up of proteins and carbohydrates, and it is a virulence factor that is necessary for the bacteria to adhere to host surface and prevent them from washing off, and helps the organism to colonize, replicate, and release the toxic substances. The adherence of pathogen to host surface is a key step in the host–pathogen interaction. All the adhesins are not virulent determinants essentially (Krogfelt 1991). There are two major groups of protein adhesins, i.e., fimbrial and afimbrial adhesins. P pilus is one of the best studied fimbrial assemblies. The tip of the fimbriae exposes the adhesin to the host receptors, and the fimbriae especially help in adherence of gram-negative bacteria like *E. coli* and *Pseudomonas* (Donnenberg 2000; Merz and So 2000; Hahn 1997). Afimbrial adhesins are adherence factors that form a close contact with host surfaces but do not have polymeric structure as that of fimbrial adhesins. *Mycobacteria* spp., *Bordetella* spp., *Streptococcus* spp., *Staphylococcus* spp., EPEC, *Neisseria* spp., and *Hemophilus* spp., have afimbrial adhesins (Donnenberg 2000; Merz and So 2000; Joh et al. 1999; Bermudez and Sangari 2000). *Bordetella pertussis* possesses four afimbrial adhesins, namely, FHA (filamentous hemagglutinin), BrkA (Bordetella resistance to killing), pertussis toxin, and pertactin (Sandros and Tuomanen 1993). FHA is homologous to the non-fimbrial adhesion of *Haemophilus influenzae* (Barenkamp and Geme 1994). FHA possesses RGD (arginyl-glycyl-aspartic acid) sequence, which is eukaryotic recognition motif that binds to CR3 (macrophage-1 antigen), and this binding facilitates the uptake of FHA by the macrophages without oxidative burst (Relman et al. 1990). Sandros and Tuomanen (1993) proposed that FHA at the molecular level mimics the host cell molecule becoming a native ligand to CR3. Thus, molecular mimicry enhances the disease pathogenesis. The polysaccharide adhesins are the components of the capsule, and they enhance the adherence of *Streptococcus* and *Staphylococcus* spp. (Walker 1998). Adherence in *Mycobacteria* is promoted by the recognition of glucan and mannan (capsule components) by the complement and mannose receptor of the host (Daffe and Etienne 1999). Surface immunoglobulin, membrane spanning proteins, glycolipids, extracellular matrix proteins, and glycoproteins are the molecules that promote the adherence of microbes to the host receptors (Finlay and Falkow 1997; Hahn 1997; Novak and Tuomanen 1999; Zhang et al. 2000). Extracellular matrix molecules such as collagen and fibronectin are exposed once the tissue gets damaged and there are several pathogens that adhere to the extracellular matrix molecules (Westerlund and Korhonen 1993). For instance, Yad A of *Yersinia* binds to fibronectin (Schulze-Koops et al. 1993; Tertti et al. 1992). Curli fibers (adhesive surface fibers) expressed by *Salmonella enterica* and *E. coli* bind contact-phase proteins and some host extracellular matrix, and help in the internalization of bacteria into host cells (Gophna et al. 2001). Rao et al. (1993) discussed that *Mycobacterium intracellulare* and *Mycobacterium avium* express integrin-like molecules that facilitate their adherence to host cell or tissues. The host–pathogen interactions (receptor–ligand) that enhance adherence may be protein–protein interactions or protein–carbohydrate interactions. It should be noted that one pathogen is able to express only one adhesin, and it is the case with both gram-positive and gram-negative bacteria (Finlay and Falkow 1997; Walker 1998). Researches in drug and vaccine development are focused to block the adherence mechanism (Flock 1999; Kelly and Younson 2000).

14.9 Invasins

Invasins are the molecules that help in the penetration of pathogens to adhere and also direct their entry into the host cells, by triggering the signal, i.e., invasion (Bliska et al. 1993; Rosenshine and Finlay 1993). Invasion can be distinguished into two categories: extracellular and intracellular. Extracellular invasive pathogens breach the host barrier; multiply and disseminate to various tissues; release toxins; and initiate host inflammatory response. It is also reported that the extracellularly surviving pathogens sometimes gain entry into the cell taking up both extracellular and intracellular pathways in infection (Cleary and Cue 2000; Dziewanowska et al. 1999; Fleiszig et al. 1997). Intracellular invasive pathogens enter into the host cell, and multiply and survive inside the cell. There are a number of gram-positive, gram-negative, and mycobacterial pathogens that tend to survive intracellularly (Finlay and Falkow 1997; Cleary and Cue 2000; Bermudez and Sangari 2000; Dehio et al. 2000). Foodborne pathogens include *Salmonella, Brucella, Yersinia,* and *Listeria.* Obligate intracellular pathogens include *Chlamydia* spp., *Rickettsia* spp., and *Mycobacterium* spp. These intracellular pathogens invade both phagocytic and non-phagocytic cells (Walker 1998). Invasion genes that encode type III secretory system are present on the bacteria that invade the non-phagocytic cells and the secretory proteins are injected into the cell that activates host cell signaling pathways and promotes the microbial entry. In *Shigella* and *Salmonella*, this mechanism of entry is well characterized (Donnenberg 2000; Sansonetti et al. 1999; Galan and Zhou 2000). This signaling leads to cytoskeletal rearrangement and cell ruffling, and the cell engulfs the pathogen. Rho GTPases, the actin regulatory proteins, are involved in cytoskeletal rearrangement pathway (Donnenberg 2000; Galan and Zhou 2000). *Mycobacterium* spp. enters into the phagocytes by binding to the complement fragments, which prevents the exposure of organisms to the reactive oxygen species. Internalin A (InlA) in *Listeria monocytogenes* helps in invasion (Gaillard et al. 1991). This InlA binds to the E-cadherin and enhances cell invasion (Mengaud et al. 1996). Bacteria such as *C. jejuni*, *Klebsiella pneumoniae*, and enteroinvasive *E. coli* (EIEC), and EPEC require cytoskeletal structures called "microtubules" for invasion (Donnenberg et al. 1990; Oelschlaeger et al. 1993). The intracellular survival of the bacteria avoiding the phagocytic mechanisms is well discussed earlier (Finlay and Falkow 1997; Hackstadt 2000; Russell 2000). Host cell cytoplasm and acidic phagolysosomes are the areas where the bacteria reside. *Coxiella burnetii* survives in the lysosomes successfully avoiding the bactericidal agents. *Coxiella burnetii* and *Salmonella* Typhimurium survive in the acidic pH to the synthesis of virulence factors necessary for their intracellular survival (Hackstadt and Williams 1981; Rathman et al. 1996). *Salmonella* survival inside a membrane-bound vacuole called "salmonella-containing vacuole" (SCV) in phagocytic and non-phagocytic cells is well established (de Chastellier et al. 1995; Russell et al. 1996; Garcia-del Portillo and Finlay 1995). *Rickettsia* and *Chlamydia* enter into the host cell, replicate, and cause lysis of host cell membrane, thus releasing the pathogenic bacteria that further attach to the adjacent cells and infect them. Apart from lysis, *Listeria* and *Shigella* directly spread from one cell to another through actin filaments without causing the cell lysis (Walker 1998). Further, the biochemical interaction between the host and the pathogen should be well studied to understand how the pathogen penetrates the cell.

14.10 Exotoxins

Exotoxins are proteinaceous in nature, and exotoxins of bacteria are divided into four groups based on their function and amino acid composition: pore-forming toxins, proteolytic toxins, A-B toxins, and other minor toxins. RTX (repeats-in-toxin) family toxins are pore-forming toxins and are found to be released by numerous gram-negative bacteria. The delivery of toxin is similar to type I secretion system (Welch 1991). Tetanus toxin (Schiavo et al. 1992), botulinum toxin, protease, and elastase of *Pseudomonas aeruginosa* (Toder et al. 1991; Engel et al. 1998) are some of the proteolytic toxins, which leads to clinical disease manifestation by breaking down the host proteins. Tetanus toxin causes spastic paralysis, and botulinum toxin causes flaccid paralysis. These two toxins are synaptobrevins, and they cause paralysis by blocking the neuromuscular junction and inhibiting the release of acetylcholine. A-B toxins are the toxins produced by *Corynebacterium diphtheria* (Fujii et al. 1991), *B. pertussis* (Stein et al. 1994), *V. cholerae* (Klose 2001), *E.coli* (Nakao and Takeda 2000) etc. Toxins also act as adhesins, which after interacting with the host cell receptors are internalized into the cell. Kesty and Kuehn (2004) described that heat-labile enterotoxin (LT, surface adhesin) of enterotoxigenic *E.coli* interacts with GM1 receptor of the host.

14.11 Innate Immunity

Once the bacteria enter the host, they are exposed to specific and nonspecific barriers of the host. The barriers of the innate immune system include skin, mucous membrane, lysozyme in the tears, low level of acid pH in the stomach, antimicrobial substances in the secretions, leukocytes, and complement proteins in blood and tissues. Bacteria fight with the host immune system to establish the infection. For the survival, the bacteria consists of molecular patterns called "PAMPs," such as LPS, cpg motifs, and lipoteichoic acids (Medzhitov and Janeway 2000). These PAMPs are recognized by PRRs (pattern recognition receptors) of host. If this PRR binds to pathogen, it regulates endocytosis and phagocytosis of the pathogen and stimulates the immunomodulatory factors to act. Signaling group of PRR includes TLRs that bind to PAMPs and stimulate the production of proinflammatory cytokines to control the infection (Aderem and Ulevitch 2000; Hacker et al. 2000). Endocytic recognition pattern receptors such as mannose receptors, macrophage scavenger receptor, and galactose receptors bind to pathogen and result in endocytosis (Neth et al. 2000; Kang and Schlesinger 1998; Kreiger and Herz 1994). Secreted pattern recognition receptors such as mannan binding lectin and surfactant protein A cause complement-mediated death of bacteria (Schagat et al. 1999; Fraser and Exekowitz 1999). Innate immunity is not only restricted to mount host inflammatory response, but also some innate immunity elements such as GTPases that control the replication of bacteria contained in the vacuole by trafficking the phagosomes in which they survive (MacMicking et al. 2003).

14.12 Subversion of Cell Signaling by Pathogens

Virulence factors of bacteria modify the host signaling pathway efficiently for their survival (Alto and Orth 2012). Bacterial virulence factors such as toxin enter the host cell by endocytosis, and they target GTPases. One example is cholera toxin—after entering into the host cell, their α-subunit is ribosylated by ADP and is not hydrolyzed to GTP, thus preventing the signaling mechanism (Henkel et al. 2010). Pathogens target MAPK (mitogen-activated protein kinase) signaling pathway that is important for innate immunity, autophagy, cell proliferation, cell migration, and apoptosis. This signaling pathway acts sequentially by phosphorylation (Morrison 2012). YopJ of *Yersinia* causes the inhibition of MAPK and NF-κB signaling pathways

(Orth et al. 1999; Hao et al. 2008; Morrison 2012; Staudt 2012). Lethal factor of *B. anthracis* causes the inhibition of MAPK by proteolysis. *Shigella* contains effector protein OspF that eliminates a phosphate group from MAPKs and inhibits the signaling pathway (Li et al. 2007). *Salmonella* effectively activates the GEF (guanine nucleotide-exchange factor) proteins such as SopE that facilitate the internalization of bacterium into the cell (Hardt et al. 1998). These bacterial GEFs are functional mimics that structurally differ from eukaryotic GEFs and subvert the small G protein signaling by molecular mimicry (Klink et al. 2010). YopE of *Yersinia paratuberculosis* mimics GAP (GTPase-activating protein) under *in vitro* conditions by stimulating the hydrolysis of GTPs and puts the molecular action of the host in switch-off mode (Black and Bliska 2000; Von Pawel-Rammingen et al. 2000; Andor et al. 2001). YopT (Shao et al. 2002, 2003) and YpkA (Juris et al. 2000; Prehna et al. 2006) are the effectors that cause subversion of G protein signaling pathway. *Clostridium botulinum* toxins modify and inactivate the small G proteins of Rho family that control the host cytoskeleton (Mohr et al. 1992). EspFu, an effector protein secreted by EHEC O157:H7, mimics the small G proteins and directly activates its substrates and facilitates for their attachment to the intestinal epithelium (Campellone et al. 2004; Garmendia et al. 2004). Another strategy used by bacteria is hijacking lipid signaling. Phosphoinositides, especially phosphatidylinositol 4,5-bisphosphate (PIP2), are involved in intracellular cell trafficking and regulation of actin cytoskeleton. Disturbance of homeostasis of phosphatidylinositol beneath the plasma membrane facilitates cell lysis by extracellular pathogens and entry of intracellular pathogens (Ham et al. 2011). Foodborne bacteria that hijack this lipid signaling pathway by targeting inositol polyphosphate 5-phosphatase include *Shigella*, *Salmonella*, and *Vibrio parahaemolyticus* with the help of IpgD, SopB, and VPA0450 effector proteins, respectively. IpgD and SopB promote bacterial invasion, whereas VPA0450 induces blebbing that hastens the lysis of the infected host cell (Norris et al. 1998; Terebiznik et al. 2002; Hernandez et al. 2004; Charras and Paluch 2008; Broberg et al. 2011). Bacterial pathogens also target host ubiquitin-mediated signal transduction. Ubiquitin regulation machinery is necessary for cell signaling, transcription, replication, immune response, and autophagy (Pickart 2004; Mukhopadhyay and Riezman 2007; Yaffe 2012). Bacteria have developed the mechanisms to evade this ubiquitination process. Many bacteria encode ubiquitin, ubiquitin ligases (E3), and deubiquitinases to modify the host ubiquitin activity posttranslationally (Randow and Lehner 2009; Collins and Brown 2010). Recent studies have reported that Q40 effectors of EPEC and *Burkholderia pseudomallei* deaminate the host ubiquitin (Cui et al. 2010; Jubelin et al. 2010; Morikawa et al. 2010). Understanding the mechanism of this ubiquitination process by the pathogen is the promising area of research.

14.13 Bacterial Metabolism: A Prerequisite for Virulence

Apart from the production of virulence factors, bacteria require nutrients such as carbon and nitrogen for their replication which is followed by colonization in the mammalian host. Hence, to understand about the host–pathogen interaction, it is essential to understand the bacterial metabolism in the host niche (Rohmer et al. 2011). Pathogenic bacteria express the virulence factors based on the concentration of nutrient at that locale. Pathogenic bacteria have to compete with the resident bacteria for the nutrient as the resident bacteria have evolved mechanisms to obtain nutrients and protect their environment being occupied by the other competing bacteria (Hibbing et al. 2010). For instance, *Staphylococcus epidermidis*, a resident flora, in the skin produces antimicrobial peptides that are toxic to the pathogenic *Streptococcus pyogenes* and *Staphylococcus aureus* (Davis 1996; Cogen et al. 2008). The intestinal bacterial flora, apart from the production of antimicrobial peptides, modulates the cytokine production by macrophages of spleen and bone marrow, and enhances the immune response against the intracellular pathogen (Lievin-Le Moa and Servin 2006; Endt et al. 2010). Unique set of metabolic genes are upregulated by the pathogenic bacteria to implement the metabolic pathway to exploit the food source in the niche. These metabolic genes are located in the pathogenicity islands (Schmidt and Hensel 2004). Examples include cluster ttrABC and ttrS metabolic genes of *Salmonella* Typhimurium that utilizes tetrathionate as electron acceptor and enhances the anaerobic growth of the bacteria in the intestine (Winter et al. 2010; Price-Carter et al. 2001). Similarly, the nan-nag gene cluster of *V. cholerae* utilizes sialic acid for their growth in the host (Almagro-Moreno and Boyd 2009). Thus, understanding the bacterial metabolism paves a way to understand their pathogenesis.

14.14 Role of Iron in Host–Pathogen Interaction

Iron acquisition is an important step required for the survival of bacteria in the host. Iron is not available to the pathogen in its free form. Iron is always present in the bound form. The iron-binding proteins are transferrin, hemoglobin, ferritin, and lactoferrin. Bacteria acquire iron from these iron-binding proteins by direct uptake or by siderophore synthesis that removes the bound iron from these proteins. Enteric bacteria like *Escherichia, Salmonella, Shigella,* and *Klebsiella* produce the iron chelators aerobactin and enterobactin, which have high attractions for ferric iron. However, *Pseudomonas* spp., hold siderophores in pyoverdin and pyochelin. All these bacteria secrete outer membrane proteins specifically involved in iron uptake. Bacterial members of the family *Neisseriaceae* and *Pasteurellaceae* employ different iron acquisition method by directly interacting with the iron-binding protein (Bullen et al. 2000). Bacteria cause changes in pH and oxidation–reduction potential, which results in an increased availability of free iron to them (Bullen et al. 2005). *Mycobacterium* produces mycobactin, a siderophore, and acids such as sialic and citric acids that acquire iron from transferrin, lactoferrin, and ferritin (Ratledge 2004). Intraphagosomal iron restriction by the host kills the intracellular pathogen by NRAMP (natural resistance-associated macrophage protein) 1, which is the member of the metal ion transporter family (Blackwell 2001). NRAMP 1, including Fe^{2+}, expressed by the endosomes of monocytes and macrophages restricts intraphagosomal environment for the survival of *Mycobacterium* (Agranoff et al. 1999). Iron also inhibits the action of IFN-γ and encourages the growth of *Mycobacterium* inside the phagosome (Gomes et al. 2001). Thus, iron is an important regulator in host–pathogen interaction. Freely obtainable iron can have disastrous effects on natural resistance to bacterial infection, and result in the elimination of defensive bactericidal systems in tissue fluids, which can lead to faster bacterial growth. It can overwhelm phagocytic mechanisms with devastating results for the host (Bullen et al. 2005).

14.15 Novel Areas of Host–Pathogen Interactions

14.15.1 MiRNAs in Host–Pathogen Interaction

Micro-RNAs are noncoding RNAs that have a main role in biological processes such as development, differentiation, proliferation, immune response, and apoptosis, and they also control the gene expression posttranscriptionally (Bartel 2004; Krol et al. 2010). They also have a crucial role during bacterial infection (Eulalio et al. 2012; Staedel and Darfeuille 2013; Cullen 2011). Expression of host miRNAs during bacterial infection is important in host response. *L. monocytogenes* is used as a model to know how gram-positive bacteria modulate the expression of host miRNA. Schnitger et al. (2011) infected the mouse bone marrow macrophages with *L. monocytogenes*, did genome-wide expression profiling, and identified the upregulation of immune-related miRNAs. Schulte et al. (2011) discussed that during *Salmonella* infection, NF-κB-dependent miRNAs were upregulated. *Mycobacterium bovis* infection has led to upregulation of NF-κB-dependent miRNAs and consequent inhibition of IL-1 production (Wu et al. 2012). Although the miRNA profiling has led to the identification of several immune-related miRNAs, the clear role of miRNAs in host–pathogen interaction is yet to be studied well.

14.15.2 Polyamines

Polyamines are the cationic molecules associated with a range of biological functions such as cell proliferation, cell differentiation, translation, and gene regulation in both eukaryotes and prokaryotes (Igarashi and Kashiwagi 2010). Shah and Swiatlo (2008) reported that polyamines are critical for virulence phenotype in many bacteria. Predominant bacterial polyamines are cadaverine, spermidine, putrescine, and spermine (Tabor and Tabor 1985). In *Salmonella* Typhimurium, polyamines promote the synthesis of type III secretion systems; in *Shigella*, polyamines help in the invasion process; and in *V. cholerae*, they are involved in the formation of biofilm. The major contribution of polyamines in bacterial pathogenesis is comprehensively discussed by Martino et al. (2013).

14.15.3 Prenylated Proteins

Prenylated proteins are the proteins to which the prenyl group is attached that facilitates protein–protein interactions, and by their increased hydrophobicity, they attach to the cell membranes (Gelb et al. 2006; Zhang and Casey 1996). These prenylated proteins are involved in the signal transduction pathway (Roberts et al. 2008), cell proliferation and differentiation, regulatory pathway, metabolism, and apoptosis (Gelb et al. 2006; Gao et al. 2009; Benetka et al. 2006; Roskoski 2003). Ras, Rac, and Rho are the well-characterized prenylated proteins in eukaryotes (Gao et al. 2009). In prokaryotes, this prenylation pathway remains undefined (Wollack et al. 2009). Intracellular bacteria manipulate the host prenylation machinery for their multiplication in the host (Price et al. 2010a, 2010b). Prenylation may be a conserved mechanism for the modification of effector in bacterial pathogens, as *in silico* analyses reveal that most bacteria possess effectors with the conserved prenylation motif. Such effector proteins include SifA protein of *Salmonella* (Reinicke et al. 2005) and AnkB protein of *Legionella* (Price et al. 2010a; Ivanov et al. 2010).

14.16 Host–Pathogen Interaction Studies

There are various methods to study the host–pathogen interaction in bacteria. Mostly, the bacterial pathogens are studied under *in vitro* simulating *in vivo* conditions. The host–pathogen interaction studies under such conditions are said to have little commonness with the natural disease conditions. Then, a new field of cellular microbiology has emerged that used immortalized cell lines as host cells. However, these cell lines lack several significant features of the diverse cell lineages that are found *in vivo* (Russell et al. 2019). Hence, appropriate laboratory animal models serve to be an essential tool in understanding the host–pathogen interaction. Another important approach is systems biology approach, which connects the wet laboratory experiments and computational analysis. Bacterial genome sequences are enormously available in the databases, and genomic approaches are promising in understanding the molecular crosstalk between the pathogen and the host. Recent genomic, transcriptomic, and proteomic approaches along with computational tools help in delineating the molecular-level interaction between the host and the pathogen (Kint et al. 2010). High-throughput techniques such as RNA sequencing, microarray, and proteomics generate large amount of data such as whole-genome sequence, whole-genome expression profile, and quantitative profile of all the proteins and metabolites (Wise et al. 2007). Microscopic and dynamic imaging techniques have provided the real-time analysis of host–pathogen interaction (Coombes and Robey 2010). Appropriate use of mathematical tools and computational techniques helps in extracting useful information from the large amount of data (Hu and Polyak 2006; Anderson et al. 2003; Liu et al. 2009). There are many web-based host–pathogen interaction databases available, and modeling strategies such as network analysis can be done using appropriate tools. Different data resources and tools used to study them are discussed in detail by Mukherjee et al. (2013).

14.17 Conclusion

The capacity of host to combat bacterial infections is constantly evolving as there is continuous exposure of different bacteria and their components. Paradoxically, bacteria are also evolving in the similar way to get adapted to the host. And it is the role immune system to protect the host from pathogenic bacteria; however, a break in the immune response or system toward the bacteria or any other pathogenic microbe results in disease. Therefore, clear knowledge on host–bacteria interaction is necessary. In the current scenario, it is mostly possible by deploying systems biology approach, i.e., mathematical modeling and high-throughput data analysis.

REFERENCES

Aderem, A., and R. Ulevitch. 2000. Toll-like receptors in the induction of the innate immune response. *Nature* 406:782–787. doi: 10.1038/35021228.

Agranoff, D., I.M. Monahan, J.A. Mangan, P.D. Butcher, and S. Krishna. 1999. Mycobacterium tuberculosis expresses a novel pH-dependent divalent cation transporter belonging to the Nramp family. *Journal of Experimental Medicine* 190:717–724. doi: 10.1084/jem.190.5.717.

Akira, S., K. Takeda, and T. Kaisho. 2001. Toll-like receptors: critical proteins linking innate and acquired immunity. *Nature Immunology* 2(8):675–680.

Alaniz, R.C., B.L. Deatherage, J.C. Lara, and B.T. Cookson. 2007. Membrane vesicles are immunogenic facsimiles of *Salmonella typhimurium* that potently activate dendritic cells, prime B and T cell responses, and stimulate protective immunity in vivo. *The Journal of Immunology* 179(11):7692–7701.

Almagro-Moreno, S., and E.F. Boyd. 2009. Sialic acid catabolism confers a competitive advantage to pathogenic *Vibrio cholerae* in the mouse intestine. *Infection and Immunity* 77:3807–3816. doi: 10.1128/IAI.00279-09.

Alto, N.M., and K. Orth. 2012. Subversion of cell signaling by pathogens. *Cold Spring Harbor Perspectives in Biology* 4(9):a006114.

Anderson, D.C., W. Li, D.G. Payan, and W.S. Noble. 2003. A new algorithm for the evaluation of shotgun peptide sequencing in proteomics: support vector machine classification of peptide MS/MS spectra and SEQUEST scores. *Journal of Proteome Research* 2(2):137–146. doi: 10.1021/pr0255654.

Andor, A., K. Trulzsch, M. Essler, et al. 2001. YopE of Yersinia, a GAP for Rho GTPases, selectively modulates Rac-dependent actin structures in endothelial cells. *Cellular Microbiology* 3(5):301–310. doi: 10.1046/j.1462 5822.2001.00114.x.

Barenkamp, S.J., and J. St. Geme. 1994. Genes encoding high-molecular weight adhesion proteins of nontypeable *Haemophilus influenzae* are part of gene clusters. *Infection and Immunity* 62:3320–3328. PMCID: PMC302962, PMID: 8039903.

Barron, R.L. 1993. Pathophysiology of septic shock and implications for therapy. *Clinical Pharmacy* 12:829–845.

Bartel, D.P. 2004. MicroRNAs: genomics, biogenesis, mechanism, and function. *Cell* 116(2):281–297. doi: 10.1016/s0092-8674(04)00045-5.

Benetka, W., M. Koranda, S. Maurer-Stroh, F. Pittner, and F. Eisenhaber. 2006. Farnesylation or geranylgeranylation? Efficient assays for testing protein prenylation in vitro and in vivo. *BMC Biochemistry* 7:6. doi: 10.1186/1471-2091-7-6.

Bergman, M.A., L.A. Cummings, S.L.R. Barrett, et al. 2005. CD4+ T cells and toll-like receptors recognize *Salmonella* antigens expressed in bacterial surface organelles. *Infection and Immunity* 73(3):1350–1356.

Bermudez, L., and F. Sangari. 2000. Mycobacterial invasion of epithelial cells. *Subcellular Biochemistry* 33:231–249.

Beveridge, T.J. 1999. Structures of gram-negative cell walls and their derived membrane vesicles. *Journal of Bacteriology* 181:4725–4733.

Black, D.S., and J.B. Bliska. 2000. The RhoGAP activity of the *Yersinia pseudo-tuberculosis* cytotoxin YopE is required for anti-phagocytic function and virulence. *Molecular Microbiology* 37:515–527. doi: 10.1046/j.1365-2958.2000.02021.x.

Blackwell, J.M. 2001. Genetics and genomics in infectious disease susceptibility. *Trends in Molecular Medicine* 7:521–526. PMID: 11689339. doi: 10.1016/s1471-4914(01)02169-4.

Blaser, M.J., J.A. Hopkins, R.M. Berka, M.L. Vasil, and W.L. Wang. 1983. Identification and characterization of *Campylobacter jejuni* outer membrane proteins. *Infection and Immunity* 42(1):276–284.

Bliska, J.B., J.E. Galan, and S. Falkow. 1993. Signal transduction in the mammalian cell during bacterial attachment and entry. *Cell* 73:903–920. doi: 10.1016/0092-8674(93)90270-z.

Borovikova, L.V., S. Ivanova, M. Zhang, et al. 2000. Vagus nerve stimulation attenuates the systemic inflammatory response to endotoxin. *Nature* 405:458–462. doi: 10.1038/35013070.

Brandtzaeg, P., P. Kierulf, P. Gaustad, et al. 1989. Plasma endotoxin as a predictor of multiple organ failure and death in systemic meningococcal disease. *Journal of Infectious Diseases* 159:195–204. doi: 10.1093/infdis/159.2.195.

Broberg, C.A., T.J. Calder, and K. Orth. 2011. *Vibrio parahaemolyticus* cell biology and pathogenicity determinants. *Microbes and Infection* 13(12–13):992–1001. doi: 10.1016/j.micinf.2011.06.013.

Bullen, J., E. Griffiths, H. Rogers, and G. Ward. 2000. Sepsis: the critical role of iron. *Microbes and Infection* 2:409–415. PMID: 10817643. doi: 10.1016/s1286-4579(00)00326-9.

Bullen, J.J., H.J. Rogers, P.B. Spalding, and C.G. Ward. 2005. Iron and infection: the heart of the matter. *FEMS Immunology and Medical Microbiology* 43:325–330. PMID: 15708305. doi: 10.1016/j.femsim.2004.11.010.

Burns, S.M., and S.I. Hull. 1998. Comparison of loss of serum resistance by defined lipopolysaccharide mutants and an acapsular mutant of uropathogenic *Escherichia coli* O75:K5. *Infection and Immunity* 66(9):4244–4253. PMCID: PMC108512.

Campellone, K.G., D. Robbins, and J.M. Leong. 2004. EspFU is a translocated EHEC effector that interacts with Tir and N-WASP and promotes Nck-independent actin assembly. *Developmental Cell* 7(2):217–228. doi: 10.1016/j.devcel.2004.07.004.

Casadevall, A., and L.A. Pirofski. 2000. Host–pathogen interactions: basic concepts of microbial commensalism, colonization, infection, and disease. *Infection and Immunity* 68(12):6511–6518. doi: 10.1128/iai.68.12.6511-6518.2000.

Charras, G., and E. Paluch. 2008. Blebs lead the way: how to migrate without lamellipodia. *Nature Reviews Molecular Cell Biology* 9(9):730–736. doi: 10.1038/nrm2453.

China, B., and F. Goffaux. 1999. Secretion of virulence factors by *Escherichia coli*. *Veterinary Research* 30:181–202. PMID: 10367354.

Cleary, P., and D. Cue. 2000. High frequency invasion of mammalian cells by β hemolytic *streptococci*. *Subcellular Biochemistry* 33:137–166.

Coburn, B., I. Sekirov, and B.B. Finlay. 2007. Type III secretion systems and disease. *Clinical Microbiology Reviews* 20(4):535–549. doi: 10.1128/CMR.00013-07.

Cogen, A.L., V. Nizet, and R.L. Gallo. 2008. Skin microbiota: a source of disease or defence? *British Journal of Dermatology* 158:442–455. doi: 10.1111/j.1365-2133.2008.08437.x.

Collins, C.A., and E. J. Brown. 2010. Cytosol as battleground: ubiquitin as a weapon for both host and pathogen. *Trends in Cell Biology* 20(4):205–213. doi: 10.1016/j.tcb.2010.01.002.

Coombes, J.L., and E.A. Robey. 2010. Dynamic imaging of host–pathogen interactions in vivo. *Nature Reviews Immunology* 10(5):353–364. doi: 10.1038/nri2746.

Coulthurst, S.J. 2013. The type VI secretion system – a widespread and versatile cell targeting system. *Research in Microbiology* 164(6):640–654. PMID: 23542428. doi: 10.1016/j.resmic.2013.03.017.

Cui, J., Q. Yao, S. Li, et al. 2010. Glutamine deamidation and dysfunction of ubiquitin/NEDD8 induced by a bacterial effector family. *Science* 329(5996):1215–1218. doi: 10.1126/science.1193844.

Cullen, B.R. 2011. Viruses and microRNAs: RISCy interactions with serious consequences. *Genes & Development* 25(18):1881–1894. doi: 10.1101/gad.17352611.

Daffe, M., and G. Etienne. 1999. The capsule of *Mycobacterium tuberculosis* and its implications for pathogenicity. *Tubercle and Lung Disease* 79:153–169. doi: 10.1054/tuld.1998.0200.

Dautin, N., and H.D. Bernstein. 2007. Protein secretion in gram-negative bacteria via the autotransporter pathway. *Annual Review of Microbiology* 61:89–112. doi: 10.1146/annurev.micro.61.080706.093233.

Davis, C.P. 1996. Normal flora. In *Medical Microbiology*, S. Baron (ed.), 4th edn. University of Texas Medical Branch at Galveston. PMID: B21413249.

de Chastellier, C., T. Lang, and L. Thilo. 1995. Phagocytic processing of the macrophage endoparasite, *Mycobacterium avium*, in comparison to phagosomes which contain *Bacillus subtilis* or latex beads. *European Journal of Cell Biology* 68:167–182. PMID: 8575463.

Dean, P., and B. Kenny. 2009. The effector repertoire of enteropathogenic *E. coli*: ganging up on the host cell. *Current Opinion in Microbiology* 12(1):101–109.

Dehio, C., S. Gray-Owen, and T. Meyer. 2000. Host cell invasion by pathogenic *Neisseriae*. *Subcellular Biochemistry* 33:61–96. doi: 10.1007/978-1-4757-4580-1_4.

Donnenberg, M.S. 2000. Pathogenic strategies of enteric bacteria. *Nature* 406:768–774. doi: 10.1038/35021212.

Donnenberg, M.S., A. Donohue-Rolfe, and G.T. Keusch. 1990. A comparison of HEp-2 cell invasion by enteropathogenic and enteroinvasive *Escherichiacoli*. *FEMS Microbiology Letters* 57:83–86. doi: 10.1111/j.1574-6968.1990.tb04179.x.

Drake, J.W., B. Charlesworth, D. Charlesworth, and J.F. Crow. 1998. Rates of spontaneous mutation. *Genetics* 148(4):1667–1686. PMCID: PMC1460098.

Dziewanowska, K., J.M. Patti, C.F. Deobald, K.W. Bayles, W.R. Trumble, and G.A. Bohach. 1999. Fibronectin binding protein and host cell tyrosine kinase are required for internalization of *Staphylococcus aureus* by epithelial cells. *Infection and Immunity* 67:4673. PMID: 10456915, PMCID: PMC96793.

Ehsani, S., C.D. Rodrigues, and J. Enninga. 2009. Turning on the spotlight—using light to monitor and characterize bacterial effector secretion and translocation. *Current Opinion in Microbiology* 12(1):24–30.

Endt, K., B. Stecher, S. Chaffron, et al. 2010. The microbiota mediates pathogen clearance from the gut lumen after non-typhoidal *Salmonella* diarrhea. *PLoS Pathogens* 6:e1001097. doi: 10.1371/journal.ppat.1001097.

Engel, L.S., J.M. Hill, A.R. Caballero, L.C. Green, and R.J.O Callaghan. 1998. Protease IV, a unique extracellular protease and virulence factor from *Pseudomonas aeruginosa*. *Journal of Biological Chemistry* 273:16792–16797. doi: 10.1074/jbc.273.27.16792.

Eulalio, A., L. Schulte, and J. Vogel. 2012. The mammalian microRNA response to bacterial infections. *RNA Biology* 9(6):742–750. doi: 10.4161/rna.20018.

Falkow, S. 1997. What is a pathogen? *ASM News* 63(7): 359–365.

Fierer, J., and D.G. Guiney. 2001. Diverse virulence traits underlying different clinical outcomes of Salmonella infection. *Journal of Clinical Investigation* 107:775–780. doi: 10.1172/JCI12561.

Finlay, B.B., and S. Falkow. 1997. Common themes in microbial pathogenicity revisited. *Microbiology and Molecular Biology Reviews* 61:136–169. PMID: 9184008. PMCID: PMC232605.

Fiocca, R., V. Necchi, P. Sommi, et al. 1999. Release of *Helicobacter pylori* vacuolating cytotoxin by both a specific secretion pathway and budding of outer membrane vesicles. Uptake of released toxin and vesicles by gastric epithelium. *Journal of Pathology* 188:220–226. doi: 10.1002/(SICI)1096-9896(199906)188:2<220::AID-PATH307>3.0.CO;2-C.

Fitzgerald, K.A., E.M. Palsson-McDermott, A.G. Bowie, et al. 2001. Mal (MyD88-adapter-like) is required for Toll-like receptor-4 signal transduction. *Nature* 413:78–83. doi: 10.1038/35092578.

Fleiszig, S.M., J.P. Wiener-Kronish, H. Miyazaki, et al. 1997. *Pseudomonas aeruginosa*-mediated cytotoxicity and invasion correlate with distinct genotypes at the loci encoding exoenzyme. *Infection and Immunity* 65:579–586. PMID: 9009316 PMCID: PMC176099.

Flock, J. 1999. Extracellular-matrix-binding proteins as targets for the prevention of *Staphylococcus aureus* infections. *Molecular Medicine Today* 5:532–537. doi: 10.1016/s1357-4310(99)01597-x.

Forst, C.V. 2006. Host–pathogen systems biology. *Drug Discovery Today* 11(5–6):220–227. doi: 10.1016/S1359-6446(05)03735-9.

Fraser, I., and R. Exekowitz. 1999. *Mannose Receptor and Phagocytosis*. *Phagocytosis and Pathogens*. Greenwich, CT: JAI Press:85–99.

Fujii, G., S.H. Choe, M.J. Bennett, and D. Eisenberg. 1991. Crystallization of diphtheria toxin. *Journal of Molecular Biology* 222:861–864. doi: 10.1016/0022-2836(91)90577-S.

Furuta, N., H. Takeuchi, and A. Amano. 2009. Entry of *Porphyromonas gingivalis* outer membrane vesicles into epithelial cells causes cellular functional impairment. *Infection and Immunity* 77(11):4761–4770.

Gaillard, J.L., P. Berche, C. Frehel, E. Gouin, and P. Cossart. 1991. Entry of *L. monocytogenes* into cells is mediated by internalin, a repeat protein reminiscent of surface antigens from gram-positive cocci. *Cell* 65:1127–1141. doi: 10.1016/0092-8674(91)90009-n.

Galan, J.E., and D. Zhou. 2000. Striking a balance: modulation of the actin cytoskeleton by salmonella. *Proceedings of the National Academy of Sciences of the United States of America* 97:8754–8761. doi: 10.1073/pnas.97.16.8754.

Galdiero, M., A. Folgore, M. Molitierno, and R. Greco. 1999. Porins and lipopolysaccharide (LPS) from *Salmonella typhimurium* induce leucocyte transmigration through human endothelial cells in vitro. *Clinical and Experimental Immunology* 116:453–461.

Gamazo, C., and I. Moriyon. 1987. Release of outer membrane fragments by exponentially growing *Brucella melitensis* cells. *Infection and Immunity* 55(3):609–615.

Gankema, H., J. Wensink, P.A. Guinée, W.H. Jansen, and B. Witholt. 1980. Some characteristics of the outer membrane material released by growing enterotoxigenic *Escherichia coli*. *Infection and Immunity* 29(2):704–713.

Gao, J., J. Liao, and G.Y. Yang. 2009. CAAX-box protein, prenylation process and carcinogenesis. *American Journal of Translational Research* 1(3):312–325. PMC2776320.

Garcia-del Portillo, F., and B.B. Finlay. 1995. Targeting of *Salmonella typhimurium* to vesicles containing lysosomal membrane glycoproteins bypasses compartments with mannose 6-phosphate receptors. *Journal of Cell Biolology* 129:81–97. doi: 10.1083/jcb.129.1.81.

Garmendia, J., A.D. Phillips, M.F. Carlier, et al. 2004. TccP is an enterohaemorrhagic *Escherichia coli* O157:H7 type III effector protein that couples Tir to the actin-cytoskeleton. *Cell Microbiology* 6(12):1167–1183. doi: 10.1111/j.1462-5822.2004.00459.x.

Gelb, M.H., L. Brunsveld, C.A. Hrycyna, et al. 2006. Therapeutic intervention based on protein prenylation and associated modifications. *Nature Chemical Biology* 2(10):518–528. doi: 10.1038/nchembio818.

Goldman, R.C., C.C. Doran, and J.O. Capobianco. 1990. Antibacterial agents which speci¢cally inhibit lipopolysaccharide synthesis. In *Cellular and Molecular Aspects of Endotoxin Reactions*, A. Nowotny, J.J. Spitzer, and E.J. Zieg-ler (eds.). Amsterdam, New York, Oxford: Elsevier Science:157–167.

Gomes, M.S., J.R. Boelaert, and R. Appelberg. 2001. Role of iron in experimental *Mycobacterium avium* infection. *Journal of Clinical Virology* 20:117–122. doi: 10.1016/s1386-6532(00)00135-9.

Gophna, U., M. Barlev, R. Seijffers, T.A. Oelschlager, J. Hacker, and E.Z. Ron. 2001. Curli Fibers mediate internalization of *Escherichia coli* by eukaryotic cells. *Infection and Immunity* 69(4):2659–2665. doi: 10.1128/IAI.69.4.2659-2665.2001.

Hacker, H., R. Vabulas, O. Takeuchi, K. Hoshino, S. Akira, and H. Wagner. 2000. Immune cell activation by bacterial CpG-DNA through myeloid differentiation marker 88 and tumor necrosis factor receptor-associated factor (TRAF) 6. *Journal of Experimental Medicine* 192:595–600. doi: 10.1084/jem.192.4.595.

Hackstadt, T. 2000. Redirection of host vesicle trafficking pathways by intracellular parasites. *Traffic* 1:93–99. doi: 10.1034/j.1600-0854.2000.010201.x.

Hackstadt, T., and J.C. Williams. 1981. Biochemical stratagem for obligate parasitism of eukaryotic cells by *Coxiella burnetii*. *Proceedings of the National Academy of Sciences of the United States of America* 78:3240–3244. doi: 10.1073/pnas.78.5.3240.

Hahn, H. 1997. The type-4 pilus is the major virulence-associated adhesin of *Pseudomonas aeruginosa*—a review. *Gene* 192:99–108. doi: 10.1016/s0378-1119(97)00116-9.

Ham, H., A. Sreelatha, and K. Orth. 2011. Manipulation of host membranes by bacterial effectors. *Nature Reviews Microbiology* 9(9):635–646. doi: 10.1038/nrmicro2602.

Hamill, R.J., and D.G. Maki. 1986. Endotoxin shock in man caused by Gram-negative bacilliçetiology, clinical features, diagnosis, natural history and prevention. In *Clinical Aspects of Endotoxin Shock*, R.A. Proctor (ed.). Amsterdam: Elsevier:55–126.

Hao Y.H., Y. Wang, D. Burdette, et al. 2008. Structural requirements for Yersinia YopJ inhibition of MAP kinase pathways. *PLoS One* 3:e1375.

Haraguchi, Y., and A. Sasaki. 1997. Evolutionary pattern of intra-host–pathogen antigenic drift: effect of cross-reactivity in immune response. *Philosophical Transactions of the Royal Society of London* 352(1349):11–20. doi: 10.1098/rstb.1997.0002.

Hardt, W.D., L.M. Chen, K.E. Schuebel, X.R. Bustelo, and J.E. Galan. 1998. *S. typhimurium* encodes an activator of Rho GTPases that induces membrane ruffling and nuclear responses in host cells. *Cell* 93(5):815–826. doi: 10.1016/s0092-8674(00)81442-7.

Hari-Dass, R., C. Shah, D.J. Meyer, and J.G. Raynes. 2005. Serum amyloid A protein binds to outer membrane protein A of Gram-negative bacteria. *Journal of Biological Chemistry* 280:18562–18567. doi: 10.1074/jbc.M500490200.

Hatfaludi, T., K. Al-Hasani, J.D. Boyce, and B. Adler. 2010. Outer membrane proteins of *Pasteurella multocida*. *Veterinary Microbiology* 144(1–2):1–17.

Henkel, J.S., M.R. Baldwin, and J.T. Barbieri. 2010. Toxins from bacteria. *EXS* 100:1–29.

Hensel, M., J.E. Shea, C. Gleeson, M.D. Jones, E. Dalton, and D.W. Holden. 1995. Simultaneous identification of bacterial virulence genes by negative selection. *Science* 269:400–403. PMID: 7618105. doi: 10.1126/science.7618105.

Hernandez, L.D., K. Hueffer, M.R. Wenk, and J.E. Galan. 2004. Salmonella modulates vesicular traffic by altering phosphoinositide metabolism. *Science* 304(5678):1805–1807. doi: 10.1126/science.1098188.

Hibbing, M.E., C. Fuqua, M.R. Parsek, and S.B. Peterson. 2010. Bacterial competition: surviving and thriving in the microbial jungle. *Nature Reviews Microbiology* 8:15–25. PMID: 19946288, PMCID: PMC2879262. doi: 10.1038/nrmicro2259.

Hodgson, J.C. 1993. Watery mouth disease in new-born lambs. In *Veterinary Annual*, M.-E. Raw, and J.J. Parkinson (eds.), Vol. 33. Oxford, London, Edinburgh, Boston: Blackwell Scienti¢c publications:102–106.

Hodgson, J.C. 1994. Diseases due to *Escherichia coli* in sheep. In *Escherichia coli in Domestic Animals and Humans*, C.L. Gyles (ed.). Wallingford: CABInternational: 135–150.

Hodgson, J.C. 2006. Endotoxin and mammalian host responses during experimental disease. *Journal of Comparative Pathology* 135;157–175. doi: 10.1016/j.jcpa.2006.09.001.

Hoekstra, D., J.W. Van Der Laan, L. De Leij, and B. Witholt. 1976. Release of outer membrane fragments from normally growing *Escherichia coli*. *Biochimica et biophysica acta* 455:889–899. doi: 10.1016/0005-2736(76)90058-4.

Horn, D.L., D.C. Morrison, S.M. Opal, R. Silverstein, K. Visvanathan, and J.B. Zabriskie. 2000. What are the microbial components implicated in the pathogenesis of sepsis? *Clinical Infectious Diseases* 31(4):851–858. doi: 10.1086/318127.

Hsing, M., K.G. Byler, and A. Cherkasov. 2008. The use of Gene Ontology terms for predicting highly-connected 'hub' nodes in protein-protein interaction networks. *BMC Systems Biology* 2(1):80.

Hu, M., and K. Polyak. 2006. Serial analysis of gene expression. *Nature Protocols* 1(4):1743–1760. doi: 10.1038/nprot.2006.269.

Igarashi, K., and K. Kashiwagi. 2010. Modulation of cellular function by polyamines. *The International Journal of Biochemistry and Cell Biology* 42(1):39–51. doi: 10.1016/j.biocel.2009.07.009.

Ismail, S., M.B. Hampton, and J.I. Keenan. 2003. *Helicobacter pylori* outer membrane vesicles modulate proliferation and interleukin-8 production by gastric epithelial cells. *Infection and Immunity* 71(10):5670–5675.

Ivanov, S.S., G. Charron, H.C. Hang, and C.R. Roy. 2010. Lipidation by the host prenyltransferase machinery facilitates membrane localization of *Legionella pneumophila* effector proteins. *The Journal of Biological Chemistry* 285(45):34686–34698. doi: 10.1074/jbc.M110.170746.

Janssens, S., and R. Beyaert. 2003. Role of Toll-like receptors in pathogen recognition. *Clinical Microbiology Reviews* 16(4):637–646. doi: 10.1128/CMR.16.4.637-646.2003.

Joanne, E., and B. Priya. 2009. Role of *Pseudomonas aeruginosa* type III effectors in disease. *Current Opinion in Microbiology* 12:61–66.

Joh, D., E. Wann, B. Kreikemeyer, P. Speziale, and M. Hook. 1999. Role of fibronectin-binding MSCRAMMs in bacterial adherence and entry into mammalian cells. *Matrix Biology* 18:211–223. doi: 10.1016/s0945-053x(99)00025-6.

Jorgensen, J.H. 1986. Clinical applications of the Limulus amebocyte lysate test. In *Clinical Aspects of Endotoxin Shock*, R.A. Proctor (ed.). Amsterdam: Elsevier:127–160.

Jubelin, G., F. Taieb, D.M. Duda, et al. 2010. Pathogenic bacteria target NEDD8-conjugated cullins to hijack host-cell signaling pathways. *PLoS Pathogens* 6(9):e1001128. doi: 10.1371/journal.ppat.1001128.

Juris, S.J., A.E. Rudolph, D. Huddler, K. Orth, and J.E. Dixon. 2000. A distinctive role for the Yersinia protein kinase: actin binding, kinase activation, and cytoskeleton disruption. *Proceedings of National Academy of Sciences* 97(17):9431–9436. doi: 10.1073/pnas.170281997.

Kadurugamuwa, J.L., and T.J. Beveridge. 1996. Bacteriolytic effect of membrane vesicles from *Pseudomonas aeruginosa* on other bacteria including pathogens: conceptually new antibiotics. *Journal of Bacteriology* 178:2767–2774. doi: 10.1128/jb.178.10.2767-2774.1996.

Kang, B.K., and L.S. Schlesinger. 1998. Characterization of mannose receptor-dependent phagocytosis mediated by *Mycobacterium tuberculosis* lipoarabinomannan. *Infection and Immunity* 66(6):2769–2777. PMCID: PMC108268.

Kelly, C., and J. Younson. 2000. Anti-adhesive strategies in the prevention of infectious disease at mucosal surfaces. *Expert Opinion on Investigational Drugs* 9:1711–1721.

Kesty, N.C., and M.J. Kuehn. 2004. Incorporation of heterologous outer membrane and periplasmic proteins into *Escherichia coli* outer membrane vesicles. *Journal of Biological Chemistry* 279(3):2069–2076.

Kim, S., A. Vela, S.M. Clohisey, et al. 2018. Host-specific differences in the response of cultured macrophages to *Campylobacter jejuni* capsule and O-methyl phosphoramidate mutants. *Veterinary Research* 49(3). doi: 10.1186/s13567-017-0501-y.

Kint, G., C. Fierro, K. Marchal, J. Vanderleyden, and S.C. De Keersmaecker. 2010. Integration of 'omics' data: does it lead to new insights into host-microbe interactions? *Future Microbiology* 5(2):313–328. doi: 10.2217/fmb.10.1.

Kitchens, R.L. 2005. Modulatory effects of sCD14 and LBP on LPS-host cell interactions. *Journal of Endotoxin Research* 11:225–229. doi: 10.1179/096805105X46565.

Klink, B.U., S. Barden, T.V. Heidler, et al. 2010. Structure of *Shigella* IpgB2 in complex with human RhoA: implications for the mechanism of bacterial guanine nucleotide exchange factor mimicry. *Journal of Biological Chemistry* 285(22):17197–17208. doi: 10.1074/jbc.M110.107953.

Klose, K.E. 2001. Regulation of virulence in *Vibrio cholerae*. *International Journal of Medical Microbiology* 291:81–88. doi: 10.1078/1438-4221-00104.

Kreiger, M., and J. Herz. 1994. Structures and functions of multiligand lipoprotein receptors: macrophage scavenger receptors and LDL receptor-related protein (LRP). *Annual Review of Biochemistry* 63:601–637. doi: 10.1146/annurev.bi.63.070194.003125. PMID: 7979249.

Krogfelt, K.A. 1991. Bacterial adhesion: genetics, biogenesis, and role in pathogenesis of fimbrial adhesins of *Escherichia coli*. *Reviews of Infectious Diseases* 13:721–735. doi: 10.1093/clinids/13.4.721.

Krol, J., I. Loedige, and W. Filipowicz. 2010. The widespread regulation of microRNA biogenesis, function and decay. *Nature Reviews Genetics* 11(9):597–610. doi: 10.1038/nrg2843.

Krug, A., A. Towarowski, S. Britsch, et al. 2001. Toll-like receptor expression reveals CpG DNA as a unique microbial stimulus for plasmacytoid dendritic cells which synergizes with CD40 ligand to induce high amounts of IL-12. *European Journal of Immunology* 31(10):3026–3037. doi: 10.1002/1521-4141(2001010)31:10%3C3026::aid-immu3026%3E3.0.co;2-h.

Lange, A., and N.M. Ferguson. 2009. Antigenic diversity, transmission mechanisms and the evolution of pathogens. *PLoS Computational Biology* 5(10):e1000536. doi: 10.1371/journal.pcbi.1000536.

Lerouge, I., and J. Vanderleyden. 2002. O-antigen structural variation: mechanisms and possible roles in animal/plant–microbe interactions. *FEMS Microbiology Reviews* 26(1):17–47.

Li, H., H. Xu, Y. Zhou, et al. 2007. The phosphothreonine lyase activity of a bacterial type III effector family. *Science* 315(5814):1000–1003. doi: 10.1126/science.1138960.

Li, Z., A.J. Clarke, and T.J. Beveridge. 1998. Gram-negative bacteria produce membrane vesicles which are capable of killing other bacteria. *Journal of Bacteriology* 180: 5478–5483.

Lievin-Le Moal, V., and A.L. Servin. 2006. The front line of enteric host defense against unwelcome intrusion of harmful microorganisms: mucins, antimicrobial peptides, and microbiota. *Clinical Microbiology Reviews* 19:315–337. doi: 10.1128/CMR.19.2.315-337.2006.

Liu, Q., A.H. Sung, M. Qiao, et al. 2009. Comparison of feature selection and classification for MALDI-MS data. *BMC Genomics* 10(Suppl 1):S3. doi: 10.1186/1471-2164-10-S1-S3.

Logan, S.M., and T.J. Trust. 1982. Outer membrane characteristics of *Campylobacter jejuni*. *Infection and Immunity* 38(3):898–906.

MacMicking, J.D., G.A. Taylor, and J.D. McKinney. 2003. Immune control of tuberculosis by IFN-γ-inducible LRG-47. *Science* 302(5645):654–659.

Mahan, M.J., J.M. Slauch, and J.J. Mekalanos. 1993. Selection of bacterial virulence genes that are specifically induced in host tissues. *Science* 259:686–688. PMID: 8430319. doi: 10.1126/science.8430319.

Maheswaran, S.K., D.J. Weiss, M.S. Kannan, et al. 1992. Effects of *Pasteurella haemolytica* A1 leukotoxin on bovine neutrophils: degranulation and generation of oxy-gen-derived free radicals. *Veterinary Immunology and Immunopatholog* 33:51–68.

Martino, M.L.D., R. Campilongo, M. Casalino, G. Micheli, B. Colonna, and G. Prosseda. 2013. Polyamines: emerging players in bacteria–host interactions. *International Journal of Medical Microbiology* 303:484–491. doi: 10.1016/j.ijmm.2013.06.008.

Matsumoto, H., and G.M. Young. 2006. Proteomic and functional analysis of the suite of Ysp proteins exported by the Ysa type III secretion system of *Yersinia enterocolitica* Biovar 1B. *Molecular Microbiology* 59(2):689–706.

Mayrand, D., and D. Grenier. 1989. Biological activities of outer membrane vesicles. *Canadian Journal of Microbiology* 35(6):607–613.

McBroom, A.J., and M.J. Kuehn. 2005. Outer membrane vesicles. *EcoSal Plus* 1(2).

McGhie, E.J., L.C. Brawn, P.J. Hume, D. Humphreys, and V. Koronakis. 2009. Salmonella takes control: effector-driven manipulation of the host. *Current Opinion in Microbiology* 12(1):117–124.

Medzhitov, R., and C. Janeway. 2000. Innate immunity. *The New England Journal of Medicine* 343(5):338–344. doi: 10.1056/nejm200008033430506.

Mengaud, J., H. Ohayon, P. Gounon, R.M. Mege, and P. Cossart. 1996. E-cadherin is the receptor for internalin, a surface protein required for entry of *L. monocytogenes* into epithelial cells. *Cell* 84:923–932. doi: 10.1016/s0092-8674(00)81070-3.

Merz, A.J., and M. So. 2000. Interactions of pathogenic *neisseriae* with epithelial cell membranes. *Annual Review of Cell and Developmental Biology* 16:423–457. doi: 10.1146/annurev.cellbio.16.1.423.

Mohr, C., G. Koch, I. Just, and K. Aktories. 1992. ADP-ribosylation by *Clostridium botulinum* C3 exoenzyme increases steady-state GTPase activities of recombinant rhoA and rhoB proteins. *FEBS Letters* 297(1–2):95–99. doi: 10.1016/0014-5793(92)80335-e.

Moran, A.P., and M.M. Prendergast. 2001. Molecular mimicry in *Campylobacter jejuni* and *Helicobacter pylori* lipopolysaccharides: contribution of gastrointestinal infections to autoimmunity. *Journal of Autoimmunity* 16(3):241–256. doi: 10.1006/jaut.2000.0490.

Morikawa, H., M. Kim, H. Mimuro, et al. 2010. The bacterial effector Cif interferes with SCF ubiquitin ligase function by inhibiting deneddylation of Cullin1. *Biochemical and Biophysical Research Communications* 401(2):268–274. doi: 10.1016/j.bbrc.2010.09.048.

Morrison, D. 2012. MAP kinase pathways. *Cold Spring Harbor Perspectives in Biology*. doi: 10.1101/cshperspect.a011254.

Morrison, D.C., and R.J. Ulevitch. 1978. The effects of bacterial endotoxins on host mediation systems: a review. *American Journal of Pathology* 93(2):526–617. PMCID: PMC2018378.

Mug-Opstelten, D., and B. Witholt. 1978. Preferential release of new outer membrane fragments by exponentially growing *Escherichia coli*. *Biochimica et Biophysica Acta (BBA)-Biomembranes* 508(2):287–295.

Mukherjee, S., A. Sambarey, K. Prashanthi, and N. Chandra. 2013. Current trends in modeling host–pathogen interactions. *Wiley Interdisciplinary Reviews-Data Mining and Knowledge Discovery* 3:109–128. doi: 10.1002/widm.1085.

Mukhopadhyay, D., and H. Riezman. 2007. Proteasome-independent functions of ubiquitin in endocytosis and signaling. *Science* 315(5809):201–205. doi: 10.1126/science.1127085.

Nakao, H., and T. Takeda. 2000. *Escherichia coli* Shiga toxin. *Journal of Natural Toxins* 9:299–313. PMID: 10994531.

Netea, M.G., M. van Deure, B.J. Kullberg, J.M. Cavaillon, and J.W. van derMeer. 2002. Does the shape of lipid A determine the interaction of LPS with Toll-like receptors? *Trends in Immunology* 23:135–139. doi: 10.1016/s1471-4906(01)02169-x.

Neth, O., D.L. Jack, A.W. Dodds, H. Holzel, N.J. Klein, and M.W. Turner. 2000. Mannose-binding lectin binds to a range of clinically relevant microorganisms and promotes complement deposition. *Infection and Immunity* 68(2):688–693. PMCID: PMC97193. doi: 10.1128/iai.68.2.688-693.2000.

Newman, R.M., P. Salunkhe, A. Godzik, and J.C. Reed. 2006. Identification and characterization of a novel bacterial virulence factor that shares homology with mammalian Toll/interleukin-1 receptor family proteins. *Infection and Immunity* 74:594–601. doi: 10.1128/IAI.74.1.594-601.2006.

Nitsche, D., C. Schulze, S. Oesser, A. Dalhoj, and M. Sack. 1994. The effects of different types of antimicrobial agents on plasma endotoxin activity in Gram-negative bacterial infection. In *Cipro£oxacin i.v.: De¢ning Its Role in Serious Infections*. International Symposium, Salzburg, September 1993, C.S. Garrard (ed.). Berlin: Springer:21–36.

Norris, F.A., M.P. Wilson, T.S. Wallis, E.E. Galyov, and P.W. Majerus. 1998. SopB, a protein required for virulence of *Salmonella dublin*, is an inositol phosphate phosphatase. *Proceedings of National Academy of Sciences* 95(24):14057–14059. doi: 10.1073/pnas.95.24.14057.

Novak, R., and E. Tuomanen. 1999. Pathogenesis of pneumococcal pneumonia. *Seminars in Respiratory Infections* 14:209–217. PMID: 10501308.

O'Neill, L.A. 2002. Toll-like receptor signal transduction and the tailoring of innate immunity: a role for Mal? *Trends in Immunology* 23:296–300. doi: 10.1016/s1471-4906(02)02222-6.

Oelschlaeger, T.A., P. Guerry, and D.J. Kopecko. 1993. Unusual microtubule-dependent endocytosis mechanisms triggered by *Campylobacter jejuni* and *Citrobacter freundii*. *Proceedings of the National Academy of Sciences of the United States of America* 90:6884–6888. doi: 10.1073/pnas.90.14.6884.

Orth, K., L.E. Palmer, Z.Q. Bao, et al. 1999. Inhibition of the mitogen-activated protein kinase kinase super-family by a Yersinia effector. *Science* 285:1920–1923.

Paape, M.J., E.M. Lilius, P.A. Wiitanen, M.P. Kontio, and R.H. Miller. 1996. Intramammary defense against infections induced by *Escherichia coli* in cows. *American Journal of Veterinary Research* 57:477–482.

Periti, P. 2000. Current treatment of sepsis and endotoxaemia. *Expert Opinion on Pharmacotherapy* 1:1203–1217. doi: 10.1517/14656566.1.6.1203.

Pettit, R.K., and R.C. Judd. 1992. The interaction of naturally elaborated blebs from serum-susceptible and serum-resistant strains of *Neisseria gonorrhoeae* with normal human serum. *Molecular Microbiology* 6(6):729–734.

Pickart, C.M. 2004. Back to the future with ubiquitin. *Cell* 116(2):181–190. PMID: 9712774. doi: 10.1016/S0092-8674(03)01074-2.

Pier, G.B. 2000. Peptides, *Pseudomonas aeruginosa*, polysaccharides and lipopolysaccharides—players in the predicament of cystic fibrosis patients. *Trends in Microbiology* 8:247–250; discussion 250–241.

Pollack, M. 1999. Biological functions of lipopolysaccharide antibodies. In *Endotoxin in Health and Disease*, H. Brade, S.M. Opal, S.N. Vogel, and D.C. Morrison (eds.). New York, Basel: Marcel Dekker:623–631.

Prehna, G., M.I. Ivanov, J.B. Bliska, and C.E. Stebbins. 2006. *Yersinia* virulence depends on mimicry of host Rho-family nucleotide dissociation in-hibitors. *Cell* 126(5):869–880. doi: 10.1016/j.cell.2006.06.056.

Price, C.T., S.C. Jones, K.E. Amundson, and Y.A. Kwaik. 2010a. Host-mediated post-translational prenylation of novel dot/icm-translocated effectors of *Legionella pneumophila*. *Frontiers in Microbiology* 1:131. doi: 10.3389/fmicb.2010.00131.

Price, C.T., T. Al-Quadan, M. Santic, S.C. Jones, and Y. Abu Kwaik. 2010b. Exploitation of conserved eukaryotic host cell farnesylation machinery by an F-box effector of *Legionella pneumophila*. *The Journal of Experimental Medicine* 207(8):1713–1726. doi: 10.1084/jem.20100771.

Price-Carter, M., J. Tingey, T.A. Bobik, and J.R. Roth. 2001. The alternative electron acceptor tetrathionate supports B12-dependent anaerobic growth of *Salmonella enterica* serovar Typhimurium on ethanolamine or 1,2-propanediol. *Journal of Bacteriology* 183(8):2463–2475. doi: 10.1128%2FJB.183.8.2463-2475.2001.

Pugin, J., D. Heumann, A. Tomasz, et al. 1994. CD14 is a pattern-recognition receptor. *Immunity* 1(6):509–516. doi: 10.1016/1074-7613(94)90093-0.

Pukatzki, S., A.T. Ma, A.T. Revel, D. Sturtevant, and J.J. Mekalanos. 2007. Type VI secretion system translocates a phage tail spike-like protein into target cells where it cross-links actin. *Proceedings of the National Academy of Sciences* 104(39):15508–15513.

Raetz, C.R.H., and C. Whitfield. 2002. Lipopolysaccharide endotoxins. *Annual Review of Biochemistry* 71:635700. doi: 10.1146/annurev.biochem.71.110601.135414.

Raetz, C.R.H., R.J. Ulevitch, S.D. Wright, C.H. Sibley, A. Ding, and C.F. Nathan. 1991. Gram-negative endotoxin: an extraordinary lipid with profound effects on eukaryotic signal transduction. *FASEB Journal* 5:2652–2660. doi: 10.1096/fasebj.5.12.1916089.

Randow, F., and P.J. Lehner. 2009. Viral avoidance and exploitation of the ubiquitin system. *Nature Cell Biology* 11(5): 527–534. doi: 10.1038/ncb0509-527.

Rao, S.P., K. Ogata, and A. Catanzaro. 1993. *Mycobacterium avium-M. intracellulare* binds to the integrin receptor alpha v beta 3 on human monocytes and monocyte-derived macrophages. *Infection and Immunity* 61:663–670. PMCID: PMC302778, PMID: 7678588.

Rathman, M., M.D. Sjaastad, and S. Falkow. 1996. Acidification of phagosomes containing *Salmonella typhimurium* in murine macrophages. *Infection and Immunity* 64:2765–2773. PMCID: PMC174137, PMID: 8698506.

Ratledge, C. 2004. Iron, mycobacteria and tuberculosis. *Tuberculosis (Edinb)* 84:110–130. PMID: 14670352. doi: 10.1016/j.tube.2003.08.012.

Reinicke, A.T., J.L. Hutchinson, A.I. Magee, P. Mastroeni, J. Trowsdale, and A.P. Kelly. 2005. A *Salmonella typhimurium* effector protein SifA is modified by host cell prenylation and S-acylation machinery. *The Journal of Biological Chemistry* 280(15):14620–14627. doi: 10.1074/jbc.M500076200.

Relman, D., E. Tuomanen, S. Falkow, D.T. Golenbock, K. Saukkonen, and S.D. Wright. 1990. Recognition of a bacterial adhesion by an integrin: macrophage CR3 (alpha M beta 2, CD11b/CD18) binds filamentous hemagglutinin of *Bordetella pertussis*. *Cell* 61:1375–1382. doi: 10.1016/0092-8674(90)90701-f.

Rietschel, E.T., T. Kirikae, F.U. Schade, et al. 1994. Bacterial endotoxin: molecular relationships of structure to activity and function. *Federation of American Societies for Experimental Biology Journal* 8:217–225. doi: 10.1096/fasebj.8.2.8119492.

Roberts, P.J., N. Mitin, P.J. Keller, et al. 2008. Rho Family GTPase modification and dependence on CAAX motif-signaled posttranslational modification. *The Journal of Biological Chemistry* 283(37):25150–25163. doi: 10.1074/jbc.M800882200.

Rohmer, L., D. Hocquet, and S.I. Miller. 2011. Are pathogenic bacteria just looking for food? Metabolism and microbial pathogenesis. *Trends in Microbiology* 19(7):341–348. doi: 10.1016/j.tim.2011.04.003.

Rosenshine, I., and B.B. Finlay. 1993. Exploitation of host signal transduction pathways and cytoskeletal functions by invasive bacteria. *Bio Essays* 15:17–24. doi: 10.1002/bies.950150104.

Roskoski Jr., R. 2003. Protein prenylation: a pivotal post-translational process. *Biochemical and Biophysical Research Communications* 303(1):1–7. doi: 10.1016/s0006 291x(03)00323-1.

Russell, D. 2000. Where to stay inside the cell: a homesteader's guide to intracellular parasitism. In *Cellular Microbiology*, P. Cossart, P. Boquet, S. Normark, et al, (eds.). Washington, DC: ASM Press.

Russell, D.G., J. Dant, and S. Sturgillkoszycki. 1996. *Mycobacterium avium*-and *Mycobacterium tuberculosis*-containing vacuoles are dynamic, fusion-competent vesicles that are accessible to glycosphingolipids from the host cell plasmalemma. *Journal of Immunology* 156:4764–4773. PMID: 8648123.

Russell, D.G., L. Huang, and B.C. VanderVen. 2019. Immunometabolism at the interface between macrophages and pathogens. *Nature Reviews Immunology* 19(5):291–304.

Salyers, A.A., and D.D. Whitt. 2002. *Bacterial Pathogenesis: A Molecular Approach*, 2nd edn. Washington, D.C.: ASM Press. ISBN: 1-55581-171-X.

Sandros, J., and E. Tuomanen. 1993. Attachment factors of *Bordetella pertussis*: mimicry of eukaryotic cell recognition molecules. *Trends in Microbiology* 1:192–196. doi: 10.1016/0966-842x(93)90090-e.

Sansonetti, P., G. Tran Van Nhieu, and C. Egile. 1999. Rupture of the intestinal epithelial barrier and mucosal invasion by *Shigella flexneri*. *Clinical Infectious Diseases* 28:466–475. doi: 10.1086/515150.

Sasaki, A. 1994. Evolution of antigen drift/switching: continuously evading pathogens. *Journal of Theoretical Biology* 168(3):291–308. doi: 10.1006/jtbi.1994.1110.

Schagat, T.L., M.J. Tino, and J.R. Wright. 1999. Regulation of protein phosporylation and pathogen phagocytosis by surfactant protein A. *Infection and Immunity* 67(9): 4693–4699. PMCID: PMC96796, PMID: 10456918.

Schiavo, G., F. Benfenati, B. Poulain, et al. 1992. Tetanus and botulinum-B neurotoxins block neurotransmitter release by proteolytic cleavage of synaptobrevin. *Nature* 359:832–835. doi: 10.1038/359832a0.

Schmidt, H., and M. Hensel. 2004. Pathogenicity islands in bacterial pathogenesis. *Clinical Microbiology Reviews* 17:14–56. doi: 10.1128/cmr.17.1.14-56.2004.

Schnitger, A.K., A. Machova, R.U. Mueller, et al. 2011. *Listeria monocytogenes* infection in macrophages induces vacuolar-dependent host miRNA response. *PLoS One* 6(11):e27435. doi: 10.1371/journal.pone.0027435.

Schulte, L.N., A. Eulalio, H.J. Mollenkopf, R. Reinhardt, and J. Vogel. 2011. Analysis of the host microRNA response to Salmonella uncovers the control of major cytokines by the let-7 family. *EMBO Journal* 30(10):1977–1989. doi: 10.1038/emboj.2011.94.

Schulze-Koops, H., H. Burkhardt, J. Heesemann, et al. 1993. Outer membrane protein YadA of enteropathogenic yersiniae mediates specific binding to cellular but notplasma fibronectin. *Infection and Immunity* 61:2513–2519. PMID: 8500887, PMCID: PMC280877.

Schwarz, S., R.D. Hood, and J.D. Mougous. 2010. What is type VI secretion doing in all those bugs? *Trends in Microbiology* 18(12):531–537. PMC 2991376. PMID: 20961764. doi: 10.1016/j.tim.2010.09.001.

Seya, T., M. Matsumoto, S. Tsuji, et al. 2001. Two receptor theory in innate immune activation: studies on the receptors for bacillus Culmet Guillen-cellwall skeleton. *Archivum Immunologiae et Therapiae Experimentalis (Warsz.)* 49(1):S13–S21.

Shah, P., and E. Swiatlo. 2008. A multifacet role for polyamines in bacterial pathogens. *Molecular Microbiology* 68:4–16. doi: 10.1111/j.1365-2958.2008.06126.x.

Shalaby, M.R., D. Pennica, and M.A. Palladino Jr. 1986. An overview of the history and biologic properties of tumor necrosis factors. *Springer Seminars in Immunopathology* 9:33–37. doi: 10.1007/bf00201903. PMID: 3523803.

Shao, F., P.M. Merritt, Z. Bao, R.W. Innes, and J.E. Dixon. 2002. A Yersinia effector and a *Pseudomonas avirulence* protein define a family of cysteine pro-teases functioning in bacterial pathogenesis. *Cell* 109(5):575–588. doi: 10.1016/s0092-8674(02)00766-3.

Shao, F., P.O. Vacratsis, Z. Bao, K.E. Bowers, C.A. Fierke, and J.E. Dixon. 2003. Biochemical characterization of the Yersinia YopT protease: cleavage site and recognition elements in Rho GTPases. *Proceedings of National Academy of Sciences* 100(3):904–909. doi: 10.1073/pnas.252770599.

Shenep, J.L., and K.A. Mogan. 1984. Kinetics of endotoxin release during antibiotic therapy for experimental gram-negative bacterial sepsis. *Journal of Infectious Diseases* 150:380–388. doi: 10.1093/infdis/150.3.380.

Shoberg, R.J., and D.D. Thomas. 1993. Specific adherence of *Borrelia burgdorferi* extracellular vesicles to human endothelial cells in culture. *Infection and Immunity* 61:3892–3900.

Shrivastava, R., and J.F. Miller. 2009. Virulence factor secretion and translocation by *Bordetella* species. *Current Opinion in Microbiology* 12(1):88–93.

Simeone, R., B. Daria, and B. Roland. 2009. ESX/type VII secretion systems and their role in host-pathogen interaction. *Current Opinion in Microbiology* 12(1):4–10.

Smith, H., J.A. Cole, and N.J. Parsons. 1992. The sialylation of gonococcal lipopolysaccharide by host factors: a major impact on pathogenicity. *FEMS Microbiology Letters* 79:287–292. doi: 10.1111/j.1574-6968.1992.tb14054.x.

Srivastava, R., V.B. Sinha, and B.S. Srivastava. 1980. Events in the pathogenesis of experimental cholera: role of bacterial adherence and multiplication. *Journal of Medical Microbiology* 13(1):1–9. doi: 10.1099/00222615-13-1-1.

Staedel, C., and F. Darfeuille. 2013. MicroRNAs and bacterial infection. *Cellular Microbiology* 15(9):1496–1507. doi: 10.1111/cmi.12159.

Staudt, L.M. 2012. TLR/NF-kBpathway. *Cold Spring Harbor Perspectives in Biology*. doi: 10.1101/cshperspect.a011247.

Stein, P.E., A. Boodhoo, G. Armstrong, et al. 1994. Structure of a pertussis toxin-sugar complex as a model for receptor binding. *Nature Structural Biology* 1:591–596. doi: 10.1038/nsb0994-591.

Tabor, C.W., and H. Tabor. 1985. Polyamines in microorganisms. *Microbiological Reviews* 49(1):81–99. PMC373019.

Takeda, K. 2005. Evolution and integration of innate immune recognition systems: the Toll-like receptors. *Journal of Endotoxin Research* 11:51–55. doi: 10.1177/09680519050110011101.

Terebiznik, M.R., O.V. Vieira, S.L. Marcus, et al. 2002. Elimination of host cell PtdIns(4,5)P(2) by bacterial SigD promotes membrane fission during invasion by *Salmonella*. *Nature Cell Biology* 4(10):766–773. doi: 10.1038/ncb854.

Tertti, R., M. Skurnik, T. Vartio, and P. Kuusela. 1992. Adhesion protein YadA of *Yersinia* species mediates binding of bacteria to fibronectin. *Infection and Immunity* 60:3021–3024. PMID: 1612772, PMCID: PMC257272.

Toder, D.S., M.J. Gambello, and B.H. Iglewski. 1991. *Pseudomonas aeruginosa* LasA: a second elastase under the transcriptional control of lasR. *Molecular Microbiology* 5:2003–2010. doi: 10.1111/j.1365-2958.1991.tb00822.x.

Triantaȼlou, M., and K. Triantaȼlou. 2005. The dynamics of LPS recognition: complex orchestration of multiple receptors. *Journal of Endotoxin Research* 11:5–11.

Van Deventer, S.J.H., H.R. Buller, J.W. tenCate, A. Sturk, and W. Pauw. 1988. Endotoxaemia: an early predictor of septicaemia in febrile patients. *Lancet* 331(8586):605–609. doi: 10.1016/s0140-6736(88)91412-2.

Vanputten, J.P.M., and B.D. Robertson. 1995. Molecular mechanisms and implications for infection of lipopolysaccharide variation in *Neisseria*. *Molecular Microbiology* 16:847–853. doi: 10.1111/j.1365-2958.1995.tb02312.x.

Verhoef, J., and E. Mattsson. 1995. The role of cytokines in gram-positive bacterial shock. *Trends in Microbiology* 3(4):136–140. doi: 10.1016/s0966-842x(00)88902-7.

Vesy, C.J., R.L. Kitchens, G. Wolfbauer, J.J. Albers, and R.S. Munford. 2000. Lipopolysaccharide-binding protein and phospholipid transfer protein release lipopolysaccharides from Gram-negative bacterial membranes. *Infection and Immunity* 68:2410–2417.

Vincent, J.L. 1996. De¢nition and pathogenesis of septic shock. In *Pathology of Septic Shock*, E.T. Rietschel, and H. Wagner (eds.). Berlin, Heidelberg, NewYork: Springer:1–13.

Viriyakosol, S., and T. Kirkland. 1995. Knowledge of cellular receptors for bacterial endotoxin. *Clinical Infectious Diseases* 21(Suppl. 2), S190–S195.

Von Pawel-Rammingen, U., M.V. Telepnev, G. Schmidt, K. Aktories, H. Wolf-Watz, and R. Rosqvist. 2000. GAP activity of the *Yersinia* YopE cytotoxin specifically targets the Rho pathway: a mechanism for disruption of actin microfilament structure. *Molecular Microbiology* 36(3):737–748. doi: 10.1046/j.1365-2958.2000.01898.x.

Vreugdenhil, A.C., M.A. Dentener, A.M. Snoek, J.W. Greve, and W.A. Buurman. 1999. Lipopolysaccharide binding protein and serum amyloid A secretion by human intestinal epithelial cells during the acute phase response. *Journal of Immunology* 163:2792–2798. PMID: 1045302.

Wai, S.N., B. Lindmark, T. Soderblom, et al. 2003. Vesicle-mediated export and assembly of pore-forming oligomers of the enterobacterial ClyA cytotoxin. *Cell* 115:25–35. doi: 10.1016/S0092-8674(03)00754-2.

Walker, T. 1998. *Microbiology*. Philadelphia: WB Saunders Company.

Welch, R.A. 1991. Pore-forming cytolysins of gram-negative bacteria. *Molecular Microbiology* 5:521–528. doi: 10.1111/j.1365-2958.1991.tb00723.x.

Werling, D., F. Sutter, M. Arnold, et al. 1996. Characterization of the acute-phase response of heifers to a pro-longed low-dose infusion of lipopolysaccharide. *Research in Veterinary Science* 61:252–257. doi: 10.1016/s0034-5288(96)90073-9.

Werling, D., and T.W. Jungi. 2003. TOLL-like receptors linking innate and adaptive immune response. *Veterinary Immunology and Immunopathology* 91:1–12. doi: 10.1016/s0165-2427(02)00228-3.

Westerlund, B., and T.K. Korhonen. 1993. Bacterial proteins binding to the mammalian extracellular matrix. *Molecular Microbiology* 9:687–694. doi: 10.1111/j.1365-2958.1993.tb01729.x.

Whiteley, L.O., S.K. Maheswaran, D.J. Weiss, and T.R. Ames. 1990. Immunohistochemical localization of *Pasteurella haemolytica* A1-derived endotoxin, leukotoxin, and capsular polysaccharide in experimental bovine Pasteurella pneumonia. *Veterinary Pathology* 27:150–161.

Whiteley, L.O., S.K. Maheswaran, D.J. Weiss, and T.R. Ames. 1991. Morphological and morphometrical analysis of the acute response of the bovine alveolar wall to *Pasteurella haemolytica* A1-derived endotoxin and leucotoxin. *Journal of Comparative Pathology* 104:23–32. doi: 10.1016/s0021-9975(08)80085-0.

Wilson, J.W., M.J. Schurr, C.L. LeBlanc, R. Ramamurthy, K.L. Buchanan, and C.A. Nickerson. 2002. Mechanisms of bacterial pathogenicity. *Postgraduate Medical Journal* 78:216–224. doi: 10.1136/pmj.78.918.216.

Winter, S.E., P. Thiennimitr, M.G. Winter, et al. 2010. Gut inflammation provides a respiratory electron acceptor for *Salmonella*. *Nature* 467:426–429. doi: 10.1038/nature09415.

Wise, R.P., M.J. Moscou, A.J. Bogdanove, and S.A. Whitham. 2007. Transcript profiling in host–pathogen interactions. *Annual Review of Phytopathology* 45(1):329–369. doi: 10.1146/annurev.phyto.45.011107.143944.

Wollack, J.W., N.A. Zeliadt, D.G. Mullen, et al. 2009. Multifunctional prenylated peptides for live cell analysis. *Journal of the American Chemical Society* 131(21):7293–7303. doi: 10.1021/ja805174z.

Wooldridge, K. 2009. *Bacterial Secreted Proteins: Secretory Mechanisms and Role in Pathogenesis*: Caister Academic Press. ISBN: 978-1-904455-42-4.

Wu, A., C.J. Hinds, and C. Thiemermann. 2004. High-density lipoproteins in sepsis and septic shock: metabolism, actions, and therapeutic applications. *Shock* 21:210–221. doi: 10.1097/01.shk.0000111661.09279.82.

Wu, Z., H. Lu, J. Sheng, and L. Li. 2012. Inductive microRNA-21 impairs anti-mycobacterial responses by targeting IL-12 and Bcl-2. *FEBS Letters* 586(16):2459–2467. doi: 10.1016/j.febslet.2012.06.004.

Yaffe, M.B. 2012. Protein regulation. *Cold Spring Harbor perspectives in biology*. doi: 10.1101/cshperspect.a005918.

Yaganza, E.S., D. Rioux, M. Simard, J. Arul, and R.J. Tweddell. 2004. Ultrastructural alterations of *Erwinia carotovora* subsp. *atroseptica* caused by treatment with aluminum chloride and sodium metabisulfite. *Applied Environmental Microbiology* 70(11):6800–6808.

Young, L.S. 1985. Gram-negative sepsis. In *Principles and Practice of Infectious Disease*, G.L. Mandell, R.G. Douglas Jr, and J.E. Bennett (eds.), 2nd edn. New York: Wiley:452–475.

Zhang, F.L., and P.J. Casey. 1996. Protein prenylation: molecular mechanisms and functional consequences. *Annual Review of Biochemistry* 65:241–269. doi: 10.1146/annurev.bi.65.070196.001325.

Zhang, J., K. Mostov, M.E. Lamm, et al. 2000. The polymeric immunoglobulin receptor translocates pneumococci across human nasopharyngeal epithelial cells. *Cell* 102:827–837. doi: 10.1016/s0092-8674(00)00071-4.

Zhou, L., R. Srisatjaluk, D.E. Justus, and R.J. Doyle. 1998. On the origin of membrane vesicles in gram-negative bacteria. *FEMS Microbiology Letters* 163(2):223–228.

Section III

Animal Microbiome

15

Reproductive Tract Microbiome in Animals: Physiological versus Pathological Condition

R. Vikram, Vivek Joshi, A. A. P. Milton, M. H. Khan, and K. P. Biam
ICAR – National Research Centre

CONTENTS

15.1 Introduction

The reproductive system consists of external genitalia as well as internal genitalia. The study of microbial community inhabiting reproductive tracts of animals is essential to understand reproductive physiology and health aiming not only towards a clinical cure but also to restore healthy microbiota status. The "one health" concept focuses on the ecological relationships between animals, humans, and environmental health (Davis et al. 2017). Hence, a better understanding of the microbiome in humans, animals, the shared environment, and their interactions could help to prevent diseases (e.g., occupational) and manage human health or various disease states.

Presently, the "sterile womb" concept, where a fetus grows up in a microbial-free (sterile environment), is under argument (Perez-Munoz et al. 2017). Recent studies have indicated that the cervical mucus plug may not be completely impermeable to ascending microbes from the vagina and may allow the transport of bacteria into the intrauterine cavity (Hansen et al. 2013; Baker et al. 2018). The reviews have been focused on correlations between commensal uterine microbial species and fertility problems, including pregnancy complications in humans (Franasiak and Scott 2017; Moreno and Franasiak 2017). Host–microbiota depends on their symbiotic relationship, thus acting as a natural barrier against colonization by pathogenic species. Hence, homeostasis is constructed and maintained. In animals particularly, determining the host–microbiome relationship within the reproductive tract may help to devise better tools for enhancing reproductive efficiency, such as treatment with probiotics to introduce microbial communities that result in positive results with regard to reproduction (Clemmons et al. 2017).

15.2 Vaginal and Uterine Microbiome in Animals

The common phyla found between the uterus and the vagina are most commonly found in many host–microbiome relationships in many species (Jami and Mizrahi 2012; Myer et al. 2015), and their function is not yet fully understood. However, relative abundance of these phyla changes according to the changes in the host physiology. The human vagina with stable bacterial communities has been related to a healthy immune state of the female reproductive tract (Gajer et al. 2012; The Human Microbiome Project Consortium et al. 2012; DiGiulio et al. 2015). Recently, the use of sequencing technologies has gained importance to characterize the microbiome. Some shared operational taxonomic units (OTUs) show there is an interaction between bacterial communities of uterus and vagina, yet the OTU differences between the vagina and uterus are due to functional differences in the tissue and microbial ecosystem niche (Clemmons et al. 2017). Both vaginal and uterine microbiota are less explored in animals, and the microbiota is composed of anaerobic, facultative anaerobic, and aerobic microorganisms (Otero et al. 2000). In postpartum cows, estrus synchronization followed by vaginal and uterine bacterial population study before insemination showed Firmicutes as the most abundant phyla in uterus and vagina (Clemmons et al. 2017), also between resulting pregnant and nonpregnant cow's uterus and vagina (Ault et al. 2019) (Table 15.1). The uterus samples of the virgin and pregnant heifers showed Firmicutes, Bacteroidetes, and Proteobacteria phyla in abundance, and this study indicates the presence of common phyla of the gastrointestinal tract inhabiting the reproductive tract (Moore et al. 2017). Microbial communities present in the uterus are also found in the vagina and gut; therefore, microbiota from the vagina and cervix migrate via intrauterine ascension, where they will subsequently colonize the uterus (Moore et al. 2017). The possible blood-borne transmission of the gut microbiome to the uterus in cows has been explored (Jeon et al. 2017). The vaginal microbiome is dominated by Proteobacteria, Bacteroidetes, and Actinobacteria in the mares. Proteobacteria, Firmicutes, Bacteroidetes, and Actinobacteria are abundant in the equine endometrial microbiome (Swartz et al. 2014). Mare endometrium during estrus showed only proteobacteria-driven microbiome (Heil et al. 2018); the estrous cycle likely influences the changes in the microbiome. During estrus, dynamic changes of the mare cervix influence communication between the vagina and the uterus. Ewe microbiota exhibited greater diversity compared to the cow with Bacteroidetes, Fusobacteria, and Proteobacteria being the dominant phyla. Archaea and lactobacilli were prevalent, but not abundant (Swartz et al. 2014) (Table 15.1). Vaginal microbiome in bitches is dominated by Bacteroidetes, Proteobacteria, Tenericutes, and Firmicutes phyla, while Proteobacteria, Firmicutes, Actinobacteria, and Bacteroidetes were the most prevalent phyla in the uterus similar to a cow (Table 15.1). At the genus level, the vaginal bacterial community of bitches was higher in richness and the uterus was higher in diversity (DeSilva et al. 2019). Similar to mares, the stages of estrous cycle influence in determining diversity and richness of microbiome in uterus and vagina in bitches.

The dominating vaginal microbiomes in the most wild primates are Firmicutes phylum, followed by Fusobacteria, Bacteroidetes, Proteobacteria, and Actinobacteria. Interestingly, the vaginal microbiome of the wild chimpanzee (the closest human relative) was dominated by Fusobacteria compared to Firmicutes and had<3.5% Lactobacilli (Yildirim et al. 2014). This is in contrast to the captured baboon, where the dominant phylum was Firmicutes but mainly consists of Clostridia genera (Uchihashi et al. 2015). The samples from wild howler, red colobus monkeys, and lemurs showed few or no reads for Lactobacilli species. The lemurs' vaginal microbiome especially showed a significantly higher proportion of unclassified taxa, suggesting possibly novel bacteria taxa not previously characterized (Yildirim et al. 2014). The Firmicutes were also the dominant phyla in guinea pig, and *Lactobacillus* spp. had a very low relative abundance with other microbes from Corynebacterium, Anaerococcus, Peptoniphilus, Aerococcus, Facklamia, and Allobaculum genera being more dominant (Neuendorf et al. 2015). Similar to mares, the semi-captive giant panda vaginal tract had Proteobacteria, Firmicutes, Actinobacteria, and Bacteroidetes as the most abundant phyla and Proteobacteria followed by Bacteroidetes in the uterus (Xia et al. 2017). The human vagina is dominated by commensal bacteria, predominantly Lactobacilli; therefore, this vaginal microbiota has an established role in female reproductive physiology, pathogen defense, and function (Wee et al. 2018). However, the Lactobacillus abundance in the reproductive tract of bovine is very low (Swartz et al. 2014; Clemmons et al. 2017). However, the reproductive tract microbiome of bovine is significantly more diverse than that of humans. The similarities exist between the uterus and the vagina of humans and bovine, where Firmicutes dominates both (Laguardia-Nascimento et al. 2015; Clemmons et al. 2017). The bacterial communities have been widely studied with less emphasis on the Archaea and Fungal communities in the vagina and uterus in animals. The vaginal, uterine, and placental microbiomes of humans and many other species have been sequenced, including the vaginal microbiomes of wild primates (chimpanzee, baboon, howler, red colobus monkey, and lemur),the guinea pig, cow, ewe, bitch, mare, and giant panda (*Ailuropoda melanoleuca*) (Figure 15.1).

15.3 Factors Affecting Reproductive Tract Microbiome

a. *The cervicovaginal microbiome:* The vaginal microbiome fluctuates according to different physiological conditions such as puberty, menstruation, sexual activity, and pregnancy (Giudice 2016; Prince et al. 2015; Song et al. 2017). In humans, the onset of puberty brings the vaginal flora under the influence of reproductive hormones, and environmental pH decreases causing predominance of *Lactobacilli* spp. (Moreno and Franasiak 2017). The microbiota development and the relative abundance of OTU greater in the vagina tract than in the uterus are strongly influenced by the proximity of the gastrointestinal tract (Clemmons et al. 2017 and Laguardia-Nascimento et al. 2015). The vaginal microbiome participates simultaneously in the host's metabolic and immune systems, thus maintaining homeostasis and contributing significantly to the reproductive success and prevention of disease (Clemmons et al. 2017; Giudice 2016).
b. *The uterine microbiome:* The comparison between uterine and vaginal microbiome showed differences between species. In healthy cows, increased richness and greater levels of diversity were found in the vagina in comparison with the uterus, which is possibly due to proximity to the external environment (Clemmons et al. 2017). In healthy women, the uterine endometrial microbiome showed lower relative abundance compared to the cervix microbiome (Verstraelen et al. 2016; Wee et al. 2018), whereas in mares, the relative abundance of the external cervical os is similar to the uterine relative abundance. The differences in the anatomy of the reproductive tract suggest the changes in the microbial communities in each species. Invasion of microbes into the

TABLE 15.1

Metagenomic Analysis of Uterine and Vaginal Microbiota in Different Species during Different Physiological Conditions

Study	Aim	Sampling and Analytical Methods	Microbiota (Taxonomic Categories)					
Clemmons et al. (2017)	Comparing the microbial communities in the bovine uterus and vagina at the time of artificial insemination.	Uterine and vaginal flush (Day -2 before estrus) Illumina MiSeq (V1–V3)	*Phyla—Uterus* *Relative abundance*		*Phyla—Vagina* *Relative abundance*		*Genera—Uterus*	*Genera—Vagina*
			Proteobacteria Actinobacteria Bacteroidetes Unassigned taxa		Bacteroidetes Proteobacteria Tenericutes Actinobacteria Unassigned taxa		Corynebacterium Ureaplasma Staphylococcus Microbacterium Butyrivibrio Helcococcus	Oscillospira Butyrivibrio Ureaplasma Campylobacter Dorea Clostridium Helcococcus Corynebacterium
Ault et al. (2019)	Characterization of the taxonomic composition of the uterine and vaginal bacterial communities during estrous synchronization up to timed artificial insemination (TAI) in bovines	Uterine and vaginal flush (days −21, −9, and −2 before estrus) Illumina MiSeq (V1–V3)	*Phyla—Uterus* *Relative abundance of significant phyla*		*Phyla—Vagina* *Relative abundance of significant phyla*		*Genera—Uterus*	
			Nonpregnant	Pregnant	Nonpregnant	Pregnant	Nonpregnant	
			Firmicutes Proteobacteria Tenericutes Verrucomicrobia Fusobacteria Thermus	Firmicutes Actinobacteria Lentisphaerae Fibrobacteres Spirochaetes Chloroflexi Armatimonadetes WPS-2	Firmicutes Proteobacteria TM7 Thermi Spirochaetes Acidobacteria Planctomycetes Verrucomicrobia Fibrobacteres	Proteobacteria TM7 Planctomycetes Verrucomicrobia SR1 Synergistetes	Corynebacterium, Staphylococcus Prevotella Microbacterium Butyrivibrio Ralstonia, Family Alcaligenaceae Family Comamonadaceae *Significantly different in nonpregnant and pregnant cows*	
Swartz et al. (2014)	Composition of sheep and cattle vaginal microbiota	Vaginal lavages were collected from 20 Rambouillet ewes and 20 crossbred beef cows of varied breeding method and pregnancy status Illumina MiSeq (V3-V4)	*Predominant phyla—Sheep and cow vagina*		*Predominant genera—Cow vagina*		*Predominant genera—Sheep vagina*	
			Bacteroidetes Fusobacteria Proteobacteria		Aggregatibacter Streptobacillus Phocoenobacter Phocoenobacter Sediminicola Sporobacter		Aggregatibacter Streptobacillus Cronobacter Phocoenobacter Psychrilyobacter	
DeSilva et al. (2019)	Characterize the normal microbiome of healthy canine vagina and endometrium and to determine the effect of the stage of estrous, on the resident microbiome.	Vaginal swab and endometrium by ovariohysterectomy Illumina MiSeq(V4)	*Phyla—Uterus, Relative abundance*	*Phyla only found in Uterus*	*Phyla—Vagina, Relative abundance*	*Phyla only found in the vagina*	*Genera*	
			Proteobacteria Firmicutes Actinobacteria Bacteroidetes	Armatimonadetes Chlamydiae Deferribacteres	Bacteroidetes Proteobacteria Tenericutes Firmicutes	Cyanobacteria Elusimicrobia Gemmatimonadetes Lentisphaerae Thermotogae	Hydrotalea, Ralstonia, Mycoplasma, Fusobacterium, and Streptococcus *predominant in the vagina.* Pseudomonas, Staphylococcus, and Corynebacterium *predominant in the uterus.*	

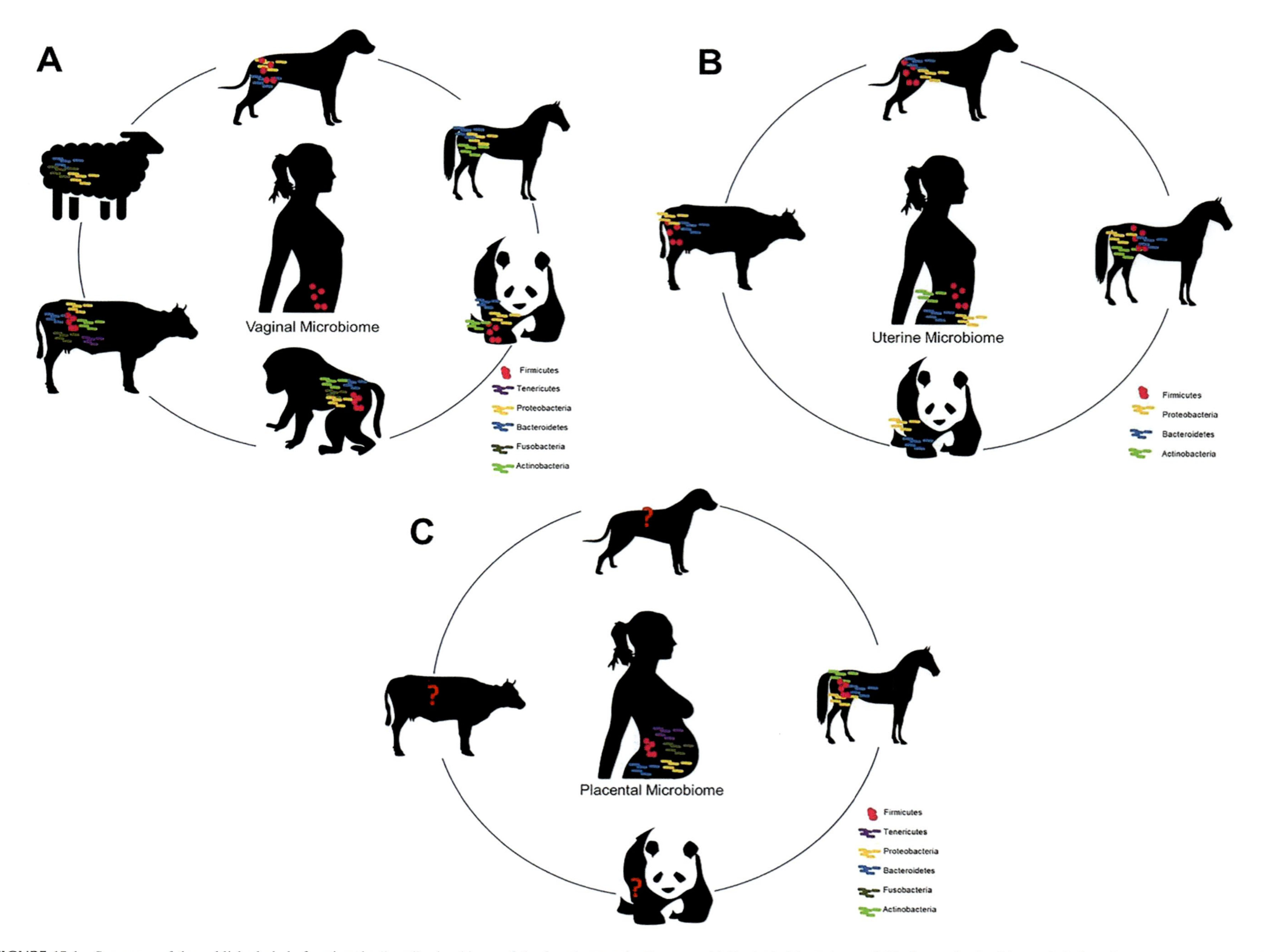

FIGURE 15.1 Summary of the published phyla forming the "core" microbiome of the female reproductive tract. (a) Vaginal, (b) uterine, and (c) placental microbiome phyla based on current metagenomic data. (Adapted from Heil et al. 2019.)

uterine cavity in humans has four proposed routes: vertical ascension from the cervicovaginal region, retrograde through the abdominal cavity, through invasive procedures, and hematogenous route (Payne and Bayatibojakhi 2014). Vertical ascension from the vagina is the major source of intrauterine colonization in humans, including mare (Macpherson 2006, Payne and Bayatibojakhi 2014).

15.4 Vaginal and Uterine Microbiome during Pregnancy in Animals

The increased presence of lactobacilli with the advancement of gestational age is due to changes in levels of estrogen and vaginal pH (Prince et al. 2015). Heifer cow vaginal microbiome during different pregnancy status showed Firmicutes as the most abundant taxa followed by Proteobacteria, Bacteroidetes, and Tenericutes, at the genera-level unclassified Enterobacteriaceae, followed by Ureaplasma and an unclassified Bacteroidaceae (Deng et al. 2019). This study used bacterial features as predictive of pregnancy; the top three bacterial features such as Clostridiaceae, *Histophilus somni*, and Campylobacter were more abundant in the unconceived cows (Deng et al. 2019). The Bacteroides, Enterobacteriaceae, and Histophilus are the dominant OTUs in the vaginal tract of cattle with reproductive disorders (Rodrigues et al. 2015). Deng et al. (2019) accurately predicted the capability of a cow to establish pregnancy after breeding by studying prebreeding fecal microbiome features. Laguardia-Nascimento et al (2015) characterized the vaginal microbiome of Nellore cattle and found the main bacterial phyla Firmicutes, Bacteroidetes, and Proteobacteria in pregnant and nonpregnant animals. There was a tendency of reduction in bacterial abundance in the vagina of pregnant animals, and an increase in Archaea (Laguardia-Nascimento et al. 2015). In fungi, the Ascomycota phylum (Mycosphaerella genus) dominated in both pregnancy and nonpregnancy animals, showing a tendency of reduction in pregnancy condition (Laguardia-Nascimento et al. 2015). There is a tendency for reduction in the microbial population soon after estrus; in the progesterone phase, the animals probably tend to have a less abundant bacterial microbiota (Otero et al. 2000). Following parturition and the return to normal postpartum estrous cycle, the equilibrium in the microbial community will be established in the vagina before pregnancy, with an increase in bacterial community and a reduction in archaea (Laguardia-Nascimento et al. 2015). Metagenetic analysis of the equine placenta and extraplacental body sites showed the presence of Firmicutes, Proteobacteria, and Bacteroidetes as the main phyla in the gravid horn of the chorioallantois, which was also the most abundant phyla within oral, fecal, and vaginal samples. The above three along with Actinobacteria were found as the main phyla in the nongravid horn of the chorioallantois (Xia et al. 2017). The most abundant bacterial phyla in the gravid and nongravid chorioallantois share a significant overlap, suggesting similar, but not identical, environments within different compartments of the chorioallantois.

15.4.1 The Influence of the Vaginal and Uterine Microbiome on Fertility and Pregnancy Outcome

The host–microbiome interaction and the existing microbiota in the reproductive tract play a significant role in the fertility and pregnancy outcome (Ault et al. 2019; Pelzer et al. 2017). Some specific microbes have been related to positive or negative pregnancy, adverse pregnancy outcomes, and preterm birth in women (Aagaard et al. 2014; Payne and Bayatibojakhi 2014). The fertile women have less vaginal dysbiosis compared to subfertile women (Uchihashi et al. 2015). Similar to humans, the chance of a stillborn puppy significantly increases by the presence of any of the genera Bibersteinia, Staphylococcus, Pasteurella, Corynebacterium, or Methylobacterium in the vagina of bitch (Cornelius et al. 2017). In cows, the taxonomic composition of the uterine and vaginal bacterial communities following estrous synchronization and timed artificial insemination (TAI) showed significant different bacterial genera in the uterus of nonpregnant and pregnant cows (Table 15.1) (Ault et al. 2019).Uterine microbiome in cows showed that *Bacteroides* spp., *Ureaplasma* spp., *Fusobacterium* spp., and *Trueperella* spp. prevalence was significantly higher in nonpregnant cows by 200 DIM (days in milking) (Ault et al. 2019).

15.5 Penile and Preputial Microbiome in Animals

The male urogenital system microbiota has not been extensively studied in all animal species. In humans, urogenital microbial communities differ between healthy, infected, males with sexually transmitted diseases or with prostatitis conditions (Wickware et al. 2020). Earlier studies of bacterial genera from penis and prepuce with bovine reproductive disease indicated Chlamydia, Campylobacter, and *H. somni* (Chaban et al. 2012; Sandal and Inzana 2010). The microbial community of seminal fluid correlates well with fertility in men (Weng et al. 2014). A recent study in postpubertal bulls identified bacteria belonging to Firmicutes, Fusobacteria, Bacteroidetes, Proteobacteria, and Actinobacteria in the prepuce (Wickware et al. 2020). Mainly two major community types were found: those with low and those with high bacterial species richness. The bull penile microbial community composition showed no differences in different breed or age groups or management practices. However, Bradyrhizobium was a distinguishing genus only found in the low-diversity samples. This study concluded that genera which were common in a cow vagina, respiratory tract, feces, and soil were the members of the bull penile microbial community (Wickware et al. 2020). The culture-dependent method identified bacteria (growth with a Total mesophilic bacterial count (TBC) ≥10,000 CFU) *Escherichia coli*, coliforms, coagulase-positive Staphylococci, and coagulase-negative Staphylococci in healthy stallion reproductive tract (Rota et al. 2011). The isolated filamentous fungi were *Aspergillus*spp.,*Penicillium*spp., *Mucoraceae*spp.,*Acremonium*spp.,*Scopulariopsis*spp.,*Chryso sporium*spp.,*Epicoccum*spp.,and *Verticillium* spp., which differed significantly between various sites of the reproductive tract in the stallion. The isolated fungal species were saprophytes commonly found in the environment. Yeasts were also isolated such as *Trichosporon* spp., *Candida famata*, and other unidentified yeast from the stallion (Rota et al. 2011). The studies on penile and preputial microbiota are minimal in other domestic animals.

15.6 The Reproductive Tract Microbiome of Animals in Pathological Condition

The microbial community in the reproductive tract is maintained in an inert state until influenced by events (e.g., parturition), in which homeostatic mechanisms of the body to maintain microbiota composition are disrupted (Moore et al. 2017). The immune function will decrease sharply during the transition to lactation in a dairy cow (Kehrli and Goff 1989). The presence of a commensal bacterial community contributes to a healthy or optimal reproductive tract, but pathogenic bacteria in the reproductive tract have negative effects (Ault et al. 2019). The uterine diseases such as metritis, clinical endometritis, subclinical endometritis, and pyometra in the dairy industry are important, as they cause economic losses due to lower conception rates, increased use of antibiotics, premature culling, reduced milk production, and increased treatment costs (Drillich et al. 2001; Sheldon et al. 2006). The phyla mainly associated with metritis, endometritis, and pyometra are Bacteroidetes, Fusobacteria, Tenericutes, and Proteobacteria (Knudsen et al. 2015).

15.6.1 Metritis

Metritis involves inflammation in all layers of the uterine wall characterized by edema, infiltration by leukocytes, and myometrial degeneration. The mucosa will be congested, and there will be the prominent leukocyte infiltration in response to the common pathogens *Fusobacterium necrophorum, A. pyogenes, Prevotella species,* and *E. coli* as studied by culture-dependent method (Sheldon et al. 2006). The shift in the uterine microbiome occurs in a metritic cow, which is characterized by a decrease in bacterial richness and a loss of heterogeneity (Galvao et al. 2019). The greater prevalence of *Trueperella pyogenes* and *E. coli* has been recorded by culture-dependent studies in the metritic cows. Additionally, it is reported that cows with metritis had a greater prevalence of gram-negative anaerobes, particularly *Bacteroides* spp. (Williams et al. 2005). Drillich et al. (2001) sampled 15 cows with metritis and reported that *F. necrophorum* and *Porphyromonaslevii* were the most prevalent (67%) bacteria isolated. The study of comparison between healthy and metritis cows postpartum showed that the metritic uterine samples predominantly contained the Bacteroidetes, Proteobacteria, Firmicutes, and Fusobacteria (Bicalhlo et al. 2017). The healthy cows showed Proteobacteria and Bacteroidetes as 40% and 10%, whereas the metritic cows showed Proteobacteria and Bacteroidetes as 10% and >45%, respectively, showing clear dysbiosis. Bacteroidetes showed that the increased relative abundance in the metritic cows was probably due to greater abundance of Bacteroides and Porphyromonas. In the metritic cow versus healthy cows, the decrease in Proteobacteria was mostly by declines in relative abundance of Shigella and Escherichia. In healthy cows, the invasion of the uterus by bacterial species that disturb homeostasis and cause dysbiosis, inflammation, and infection can be prevented by the presence of tolerant colicin E2 commensal bacteria within the uterine microbiota (Bicalhlo et al. 2017). This dysbiosis favors the growth of strict gram-negative anaerobes, such as *F. necrophorum* and *Bacteroides* spp. Additionally, the microbiota of the metritic cows express genes in high levels that are involved in "protein translocation across membranes" followed by "secretion"; this process aids in uterine epithelial colonization and invasion of the mucosal surface by bacteria (Bicalhlo et al. 2017). There will be a predominant microbial shift at the phylum level from calving until the establishment of metritis in dairy cows (Figure 15.2). Another study reported that Bacteroidetes, Peptostreptococcus, and Fusobacterium were more abundant in the metritic cows than in healthy cows on 10 days postpartum (Peng et al. 2013). The examination of uterine microbiota from the start of calving until the initiation of metritis using sequencing of 16S rRNA genes showed the identical uterine microbiota in healthy and metritic cows until day 2 postpartum (Jeon et al. 2015). Later on, the bacterial population diverged in favor of the relative abundance of Bacteroides, Porphyromonas, and Fusobacterium in the metritic cows (Jeon et al. 2015). Additionally, metritis cured cows either naturally or with antibiotic treatment displayed a decrease in abundance of Bacteroides, Porphyromonas, and Fusobacterium (Jeon and Galvao 2018). In conclusion, Bacteroides, Porphyromonas, and Fusobacterium play a key role in the development of metritis (Table 15.2). At the species level, *Bacteroides pyogenes, Por. levii*, and *Helcococcus ovis* are potential emerging uterine pathogens (Galvao et al. 2019).

15.6.2 Endometritis

Endometritis is a superficial inflammation of the endometrium, with no involvement deeper than stratum spongiosum (Sheldon et al. 2006). Clinical endometritis is characterized by the presence of purulent or mucopurulent vaginal discharge at 21 or more days postpartum along with prominent leukocyte infiltration into the uterine lumen. Subclinical endometritis exhibits no signs of clinical endometritis and is characterized by an increased proportion of polymorphonuclear neutrophil (PMN) cells in the endometrium (Kasimanickam et al. 2004). The most common microbes responsible for chronic endometritis in humans include Enterobacteriaceae, *Enterococcus faecalis, Gardnerella vaginalis, Streptococcus* spp., *Staphylococcus* spp., *Mycoplasma* spp., *Ureaplasma urealyticum, Chlamydia trachomatis*, and *Neisseria gonorrhoeae* (Moreno et al. 2018). Williams et al. (2005) sampled cows from 21 to 28 days postpartum, and culture studies reported that *E. coli* and *T. pyogenes* were the most prevalent recognized uterine pathogens isolated from the uterus, followed by *Prevotella melaninogenica* and *F. necrophorum. T. pyogenes* is a critical pathogen involved in the development of clinical endometritis along with gram-negative anaerobes such as *F. necrophorum, Por. levii,* and *Prev. melaninogenica*, which may act synergistically with *T. pyogenes* to cause clinical endometritis (Galvao et al. 2019). Apart from above, members of the genera Bacillus, Streptococcus, Enterococcus,

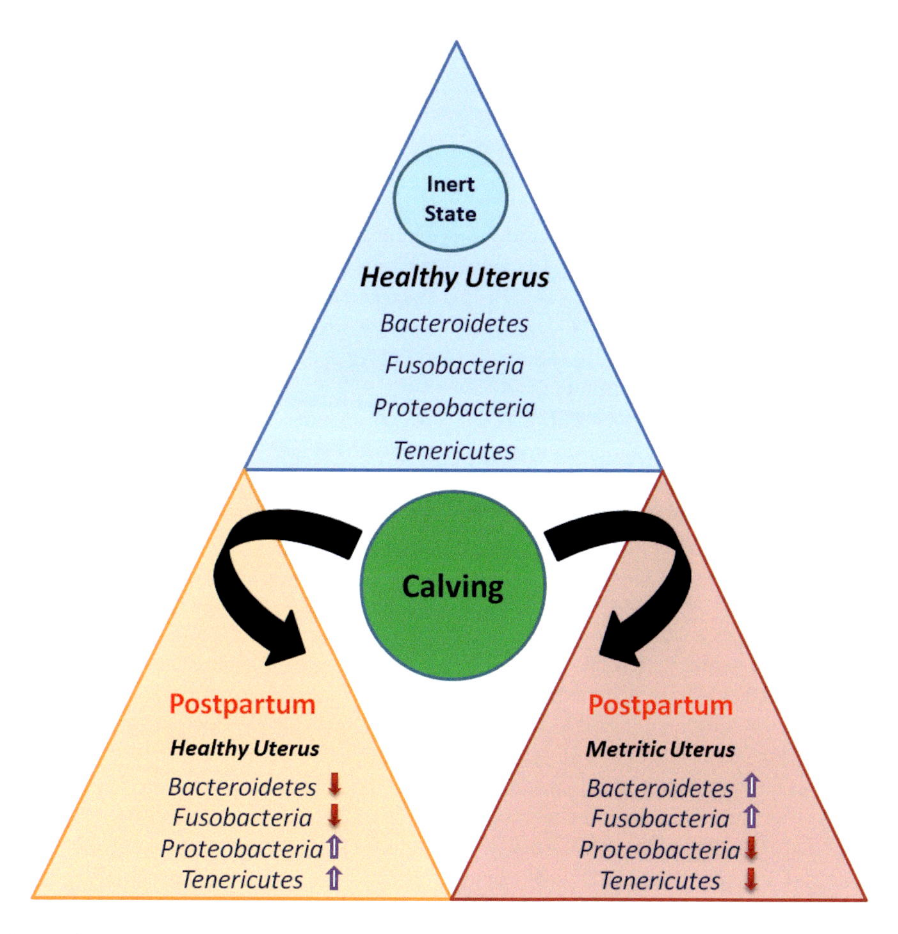

FIGURE 15.2 Microbial shift at the phylum level in healthy and metritic uterus of dairy cows post calving. Arrows following phylum names indicate the relative abundance comparison among healthy amd metritic uterus. .)

TABLE 15.2

Metagenomic Analysis of Uterine Microbiota Indicating Pathogenic Bacteria (Taxonomic Categories) Associated with Metritis in Postpartum Dairy Cows

Study	Bacteria (Taxonomic Categories)	Sample (Day Postpartum)	Analytical Methods
Santos et al. (2011)	Fusobacteria (phylum)	Uterine fluid	DGGE and 16S clone library sequencing
Machado et al. (2012)	Ureaplasma (genus), Bacteroides (genus	Uterine fluid (35±3)	Pyrosequencing (V1-2)
Peng et al. 2013	Bacteroidetes (phylum) Peptostreptococcus (genus) Fusobacterium (genus)	Uterine fluid (10)	qPCR and 16S clone library sequencing
Jeon et al. (2015, 2016, 2018)	Bacteroides (genus) Porphyromonas (genus) Fusobacterium (genus)	Swab (6±2)	Illumina MiSeq (V4)
Knudsen et al. (2016)	Porphyromonadaceae (family) Fusobacteriaceae (family)	Uterine flush and endometrial biopsy (4–12)	Illumina MiSeq (V1-2)
Bicalho et al. (2017)	Bacteroidetes (phylum) Fusobacteria (phylum)	Swab (3–12)	Illumina MiSeq (shotgun)
Sicsic et al. (2018)	Bacteroides (genus) Porphyromonas (genus) Fusobacterium (genus)	Swab and biopsy (5–10)	Pyrosequencing (V1-3)

DGGE, denaturing gradient gel electrophoresis; qPCR, quantitative polymerase chain reaction. Adapted from Jeon and Galvao (2018).

and coagulase-negative Staphylococci are among the most frequently isolated intrauterine bacteria and have been described as potential or opportunistic pathogens (Carneiro et al. 2016). Studies of identifying bacteria via 16S rRNA gene profiling by high-throughput sequencing in healthy, subclinical endometritis and clinical endometritis showed that Lactococcus, Bacillus, Solibacillus, Pseudomonas, and Arthrobacter were the five most abundant genera with little variations in the uterus of healthy and subclinical endometritis cows (Wang et al. 2018). In clinical endometritis cows, the Bacillus, Lactococcus, Solibacillus, and Fusobacterium were the top five abundant genera. The genus-level heat map analysis showed a shift in the uterine microbiota of clinical endometritis cows, with an increase in the genera Fusobacterium,

Parvimonas, Peptoniphilus, Porphyromonas, Trueperella, and Helcococcus compared with healthy and subclinical endometritis cows (Wang et al. 2018). Fusobacterium acts synergistically with Trueperella, Parvimonas, Porphyromonas, and other bacteria, to cause dysbiosis in the uterine microbiota of clinical endometritis cows. Other studies showed that Trueperella and the Gram-negative anaerobes Fusobacterium, Bacteroides, and Porphyromonas act synergistically to cause metritis and endometritis in the uterus (Bicalho et al. 2012; Prunner et al. 2014). The pathogenic bacteria (*Trueperella* spp. and *Fusobacterium* spp.) were also present in the uterus of virgin heifers and pregnant cows (Karstrup et al. 2017; Moore et al. 2017), but the co-occurrence of uterine pathogens could be considered as important in the development of uterine infection. The study of comparison of both vaginal and uterine microbiomes during the postpartum period in the dairy animals showed a shared community in the vagina and uterus. The changes associated with the development of endometritis were observed as early as 7 days postpartum, a time when vaginal and uterine microbiomes were most similar. The 16S rRNA pyrosequencing of the vaginal microbiome as early as 7 days postpartum showed at least three different microbiome types that were associated with the later development of postpartum endometritis (Miranda-Casoluengo et al. 2019). In mares, the "gold standard" to identify bacterial endometritis is tissue culture from an endometrial biopsy, but this technique is not routinely performed (Ferris 2016). More commonly, a double-guarded endometrial swab technique is performed (Ferris 2016). The most common bacteria identified by these techniques in clinical cases of bacterial endometritis in mares include *E. coli, Klebsiella pneumonia, Streptococcus equi* subsp. *zooepidemicus,* and *Pseudomonas aeruginosa* (Ferris 2016). The relationship between endometritis, successful conception, implantation, or pregnancy failure remains unclear in bitches with endometritis (Fontaine et al. 2009) as the collection of samples is a major difficulty. The study in infertile bitches found 70% to have heavy bacterial growth by *in vitro* culture, and the most common bacteria found include Group G *Streptococcus, Staphylococcus intermedius, Pasteurella multocida, E. coli,* and *Proteus mirabilis* (Fontaine et al. 2009). Although fungal endometritis is rare in mare, the most common causes of fungal endometritis in the mare are yeasts (*Candida* spp.) and molds with septated hyphae (*Aspergillus* spp.) (Scott 2018).

15.6.3 Pyometra

Pyometra is a disease in the uterus, where purulent exudate accumulates in the uterine lumen with persistent corpus luteum (PCL) (Sheldon et al. 2006). After calving, the uterine lumen is usually contaminated by the external environment bacteria, and cows that fail to clear contamination will develop various uterine diseases, in which 5% develop pyometra (Sheldon et al. 2008). Unlike cows, in bitches, it is proposed that bacteria ascend from the vagina during pro-estrus and estrus, and induce the pyometra during metoestrus by acting on the progesterone-primed endometrium (Noakes et al. 2001). It is well known that the parasite *Tritrichomonas foetus* is associated with pyometra in cattle (Rae and Crews 2006). The most frequently isolated bacteria by culture method from the cases of pyometra in cattle are *T. pyogenes* and Gram-negative bacteria, especially *F. necrophorum, Prev. melaninogenica.* Other unidentified anaerobic gram-negative bacteria act synergistically to cause pyometra in the uterus (Ruder et al. 1981). The study of microbiota in the uterus of 21 cows with pyometra targeting the V1 and V2 hypervariable regions of the 16S rRNA gene found that Fusobacteriaceae, Bacteroidaceae, Mycoplasmataceae, Pasteurellaceae, and Porphyromonadaceae accounted for approximately 85% of the OTUs observed in the uterine fluid samples (Knudsen et al. 2015). They concluded that the Actinomycetaceae family, which includes *T. pyogenes,* constituted only 1% of the total number of reads and suggested that pathogens in addition to *F. necrophorum* may be involved in the pathogenesis of pyometra (Knudsen et al. 2015). Bacterial invasion of the uterus and oviducts was studied in 21 cows with pyometra by fluorescence *in situ* hybridization (FISH) using probes targeting 16S ribosomal RNA of *F. necrophorum, Por. levii, Trueperella pyogenes,* and the overall bacterial domain. Results depicted that the *F. necrophorum* and *Por. levii* were found to invade the endometrium, especially if the endometrium was ulcerated and *T. pyogenes* did not invade the uterine tissue (Karstrup et al. 2017).

Pyometra in bitches is an important disease as it causes fatal consequences. In contrary to bovine pyometra where a reduced bacterial diversity is seen in the uterus, a higher bacterial diversity is seen in canine pyometra when compared to the healthy bitches. The 16S rRNA gene analysis showed that Proteobacteria, Pasteurellaceae, Porphyromonadaceae, and Fusobacteriaceae families were abundant in bitches with pyometra in comparison with healthy bitches (Song et al. 2017). However, culture-dependent method isolated the important pathogen of pyometra in bitches as *E. coli,* which accounted for 76.6% of the total isolates (Coggan et al. 2008). Other bacteria that have been isolated were *P. aeruginosa, Staphylococcus kloosii, Salmonella* spp., *Citrobacter diversus, Klebsiella pneumoniae* subsp. *pneumoniae, Proteus mirabilis, Streptococcus* spp., *Morganella morganii Klebsiella pneumoniae* subsp. *azanae, Staphylococcus schleiferi* subsp. *coagulans, Staphylococcus intermedius, Staphylococcus epidermidis, Streptococcus canis,* and *Corynebacterium jeikeium* (Coggan et al. 2008). The Haemophilus, Porphyromonas, and Fusobacterium could be potentially detrimental bacteria in dogs with pyometra. This group of bacteria has not been previously reported in dogs with pyometra and may have been underestimated in culture studies. Besides, the metagenomics studies along with clinical investigations will help to elucidate the pathogenesis of canine pyometra.

15.7 Conclusions

The comparison between different microbiome studies is always difficult given the different methodologies used. The microbial environment in the reproductive tract varies within and among individuals in both healthy and diseased states. The study of the reproductive microbiome is important as a small shift in the microbiome can be identified, which may not be clinically apparent but clinically important. Therefore, it is beneficial to have an understanding of the composition of the microbiome of the healthy and pathogenic reproductive tract. This will help to devise better tools for treating reproductive tract infections and enhancing reproductive efficiency.

REFERENCES

Aagaard, Kjersti, Jun Ma, Kathleen M. Antony, Radhika Ganu, Joseph Petrosino, and James Versalovic. 2014. "The Placenta Harbors a Unique Microbiome." *Science Translational Medicine* 6 (237). Doi:10.1126/scitranslmed.3008599.

Ault, Taylor B., Brooke A. Clemmons, Sydney T. Reese, Felipe G. Dantas, Gessica A. Franco, Tim P.L. Smith, J. Lannett Edwards, Phillip R. Myer, and Ky G. Pohler. 2019. "Bacterial Taxonomic Composition of the Postpartum Cow Uterus and Vagina Prior to Artificial Insemination." *Journal of Animal Science* 97 (10): 4305–13.Doi:10.1093/jas/skz212.

Baker, James M., Dana M. Chase, and Melissa M. Herbst-Kralovetz. 2018. "Uterine Microbiota: Residents, Tourists, or Invaders?" *Frontiers in Immunology* 9 (MAR). Doi:10.3389/fimmu.2018.00208.

Bicalho, M. L.S., V. S. Machado, C. H. Higgins, F. S. Lima, and R. C. Bicalho. 2017. "Genetic and Functional Analysis of the Bovine Uterine Microbiota. Part I: Metritis versus Healthy Cows." *Journal of Dairy Science* 100 (5): 3850–62. Doi:10.3168/jds.2016–12058.

Carneiro, Luísa Cunha, James Graham Cronin, and Iain Martin Sheldon. 2016. "Mechanisms Linking Bacterial Infections of the Bovine Endometrium to Disease and Infertility." *Reproductive Biology* 16 (1): 1–7. Doi:10.1016/j.repbio.2015.12.002.

Chaban, Bonnie, Shirley Chu, Steven Hendrick, Cheryl Waldner, and Janet E. Hill. 2012. "Evaluation of a Campylobacter Fetus Subspecies Venerealis Real-Time Quantitative Polymerase Chain Reaction for Direct Analysis of Bovine Preputial Samples." *Canadian Journal of Veterinary Research* 76(3): 166–73.

Clemmons, Brooke A., Sydney T. Reese, Felipe G. Dantas, Gessica A. Franco, Timothy P.L. Smith, Olusoji I. Adeyosoye, Ky G. Pohler, and Phillip R. Myer. 2017. "Vaginal and Uterine Bacterial Communities in Postpartum Lactating Cows." *Frontiers in Microbiology* 8 (JUN): 1–10. Doi:10.3389/fmicb.2017.01047.

Coggan, Jennifer Anne, Priscilla Anne Melville, Clair Motos De Oliveira, Marcelo Faustino, Andréa Micke Moreno, and Nilson Roberti Benites. 2008. "Microbiological and Histopathological Aspects of Canine Pyometra." *Brazilian Journal of Microbiology* 39 (3): 477–83. Doi:10.1590/S1517–83822008000300012.

Cornelius, A. J., Bichalho, R.C., and Cheong, S. H. 2017. "The Canine Vaginal Microbiome and Associations with Puppy Survival." *Clincal Theriogenology* 3: 424.

Davis, Meghan F., Shelley C. Rankin, Janna M. Schurer, Stephen Cole, Lisa Conti, Peter Rabinowitz, Gregory Gray, et al. 2017. "Checklist for One Health Epidemiological Reporting of Evidence (COHERE)." *One Health* 4: 14–21. Doi:10.1016/j.onehlt.2017.07.001.

Deng, Feilong, Maryanna McClure, Rick Rorie, Xiaofan Wang, Jianmin Chai, Xiaoyuan Wei, Songjia Lai, and Jiangchao Zhao. 2019. "The Vaginal and Fecal Microbiomes Are Related to Pregnancy Status in Beef Heifers." *Journal of Animal Science and Biotechnology* 10 (1): 1–13. Doi:10.1186/s40104-019-0401-2.

DeSilva, U., C. C. Lyman, G. R. Holyoak, K. Meinkoth, X. Wieneke, and K. A. Chillemi. 2019. "Canine Endometrial and Vaginal Microbiomes Reveal Distinct and Complex Ecosystems." *PLoS ONE* 14 (1): 1–17. Doi:10.1371/journal.pone.0210157.

DiGiulio, Daniel B., Benjamin J. Callahan, Paul J. McMurdie, Elizabeth K. Costello, Deirdre J. Lyell, Anna Robaczewska, Christine L. Sun, et al. 2015. "Temporal and Spatial Variation of the Human Microbiota during Pregnancy." *Proceedings of the National Academy of Sciences of the United States of America* 112 (35): 11060–65. Doi:10.1073/pnas.1502875112.

Drillich, M., O. Beetz, A. Pfützner, M. Sabin, H. J. Sabin, P. Kutzer, H. Nattermann, and W. Heuwieser. 2001. "Evaluation of a Systemic Antibiotic Treatment of Toxic Puerperal Metritis in Dairy Cows." *Journal of Dairy Science* 84 (9): 2010–17. Doi:10.3168/jds.S0022–0302(01)74644-9.

Ferris, Ryan A. 2016. "Endometritis: Diagnostic Tools for Infectious Endometritis." *Veterinary Clinics of North America - Equine Practice* 32 (3): 481–98. Doi:10.1016/j.cveq.2016.08.001.

Fontaine, E., X. Levy, A. Grellet, A. Luc, F. Bernex, H. J. Boulouis, and A. Fontbonne. 2009. "Diagnosis of Endometritis in the Bitch: A New Approach." *Reproduction in Domestic Animals* 44 (SUPPL. 2): 196–99. Doi:10.1111/j.1439-0531.2009.01376.x.

Franasiak, Jason M., and Richard T. Scott. 2017. "Endometrial Microbiome." *Current Opinion in Obstetrics and Gynecology* 29 (3): 146–52. Doi:10.1097/GCO.0000000000000357.

Gajer, Pawel, Rebecca M. Brotman, Guoyun Bai, Joyce Sakamoto, Ursel M.E. Schütte, Xue Zhong, Sara S.K. Koenig, et al. 2012. "Temporal Dynamics of the Human Vaginal Microbiota." *Science Translational Medicine* 4 (132). Doi:10.1126/scitranslmed.3003605.

Galvão, Klibs N., Rodrigo C. Bicalho, and Soo Jin Jeon. 2019. "Symposium Review: The Uterine Microbiome Associated with the Development of Uterine Disease in Dairy Cows." *Journal of Dairy Science* 102 (12): 11786–97. Doi:10.3168/jds.2019–17106.

Giudice, Linda C. 2016. "Challenging Dogma: The Endometrium Has a Microbiome with Functional Consequences!" *American Journal of Obstetrics and Gynecology* 215 (6): 682–83. Doi:10.1016/j.ajog.2016.09.085.

Hansen, Lea K., Naja Becher, Sara Bastholm, Julie Glavind, Mette Ramsing, Chong J. Kim, Roberto Romero, Jørgen S. Jensen, and NielsUldbjerg. 2014. "The Cervical Mucus Plug Inhibits, but Does Not Block, the Passage of Ascending Bacteria from the Vagina during Pregnancy." *Acta Obstetricia et Gynecologica Scandinavica* 93 (1): 102–8. Doi:10.1111/aogs.12296.

Heil, B.A., Paccamonti, D.L., and J.L. Sones. 2019. "A Role for the Mammalian Female Reproductive Tract Microbiome in Pregnancy Outcomes." *Physiol Genomics* 51(8): 390–399. Doi: 10.1152/physiolgenomics.00045.2019.

Heil, B.A., S.K. Thompson, T.A. Kearns, G.M. Davolli, G. King, and J.L. Sones. 2018. "Metagenetic Characterization of the Resident Equine Uterine Microbiome Using Multiple Techniques." *Journal of Equine Veterinary Science* 66: 111. Doi:10.1016/j.jevs.2018.05.156.

Jami, Elie, and Itzhak Mizrahi. 2012. "Composition and Similarity of Bovine Rumen Microbiota across Individual Animals." *PLoS ONE* 7 (3): 1–8. Doi:10.1371/journal.pone.0033306.

Jeon, Soo Jin, Federico Cunha, Achilles Vieira-Neto, Rodrigo C. Bicalho, Svetlana Lima, Marcela L. Bicalho, and Klibs N. Galvão.2017."Blood as a Route of Transmission of Uterine Pathogens from the Gut to the Uterus in Cows." *Microbiome* 5 (1): 109. Doi: 10.1186/s40168-017-0328-9.

Jeon, Soo Jin, and Klibs N. Galvão. 2018. "An Advanced Understanding of Uterine Microbial Ecology Associated with Metritis in Dairy Cows." *Genomics & Informatics* 16 (4): e21. Doi: 10.5808/gi.2018.16.4.e21.

Jeon, Soo Jin, Fabio S. Lima, Achilles Vieira-Neto, Vinicius S. Machado, Svetlana F. Lima, Rodrigo C. Bicalho, Jose Eduardo P. Santos, and Klibs N. Galvão. 2018. "Shift of Uterine Microbiota Associated with Antibiotic Treatment and Cure of Metritis in Dairy Cows." *Veterinary Microbiology* 214 (December 2017): 132–39. Doi: 10.1016/j.vetmic.2017.12.022.

Jeon, Soo Jin, Zhengxin Ma, Minyoung Kang, Klibs N. Galvão, and Kwangcheol Casey Jeong. 2016. "Application of Chitosan Microparticles for Treatment of Metritis and in Vivo Evaluation of Broad Spectrum Antimicrobial Activity in Cow Uteri." *Biomaterials* 110: 71–80. Doi: 10.1016/j.biomaterials.2016.09.016.

Jeon, Soo Jin, Achilles Vieira-Neto, Mohanathas Gobikrushanth, Rodolfo Daetz, Rodolfo D. Mingoti, Ana Carolina Brigolin Parize, Sabrina Lucas de Freitas, et al. 2015. "Uterine Microbiota Progression from Calving until Establishment of Metritis in Dairy Cows."*Applied and Environmental Microbiology* 81 (18): 6324–32. Doi: 10.1128/AEM.01753-15.

Karstrup, C. C., Pedersen, H. G., Jensen, T. K., and Agerholm. J. S. 2017. "Bacterial Invasion of the Uterus and Oviducts in Bovine Pyometra." *Theriogenology* 15: 93–98.

Kasimanickam, R., T. F. Duffield, R. A. Foster, C. J. Gartley, K. E. Leslie, J. S. Walton, and W. H. Johnson. 2004. "Endometrial Cytology and Ultrasonography for the Detection of Subclinical Endometritis in Postpartum Dairy Cows." *Theriogenology* 62 (1–2): 9–23. Doi:10.1016/j.theriogenology.2003.03.001.

Kehrli, Marcus E., and Jesse P. Goff. 1989. "Periparturient Hypocalcemia in Cows: Effects on Peripheral Blood Neutrophil and Lymphocyte Function." *Journal of Dairy Science* 72 (5): 1188–96. Doi:10.3168/jds.S0022–0302(89)79223-7.

Knudsen, Lif Rødtness Vesterby, Cecilia Christensen Karstrup, Hanne Gervi Pedersen, Jørgen Steen Agerholm, Tim Kåre Jensen, and Kirstine Klitgaard. 2015. "Revisiting Bovine Pyometra-New Insights into the Disease Using a Culture-Independent Deep Sequencing Approach." *Veterinary Microbiology* 175 (2–4): 319–24. Doi:10.1016/j.vetmic.2014.12.006.

Knudsen, L. R. V., Karstrup, C. C., Pedersen, H. G., Angen, Ø., Agerholm, J. S., Rasmussen, E. L.,… and Klitgaard, K. 2016. An investigation of the microbiota in uterine flush samples and endometrial biopsies from dairy cows during the first 7 weeks postpartum. *Theriogenology* 86 (2): 642–650.

Laguardia-Nascimento, Mateus, Kelly Moreira Grillo Ribeiro Branco, Marcela Ribeiro Gasparini, Silvia Giannattasio-Ferraz, Laura Rabelo Leite, Flávio Marcos Gomes Araujo, Anna Christina De Matos Salim, Jacques Robert Nicoli, Guilherme Corrêa De Oliveira, and Edel Figueiredo Barbosa-Stancioli. 2015. "Vaginal Microbiome Characterization of Nellore Cattle Using Metagenomic Analysis." *PLoS ONE* 10 (11):1–19. Doi:10.1371/journal.pone.0143294.

Machado, V. S., G. Oikonomou, M. L.S. Bicalho, W. A. Knauer, R. Gilbert, and R. C.Bicalho. 2012. "Investigation of Postpartum Dairy Cows' Uterine Microbial Diversity Using Metagenomic Pyrosequencing of the 16S RRNA Gene." *Veterinary Microbiology* 159 (3–4): 460–69. Doi:10.1016/j.vetmic.2012.04.033.

Macpherson, Margo L. 2006. "Diagnosis and Treatment of Equine Placentitis." *Veterinary Clinics of North America - Equine Practice* 22 (3): 763–76. Doi:10.1016/j.cveq.2006.08.005.

Miranda-CasoLuengo, Raúl, Junnan Lu, Erin J. Williams, Aleksandra A. Miranda-CasoLuengo, Stephen D. Carrington, Alexander C.O. Evans, and Wim G. Meijer. 2019. "Delayed Differentiation of Vaginal and Uterine Microbiomes in Dairy Cows Developing Postpartum Endometritis." *PLoS One* 14 (1): 1–23. Doi:10.1371/journal.pone.0200974.

Moore, S. G., A. C. Ericsson, S. E. Poock, P. Melendez, and M. C. Lucy. 2017. "Hot Topic: 16S RRNA Gene Sequencing Reveals the Microbiome of the Virgin and Pregnant Bovine Uterus." *Journal of Dairy Science* 100 (6): 4953–60. Doi:10.3168/jds.2017–12592.

Moreno, Inmaculada, Ettore Cicinelli, Iolanda Garcia-Grau, Marta Gonzalez-Monfort, Davide Bau, Felipe Vilella, Dominique De Ziegler, Leonardo Resta, Diana Valbuena, and Carlos Simon. 2018. "The Diagnosis of Chronic Endometritis in Infertile Asymptomatic Women: A Comparative Study of Histology, Microbial Cultures, Hysteroscopy, and Molecular Microbiology." *American Journal of Obstetrics and Gynecology* 218 (6): 602.e1–602.e16. Doi:10.1016/j.ajog.2018.02.012.

Moreno, Inmaculada, and Jason M. Franasiak. 2017. "Endometrial Microbiota—New Player in Town." *Fertility and Sterility* 108 (1): 32–39. Doi: 10.1016/j.fertnstert.2017.05.034.

Myer, Phillip R., Timothy P.L. Smith, James E. Wells, Larry A. Kuehn, and Harvey C. Freetly. 2015. "Rumen Microbiome from Steers Differing in Feed Efficiency." *PLoS One* 10 (6): 1–17. Doi:10.1371/journal.pone.0129174.

Neuendorf, Elizabeth, Pawel Gajer, Anne K. Bowlin, Patricia X. Marques, Bing Ma, Hongqiu Yang, Li Fu, et al. 2015. "Chlamydia Caviae Infection Alters Abundance but Not Composition of the Guinea Pig Vaginal Microbiota." *Pathogens and Disease* 73 (4): 1–12. Doi:10.1093/femspd/ftv019.

Noakes, D. E., Dhaliwal, G. K., and England G. C. 2001. "Cystic Endometrial Hyperplasia/Pyometra in Dogs: A Review of the Causes and Pathogenesis." *Journal of Reproduction and Fertility* 57: 395–406.

Otero, C., L. Saavedra, C. SilvaDe Ruiz, O. Wilde, A. R. Holgado, and M. E. Nader-Macías. 2000. "Vaginal Bacterial Microflora Modifications during the Growth of Healthy Cows." *Letters in Applied Microbiology* 31 (3): 251–54. Doi:10.1046/j.1365-2672.2000.00809.x.

Payne, Matthew S., and Sara Bayatibojakhi. 2014. "Exploring Preterm Birth as a Polymicrobial Disease: An Overview of the Uterine Microbiome." *Frontiers in Immunology* 5 (NOV). Doi:10.3389/fimmu.2014.00595.

Pelzer, Elise, Luisa F. Gomez-Arango, Helen L. Barrett, and Marloes Dekker Nitert.2017. "Review: Maternal Health and the Placental Microbiome." *Placenta* 54: 30–37. Doi:10.1016/j.placenta.2016.12.003.

Peng, Yu, Yi Hao Wang, Su Qin Hang, and Wei Yun Zhu. 2013. "Microbial Diversity in Uterus of Healthy and Metritic Postpartum Holstein Dairy Cows." *Folia Microbiologica* 58 (6):593–600. Doi:10.1007/s12223-013-0238-6.

Perez-Muñoz, Maria Elisa, Marie Claire Arrieta, Amanda E. Ramer-Tait, and Jens Walter. 2017. "A Critical Assessment of the 'Sterile Womb' and 'in Utero Colonization' Hypotheses: Implications for Research on the Pioneer Infant Microbiome." *Microbiome* 5 (1): 1–19.Doi:10.1186/s40168-017-0268-4.

Prince, Amanda L., Derrick M.Chu, Maxim D.Seferovic, Kathleen M.Antony, JunMa, and Kjersti M.Aagaard.2015."The Perinatal Microbiome and Pregnancy: Moving beyond the Vaginal Microbiome."*Cold Spring Harbor Perspectives in Medicine*5 (6):1–23.Doi:10.1101/cshperspect.a023051.

Prunner, Isabella, Karen Wagener, Harald Pothmann, Monika Ehling-Schulz, and Marc Drillich. 2014. "Risk Factors for Uterine Diseases on Small- and Medium-Sized Dairy Farms Determined by Clinical, Bacteriological, and Cytological Examinations." *Theriogenology* 82 (6): 857–65. Doi:10.1016/j.theriogenology.2014.06.015.

Rae, D. Owen, and John E. Crews. 2006. "Tritrichomonas Foetus." *Veterinary Clinics of North America - Food Animal Practice* 22 (3): 595–611. Doi:10.1016/j.cvfa.2006.07.001.

Rodrigues, N. F., J. Kästle, T. J. D. Coutinho, A. T. Amorim, G. B. Campos, V. M. Santos, L. M. Marques, J. Timenetsky, and S. T. de Farias. 2015. "Qualitative Analysis of the Vaginal Microbiota of Healthy Cattle and Cattle with Genital-Tract Disease." *Genetics and Molecular Research* 14 (2): 6518–28. Doi:10.4238/2015.June.12.4.

Rota, A., E. Calicchio, S. Nardoni, F. Fratini, V. V. Ebani, M. Sgorbini, D. Panzani, F. Camillo, and F. Mancianti. 2011. "Presence and Distribution of Fungi and Bacteria in the Reproductive Tract of Healthy Stallions." *Theriogenology* 76 (3): 464–70. Doi:10.1016/j.theriogenology.2011.02.023.

Ruder, C. A., R. G. Sasser, R. J. Williams, J. K. Ely, R. C. Bull, and J. E. Butler. 1981. "Uterine Infections in the Postpartum Cow. II. Possible Synergistic Effect of Fusobacterium Necrophorum and Corynebacterium Pyogenes." *Theriogenology* 15 (6): 573–80. Doi:10.1016/0093-691X(81)90060-1.

Sandal, Indra, and Thomas J. Inzana. 2010. "A Genomic Window into the Virulence of Histophilus Somni." *Trends in Microbiology* 18 (2): 90–99. Doi:10.1016/j.tim.2009.11.006.

Santos, T. M.A., R. Gilbert, and R. C. Bicalho. 2011. "Metagenomic Analysis of the Uterine Bacterial Microbiota in Healthy and Metritic Postpartum Dairy Cows." *Journal of Dairy Science* 94 (1): 291–302. Doi:10.3168/jds.2010-3668.

Scott, C. J. 2018. "A Review of Fungal Endometritis in the Mare." *Equine Veterinary Education*, 1–5. Doi:10.1111/eve.13010.

Sheldon, I. Martin, Gregory S. Lewis, Stephen LeBlanc, and Robert O. Gilbert. 2006. "Defining Postpartum Uterine Disease in Cattle." *Theriogenology* 65 (8):1516–30. Doi: 10.1016/j.theriogenology.2005.08.021.

Sheldon, I. Martin, Erin J. Williams, Aleisha N.A. Miller, Deborah M. Nash, and Shan Herath. 2008. "Uterine Diseases in Cattle after Parturition." *Veterinary Journal* 176 (1): 115–21. Doi:10.1016/j.tvjl.2007.12.031.

Sicsic, Ron, Tamir Goshen, Rahul Dutta, Noa Kedem-Vaanunu, Veronica Kaplan-Shabtai, Zohar Pasternak, Yuval Gottlieb, Nahum Y. Shpigel, and Tal Raz. 2018. "Microbial Communities and Inflammatory Response in the Endometrium Differ between Normal and Metritic Dairy Cows at 5–10 Days Post-Partum." *Veterinary Research* 49 (1): 1–15. Doi:10.1186/s13567-018-0570-6.

Song, Young Gang, Robin B. Guevarra, Jun Hyung Lee, Suphot Wattanaphansak, Bit Na Kang, Hyeun Bum Kim, and Kun Ho Song. 2017. "Comparative Analysis of the Reproductive Tract Microbial Communities in Female Dogs with and without Pyometra through the 16S RRNA Gene Pyrosequencing." *Japanese Journal of Veterinary Research* 65 (4): 193–200. Doi:10.14943/jjvr.65.4.193.

Swartz, Jeffrey D., Medora Lachman, Kelsey Westveer, Thomas O'Neill, Thomas Geary, Rodney W. Kott, James G. Berardinelli, et al. 2014. "Characterization of the Vaginal Microbiota of Ewes and Cows Reveals a Unique Microbiota with Low Levels of Lactobacilli and Near-Neutral PH." *Frontiers in Veterinary Science* 1 (OCT): 1–10. Doi:10.3389/fvets.2014.00019.

The Human Microbiome Project Consortium, Huttenhower, C., Gevers, D. et al. 2012. Structure, function and diversity of the healthy human microbiome. *Nature* 486:207–214.

Uchihashi, M., I. L. Bergin, C. M. Bassis, S. A. Hashway, D. Chai, and J. D. Bell. 2015. "Influence of Age, Reproductive Cycling Status, and Menstruation on the Vaginal Microbiome in Baboons (Papio Anubis)." *American Journal of Primatology* 77 (5): 563–78. Doi:10.1002/ajp.22378.

Verstraelen, Hans, Ramiro Vilchez-vargas, Fabian Desimpel, Ruy Jauregui, Nele Vankeirsbilck, Steven Weyers, Rita Verhelst, Petra De Sutter, Dietmar H. Pieper, and Tom Van De Wiele. 2016. "Characterisation of the Human Uterine Microbiome in Non-Pregnant Women through Deep Sequencing of the V1–2 Region of the 16S RRNA Gene." 1–23. Doi:10.7717/peerj.1602.

Wang, Meng Ling, Ming Chao Liu, Jin Xu, Li Gang An, Jiu Feng Wang, and Yao Hong Zhu. 2018. "Uterine Microbiota of Dairy Cows with Clinical and Subclinical Endometritis." *Frontiers in Microbiology* 9 (NOV): 1–11. Doi:10.3389/fmicb.2018.02691.

Wee, Bryan A., Mark Thomas, Emma Louise Sweeney, Francesca D. Frentiu, Melanie Samios, Jacques Ravel, Pawel Gajer, et al. 2018. "A Retrospective Pilot Study to Determine Whether the Reproductive Tract Microbiota Differs between Women with a History of Infertility and Fertile Women." *Australian and New Zealand Journal of Obstetrics and Gynaecology* 58 (3): 341–48. Doi:10.1111/ajo.12754.

Weng, Shun-long, Chih-min Chiu, Feng-mao Lin, Wei-chih Huang, Chao Liang, Ting Yang, Tzu-ling Yang, et al. 2014. "Bacterial Communities in Semen from Men of Infertile Couples : Metagenomic Sequencing Reveals Relationships of Seminal Microbiota to Semen Quality."9 (10). Doi:10.1371/journal.pone.0110152.

Wickware, Carmen L., Timothy A. Johnson, and Jennifer H. Koziol. 2020. "Composition and Diversity of the Preputial Microbiota in Healthy Bulls." *Theriogenology* 145: 231–37. Doi:10.1016/j.theriogenology.2019.11.002.

Williams, Erin J., Deborah P. Fischer, Dirk U. Pfeiffer, Gary C. W. England, David E. Noakes, Hilary Dobson, and I. Martin Sheldon. 2005. "Clinical Evaluation of Postpartum

Vaginal Mucus Reflects Uterine Bacterial Infection and the Immune Response in Cattle." *Theriogenology* 63: 102–17. Doi:10.1016/j.theriogenology.2004.03.017.

Xia, Y. X., Cornelius, A. J., Donnelly, C. G., Bicalho, R. C., Cheong, S. H., and Sones, J. L. 2017. "Metagenomic Analysis of the Equine Placental Microbiome." *Clinical Theriogenology* 9: 452.

Yildirim, Suleyman, Carl J Yeoman, Sarath Chandra Janga, Susan M. Thomas, Mengfei Ho, Steven R. Leigh, Primate Microbiome Consortium, Bryan A. White, Brenda A. Wilson, and Rebecca M. Stumpf. 2014. "Primate Vaginal Microbiomes Exhibit Species Specificity without Universal Lactobacillus Dominance." *The ISME Journal* 1–14. Doi:10.1038/ismej.2014.90.

16

Community Structures of Fecal Actinobacteria in Animal Gastrointestinal System

Selvanathan Latha and D. Dhanasekaran
Bharathidasan University

CONTENTS

16.1 Actinobacteria as Typical Inhabitants of Animal GI Tract

Animals have coevolved with microbial components that outnumber their own body cells and are known to provide genes which contribute significant functions in the colonized organs (De Jesus-Laboy et al., 2011). Primarily, the microbiota of gastrointestinal (GI) tract comprising resident and transient floras play a crucial role in metabolic, nutritional, physiological, and immunological processes related to the well-being of animals (Figure 16.1). Among many important functions, the gut microbiota can convert feedstuffs into microbial biomass and the fermentation end products can be utilized by the animal host (Kong et al., 2010). A huge number of microbial types are prevailing in animal gut; however, most of them are uncultivable; and also, the relationship between the gut microbiota and the host is remaining unclaimed due to the complexity of internal ecological systems. Moreover, the number and dominance of microbial species that persist and colonize the GI tract usually fluctuate between individuals and are determined by a combination of host-specific and environmental factors such as age, gender, diet, and genotype.

The composition of animal gut microbiota has been extensively studied over recent decades, which reveals the predominance of bacterial genera such as *Lactobacillus* (*L. acidophilus*, *L. rhamnosus*, and *L. salivarius*), *Lactococcus*, and *Bacillus*. Next to bacteria, actinobacteria are one of the imperative gut floras of animals and are well known to form an intimate association with vertebrates and invertebrates. Most of the animal gut actinobacteria (e.g., *Bifidobacterium*, *Streptomyces*, and *Rhodococcus*) are beneficial and have been reported markedly in the literature (Kim et al., 2005; Tan et al., 2009; Wu et al., 2010). The symbiotic interactions of actinobacteria with animals are indispensable for their survival and reproduction because they play a decisive role in nutrition, detoxification of certain compounds, growth performance, and defense against pathogenic bacteria (Scott et al., 2008; Visser et al., 2012).

The protection of European beewolf (*Philanthus triangulum*, Hymenoptera, Crabronidae) offspring by *Streptomyces* strains against pathogens (Kaltenpoth et al., 2006) and growth suppression of parasitic fungus *Escovopsis* sp. through the production of specific antibiotics by ant-associated actinobacterial strains (*Streptomyces* from leaf-cutting ant *Acromyrmex octospinosus* and *Pseudonocardia* from attine ant *Apterostigma dentigerum*) are some of the convincing evidences for the mutualistic association of actinobacteria (Haeder et al., 2009; Oh et al., 2009). In this view, knowledge on actinobacterial genera and species as well as their frequency in animal GI tract is very essential in understanding of symbiosis between actinobacteria and their respective host animals. Furthermore, the enumeration of diverse and novel actinobacterial strains propounds a theoretical guide for exploitation and utilization of beneficial actinobacterial resources at an elevated level.

In general, culture-dependent methods, i.e., the traditional microbiological approaches applying selective media and isolation practices, are much suited for the investigation of actinobacteria in animal GI tract. For instance, a study conducted by Mead (1989) on the bacterial flora of chicken's ceca demonstrates that 9% of their total culturable bacteria were composed of *Bifidobacterium* spp. Although to date a vast number of animal intestinal actinobacteria have been identified by means of culture-dependent methods, the rate of uncultivable gut actinobacteria remains high. Hence, researchers look forward in the identification of uncultured gut actinobacteria since they might offer novel strains with profitable biological potentials. As a result, molecular approaches, including hybridization and G+C% profiling of 16S rDNA sequences,

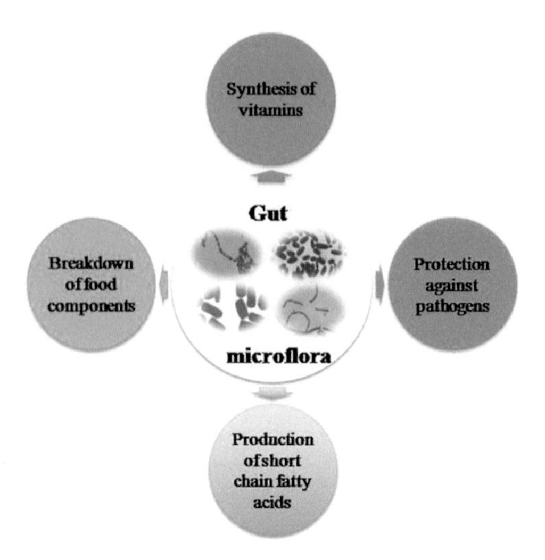

FIGURE 16.1 Key functions of gut microflora in animal system.

have been recently commenced to scrutinize the actinobacterial diversity in animal gut.

The retrieval of 16S rRNA gene sequences from DNA isolated from the cecal mucosa of broiler chickens represents the existence of a typical gut inhabitant *Bifidobacterium infantis* (Zhu et al., 2002). Similarly, analysis of bacterial rDNA sequences from the intestinal contents of Ross-hybrid chickens implies that their cecum and ileum were inhabited by *Bifidobacterium* spp. with high G+C content. Further, a study on 16S rRNA gene sequence-based comparison of bacterial communities in grass carp (*Ctenopharyngodon idellus*) intestinal contents and fish culture-associated environments exemplifies that actinobacteria were the prevalent members among their intestinal bacterial communities (Wu et al., 2012). Also, a combination of culture-based techniques and 16S rDNA sequencing applied by Wu et al. (2010) for the investigation of microbial communities in intestinal contents and mucosal layers of yellow catfish (*Pelteobagrus fulvidraco*) describes the inhabitancy of *Microbacterium lacticum* and *Corynebacterium* sp.

16.2 Diversity and Biological Attributes of Animal Fecal Actinobacteria

"Feces," also known as stool, manure, dung, litter, or dropping, is a solid waste product of an animal digestive tract discharged through the anus or cloaca during a process called "defecation." Since animal feces is a good source of plant nutrients and organic matters, it stimulates the growth of crops and grasses, which can be further used as a crop fertilizer and soil amendment in agriculture as well as feedstock for energy production. It also encourages the soil microbial activity by improving plant nutrition and by promoting trace minerals supply. The macro- and micronutrients of feces such as nitrogen, phosphate, and potassium are most essential to sustain plant growth and agriculture—especially, nitrogen assists the plants to grow greener and stronger.

Each animal feces has its own qualities and so requires different application rates while using as a fertilizer. For example, sheep manure is high in nitrogen and potash, whereas pig manure is comparatively low in both. The chicken litter is concentrated with nitrogen and phosphate and is valued for both properties. Similar to nutrient levels, every animal feces has its unique and complex fecal microbiota (i.e., healthy bacterial flora habitually residing the GI tract of animal by digesting the feed and excreted via feces), and it is well known that fecal microbiome are a good reflection of intestinal microflora (Campbell et al., 1997). Moreover, earlier studies on fecal microbes are mostly associated with diversity analysis and verification of gut microbial changes due to the imbalance between beneficial and pathogenic bacteria.

As actinobacteria are one of the animal intestinal microbes, a number of studies have been carried out to characterize the cultivable animal fecal actinobacteria in order to measure the diversity of gut actinobacteria and their biological traits. In 1981, Mara and Oragui isolated *Micromonospora* spp. and *Streptomyces* spp. from feces of domestic avian species (ducks, geese, and turkeys) and seabird (seagulls), whereas a novel actinobacterium *Brevibacterium yomogidense* MN-6-a^T was recovered by Tonouchi et al. (2013) from the soil conditioner made by poultry manure in Yomogida village, Aomori, Japan. Alike, Chen et al. (2014) obtained 28 genera of actinobacteria from the feces of six bird species, namely, *Larus ridibundus*, *Pavo cristatus*, *Aceros undulatus*, *Grus japonensis*, *Cygnus*, and *Struthio camelus* (Table 16.1). Among them, *Streptomyces*, *Rhodococcus*, and *Arthrobacter* were the prime genera with the widest distribution. Further, their antimicrobial activity results indicate that not only the diversity of bird fecal actinobacteria was very complex but also the bioactivity was wide and high.

In 2012, Gouliamova et al. identified three *Thermoactinomyces sacchari* strains (IMA Z3, IMA Z9, and IMA Z10) with a strong proteolytic activity from the feces of mammals hippopotamus, and elephant. At the same time, Cao et al. (2012) isolated 110 cultural actinobacteria from the feces of *Panthera tigris tigris* living in Yunnan Safari Park, Kunming, China, which mainly belonged to 12 genera and 10 different families. A similar study of Jiang et al. (2013a) reveals the detection of 31 genera from the feces of carnivorous, omnivorous, and phytophagous animals, which live in Yunnan Wild Animal Park,

Kunming, China (Table 16.2). Comparable number of actinobacterial genera (29 genera) were recorded in the feces of six herbivore species, namely, *Cervus nippon*, *Rhizomys sinensis*, *Rhinoceros sondaicus*, *Vicugna pacos*, *Connochaetes taurinus*, and *Elephas maximus*(Table 16.3), in which the members of three genera, i.e., *Streptomyces*, *Rhodococcus*, and *Microbacterium*, and the order Micrococcales were dominant (Jiang et al., 2013b).

A study on diversity and bioactivity of cultivable fecal actinobacteria from 31 species of animals comprising primate, mammality, birds, amphibian, and insect; perissodactyla, artiodactyla, and ruminant; and carnivore, herbivore, and omnivore discloses the identification of 35 actinobacterial genera, including a new genus *Enteractinococcus*. The members of *Rhodococcus* were predominant next to *Streptomyces*, and a few of them had high antitumor and antimicrobial activities. Besides, the recovery of more than 50 secondary metabolites from the fecal actinobacteria illustrates that they could possibly be a new source for discovering drug leader, agricultural chemicals, and other industrially useful molecules (Jiang et al., 2013c).

In 2013, Lu et al. isolated an oil-producing actinobacterium *Streptomyces* sp. S161 from the feces of sheeps (*Ovis aries*) and reported it as the first strain producing biodiesel directly from starch. They also stated that the strain would be useful for the direct conversion of renewable lignocellulose into biodiesel due to its cellulase and xylanase activities. Moreover, a study performed by Wang et al. (2014) with the same strain shows that it has the ability to produce ionophores antibiotics nactins (nonactin, monactin, dinactin, trinactin, and tetranactin),which are commonly fed to ruminant animals for the improvement of feed conversion efficiency (FCE). In 2014, Tan et al. made an attempt to isolate actinobacteria from ruminant feces, specifically from sheeps (*O. aries*) and cows (*Bos taurus*) in which most of the isolated actinobacteria belonged to the genera *Streptomyces*, *Amycolatopsis*, *Micromonospora*, and *Cellulosimicrobium*. Further, all the strains displayed

TABLE 16.1

Composition of Fecal Actinobacteria in Birds

S. No.	Scientific Name	Common Name	Generic Abundance	Name of the Actinobacteria
1.	*A. undulates*	Wreathed hornbill	8	*Arthrobacter*, *Corynebacterium*, *Dietzia*, *Gordonia*, *Nocardia*, *Rhodococcus*, *Streptomyces*, and *Yaniella*
2.	*S. camelus*	Ostrich	4	*Arthrobacter*, *Oerskovia*, *Rhodococcus*, and *Streptomyces*
3.	*C. cygnus*	Duck	5	*Cellulosimicrobium*, *Micromonospora*, *Rhodococcus*, *Streptomyces*, and *Verrucosispora*
4.	*P. cristatus*	Peafowl/peacock	18	*Arthrobacter*, *Brevibacterium*, *Cellulosimicrobium*, *Curtobacterium*, *Dietzia*, *Gordonia*, *Isoptericola*, *Janibacter*, *Kineococcus*, *Kocuria*, *Leucobacter*, *Microbacterium*, *Nocardiopsis*, *Oerskovia*, *Pseudonocardia*, *Rhodococcus*, *Sanguibacter*, and *Streptomyces*
5.	*L. ridibundus*	Black-headed gull	8	*Arthrobacter*, *Blastococcus*, *Microbacterium*, *Oerskovia*, *Plantibacter*, *Promicromonospora*, *Pseudoclavibacter*, and *Streptomyces*
6.	*G. japonensis*	Red-crowned crane	9	*Arthrobacter*, *Blastococcus*, *Cellulosimicrobium*, *Microbacterium*, *Mycobacterium*, *Nocardia*, *Rhodococcus*, *Streptomyces*, and *Yaniella*

Source: Chen et al. (2014).

TABLE 16.2

Composition of Fecal Actinobacteria in Carnivorous, Omnivorous, and Phytophagous Animals

S. No.	Scientific Name	Common Name	Generic Abundance	Name of the Actinobacteria
1.	*P. t. altaica*	Manchurian tiger	9	*Arthrobacter*, *Enteractinococcus*, *Microbacterium*, *Nocardia*, *Oerskovia*, *Promicromonospora*, *Saccharomonospora*, *Streptomyces*, and *Yaniella*
2.	*P. tigris*	Bengal tiger	12	*Arthrobacter*, *Corynebacterium*, *Dietzia*, *Enteractinococcus*, *Kocuria*, *Microbacterium*, *Nocardia*, *Nocardiopsis*, *Oerskovia*, *Promicromonospora*, *Saccharomonospora*, and *Streptomyces*
3.	*A. melanoleuca*	Giant panda	13	*Agrococcus*, *Arthrobacter*, *Cellulomonas*, *Cellulosimicrobium*, *Janibacter*, *Micrococcus*, *Micromonospora*, *Mycobacterium*, *Oerskovia*, *Patulibacter*, *Rhodococcus*, *Streptomyces*, and *Verrucosispora*
4.	*V. zibetha*	Zibet	10	*Arthrobacter*, *Cellulosimicrobium*, *Curtobacterium*, *Isoptericola*, *Kocuria*, *Microbacterium*, *Micrococcus*, *Rhodococcus*, *Sanguibacter*, and *Streptomyces*
5.	*U. thibetanus*	Asiatic black bear	8	*Dietzia*, *Gordonia*, *Microbacterium*, *Nocardiopsis*, *Promicromonospora*, *Rhodococcus*, *Saccharomonospora*, and *Streptomyces*
6.	*C. nippon*	Shansi sika	11	*Agrococcus*, *Arthrobacter*, *Citricoccus*, *Kocuria*, *Leucobacter*, *Microbacterium*, *Nocardiopsis*, *Rhodococcus*, *Salinibacterium*, *Streptomyces*, and *Tsukamurella*
7.	*V. pacos*	Vicuna	9	*Arthrobacter*, *Cellulosimicrobium*, *Dietzia*, *Isoptericola*, *Kocuria*, *Nocardiopsis*, *Rhodococcus*, *Saccharomonospora*, and *Streptomyces*

Source: Jiang et al. (2013a).

TABLE 16.3

Composition of Fecal Actinobacteria in Herbivorous Animals

S. No.	Scientific Name	Common Name	Generic Abundance	Name of the Actinobacteria
1.	*V. pacos*	Alpaca	11	*Arthrobacter*, *Cellulosimicrobium*, *Dietzia*, *Gordonia*, *Isoptericola*, *Kocuria*, *Microbacterium*, *Nocardiopsis*, *Rhodococcus*, *Saccharomonospora*, and *Streptomyces*
2.	*C. taurinus*	Common wildebeest	5	*Citricoccus*, *Microbacterium*, *Micrococcus*, *Rhodococcus*, and *Streptomyces*
3.	*C. Nippon*	Sika deer	15	*Actinocorallia*, *Agrococcus*, *Arthrobacter*, *Citricoccus*, *Isoptericola*, *Kocuria*, *Leucobacter*, *Microbacterium*, *Mycobacterium*, *Nocardiopsis*, *Promicromonospora*, *Rhodococcus*, *Salinibacterium*, *Streptomyces*, and *Tsukamurella*
4.	*R. sondaicus*	Rhino	6	*Dietzia*, *Nocardiopsis*, *Promicromonospora*, *Rhodococcus*, *Saccharomonospora*, and *Streptomyces*
5.	*E. maximus*	Indian elephant	10	*Arthrobacter*, *Cellulomonas*, *Cellulosimicrobium*, *Leucobacter*, *Microbacterium*, *Micromonospora*, *Promicromonospora*, *Rhodococcus*, *Streptomyces*, and *Verrucosispora*
6.	*R. sinensis*	Chinese bamboo rat	13	*Agrococcus*, *Arthrobacter*, *Brachybacterium*, *Corynebacterium*, *Dietzia*, *Gordonia*, *Labedella*, *Microbacterium*, *Oerskovia*, *Rhodococcus*, *Sanguibacter*, *Streptomyces*, and *Williamsia*

Source: Jiang et al. (2013b).

growth on medium containing pectin, cellulose, or xylan by using them as sole carbon sources; however, strains belonging to the genus *Streptomyces* only exhibited antibacterial and antifungal activities.

Previous reports on animal fecal actinobacteria also highlighted several key points to the researchers which are as follows: (1) Diversity of cultivable actinobacteria is rich in animal feces like those in soils, oceans, extreme environments, and plants but differed in their composition, (2) members of the genus *Streptomyces* are preponderant microbes present almost in all animal feces, (3) functions of actinobacteria in GI tract of animals may be determined by their bioactive compounds production, (4) secondary metabolites with bioactivities produced by fecal actinobacteria (except pathogens) are normally not toxic or less toxic to their hosts, and (5) fresh feces has to be collected from healthy animals and the entire research work should be carried out under the strict sterile conditions to avoid any possibility of pathogen interference. The above-mentioned key points are the distinct features of fecal actinobacterial communities, which could make them unique from actinobacteria of different ecological habitats.

16.3 Community Structures of Fecal Actinobacteria in Chicken and Goat

Though actinobacteria are rich in animals, only a very few reports are available on the diversity of gut actinobacteria in feces of chicken and goat. Mara and Oragui (1981) reported the incidence of *Streptomyces* spp. in the feces of chicks, and *R. coprophilus*, *Micromonospora* spp., and *Streptomyces* spp. in hens. Furthermore, they stated that the ratio of *R. coprophilus* to other actinobacteria (*Micromonospora* spp. and *Streptomyces* spp.) would offer a positive index to discriminate dairy farm effluents from nonanimal fecal contamination. Likewise, Tan et al. (2009) described the widest distribution of *Oerskovia* spp. with the dominance of *O. turbata*-like isolates as well as *Nocardiopsis* spp. and *Streptomyces* spp. in feces of goats, and suggested that their physiological characteristics play a significant role in GI ecology of goats.

Aside from traditional culture methods, the advanced culture-independent molecular analysis provides a compelling support for the presence of gut actinobacteria in feces of chicken and goat. In 2011, De Jesus-Laboy et al. depicted the existence of actinobacteria belonging to seven different families in the feces of domestic goats using high-density universal 16S rRNA microarray analysis. Besides, a study on assessment of fecal microbiota composition of broilers and hens by 16S rRNA gene pyrosequencing represents the occurrence of *Corynebacterium* sp. and *Brevibacterium* sp. as a part of core chicken fecal microbiome (Videnska et al., 2014).

Taken as a whole, the critical analyses on culture-dependent and culture-independent recognition of gut or fecal microbiota of animals undeniably specify the extensive assemblage of a wide variety of actinobacteria within different animal species. Hence, the animals could also be used as a prominent resource for the isolation of novel as well as the existing actinobacterial strains and eventual screening for proficient probiotic properties. However, proper isolation and identification are constantly needed to achieve the revenue of probiotic actinobacteria originated from animals.

16.4 Population Density of Cultivable Actinobacteria in Animal Feces

The GI tract of production animal is a complex and dynamic ecosystem comprising a diverse collection of microorganisms known as gut flora, gut microbiota, or intestinal microbiota which include resident members as well as transient passengers introduced from the environment. Most of them coevolved with the host animal in symbiotic relationship to establish and ensure gut homeostasis. Moreover, they improve the health and welfare of animals by contributing nutrient digestion and absorption, protection against pathogens, and development of immune system (De Jesus-Laboyet al., 2011; Kamada et al., 2013). The dominant bacterial genera inhabiting the GI tract of different animal species are *Lactobacillus*, *Streptococcus*, *Bacteroides*, *Enterococcus*, *Clostridia*, *Faecalibacterium*, *Eubacterium*, *Ruminococcus*, *Peptococcus*, *Peptostreptococcus*, *Bifidobacterium*, *Streptomyces*, and *Nocardiopsis* belonging to the phyla Firmicutes, Bacteroidetes, Proteobacteria, and Actinobacteria.

In general, the gut microbial communities of animals are host-specific evolving throughout an individual's lifetime, and susceptible to both exogenous and endogenous influential factors, including age, diet, state of health, and antibiotic use (Sekirov et al., 2010). It is well known that animal gut microbial composition could be easily encountered by their fecal microbiome, since the gut flora are regularly shed in feces and occupy 60% of its dry mass. Also, the critical analyses of animal fecal microbiota using integrated molecular and cultivation-based approaches proved the feasible recovery of gut microbes, especially actinobacteria in feces (De Jesus-Laboy et al., 2011; Gouliamova et al., 2012; Lu et al., 2013; Tan et al., 2014). However, understanding of animal gut-associated actinobacterial communities is still inadequate. In this perspective, the present study has been initiated with isolation and identification of actinobacteria from feces of chickens and goats in order to explore the population density and diversity of cultivable gut actinobacteria in domestic animals using cultivation-based approaches.

The study results reveal that actinobacterial density of each fecal sample was different from one another with the range of 2.48–6.11 log CFU gfw^{-1}. Maximum actinobacterial population was registered in the feces of indigenous chickens from Sooriyur, broiler chickens from Gundur, indigenous goats from Nallur, and farm goats from S.M.S. farm-I, while no actinobacteria were recorded in broiler chickens from Sooriyur, Mandaiyur, and Puthutheru as well as in farm goats from Nitherson farm-II,III and S.M.S. farm-II,III. The results further disclose the abundance of cultivable animal fecal actinobacteria in indigenous chickens and goats as compared to broiler chickens and farm goats (Figures16.2 and 16.3). The major difference in actinobacterial count between indigenous and farm animals could be due to a variability in gut microbial determinants such as mode of delivery, initial microbial communities, nutritional intake and food habits, host genetics, and immune system together with excessive hygienic conditions, use of disinfectants, and continuous nontherapeutic applications of antibiotics in farm animals.

A number of reports are available on the isolation of actinobacteria from the feces of different kinds of animals (including species of herbivorous, carnivorous, and omnivorous animals) with reference to their gut actinobacterial assessment (Cao et al., 2012; Lu et al., 2013; Jiang et al., 2013a, 2013b, 2013c, 2014; Chen et al., 2014; Wang et al., 2014); however, information about actinobacterial isolation from chicken and goat feces is very minimal. In 1981, Mara and Oragui investigated the feces of 14-day-old chicks obtained from the animal house of Department of Agricultural Sciences, Leeds University, West Yorkshire, for the occurrence of *R. coprophilus* and associated actinobacteria, where they found *Streptomyces* spp. alone with 2.65 log CFU gfw^{-1} (4.5×10^{2} CFU gfw^{-1}). At the same time, their extensive evaluation on feces of hens collected from the Brookland farm, West Yorkshire, England, illustrates the presence of *Micromonospora* spp. (4.11 log CFU gfw^{-1}, i.e., 1.3×10^{4} CFU gfw^{-1}), *R. coprophilus* (3.80 log CFU gfw^{-1}, i.e., 6.3×10^{3} CFU gfw^{-1}), and *Streptomyces* spp. (3.11 log CFU gfw^{-1}, i.e., 1.3×10^{3} CFU gfw^{-1}). Alike, Tan et al. (2009) endeavored to characterize the culturable actinobacterial communities in the feces of healthy goats roaming the pastures in Runan Town, Henan Province, China, and reported a total actinobacterial population of 7.15 log CFU gfw^{-1} (1.4×10^{7} CFU gfw^{-1}), in which 3.52 log CFU gfw^{-1} (3.3×10^{3} CFU gfw^{-1}) was occupied by streptomycete-like strains.

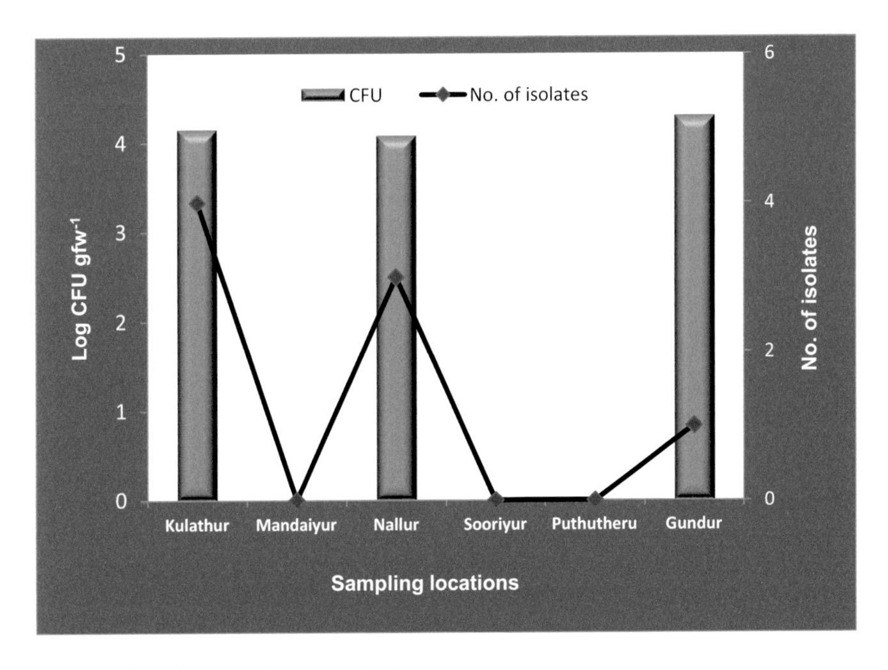

FIGURE 16.2 Total actinobacterial population in feces of broiler chickens.

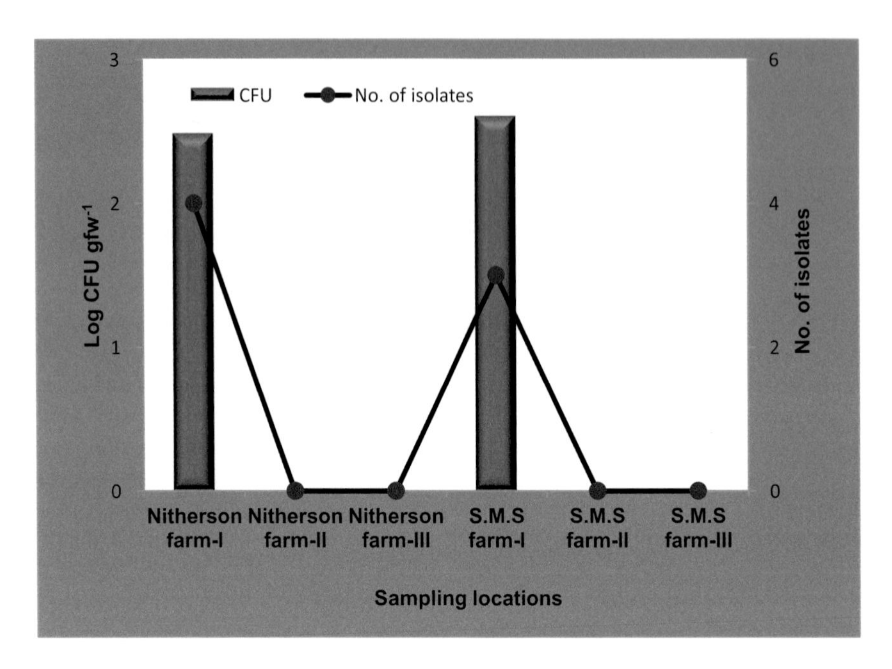

FIGURE 16.3 Total actinobacterial population in feces of farm goats.

16.5 Diversity and Distribution Pattern of Animal Fecal Actinobacteria

The feces of domestic animals examined in the present study harbor taxonomically diverse actinobacterial communities. In particular, highest diversity of four actinobacterial genera (i.e., *Streptomyces*, *Nocardiopsis*, *Kitasatospora*, and *Actinomadura*) was found in chickens, while three different genera (i.e., *Streptomyces*, *Nocardiopsis*, and *Saccharopolyspora*) were obtained from goats mainly belonging to the families Streptomycetaceae, Nocardiopsaceae, Thermomonosporaceae, and Pseudonocardiaceae. Although the genera *Streptomyces* and *Nocardiopsis* were recovered from the feces of both animals, *Streptomyces* was the most frequently isolated genus with different morphology that might be due to its ubiquitous nature, and was corroborated by previous investigations. A study conducted by Mara and Oragui (1981) demonstrates the occurrence of actinobacteria in animal feces, especially *Streptomyces* spp. (Streptomycetaceae) in chicks, and *Micromonospora* spp. (Micromonosporaceae), *R. coprophilus* (Nocardiaceae), and *Streptomyces* spp. (Streptomycetaceae) in hens. Similarly, Tan et al. (2009) reported the extensive distribution of *Oerskovia* spp. along with *Nocardiopsis* spp. and *Streptomyces* spp. in feces of goats belonging to the families Cellulomonadaceae, Nocardiaceae, and Streptomycetaceae, respectively.

Additionally, the culture-independent molecular analyses on fecal microbial constituents of domestic animals confirm the existence of diverse as well as novel or rare actinobacteria. For instance, Videnska et al. (2014) established the incidence of *Corynebacterium* sp. (Corynebacteriaceae) and *Brevibacterium* sp. (Brevibacteriaceae) in the feces of egg-laying hens and broilers originating from four different Central European countries by pyrosequencing the V3 and V4 variable region of 16S rRNA genes. Further, an investigation on fecal bacterial community structures of domestic and feral goats via high-density microarray (16S rDNA PhyloChip) analysis provides evidence for the presence of actinobacteria in domestic goats belonging to the families Acidimicrobiaceae,Micromonosporaceae,Microthrixaceae,Promicromonosporaceae,Gordoniaceae,Kineosporiaceae, and Thermomonosporaceae (De Jesus-Laboyet al., 2011). The above manifestations undoubtedly signify that the chicken and goat feces are one of the vital resources exhibiting a great pool of actinobacterial diversity with potential interest for probiotics and drug discovery.

Though direct comparison of cultivable fecal actinobacteria reveals their occurrence in all the indigenous animals, in few farm animals, they greatly differ in abundance and diversity. In general, the actinobacterial abundance does not necessarily correspond to the genus diversity, which can be substantiated from the present observations as the actinobacterial communities of chicken feces contributed highest diversity with the lowest number of total isolates. Moreover, the dissimilarity in actinobacterial abundance and diversity of the tested animal feces may be viewed in the background that various physical, chemical, and geographical factors—including animal species and their GI tract length, habitats, geographical areas, sources of diet, culture media, chemical and physical pretreatment of the samples, use of chemical inhibitors or antibiotics, growth conditions, and recognition of candidate colonies on primary isolation plate—can possibly affect the recovery of animal gut actinobacteria in feces.

Certainly, the cultivation-based methods are highly selective due to the choice of media and culture conditions (Imhoff and Stohr, 2003). The SCA medium used in the present study has been reported as a suitable medium for the isolation of actinobacteria from various natural environments (e.g., soil, water, and organic matter as well as less-explored ecosystems such as mountain, forest, and desert) because the development of bacterial and fungal colonies was usually very much suppressed that allows only the growth of actinobacteria (Meena et al., 2013; Leeet al., 2014). However, the reports on enumeration of fecal actinobacteria using different media such as SCA,CM medium,M3, MM3, M4, M5, M6, YIM 7, YIM 47, YIM 171, YIM 212, YIM 213, and YIM 601 agar illustrate the fact that nutrients present in the individual medium may have multifarious effects on the isolation of animal-associated actinobacteria, and thus, various culture media may therefore be appropriate when examining different types of animal feces (Mara and Oragui, 1981; Tan et al., 2009; Jiang et al., 2013a).

Nevertheless, the total actinobacterial density and genera recorded in this study were comparable with those inhabiting the soil, lake sediment, plant interior, and animal gut (Xu et al., 1996; Tian et al., 2007). Also, it was hypothesized that the indigenous chickens and goats would have obtained their remarkable gut actinobacterial composition from their mothers, natural environmental factors (e.g., air, soil, and water) as well as through the ingestion of herbal feeds (which are known to be naturally occurring niches of actinobacteria),and adapted physiologically becoming biologically active with eventual colonization in the gut during their lifetime to reach high proportions encountered in fecal droppings. Moreover, the nonexistence of pathogenic actinobacteria (e.g., *Corynebacterium*, *Mycobacterium*, and *Nocardia*) in feces of chickens and goats indicates that the actinobacteria recovered in the study were normal and nonpathogenic, which might have beneficial associations with their GI tract; however, it has to be ascertained using a range of *in vitro* and *in vivo* analyses.

16.6 Conclusions

The feces of chicken, goat, and other animals act as a richest source of diverse cultivable actinobacterial genera similar to soil, water, plant, and other ecosystems. Like traditional probiotics, most actinobacteria of chicken and goat origin fulfilled many of the *in vitro* probiotic selection criteria.

Acknowledgments

We acknowledge the financial assistance of the Department of Science and Technology, Government of India, under DST-Promotion of University Research and Scientific Excellence (PURSE) scheme-Phase II, Rashtriya Uchchatar Shiksha Abhiyan (RUSA)-2.O.

REFERENCES

Cao, Y., Jiang, Y., Li, Y., Chen, X., Jin, R., & He, W. (2012). Isolation methods and diversity of culturable fecal actinobacteria associated with *Panthera tigris* tigris in Yunnan Safari Park. *Acta Microbiologica Sinica*, *52*(7), 816–824.

Campbell, J. M., Fahey, G. C., & Wolf, B. W. (1997). Selected indigestible oligosaccharides affect large bowel mass, cecal and fecal short-chain fatty acids, pH and microflora in rats. *The Journal of Nutrition*, *127*(1), 130–136.

Chen, X., Qiu, S. M., Jiang, Y., Han, L., Huang, X. S., & Jiang, C. L. (2014). Diversity, bioactivity and drug development of cultivable actinobacteria in six species of bird feces. *American Journal of Agricultural and Biological Sciences*, *2*(1), 13–18.

De Jesus-Laboy, K. M., Godoy-Vitorino, F., Piceno, Y. M., Tom, L. M., Pantoja-Feliciano, I. G., Rivera-Rivera, M. J., Andersen, G. L., & Dominguez-Bello, M. G. (2011). Comparison of the fecal microbiota in feral and domestic goats. *Genes*, *3*(1), 1–18.

Gouliamova, D. E., Stoilova-Disheva, M. M., Dimitrov, R. A., Gushterova, A. G., Vasileva-Tonkova, E. S., Paskaleva, D. A., & Stoyanova, P. E. (2012). Preliminary characterization of yeasts and actinomycetes isolated from mammalian feces. *Biotechnology & Biotechnological Equipment*, *26*(1), 1–4.

Haeder, S., Wirth, R., Herz, H., & Spiteller, D. (2009). Candicidin-producing *Streptomyces* support leaf-cutting ants to protect their fungus garden against the pathogenic fungus *Escovopsis*. *Proceedings of the National Academy of Sciences*, *106*(12), 4742–4746.

Imhoff, J. F., & Stohr, R. (2003). Sponge-associated bacteria: General overview and special aspects of bacteria associated with *Halichondria panicea*. In: Muller, W. E. G. (Eds.), *Sponges(Porifera)*, Springer-Verlag, Heidelberg, pp 35–57.

Jiang, Y., Chen, X., Han, L., Huang, X., Qiu, S., & Jiang, C. (2013a). Cultivable actinomycete communities in mammal feces of three diet habits. *International Journal of Microbiology Research and Reviews*, *1*(4), 76–84.

Jiang, Y., Chen, X., Han, L., Li, Q., Huang, X., Qiu, S., Ding, Z., &Jiang, C. (2013b). Diversity of cultivable actinomycetes in 6 species of herbivore feces. *International Journal of Microbiology Research and Reviews*, *2*(6), 110–117.

Jiang, Y., Chen, X., Han, L., Huang, X., &Jiang, C. (2014). Community of actinomycetes in 42 species of animal feces. *Online International Interdisciplinary Research Journal*, *4*, 23–38.

Jiang, Y., Han, L., Chen, X., Yin, M., Zheng, D., Wang, Y., Qiu, S., & Huang, X. (2013c). Diversity and bioactivity of cultivable animal fecal actinobacteria. *Advances in Microbiology*, *3*, 1–13.

Kamada, N., Seo, S. U., Chen, G. Y., & Nunez, G. (2013). Role of the gut microbiota in immunity and inflammatory disease. *Nature Reviews Immunology*, *13*(5), 321–335.

Kaltenpoth, M., Goettler, W., Dale, C., Stubblefield, J. W., Herzner, G., Roeser-Mueller, K., & Strohm, E. (2006). *Candidatus Streptomyces philanthi*, an endosymbiotic streptomycete in the antennae of *Philanthus* digger wasps. *International Journal of Systematic and Evolutionary Microbiology*, *56*(6), 1403–1411.

Kim, T. K., Garson, M. J., & Fuerst, J. A. (2005). Marine actinomycetes related to the '*Salinospora*' group from the Great Barrier Reef sponge *Pseudoceratina clavata*. *Environmental Microbiology*, *7*(4), 509–518.

Kong, Y., Teather, R., & Forster, R. (2010). Composition, spatial distribution, and diversity of the bacterial communities in the rumen of cows fed different forages. *FEMS Microbiology Ecology*, *74*(3), 612–622.

Lee, L. H., Zainal, N., Azman, A. S., Eng, S. K., Goh, B. H., Yin, W. F., Ab Mutalib, N. S., Chan, K. G. (2014a). Diversity and antimicrobial activities of actinobacteria isolated from tropical mangrove sediments in Malaysia. *The Scientific World Journal*, *2014*, 1–14.

Lu, Y., Wang, J., Deng, Z., Wu, H., Deng, Q., Tan, H., & Cao, L. (2013). Isolation and characterization of fatty acid methyl ester (FAME)-producing *Streptomyces* sp. S161 from sheep (*Ovis aries*) faeces. *Letters in Applied Microbiology*, *57*(3), 200–205.

Mara, D. D., & Oragui, J. I. (1981). Occurrence of *Rhodococcuscoprophilus* and associated actinomycetes in feces, sewage, and freshwater. *Applied and Environmental Microbiology*, *42*(6), 1037–1042.

Mead, G. C. (1989). Microbes of the avian cecum: Types present and substrates utilized. *Journal of Experimental Zoology*, *252*(S3), 48–54.

Meena, B., Rajan, L. A., Vinithkumar, N. V., & Kirubagaran, R. (2013). Novel marine actinobacteria from emerald Andaman & Nicobar Islands: A prospective source for industrial and pharmaceutical byproducts. *BMC Microbiology*, *13*, 145.

Oh, D. C., Poulsen, M., Currie, C. R., & Clardy, J. (2009). Dentigerumycin: A bacterial mediator of an ant-fungus symbiosis. *Nature Chemical Biology*, *5*(6), 391–393.

Scott, J. J., Oh, D. C., Yuceer, M. C., Klepzig, K. D., Clardy, J., & Currie, C. R. (2008). Bacterial protection of beetle-fungus mutualism. *Science*, *322*(5898), 63–63.

Sekirov, I., Russell, S. L., Antunes, L. C. M., & Finlay, B. B. (2010). Gut microbiota in health and disease. *Physiological Reviews*, *90*(3), 859–904.

Tan, H., Deng, Z., & Cao, L. (2009). Isolation and characterization of actinomycetes from healthy goat faeces. *Letters in Applied Microbiology*, *49*(2), 248–253.

Tan, H., Deng, Q., & Cao, L. (2014). Ruminant feces harbor diverse uncultured symbiotic actinobacteria. *World Journal of Microbiology and Biotechnology*, *30*(3), 1093–1100.

Tian, X., Cao, L., Tan, H., Han, W., Chen, M., Liu, Y., & Zhou, S. (2007). Diversity of cultivated and uncultivated actinobacterial endophytes in the stems and roots of rice. *Microbial Ecology*, *53*(4), 700–707.

Tonouchi, A., Kitamura, K., & Fujita, T. (2013). *Brevibacteriumyomogidense* sp. nov., isolated from a soil conditioner made from poultry manure. *International Journal of Systematic and Evolutionary Microbiology*, *63*(2), 516–520.

Videnska, P., Rahman, M. M., Faldynova, M., Babak, V., Matulova, M. E., Prukner-Radovcic, E., Krizek, I., Smole-Mozina, S., Kovac, J., Szmolka, A., Nagy, B., Sedlar, K., Cejkova, D., & Rychlik, I. (2014). Characterization of egg laying hen and broiler fecal microbiota in poultry farms in Croatia, Czech Republic, Hungary and Slovenia. *Plos One*, *9*(10), e110076.

Visser, A. A., Nobre, T., Currie, C. R., Aanen, D. K., & Poulsen, M. (2012). Exploring the potential for actinobacteria as defensive symbionts in fungus-growing termites. *Microbial Ecology*, *63*(4), 975–985.

Wang, J., Tan, H., Lu, Y., & Cao, L. (2014). Determination of ionophore antibiotics nactins produced by fecal *Streptomyces* from sheep. *Biometals*, *27*(2), 403–407.

Wu, S., Gao, T., Zheng, Y., Wang, W., Cheng, Y., & Wang, G. (2010). Microbial diversity of intestinal contents and mucus in yellow catfish (*Pelteobagrusfulvidraco*). *Aquaculture*, *303*(1), 1–7.

Wu, S., Wang, G., Angert, E. R., Wang, W., Li, W., & Zou, H. (2012). Composition, diversity, and origin of the bacterial community in grass carp intestine. *Plos One*, *7*(2), e30440.

Xu, L., Li, Q., & Jiang, C. (1996). Diversity of soil actinomycetes in Yunnan, China. *Applied and Environmental Microbiology*, *62*(1), 244–248.

Zhu, X. Y., Zhong, T., Pandya, Y., & Joerger, R. D. (2002). 16S rRNA-based analysis of microbiota from the cecum of broiler chickens. *Applied and Environmental Microbiology*, *68*(1), 124–137.

17

Microbiota Functions in Caenorhabditis elegans

Arun Kumar
Institute of Advanced Study in Science and Technology

Somarani Dash
Institute of Advanced Study in Science and Technology

Mojibur R. Khan
Institute of Advanced Study in Science and Technology

CONTENTS

17.1 Introduction

The colonization of microorganisms in the gut has always been an important factor in determining the evolutionary events of all multicellular organisms. This type of interaction between microbiota and their hosts is highly interdependent, together known as "holobiont." In the past two decades, there are numerous studies that suggest the role of colonized gut microbiota in affecting their host health. The alteration in the composition of gut microbiota is associated with several disease conditions, termed as "dysbiosis" (Adak and Khan 2019). This alteration or dysbiosis makes the host more susceptible towards a disease progression or pathobiont. For example, a reduction in the diversity of gut microbiota is found in several human diseases such as inflammatory bowel disease (IBD) and Crohn's disease (CD) (Manichanh et al. 2012; Jansson et al. 2009). However, the studies on investigating mechanisms by which these specific microbial alterations impact the host health are always challenging in humans. Other important question is how gut microbiota serves as a universal factor in the host physiological traits, such as development, immunity, reproduction, behavior, and lifespan.

To understand these complex phenomena, the genetic tractability of model organisms such as *Caenorhabditis elegans*, *Drosophila melanogaster*, and *Mus musculus* has provided us a platform. Compared to other models, *C. elegans* is self-fertilizing and maintains genetically homogenous populations. This advantage strengthens its candidature to study how genetics plays an important role in host–microbiota communication (Han et al. 2017). In support, many reports have suggested that most signaling pathways of *C. elegans* are conserved in different other organisms, including humans (Lai et al. 2000; Kim et al. 2002; Shapira et al. 2006; Walker et al. 2011; Tenor and Aballay 2008; Jose et al. 2012; Melo and Ruvkun 2012; Paek et al. 2012). Thus, understanding these molecular pathways in *C. elegans* may aid in developing interventions towards treating human diseases. In this chapter, we have summarized the role of microbiota composition in determining the life history traits of *C. elegans*. In addition, we discuss the use of non-native microbes to study the underlying mechanisms of host–microbe cometabolism and their influence on host physiology.

17.2 *Caenorhabditis elegans*

C. elegans, an invertebrate microbivore (i.e., feed on microbes) organism, is commonly found on the decaying fruits and rotting vegetations (Frézal and Félix 2015). *C. elegans* has been in use in the laboratories since Sydney Brenner first used this

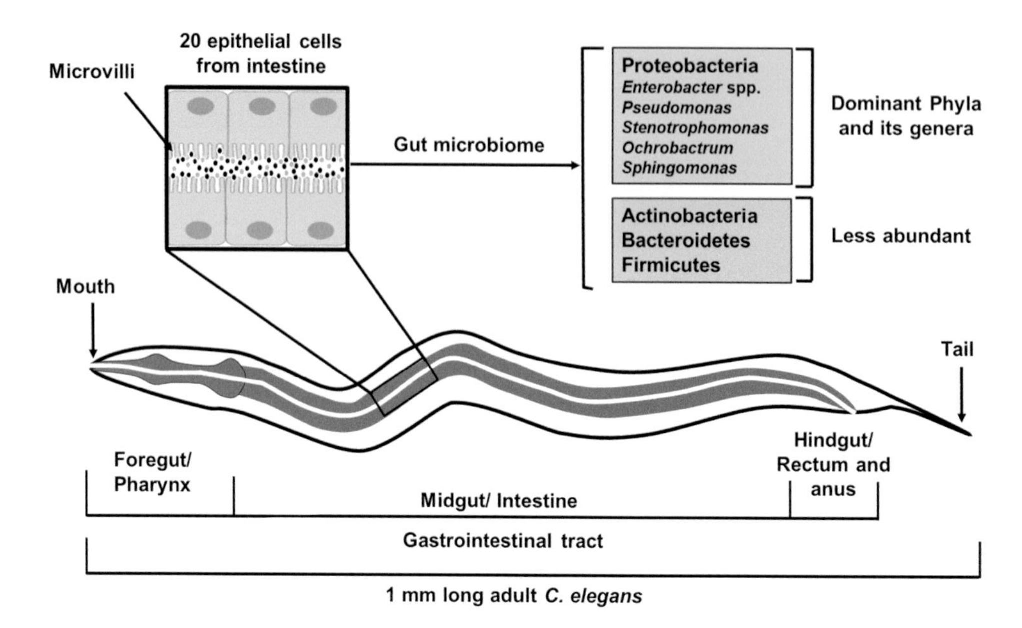

FIGURE 17.1 Schematic diagram showing the gastrointestinal tract of *C. elegans* and its microbiota.

organism as model to study neurobiology and developmental aspects in the early 1960s (Brenner 1974). *C. elegans* body consists of a mouth, pharynx, intestine, collagenous cuticle, and gonad. The gastrointestinal tract of *C. elegans* starts with its mouth, which has an opening to feed on microbes. After feeding, the microbe is taken to a tube-like structure where strong muscles grind it at a rate of 200 per minute, which is called as its "pharyngeal pumping." The grinded microbial material then goes to their intestinal lumen; there a pair of 20 epithelial cells absorb the available nutrients through their fast contraction of surrounding muscles every 50 seconds and the rest unabsorbed content is passed through its anus (McGhee 2013) (Figure 17.1). Despite their efficient rate of pharyngeal grinding in young-aged nematodes, a lower number of microbes are engulfed without grinding, which accumulate in their intestine and considered as their microbiota (Figure 17.1). In aged nematodes, reports suggested the accumulation of higher microbial load in their intestine, which may lead to a decrease in pharyngeal activity, and defecation rate, and lose in the intestinal structure and function of *C. elegans* (McGee et al. 2011; Rae et al. 2012; Gomez et al. 2012).

Under laboratory conditions, the wild-type N2 nematode is considered as reference strain for all studies and grown on nematode growth medium (NGM) seeded with nonpathogenic *Escherichia coli* OP50, a standard bacterial source to feed these nematodes. *E. coli* OP50 is an uracil auxotroph which limits their growth on NGM in the absence of uracil. The axenic condition for hermaphrodite nematodes is maintained by bleaching with sodium hypochlorite once the nematodes lay eggs, and again propagated on a monoaxenic culture of *E. coli* OP50 (Stiernagle 1999). Notably, *E. coli* OP50 helps in forming thinner bacterial lawns for easier visualization of these nematodes under microscopes. Starting from the larval stages (L1-L4 stages) to adulthood (post-L4 stages), the *E. coli* OP50 colonize the intestinal lumen of *C. elegans* and is considered as its entire microbiota. One important point is that *E. coli* OP50 does not have any resemblance to the native gut microbiota found in *C. elegans* in their natural habitat (Schulenburg and Félix 2017). This artificial culturing of nematodes somehow hinders the importance of their native microbiota, as they are cultured without its native microbiota. In this regard, the recent studies have characterized the composition of gut microbiota of *C. elegans* in their natural habitat and its role in the physiology of *C. elegans*, such as growth, lifespan, fecundity, pathogen-associated avoidance behavior, and immune responses (Dirksen et al. 2016; Berg et al. 2016; Samuel et al. 2016). Moreover, *C. elegans* has also been grown in artificial environment mimicking their natural habitat under laboratory conditions. These investigations have suggested the role of gut microbiota in modulating the host factors at the gene and transcript level.

17.3 Microbiome Composition and Diversity

Our current knowledge on the composition and functionality of gut microbiota in *C. elegans* is a result of three studies, i.e., Dirksen et al. (2016), Berg et al. (2016), and Samuel et al. (2016). The next-generation sequencing was used to characterize their native microbiota. Dirksen et al. (2016) and Samuel et al. (2016) studied *C. elegans* in their natural habitats (composting fields, rotting fruits, and their vectors, e.g., snails) of different countries. Another study by Berg et al. (2016) investigated the composition of gut microbiota in artificially cultured *C. elegans* larvae (L1) on autoclaved soil supplemented with decaying fruits. All three studies collected samples from several geographical location or followed different approaches to characterize the gut microbiota of *C. elegans*, but suggested that their gut is rich in diverse microbial communities, majorly dominated by the phylum Proteobacteria, including the most abundant Gammaproteobacteria (*Enterobacteriaceae*, *Pseudomonadaceae*, and *Xanthomonadaceae*), which constitute more than 80% of total microbiome, and lower abundance of Alphaproteobacteria, as well as the member of the other phyla Actinobacteria, Firmicutes, and Bacteroidetes (Figure 17.1) (Berg et al. 2016a; Samuel et al. 2016; Dirksen et al. 2016). Although nematodes were isolated from different geographical locations or cultured in artificial laboratory conditions, these studies confirmed that their gut shared a *C. elegans*-restricted microbiota known as "core microbiota," and there were differences at the genera and species level (Berg et al. 2016a; Samuel et al. 2016; Dirksen et al. 2016).

Several conclusions can be drawn from both of these studies. First, the composition of gut microbiota in *C. elegans* differs from the composition of soil microbiota, as the nematodes were collected from different geographical sampling sites. This suggests that host factors greatly affect the microbiota composition in *C. elegans*, which is a defined and nonrandom microbial colonization (Dirksen et al. 2016; Berg et al. 2016a). Second, the nematodes were also cultured on the same soil but grown at different temperature (a gradient of 15°C–25°C) to find the role of temperature in the gut microbiome of *C. elegans*. As expected, the temperature differences affected the microbiota of both soil and the nematode's gut. The increase in temperature from 15°C to 25°C increased the abundance of Sphingobacterium in soil communities, but no such difference was observed in the nematodes. On the contrary, the abundance of Agrobacterium was increased in the nematode's guts, the temperature increased, but no such differences were observed in the soil communities. Importantly, there were also no significant temperature-dependent changes in the core microbiota (i.e., *Enterobacteriaceae*, *Pseudomonadaceae*, and *Xanthomonadaceae*) of *C. elegans* (Berg et al. 2016a). These interpretations hint that temperature-dependent alteration in soil microbial communities was not universal to the nematode's gut microbiota, which indicates that host-associated processes determine the composition of gut microbiota. Third, the family of Rhabditidae nematodes, including *C. elegans*, also grouped with other microbivore nematodes sharing the same soil environment. One interesting observation was made by Ladygina et al. (2009) that all the members of this family Rhabditidae host *Pseudomonas* sp. in their gut. In addition, the studies on an insect parasitic nematode *Pristionchus* sp. belonging to the family Diplogastridae also found *Pseudomonas* sp. as their gut commensal (Rae et al. 2008). These studies suggest a long-standing interaction of core microbiota with the evolutionary history of families Rhabditidae and Diplogastridae for more than approximately 200–300 million years ago.

17.4 Genetics versus Environmental Factors on Gut Microbiome

The composition of gut microbiota in *C. elegans* largely depends on the environmental factors and host genetics (Berg et al. 2016a; Dirksen et al. 2016; Samuel et al. 2016). The culturing of nematodes in the same environmental conditions consisting of distinct genotypes, such as different species and strains, showed a significant difference in the composition of their gut microbiota. Berg et al. (2016a) isolated and characterized the gut isolates of nematodes *C. elegans* and *C. briggsae*, and investigated their role in their health. Their study showed that composition of gut microbiota was similar in the same genotypic nematodes, but differed in other genotypes. Mostly, the genotypically distinct nematodes showed the enrichment of few genotype-specific indicator species (e.g., members of Enterobacteriaceae, but distinct in specific operational taxonomic units (OTUs)) in both *C. elegans* and *C. briggsae* (Berg et al. 2016a). In contrast, the culturing of nematodes on different soil or their combinations suggested a large variation of OTUs in genotypically distinct nematodes, but this variation was significantly reduced if cultured on the same soil. This suggests that environment is the dominant contributing factor in shaping the composition of gut microbiota in *C. elegans* (Berg et al. 2016a).

The competition between colonizing microbes is one of the driving forces to shape microbiota composition within the gut. This competition somehow states the mystery of characteristic core microbiota in nematodes. For example, the gut microbes under Enterobacteriaceae, Pseudomonadaceae, and Xanthomonadaceae are known as "early colonizer" of their gut, because all of these have flexible metabolic pathways and faster growth compared to other gut colonizers (Figure 17.1) (Samuel et al. 2016). In addition, the members of the family Comamonadaceae were found to negatively interact with other core microbiota of *C. elegans* and reduced the abundance of certain core microbiota, but this reduction was observed in only few samples. Therefore, the Comamonadaceae was not considered as a member of their core microbiota (Figure 17.1) (Berg, Zhou, and Shapira 2016).

The pathogenic infections also modulate the gut microbiome of the host. A study focused on how a pathogen *Bacillus nematocida* B16 infection modulated the composition of gut microbiota and killed the *C. elegans*. To answer this question, the nematode *C. elegans* were precultured on soil or rotten fruits to study their native microbiota under laboratory conditions. Later, these nematodes were exposed to pathogen *B. nematocida* B16 for 24 hours, and found a significant reduction in the gut microbial diversity after infection. These changes in microbiota were consistent in all the nematodes which were cultured on soil or rotten fruits. The results showed the increase in abundance of Firmicutes, but the reduction in the richness of microbial communities belonging to Proteobacteria, Acidobacteria, Actinobacteria, and Cyanobacteria compared to uninfected nematodes (Niu et al. 2016).

The microbiota composition is not only influenced by microbial pathogens, but also influenced by viral infections. A study investigated the effect of naturally infecting virus of *C. elegans*, i.e., Orsay virus, on the gut microbial composition of wild-type strain N2 and its mutant *rde-1* (ne219) deficient in the innate immune system. The important function of rde-1 gene is germline development and maintenance, which is conserved from invertebrate to vertebrates. The previous researches on Orsay virus have suggested that its infection causes intestinal abnormalities in *C. elegans*, which not only affects their survival but also is horizontally transmitted to their offspring (Franz et al. 2014). Their results showed that Orsay virus multiplies 100-fold in rde-1 mutants compared to wild-type N2. While the gut microbiome comparison showed a significant decrease in the diversity of gut microbiota in *rde-1* mutants, there was an increase in the abundance of Sphingobacterium, Brevundimonas, and Trabulsiella. In addition, the transcriptomics results showed an upregulation of C-type lectin (CLEC) in *rde-1* nematodes (i.e., *clec*-70, *clec*-71, and *clec*-72), which suggests that CLEC genes are involved in the modulation of gut microbiota in immune-deficient *rde-1* mutants (Guo et al. 2017). Therefore, the *C. elegans* provides a platform to study the role of influencing factors in the gut microbiome.

17.5 Functionality of Gut Microbiota

Taking the advantage of artificial laboratory conditions, Samuel et al. (2016) cultured over 500 native gut microbial strains of *C. elegans* and studied their effect on growth rate, induction of immune and stress reporters. These reporters were used to determine how culturing of *C. elegans* on their native microbiota may affect their physiology. Out of these, 78% showed beneficial effects and 22% negatively affected the nematode's physiology compared to control *E. coli* OP50. Their results suggested that the effects of these microbial isolates on nematodes were genera-specific. Microbial strains that fall into Enterobacteriaceae, *Enterobacter*, *Gluconobacter*, *Lactococcus*, and *Providencia* were beneficial, while Gammaproteobacteria (*Stenotrophomonas* and *Xanthomonas*) and Bacteroidetes (*Sphingobacterium* and *Chryseobacterium*) were found to negatively affect their physiology (Samuel et al. 2016). They expected that the detrimental microbes may serve as poor nutritional source and show potent antagonistic activities which may further affect the nematode's growth and survival. To support this hypothesis, the binary mixture of both beneficial (*Enterobacter* sp., *Gluconobacter* sp., and *Providencia* sp.) and detrimental bacteria (*Pseudomonas* sp., *Serratia* sp., and *Chryseobacterium* sp.) was used to culture the *C. elegans*. Their results suggested that higher or equal levels of beneficial microbes neutralized the consequences caused by detrimental microbes, as the higher level may supply sufficient quantity of nutrition and consists of antipathogenic potential to support the growth and survival of nematodes. In contrast, the *E. coli* OP50 in a binary mixture with detrimental microbes were unable to support the growth of *C. elegans* (Samuel et al. 2016).

Similar to mammalian host, the native microbiota of *C. elegans* has shown to protect them from pathogens. This may be done by limiting the colonization of pathogen to *C. elegans* intestine or activating the innate immune responses of *C. elegans*. A commensal microbe *Bacillus subtilis* GS67 inhibited the colonization of pathogen *Bacillus thuringiensis* DB27 in the gut of *C. elegans*. However, further investigation suggested that this commensal releases fengycin to protect *C. elegans* from infection by *B. thuringiensis* DB27 (Figure 17.2) (Iatsenko et al. 2014). The other microbiota-associated metabolites have also shown to protect *C. elegans* from pathogenic infections. *Pseudomonas aeruginosa* produced two secondary metabolites, i.e., pyochelin and phenazine-1-carboxamide, which promoted the pathogenic-avoidance behavior of nematodes though the activation of TGF-β signaling pathway in their interneurons (Kirienko et al. 2013). In support, Berg et al. (2016a) also showed that the gut commensal *Pseudomonas mendocina* protected their native host *C. elegans* from pathogenic infection by *P. aeruginosa*, but did not protect *C. briggsae*. Further, investigations suggested that *P. mendocina* activates p38-MAPK pathway of host to protect them from subsequent infection by *P. aeruginosa* (Berg, Zhou, and Shapira 2016). Their study found that *P. mendocina* was successful in inhibiting the fungal growth

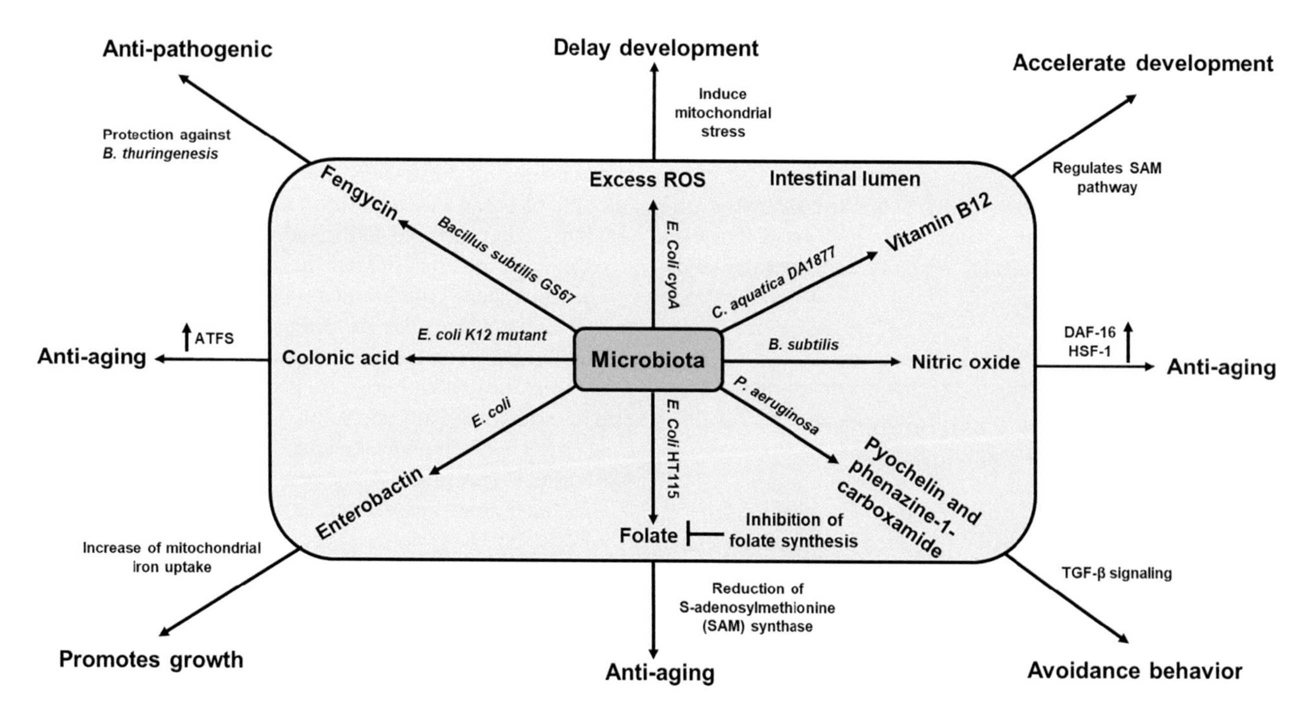

FIGURE 17.2 Microbiota-released metabolites and their impact on the physiology of *C. elegans*.

under both *in vitro* conditions and in their native host *C. elegans* (Figure 17.2). Similar results were shown by another gut microbe *Enterobacter cloacae* isolated from *C. elegans* and *C. briggsae*, which provided protection to only their native host from pathogenic infections, but not the non-native host (Berg, Zhou, and Shapira 2016). These adaptations of gut commensals suggested their host-restricted role at the functional level, as the nematodes are exposed to a wide array of pathogens in their natural environment.

On the other hand, Dirksen et al. (2016) cultured *C. elegans* on their common 14 gut isolates to study their role in population size and stress resistance compared to its laboratory bacterial food *E. coli* OP50. They found that the gut isolates under Beta- and Gammaproteobacteria were beneficial in increasing the population size and fitness of nematodes, while Actinobacteria, Alphaproteobacteria, Bacteroidetes, and Firmicutes were found to reduce the nematode's fitness. Their study also suggested that even though the abundance of gut microbes belonging to the Phylum such as Ochrobactrum and Stenotrophomonas was lower on petri plates, both contribute up to 20% to their total gut microbiota (Berg, Zhou, and Shapira 2016). Further investigation focused in culturing nematodes on native gut microbe *Ochrobactrum* spp. MYb71, which suggested that host signaling pathways upregulated the level of foreign entity-degrading enzymes such as proteases, lipases, and glutathione metabolism (Cassidy et al. 2018). The recent publication from this laboratory group suggested that *Ochrobactrum* spp. MYb71 and MYb237 influence the development, fertility, immunity, and energy metabolism of *C. elegans*. In addition, the transcriptomic analysis showed that these responses are mediated through upregulation of ELT-2 (a GATA transcription factor). Moreover, CLEC genes were also found to be involved in their energy metabolism or fecundity throughout their development and adulthood (Yang et al. 2019).

In addition, Proteobacteria is a common commensal of different organisms. In zebrafish *Danio rerio*, the laboratory-reared and zebrafish collected from their natural environment is dominated by Proteobacteria (mainly Gammaproteobacteria) at all life stages, and other less-abundant genera include Fusobacteria, Firmicutes, Bacteroidetes, Verrucomicrobia, and Cyanobacteria. The dominated Gammaproteobacteria are found to regulate their innate immunity (Roeselers et al. 2011). In fruit fly *Drosophila melanogaster*, Proteobacteria constitutes about 30% of its core microbiota and affects almost all aspects of their physiology, including oogenesis, longevity, and immunity. However, the alteration of the members of Enterobacteriaceae is correlated with their gut dysbiosis in aged fruit flies (Lizé, McKay, and Lewis 2014). In vertebrates such as humans and mice, the Proteobacteria is present as minor phyla, of which commensal *E. coli* activates the signaling cascades to protect gut barrier of host against pathogenic infections. On the other hand, the higher abundance of Proteobacteria has been correlated with severity of several diseases (e.g., IBD and metabolic disorders) associated with gut dysbiosis (Sommer and Bäckhed 2013). Thus, further investigation on the role of Proteobacteria would help us to understand the complex host–microbiota interaction.

17.6 A Gateway for Understanding Host–Microbe Cometabolism

Recent advances have been made in understanding the nematode's metabolism by culturing them on non-native microbes. This may help to establish the fundamental role of gut microbiota in determining the host metabolism. For example, there was always a mystery how microbiota and their host compete for the available mitochondrial uptake of iron. A study focused on the role of microbe-secreted enterobactin (Ent) and their effect on host health. They found that Ent-secreting microbe unexpectedly promoted the mitochondrial iron uptake in *C. elegans* and promoted their growth (Qi and Han 2018). The similar mechanisms may also be conserved in mammals. The microbiota-released metabolites were also reported to affect the nematode's physiology, including folate, vitamin B12, nitric oxide, and colonic acid.

17.6.1 Vitamin B12

All organisms require a small proportion of micronutrients in their diet as humans consume these micronutrients from their diet or secretion by the colonized gut microbiota, such as vitamin K and vitamin B12. Similarly, the reports have suggested the role of micronutrients in the physiology and metabolism of *C. elegans*. *Comamonas aquatica* have found to accelerate the developmental process in *C. elegans* and resulted in their shortened lifespan, which came at the expense of reducing their fecundity ability. Further detailed investigation has suggested that this microbe releases vitamin B12, which further affects the physiology (i.e., development and fecundity) of nematodes through S-adenosyl methionine (SAM) regulatory pathways and breaks down the large amount of accumulation of propionic acid (Figure 17.2) (Watson et al. 2014). Vitamin B12 is considered an important essential micronutrient from yeast to plants and animals (Scott 1999). Thus, the research of the last few decades suggests that *C. elegans* can be maintained at low level of vitamin B12 (i.e., by feeding them with *E. coli* OP50). However, the nematodes can activate a propionic shunt to prevent the buildup of this metabolite (Watson et al. 2014).

17.6.2 Folate

Several bacterial genetic variants of *E. coli* OP50 and nematode's mutants provide a platform to study how bacterial genetics affects the host factors, or vice versa. A bacterial mutant *E. coli* HT115 lacking *aroD* gene (encodes for an enzyme 3-dehydroquinate dehydratase) is found to enhance the lifespan of *C. elegans*. Interestingly, the *aroD*-encoded enzyme, i.e., 3-dehydroquinate dehydratase, synthesizes the aromatic compound folate or vitamin B9 (Figure 17.2) (Virk et al. 2016, 2012). The further culturing of *C. elegans* on a combination of *E. coli* HT115 and folate precursors (e.g., para-aminobenzoic acid (PABA)) shortened the lifespan of *C. elegans*. The addition of drugs inhibiting the folate synthesis in *E. coli* also extended the nematode's lifespan. Interestingly, it should be noted that folate is not required by *C. elegans*, although it is released by *E. coli* HT115 to support its growth, which in turn modulates the longevity in *C. elegans* (Virk et al. 2016). Another follow-up study also found that the effect of few folate derivatives is dependent on folate receptor homolog folr-1 of *C. elegans*, which may further induce the proliferation of germ cell (Chaudhari et al. 2016).

Nowadays, the recent focus is to repurpose the available drugs and study their interaction with the resident microbiota of host. Metformin, a drug used for diabetes, has been found to extend longevity in *C. elegans*. The further investigations suggested that metformin modulates the commensal (i.e., *E. coli* OP50) folate metabolism and leads to extension in the lifespan of the host (Cabreiro et al. 2013). In support, a metagenomic study was conducted to compare the functionality of microbiome in children and adults. In the microbiome of babies, the genes involved in the folate synthesis were highly upregulated compared to adults. However, the genes for folate salvage pathway were upregulated in adult's microbiome (Engevik et al. 2019). These studies hint that controlling the folate synthesis of core microbiota can be a route to healthy aging.

17.6.3 Colonic Acid

A study conducted a bacterial mutant screening on *C. elegans*—out of 4,000, only 29 bacterial mutants showed enhanced longevity. These mutants were found to affect aging of *C. elegans* through mTOR, insulin/IGF signaling, JNK, and caloric restriction (Han et al. 2017). In addition, Han et al. studied two bacterial mutants (Dlon and Dhns) which suppressed the secretion of polysaccharide colonic acid and reduced the longevity of *C. elegans*. However, the supplementation of colonic acid increased the lifespan of *C. elegans* through upregulation of activating transcription factor associated with stress (ATFS), which is involved in regulating unfolded protein response (UPRmt) and mitochondrial dynamics of the *C. elegans* (Figure 17.2) (Han et al. 2017). These types of microbe-associated effects on the host factor are conserved across all organisms.

17.6.4 Nitric Oxide

The microbiota-released nitric oxide, a diffusible signaling molecule, was found to enhance the lifespan of *C. elegans*. However, the microbes deficient in producing nitric oxide decreased the nematode's lifespan. In this regard, the supplementation of exogenous nitric oxide also resulted in the extension of lifespan through activating DAF-16 and HSF-1 (Figure 17.2) (Donato et al. 2017). These pathways are known to be involved in antiaging and stress resistance. On the other hand, the microbiota-released olfactory signals also modulate the nematode's physiology, prior to any nutrient-sensing metabolic processes. A microbe *B. subtilis* releases nitric oxide which activates soluble guanylyl cyclase of neuron in *C. elegans*, thus extending the lifespan of these nematodes through upregulating the expression of heat shock proteins (Gusarov et al. 2013).

17.6.5 Reactive Oxygen Species

The reactive oxygen species (ROS) produced by gut commensals affect the physiology of nematodes. The increased level of microbiota-derived ROS (e.g., *E. coli* cyo A mutant) may activate the mitochondrial stress responses in *C. elegans*, which further may delay the developmental process (Figure 17.2) (Saiki et al. 2008; Larsen and Clarke 2002). In contrast, the bacterial mutations with increased ROS detoxifications may prevent this detrimental effect. Further studies showed that bacterial ROS might be transported (e.g., as carbonylated peptides) in the gut through pept-1 intestinal peptide transporter (Brooks, Liang, and Watts 2009).

17.6.6 A Metabolic Communication between Pathogen and the Commensal

Gut commensals of *C. elegans* may not be identical to those of humans, but their effects on the host physiology are likely universal. One interesting study was conducted to find out how *Giardia duodenalis* induces the functional changes in host commensals. *G. duodenalis* is the causative agent of parasitic diarrhea and associated with higher risk of irritable bowel syndrome (Gerbaba et al. 2015). Surprisingly, the culturing of *C. elegans* on pathogen *Giardia* or non-native commensal *E. coli* OP50 alone did not affect their viability, but a combination of both becomes lethal to the host. This suggests that the pathogen makes the commensal a pathobiont. To support this study, the nematodes were cultured on the microbiota of noninflamed colonic sites from healthy human donor and microbiota from the inflamed sites of ulcerative colitis (UC) patients. Similar to previous results, the combination of *Giardia* with noninflamed site microbiota did not affect nematode's survival, but *Giardia* with microbiota from inflamed colonic sites was found to be lethal. The first important observation was that the combination of pathogen and commensals has facilitated the colonization of *Giardia* to the nematode's gut. The transcriptomic analyses suggested that their synergism alters the expression of commensal genes involved in hydrogen sulfide biosynthesis (HSB). To further characterize the role of HSB in worm survival, its positive regulator was deleted and found that it was sufficient to kill *C. elegans* even in the absence of *Giardia* (Gerbaba et al. 2015). The study states that pathogen *Giardia* makes the commensal *E. coli* OP50 an opportunistic pathogen by altering their metabolism, which may further modulate the important host–microbe cometabolism pathways deleterious to the host.

17.7 Future Perspective

The research of the past decade suggests that gut microbes affect the physiology and metabolism of the host. It is important to understand their underlying mechanisms by which they affect the host health. For example, the research has shown that 71% of the total human gut-associated Proteobacteria synthesizes folate, and their higher levels have been associated to induce inflammation in UC and CD (Vester-Andersen et al. 2019). Due to higher genetic homology and metabolic similarities with the humans, *C. elegans* can be used for understanding folate-associated inflammation disorders and aging. Thus, a high-throughput investigation on bacterial genes and their metabolites may enrich our knowledge towards microbiota-induced impact on host health. Further investigations are needed to understand the effect of these microbiota-released metabolites with complex microbiota to characterize their effect at a community level on their host physiology.

REFERENCES

Adak A, Khan MR. 2019. An insight into gut microbiota and its functionalities. *Cellular and Molecular Life Sciences* 76 (3):473–493.

Berg M, Stenuit B, Ho J, et al. 2016. Assembly of the *Caenorhabditis elegans* gut microbiota from diverse soil microbial environments. *The ISME Journal* 10 (8):1998.

Berg M, Zhou XY, and Shapira M. 2016. Host-specific functional significance of *Caenorhabditis* gut commensals. *Frontiers in Microbiology* 7: 1622.

Brenner S. 1974. The genetics of *Caenorhabditis elegans*. *Genetics* 77 (1):71–94.

Brooks KK, Liang B, and Watts JL. 2009. The influence of bacterial diet on fat storage in *C. elegans*. *PloS One* 4 (10):e7545.

Cabreiro F, Au C, Leung K, et al. 2013. Metformin retards aging in *C. elegans* by altering microbial folate and methionine metabolism. *Cell* 153 (1):228–239.

Cassidy L, Petersen C, Treitz C, et al. 2018. The *Caenorhabditis elegans* proteome response to naturally associated microbiome members of the genus Ochrobactrum. *Proteomics* 18 (8):1700426.

Chaudhari SN, Mukherjee M, Vagasi AS, et al. 2016. Bacterial folates provide an exogenous signal for *C. elegans* germline stem cell proliferation. *Developmental Cell* 38 (1):33–46.

Dirksen P, Marsh SA, Braker I, et al. 2016. The native microbiome of the nematode *Caenorhabditis elegans*: gateway to a new host-microbiome model. *BMC Biology* 14 (1):38.

Donato V, Ayala FR, Cogliati S, et al. 2017. *Bacillus subtilis* biofilm extends *Caenorhabditis elegans* longevity through downregulation of the insulin-like signalling pathway. *Nature Communications* 8:14332.

Engevik MA, Morra CN, Röth D, et al. 2019. Microbial metabolic capacity for intestinal folate production and modulation of host folate receptors. *Frontiers in Microbiology* 10:2305.

Franz CJ, Carl J., Renshaw H, et al. 2014. Orsay, Santeuil and Le Blanc viruses primarily infect intestinal cells in *Caenorhabditis nematodes*. *Virology* 448:255–264.

Frézal L, and Félix MA. 2015. The natural history of model organisms: *C. elegans* outside the Petri dish. *Elife* 4:e05849.

Gerbaba TK, Gupta P, Rioux K, et al. 2015. Giardia duodenalis-induced alterations of commensal bacteria kill *Caenorhabditis elegans*: a new model to study microbial-microbial interactions in the gut. *American Journal of Physiology-Gastrointestinal and Liver Physiology* 308 (6):G550–G561.

Gomez F, Monsalve GC, Tse V, et al. 2012. Delayed accumulation of intestinal coliform bacteria enhances life span and stress resistance in *Caenorhabditis elegans* fed respiratory deficient *E. coli*. *BMC Microbiology* 12 (1):300.

Guo Y, Xun Z, Coffman SR, et al. 2017. The Shift of the Intestinal Microbiome in the Innate Immunity-Deficient Mutant rde-1 Strain of *C. elegans* upon Orsay Virus Infection. *Frontiers in Microbiology* 8:933.

Gusarov I, Gautier L, Smolentseva O, et al. 2013. Bacterial nitric oxide extends the lifespan of *C. elegans*. *Cell* 152 (4):818–830.

Han B, Sivaramakrishnan P, Lin CJ, et al. 2017. Microbial genetic composition tunes host longevity. *Cell* 169 (7):1249–1262. e13.

Iatsenko I, Yim JJ, Schroeder FC, et al. 2014. B. subtilis GS67 protects *C. elegans* from Gram-positive pathogens via fengycin-mediated microbial antagonism. *Current Biology* 24 (22):2720–2727.

Jansson J, Willing B, Lucio M, et al. 2009. Metabolomics reveals metabolic biomarkers of Crohn's disease. *PloS One* 4 (7):e6386.

Jose AM, Kim YA, Leal-Ekman S, et al. 2012. Conserved tyrosine kinase promotes the import of silencing RNA into *Caenorhabditis elegans* cells. *Proceedings of the National Academy of Sciences* 109 (36):14520–14525.

Kim DH, Feinbaum R, Alloing G, et al. 2002. A conserved p38 MAP kinase pathway in *Caenorhabditis elegans* innate immunity. *Science* 297 (5581):623–626.

Kirienko NV, Kirienko DR, Larkins-Ford J, et al. 2013. Pseudomonas aeruginosa disrupts *Caenorhabditis elegans* iron homeostasis, causing a hypoxic response and death. *Cell Host & Microbe* 13 (4):406–416.

Ladygina N, Johansson T, Canbäck B, et al. 2009. Diversity of bacteria associated with grassland soil nematodes of different feeding groups. *FEMS Microbiology Ecology* 69 (1):53–61.

Lai CH, Chou CY, Ch'ang LY, et al. 2000. Identification of novel human genes evolutionarily conserved in *Caenorhabditis elegans* by comparative proteomics. *Genome Research* 10 (5):703–713.

Larsen PL, and Clarke CF. 2002. Extension of life-span in *Caenorhabditis elegans* by a diet lacking coenzyme Q. *Science* 295 (5552):120–123.

Lizé A, McKay R, and Lewis Z. 2014. Kin recognition in Drosophila: the importance of ecology and gut microbiota. *The ISME Journal* 8 (2):469–477.

Manichanh C, Borruel N, Casellas F, et al. 2012. The gut microbiota in IBD. *Nature Reviews Gastroenterology and Hepatology* 9 (10):599.

McGee MD, Weber D, Day N, et al. 2011. Loss of intestinal nuclei and intestinal integrity in aging *C. elegans*. *Aging Cell* 10 (4):699–710.

McGhee JD. 2013. The *Caenorhabditis elegans* intestine. *Wiley Interdisciplinary Reviews: Developmental Biology* 2 (3):347–367.

Melo JA, and Ruvkun G. 2012. Inactivation of conserved *C. elegans* genes engages pathogen-and xenobiotic-associated defenses. *Cell* 149 (2):452–466.

Niu Q, Zhang L, Zhang K, et al. 2016. Changes in intestinal microflora of *Caenorhabditis elegans* following *Bacillus nematocida* B16 infection. *Scientific Reports* 6:20178.

Paek J, Lo JY, Narasimhan SD, et al. 2012. Mitochondrial SKN-1/Nrf mediates a conserved starvation response. *Cell Metabolism* 16 (4):526–537.

Qi B, and Han M. 2018. Microbial siderophore enterobactin promotes mitochondrial iron uptake and development of the host via interaction with ATP synthase. *Cell* 175 (2):571–582. e11.

Rae R, Riebesell M, Dinkelacker I, et al. 2008. Isolation of naturally associated bacteria of necromenic *Pristionchus nematodes* and fitness consequences. *Journal of Experimental Biology* 211 (12):1927–1936.

Rae R, Witte H, Rödelsperger C, et al. 2012. The importance of being regular: *Caenorhabditis elegans* and *Pristionchus pacificus* defecation mutants are hypersusceptible to bacterial pathogens. *International Journal for Parasitology* 42 (8):747–753.

Roeselers G, Mittge EK, Stephens WZ, et al. 2011. Evidence for a core gut microbiota in the zebrafish. *The ISME Journal* 5 (10):1595–1608.

Saiki R, Lunceford AL, Bixler T, et al. 2008. Altered bacterial metabolism, not coenzyme Q content, is responsible for the lifespan extension in *Caenorhabditis elegans* fed an *Escherichia coli* diet lacking coenzyme Q. *Aging Cell* 7 (3):291–304.

Samuel BS, Rowedder H, Braendle C, et al. 2016. *Caenorhabditis elegans* responses to bacteria from its natural habitats. *Proceedings of the National Academy of Sciences* 113 (27):E3941–E3949.

Schulenburg H, and Félix MA. 2017. The natural biotic environment of *Caenorhabditis elegans*. *Genetics* 206 (1):55–86.

Scott JM. 1999. Folate and vitamin B 12. *Proceedings of the Nutrition Society* 58 (2):441–448.

Shapira M, Hamlin BJ, Rong J, et al. 2006. A conserved role for a GATA transcription factor in regulating epithelial innate immune responses. *Proceedings of the National Academy of Sciences* 103 (38):14086–14091.

Sommer F, and Bäckhed F. 2013. The gut microbiota—masters of host development and physiology. *Nature Reviews Microbiology* 11 (4):227–238.

Stiernagle T. 1999. Maintenance of *C. elegans*. *C. elegans* 2:51–67.

Tenor JL, and Aballay A. 2008. A conserved toll-like receptor is required for *Caenorhabditis elegans* innate immunity. *EMBO Reports* 9 (1):103–109.

Vester-Andersen MK, Mirsepasi-Lauridsen HC, Prosberg MV, et al. 2019. Increased abundance of proteobacteria in aggressive Crohn's disease seven years after diagnosis. *Scientific Reports* 9 (1):1–10.

Virk B, Correia G, Dixon DP, et al. 2012. Excessive folate synthesis limits lifespan in the *C. elegans: E. coli* aging model. *BMC Biology* 10 (1):67.

Virk B, Jia J, Maynard CA, et al. 2016. Folate acts in *E. coli* to accelerate *C. elegans* aging independently of bacterial biosynthesis. *Cell Reports* 14 (7):1611–1620.

Walker AK, Jacobs RL, Watts JL, et al. 2011. A conserved SREBP-1/phosphatidylcholine feedback circuit regulates lipogenesis in metazoans. *Cell* 147 (4):840–852.

Watson E, MacNeil LT, Ritter AD, et al. 2014. Interspecies systems biology uncovers metabolites affecting *C. elegans* gene expression and life history traits. *Cell* 156 (4):759–770.

Yang W, Petersen C, Pees B, et al. 2019. The inducible response of the nematode *Caenorhabditis elegans* to members of its natural microbiome across development and adult life. *Frontiers in Microbiology* 10:1793.

18

Impact of Microbial Communities on the Female Reproductive Tract of Bovine

M. Srinivasan
Bharathidasan University

J. Helan Chandra
Ampersand Academy

M.S. Murugan
Veterinary University Training and Research Centre

C. Manikkaraja, D. Dhanasekaran, and G. Archunan
Bharathidasan University

CONTENTS

18.1 Introduction

Animals and humans are exposed to microbes which colonize with diversity in/on the body sites such as moisture places, gut, skin, reproductive tract, faces, and urine. Vagaries of microbial diversity perturbs the animals and humans life by interacting with mucosal natural orifices such as reproductive tract, alimentary tract, urinary tract and integuments, which leads to colonization. Among these, the female reproductive tract (FRT) microbiome appears to play an important role in significant aspects of reproduction since its composition is associated with the health, physiology, development, and behavior of the host (Parfrey et al. 2018). The vaginal microbiome influences the host signaling within the reproductive tract during proinflammatory signals, which may play an important role in pregnancy and estrus. Therefore, the characterization of the reproductive microbiome is a considerable task in the host-related microbial communities. Current DNA sequencing analysis also suggests the presence of dynamic microbial communities in the vagina (Chen et al. 2017; Mahalingam et al. 2019). Human Microbiome Project (HMP), in healthy people, has increasingly gained attention representing the colonized unique microbiome at various body sites (Moreno and Simon 2019). Increasing evidences suggests that delineating the normal microbiota of FRT in humans and animals will provide information on strategies to optimize reproductive health in humans and animals.

Efficient reproduction is necessary for the survival of any species. Subfertility and infertility are the prime causes of failure to reproduce in mammals, including humans, exotic as well as endangered species and domestic livestock (Clemmons et al. 2017). Studies had demonstrated the various strategies to curb microbial invasion and interaction with reproduction in females in order to improve the efficiency of cattle reproduction and its health (Santos and Bicalho 2012; Stevenson et al. 2015). Studies pertaining to FRT microbiota is still its infancy, particularly in reference with the change in hormonal mileu towards pregnancy and estrous cycle. Very few reports have evaluated the cow vaginal microbial community through the metagenomics approach (Laguardia-Nascimento et al. 2015; Mahalingam et al. 2019). There is little information on the contribution of the vaginal microbial community with respect to reproductive physiology. Although the microbiome influences host biology significantly, knowledge about microbial communities in the reproductive tract of dairy and beef cattle is meager. Hence, this review explains about the functional role of reproductive tract microflora and possible source in cattle reproduction.

18.2 Vaginal Microbiota

The vaginal microbiota of cattle has been assessed based on culture-dependent and advanced technologies such as metagenomics techniques in disease condition (Wang et al. 2016). However, the vaginal bacteria of Nellore, Holstein, and Fleckvieh cattle comprised the following phylum: Firmicutes, Bacteroidetes, and Proteobacteria (Laguardia-Nascimento et al. 2015; Nesengani et al. 2017) in the existence of vaginal microbiota. Major bacterial genus similarity between the Nellore cattle and Gyr breed was mainly *Aeribacillus*, *Bacillus*, *Clostridium*, *Bacteroides*, and *Ruminococcus* (Giannattasio-Ferraz et al. 2019). This information begins to unravel the FRT microbiota and also defines the existence of the primary microbiota. The existence of microbial communities in the reproductive tract of buffalo and cow is similar in populations (Figure 18.1), as well as the colonization of bacterial flora in the reproductive tract (Table 18.1). Besides, similar functional opportunistic pathogenic groups of microbiota are reported in different breeds of bovine; the cervical–vaginal fluids/mucus which play a significant role in the various aspects of theriogenology require further investigation.

18.2.1 Estrus and Pregnancy Microbiota

The healthy microbial flora of the FRT consists of aerobic, anaerobic, and facultative anaerobic bacteria. Microbes such as *Streptococcus* spp., *Staphylococcus* spp., *Enterococci*, and Enterobacteriaceae are repeatedly introduced into the dynamic ecosystem of the FRT (Otero et al. 1999; Hafez and Hafez 2000; Otero et al. 2000; Jainudeen and Hafez 2000). Bacterial profile in the genital tract of female buffalos had reported in various stages of the reproductive cycle (El-Jakee et al. 2008). During a normal estrous cycle in

TABLE 18.1

Wide Variety of Bacterial Genera Colonizing in the FRT in Cows and Buffaloes

Bovine Breed	Source	Method	Genera	Phylum	Reference
Nellore	Vaginal fluid	Culture independent	*Aeribacillus, Bacteroides, Clostridium, Ruminococcus, Rikenella, Alistipes, Bacillus, Eubacterium, Prevotella*	Firmicutes, Bacteroidetes, Proteobacteria	Laguardia-Nascimento et al. (2015)
Criolla	Vaginal tract	Culture dependent	*Enterococci, Staphylococci,* and *Lactobacilli*	Firmicutes	Otero et al. (2000)
Cross bred *(Nelore x Hereford)*	Vagina	Culture dependent	*Enterococci, Lactobacilli*	Firmicutes	Otero et al. (1999)
Cross bred	Vaginal fluid	Culture independent	*Aggregatibacter, Streptobacillus, Phocoenobacter, Sediminicola, Sporobacter, Lactobacilli*	Proteobacteria, Fusobacteria, Bacteroidetes, Firmicutes, Actinobacteria, Tenericutes	Swartz et al. (2014)
	Vaginal fluid	Culture dependent	*Bacillus, Staphylococcus, Streptococcus, Enterococcus*	Firmicutes	Swartz et al. (2014)
Criollo Limonero	Vaginal swab	Culture dependent	*Arcanobacterium, Staphylococcus, Erysipelothrix rhusiopathiae, Bacteroides, Peptostreptococcus*	Actinobacteria Firmicutes Bacteroidetes	Zambrano-Nava et al. (2011)
Holstein	Vaginal fluid	Culture independent	*Bacteroides, Ureaplasma, Fusobacterium, Peptostreptococcus, Sneathia., Prevotella, Arcanobacterium, Anaerococcus, Parabacteroide, Propionibacterium*	Actinobacteria, Bacteroidetes, Firmicutes, Fusobacteria, Proteobacteria, Tenericutes	Machado et al. (2012)
Holstein	Vaginal swab	Culture independent	*Ureaplasma, Bacteroides, Fusobacterium, Sneathia, Porphyromonas*	Bacteroidetes, Proteobacteria, Firmicutes, Actinobacteria, Fusobacteria, Spirochaetes	Bicalho, Machado, et al. (2017), Bicalho, Santin, et al. (2017), Bicalho, Lima, et al. (2017)

(*Continued*)

TABLE 18.1 (*Continued*)

Wide Variety of Bacterial Genera Colonizing in the FRT in Cows and Buffaloes

Bovine Breed	Source	Method	Genera	Phylum	Reference
Gyr	Vaginal mucus	Culture independent	*Aeribacillus, Bacillus*	Firmicutes, Bacteroidetes, Proteobacteria, Actinobacteria. Unclassified bacteria	Giannattasio-Ferraz et al. (2019)
Holstein	Vaginal mucus	Culture independent	*Ruminococcaceae, Streptococcus, Fusobacterium, Porphyromonas, Aggregatibacter, Helcococcus, Corynebacterium, Bacteroides*	Firmicutes, Bacteroidetes, Fusobacteria, Proteobacteria, Tenericutes, Actinobacteria	Galvao et al. (2019)
Holstein	Vaginal swab	Culture independent	*Mannheimia, Moraxella, Bacteroides, Streptococcus, Pseudomonas*	Proteobacteria Firmicutes Bacteroides	Lima et al. (2019)
Holstein	Uterine fluid	Culture-independent	*Porphyromonas Mycoplasma, Mannheimia, Pasteurella, Fusobacterium*	Proteobacteria, Firmicutes, Fusobacteria, Bacteroidetes, Tenericutes, Actinobacteria.	Santos et al. (2011)
Brangus	Vaginal Swab	Culture independent	*Pasteurella, Fusobacterium*	Tenericutes, Proteobacteria, Fusobacteria, Firmicutes	Messman et al. (2020)
Holstein Friesian	Vaginal sample	Culture dependent/Culture independent	*Ruminococcus, Dialister, Escherichia /Shigella, Virgibacillus, Campylobacter, Helcococcus, Staphylococcus, Bacillus, Actinopolymorpha, Exiguobacterium, Haemophilus /Histophilus, Aeribacillus, Porphyromonas, Lactobacillus, Clostridium*	Bacteroidetes, Proteobacteria, Firmicutes	Gonzalez Moreno et al. (2016)
Murrah buffalo	Cervicovaginal mucus	Culture independent	*Burkholderia-Paraburkholderia, Stenotrophomonas, Corynebacterium_1, Porphyromonas, Helcococcus*	Proteobacteria, Firmicutes, Actinobacteria, Bacteroidetes, Tenericutes	Mahalingam et al. (2019)
Buffalo	Vaginal swab	Culture independent	*E. coli, Proteus, Klebsiella, Staphylococcus, Streptococcus, Bacillus*	Proteobacteria Firmicutes	Kumar et al. (2019)
Buffalo	Vaginal fluid	Culture independent	*Salmonella, Sporosarcina, Microlunatus, Syntrophococcus, Arcobacter, Shigella, Kocuria, Paracoccus, Acetobacter, Lechevalieria, Hyphomicrobium, Chromohalobacter, Bordetella, Peptostreptococcus, Brevundimonas, Moraxella, Microcoleus, Thiomonas, Weissella*	Proteobacteria Firmicutes Actinobacteria	Mahesh et al. (2020)

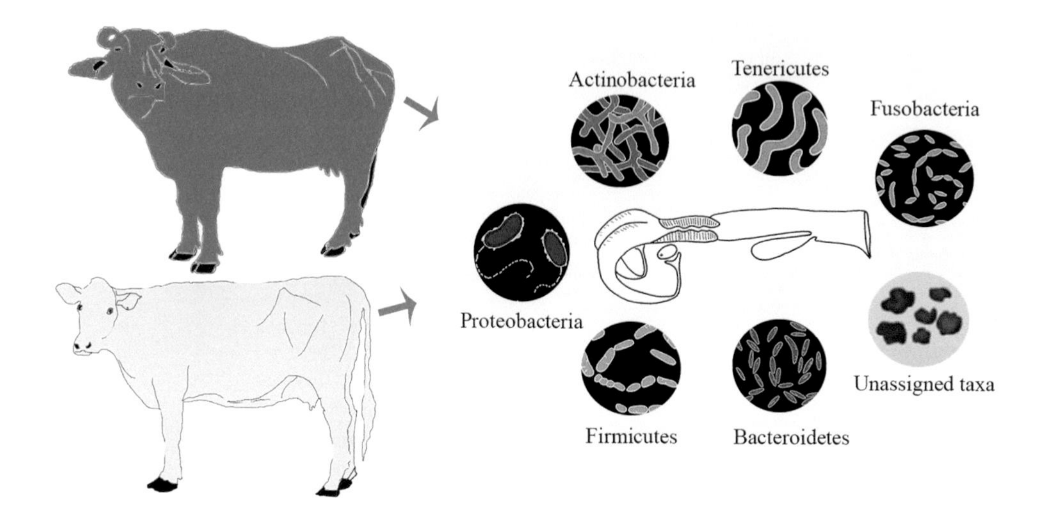

FIGURE 18.1 Microflora compositions in the reproductive tract of cow and buffalo. Proportions of main bacterial phyla are estimated based on studies that used 16S rRNA gene sequencing.

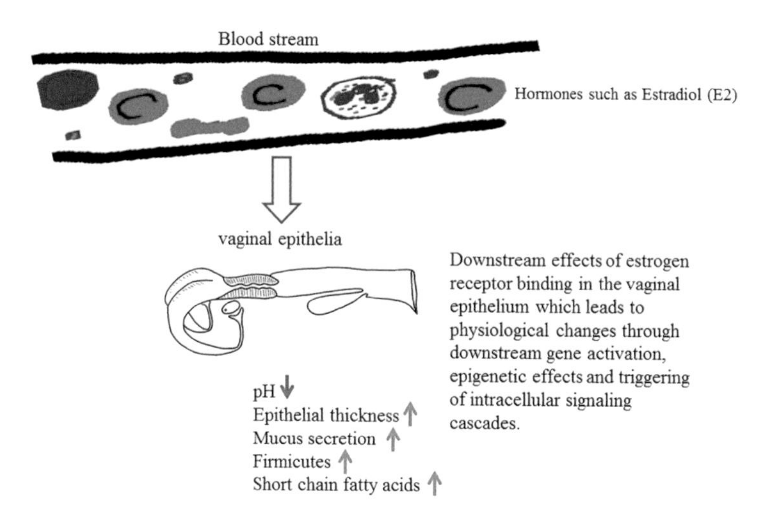

FIGURE 18.2 Possible sources of microflora to the reproductive tract during estrus: the hematogenous pathway (via blood) of bacterial transmission from the gut to the vaginal tract.

buffaloes, microbes such as *Escherichia coli*, *Enterococcus faecalis*, *Yersinia enterocolitica*, *Micrococcus* spp., *Citrobacter diversus*, *Corynebacterium bovis*, *Klebsiella* spp., and *S. epidermidis* were isolated. It was observed from the various studies that mucosa of FRT harbors both aerobic and anaerobic bacteria of pathogenic and commensals such as Gram-positive cocci and *bacilli* as well as *Enterobacteria* (Otero et al. 2000; Wang et al. 2013). Otero and colleagues described the dynamics of the cultured microbial population during the estrous cycle in cattle (Otero et al. 2000). Thus, in the estrus period dominated by estradiol (E2) release, the animals probably tend to present a high abundant microbiota (Figure 18.2). Vaginal bacterial microbiota could tend to be limited in pregnant females due to the presence of progesterone at a dominant level. After parturition, the microbial population in the vagina would return to the equilibrium as prior to pregnancy during the normal estrous cycle with an increment in bacterial population and a decrement in archaea populations (Laguardia-Nascimento et al. 2015).

Presence or absence of certain bacteria might serve as biomarkers for both optimal reproduction and failure (Deng et al. 2019). Research in this field is highly imperative to have preventive and curative strategies. Moreover, the phylogenetic relationship diversity of the vaginal microbial communities shift during the breeding period can lead to a successful pregnancy in cattle (Ault et al. 2019; Serrano et al. 2019). The vaginal fluids of cows contain the most abundant bacterial phyla, such as Firmicutes, Proteobacteria, Bacteroidetes, and Actinobacteria (Chen et al. 2020). The presence of colonized microbial flora in the vaginal tract with interspecies difference leads to comprehending animal physiology as well as health in a better way.

Studies described that the bovine reproductive tract microflora varies between the luteal and follicular stages (Wang et al. 2019). *E. coli* was predominately found during the assessment of the vaginal microflora during follicular and luteal phases, followed by *Aerococcus vaginalis*, *Aerococcus viridans*, *Haemophilus somnus*, *Streptococcus pluranimalium*, *Sphingomonas roseiflava*, *Psychrobacter marincola*, and *Lactobacillus* spp. in beef cattle (Wang et al. 2019). However, these bacterial populations varied in dairy cows, and this mismatch in the presence of microbial diversity may be due to the geological area, the methods used, and the types of of breed and immunity status of the animals. Noticeably, the vaginal *Streptococcus* spp. population in the luteal phase was significantly higher than that in the follicular phase, while the vaginal *Lactobacillus* spp. population in the follicular phase was considerably higher than in the luteal phase. However, the contrasts between the vaginal microbiota during follicular and luteal phasesare to be assessed, and it is suggested that future examinations should be conducted using a metagenomic investigation, which would be the better analysis of the bacteria within the genome community (Fettweis et al. 2019; Goltsman et al. 2018).

Studying the changes in the microbiota is significant during pregnancy and throughout the estrous cycle will demonstrate the potential influence of up and down regulation of ovarian steroids in growth of microbes. Our latest study revealed that the presence of the buffalo vaginal microbiota varies during the different phases of estrous cycle (Mahalingam et al. 2019). The prominent operational taxonomic unit (OTU) of 16S rRNA gene sequences, clearly depicted that the buffalo vaginal fluid bacteria were closely related to Corynebacterium, Porphyromonas, Helcococcus, Anaerococcus, and Fastidiosipila. The Firmicutes phylum was found particularly during the estrus phase. Moreover, the particular genera—including *Campylobacter*, *Porphyromonas*, and *Corynebacterium*—are found to be specifically increased at the time of estrus. Details of the taxonomic profile of buffalo vaginal microbiome during estrous cycle in STAT (**S**RA **T**axonomy **A**nalysis **T**ool, NCBI) are represented in Figure 18.3. Identifying bacterial communities at various phases of estrus cycle leads to the development of new approaches in estrus detection and also facilitates the buffalo fertility.

18.2.2 Cervico-Vaginal Microbiota

Cervix is protected by the significant physiological and anatomical barriers, which is the series of mucous-lined collagenous rings from the entry of environmental pathogens (Sheldon and Dobson 2004; Azawi 2008). Apart from that, cervical mucus also functions as a biological/physical barrier that prevents the invasion of microbes from the lower genital tract (Sheldon and Dobson 2004). The cervix must be assessed individually during the postpartum period to predict subsequent reproductive success (Deguillaume et al. 2012). The "uterine health" includes not only inflammation in uterine compartments but also inflammation of the cervix as well. These mechanisms would correlate link endocervical inflammation with delayed conception could be interest to therapy, and it should further investigated in future studies. Usually, cervix bacterial diversity

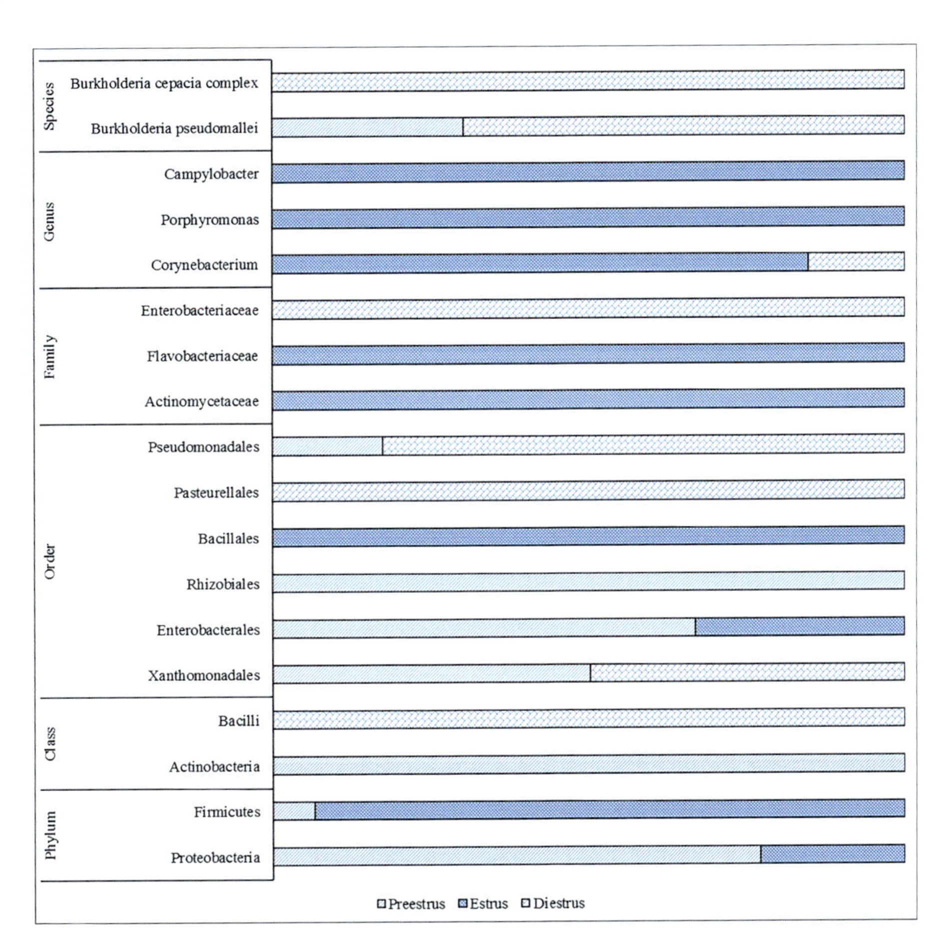

FIGURE 18.3 The taxonomic distribution from the next-generation sequencing runs of buffalo vaginal microbiome during estrous cycle. Distribution of reads mapping to specific taxonomy nodes as a percentage of total reads within the analyzed run.

was observed in cattle by traditional culture-dependent methods (Singer et al. 2016). Furthermore, next-generation sequencing is found to assist in studying and identifying uncultivated cervical bacteria in vaginal swab which conformed the most commonly found bacterial pathogens. Notably, the occurrence of *Staphylococcus aureus* from cervical swabs was higher in the case of animals with the record of abortion (Galvão 2018). Similar to vaginal flora, cervix is mainly comprised of Proteobacteria, Bacteroidetes, and Firmicutes during clinical formative (CF), clinical gestation (CG), and clinical postpartum (CP) phases of dairy cows (Wang et al. 2018). Conversely, the cervical bacterial community of metritis cows has a higher level of Bacteroidetes and Fusobacteria than in dairy cows during CF, CG, and CP phases. The findings showed that the presence of the microbial community in the cervical region alters the metritis in the infected cows.

Advanced sequencing technology confirmed the presence of cervical microbiome population in humans, and its fluctuations could be used as a potential clinical biomarker towards cervical intraepithelial neoplasia (CIN) and invasive cervical cancer associated with human papillomavirus (HPV) infection (Curty et al. 2019). Analysis of Cervicovaginal microbiome could be used as inflammatory biomarkers to diagnosis the genital pathology in both dairy and beef cows (Adnane et al. 2018; Chen et al. 2017).

18.3 Uterine Microbiota

Fluorescent probes that image bacteria and 16S ribosomal RNA sequencing confirmed the existence of microbiota in uterus even during pregnancy. But the microbial composition of uterine microbiota is less abundant than that of gut or vagina, whereas only a few parts of the bacterial loads are responsible for the postpartum uterine disease (Moore et al. 2017; Karstrup et al. 2017). However, other studies disagree with this information, and the difference in the methods used and the breeds of cattle could explain these conflicting results (Santos and Bicalho 2012; Santos et al. 2011). Irrespective of the cow's health, presence of bacterial diversity in the uterus is greater at the phylum levels, such as Proteobacteria, Tenericutes, Firmicutes, Bacteroidetes, Fusobacteria, Actinobacteria, and unassigned taxa (Santos and Bicalho 2012), as compared to that of vagina and cervical niche.

After parturition, pathogenic bacteria from the uterine microbiota could cause disease and reduce uterine health (Ryan et al. 2020). A recent study on lactating dairy cows clearly emphasizes the state of association among uterine health, corpus luteum (CL), fertility subindex, and reproductive function (Canadas et al. 2020). Interestingly, similar results were not observed in the recent study in lactating cows (Canadas et al. 2020). These altered results might be due to the methods adopted to study genetic factors of the host or the type of breed. Most importantly, the microbes like *Streptococcus pyogenes* and *E. coli* have been observed to be the major regulators of uterine diseases, and specifically, the culture-independent methods have been identified without genetic sequences associated with these organisms. Bacterial pathogens could cause or contribute to endometrial pathology. But bacteria such as *Peptostreptococcus* and *Propionibacterium* associate with the health status of the uterus (Sheldon et al. 2019). These results concluded that the bacterial community varies from animal to animal, between diseases, and with time postpartum. However, there is a lack of knowledge about pathogenic bacteria of the uterine disease. The presence of alpha-hemolytic streptococci at seven days postpartum by culture methods amplified the fertility yields in cows and improved the uterine health. Detection of *E. coli* in lactating cattle might promote the uterine health (LeBlanc et al. 2002). But these results are variable (de Boer et al. 2015). Therefore, it is concluded that the presence of bacteria species in the reproductive tract does not always cause infection/disease in cattle. Still, it is important to note that they can also improve uterine health in cattle. It is supported by a recent study which revealed that microbes contribute to health status of the uterus (Sheldon et al. 2019). Innate and adaptive immune responses during inflammation support the healthy uterus by eradicating pathogenic bacteria (Sheldon et al. 2020), and innate immune response also supports uterus during postpartum (Machado and Silva 2020; Pascottini and LeBlanc 2020).

18.4 Functional Role of Bacterial Flora in Reproductive Disorders and Failures

Microbial invasion during postpartum period in dairy and beef cows often impedes follicular growth by poor follicular growth and anovulatory follicles (Gilbert 2019). Understanding the functional role of these bacterial communities in the reproductive tracts of beef and dairy cattle provides a vital data to determine how these factors are linked with reproductive success.

It is ambiguous with respect to the pathogen, causing infection, and supports the uterine health status, while others are ubiquitous. Dysbiosis, which increases inflammation, might be due to disruption in the microbiome (Chase and Kaushik 2019). Both welfare and economic reasons made farmers be concern about postpartum uterine diseases such as endometritis, metritis, and pyometra occurs during and after calving (Galvão 2018). Uterine inflammation in dairy and beef cattle is one of the most important causes of infertility and predisposes the animals for reproductive culling (Helfrich et al. 2020; Sheldon and Owens 2017). The summary of the reproductive tract microflora associated with infection is depicted in Figure 18.4.

18.4.1 Metritis Microbiome Biomarker

Metritis and endometritis were not considered as two distinct clinical diseases (Bruun et al. 2002), and it would be unbearable to differentiate between these two diseases associated with the uterus (Sheldon et al. 2009). Recent discoveries revealed that uterine microbiota could indicate the metritis with a dysbiosis by reducing the richness and expansion of Bacteroidetes and Fusobacteria, particularly Bacteroides, Porphyromonas, and Fusobacterium (Galvao et al. 2019). However, in other studies, *Helcococcus ovis* was rather recently revealed as an emerging bacteria implicated in the pathogenesis of metritis in Holstein dairy cows (Cunha et al. 2019).

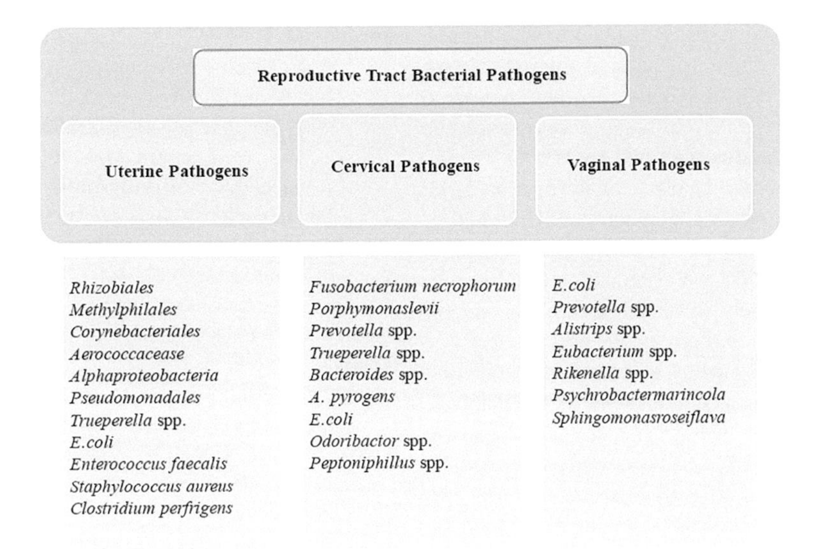

FIGURE 18.4 Microflora interactions in the reproductive tract of infected animals. Isolates of microbial flora from uterine, cervix, and vaginal flora and their interactions with the host animal and environmental factors (Appiah et al. 2020).

Investigations further focused on the incidence of a high comparative frequency of specific genera that are low in abundance, as it has been reported in a group infected with metritis, such as *Sneathia* and *Peptostreptococcus*; meanwhile, *Peptostreptococcus* has already been linked to uterine infections (Peng et al. 2013). Conversely, a recent study reported an association existing between *Sneathia* and metritis (Sicsic et al. 2018). However, these results are in disagreement with the previous investigation (Jeon et al. 2015), where *Sneathia* associates with the health of the uterus. The inconsistent results of each study revealed the fact that environmental bacteria contributed to the populations of microflora in the uterus of postpartum cattle and played a vital role in the frequency of metritis compared to the predicted relative incidence.

18.4.2 Clinical and Subclinical Endometritis Microbiome Biomarker

In regard to postpartum complications, it is challenging to relate decrement infertility to endometritis caused by *E. coli* and *S. pyogenes in vitro* to analyze reproductive failure associated with endometritis in Holstein heifers (Piersanti et al. 2019). Besides, most of the results exhibited *E. coli* as the major endometrial pathogens in beef cattle (Salah and Yimer 2017) and dual-purpose cattle (Ricci et al. 2015; de Cássia Bicudo et al. 2019). However, in other study on subclinical endometritis (ScE) was regarded as a multifactorial disease where *E. coli* pathogen was viewed as a causal agent and no other bacteria were isolated (Madoz et al. 2013); rather, coagulase-negative *staphylococci* (CoNS) were identified in healthy cows, and also, *S. pyogenes* were isolated from cows with clinical endometritis (CE) (Werner et al. 2012). Furthermore, *Corynebacterium endometrii sp.* were observed in the uterus of a cow with endometritis (Mogheiseh et al. 2020), whereas the abundance of Bacteroidetes and Fusobacterium was confirmed in cows having CE (Miranda-CasoLuengo et al. 2019).

The interaction exists between endometritis and cystic ovarian disease (COD) in Japanese black cattle due to the presence of *E. coli* at 40–60 days postpartum when compared with the normal ovarian cycle (Yamamoto et al. 2020). Endometritis exhibits adverse effects during the postpartum ovarian cycle of Holstein dairy cows (Mohammed et al. 2019). In particular, endometritis increases the chance of prolonged CL activity that might extend the interval from calving to conception and affects the economic status of farmers.

Unusual vaginal discharge is the early symptom which highlights the importance of monitoring cows with endometritis to proceed for treatment. The recent data also suggested that the development of postpartum endometritis could be associated with a delayed differentiation of vaginal and uterine microbiomes at the beginning of the postpartum period (Miranda-CasoLuengo et al. 2019). At the same time, alteration that happened in the histopathological and hematological profiles of endometrium could be associated with endometritis. About 90% of repeat breeder buffaloes exhibited signs of moderate or severe endometritis that could contribute to the low fertility recounted in repeat breeder buffaloes (Salzano et al. 2020; Dash et al. 2019). Ultimately, future studies are required to explore its underlying mechanism of action of endometritis depicted in the complex nature of the disease.

18.5 Conclusions and Future Recommendations

Overall, the microbiota of the bovine reproductive tract is a rapidly developing area, and its physiological role in bovine reproduction remains to be understood. The importance of the interaction between the host and the supporting microbial flora need to be studied exhaustively . However, the difference between the microbial ecosystem of cattle under healthy and diseased conditions has still to be unveiled as it causes pathological conditions in certain species of healthy cattle under various factors. The current review focuses on the varied composition of the microbial communities in the reproductive tract, and it might be due to contingent factors. The healthy cows have a complex and dynamic bacterial flora consisting of various anaerobic and fastidious bacteria when compared with metritis and endometritis. Even though variations in the microbiome of the reproductive tract and molecular characterization indicate the specific changes of normal bacterial flora, it could be used as a tool to understand the physiological changes in livestock and to monitor the reproductive health. Knowing the basic information and establishing baselines for vaginal microbiomes will help in understanding the importance of microbes in relation to reproductive efficiency.

In the past two decades, the use of advanced metagenomics has led to an explosion in understanding the bacterial microflora compositions in various ecological niches. Additionally, the recent cultivation techniques have been used as a molecular method to isolate specific and different forms of bacterial species in the vagina, cervix, and uterus. Identification and comparison of bacterial communities present in cattle reproductive tract (vaginal, uterine, and cervical microbiomes) during various physiological conditions by combining culture-dependent and culture-independent methods might lead to the development of new approaches towards treatment with probiotics, which would have a positive effect on reproductive efficiency. Determination of microbiome biomarkers to increase the success of pregnancy and enhance fertility in livestock would result in saving millions of dollars annually to the farmers. Nevertheless, it is recommended that enthusiastic work be continued in the future that aims at understanding the role of the microbiome in the reproductive tract of bovine in reference to health aspect.

Acknowledgments

GA acknowledges with thanks UGC, New Delhi, for the award of UGC-BSR Faculty Fellowship (No.F.18-1/2011 (BSR) dt.04.01.2017). The facility is availed from DST-FIST-II, and DST-PURSE phase-II, Rashtriya Uchchatar Shiksha Abhiyan (RUSA) 2.0 is gratefully acknowledged.

REFERENCES

Adnane, M., K. G. Meade, and C. O'Farrelly. 2018. Cervicovaginal mucus (CVM) - an accessible source of immunologically informative biomolecules. *Vet Res Commun* 42 (4):255–263.

Appiah, M. O., J. Wang, and W. Lu. 2020. Microflora in the reproductive tract of cattle: A review (running title: The microflora and bovine reproductive tract). *Agriculture-Basel* 10 (6):232.

Ault, T. B., B. A. Clemmons, S. T. Reese, F. G. Dantas, G. A. Franco, T. P. Smith, J. L. Edwards, P. R. Myer, and K. G. Pohler. 2019. Uterine and vaginal bacterial community diversity prior to artificial insemination between pregnant and nonpregnant postpartum cows. *J Anim Sci* 97 (10):4298–4304.

Azawi, O. I. 2008. Postpartum uterine infection in cattle. *Anim Reprod Sci* 105 (3–4): 187–208.

Bicalho, M. L. S., S. Lima, C. H. Higgins, V. S. Machado, F. S. Lima, and R. C. Bicalho. 2017. Genetic and functional analysis of the bovine uterine microbiota. Part II: Purulent vaginal discharge versus healthy cows. *J Dairy Sci* 100 (5):3863–3874.

Bicalho, M. L. S., V. S. Machado, C. H. Higgins, F. S. Lima, and R. C. Bicalho. 2017. Genetic and functional analysis of the bovine uterine microbiota. Part I: Metritis versus healthy cows. *J Dairy Sci* 100 (5):3850–3862.

Bicalho, M. L. S., T. Santin, M. X. Rodrigues, C. E. Marques, S. F. Lima, and R. C. Bicalho. 2017. Dynamics of the microbiota found in the vaginas of dairy cows during the transition period: Associations with uterine diseases and reproductive outcome. *J Dairy Sci* 100 (4):3043–3058.

Bruun, J., A. K. Ersboll, and L. Alban. 2002. Risk factors for metritis in Danish dairy cows. *Prev Vet Med* 54 (2):179–190.

Chase, C., and R. S. Kaushik. 2019. Mucosal immune system of cattle: All immune responses begin here. *Vet Clin North Am Food Anim Pract* 35 (3):431–451.

Chen, S. Y., F. Deng, M. Zhang, X. Jia, and S. J. Lai. 2020. Characterization of vaginal microbiota associated with pregnancy outcomes of artificial insemination in dairy cows. *J Microbiol Biotechnol* 30 (6):804–810.

Chen, C., X. Song, W. Wei, H. Zhong, J. Dai, Z. Lan, F. Li, X. Yu, Q. Feng, Z. Wang, and H. Xie. 2017. The microbiota continuum along the female reproductive tract and its relation to uterine-related diseases. *Nat Commun* 8 (1):875.

Clemmons, B. A., S. T. Reese, F. G. Dantas, G. A. Franco, T. P. Smith, O. I. Adeyosoye, K. G. Pohler, and P. R. Myer. 2017. Vaginal and uterine bacterial communities in postpartum lactating cows. *Front Microbiol* 8:1047.

Cunha, F., S. J. Jeon, P. Kutzer, K. C. Jeong, and K. N. Galvao. 2019. Draft genome sequences of *Helcococcus ovis* strains isolated at time of metritis diagnosis from the uterus of holstein dairy cows. *Microbiol Resour Announc* 8 (22):e00402-19.

Curty, G., P. S. de Carvalho, and M. A. Soares. 2019. The role of the cervicovaginal microbiome on the genesis and as a biomarker of premalignant cervical intraepithelial neoplasia and invasive cervical cancer. *Int J Mol Sci* 21 (1):222.

Dash, S., S. Basu, P. C. Mishra, and K. Ray. 2019. Effect of immune modulators on certain haematological parameters in ameliorating bovine endometritis. *Int J Chem Studies* 7 (2):1736–1739.

de Boer, M., B. M. Buddle, C. Heuer, H. Hussein, T. Zheng, S. J. LeBlanc, and S. McDougall. 2015. Associations between intrauterine bacterial infection, reproductive tract inflammation, and reproductive performance in pasture-based dairy cows. *Theriogenology* 83 (9):1514–1524.

de Cássia Bicudo, L., E. Oba, S. D. Bicudo, D. da Silva Leite, A. K. Siqueira, M. M. de Souza Monobe, M. Nogueira, J. C. de Figueiredo Pantoja, F. J. P. Listoni, and M. G. Ribeiro. 2019. Virulence factors and phylogenetic group profile of uterine Escherichia coli in early postpartum of high-producing dairy cows. *Anim Prod Sci* 59 (10):1898–1905.

Deguillaume, L., A. Geffre, L. Desquilbet, A. Dizien, S. Thoumire, C. Vornière, F. Constant, R. Fournier, and S. Chastant-Maillard. 2012. Effect of endocervical inflammation on days to conception in dairy cows. *J Dairy Sci* 95 (4):1776–1783.

Deng, F., M. McClure, R. Rorie, X. Wang, J. Chai, X. Wei, S. Lai, and J. Zhao. 2019. The vaginal and fecal microbiomes are related to pregnancy status in beef heifers. *J Anim Sci Biotechnol* 10 (1):92.

El-Jakee, J. A., W. M. Ahmed, F. R. El-Seedy, and S. I. Abd El-Moez. 2008. Bacterial profile of the genital tract in female buffaloes during different reproductive stages. *Glob Vet* 2 (1):7–14.

Fettweis, J. M., M. G. Serrano, J. P. Brooks, D. J. Edwards, P. H. Girerd, H. I. Parikh, B. Huang, T. J. Arodz, L. Edupuganti, A. L. Glascock, and J. Xu. 2019. The vaginal microbiome and preterm birth. *Nat Med* 25 (6):1012–1021.

Galvão, K. N. 2018. Postpartum uterine diseases in dairy cows. *Anim Reprod* 9 (3):290–296.

Galvao, K. N., R. C. Bicalho, and S. J. Jeon. 2019. Symposium review: The uterine microbiome associated with the development of uterine disease in dairy cows. *J Dairy Sci* 102 (12):11786–11797.

Giannattasio-Ferraz, S., M. Laguardia-Nascimento, M. R. Gasparini, L. R. Leite, F. M. G. Araujo, A. C. de Matos Salim, A. P. de Oliveira, J. R. Nicoli, G. C. de Oliveira, F. G. da Fonseca, and E. F. Barbosa-Stancioli. 2019. A common vaginal microbiota composition among breeds of Bos taurus indicus (Gyr and Nellore). *Braz J Microbiol* 50 (4):1115–1124.

Gilbert, R. O. 2019. Symposium review: Mechanisms of disruption of fertility by infectious diseases of the reproductive tract. *J Dairy Sci* 102 (4):3754–3765.

Goltsman, D. S. A., C. L. Sun, D. M. Proctor, D. B. DiGiulio, A. Robaczewska, B. C. Thomas, G. M. Shaw, D. K. Stevenson, S. P. Holmes, J. F. Banfield, and D. A. Relman. 2018. Metagenomic analysis with strain-level resolution reveals fine-scale variation in the human pregnancy microbiome. *Genome Res* 28 (10):1467–1480.

Gonzalez Moreno, C., C. Fontana, P. S. Cocconcelli, M. L. Callegari, and M. C. Otero. 2016. Vaginal microbial communities from synchronized heifers and cows with reproductive disorders. *J Appl Microbiol* 121 (5):1232–1241.

Hafez, B., and E. S. E. Hafez. 2000. Anatomy of female reproduction. In *Reproduction in Farm Animals*, edited by B. Hafez, and E. S. E. Hafez. Baltimore, MD: Lippincott Williams & Wilkins. pp: 13–29.

Helfrich, A. L., H. D. Reichenbach, M. M. Meyerholz, H. A. Schoon, G. J. Arnold, T. Fröhlich, F. Weber, and H. Zerbe. 2020. Novel sampling procedure to characterize bovine subclinical endometritis by uterine secretions and tissue. *Theriogenology* 141: 186–196.

Jainudeen, M. R., and E. S. E. Hafez. 2000. Reproductive failure in females. In *Reproduction in Farm Animals*, edited by B. Hafez, and E. S. E. Hafez. Baltimore, MD: Lippincott Williams & Wilkins. pp: 259–278.

Jeon, S. J., A. Vieira-Neto, M. Gobikrushanth, et al. 2015. Uterine microbiota progression from calving until establishment of metritis in dairy cows. *Appl Environ Microbiol* 81 (18):6324–6332.

Karstrup, C. C., K. Klitgaard, T. K. Jensen, J. S. Agerholm, and H. G. Pedersen. 2017. Presence of bacteria in the endometrium and placentomes of pregnant cows. *Theriogenology* 99:41–47.

Kumar, S., U. Sharma, M. A. Malik, and S Kumar. 2019. Vaginal bacterial profile in buffaloes following treatment with progesterone insert. *J Anim Res* 9 (2):359–361.

Laguardia-Nascimento, M., K. M. Branco, M. R. Gasparini, et al. 2015. Vaginal microbiome characterization of Nellore cattle using metagenomic analysis. *PLoS One* 10 (11):e0143294.

LeBlanc, S. J., T. F. Duffield, K. E. Leslie, K. G. Bateman, G. P. Keefe, J. S. Walton, and W. H. Johnson. 2002. Defining and diagnosing postpartum clinical endometritis and its impact on reproductive performance in dairy cows. *J Dairy Sci* 85 (9):2223–2236.

Lima, S. F., M. L. S. Bicalho, and R. C. Bicalho. 2019. The Bos taurus maternal microbiome: Role in determining the progeny early-life upper respiratory tract microbiome and health. *PLoS One* 14 (3):e0208014.

Machado, V. S., G. Oikonomou, M. L. Bicalho, W. A. Knauer, R. Gilbert, and R. C. Bicalho. 2012. Investigation of postpartum dairy cows' uterine microbial diversity using metagenomic pyrosequencing of the 16S rRNA gene. *Vet Microbiol* 159 (3–4):460–469.

Machado, V. S., and T. H. Silva. 2020. Adaptive immunity in the postpartum uterus: Potential use of vaccines to control metritis. *Theriogenology* 150: 201–209.

Madoz, L. V., M. J. Giuliodori, M. Jaureguiberry, J. Plontzke, M. Drillich, and R. L. de la Sota. 2013. The relationship between endometrial cytology during estrous cycle and cutoff points for the diagnosis of subclinical endometritis in grazing dairy cows. *J Dairy Sci* 96 (7):4333–4339.

Mahalingam, S., D. Dharumadurai, and G. Archunan. 2019. Vaginal microbiome analysis of buffalo (*Bubalus bubalis*) during estrous cycle using high-throughput amplicon sequence of 16S rRNA gene. *Symbiosis* 78 (1):97–106.

Mahesh, P, V. S. Suthar, D. B. Patil, M. Joshi, S. Bagatharia, and C. Joshi.. 2020. Vaginal microbiota during estrous cycle and its plausible association with certain hematological parameters in *Bubalus bubalis*. *Indian J Vet Sci Biotech* 15 (04):54–58.

Messman, R. D., Z. E. Contreras-Correa, H. A. Paz, G. Perry, and C. O. Lemley. 2020. Vaginal bacterial community composition and concentrations of estradiol at the time of artificial insemination in Brangus heifers. *J Anim Sci*98 (6): 1-9.

Miranda-CasoLuengo, R., J. Lu, E. J. Williams, A. A. Miranda-CasoLuengo, S. D. Carrington, A. C. Evans, and W. G. Meijer. 2019. Delayed differentiation of vaginal and uterine microbiomes in dairy cows developing postpartum endometritis. *PLoS One* 14 (1):e0200974.

Mohammed, Z. A., G. E. Mann, and R. S. Robinson. 2019. Impact of endometritis on post-partum ovarian cyclicity in dairy cows. *Vet J* 248:8–13.

Moore, S. G., A. C. Ericsson, S. E. Poock, P. Melendez, and M. C. Lucy. 2017. Hot topic: 16S rRNA gene sequencing reveals the microbiome of the virgin and pregnant bovine uterus. *J Dairy Sci* 100 (6):4953–4960.

Moreno, I., and C. Simon. 2019. Deciphering the effect of reproductive tract microbiota on human reproduction. *Reprod Med Biol* 18 (1):40–50.

Nesengani, L. T., J. Wang, Y. J. Yang, L. Y. Yang, and W. F. Lu. 2017. Unravelling vaginal microbial genetic diversity and abundance between Holstein and Fleckvieh cattle. *Rsc Advances* 7 (88):56137–56143.

Otero, C., C. S. de Ruiz, R. Ibanez, O. R. Wilde, A. A. P. D. Holgado, and M. E. Nader-Macias. 1999. Lactobacilli and enterococci isolated from the bovine vagina during the estrous cycle. *Anaerobe* 5 (3–4):305–307.

Otero, C., L. Saavedra, C. Silva de Ruiz, O. Wilde, A. R. Holgado, and M. E. Nader-Macias. 2000. Vaginal bacterial microflora modifications during the growth of healthy cows. *Lett Appl Microbiol* 31 (3):251–254.

Parfrey, L. W., C. S. Moreau, and J. A. Russell. 2018. Introduction: The host-associated microbiome: Pattern, process and function. *Mol Ecol* 27 (8):1749–1765.

Pascottini, O. B., and S. J. LeBlanc. 2020. Modulation of immune function in the bovine uterus peripartum. *Theriogenology* 150: 193–200.

Peng, Y., Y. Wang, S. Hang, and W. Zhu. 2013. Microbial diversity in uterus of healthy and metritic postpartum Holstein dairy cows. *Folia Microbiol (Praha)* 58 (6):593–600.

Piersanti, R. L., R. Zimpel, P. C. Molinari, M. J. Dickson, Z. Ma, K. C. Jeong, J. E. Santos, I. M. Sheldon, and J. J. Bromfield. 2019. A model of clinical endometritis in Holstein heifers using pathogenic *Escherichia coli* and *Trueperella pyogenes*. *J Dairy Sci* 102 (3):2686–2697.

Ricci, A., S. Gallo, F. Molinaro, A. Dondo, S. Zoppi, and L. Vincenti. 2015. Evaluation of subclinical endometritis and consequences on fertility in piedmontese beef cows. *Reprod Domest Anim* 50 (1):142–148.

Canadas, E. R., M. M. Herlihy, J. Kenneally, J. Grant, F. Kearney, P. Lonergan, and S. T. Butler. 2020. Associations between postpartum phenotypes, cow factors, genetic traits, and reproductive performance in seasonal-calving, pasture-based lactating dairy cows. *J Dairy Sci* 103 (1):1016–1030.

Ryan, N. J., K. G. Meade, E. J. Williams, et al. 2020. Purulent vaginal discharge diagnosed in pasture-based Holstein-Friesian cows at 21 days postpartum is influenced by previous lactation milk yield and results in diminished fertility. *J Dairy Sci* 103 (1):666–675.

Salah, N., and N. Yimer. 2017. Cytological endometritis and its agreement with ultrasound examination in postpartum beef cows. *Vet World* 10 (6):605–609.

Salzano, A., A. Pesce, L. D'Andrea, O. Paciello, F. della Ragione, P. Ciaramella, C. Salzano, A. Costagliola, F. Licitra, and G. Neglia. 2020. Inflammatory response in repeat breeder buffaloes. *Theriogenology* 145: 31–38.

Santos, T. M., and R. C. Bicalho. 2012. Diversity and succession of bacterial communities in the uterine fluid of postpartum metritic, endometritic and healthy dairy cows. *PLoS One* 7 (12):e53048.

Santos, T. M., R. O. Gilbert, and R. C. Bicalho. 2011. Metagenomic analysis of the uterine bacterial microbiota in healthy and metritic postpartum dairy cows. *J Dairy Sci* 94 (1):291–302.

Serrano, M. G., H. I. Parikh, J. P. Brooks, D. J. Edwards, T. J. Arodz, L. Edupuganti, B. Huang, P. H. Girerd, Y. A. Bokhari, S. P. Bradley, and J. L. Brooks. 2019. Racioethnic diversity in the dynamics of the vaginal microbiome during pregnancy. *Nat Med* 25 (6):1001–1011.

Sheldon, I. M., J. G. Cronin, and J. J. Bromfield. 2019. Tolerance and innate immunity shape the development of postpartum uterine disease and the impact of endometritis in dairy cattle. *Annu Rev Anim Biosci* 7 (1):361–384.

Sheldon, I. M., J. Cronin, L. Goetze, G. Donofrio, and H. J. Schuberth. 2009. Defining postpartum uterine disease and the mechanisms of infection and immunity in the female reproductive tract in cattle. *Biol Reprod* 81 (6):1025–1032.

Sheldon, I. M., and H. Dobson. 2004. Postpartum uterine health in cattle. *Anim Reprod Sci* 82–83:295–306.

Sheldon, I. M., P. C. C. Molinari, T. J. R. Ormsby, and J. J. Bromfield. 2020. Preventing postpartum uterine disease in dairy cattle depends on avoiding, tolerating and resisting pathogenic bacteria. *Theriogenology* 150: 158–165.

Sheldon, I. M., and S. E. Owens. 2017. Postpartum uterine infection and endometritis in dairy cattle. *Anim Reprod* 14 (3):622–629.

Sicsic, R., T. Goshen, R. Dutta, et al. 2018. Microbial communities and inflammatory response in the endometrium differ between normal and metritic dairy cows at 5–10 days postpartum. *Vet Res* 49 (1):77.

Singer, E., B. Bushnell, D. Coleman-Derr, B. Bowman, R. M. Bowers, A. Levy, E. A. Gies, J. F. Cheng, A. Copeland, H. P. Klenk, and S. J. Hallam. 2016. High-resolution phylogenetic microbial community profiling. *ISME J* 10 (8):2020–2032.

Stevenson, J. S., S. L. Hill, G. A. Bridges, J. E. Larson, and G. C. Lamb. 2015. Progesterone status, parity, body condition, and days postpartum before estrus or ovulation synchronization in suckled beef cattle influence artificial insemination pregnancy outcomes. *J Anim Sci* 93 (5):2111–2123.

Swartz, J. D., M. Lachman, K. Westveer, T. O'Neill, T. Geary, R. W. Kott, J. G. Berardinelli, P. G. Hatfield, J. M. Thomson, A. Roberts, and C. J. Yeoman. 2014. Characterization of the vaginal microbiota of ewes and cows reveals a unique microbiota with low levels of lactobacilli and near-neutral pH. *Front Vet Sci* 1:19.

Wang, Y., B. N. Ametaj, D. J. Ambrose, and M. G. Ganzle. 2013. Characterisation of the bacterial microbiota of the vagina of dairy cows and isolation of pediocin-producing Pediococcus acidilactici. *BMC Microbiol* 13 (1):19.

Wang, Jun, Chang Liu, Lucky T. Nesengani, Y. Gong, Y. Yang, L. Yang, and W. Lu. 2019. Comparison of vaginal microbial community structure of beef cattle between luteal phase and follicular phase. *Indian J Anim Res* 53 (10):1298–1303.

Wang, J., C. Sun, C. Liu, Y. Yang, and W. Lu. 2016. Comparison of vaginal microbial community structure in healthy and endometritis dairy cows by PCR-DGGE and real-time PCR. *Anaerobe* 38:1–6.

Wang, Y., J. Wang, H. Li, K. Fu, B. Pang, Y. Yang, Y. Liu, W. Tian, and R. Cao. 2018. Characterization of the cervical bacterial community in dairy cows with metritis and during different physiological phases. *Theriogenology* 108:306–313.

Werner, A., V. Suthar, J. Plontzke, and W. Heuwieser. 2012. Relationship between bacteriological findings in the second and fourth weeks postpartum and uterine infection in dairy cows considering bacteriological results. *J Dairy Sci* 95 (12):7105–7114.

Yamamoto, N., R. Nishimura, Y. Gunji, and M. Hishinuma. 2020. Research of postpartum endometritis in Japanese Black cattle with cystic ovarian disease by vaginal mucus test and endometrial cytology. *Arch Anim Breed* 63 (1):1–8.

Zambrano-Nava, S., J. Boscan-Ocando, and J. Nava. 2011. Normal bacterial flora from vaginas of Criollo Limonero cows. *Trop Anim Health Prod* 43 (2):291–294.

Section IV

Plant Microbiome

19

Insights into the Structure, Function, and Dynamics of Rice Root and Rhizosphere-Associated Microbiome

Ekramul Islam and Kiron Bhakat
University of Kalyani

CONTENTS

19.1 Introduction

Each plant harbors a diverse group of microorganisms in their body parts. A system consisting of plants, their environments, and all microorganisms in that environment to which the plant interacts is known as "phytobiome" (Leach et al., 2017). Dynamic interactions between the components of phytobiome tightly regulate the agroecosystem function, and thus, sustainable agriculture is largely dependent on this tripartite interaction (Kim and Lee, 2020). Plant-associated microbial community is the vital biotic component referred to as "microbiome." Soil is the greatest reservoir of microorganisms, which serves as the source of plant microbiome. Terrestrial plants are directly exposed to the soil microbial reservoir through their root. The bulk soil microorganisms are found to colonize the various compartments of the root system (the root microbiome) that include the endosphere (inside the root), rhizoplane (root surface), and rhizosphere (soil surrounding the root) (Ding et al., 2019). Each plant and the below-ground microorganisms are interdependent entities. Plants are well known to secrete organic compounds through their root that attract the microorganisms in surrounding soil and support their growth and activities (Hussain et al., 2012; Zhu et al., 2017, 2018). Conversely, soil and root-associated microorganisms help the plant in many aspects, like plant growth promotion by supplying growth hormones, mobilizing micronutrients, inhibiting phytopathogens, and enhancing tolerance to environmental stress (Kim and Lee, 2020). Microorganisms do these beneficial jobs as a part of complex microbial community. Directed functions of microbial community largely depend on the abundance, diversity, and composition of the participating microbial species (Escalas et al., 2019). Some part of the microbial community remains static, while the other part is highly dynamic during the growing cycle of the crops (Hussain et al., 2012). Any perturbation to these attributes of the microbial community due to different reasons might lead to the loss of beneficial outcome imparted by the microbes to the plant (Kim and Lee, 2020). As the below-ground microorganisms play a critical role in maintaining plant health, one should carefully look after the root-associated microbiome also.

Rice is the staple food of more than 50% of the world population. Unlike other plants, rice is generally cultivated under deep-irrigated flooded paddy soil, which creates anaerobic condition in the soil. In this condition, microbial reduction processes get enhanced, which allows the development of various functional microbial guilds that sequentially use NO_3^-, Mn^{4+}, Fe^{3+}, and SO_4^{2-} as electron acceptors, where trace gases such as N_2O, N_2, H_2S, and CH_4 are found to be released (Kögel-Knabner et al., 2010). On the other hand, the rice plant is well known to release O_2 into the soil through root aerenchyma, which creates oxygenated environment around the root that is surrounded by the anoxic bulk soil (Colmer et al., 2019). This oxic–anoxic zone develops a redox gradient around the root, which is very much conducive for various microbial groups with diverse physiology, making the rhizosphere microbial hotspot (Zhang et al., 2020). Disruption of anaerobic condition by radial oxygen loss (ROL) is found to inhibit various microbial guilds that facilitate biogeochemical cycle of key elements like C, N, P, and Fe (Wei et al., 2019). Oxic–anoxic condition also allows the formation of root plaque, a brown precipitate of Fe and Mn oxide, on the rice root surface. Investigation showed that microorganisms in root plaque play an important role in uptake of ions by plants from soil (Liu et al., 2019). Among various factors, soil health and rice genotype largely influence the assemblage of root microbiota (Kim and Lee, 2020). It has been demonstrated that rice plants those have unique nitrogen-use efficiency gene, could allow specific microbial groups to colonize with their root (Zhang et al., 2019).

Besides, feeding the world's growing population, rice cultivation is well known to contribute to substantial amount of greenhouse gases such as methane (CH_4) and nitrous oxide (N_2O). Multiple microbial pathways in soil can lead to the production of N_2O, including nitrification and heterotrophic denitrification (Liu et al., 2019). Enriched methanogens under the flooded condition in rice rhizosphere and bulk soil increased CH_4 release (Liechty et al., 2020). Given that microbiome in soils greatly influences the plant productivity, comprehensive understanding and exploitation of the rice root-associated microbiome are potentially beneficial to the promotion of crop health and sustainable productivity of paddy ecosystems (Ding et al., 2019). However, to date, there are lacks of studies which provide a summary of literature focused on the microbiomes residing in the rice root-related compartments.

In this chapter, first, microbial diversity and community structure of the rice root system have been described based on the published literatures. Then, functional diversity of microbial groups is discussed. Characteristics of rhizoplane-, endosphere-, and rhizosphere-associated microbiomes have been discussed in relation to paddy ecosystem functioning. Finally, we tried to find the gaps in knowledge and provide a glimpse on the future research perspectives.

19.2 Global Importance and Cultivation Strategies of Rice

Rice, *Oryza sativa*, is the second largest produced cereal in the world belonging to the grass family Poaceae of the plant kingdom. Asia contributes about 90% of the world rice production and is the largest rice consumer where it is taken as principal staple food. India ranks as the second largest producer and exporter of rice after China. Rice cultivation generally requires deep irrigation and is often grown in monoculture with two to three crops a year depending upon water availability. The demand for rice is increasing day by day to feed world increasing population. The production of rice can be increased by increasing the area of rice field, increasing rice yields, and increasing cropping intensity. But the scope for expansion of rice field is limited. Therefore, future increase in rice supply must come from increased yields and intensified cropping, particularly in the irrigated rice ecosystem.

19.3 Strategies of Microbiome Profiling

Microbiota of plants is generally profiled by analyzing extracted DNA from various plant compartments such as rhizosphere, endosphere, and rhizoplane using marker genes such as ribosomal RNA genes or internal transcribed spacer (ITS) region. Before the advent of the –"omics" technologies, terminal restriction fragment length polymorphism (T-RFLP), amplified ribosomal DNA restriction analysis (ARDRA), PCR-denaturing gradient gel electrophoresis (PCR-DGGE), and clone library analysis of 16S rRNA genes were used to explore the microbiome. With these approaches, however, only information regarding the composition of microbial communities could be obtained. But the term "microbiome" does not only refer to the composition but also their niche and collective genome (Bäckhed et al., 2005, Ursell et al., 2012). In recent years, high-throughput sequencing and –omics technologies (metagenomics, metaproteomics, metatranscriptomics, and metabolomics) have greatly enhanced the microbiome analysis. Various techniques used to decipher the microbial community are presented in Table 19.1. These technologies have been widely used in human and animal microbiome analyses that greatly facilitated to build functional relationship between microbiome and host (Arnold et al., 2016). Dysbiosis of the microbial communities is found related to various disease conditions like obesity and inflammatory bowel disease (Carding et al., 2015). However, little attention has been paid to microbiome of plants—dysbiosis in which might lead to a reduction in soil fertility, reduced plant growth, and loss of resistance to disease. Microbial community helps the plant to cope up the environmental and nutritional stresses as well.

19.4 Rice Root Anatomy and Microbial Distribution Therein

Rice root together with soil can be divided into various compartments. From bulk soil, these compartments are rhizosphere, root plaque, rhizoplane, and endosphere (Figure 19.1). Microorganisms are colonized in these compartments and performed distinct function being the part of microbial community. Recent studies related to the structure and function of microbial communities in various compartments and their exploitation in sustainable rice cultivation are presented in Table 19.1.

19.4.1 Rhizosphere

Rhizosphere is the small compartment of soil surrounding the plant root that extends from root surface to bulk soil. Anaerobic condition due to flooded irrigation during rice cultivation and ROL from rice root makes this region a unique habitat for microorganisms (Kögel –Knabner et al., 2010; Wu et al., 2012). ROL creates a redox gradient, which is gradually decreased from the root surface to the bulk soil. The width of the oxic zone around the root depends on the magnitude of O_2 loss, root permeability, and the root and soil respiration

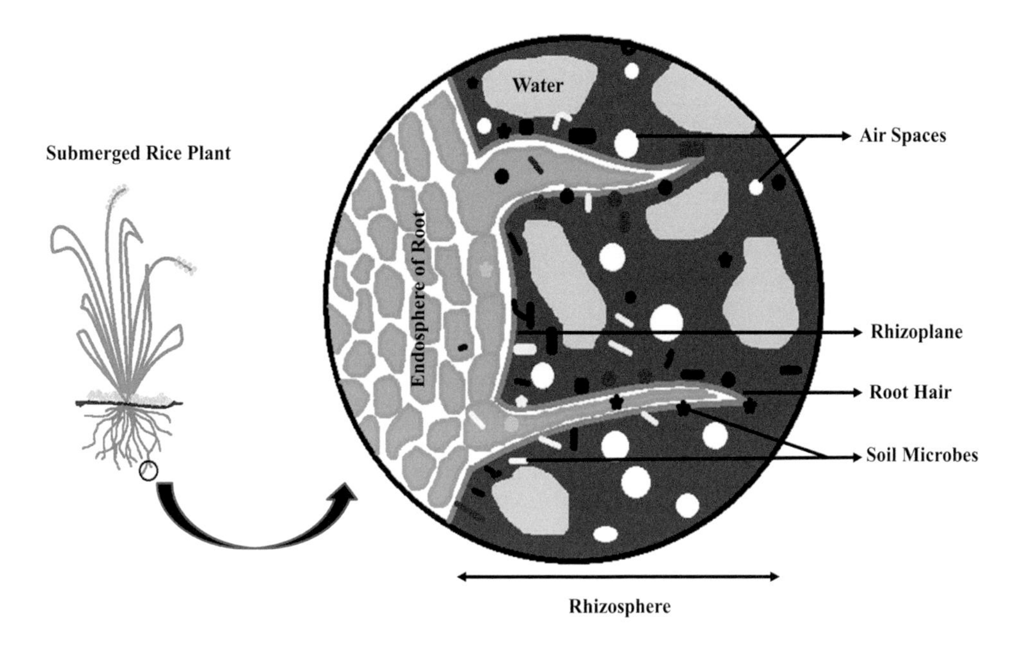

FIGURE 19.1 Schematic representation of various compartments (rhizosphere, rhizoplane, and endosphere) of rice root system under flooded condition.

TABLE 19.1

Diversity and Distribution of Root-Associated Microbiome and Techniques Used

Compartment Studied	Major Findings	Tools Used for Deciphering the Microbiome	References
Rhizosphere	Dynamics of the soil microbial community of rice	Extraction of soil genomic DNA followed by PCR amplification and construction of sequencing library using Illumina MiSeq system	Li et al. (2019)
	Investigation of various soil microbial community of rice rhizosphere	Culture-independent techniques: PCR-DGGE (denaturing gradient gel electrophoresis)	Hussain et al. (2011, 2012)
	Study of the composition and diversity of rhizosphere microbiota (bacteria and archaea)	Meta-omics technology	Breidenbach et al. (2016)
	Distribution of anaerobic bacterial community coexisting under oxic and anoxic conditions of rice fields	Gene marker-based community study; meta-omics approach	Ding et al. (2019)
	Comparative study of the rice rhizosphere microbial diversity with other common plants/crops and their influence in nutrient cycling	High-throughput sequencing (HTS) studies; metaproteomics and metagenomics approaches	Ding et al. (2019)
	Insight study of rhizomicrobiome of rice cultivation and their responses to environmental shifts	Soil DNA extraction followed by PCR amplification and HTS using Illumina HiSeq 2500 PE250 platform; OTU cluster using UCLUST in QIIME; taxonomic classification using RDP classifier based on SILVA and UNITE database for bacteria and fungi, respectively	Xu et al. (2019b)
Rhizoplane	Alteration of community diversity of endophyte-enriched rice root system using different agricultural strategies	HTS and pyrosequencing approaches	Jha et al. (2020)
	Functional characteristics of the endophytes colonizing the rice root system	Metagenomic techniques	Sessitsch et al. (2012)
	Microbial community study of iron plaque of rice roots and their significance on arsenic metabolism	16S rRNA extraction followed by sequencing using Illumina and clustering of sequence into operational taxonomic units (OTUs) using UCLUST tool	Hu et al. (2019)
	Study of the total and group-specific bacterial community associated with the roots of different rice cultivars	PCR amplification following DNA extraction and DGGE analysis	Hardoim et al (2011)

(*Continued*)

TABLE 19.1 (*Continued*)

Diversity and Distribution of Root-Associated Microbiome and Techniques Used

Compartment Studied	Major Findings	Tools Used for Deciphering the Microbiome	References
Endosphere	Contribution of soil microbiota in ultrahigh yield of rice	Root RNA and DNA extraction followed by PCR and sequencing using Illumina HiSeq 2,500 platform. Metagenomic analysis	Zhong et al. (2020)
	Interrelationship study between root microbiology, host genotype regulation, and nutrient cycling	Culture-independent analysis: DNA extraction, PCR amplification and sequencing using Illumina HiSeq 2,500 platform; metagenomic study using MetaGeneMark_v1_mod Culture-dependent analysis: inoculating the tryptic soy broth on microtiter plate and incubation at room temperature	Zhang et al. (2019)
	A comparative study of the endophytic bacterial community of aerobic rice under different water regimes	Total DNA extraction of microbial community followed by cloning and sequencing; restriction fragment length polymorphism (RFLP) and phylogenetic analysis using MEGA 4.0 software: Diversity and coverage analysis using MOTHUR	Vishwakarma and Dubey (2019)
Bulk soil	Abundance of bacterial and archaeal communities of rice fields with the effects of nitrogen fertilization and cropping season on the community	Culture-independent techniques like quantitative PCR (qPCR), terminal restriction fragment length polymorphism (T-RFLP), pyrosequencing, nonmetric multidimensional scaling (NDMS) analysis for archaea	Ji et al. (2020)
Rhizoplane/rhizosphere	Dynamics of the root-associated microbiomes over the life cycle of paddy	16S rRNA extraction followed by sequencing using Illumina MiSeq system and further processing of the sequence by PYTHON sequence clustering and filtering into operational taxonomic units (OUTs) using Ninja-OPS pipeline	Edwards et al. (2018)
Rhizoplane/rhizosphere/ bulk soil	Composition, structure, and role of microbiomes associated with the roots of rice plant	16S rRNA gene amplification and sequencing by Illumina MiSeq platform, clustering of sequence into OUT using UCLUST, taxonomic classification of RDP using QIIME version	Edwards et al. (2015)
Rhizosphere/bulk soil	Study of core microbial community associated with rice root system	PCR amplification of bacterial 16S rRNA genes was followed by sequencing using Illumina MiSeq platform; pyrosequencing using QIIME pipeline; OTU classification	Xu et al. (2019a)
Rhizosphere/rhizoplane/ endosphere	Recognizing the association of root microbiota and transcriptomes of wild and cultivable rice varieties	DNA extraction and purification study of 16S rRNA followed by sequencing using Illumina MiSeq platform; sequence processing using QIIME and RDP classifier: OTU constructed using USEARCH	Tian et al. (2018)
	Study of different methane-emitting microbial species associated with rice	16S rRNA gene sequencing followed by amplicon library preparation; OTU clustering analysis using UCLUST	Liechty et al. (2020)
Bulk soil/rhizosphere/ rhizoplane/endosphere	Impact of drought on the compositional shift of rice root microbiota	DNA extraction followed by the construction of 16S rRNA gene library; OTU clustering using QIIME and phylogenetic tree construction using Fast Tree	Santos-Medellin et al. (2017)
Bulk soil/rhizoplane/ endosphere	Study of the effects of phosphorus application on species richness and diversity-associated microbial communities	Genomic DNA extraction by CTAB protocol followed by HTS using Illumina MiSeq PE 300 platform; gene analysis of microbiome using QIIME platform; fungal nuclear ribosome was assigned using RDP classifier, and ITS (internal transcribed spacer) gene was analyzed using OTU clustering	Long and Yao (2020)

(Larsen et al., 2015). This redox gradient of oxic–anoxic interface favors certain groups of microorganisms such as methanogens and methanotrophs, which are adapted to this unique niche. Rhizodeposition and sloughed-off cells from root also greatly influence the distribution and diversity of microorganisms in the rhizosphere (Wu et al., 2017; Sasse et al., 2018).

Microbial communities in the rice rhizosphere in terms of diversity and composition of bacteria, archaea, and fungi are widely studied. Rhizosphere is dominated by bacteria. Although the growth stages of rice plant and soil depth greatly affect the microorganisms, in general, rhizospheric bacteria, archaea, and fungi are more abundant compared to those of bulk soil (Hussain et al., 2012; Lee et al., 2015; Breidenbach et al., 2016). This might be due to the deposition of carbon metabolite in the rhizosphere that influences the microbial growth (Ding et al., 2019). Alpha diversities of rice rhizosphere microbial communities are found to be close or even significantly less than in the bulk soil (Edwards et al., 2015).

Microbial communities associated with rice rhizosphere are found to be very diverse and dynamic, especially for bacteria, which is largely influenced by geographic location of rice field, soil type, genotype of rice, and agricultural management (Edwards et al., 2015; Santos-Medellin et al., 2017). Based on the published literature, Ding et al. (2019) summarize the information regarding assemblage of bacterial communities in rice rhizosphere at different taxonomic level, which indicates larger occupancy of *Proteobacteria* (mainly *Alpha-*, *Beta-*, and *Deltaproteobacteria* classes), *Acidobacteria*, *Actinobacteria*, and *Chloroflexi* phyla. It has been observed that rice rhizosphere microbiome is significantly different from the rhizosphere microbiome of other crops, while bacteria within the class *Deltaproteobacteria* are found to be predominated in rice rhizosphere. Particularly, *Geobacter* and *Desulfococcus* genera of this class are notable (Sun et al., 2015; Ding et al., 2019). Other abundant groups *Alpha-* and *Betaproteobacteria* are involved in the regulation of nutrient cycle, production of phytohormones, and synthesis of antibiotics that inhibit the growth of phytopathogen in the rhizosphere (Ding et al., 2019).

Rice rhizosphere archaeal communities are commonly composed of *Crenarchaeota*, *Thaumarchaeota*, and *Euryarchaeota* (Ding et al., 2019). Similar to bacteria, archaeal population of rice rhizosphere also markedly differs from that of other crops where dominance of *Thaumarchaeota* is noticed. The population of *Euryarchaeota* is found to be quite enriched in the rhizosphere because of abundance of methanogen genera like *Methanosarcina* and *Methanosaeta* (Lee et al., 2014). Compared to bacterial community, archaeal community is relatively stable in respect to irrigation, soil horizon, and growth season (Ding et al., 2019).

Among fungi, mostly aquatic species are found to associate with rice rhizosphere microbial community (Barr, 2001). *Rhizophlyctis* and *Cladochytrium* genera are frequently noticed in the rice rhizosphere that might be involved in the decomposition of cellulose, facilitating C cycle (Gleason et al., 2011; Eichorst and Kuske, 2012). Members of the genus *Aspergillus* are well known to be involved in P solubilization, while arbuscular mycorrhizal fungi (AMF) facilitate P transport and uptake (Mendes et al., 2013). Rice genotypes, crop management practices, and soil types affect the overall diversity of fungi as well as an effective root colonization of AMF (Diedhiou et al., 2016; Mbodj et al., 2018).

19.4.2 Iron Root Plaque

A thin layer of reddish-brown color deposition on the root surface of rice plant is generally observed, which is known as "root iron plaque" (Khan et al., 2016) (Figure 19.2). This deposition mainly contains iron and manganese. Root plaque is formed when rice plant is cultivated in weathered soil with high iron. Chemical oxidation Fe(II) by O_2 has been considered as the main driver for iron root plaque formation; however, recently, it has been accepted that microorganisms are also involved in Fe oxidation in the rhizosphere. In microaerophilic conditions that coincide 0.5–1.5 mm thick niche from the root surface, Fe-oxidizing bacteria are found to contribute to 18%–62% of the total Fe^{2+} oxidation (Chen et al., 2017; Dubinina and Sorokind, 2014; Han et al., 2016). Recent reports also suggest the importance of Fe-reducing bacteria in root plaque formation in submerged condition as they contribute to more than 12% of the bacteria in the rhizosphere, but only <1% in the bulk soil. The microorganisms responsible for iron reduction in rice rhizosphere are affiliated with *Geobacter* sp. and *Shewanella* sp. (Wang et al., 2009; Zecchin et al., 2017).

19.4.3 Rhizoplane

The root surface of plants that is in direct contact with rhizosphere soils and drives nutrient exchange and transformation in the soil–plant systems is known as "rhizoplane" (Sasse et al., 2018). Rhizoplane serves as a barrier that selectively allows rhizosphere microbiome to attach the rhizoplane and enter into the endosphere. Among the bacteria, members of the genera *Bacillus* and *Pseudomonas* are found to be typical inhabitant of the rhizoplane of rice (Hwangbo et al., 2016). The inhabitant of rhizoplane microorganism is involved in P solubilization (*Bacillus velezensis*), production of cyanide or siderophore that inhibits the growth of root pathogen (*Pseudomonas fluorescens*), N_2 fixation (e.g., *Azospirillum*

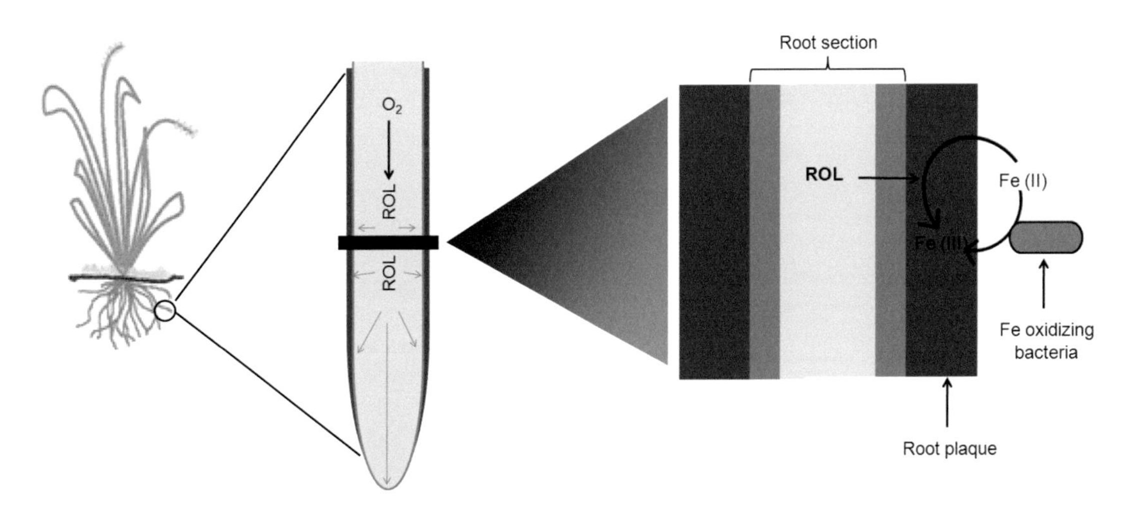

FIGURE 19.2 Schematic diagram of formation of root plaque on root surface.

spp.), and nitrification (e.g., *Nitrobacter* and *Nitrosospira* spp.) (Hwangbo et al., 2016; Banik et al., 2016). Using metaproteomic analyses, Knief et al. (2012) found the predominance of *Alpha-*, *Beta-*, and *Deltaproteobacteria* in the rice rhizoplane. The latter research group found the relatedness of the proteins recovered to *Azospirillum* and *Bradyrhizobium* for N_2 fixation, *Methylosinus* for CH_4 oxidation, and *Anaeromyxobacter* and *Geobacter* for Fe(III) reduction (Knief et al., 2012; Ding et al., 2019). Edwards et al. (2015) noticed that members of *Fibrobacteres* and *Spirochetes* from rice rhizoplane could degrade cellulose. Plant species and agricultural practice have been shown to contribute to variations in the diversity, abundance, and composition of rice rhizoplane microbiome. However, the decisive factor for shaping the rice rhizoplane microbiome is still unclear and requires further investigation (Ding et al., 2019).

19.4.4 Endosphere

Internal regions of plant parts are known as "endosphere." Microorganisms that are found inside the surface-sterilized endosphere are known as "endophytes" (Mano and Morisaki, 2008). Microorganisms can internally colonize the rice endosphere via either vertical or horizontal transmission routes. Microbes colonized the seeds might be transmitted vertically from mother generation to daughter during germination of seeds (Hardoim et al., 2012). In horizontal transmission, rhizosphere microbes colonize the endosphere by cell wall-degrading enzymes like cellulase and pectinase or by accidentally entering through the root wounds (Hardoim et al., 2015). Among bacteria, the common inhabitant microorganisms of the rice root endosphere are affiliated to *Gammaproteobacteria* (*Enterobacteriaceae* and *Pseudomonadaceae*), *Firmicutes* (*Bacillus*), *Actinobacteria*, and *Bacteroidetes* (Reinhold-Hurek and Hurek, 2011; Sessitsch et al., 2012). Compared to other compartment, reduced alpha-diversity is observed in rice endosphere, but it harbors considerably diverse microbial community (Edwards et al., 2015). A wide distribution but with small-sized population of archaea and fungi are also observed in rice root endosphere (Santos-Medellin et al., 2017). Among archaea, members of the methanogen like *Methanospirillum* and *Methanobacterium* are frequently observed (Sun et al., 2008; Edwards et al., 2015). Among fungi, prominence of the members of the genera *Aspergillus*, *Fusarium*, *Penicillium*, and *Trichoderma* within the phylum *Ascomycota* is noticed (Potshangbam et al., 2017; Santos-Medellin et al., 2017). Endosphere community is a stable community; however, it might be indirectly affected by environmental factors, such as in upland rice ecosystems; precipitation indirectly influences the endosphere microbial community diversity by modulating root growth and the rhizosphere microbiome (Crusciol et al., 2013).

19.4.5 Bulk Soil

In paddy field, bulk soil or upland soil is the most widely studied subject for different agricultural parameters and microorganisms. Among microorganisms, *Proteobacteria*, *Chloroflexi*, *Actinobacteria*, and *Acidobacteria* are the bacterial phyla and *Ascomycota*, *Basidiomycota*, and *Glomeromycota* are the fungal group typically found to dominate the bulk soil (Kim and Lee, 2020). Regardless of the geographical location of paddy field, at phylum level the consistent pattern of abundance of the microorganisms suggests the effect of environmental commonality during rice cultivation (Jiang et al., 2016; Yuan et al., 2018). But at lower taxonomic level, high heterogeneity in the abundance of the microorganisms is observed possibly correlated with the chemical heterogeneity of the rice field. This heterogeneity is linked to various factors such as edaphic factors, irrigation and cultivation regime, cultivation period, soil depth, and pH (Kim and Lee, 2020). Liu et al. (2016) noticed that an increased microbial abundance is correlated with the increased cultivation period that might be due to the accumulation of organic carbon and nitrogen. Fungal and bacterial community composition in bulk soil also varied with the activity of the soil enzymes such as phosphatases, ureases, and invertases (Liu et al., 2016). Soil pH is one of the key factors that affect diversity and function of microbial community in bulk soil by regulating the availability of electron donors and acceptors (Yuan et al., 2019).

19.5 Functional Characterization of Rice Rhizosphere Microbiome

Characterization of function of microbial communities in the rice root system is the important task of rice microbiome research. To decipher the function of microbiome in rice, metagenomic, metatranscriptomic, and metaproteomic approaches are utilized (Ding et al., 2019; Kim and Lee, 2020). By analyzing rice root microbiome through metagenomic approach, Sessitsch et al. (2012) found that endophytic community is represented by flagellated bacteria with plant polymer-degrading enzymes; protein secretion systems; iron acquisition and storage mechanisms; quorum-sensing systems; reactive oxygen species detoxification mechanisms; and proteins for nitrogen fixation, denitrification, and nitrification. Knief et al. (2012) compared the function of rice phyllosphere and rhizosphere microbiome using metagenomics and metaproteomics. The latter author observed the prevalence of proteins related to methanogenesis, methanotrophy, chemotaxis, and nitrogen metabolism in the rhizosphere but proteins associated with methanol-based methylotrophy, substrate uptake, and stress responses are more frequently found in the phyllosphere. Using mutant genotype on NRT1.1B (rice nitrate transporter and sensor), Zhang et al. (2019) showed that the superior nitrogen-use efficiency of *indica* varieties to that of *japonica* varieties is related to the compositional and functional differences in bacterial communities. Although some studies investigated the composition and diversity of endophytic fungal communities associated with rice root, understanding the functional and ecological roles of endophytic fungal communities is lacking (Kim and Lee, 2020). Rice rhizosphere is characterized by the presence of oxic–anoxic interface with substantial rhizodeposition of carbon, which made this region a unique habitat for various microbial functional groups involving biogeochemical cycling of essential elements.

19.5.1 Cycling of Methane

Cultivation of rice is one of the major contributors of global greenhouse gas such as CH_4 emission. Methane emission from rice field is the net outcome of many chemical and biological processes that include CH_4 formation and oxidation. From soil to atmosphere, CH_4 is released through the rice vascular system. A net balance of complex microbial consortia of methanogens and methanotrophs regulates the CH_4 emission (Conrad, 2009). Under submerged oxygen limited condition, methanogenic archaea of the phylum *Euryarchaeota* functions in the bulk soil of rice field. However, several investigations reported their assemblage in rice rhizosphere and explained that detected members of the *Methanosarcinaceae* and *Methanocellaceae* could tolerate O_2 toxicity as well as could adopt strategies to dwell old root segment where O_2 is limited (Angel et al., 2011; Lee et al., 2015). Other microorganisms such as hydrogenotrophic *Methanocellales*, aceticlastic *Methanosaetaceae*, *Methanomicrobiaceae*, and *Methanosarcinaceae* are the active rhizosphere methanogens (Zhu et al., 2014). It has been observed that community composition of methanogens in rhizosphere is relatively stable under field conditions; however, their abundance and activities vary with rice growth (Lee et al., 2014, 2015). Anaerobic CH_4 oxidation is a very slow process. It is estimated that 10%–30% of the CH_4 produced by the methanogenic archaea is consumed by the aerobic methanotrophic bacteria inhabiting the oxic–anoxic interface of rice roots (Shrestha et al., 2010). Aerobic methanotrophic bacteria are affiliated to *Proteobacteria* (e.g., type I and type II methanotrophs) and Verrucomicrobia phyla (Lee et al., 2014; Shen et al., 2014). Under *in situ* condition in the rhizosphere, type I methanotrophs (e.g., *Methylomonas*, *Methylobacteria*, *Methylosarcina*, and *Methylomicrobium*) performed well than type II methanotrophs (e.g., *Methylocystis*) (Qiu et al., 2008). There composition, abundance, and activities are dynamic and changed under various environmental conditions such as availability of O_2 and rice growth stage (Reim et al., 2012; Lee et al., 2014, 2015).

19.5.2 Nitrogen Cycling

Oxic–anoxic interface of the rice rhizosphere offers a favorable microhabitat for coupled nitrification–denitrification (Wei et al., 2017). ROL in the rhizosphere supports aerobic nitrifiers that drive the oxidation of NH_3 to NO_2^-/NO_3^- via nitrification. The produced NO_2^-/NO_3^- diffuses to the surrounding anoxic sites where heterotrophic denitrifiers via denitrification reduce them stepwise to nitric oxide (NO), nitrous oxide (N_2O), and dinitrogen gas (N_2). The aerobic oxidation of NH_3 to NO_2 is carried out by ammonia-oxidizing archaea (AOA) and ammonia-oxidizing bacteria (AOB) (Chen et al., 2010a). Ammonia monooxygenase gene (*amoA*) analysis revealed that rice rhizosphere AOA are affiliated with *Thaumarchaeota*, whereas AOB with *Nitrosospira* and *Nitrosomonas* of *Betaproteobacteria* (Hussain et al., 2011; Li et al., 2018). Communities of AOB are more responsive to rice genotypes, soil depth, rice growth stages, N_2 fertilization regime, and soil pH than those of AOA (Hussain et al., 2011; Chen et al., 2011; Yao et al., 2011). Jiang et al. (2015), based on $^{13}CO_2$-based stable isotope probing (SIP) of DNA/RNA in combination with functional gene analysis, showed that AOA drives the autotrophic nitrification in acidic paddy soils, while AOB performs in alkaline paddy soils.

The process of reduction of NO_2^- to NO in rice rhizosphere is carried out by denitrifying bacterial community that either contains the reductase gene *nirK* (related to Rhizobiales) or contains the *nirS* (related to Burkholderiales and Rhodocyclales) (Yoshida et al., 2009; Chen et al., 2010b). It is observed that the community composition of nirK-bearing denitrifiers is largely represented by unclassified bacteria which dominate the ammonia oxidizers (Hussain et al., 2011). Ammonia oxidation is also carried out in the rice rhizosphere under anoxic condition by anaerobic ammonia oxidation (anammox) by anammox bacteria of the order Brocadiale within the phylum Planctomycetes (Nie et al., 2015; Zhou et al., 2017).

19.6 Community Dynamics of Rice Microbiome

Edwards et al. (2018) studied the dynamics of microbial assemblies over the plant life cycle based on the samples collected from root spatial compartments of field-grown rice (*O. sativa*) over the course of three consecutive growing seasons, as well as two sites in diverse geographic regions. The root microbiota was found to be highly dynamic during the vegetative phase of plant growth and then stabilized compositionally for the remainder of the life cycle. Hu et al. (2019) performed metagenomic analysis of microbial community in iron root plaque of paddy rice to understand the influence of arsenic contamination on community structure and putative arsenic metabolism gene abundance. Analysis of 454 amplicon pyrosequencing data of 16S ribosomal DNA by Jha et al. (2020) showed alterations in the structure of root-associated microbiome of rice receiving growth-promoting treatments of urea fertilizer and rhizobium biofertilizer. Li et al. (2019), using MiSeq sequencing platform, investigated the dynamics between the rhizosphere microbial community and heavy metal ions under continuous flooding (CF) and intermittent flooding (IF) conditions to decipher the relationship between microbial community and environmental factors. Long-term flooding resulted in an abundance of *Anaeromyxobacter* sp., *Geobacter*, and *Desulfovibrio*, and the abundance of these taxa displayed a significant relationship to Pb and Zn contents of rice roots. Liechty et al. (2020) performed a comparative analysis of root microbiomes of rice cultivars with high and low methane emissions and found differences in abundance of methanogenic archaea and putative upstream fermenters. Root-associated community assembly is also investigated under different phosphate input levels in phosphorus (P)-deficient paddy soil (Long and Yao, 2020). Sequencing of DNA on Illumina MiSeq PE300 platform showed that soil P application affected both the rice root endosphere and soil rhizosphere microbial community and interaction between rice root endophytic bacteria and fungi, especially species related to P cycling. Transcriptome data of metabolic pathways and 16S rRNA and ITS amplicon data of root-associated microbiomes obtained from rice indicated coevolutionary associations between root-associated microbiomes and root transcriptomes in wild and cultivated rice varieties (Tian et al., 2018). It has been observed that cultivated rice rhizomicrobiome is more sensitive to environmental shifts than that of wild rice in natural environments (Xu et al., 2019b). Xu et al. (2019a) used deep pyrosequencing of bacterial 16S rRNA to identify and characterize the root-associated microbial community of three traditional rice cultivars, and showed that the assembly of root-associated microbial community of typical rice cultivars is strongly influenced by soil type. A recent study identifies soil biotic factors affecting ultrahigh yield, considering soil microbial community structure and metagenomic functions during four key rice growth stages, together with results from nitrogen enrichment experiments and rice root transcriptome analysis (Zhong et al., 2020). The latter study found strong soil microbial mechanisms for the promotion of ultrahigh rice yield.

REFERENCES

Angel R, Matthies D, Conrad R. Activation of methanogenesis in arid biological soil crusts despite the presence of oxygen. *PLoS One*. 2011; 6:e20453.

Arnold JW, Roach J, Azcarate-Peril MA. Emerging technologies for gut microbiome research. *Trends Microbiol*. 2016; 24(11):887–901. Doi: 10.1016/j.tim.2016.06.008.

Bäckhed F, Ley RE, Sonnenburg JL, Peterson DA, Gordon JI. Host-bacterial mutualism in the human intestine. *Science*. 2005; 307:1915–1920.

Banik A, Mukhopadhaya SK, Dangar TK. Characterization of N_2-fixing plant growth promoting endophytic and epiphytic bacterial community of Indian cultivated and wild rice (Oryza spp.) genotypes. *Planta* 2016; 243: 799–812.

Barr DJS. Chytridiomycota. In: Esser K, Lemke PA (eds). *The Mycota VII: Systematics and Evolution Part A*. Heidelberg: Springer-Verlag Berlin Heidelberg GmbH. 2001; 93–112.

Breidenbach B, Pump J, Dumont MG. Microbial community structure in the rhizosphere of rice plants. *Front Microbiol*. 2016; 6:1537.

Carding S, Verbeke K, Vipond DT, Corfe BM, Owen LJ. Dysbiosis of the gut microbiota in disease. *Microb Ecol Health Dis*. 2015; 26:26191. Doi: 10.3402/mehd.v26.26191.

Chen YT, Li XM, Liu TX, Li FB. Microaerobic iron oxidation and carbon assimilation and associated microbial community in paddy soil. *Acta Geochim*. 2017; 36:502–05.

Chen Z, Luo X, Hu R et al. Impact of long-term fertilization on the composition of denitrifier communities based on nitrite reductase analyses in a paddy soil. *Microb Ecol*. 2010b; 60:850–61.

Chen X, Zhang LM, Shen JP et al. Soil type determines the abundance and community structure of ammonia-oxidizing bacteria and archaea in flooded paddy soils. *J Soils Sediments*. 2010a; 10:1510–6.

Chen X, Zhang LM, Shen JP et al. Abundance and community structure of ammonia-oxidizing archaea and bacteria in an acid paddy soil. *Biol Fertil Soils*. 2011; 47:323–31.

Colmer TD, Kotula L, Malik AI, Takahashi H, Konnerup D, Nakazono M, Pedersen O. Rice acclimation to soil flooding: Low concentrations of organic acids can trigger a barrier to radial oxygen loss in roots. *Plant Cell Environ*. 2019; 42(7):2183–97.

Conrad R. The global methane cycle: Recent advances in understanding the microbial processes involved. *Environ Microbiol Rep*. 2009; 1:285–92.

Crusciol CAC, Soratto RP, Nascente AS et al. Root distribution, nutrient uptake, and yield of two upland rice cultivars under two water regimes. *Agron J*. 2013; 105:237–47.

Diedhiou AG, Mbaye FK, Mbodj D et al. Field trials reveal ecotype specific responses to mycorrhizal inoculation in rice. *PLoS One*. 2016; 11:e0167014.

Ding LJ, Cui HL, Nie SA, Long XE, Duan GL, Zhu YG. Microbiomes inhabiting rice roots and rhizosphere. *FEMS Microbiol Ecol*. 2019; 95(5):fiz040.

Dubinina GA, Sorokina AY. Neutrophilic lithotrophic iron-oxidizing prokaryotes and their role in the biogeochemical processes of the iron cycle. *Microbiology*. 2014; 83:1–14.

Edwards J, Johnson C, Santos-Medellin C et al. Structure, variation, and assembly of the root-associated microbiomes of rice. *Proc Natl Acad Sci USA*. 2015; 112(9):11–20.

Edwards JA, Santos-Medellín CM, Liechty ZS, et al. Compositional shifts in root-associated bacterial and archaeal microbiota track the plant life cycle in field-grown rice. *PLoS Biol*. 2018; 16(2):e2003862. Doi:10.1371/journal.pbio.2003862.

Eichorst SA, Kuske CR. Cellulose-responsive bacterial and fungal communities in geographically and edaphically different soils identified using stable isotope probing. *Appl Environ Microbiol* 2012; 78:2316–27.

Escalas A, Hale L, Voordeckers JW, Yang Y, Firestone MK, Alvarez-Cohen L, Zhou J. Microbial functional diversity: From concepts to applications. *Ecol Evol*. 2019; 9(20):12000–16.

Gleason FH, Marano AV, Digby AL et al. Patterns of utilization of different carbon sources by Chytridiomycota. *Hydrobiologia*. 2011; 659:55–64.

Han C, Ren JH, Tang H, Xua D, Xie XC. Quantitative imaging of radial oxygen loss from Valisneria spiralis roots with a fluorescent planar optode. *Sci Total Environ*. 2016; 569:1232–40.

Hardoim PR, Andreote FD, Reinhold-Hurek B, Sessitsch A, Simon van Overbeek L, Dirk van Elsas J. Rice root-associated bacteria: Insights into community structures across 10 cultivars. *FEMS Microbiol Ecol*. 2011; 77:154–64.

Hardoim PR, Hardoim CCP, van Overbeek LS et al. Dynamics of seed-borne rice endophytes on early plant growth stages. *PLoS One*. 2012; 7:e30438.

Hardoim PR, van Overbeek LS, Berg G et al. The hidden world within plants: Ecological and evolutionary considerations for defining functioning of microbial endophytes. *Microbiol Mol Biol Rev*. 2015; 79:293–320.

Hu M, Sun W, Krumins V, Li F. Arsenic contamination influences microbial community structure and putative arsenic metabolism gene abundance in iron plaque on paddy rice root. *Sci Total Environ*. 2019; 649:405–12. Doi: 10.1016/j.scitotenv.2018.08.388.

Hussain Q, Liu Y, Jin Z et al. Temporal dynamics of ammonia oxidizer (*amoA*) and denitrifier (*nirK*) communities in the rhizosphere of a rice ecosystem from Tai Lake region, China. *Appl Soil Ecol*. 2011; 48:210–8.

Hussain Q, Pan GX, Liu YZ et al. A Microbial community dynamics and function associated with rhizosphere over periods of rice growth. *Plant Soil Environ*. 2012; 58:55–61.

Hwangbo K, Um Y, Kim KY et al. Complete genome sequence of *Bacillus velezensis* CBMB205, a phosphate-solubilizing bacterium isolated from the rhizoplane of rice in the republic of Korea. *Genome Announc*. 2016; 4:e00654–16.

Jha PN, Gomaa AB, Yanni YG, et al. Alterations in the endophyte-enriched root-associated microbiome of rice receiving growth-promoting treatments of urea fertilizer and rhizobium biofertilizer. Microb Ecol. 2020; 79(2):367–82. Doi: 10.1007/s00248-019-01406-7

Ji Y, Conrad R, Xu H. Responses of archaeal, bacterial, and functional microbial communities to growth season and nitrogen fertilization in rice fields. *Biol Fertil Soils*. 2020; 56:81–95 Doi: 10.1007/s00374-019-01404-4

Jiang X, Hou X, Zhou X et al. pH regulates key players of nitrification in paddy soils. *Soil Biol Biochem*. 2015; 81:9–16.

Jiang YJ, Liang YT, Li CM, Wang F, Sui YY, Suvannang N, Zhou JZ, Sun B. Crop rotations alter bacterial and fungal diversity in paddy soils across East Asia. *Soil Biol Biochem*. 2016; 95:250–61.

Khan N, Seshadri B, Bolan N, Saint CP, Kirkham MB, Chowdhury S, Yamaguchi N, Lee DY, Li G, Kunhikrishnan A, Qi F, Karunanithi R, Qiu R, Zhu YG, Syu CH. Root iron plaque on wetland plants as a dynamic pool of nutrients and contaminants. *Adv Agron*. 2016; 138:1–96.

Kim H, Lee YH. The rice microbiome: A model platform for crop holobiome. *Phytobiomes J*. 2020; 4(1):5–18.

Knief C, Delmotte N, Chaffron S, Stark M, Innerebner G, Wassmann R, von Mering C, Vorholt JA. Metaproteogenomic analysis of microbial communities in the phyllosphere and rhizosphere of rice. *ISME J*. 2012; 6:1378–90.

Kögel-Knabner I, Amelung W, Cao Z, Fiedler S, Frenzel P, Jahn R, Kalbitz K, Kölbl A, Schloter M. Biogeochemistry of paddy soils. *Geoderma*. 2010; 157(1–2):1–14.

Larsen M, Santner J, Oburger E, Wenzel WW, Glud RN. O_2 dynamics in the rhizosphere of young rice plants (*Oryza sativa L.*) as studied by planar optodes. *Plant Soil*. 2015; 390:279–92.

Leach JE, Triplett LR, Argueso CT, Trivedi P. Communication in the Phytobiome. *Cell*. 2017; 169(4):587–596. Doi: 10.1016/j.cell.2017.04.025.

Lee HJ, Jeong SE, Kim PJ et al. High resolution depth distribution of bacteria, archaea, methanotrophs, and methanogens in the bulk and rhizosphere soils of a flooded rice paddy. *Front Microbiol*. 2015; 6:639.

Lee HJ, Kim SY, Kim PJ et al. Methane emission and dynamics of methanotrophic and methanogenic communities in a flooded rice field ecosystem. *FEMS Microbiol Ecol*. 2014; 88:195–212.

Li Y, Chapman S, Nicol GW et al. Nitrification and nitrifiers in acidic soils. *Soil Biol Biochem*. 2018; 116:290–301.

Li H, Yu Y, Guo J, Li X, Rensing C, Wang G. Dynamics of the rice rhizosphere microbial community under continuous and intermittent flooding treatment. *J Environ Manage*. 2019; 249:109326. Doi:10.1016/j.jenvman.2019.109326.

Liechty Z, Santos-Medellín C, Edwards J, et al. Comparative analysis of root microbiomes of rice cultivars with high and low methane emissions reveals differences in abundance of methanogenic archaea and putative upstream fermenters. *mSystems*. 2020; 5(1):e00897–19. Doi:10.1128/mSystems.00897-19.

Liu H, Ding Y, Zhang Q, Liu X, Xu J, Li Y, Di H. Heterotrophic nitrification and denitrification are the main sources of nitrous oxide in two paddy soils. *Plant Soil*. 2019, 445(1):39–53.

Liu Y, Lv H, Yang N, Li Y, Liu B, Rensing C, Dai J, Fekih IB, Wang L, Mazhar SH, Kehinde SB. Roles of root cell wall components and root plaques in regulating elemental uptake in rice subjected to selenite and different speciation of antimony. *Environ Exp Bot*. 2019; 163:36–44.

Liu Y, Wang P, Pan G, Crowley D, Li L, Zheng J, Zhang X, Zheng J. Functional and structural responses of bacterial and fungal communities from paddy fields following long-term rice cultivation. *J. Soils Sediments*. 2016; 16:1460–71.

Long XE, Yao H. Phosphorus input alters the assembly of rice (*Oryza sativa* L.) root-associated communities. *Microb Ecol*. 2020; 79(2):357–66. Doi:10.1007/s00248-019-01407-6

Mano H, Morisaki H. Endophytic bacteria in the rice plant. *Microbes Environ*. 2008; 23:109–17.

Mbodj D, Effa-Effa B, Kane A et al. Arbuscular mycorrhizal symbiosis in rice: Establishment, environmental control and impact on plant growth and resistance to abiotic stresses. *Rhizosphere*. 2018; 8:12–26.

Mendes R, Paolina G, Jos MR. The rhizosphere microbiome: Significance of plant beneficial, plant pathogenic, and human pathogenic microorganisms. *FEMS Microbiol Rev*. 2013; 37:634–63.

Nie SA, Li H, Yang XR et al. Nitrogen loss by anaerobic oxidation of ammonium in rice rhizosphere. *ISME J*. 2015; 9:2059–67.

Potshangbam M, Devi SI, Sahoo D et al. Functional characterization of endophytic fungal community associated with *Oryza sativa* L. and *Zea mays* L. *Front Microbiol*. 2017; 8: 325.

Qiu Q, Noll M, Abraham WR et al. Applying stable isotope probing of phospholipid fatty acids and rRNA in a Chinese rice field to study activity and composition of the methanotrophic bacterial communities in situ. *ISME J*. 2008; 2:602–14.

Reim A, Lüke C, Krause S et al. One millimetre makes the difference: High-resolution analysis of methane-oxidizing bacteria and their specific activity at the oxic-anoxic interface in a flooded paddy soil. *ISME J*. 2012; 6:2128–39.

Reinhold-Hurek B, Hurek T. Living inside plants: Bacterial endophytes. *Curr Opin Plant Biol*. 2011; 14:435–43.

Santos-Medellín C, Edwards J, Liechty Z, Nguyen B, Sundaresan V. Drought stress results in a compartment-specific restructuring of the rice root-associated microbiomes. *mBio*. 2017; 8(4):e00764–17. Doi:10.1128/mBio.00764-17.

Sasse J, Martinoia E, Northen T. Feed your friends: Do plant exudates shape the root microbiome? *Trends Plant Sci*. 2018; 23:25–41.

Sessitsch A., Hardoim P, Döring J, Weilharter A, Krause A, Woyke T, Mitter B, Hauberg-Lotte L, Friedrich F, Rahalkar M. Functional characteristics of an endophyte community colonizing rice roots as revealed by metagenomic analysis. *Mol Plant-Microbe Interact*. 2012; 25:28–36.

Shen LD, Liu S, Huang Q et al. Evidence for the co-occurrence of nitrite-dependent anaerobic ammonium and methane oxidation processes in a flooded paddy field. *Appl Environ Microbiol*. 2014; 80:7611–9.

Shrestha M, Shrestha PM, Frenzel P et al. Effect of nitrogen fertilization on methane oxidation, abundance, community structure, and gene expression of methanotrophs in the rice rhizosphere. *ISME J.* 2010; 4:1545–56.

Sun L, Qiu F, Zhang X et al. Endophytic bacterial diversity in rice (*Oryza sativa* L.) roots estimated by 16S rDNA sequence analysis. *Microb Ecol.* 2008; 55:415–24.

Sun M, Xiao T, Ning Z et al. Microbial community analysis in rice paddy soils irrigated by acid mine drainage contaminated water. *Appl Microbiol Biotechnol.* 2015; 99:2911–22.

Tian L, Shi S, Ma L, et al. Co-evolutionary associations between root-associated microbiomes and root transcriptomes in wild and cultivated rice varieties. *Plant Physiol Biochem.* 2018; 128:134–41. Doi: 10.1016/j.plaphy.2018.04.009.

Ursell LK, Metcalf JL, Parfrey LW, Knight R. Defining the human microbiome. *Nutr Rev.* 2012; 70:S38–44.

Vishwakarma P, Dubey SK. Diversity of endophytic bacterial community inhabiting in tropical aerobic rice under aerobic and flooded condition. *Arch Microbiol.* 2020; 202(1):17–29. Doi:10.1007/s00203-019-01715-y.

Wang XJ, Yang J, Chen XP, Sun GX, Zhu YG. Phylogenetic diversity of dissimilatory ferric iron reducers in paddy soil of Hunan, South China. *J Soils Sediments.* 2009; 9:568–77.

Wei X, Hu Y, Peng P et al. Effect of P stoichiometry on the abundance of nitrogen-cycle genes in phosphorus-limited paddy soil. *Biol Fertil Soils.* 2017; 53:767–76.

Wei X, Zhu Z, Wei L, Wu J, Ge T. Biogeochemical cycles of key elements in the paddy-rice rhizosphere: Microbial mechanisms and coupling processes. *Rhizosphere.* 2019; 10:100145. Doi: 10.1016/j.rhisph.2019.100145.

Wu X, Ge T, Yan W et al. Irrigation management and phosphorus addition alter the abundance of carbon dioxide-fixing autotrophs in phosphorus-limited paddy soil. *FEMS Microbiol Ecol.* 2017; 93. Doi: 10.1093/femsec/fix154.

Wu C, Ye Z, Li H, Wu S, Deng D, Zhu Y, Wong M. Do radial oxygen loss and external aeration affect iron plaque formation and arsenic accumulation and speciation in rice? *J Exp Bot.* 2012; 63:2961–70.

Xu Y, Ge Y, Song J, Rensing, C. Assembly of root-associated microbial community of typical rice cultivars in different soil types. *Biol Fert Soils.* 2019a; 56:249–60.

Xu S, Tian L, Chang C, Li X, Tian C. Cultivated rice rhizomicrobiome is more sensitive to environmental shifts than that of wild rice in natural environments. *Appl Soil Ecol.* 2019b; 140:68–77.

Yao H, Gao Y, Nicol GW et al. Links between ammonia oxidizer community structure, abundance and nitrification potential in acidic soils. *Appl Environ Microbiol.* 2011; 77:4618–25.

Yoshida M, Ishii S, Otsuka S et al. Temporal shifts in diversity and quantity of *nirS* and *nirK* in a rice paddy field soil. *Soil Biol Biochem.* 2009; 41:2044–51.

Yuan C, Zhang L, Hu H, Wang J, Shen J, He J. The biogeography of fungal communities in paddy soils is mainly driven by geographic distance. *J Soils Sediments.* 2018; 18:1795–805.

Yuan CL, Zhang LM, Wang JT, Hu HW, Shen JP, Cao P, He JZ. Distributions and environmental drivers of archaea and bacteria in paddy soils. *J Soils Sediments.* 2019; 19:23–37.

Zecchin S, Corsini A, Martin M, Romani M, Beone GM, Zanchi R, Zanzo E, Tenni D, Fontanella MC, Cavalca L. Rhizospheric iron and arsenic bacteria affected by water regime: Implications for metalloid uptake by rice. *Soil Biol Biochem.* 2017; 106:129–37.

Zhang X, Kuzyakov Y, Zang H, Dippold MA, Shi L, Spielvogel S, Razavi BS. Rhizosphere hotspots: Root hairs and warming control microbial efficiency, carbon utilization and energy production. *Soil Biol Biochem.* 2020; 107872. Doi: 10.1016/j.soilbio.2020.107872.

Zhang J, Liu YX, Zhang N, Hu B, Jin T, Xu H, Qin Y, Yan P, Zhang X, Guo X, Hui J, Cao S, Wang X, Wang C, Wang H, Qu B, Fan G, Yuan L, Garrido-Oter R, Chu C, Bai Y. NRT1.1B is associated with root microbiota composition and nitrogen use in field-grown rice. *Nat Biotechnol.* 2019; 37:676–84.

Zhong Y, Hu J, Xia Q, Zhang S, Li X, Pan X, Zhao R, Wang R, Yan W, Shangguan Z, Hu F, Yang C, Wang W. Soil microbial mechanisms promoting ultrahigh rice yield. *Soil Biol Biochem.* 2020; 143: 107741.

Zhou XH, Zhang JP, Wen CZ. Community composition and abundance of anammox bacteria in cattail rhizosphere sediments at three phenological stages. *Curr Microbiol.* 2017; 74:1349–57.

Zhu Z, Ge T, Hu Y et al. Fate of rice shoot and root residues, rhizo-deposits, and microbial assimilated carbon in paddy soil-part 2: Turnover and microbial utilization. *Plant Soil.* 2017; 416:243–57.

Zhu Z, Ge T, Liu S et al. Rice rhizo-deposits affect organic matter priming in paddy soil: The role of N fertilization and plant growth for enzyme activities, CO_2 and CH_4 emissions. *Soil Biol Biochem.* 2018; 116:369–77.

Zhu W, Lu H, Hill J et al. ^{13}C pulse-chase labelling comparative assessment of the active methanogenic archaeal community composition in the transgenic and non-transgenic parental rice rhizospheres. *FEMS Microbiol Ecol.* 2014; 87:746–56.

20

Mangrove Ecosystem and Microbiome

Snehal O. Kulkarni and Yogesh S. Shouche
National Centre for Cell Science

CONTENTS

20.1 Introduction

20.1.1 Mangrove Forests

Mangrove forests occurring at the interface of terrestrial and marine ecosystems make up one of the most productive and biologically diverse ecosystems on the planet. They support a diverse group of aquatic animals and microorganisms. They are coastal wetland forests mainly found at the intertidal zones of estuaries, backwaters, deltas, creeks, lagoons, marshes, and mudflats of tropical and subtropical latitudes. Nearly 60%–70% of the world's tropical and subtropical coastlines are covered with mangroves (Thatoi et al., 2013). Despite being fragile and sparsely distributed, they are unique intertidal ecosystems with the highest productivity all over the world. Mangroves are a group of trees or shrubs that are present in the coastal intertidal zone. There are about 110 different species of mangrove trees. They all belong to the families *Rhizophoraceae*, *Acanthaceae*, *Lythraceae*, *Combretaceae*, and *Arecaceae*. Mangroves are found mainly in Central America, South Africa, northern South America, and Southeast Asia. Common mangrove species include black mangrove, red mangrove, white mangrove, buttonwood mangrove, gray mangrove, mangrove palm, and mangrove apple. Mangrove forests in Southeast Asia include *Sonneratia* of the family Lythraceae and nipa palm of the family Arecaceae. Mangrove flora along the coast of tropical America and along the coast of Gulf of Mexico to Florida consists of red mangrove, i.e., *Rhizophora mangle* of the family *Rhizophoraceae*, and black mangrove, i.e., *Avicennia nitida* or *A. marina* of the family *Acanthaceae*. In the United States, mangroves are mainly found in Florida and South Texas. Only four species of mangrove are found in the Unites States, i.e., red, black, white, and buttonwood mangroves. The highest mangrove forests are found in Indonesia followed by Brazil, Malaysia, New Guinea, Australia, Mexico, Nigeria, and Myanmar. In Africa, the important mangrove swamps are found in Kenya, Tanzania, and Madagascar. In Asia, mangroves are present in India, Malaysia, Vietnam, Indonesia, Taiwan, and Japan. In India, the highest area covered by mangrove forests is seen in West Bengal, which is 2,106 km^2, followed by Gujarat and Andaman and Nicobar Islands. Australia and New Guinea both rank in top five mangrove-holding nations globally.

Mangrove plant species grow in the areas with low-oxygen soil where slow-moving seawater allows fine sediments to settle. Mangrove forests only grow at tropical and subtropical latitudes near the equator because they can not withstand freezing temperatures. Mangrove forests stabilize the coastline and reduce the erosion from storm, surge, current waves, tides, and especially tsunamis (Prance, 1998). They act as a physical barrier to mitigate the effects of coastal disasters such as hurricanes.

Mangroves are salt-tolerant trees, also called halophytes, and adapt themselves to life under harsh coastal conditions. They are the only species of trees in the world that can tolerate saline conditions. They contain a complex salt filtration system and a complex root system to cope with sea water immersion and wave action. They also adapt themselves to the low-oxygen and waterlogged mud conditions. The height of the mangrove plant species ranges from 2 to 10 m. All mangrove species feature oblong- or oval-shaped leaves. According to NASA, mangrove ecosystems move carbon dioxide from the atmosphere into long-term storage in greater quantities than any other type of ecosystem. It is reported that they have the ability to absorb up to four times carbon dioxide by the area than upland terrestrial forests (Donato et al., 2011). Thus, they are called the planet's best carbon scrubbers.

Mangrove plants have to tolerate broad ranges of salinity, temperature, and moisture, and thus, only a selective species of plants make up the mangrove community. The mangrove biome or "mangal" is a distinct saline woodland or shrubland habitat that grows in coastal saline or brackish water where high organic fine sediment collects, and salinity ranges from 3%–4% to 9% (w/v) due to evaporation of sea water. The term "mangrove" has come to English probably from Spanish.

The unique ecosystem found in the intricate mesh of mangrove roots offers an ideal habitat for many organisms. In the areas where roots are permanently submerged, organisms like algae, barnacles, oysters, and sponges, which generally require hard surface to adhere, are present. Shrimps and lobsters use muddy bottoms as their homes. Mangrove crabs munch the mangrove leaves, thereby adding the nutrients to the mangrove mud ecosystem for other bottom feeders.

Of the recognized 110 mangrove species, only about 54 species and 20 genera from 16 families constitute the "true mangroves", because only these species occur almost exclusively in all mangrove habitats. The great biodiversity in this mangrove ecosystem is found only in few countries like New Guinea, Indonesia, and Malaysia.

20.1.2 Adaptation Strategies of Mangrove Plant Species

The mangrove plant species adapt themselves to the low-oxygen environment by making their roots stilt that can absorb air through pores in their bark or lenticels in the case of red mangroves, while in the case of black mangroves, which live on higher ground, they make many pneumatophores that are specialized root-like structures, which come out of the soil like straws for breathing. Pneumatophores are the breathing tubes that typically reach heights up to 30 cm, while in some species, the height of the pneumatophores reaches up to 3 m. These pneumatophores or aerial roots of the mangroves absorb gases directly from the atmosphere and micronutrients like iron from the soil. These mangrove plant species adapt themselves to the saline conditions by excreting salts outside the plants. In the case of red mangroves, the plant species excrete salts by having significantly impermeable roots that act as an ultrafiltration mechanism to exclude NaCl from the rest of the plant. These plant species store NaCl in cell vacuoles. White and gray mangroves excrete salt directly via their two salt glands present at each leaf base. Thus, their leaves are covered with salt crystals.

20.1.3 Mangroves around the World

Mangroves are found in over 118 countries in the tropical and subtropical regions of the world spanning an area of approximately 137,800 km^2. Asia has the highest percentage, i.e., 42%, of the world's mangrove followed by Africa 21%, North Central America 12%, and South America 11%. The largest amount of mangrove forests is found in Indonesia covering approximately 42,500 km^2 of the world.

20.1.4 Mangroves in Asia

In Asia, the highest mangrove vegetation is found in Indonesia. The mangroves are distributed in Indonesia in 42,500 km^2. This luxuriant growth of mangroves could be due to high rainfall and high humidity. Other than Indonesia, in Asia, mangrove forests can be found in India, Pakistan, Bangladesh, China, Sri Lanka, Maldives, Malaysia, Vietnam, Taiwan, Iran, Oman, etc. (Kathiresan and Rajendran, 2005).

20.1.4.1 Mangrove Ecosystems in India

India has a total of 6,740 km^2 mangrove forests, which is 0.1% of the country's total geographical area and 7% of the world's mangrove vegetation (Sahoo et al., 2017). The east coast of India has larger mangroves, i.e., 80%, than the west coast (20%) due to the river deltas of Ganges, Krishna, Godavari, Kaveri, Brahmaputra, and Mahanadi. Sundarban mangrove forest situated in the joint delta of Ganges, Brahmaputra, and Meghana Rivers in West Bengal, India is one of the largest mangrove forests in the globe. It covers the area from Hooghly River in India to Baleswar River in Bangladesh. Sundarban mangrove forest covers a total area of 10,000 km^2 of which Bangladesh covers approximately 6,000 km^2, and West Bengal covers approximately 4,260 km^2. It is comprised of mangrove forests, agricultural land mudflats, and barren land. Four protected areas in the Sundarban mangrove site are enlisted as UNESCO World Heritage sites viz. Sundarban National Park, Sundarbans West, Sundarbans South, and Sundarban East Wildlife Sanctuaries (Giri et al., 2007). The dominant mangrove plant species found in Sundarban is *Heritiera fomes* locally known as *Sundari*. Apart from *Sundari*, 26 different mangrove species grow well in Sundarban. A large biodiversity exists in the Sundarban mangroves including 40 mammals, 35 reptiles, and 260 bird species. Most importantly, Sundarban mangrove is known for its tiger reserve. It is an important habitat for endangered tigers such as the Royal Bengal Tiger. It is estimated that there are approximately 200 Bengal Tigers present in the area of Sundarban mangroves. The forest also provides habitat for jungle cats, fishing cats, and leopard cats. The Sundarbans National Park is a home of Olive Ridley turtles, green turtles, sea snakes, estuarine crocodiles, chameleons, king cobras, pythons, etc. Fishes and amphibians found in Sundarbans include starfish, king crab, hermit crab, prawns, tree frogs, Gangetic dolphins, and mudskippers. Bhitarkanika mangrove ecosystem is India's second largest mangrove ecosystem located in the state of Odisha, India. It is created by two river deltas viz. Brahmani and Baitarani. It is considered as one of the top-ranking mangrove forests in the world in terms of its rich biodiversity (Thatoi et al., 1999). It is the home of saltwater crocodiles and nesting of Olive Ridley sea turtles. Few studies have been undertaken till now to understand the flora, fauna, and microbial diversity of the Bhitarkanika mangrove ecosystem (Mishra et al., 1995; Gupta et al., 2007; Thatoi and Biswal, 2008; Mishra et al., 2011). The other mangroves from India include the Godavari–Krishna mangroves in the state of Andhra Pradesh, Pichavaram mangroves in the state of Tamil Nadu, mangroves in Mumbai, and Baratang Island mangroves located within the Andaman and Nicobar Islands.

20.1.4.2 Mangroves in Indonesia

Indonesia contains the largest area of mangrove forests in the world. About 3 million hectares of mangrove forests grow along Indonesia's 95,000 km^2 coastline, which accounts for approximately 23% of mangrove ecosystems all over the world (Giri et al., 2011). In Indonesia, mangrove forests are found mainly in Papua, Kalimantan, and Sumatra. Mangrove trees in Indonesia can reach up to the height of 50 m, and they are densely packed with intertwined roots extending from the tree trunks. Indonesia's mangroves are among the most carbon-rich forests in the world. But over the past three decades, Indonesia has lost 40% of its mangroves (Campbell and Brown, 2015). The main cause of mangrove loss in Indonesia includes conversion of mangrove forests to shrimp ponds, which is also known as "blue revolution", conversion of mangrove forests into agricultural land and salt pans, etc. The degradation of mangrove ecosystems due to oil spills and pollution also resulted in mangrove loss in Indonesia (FAO, 2007).

20.1.5 Mangroves in Africa

In Africa, mangrove forests are observed mainly in Kenya, Tanzania, Democratic Republic of Congo, and Madagascar. Nigeria experiences the highest mangrove forest covering an area of 36,000 km^2.

20.1.6 Mangrove in America

In America, mangroves are found in the tropical and subtropical zones of North and South America. In the United States, mangroves are mainly found in Florida and South Texas. Mangroves are also present in Mexico, Central America, and Caribbean. In South America, Brazil contains approximately 26,000 km^2 of mangrove forest, which accounts for 15% of the total world's mangrove area making it the second highest mangrove vegetation-bearing nation.

20.2 Mangroves as an Ecosystem

Since mangroves are mostly situated tropically, they receive a lot of solar energy, and thus, abundant sunlight, water, and organic matter-rich sediment make the mangrove ecosystem as among the most productive and biologically complex ecosystems on the earth with high rates of biomass production (Ghizelini et al., 2012). It acts as a bridge connecting the land and the sea (Figure 20.1). They are foundations of coastal food web. Mangrove forests are important feeding grounds to thousands of species of aquatic animals. Few animals like crabs and insects directly eat mangrove leaves, while decomposers wait for mangrove leaves to fall to the ground and consume the decaying material. Microorganisms including bacteria and fungi use this decaying material as a fuel, and in return, they recycle the nutrients like nitrogen, phosphorus, sulfur, and iron for the mangrove ecosystem. Various groups of bacteria such as nitrogen-fixing bacteria, phosphate-solubilizing bacteria and fungi, cellulose decomposers, chitin degraders, pectin degraders, anaerobic microorganisms such as methanogens are prevalent in this ecosystem (Das et al., 2006). Sea anemones, brittle stars, underwater sponges, snails, worms,

FIGURE 20.1 Mangrove ecosystem in Kolmandala, District Raigad, west coast of India showing pneumatophres.

barnacles, oysters, and sea urchins make mangrove roots their home. Monkeys, birds, insects, and other plants live on mangrove branches. Many crabs, shrimps, and fishes spend their early stages of life within a safety environment of underwater mangrove roots before making their way out into the open ocean as adults. Thus, mangrove forests also serve as "nursery habitats". The endangered species such as rainbow parrotfish, Goliath grouper, mangrove hummingbird, etc. rely on this mangrove ecosystem for their survival, protection, and food.

20.2.1 Creatures in the Mangrove Ecosystem

Not many large animals can navigate the muddy, underwater mangrove forest, but the Royal Bengal Tiger can. Sundarban mangrove forest located in the Ganges–Brahmaputra delta in West Bengal, India is a home of approximately 200 Royal Bengal Tigers. Mudskippers are fishes that are found in some mangrove forests of the Indo-Pacific region. They spend the majority of their time out of the water and use their pectoral fins to climb mangrove trees. They breathe by storing water in their mouth and gill chamber. Mud lobsters are another important aquatic animal found in the Indo-Pacific mangrove ecosystem. They burrow the mangrove sediment down up to 2 m deep with the help of their claws. This excavated mangrove sediment is rich in nutrients from decaying matter from the deep underground. Due to this procedure, aeration of the soil takes place, and the fertility of the sediment increases. This fertile sediment becomes food for the other species of microorganisms and insects. The mangrove forests from the tip of Florida to the Caribbean are the home of the American crocodile, which is one of the endangered species, while in Australia, mangrove forests are the home of massive saltwater crocodiles.

20.2.2 Detritus as an Important Element in the Mangrove Ecosystem

Degradation of mangrove vegetative materials viz. mangrove leaves and wood produces "detritus". "Detritus" can be defined as organic matter in the active process of decomposition. Degradation of fallen mangrove vegetation starts immediately by bacteria and fungi residing in the mangrove sediment having cellulolytic, pectinolytic, amylolytic, proteolytic, and lignocellulolytic activities (Findlay et al., 1986). The undecomposed leaves are poor in nutrients, and they become nutritious when microorganisms start decomposing them (Odum, 1971). The detritus is rich in protein and energy, and contains a large population of microorganisms (Odum and Herald, 1975a; Bano et al., 1997). It serves as a nutrient for many microorganisms living in the mangrove ecosystem as well as for mangrove vegetation, and thus, microbes develop a link between themselves and the mangrove plants species. Crustaceans, mollusks, insect larvae, nematodes, polychaetes, and few fishes consume detritus, and thus, they are called detritus feeders (Odum and Herald, 1975b). Fungi are particularly important in the decomposition of dead organic matter (Kohlmeyer and Kohlmeyer, 1979). The microbial decomposition of mangrove litter is studied by few researchers over the globe (Fell et al., 1984; Raghukumar et al., 1994; Rajendran and Kathiresan, 2004). Rajendran and Kathiresan (2006) extensively studied the microbial flora associated with submerged mangrove leaf litter in India in relation to the changes in nitrogen and tannin levels and juvenile prawns' assemblage. The total heterotrophic bacterial counts were high (5.58×10^5 cfu g^{-1} wet tissue) in decomposed leaves than in undecomposed leaves. Six genera of heterotrophic bacteria viz. *Flavobacterium*, *Vibrio*, *Pseudomonas*, *Acinetobacter*, *Corynebacterium*, and *Azotobacter* and 19 species of fungi viz. *Aspergillus niger*, *A. glaucus*, *A. fumigatus*, *A. candidus*, *Alternaria alternata*, *Halosarphia fibrosa*, *Ophiobolus littoralis*, *Spathulospora lanata*, *Pontoporeia biturbinata*, *Fusarium* sp., *Mucor* sp., *Penicillium* sp., and *Curvularia* sp. have been identified from decomposed leaves of mangroves. The total leaf nitrogen levels increased during the decomposition of leaves, while the total tannin levels decreased. The juvenile prawns' assemblage around the decomposing leaves was also found to be increased during the decomposition of mangrove leaves. The high nitrogen content in the decomposing leaves of mangroves could be partially due to nitrogen-fixing bacteria such as *Azotobacter* sp. It was also observed from this study that *Azotobacter* sp. might be increasing the level of nitrogen in the decomposing leaves of mangroves, which attracts juvenile prawns as there was an increase in the number of juvenile prawns after 40–50 days of decomposition. It was also noted that the fungal species colonizes on the tannin-rich mangrove leaves during early decomposition, which might be due to the presence of tannase enzymes in the *Aspergillus* species, which helps in degrading tannins in the leaves, and then when the tannin content decreases, bacterial population was found to increase on the decomposing leaves of mangroves. Zhuang and Lin (1993) have measured the bacterial densities as 2×10^5–10×10^5cfu g^{-1} of wet tissue of *Kandelia candel* mangrove plant leaves that decomposed after 2–4 weeks in the sediment.

20.3 Physicochemical Parameters of the Mangrove Ecosystem

The physicochemical characteristics of a particular ecosystem play an important role in the microbiome of that ecosystem. Various physicochemical and biological processes in the mangrove ecosystem make it an ideal habitat for a vast number of microorganisms leading to a rich biodiversity (Srilatha et al., 2013). It is said that seasonal variation and anthropogenic activities change the physicochemical parameters of the mangrove ecosystem, which ultimately affects its microbiome. The most important variables that influence the microbiota of the mangrove ecosystem are temperature, pH, redox potential, salinity, soil texture and soil nutrients, organic matter present in the sediment, metal concentration of sediment, etc. (Holguin et al., 2001; Mishra et al., 2011). In mangrove ecosystem, sediment and vegetation have a strong relationship with each other. It is reported that the soil particle size not only affects the bacterial biomass but also determines the structure of bacterial communities in the mangrove ecosystem. It is also reported that mangrove soil composed of clay and fine silt particles showed a high diversity of bacteria than soil with large particles (Sessitsch et al., 2001). Thus, a study of physicochemical parameters of the mangrove ecosystem is a prerequisite in understanding that habitat and the microorganisms inhabiting that habitat. Similarly, studying the physicochemical parameters of mangrove sediment and water helps in understanding the nutrient cycling occurring in the mangrove ecosystem. The mangrove ecosystems are characterized by periodic flooding; thus, environmental factors such as salinity and nutrient availability of the ecosystem are highly variable (Thatoi et al., 2013). Mangroves provide a unique ecological niche for a diverse group of bacterial communities, and these bacterial communities then largely control the carbon, nitrogen, phosphorus, and sulfur dynamics in the mangrove sediment. Few researchers have described the physicochemical characters of mangrove sediment and water present in the mangrove ecosystem present in different parts of the world. Ramanathan et al. (2008) reported the relationship between mangrove sediment nutrients and microbial flora present in the sediments of the Sundarban mangroves, India. The mangrove sediment was studied for its physicochemical parameters such as pH, electrical conductivity, total dissolved solids, total nitrogen, carbon, and phosphorus, redox potential salinity, nitrate, bicarbonate, and sulfate concentration, and chloride, ammonium, sodium, potassium, calcium, and magnesium ions. The sediment analysis of the Sundarban mangrove ecosystem showed that it was clayey in nature, which might be due to the dense network of mangrove troops acting as a trap for fine particles. The sediment was higher in organic matter. The phosphorus concentration in the sediment was high (795 $\mu g\ g^{-1}$), which was again due to high microbial activity, mainly phosphorus-solubilizing bacterial activity of the pristine mangroves of Sundarban. Goutam and Ramanathan (2013) also studied the microbial diversity in the surface sediments of mangroves of Gulf of Kutch, Gujarat and its interactions with the nutrients of the mangrove sediment. The nutrients present in the mangrove water such as phosphates, sulfates, dissolved silica, salinity, etc. were reported. From their study, a strong correlation was observed between nitrate concentration, fungal and bacterial loads, and phosphate concentration, which indicates that these factors significantly influence the microbial population in mangrove sediment. Gonzalez-Acosta et al. (2005) studied the effect of environmental factors on the population of cultivable heterotrophic bacteria and *Vibrio* spp. in the mangrove sediment of Baja, California. The physicochemical parameters studied were seawater temperature, salinity, concentration of dissolved oxygen, sediment nutrient concentration, pH, and sediment texture. The results of general cluster analyses and several principal component analyses showed that seawater temperature was the principal determinant of the seasonal distribution of cultivable heterotrophic bacteria and *Vibrio* spp. in the mangrove ecosystem of Baja, California. When the sea water temperature increased, the heterotrophic bacterial population decreased. The yearly mean population of heterotrophic bacteria in the mangrove ecosystem of Baja, California was 1.64×10^7 cfu g^{-1} of sediment, while the yearly population of *Vibrio* spp. was 2.12×10^4 cfu g^{-1} of sediment. Mishra et al. (2011) analyzed the microbial community structure of Bhitarkanika mangroves, Odisha, India in relation to soil physicochemical properties and seasonal fluctuations. From their study, it was observed that soil nutrients like N, P, and K and total carbon content was higher in the rainy season. Thus, heterotrophic, phosphate-solubilizing, sulfur-oxidizing bacterial population and fungi were maximum in the rainy season while nitrifying, denitrifying cellulose degraders and actinomycetes were higher in the winter season. Soil pH and salinity varied from 6 to 8 and 6.4 to 19.5 ppt, respectively, and they affected the microbial population in the mangrove sediment. Organic matter present in the mangrove sediment, pH, and available soil phosphorus accounted for the significant amount of variability in the bacterial community composition of Bhitarkanika mangrove ecosystem. Saravanakumar et al. (2016) studied the ecology of sol microbes present in the Pichavaram mangrove forest located in the Bay of Bengal on the southeast coast of India. They enumerated nine groups of microorganisms namely *Cyanobacteria*, total heterotrophic bacteria (THB), *Lactobacilli*, *Azotobacters*, *Actinobacteria*, *fungi*, *yeasts*, *thraustrochytrids*, and *Trichoderma* from mangrove sediments drawn from two different mangrove sites (dense mangrove and sparse mangrove) and at 10 different depths (10–100 cm) for four seasons (pre monsoon, monsoon, post monsoon, and summer) and compared the microbial population with 23 different physicochemical parameters mainly temperature, pH, redox potential, salinity, soil texture, soil composition, total organic matter, major elements and trace elements, etc. of the mangrove sediments. They reported that the microbial density was higher in the dense mangrove site than in the sparse one. The microbial counts decreased with increase in depth from 10 to 100 cm. The physicochemical parameters of Pichavaram mangrove ecosystem were described as follows: soil temperature varied from 26°C to 28.50°C, soil pH varied from 7.38 to 8.13, and redox potential was 29.81 mV, while the salinity was 30.48 ppt. Silt and clay were higher in Pichavaram's dense mangrove sediment than in the sparse one. Soil nitrogen was 10.17 g m^{-2}, phosphorus was 5.71 g m^{-2}, potassium was 115 g m^{-2}, and total organic carbon was 2.89 mgC g^{-1}. In general, wet seasons were found to be favorable for the microbial colonizations in the Pichavaram mangrove ecosystem. The physical and chemical factors in the mangrove ecosystem of Pichavaram exhibited a profound effect on the microbial population residing in the mangrove sediment. Microbial density increased with soil temperature, pH, salinity, and soil composition (silt and clay), and this revealed the tolerance of microbes to the changes in pH, temperature, salinity, and sol composition.

20.4 Microbial Diversity in the Mangrove Ecosystem

The mangrove ecosystem is a unique environment harboring a diverse group of microorganisms that perform an important role of nutrient cycling in the ecosystem. They play an important role in the biochemical cycle of the ecosystem. The mangrove ecosystem is very rich in organic matter, and microbes are active participants in this ecosystem. Since the ecosystem is saline in nature, halophiles are believed to be predominant in this ecosystem. According to Alongi (1988), in tropical mangrove, bacteria and fungi make up approximately 91% of the total biomass, while algae and protozoa represent a very low percentage. The mangrove ecosystem provides a unique environment having abundance of carbon and other nutrients to the large microbial communities like bacteria, fungi, cyanobacteria, microalgae, fungi, etc. The common bacteria present in the mangrove ecosystem includes sulfate-reducing bacteria viz. *Desulfovibrio*, *Desulfotomaculum*, *Desulfococcus*, etc.; phosphate-solubilizing bacteria such as *Bacillus*, *Vibrio*, *Xanthobacter*, etc.; anoxygenic photosynthetic bacteria such as *Chloronema*, *Beggiatoa*, etc.; and methanogens such as *Methanococcoides*, etc. (Thatoi et al., 2013). The total bacterial count in the pristine and dense mangrove of Sundarban, India is reported in the range of 4.71×10^6–2.98×10^7 cfu g^{-1} of soil (Ramanathan et al., 2008), while that of Bhitarkanika mangrove ecosystem is in the range of 1.8–2.1×10^6cfu g^{-1} of soil with the predominance of bacterial genera such as *Bacillus*, *Pseudomonas*, *Desulfotomaculum*, *Desulfovibrio*, *Desulfomonas*, *Methylococcus*, *Vibrio*, *Micrococcus*, *Klebsiella*, and *Azotobacter* (Mishra et al., 2011). Mangrove forests are believed to be "hotspots" for marine fungi belonging to *Ascomycetes*, *Deuteromycetes*, and *Basidiomycetes*. Marine, terrestrial, and endophytic fungi such as *Alternaria*, *Aspergillus*, *Fusarium*, *Penicillium*, etc. are reported in high number from mangrove sediments globally. Ramanathan et al. (2008) reported the total fungal count in the mangrove sediments of Sundarban, India in the range of 5×10^3–4×10^4 cfu g^{-1} of soil. Saravanakumar et al. (2016) reported that the total heterotrophic bacterial count in the Pichavaram dense mangrove ecosystem was 1.94×10^4cfu g^{-1} of soil. The counts were higher at 10 cm depth mangrove sediment and lowest in 100 cm depth mangrove sediment, while the highest count was observed post monsoon and lowest in monsoon.

The study of microbial diversity and distribution in the mangrove ecosystem would improve our understanding of mangrove functionality and interactions of different microorganisms in it. In the following paragraphs, we will discuss in depth about the role of different microorganisms in the mangrove ecosystem.

20.4.1 Culturable Diversity

Several researchers viz. Alongi (1988), Alongi et al. (1992, 1993, 1998), Holguin et al. (1992, 1999, 2001), Das et al. (2006), Mishra et al. (2011), etc. have reported the culturable diversity of the mangrove ecosystem globally. Few reviews viz. Holguin et al. (2001), Sahoo and Dhal (2009), and Thatoi et al. (2013) explained in detail about the microbial diversity of mangrove ecosystems all over the globe.

20.4.1.1 Bacterial Diversity

According to Bano et al. (1997), microbial-generated detritus plays an important role in the bacterial growth in any mangrove ecosystem. Although mangrove ecosystems are rich in organic matter, they are generally nutrient-deficient ecosystems (Holguin et al., 1992). It is only the efficient recycling of nutrients by bacterial population present in the mangrove ecosystem that makes it highly productive (Alongi et al., 1993). Bacteria are responsible for most of the carbon flux in the mangrove sediments. They are believed to be major participants in the nitrogen and phosphorus cycles (Toledo et al., 1995). Complex interactions among these microorganisms maintain the nutritional status and ecological balance of the mangroves (Holguin et al., 2006). It is observed that tree exudates fuel many bacterial activities in mangrove sediment (Alongi et al., 1993). There is a close relationship between microorganisms present in the mangrove ecosystem, nutrients present in the mangrove sediment, and mangrove tree parts like aerial roots, barks stem, submerged parts, etc. The recycling of nutrients takes place over here, and most of the necessary nutrients for the natural sustainability of this ecosystem (Alongi, 1994) are preserved here. In the following sections, we will discuss about different bacterial groups present in the mangrove ecosystem and their contribution toward the productivity of the mangrove ecosystem.

20.4.1.2 Sulfur-Oxidizing and Sulfate-Reducing Bacteria

Mangrove sediments are mostly anaerobic with an overlying thin aerobic sediment layer. The decomposition of the anaerobic layer occurs mainly through sulfate reduction. According to Howarth (1984), in anoxic mangrove sediment layers, 70%–90% of the total respiration is by sulfate reduction. In mangrove anaerobic sediments, sulfate-reducing bacteria (SRBs) appear to be the main decomposers of organic matter and play an important role in the mineralization of organic sulfur and in the production of soluble iron and phosphorus, which are then utilized by other microorganisms present in the mangrove ecosystem (Thatoi et al., 2013). SRBs have been isolated from many mangrove sediments of Florida (Zuberer and Silver, 1978), Goa (Saxena et al., 1988), Denmark (Jørgensen, 1977), etc. From mangroves of Goa, India, SRBs such as *Desulfovibrio sapovorans*, *Desulfovibrio desulfuricans*, *Desulfovibrio salexigens*, *Desulfotomaculum acetoxidans*, etc. are isolated (Loka Bharathi et al., 1991). In anoxic mangrove sediments, sulfur-oxidizing bacteria play an important role in the generation of sulfate, which is utilized by SRBs as an alternative electron acceptor in an anaerobic respiration to create hydrogen sulfide. Sulfur-oxidizing bacteria, e.g., *Chromatiaceae* and *Desulfobacterales*, were reported from Brazilian mangrove sediments. Some free-living and symbiotic sulfur oxidizers are reported from mangrove swamps of China (Liang et al., 2006).

20.4.1.3 Phosphate-Solubilizing Bacteria

Phosphorus is one of the major plant nutrients. Mangrove sediments have a strong capability to absorb phosphates carried by the tides, but in mangrove sediments, inorganic and insoluble phosphates are present. Solubilization of this inorganic and insoluble phosphate is carried out by fungi and phosphate-solubilizing bacteria residing in the mangrove sediments. Thus, phosphate-solubilizing microorganisms play an important role in supplementing phosphorus to the plants. Soil phosphate-solubilizing bacteria are associated with black mangrove root system and are involved in carbon recycling under both aerobic and anaerobic conditions (Das et al., 2012). The mangrove soil-associated microbe *Trichoderma* sp. is known to be involved in soil phosphate solubilization and in the improvement of growth of mangrove plant species *Avicennia marina*

(Saravanakumar et al., 2016). Phosphate-solubilizing bacteria such as *Bacillus amyloliquefaciens*, *Paenibacillus macerans*, *Xanthobacter agilis*, *Vibrio proteolyticus*, *Enterobacter aerogenes*, etc. are reported from mangrove roots (Vazquez et al., 2000). Few studies from Indian mangrove ecosystem, i.e., Bhitarkanika, Sundarban, Pichavaram, Tamil Nadu, and Great Nicobar also reported the isolation of phosphate-solubilizing bacterial genera like *Pseudomonas*, *Bacillus*, *Vibrio*, *Micrococcus*, *Alcaligenes*, etc. (Thatoi et al., 2013). Ramanathan et al. (2008) reported high counts of phosphate-solubilizing bacteria, i.e., 1.40×10^5cfu g^{-1} soil from mangrove sediment associated with dense and pristine mangroves of Sundarban, India. Mishra et al. (2011) reported phosphate-solubilizing bacterial population in the Bhitarkanika mangrove ecosystem in the range of $2.2–15.7 \times 10^5$cfu g^{-1} soil. High counts of phosphate-solubilizing bacteria (17×10^3cfu g^{-1} soil) are reported in the mangrove sediment of Gulf of Kuchh, India (Gautam and Ramanathan, 2013).

20.4.1.4 Nitrogen-Fixing Bacteria

In mangrove ecosystem, nitrogen fixation is the second major activity observed after carbon decomposition of detritus by SRBs. High rates of nitrogen fixation are observed with dead and decaying leaves, pneumatophores, rhizosphere soil, etc. Nitrogen-fixing bacteria such as members of genera *Azospirillum*, *Azotobacter*, *Rhizobium*, *Clostridium*, and *Klebsiella* have been isolated from mangrove sediments, rhizosphere root surfaces of many mangrove plant species present in Mexico (Holguin et al., 1992), Pichavaram mangroves, India (Lakshmanaperumalsamy, 1987, Ravikumar, 1995), and mangrove swamps of Sundarban, India (Sengupta and Chaudhary, 1990). Mangrove ecosystems are actually nitrogen-deficient ecosystems as salinity present in the mangrove sediment decreases the availability of nitrogen (Kathiresan, 2000), and hence, the mangrove ecosystem is dependent upon the nitrogen-fixing microorganisms such as *cyanobacteria*, *Azotobacter*, etc. (Alongi et al., 1993). Gautam and Ramanathan (2013) reported high counts of nitrogen-fixing bacteria, i.e., 4×10^3cfu g^{-1} soil, in the mangrove sediment of Gulf of Kuchh, Gujarat, which indicated the dominance of the nitrogen-fixing process. Ramanathan et al. (2008) also reported that the free-living nitrogen fixers were high in count (1.36×10^5cfu g^{-1} soil) in the mangrove sediments of Sundarban, India. Mishra et al. (2011) reported that the nitrogen-fixing bacterial population in the Bhitarkanika mangrove ecosystem increased to a great extent (1.35×10^7 cfu g^{-1} soil) in the rainy season. Saravanakumar et al. (2016) reported that *Azotobacter*'s count in the Pichavaram dense mangrove ecosystem was 4.04×10^3 cfu g^{-1} of soil. The counts were higher in 10 cm depth mangrove sediment and lowest in 100 cm depth mangrove sediment, while the highest count was observed in monsoon and lowest in the premonsoon period.

20.4.1.5 Methanogenic Bacteria

An important characteristic of the mangrove ecosystem is the absence of oxygen in deeper soil, which makes it an ideal environment for methanogens (Dar et al., 2008). Mobanraju et al. (1997) and Lyimo et al. (2008) reported the presence of *Methanococcoides methylutens* and *Methanosarcina semesiae* in the mangrove sediments of Pichavaram, Southeast India and Tanzanian mangrove sediment, respectively. Taketani et al. (2010) reported the occurrence of *Methanopyrus kandleri* and *Methnothermococcus thermolithotrophicus* from the mangrove soil of Brazil.

20.4.1.6 Photosynthetic Anoxygenic Bacteria

The group photosynthetic anoxygenic bacteria include purple sulfur bacteria and green and purple nonsulfur bacteria. Mangrove sediments are generally rich in sulfur and are anaerobic, which provides an ideal habitat for photosynthetic anoxygenic bacteria. There are few reports of isolation of representatives of family *Chromatiaceae* (purple sulfur bacteria) and *Rhodospirillaceae* (purple nonsulfur bacteria) from Indian mangrove sediments (Vethanayagam, 1991). Vethanayagam and Krishnamurthy (1995) reported the presence of genera like *Chloronema*, *Chromatium*, *Beggiatoa*, *Thiopedia*, and *Leucothiobacteria* from the mangrove ecosystem of Cochin, India. In the mangroves of coast of Red Sea in Egypt, 225 isolates of purple nonsulfur bacteria belonging to the ten species of four different genera were isolated from water, mud, and roots of *A. marina* samples (Thatoi et al., 2013). Holguin et al. (2001) reported the isolation of purple sulfur bacteria from the submerged part of the pneumatophores of *A. germinans* in the mangroves of Baja California, Mexico. Purple nonsulfur bacteria of the family *Rhodospirillaceae* and nonheterocystous cyanobacteria are responsible for two-thirds of nitrogen fixation association with decomposing leaves of *R. mangle* (Gotto and Taylor, 1976). Thus, one could say that photosynthetic anoxygenic bacteria contribute significantly to the productivity of mangroves.

20.4.1.7 Algae

According to the reports of Kathiresan and Qasim (2005), in the mangrove ecosystem, algal species are found on the pneumatophores, i.e., roots of the mangrove trees and in soft mud of mangroves. Algal species belonging to the genera *Bostrychia*, *Caloglossa*, and *Catenella* are commonly associated with the roots and trunks of the mangrove, while species of the genera *Rhizoclonium*, *Enteromorpha*, and *Cladophora* normally present in the sediments of the mangrove. Most of the algal species form small filaments, which are resistant to desiccation and saline environment of the mangrove ecosystem. It is reported that along with cyanobacteria and SRBs, these algae form a biofilm which helps them to adhere to the trees of mangroves. There are many reports of different algal taxa from different mangrove ecosystems all over the world. A total of 150 algal taxa are reported from New World mangroves, Latin America (Cordeiro-Marino et al., 1992), and 109 algal species from Caribbean Coast mangroves (Kathiresan and Qasim, 2005). From Indian mangrove ecosystem, 558 algal species are recorded, while from Andaman and Nicobar mangroves, 71 species are recorded (Kathiresan and Qasim, 2005). A total of 150 different species of algae belonging to the group *Chlorophyta*, *Chrysophyta*, and *Phaeophyta* are reported from

Sundarban mangroves of West Bengal, India (Sen and Naskar, 2003). From mangroves of Bhitarkanika, Orissa, India, algal species such as *Gloeocapsa*, *Chlorella*, *Ulva*, *Anabaena*, and *Oscillotoria* were identified (Mishra, 2010). Saravanakumar et al. (2016) reported 63 species of cyanobacteria, and the cyanobacterial count in the Pichavaram dense mangrove forest was 1.32×10^3cfu g^{-1} of mangrove sediment. They also reported that the cyanobacterial count was higher post monsoon and minimum in monsoon. The count was higher in upper mangrove sediment and minimum at 100cm depth.

20.4.1.8 Fungi and Actinomycetes

Mangrove ecosystems are called as the "home" of manglicolous fungi and marine fungi (Sahoo and Dhal, 2009). The high fungal diversity in the mangrove sediments is because of the favorable environment, i.e., moist conditions, high organic matter in the sediment, aeration, and low pH (Ghizelini et al., 2012). The terrestrial fungi and lichen occupy the upper part of the mangrove trees, while marine fungi occupy the lower and submerged parts of the mangrove trees such as aerating roots, etc. (Thatoi et al., 2013). Since fungi are able to synthesize all types of degrading enzymes that degrade lignin, cellulose, and other plant components, they are important for nutrient cycling of this ecosystem (Sahoo and Dhal, 2009). Thus, fungi are known to be a prerequisite for the decomposition of the mangrove litter and nutrient conservation in the mangrove ecosystem (Saravanakumar et al., 2016).The primary role of fungi in the mangrove sediment is considered to be the mineralization of organic matter. Fungi also serve as a food source for benthic fauna. A coarser sandy mangrove sediment has a lower number of fungi than sediments with finer texture (Ghizelini et al., 2012). Liu et al. (2007) reported more than 200 species of endophytic fungi mainly *Alternaria*, *Aspergillus*, *Cladosporium*, *Fusarium*, *Penicillium*, *Trichoderma*, etc. from the mangrove environment. According to Hyde et al. (1998), there are 1,500 fungal species present in the mangrove ecosystem. Thatoi et al. (2013) in their review of biodiversity of microorganisms from the mangrove ecosystem described at length about fungal and actinomycetes diversity in different mangrove ecosystems present around the world. The details are described in Table 20.1. In addition to the role of degradation of lignin and cellulose from mangrove trees, few fungi like *Cladosporium*, *Fusarium*, etc. are found on dead leaves of mangroves showing pectinolytic, proteolytic, and amylolytic activities (Raghukumar et al., 1994). According to the study of Raghukumar et al. (1995), in Indian mangrove, first colonizers on the fallen mangrove trees were fungi; thus, it is possible that these fungi might be tolerating high levels of phenolic compounds present in the mangrove leaves. Saravanakumar et al. (2016) reported the actinobacterial and fungal count in the Pichavaram dense mangrove forest as 6.35×10^3cfu g^{-1} of soil for actinobacteria and 7.17×10^3cfu g^{-1} of soil for fungi, respectively. They also reported that the actinobacterial count was higher pre monsoon and minimum in monsoon. The fungal count was more pre monsoon and minimum post monsoon. Both actinobacterial and fungal counts were higher in upper mangrove sediment and minimum at 100cm depth. Saravanakumar et al. (2016) also reported the *Trichoderma* counts in the Pichavaram dense mangrove forest as 6.35×10^3cfu g^{-1} of soil. The count was higher in upper mangrove sediment (10cm depth) and in the summer season, while it was lower in the postmonsoon season.

20.4.1.9 Cellulose Degraders

Thatoi et al. (2013) reviewed the occurrence of cellulose-degrading bacteria in various mangrove ecosystems all over the world. Cellulose degradation by microorganisms is one of the important processes in any mangrove ecosystem because cellulase enzyme produced by the microorganisms represents a major flow of carbon from fixed carbon sink to the atmospheric carbon dioxide. As cellulase is insoluble, the bacteria and fungi capable of producing cellulose degrade cellulose into carbon and energy sources, which are required by other microorganisms present in the mangrove ecosystem. Compared to other microbial groups, very less information is available of the diversity of cellulose degraders from different mangrove ecosystems. Gautam and Ramanathan (2013) reported high counts of cellulose degraders, i.e., $5–10\times10^3$cfu g^{-1} soil, from the mangrove sediment of Gulf of Kuchh, Gujarat, India. Ramanathan et al. (2008) also reported high counts of cellulose-degrading bacteria, i.e., 4.51×10^5cfu g^{-1} soil, from mangrove sediment associated with dense and pristine mangroves of Sundarban, India. The excessive litter fall, optimum temperature, and redox conditions in the dense mangroves of Sundarban might have created an ideal environment for cellulose degraders.

20.4.2 Unculturable Diversity

Knowledge of the microbial diversity and the activities of microorganisms in mangrove sediments is important for understanding how the mangrove ecosystems function. But

TABLE 20.1
Diversity of Fungi and Actinomycetes in Different Mangrove Ecosystems Present All over the World

S. No.	Source of Isolation of Fungi/Actinomycetes	References
1.	169 fungal species from mangroves of Malaysia	Alias et al. (1995)
2.	44 fungal species from mangrove of Hong Kong	Sadaba et al. (1995)
3.	91 fungal species from mangrove of Egyptian Red Sea	Abdel-Wahab (2005)
4.	112 fungal species from mangrove of Bahamas islands	Jones and Abdel-Wahab (2005)
5.	48 fungal species from mangrove of Pichavaram, India	Ravikumar and Vittal (1996)
6.	31 fungal species have been studied from sediment and 27 species from decaying leaves, stems, roots and pneumatophores of an estuarine mangrove ecosystem in Cochin, India	Prabhakaran et al. (1990)

our knowledge about the microbiome of the mangrove ecosystem is limited because several microbes are unculturable, and therefore, their abundance and diversity cannot be assessed through use of traditional culture-based methods. During past decades, the development of molecular techniques using nucleic acids has led to many new findings in the field of microbial ecology, and numerous new members belonging to known or new lineages of the bacteria and Archaea have been detected (Liang et al., 2006). Next-generation sequencing such as pyrosequencing and Illumina sequencing provides a relatively detailed picture of microbial communities in a particular ecosystem as compared to other methods (Liu et al., 2019). This approach avoids the limitation of traditional culture-based techniques for assessing microbial diversity in the natural environment. In-depth analysis of prokaryotic communities is crucial in understanding any ecosystem's function. So far, few culture-independent studies either using clone library and sequencing approach (Liang et al., 2006) or using new technology such as pyrosequencing of the hypervariable region of 16S rRNA gene (Fernandes et al., 2014) have been undertaken to assess the taxonomic diversity of bacteria in the mangrove ecosystem. Bacterial diversity data obtained from unculturable studies help in understanding the ecological roles that microbes play in an environment.

Liang et al. (2006) investigated the bacterial diversity present in the surface sediments of the mangrove ecosystem of China using a culture-independent molecular approach such as clone library preparation and phylogenetic analysis. Phylogenetic analysis of nearly full length 16S rRNA genes revealed that 148 bacterial clones fell into at least 13 major lineages of the bacteria domain: *Alpha-*, *Beta-*, *Gamma-*, *Delta-*, and *Epsilonproteobacteria*, the *Cytophaga–Flexibacter–Bacteroides* group, *Actinobacteria*, *Chloroflexi*, *Firmicutes*, *Fusobacteria*, *Chlamydiae/Verrucomicrobia* group, *Fibrobacteres/Acidobacteria* group, and Planctomycetes. Sixty-seven percentage of the clones of the gene library were grouped within the five subdivisions of Proteobacteria. The *Gammaproteobacteria* represented the most abundant proteobacterial group. Many of the sequences were phylogenetically associated with cultivated organisms involved in sulfur and nitrogen cycles, e.g., *Thioalkalivibrio nitratireducens*, *Thioalkalivibrio denitrificans*, *Rhabdochromatium marinum*, and *Thiococcus* sp. A substantial portion (8%) of the clones were grouped within the *Deltaproteobacteria*, a subdivision with anaerobic sulfate or metal reduction as the predominant metabolic trait of its members. In this way, the results significantly expanded the knowledge of the bacterial diversity of the unique mangrove environment.

Dias et al. (2010) studied the bacterial diversity of Brazilian well-preserved mangrove using another culture-independent approach like denaturing gradient gel electrophoresis. The data revealed that the Brazilian mangrove bacterial community was dominated by *Alphaproteobacteria* (40% of clones), *Gammaproteobacteria* (19% of clones), and *Acidobacteria* (28% of clones), while minor components of the assemblage were affiliated to *Betaproteobacteria, Deltaproteobacteria, Firmicutes, Actinobacteria,* and *Bacteroidetes*.

Fernandes et al. (2014) studied the bacterial community composition in the relatively pristine Tuvem mangrove site situated in Goa, India and compared the results with the anthropogenically influenced Divar mangrove site in the same region using the pyrosequencing approach. Their observations revealed that phylum *Proteobacteria* was dominant at both pristine as well as anthropogenically influenced mangroves of Goa comprising 43% and 46% of the total tags, respectively. *Deltaproteobacteria* was the next most abundant class at the pristine Tuvem mangrove ecosystem comprising 21% of the total tag sequences, while at Divar, their abundance was 15%. Within class *Deltaproteobacteria*, up to 53% tags were comprised of order *Desulfobacterales*, and others were belonging to the *Desulfuromonadales*, *Myxococcales*, and *Synthrophobacterales*. The bacteria belonging to *Desulfobacterales* have been involved in sulfur cycling in particular sulfate reduction. Other important classes recorded in both mangrove ecosystems were *Actinobacteria*, *Gammaproteobacteria*, and *Alphaproteobacteria*. *Actinobacteria* were found to be the next most abundant phylum following *Proteobacteria*. Many *Actinobacteria* play an important role in the mangrove ecosystem such as cellulose degradation, hydrocarbon degradation, nitrate reduction, etc. Finally, due to high inputs of anthropogenically derived organic and inorganic material, the bacterial diversity in the anthropogenically influenced Divar mangrove site was relatively more diverse than that of the pristine Tuvem mangrove site.

Metagenomic analysis provides a method to evaluate the potential microbial pathways occurring in the environment in which these microbes inhabit representing a single snapshot where DNA present in the environment can be sequenced to provide a wide view of microbial community present in that ecosystem and potential functioning of that ecosystem. Such an approach provides a relatively unbiased view of microbial diversity present in an ecosystem (Andreote et al., 2012). For the first time, Andreote et al. (2012) revealed the microbiome of Brazilian mangrove sediment using metagenomic data obtained by direct 454-pyrosequencing of DNA collected from the mangrove sediment environment. This study described the microbial groups present in the Brazilian mangrove ecosystem, the preferential metabolic processes that might be occurring in the mangrove ecosystem, and the biochemical cycles like carbon, nitrogen, and sulfur occurring in the mangrove ecosystem. The major abundance of *Proteobacteria* (47%–56%), *Firmicutes* (10%–13%), *Actinobacteria* (5%–12%), *Bacteroidetes* (3%–11%), and *Chloroflexi* (1%–5%) was observed in Brazilian mangrove ecosystem. Minor groups detected were *Planctomycetes* (1%–4%), *Cyanobacteria* (1%–3.5%), *Acidobacteria* (2.7%), and *Archaea* (3.4%). Within *Proteobacteria*, the most frequent class detected was *Gammaproteobacteria* (32%–42%), followed by *Deltaproteobacteria* (30%–40%), *Alphaproteobacteria* (7.5%–18%), *Betaproteobacteria* (2%–9%), and *Epsilonproteobacteria* (2%–20%). The high occurrence of SSU sequences affiliated to *Deltaproteobacteria* might be related to the mangrove ecosystem for specific microbial groups such as sulfate-reducing bacteria. The affiliation of sequences in the KEGG database allowed to map the biogeochemical transformations that are possibly performed by microbes present in the mangrove sediment. In terms of carbon cycle, the sequences indicated the prevalence of genes involved in the metabolism of methane, formaldehyde, and carbon dioxide. With respect to nitrogen cycle, the sequences related to dissimilatory reduction of nitrate, nitrogen immobilization, and denitrification were detected, while in the case of sulfur cycle, sequences related to adenylyl sulfate, sulfite, and hydrogen sulfide were detected. Thus, this study tentatively described the "microbial core" for mangrove functioning, which was mainly composed of *Desulfobacteraceae* and the other groups involved in carbon and nitrogen cycle such as *Rhodobacteraceae*, *Planctomycetes*, and *Burkholderiaceae*.

Paingankar and Deobagkar (2018) also studied the microbial community structure of mangroves of west coast of Maharashtra, India using the Illumina MiSeq sequencing platform and metagenomic analyses. From the results, it could be seen that in mangroves of west coast of Maharashtra, India, *Proteobacteria* and *Bacteroidetes* were most common, followed by taxon such as *Firmicutes*, *Spirochaetes*, *Chloroflexi*, and *Verrucomicrobia*. In *Proteobacteria* group, *Gammaproteobacteria*, *Alphaproteobacteria*, and *Deltaproteobacteria* were most abundant. Interestingly, bacteria having the capacity to utilize sulfate were present along with methanogens in mangrove sediment samples, suggesting that anaerobic and sulfur-based metabolic pathways play an important role in these mangrove ecosystems. Sulfate-reducing bacteria, such as *Desulfobacca acetoxidans*, *Desulfotalea psychrophila*, and *Desulfobulbus propionicus*, were present in high numbers in all mangrove sediment samples analyzed in this study from the west coast of Maharashtra, India.

The rapid development of high-throughput sequencing technology provides a great quantity of microbial sequence data. The Earth Microbiome Project (EMP) has assembled a large set of samples submitted by scientists around the world, which provides a reference set against which a new data set can be compared. This offers an opportunity to compare the different ecologies present on the earth (Thompson et al., 2017). Zhang et al. (2019) recently revealed the prokaryotic diversity in the mangrove sediments across Southeastern China using high-throughput sequencing technology. The data obtained from this study was compared with the data of 1,370 sediment samples collected from the EMP to compare the microbial diversity of mangroves with other biomes such as fresh water lakes, salt lakes, hot springs, freshwater rivers, etc. The results showed that the prokaryotic alpha diversity as measured by Shannon index was significantly higher in mangroves of Southeastern China than other biomes. Principal coordinates analysis (PCoA) revealed that the prokaryotic beta diversity in the mangrove ecosystem showed a distinct cluster when compared with other biomes. The core operational taxonomic units (OTUs) in the mangroves of Southeastern China were mostly assigned to *Gammaproteobacteria*, *Deltaproteobacteria*, *Chloroflexi*, and *Euryarchaeota*.

The anaerobic and high-salinity environment of the mangrove ecosystem provides conditions for archaea to survive. So, domain Archaea is also an important part of microbial communities of the mangrove ecosystem. However, previous studies till now have been mainly focused on bacterial communities present in the mangrove sediments, and thus, there is lack of information regarding the archaeal communities in the mangrove ecosystem except studies of Yan et al. (2006), Lyimo et al. (2009), Pires et al. (2012), and Liu et al. (2019). For the first time, Bhattacharyya et al. (2015) described the diversity and distribution of archaea in the tropical mangrove sediments of Sundarbans, India using 16S rRNA gene amplicon sequencing. The taxonomic analysis revealed the dominance of phyla *Euryarchaeota* and *Thomarchaeota* within the dataset. Within euryarchaeal sequences, a number of members of class *Halobacteria*, e.g. *Halosarcina*, *Halorientalis*, *Halolamina*, *Halorhabdus*, *Halogranum*, *Haloferax*, *Halomarina*, *Halorussus*, *Haloplanus*, and *Halarchaeum*, and of the class *Methanomicrobia*, for example, *Methanosarcina*, *Methermicoccus*, *Methanocella*, *Methanococcoides*, *Methanosalsum*, *Methanolobus*, and *Methanogenium* were detected. Thus, the study indicated the presence and active participation of archaea in the mangrove ecosystem of Sundarban. Basak et al. (2014) and Chakraborty et al. (2015) also described the community composition of the surface sediments and water sample of the Indian Sundarban mangrove ecosystem. Their results revealed that Sundarban mangrove sediments are dominated by *Deltaproteobacteria* followed by *Gammaproteobacteria*, *Alphaproteobacteria*, *Betaproteobacteria*, and *Epsilonproteobacteria* under phylum *Proteobacteria*. Abundant bacterial orders are *Desulfobacterales*, *Desulfuromonadales*, *Myxococcales*, and *Bdellovibrionales*. Recently, Dhal et al. (2020) investigated the aquatic microbiome of Sundarban mangrove ecosystem using the 16S rRNA gene-based amplicon approach. The study reports bacterial and archaeal community composition of Sundarban mangroves. The results revealed that the microbial community of marine estuary water from Sundarbans was dominated with bacteria occupying more than 96% of the total community, while archaea represented only 4%. The dominant bacterial groups were *Flavobacteria*, *alphaproteobacteria*, and *Acidimicrobiia*. The most dominant bacterial family was *Rhodobacteraceae*. The dominancy of *Thaumarchaeota* and *Euryarchaeaota* in archaeal community assemblage was seen.

Liu et al. (2019) described the bacterial and archaeal community composition of the Bamenwan mangrove wetland soil in China using pyrosequencing. The top 10 phyla present in the mangrove sediment were *Proteobacteria*, *Actinobacteria*, *Chloroflexi*, *Acidobacteria*, *Firmicutes*, *Bacteroidetes*, *Cyanobacteria*, *Nitrospirae*, and *Chlorobi*. *Proteobacteria* (53%) was a dominant phylum followed by *Actinobacteria* (48.40%). In *Proteobacteria* phylum, *Alphaproteobacteria* (40%), *Deltaproteobacteria* (33%), and *Gammaproteobacteria* (23%) were three main classes detected. Within *Alphaproteobacteria*, dominant orders were *Rhizobiales* and *Rhodospirillales*. The majority of *Deltaproteobacteria* sequences were belonging to *Desulfobacterales* and *Syntrophobacterales*, while within *Gammaproteobacteria*, the dominant orders were *Xanthomonadales* and *Chromatiales*. Three different archaeal phyla viz. *Crenarchaeota*, *Euryarchaeota*, and *Thomarchaeota* were detected with dominance of *Crenarchaeota* and *Euryarchaeota*. Within *Euryarchaeota*, *Halobacteriales*, *Methanosarcinales*, and *Thermoplasmatales* orders were dominant.

Thus, in conclusion, we can state that *Proteobacteria*, well known for their role in nitrogen fixation, was dominant in almost all mangrove sediments around the world. The difference in community compositions can be attributed to different geographical locations, mangrove plant species, and physicochemical parameters. Additional divergence could also be attributed to anthropogenic stressors acting on mangrove areas, chemical composition of the area, and pollution. The higher abundance of sulfate reducers in almost all mangrove sediments clearly suggests that these organisms are important players in mangrove ecology. The high concentration of sulfate in mangroves, the anaerobic conditions, and very low redox potential provide an ideal environment in mangroves for sulfate reducers and methanogens. In the mangrove ecosystem, sulfate reducers play an important role in the oxidation of organic matter, degradation of long-chain and aromatic compounds, production of H_2S, and release of phosphate and other ions that are essential for plant growth (Paingankar and Deobagkar, 2018).

Even though mangrove ecosystems are rich in microbial diversity, less than 5% of the mangrove species have been described till now. Thus, recently developed technologies in molecular biology such as next-generation sequencing, metagenomics, etc. offer a great opportunity for ecologists all over the world to explore this rich microbial diversity.

20.5 Biotechnological Potential of Mangrove Microbiome

Mangrove microorganisms have proven to be an important source of food, feed, medicine, enzymes, and antimicrobial substances. Thatoi et al. (2013) have extensively reviewed the biotechnological application of mangrove isolates. Thus, here the biotechnological applications of mangrove isolates are summarized in Table 20.2. The halophilic and halotolerant bacterial communities present in the mangrove ecosystem are capable of producing unique enzymes, biosurfactants, biodegradable plastics, and compatible solutes, which are commercially important. Fungi that are the main components of

TABLE 20.2

Biotechnological Potential of Microorganisms Isolated from Mangroves All over the World

Sr. No.	Microorganisms Isolated from Mangroves	Source of Isolation of Microorganisms	Biotechnological Potential	Reference
1.	Fungi	Mangrove sediment from China	Biopesticide	Xiao et al. (2005)
2..	*Vibrio fluvialis*	Mangrove sediment	Alkaline extracellular protease used in the detergent industry	Venugopal and Saramma (2006)
3.	*Aspergillus niger*	Mangrove sediment	Thermostable alkaline xylanase used in biobleaching of paper pulp	Raghukumar et al. (2004)
4.	*Leucobacter komagatae*	Mangrove sediment of Southern Thailand	Biosurfactant production	Saimmai et al. (2011)
5.	Fungal strain *Preussia aurantiaca*	Mangrove sediment	Antimicrobial compounds such as Auranticins (A and B)	Poch and Gloer 1991)
6.	*Fusarium* sp.	Mangrove forest	Enniatin G—a novel compound with a structure of cyclohexapeptide, which is having antitumor, antibiotic, insecticidal, and phytotoxic activity	Lin and Zhou (2003)
7.	Ascomycetes, *Verruculina enalia*	Mangrove wood	New phenolic compounds, enalin A and B, with hydroxymethyl furfural and three cyclopeptides with antimicrobial, antifungal, phytotoxic, and antidiabetic activities	Lin et al. (2002)
8.	*Streptomyces* sp.	Mangrove soil in Indonesia	Series of antibiotics that strongly inhibit the growth of Gram-positive and Gram-negative bacteria	Wiwin (2010)
9.	*Streptomyces* sp.	Mangrove sediment of Manakudy estuary, India	Potent antimicrobial effective against methicillin-resistant *Staphylococcus aureus* (clinical isolate)	Santhi and Jebakumar (2011)
10.	2,000 species of *actinomycetes*	Mangrove sediment of China	Bioactive compounds having anti-tumor, anti-cancer, and anti-infection properties	Hong et al. (2009)
11.	*Streptomyces albidoflavus*	Mangroves of Pichavaram, India	Bioactive compound having anti-tumor activity	Sivakumar et al. (2005)
12.	*Fungi*	Mangroves of South China Sea	Isoflavone and prostaglandin analog compounds that appeared to be promising for treating cancer patients with multidrug resistance	Tao et al. (2010)
13.	White-rot basidiomycetous fungus	Mangrove sediment	Enzyme laccase having activity to decolorize colored effluents and synthetic dyes	D'Souza et al. (2006)

the mangrove ecosystem are reported to produce xylanolytic enzymes, which have direct industrial applications in paper manufacturing, bread making, animal feed making, juice preparation, wine industry, etc. Marine algae which are also an important component of the mangrove ecosystem are the sole producers of industrially important phycocolloids such as agar-agar, alginate, etc. Actinomycetes isolated from the mangrove ecosystem are a rich source of anti-tumor, anti-infection, and anti-diabetic compounds. It is reported that mangrove microbial strains are beneficial for agriculture. The mangrove isolates have the ability to fix nitrogen, solubilize phosphates, and produce ammonia and plant growth-promoting hormones like IAA. Thus, these mangrove isolates can be inoculated in crops for high yields. The halophilic microbes from mangroves can be used in bioremediation of salt-affected soil or hypersaline soil for crop productivity. It is also reported that the mangrove fungi are capable of producing insecticidal metabolites, which can be explored as biopesticides. Dias et al. (2009) and Mishra (2010) reported the activity of enzymes such as amylase, protease, esterase, and lipase from bacteria isolated from Brazilian mangrove ecosystem and catalase, peroxidase, oxidase, polyphenol oxidase, and ascorbic acid oxidase from bacteria isolated from mangroves of Bhitarkanika, Orissa, India, respectively. Phytase- and tannase-producing microbial strains were reported from mangrove ecosystems of Kerala, India. A significant number of reports have focused on antimicrobial metabolites and bioactive compounds like anti-tumor, anti-cancer, anti-infection isolated from mangrove saprophytic fungi and actinomycetes, respectively. Thus, mangrove ecosystems are called "hot spots" for newer drugs naturally produced by mangrove isolates. The mangrove bacterial isolates such as *Pseudomonas*, *Marinobacter*, *Alcanivorax*, *Microbulbifer*, *Sphingomonas*, *Micrococcus*, *Cellulomonas*, *Dietzia*, and *Gordonia* are reported to degrade hydrocarbons and thus could be explored for bioremediation of oil-contaminated mangrove soil. Many mangrove bacterial isolates are reported to degrade polyaromatic hydrocarbons (PAHs) and utilized for bioremediation of PAH-contaminated areas. Table 20.2 describes the biotechnological potential of microorganisms isolated from mangroves all over the world.

20.6 Need for Conservation of Mangrove Ecosystem

Mangrove ecosystems are extremely important for offering protection against destructive ocean events like tsunamis and tropical cyclones. But they are not always valued for that function. Today, mangroves are found in or near the urban areas, and thus, they are under the constant impact of anthropogenic activities. Currently, mangroves are one of the most threatened habitats on the earth. Currently, mangroves can be seen as a thin green line of vegetation around the coasts and estuaries because they account for less than 1% of the world's tropical forest. In many regions, mangrove ecosystems are shrinking at an alarming rate because of urbanization and agricultural expansion, reclaiming mangrove area for industry, housing, coastal road construction, tourism and ports, aquaculture, mainly shrimp cultivation, charcoal production, salt pan production, and most importantly discharge of pollutants, sewage and waste water from various industries, etc. In few countries, mangrove wood is used for cooking. Oil spills, which occur frequently in Panama, the Persian Gulf, and southern Mexico, are particularly damaging for mangrove ecosystems because they block the diffusion of gases in plants and soil. Wastewater contaminates the mangrove sediments and detritus food web and introduces the heavy-metal residues in the mangrove ecosystem. This ultimately disturbs the microbial ecology of the mangrove ecosystem resulting in species extinction. According to Duke et al. (2007), many mangroves are on the verge of extinction and are expected to disappear from at least 26 of the 120 countries in which they are currently found. Rates of destruction are currently measured to be between 1% and 2% per year. According to Ghizelini et al. (2012), the importance of mangroves to tropical coastal regions has been a continual source of scientific discussion. American Institute for Global Change Research has described mangroves as one of the most critical ecosystems in the tropical region and vulnerable to global climate change. In 1965, Brazilian authorities recognized the importance of mangroves and gave them permanent protection in law. Still, the protection of the mangrove ecosystem remains a challenge for all countries.

July 26 is celebrated all over the world as "International Mangrove Day", which is dedicated to the unique forests of mangroves that survive at the interface of land, river, and sea. The tsunami that struck Southeast and South Asia in 2004 destroyed many villages, towns, and cities across 14 countries. But at the same time, some affected areas escaped the massive destructions due to healthy mangroves present at their coastlines acting as a protective barrier. Mangrove forests are also skilled at filtering the pollutants from river water. However, mangroves are declined rapidly around the world, and without mangroves, sea coasts will be unproductive and barren and fisheries will be collapsed. The sea level will increase, and there will be ultimate loss of sea animals and plants and microbes residing in the mangrove ecosystem. Loss of mangroves will release a large amount of carbon dioxide into the atmosphere, which ultimately contributes to global warming. Thus, protection and restoration of mangroves should have the highest priority for the benefit of people, biodiversity, and marine life and finally our planet. Mangrove reforestation using plant growth-promoting bacteria (PGPB) is also one of the remedies to the extinction of the mangrove ecosystem. Holguin et al. (2001) and Sahoo and Dhal (2009) shed light on the rehabilitation of the mangrove ecosystem through use of PGPB. Inoculation of plants with plant growth-promoting bacteria has been proposed as a useful agricultural tool to enhance crop yields. PGPB promote plant growth by mechanisms such as N_2 fixation, phosphate solubilization, phytohormone production, siderophore synthesis, or biocontrol of phytopathogens (Holguin et al., 2001). Thus, these PGPB will speed up the development of mangrove plantlets for the reforestation of the damaged and degraded mangrove areas all over the world. Currently, PGPB specific to mangrove plant species are not known, but attempts are made to colonize the black mangrove roots with cyanobacterial or *Azospirillum* sp., which are common terrestrial PGPB, and the results are encouraging. The dense population of *Azospirillum* sp. colonized on the black mangrove roots within four days of inoculation. The studies also showed that coinoculation of PGPB and phosphate-solubilizing bacteria like *Bacillus licheniformis* to the mangrove plantlets increased the N_2 fixation rate drastically. So, in all, it can be concluded that PGPB will effectively promote the growth of mangrove plantlets and help in the reforestation process.

Acknowledgment

SOK would like to acknowledge DBT BIOCARe (BT/PR19641/BIC/101/465/2016) for providing fellowship and National Center for Cell Science (NCCS), Pune for providing infrastructural facilities.

REFERENCES

Abdel-Wahab MAA (2005) Diversity of marine fungi from Egyptian red sea mangroves. *Bot Mar* 48:248–355.

Alias SA, Kuthubutheen AJ, Jones EBG (1995) Frequency of occurrence of fungi on wood in Malaysian mangroves. *Hydrobiologia* 295:97–106.

Alongi DM (1988) Bacterial productivity and microbial biomass in tropical mangrove sediments. *Microb Ecol* 15:59–79.

Alongi DM, Boto KG, Robertson AI (1992) Nitrogen and phosphorus cycles in tropical mangrove ecosystems. *Washington DC: AmGeophys Univ* 41:251–292.

Alongi DM, Christoffersen P, Tirendi F (1993) The influence of forest type on microbial-nutrient relationships in tropical mangrove sediments. *J Exp Mar Biol Ecol* 171:201–223.

Alongi DM, Sasekumar A, Tirendi F, Dixon P (1998) The influence of stand age on benthic decomposition and recycling of organic matter in managed mangrove forests of Malaysia. *J Exp Mar Biol Ecol* 225:197–218.

Andreote FD, Jime´nez DJ, Chaves D, et al. (2012) The microbiome of Brazilian Mangrove sediments as revealed by metagenomics. *PLoS ONE* 7(6):e38600. doi:10.1371/journal.pone.003860.

Bano N, Nisa MU, Khan N, Saleem M, Harrison PJ, Ahmed SI, Azam F (1997) Significance of bacteria in the flux of organic matter in the tidal creeks of the mangrove ecosystem of the Indus River Delta, Pakistan. *Mar Ecol Prog Ser* 157:1–12.

Basak NS, Majumder S, Nag et al. (2014), Spatiotemporal analysis of bacterial diversity in sediments of Sundarbans using parallel 16S rRNA gene tag sequencing, *Microb Ecol* 69:500–511.

Bhattacharyya, A, Majumder NS, Basak P, et al. (2015) Diversity and distribution of Archaea in mangrove sediments of Sundarbanss. *Archaea*, Volume 2015, Article ID 968582, 14 pages

Campbell A, Brown B (2015) Indonesia's vast mangroves are a treasure worth saving. The Conversation. From http://thconversation.com/indonesia-vast-mangroves-are-a-treasure-worth-saving-39367.

Chakraborty A, Bera A, Mukherjee A, et al. (2015) Changing bacterial profile of Sundarbans, the world eritagemangrove: impact of anthropogenic interventions. *World J Microbiol Biotechnol* 31:593–610.

Cordeiro-Marino M, Braga MRA, Eston VR, Fujii MT, Yokoya NA (1992) Mangrove macro algal communities in Latin America. The state of the art and perspectives. In: Seeliger U (ed) *Coastal Plant Communities of Latin America*. Academic, San Diego, CA, pp 51–64.

D'Souza DT, Tiwari R, Sah AK, Raghukumar C (2006) Enhanced production of laccase by a marine fungus during treatment of colored effluents and synthetic dyes. *Enzym Microb Technol* 38:504–511.

Dar SA, Kleerebezem R, Stams AJM, Kuenen JG, Muyzer G (2008) Competition and coexistence of sulfate-reducing bacteria, acetogens and methanogens in a lab-scale anaerobic bioreactor as affected by changing substrate to sulfate ratio. *Appl Environ Microbiol* 78:1045–1055.

Das S, Lyla PS, Khan SA (2006) Spatial variation of aerobic culturable heterotrophic bacterial population in sediment of the Continental slope of western Bay of Bengal. *Indian J Mar Sci* 36(1):51–58.

Dhal PK, Kopprio GA, Ga¨rdes A (2020) Insights on aquatic microbiome of the Indian Sundarbans mangrove areas. *PLoS ONE* 15(2):e0221543.

Dias ACF, Andreote FD, Dini-Andreote F, Lacava PT, Sa ALB, Melo IS, Azevedo JL, Araujo WL (2009) Diversity and biotechnological potential of culturable bacteria from Brazilian mangrove sediment. *World J Microbiol Biotechnol* 25:1305–1311.

Dias ACF, Andreote FD, Rigonato J, Fiore MF, Melo IS, Araujo WL (2010) The bacterial diversity in Brazilian non disturbed mangrove sediment. *Antonie Van Leeuwenhoek* 98:541–555.

Donato DC, Kauffman JB, Murdiyarso D, Kurnianto S, Stidham M, Kanninen M (2011) Mangroves among the most carbon rich forests in the tropics. *Nat Geosci* 4:293–297.

Duke NC, Meynecke JO, Dittmann S, et al. (2007). A world without mangroves? *Science* 317:41–42.

FAO (2007) *The World's Mangroves 1980–2005*. Rome: Food and Agricultural Organization of the United Nations.

Fell JW, Master IM, Wiegert RG (1984) Litter decomposition and nutrient enrichment. In: Snedaker SC, Snedaker JG (eds) *The Mangrove Ecosystem: Research Methods. (Monograph on Oceanographic Methodology, no 8)* UNESCO, Paris, pp 239–251.

Fernandes SO, Kirchman DL, Michotey VD, Bonin PC, Loka Bharathi PA (2014) Bacterial diversity in relatively pristine and anthropogenically-influenced mangrove ecosystems (Goa, India). *Brazilian J Microbiol* 45:1161–1171.

Findlay RH, Fell JW, Coleman NK, Vestal JR (1986) Biochemical indicators of the role of fungi and thraustochytrids in mangrove detrital systems. In: Moss ST (ed) *The Biology of Marine Fungi*. Cambridge University Press, Cambridge, pp 91–104.

Ghizelini AM, Mendonça-Hagler LCS, Macrae A (2012) Microbial diversity in Brazilian mangrove sediment. *Brazilian J Microbiol* 2012:1242–1254. ISSN 1517-8382

Giri C, Ochieng E, Tieszen LL, et al. (2011) Status and distribution of mangrove forests of the world using Earth observation satellite data. *Global Ecol Biogeography* 20:154–159.

Giri, C, Pengra B, Zhu Z, Singh A, Tieszen LL (2007) Monitoring mangrove forest dynamics of the Sundarbans in Bangladesh and India using multi-temporal satellite data from 1973 to 2000. *Estuarine, Coastal Shelf Sci* 73(1–2):91–100

Gonzalez-Acosta GB, Bashan Y, Hernandez-Saavedra NY, Ascencio F, Cruz-Aguero G (2005) Seasonal seawater temperature as the major determinant for populations of culturable bacteria in the sediments of an intact mangrove in an arid region. *FEMS Microbiol Ecol* 55:311–321.

Gotto JW, Taylor BF (1976) N2 fixation associated with decaying leaves of the red mangrove (*Rhizophora mangle*). *Appl Environ Microbiol* 31:781–783.

Goutam K and Ramanathan AI (2013) Microbial diversity in the surface sediments and its interaction with nutrients of mangroves of Gulf of Kachchh, Gujarat, India. *Int Res J Environ Sci* 2:25–30.

Gupta N, Das S, Basak UC (2007) Useful extracellular activity of bacteria isolated from Bhitrkanika mangrove ecosystem of Odisha coast. *Malaysian J. Microbiol* 3:15–18.

Holguin G, Guzman MA, Bashan Y (1992) Two new nitrogen fixing bacteria from the rhizosphere of mangrove trees: their isolation, identification and in vitro interaction with rhizosphere staphylococcus sp. *FEMS Microbiol* 101:207–216.

Holguin G, Vazquez P, Bashan Y (2001) The role of sediment microorganisms in the productivity, conservation, and rehabitation of mangrove ecosystems: an overview. *Biol Fertil Soil* 33:265–278.

Hong K, Gao AH, Xie QY, et al. (2009) Actinomycetes for marine drug discovery isolated from mangrove soils and plants in China. *Mar Drugs* 7:24–44.

Howarth RW (1984) The ecological significance of sulfur in the energy dynamics of salt marsh and coastal marine sediments. *Biogeochemistry* 1:5–27.

Hyde KD, Jones EBG, Leano E, Pointing SB, Poonyth AD, Vrijmoed LLP (1998) Role of fungi in marine ecosystems. *Biodiv Conserv* 7:1147–1161.

Jones EBG, Abdel-Wahab MA (2005) Marine fungi from the Bahamas Islands. *Bot Mar* 48:356–364.

Jørgensen BB (1977) The sulfur cycle of a coastal marine sediment (Limfjorden, Denmark). *Limnol Oceanogr* 22:814–832.

Kathiresan K (2000) A review of studies on Pichavaram mangrove, Southeast India. *Hydrobiologia* 430:185–205.

Kathiresan K, Qasim SZ (2005) *Biodiversity of Mangrove Ecosystems*. New Delhi: Hindustan, P 51.

Kathiresan K, Rajendran N (2005) Mangrove ecosystem of Indian Ocean region. *Indian J Mar Sci* 34:104–113.

Kohlmeyer J, Kohlmeyer E. (1979) *Marine mycology. Then higher fungi*. New York: Academic, 690 p.

Lakshmanaperumalsamy P (1987) Nitrogen fixing bacteria, Azotobacter sp. in aquatic sediment. *Fish Technol Soc Fish Technol* 24(2):126–128.

Liang JB, Chen YQ, Lan CY, Tam NFY, Zan QJ, Huang LN (2006) Recovery of novel bacterial diversity from mangrove sediment. *Mar Biol* 150:739–747 DOI 10.1007/s00227-006-0377-2.

Lin YC, Wu XY, Deng ZJ, Wang J, Zhou SN, Vrijmoed LLP, Jones EBG (2002) The metabolites of the mangrove fungus Verruculina enalia No. 2606 from a salt lake in the Bahamas. *Phytochemistry* 59:469–471.

Liu M, Huang H, Bao S, Tong Y (2019) Microbial community structure of soils in Bamenwan mangrove wetland. *Sci Rep* 9:8406.

Liu A, Wu X, Xu T (2007) Research advances in endophytic fungi of mangrove. *Chin J Appl Ecol* 18(4):912–918.

Lin YC, Zhou SN (2003) *Marine Microorganism and its Metabolites*. Beijing: Chemical Industry, pp 426–427.

Loka Bharathi PA, Oak S, Chandramohan D (1991) Sulfate-reducing bacteria from mangrove swamps II: their ecology and physiology. *Oceanol Acta* 14:163–171.

Lyimo TJ, Pol A, Jetten SMM, Op Den Camp HJM (2008) Diversity of methanogenic archaea in a mangrove sediment and isolation of a new Methanococcoides strain. *FEMS Microbiol Lett* 291:247–253.

Lyimo TJ, Pol A, Jetten MSM, Op Den Camp HJM (2009) Diversity of methanogenic archaea in a mangrove sediment and isolation of a new Methanococcoides strain. *FEMS Microbiol Lett* 291(2):247–253.

Mishra, RR (2010) *Microbial Biodiversity in Mangroves of Bhitarakanika, Orissa—A Study on Genotypic, Phenotypic and Proteomic Characterisation of Some Predominant Bacteria*. PhD thesis submitted to North Orissa University, Orissa, India.

Mishra RR, Prajapati S, Das J, Dangar TK, Das N, Thatoi HN (2011) Reduction of selenite to red elemental selenium by moderatelyhalotolerant *Bacillus megaterium* strains isolated from Bhitarkanika mangrove soil and characterization of reduced product. *Chemosphere* 84(9):1231–1237.

Mishra, AK, Routray TK, Satapathy GC (1995) Ecological study on soil microflora of bhitarkanika mangroves in relation to vegetational patterns, pp. 46–51. In RC Mohanty (ed.). *Environment: Change and Management*. Delhi: Kamla-Raj.

Mobanraju R, Rajgopal BS, Daniels L, Natrajan R (1997) Isolation and characterisation of methanogenic bacteria from mangrove sediment. *J Mar Biotechnol* 5:147–152.

Odum EP (1971) *Fundamentals of Ecology*. Philadelphia, PA: WB Saunders, 574 p.

Odum WE, Heald EJ (1975a) Mangrove forests and aquatic productivity. In: Hasler AD (ed) *Coupling of Land and Water Systems. Ecological Studies Series*. Springer, Berlin, pp 129–136.

Odum WE, Heald EJ (1975b) The detritus-based food web of an estuarine mangrove community. In: Ronin LT (ed) *Estuarine Research*. Academic, New York, pp 265–286.

Paingankar MS and Deobagkar DD (2018) Pollution and environmental stressors modulate the microbiome in estuarine mangroves: a metagenome analysis. *Current Science* 115:1525–1535.

Pires, ACC, Cleary DFR, Almeida A. et al. (2012) Denaturing gradient gel electrophoresis and barcoded pyrosequencing reveal unprecedented archaeal diversity in mangrove sediment and rhizosphere samples. *Appl Environ Microbiol* 78:5520–5528.

Poch GK, Gloer JB (1991) Auranticins A and B: two depsidones from a mangrove isolate of the fungus Preussia aurantiaca. *J Nat Prod* 54:213–217.

Prabhakaran NR, Gupta N, Krishnankutty M (1990) Fungal activity in Mangalvan—an estuarine mangrove ecosystem. In: Nair NB (ed) *Proceedings of the National Seminar on Estuarine Management, Trivandrum*. Academic, New York, pp 458–463.

Prance GT (1998) *Rainforests of the World*. New York: Crown Publishers.

Raghukumar S, Sathe-Pathak V, Sharma S, Raghukumar C (1995) Thraustochytrid and fungal component of marine detritus. Field studies on decomposition of leaves of the mangrove Rhizophora apiculata. *Aquat Microb Ecol* 9:117–125.

Raghukumar SS, Sharma C, Raghukumar C, Satha Pathak, V Chandramohan D (1994) Thraustochytrid and fungal component of marine detritus. 4. Laboratory studies on decomposition of leaves of the mangrove *Rhizophora apiculata* Blume. *J Exp Mar Biol Ecol* 183:113–131.

Rajendran N, Kathiresan K (2004) How to increase juvenile shrimps in mangrove waters? *Wetl Ecol Mang* 12:179–188.

Rajendran N, Kathiresan K (2006) Microbial flora associated with submerged mangrove leaf litter in India. *Rev Biol Trop* (Int J Trop Biol. ISSN-0034-7744) 55:393–400.

Ramanathan AL, Singh G, Majumdar J, Samal AC, Chauhan R, Ranjan RK, Rajkumar K, Santra SC (2008) A study of microbial diversity and its interaction with nutrients in the sediments of Sundarban mangroves. *Indian J Mar Sci* 37(2):159–165.

Ravikumar S (1995) *Nitrogen fixing Azotobacters from the mangrove habitat and their utility as biofertilizers*. PhD thesis, Annamalai University., Chidambaram, Tamil Nadu, India.

Ravikumar DR, Vittal BPR (1996) Fungal diversity on decomposing biomass of mangrove plant Rhizophora in Pichagram estuary, east coast of India. *Indian J Mar Sci* 21(1):64–66.

Sadaba RB, Vrijmoed LLP, Jones EBG, Hodgkiss IJ (1995) Observations on vertical distribution of fungi associated with standing senescent Acanthus ilicifolius stems at Mai Po mangrove, Hong Kong. *Hydrobiologia* 295:119–126.

Sahoo K, Dhal NK (2009) Potential microbial diversity in mangrove ecosystem: a review. *Indian J Mar Sci* 38(2):249–256.

Sahoo K, Jee PK, Dhal NK, Ritarani Das R (2017) Physicochemical sediment properties of mangroves of odisha, India. *J Oceonograp Marine Res* 5:2. DOI: 10.4172/2572-3103.1000162.

Saimmai A, Sobhon V, Maneerat S (2011) Production of biosurfactants from a new and promising strain of Leucobacter komagatae 183. *Ann Microbiol* 62(1):391–402.

Santhi VS, Jebakumar SRD (2011) Phylogenetic analysis and antimicrobial activities of Streptomyces isolates from mangrove sediment. *J Basic Microbiol* 51:71–79.

Saravanakumar K, Anburaj R, Gomathi V, Kathiresan K (2016) Ecology of soil microbes in a tropical mangrove forest of south east coast of India. *Biocatalysis and Agri Biotechnol* 8:73–85.

Saxena D, Loka-Bharathi PA, Chandramohan D (1988) Sulfate reducing bacteria from mangrove swamps of Goa, central west coast of India. *Indian J Mar Sci* 17:153–157.

Sen N, Naskar K (2003) *Algal Flora of Sundarbans Mangal*. New Delhi: Daya.

Sengupta A, Chaudhuri S (1990) Halotolerant *Rhizobium* strains from mangrove swamps of the Ganges River Delta. *Indian J Microbiol* 30:483–484.

Sessitsch A, Weilharter A, Gerzabek MH, Kirchman H, Kandeler E (2001) Microbial population structures in soil particle size fractions of a long-term fertilizer field experiment. *Appl Environ Microbiol* 67:4215–4224.

Sivakumar K, Sahu MK, Kathiresan K (2005) An antibiotic producing marine Streptomyces from the Pichavaram mangrove environment. *J Annamalai Univ, Part-B* XLI:9–18.

Srilatha G, Varadharajan D, Chamundeeswari K and Mayavu P (2013) Study on physic-chemical parameters in different mangrove regions, Southeast Coast of India. *J Environ Anal Toxicol* 3:5.

Taketani GR, Yoshiura AC, Dias FCA, Andreote DF, Tsai MS (2010) Diversity and identification of methanogenic archaea and sulphate-reducing bacteria in sediments from a pristine tropical mangrove. *Antonie van Leeuwenhoek* 97:401–411.

Tao L, Zhang JY, Liang YJ, et al. (2010) Anticancer effect and structureactivity analysis of marine products isolated from metabolites of mangrove fungi in the South China Sea. *Mar Drugs* 8:1094–1105.

Thatoi H, Behera BC, Mishra RR, Dutta SK (2013) Biodiversity and biotechnological potential of microorganisms from mangrove ecosystems: a review. *Ann Microbiol* 63:1–19.

Thatoi HN, Biswal AK (2008) Mangroves of Orissa coast: floral diversity and conservation status. Special habitats and threatenedplants of India. *ENVIS Wild Life Protected Area* 11 (1):201–207.

Thatoi HN, Ouseph A, Mishra PK, Mohanty JR, Acharjyo LN, Ranjit Daniels J (1999) Mangrove restoration in Odisha, South India: an experiment. *Tiger paper* 26: 23–26.

Thompson LR, Sanders JG, McDonald D, et al. (2017) A communal catalogue reveals earth's multiscale microbial diversity. *Nature* 551:457–463.

Toledo G, Bashan Y, Soeldner A (1995) Cyanobacteria and black Mangrooves in North Western Mexico. Colonization and diurnal and seasonal nitrogen fixation on aerial roots. *Can J Microbiol* 41:999–1011.

Vazquez P, Holguin G, Puente ME, Lopez-Cortes A, Bashan Y (2000) Phosphate-solubilizing microorganisms associated with the rhizosphere of mangroves in a semiarid coastal lagoon. *Biol Fertil Soils* 30:460–468.

Venugopal M, Saramma AV (2006) Characterization of alkaline protease from Vibrio fluvialis strain VM10 isolated from a mangrove sediment sample and its application as a laundry detergent additive. *Process Biochem* 41:1239–1243.

Vethanayagam RR (1991) Purple photosynthetic bacteria from a tropical mangrove environment. *Mar Biol* 110:161–163.

Vethanayagam RR, Krishnamurthy K (1995) Studies on an oxygenic photosynthetic bacterium *Rhodopseudomonas* sp. from the tropical mangrove environment. *Indian J Mar Sci* 24:19–23.

Wiwin R (2010) Identification of Streptomyces sp-MWS1 producing antibacterial compounds. *Indonesian J Trop Infect Dis* 1(2):80–85.

Xiao YT, Zheng ZH, Huang YJ, Xu QY, Su WJ, Song SY (2005) Nematicidal and brine shrimp lethality of secondary metabolites from marine-derived fungi. *J Xiamen Univ (Nat Sci)* 44(6):847–850.

Yan B, Hong K, Yu ZN (2006) Archaeal communities in mangrove soil characterized by 16S rRNA gene clones. *J Microbiol* 44:566–571.

Zhang C-J, Pan J, Duan C-H, et al. 2019.Prokaryotic diversity in mangrove sediments across southeastern China fundamentally differs from that in other biomes. *mSystems* 4:e00442–19.

Zhuang T, Lin P (1993).Soil microbial amount variations of mangroves (Kandelia candel) in process of natural decomposition of litter leaves. *J. Xiamen Univ. Nat. Sci.* 32(3):365–370.

Zuberer DA, Silver WS (1978) Biological dinitrogen fixation (Acetylene reduction) associated with Florida mangroves. *Appl Environ Microbiol* 35:567–575.

21

Role of the Mycobiome in Agroecosystems

Ahmed Abdul Haleem Khan
Telangana University

CONTENTS

21.1 Introduction

Fungi are well-known eukaryotic organisms that inhabit a variety of habitats and adapt to typical ecosystems on the planet Earth. Though the systematic position enabled fungi among lower plants without chlorophyll and heterotrophic nutrition, a separate phylum was established. The habit of these organisms is microscopic and macroscopic with the ability to decay and decompose the organic matter in the environment. The initial phases in studies on, fungi were reported to attack cereal crop, rice that resulted in diseases i.e., Bengal famine and Irish famine that attacked potato crop and Salem witch trials due ergot toxins by *Claviceps*. The episodes of dreadful diseases by different forms of fungi in history made fungi as culprit in society. The discovery of miracle drug 'penicillin' from *Penicillium* sp. revolutionized the focus on fungi, though from a long time it was an inoculum in making wine, beer, and bread.

The dense population in urban settings resulted in pollution (air, water, and soil) and increased the health risks. The industrial and green revolution resulted in increased levels of recalcitrant and xenobiotics with different trophic levels of the food chain. The hazards in the farming system were common in conventional practices to cultivate a variety of crops around the globe. The present generation is turning toward organic, precision, and sustainable systems of cropping to cut the burden of fertilizers, pesticides, herbicides, and insecticides of fossil fuel refinery origin (Khan, 2020, 2019a, 2019b).

Fungi are drivers of evolution by colonizing the higher living organisms (plants/animals) that improved the fitness and adaption to changing abiotic and biotic factors in the ecosystem. Plants exist in diverse habits and habitats with different climates by effective mutualistic or symbiotic associations with microorganisms (rhizosphere, phyllosphere, phylloplane, and spermosphere) (Khan et al., 2010, 2011, 2014, 2015, 2016, 2017). The colonization of fungi in a variety of plant tissues creates a mycobiome that benefits the host by exchange of metabolites (Table 21.1). The abundance of fungal partners depends on the type of tissue, maturity, chemical composition, temperature, and humidity. The beneficial fungi coexisting with plants could be endophytes, dark septate endophytes (DSE), and arbuscular mycorrhizal fungi (AMF) (Table 21.2). The interactions between plants and fungi promote growth and nutrient uptake, and limit pathogen-specific defense in extreme environments (Table 21.3). The symbiotic associations of different fungi explored in the agroecosystems are enlisted to encompass the reports on plant–fungi relations that revoke the conventional farming and popularize the sustainable system for a safe ecosystem.

21.1.1 *Achnatherum inebrians* (Drunken Horse Grass)

The comparison study was reported to evaluate the potential of an endophyte (*Epichloe gansuensis*) for physiological barriers (salinity, water pH variability, light exposure, and temperature regimes) in endophyte-infected and noninfected seed germination percentage of *A. inebrians*. The infected seeds exhibited improved germination and survival compared to noninfected seeds under a varying range of salinity, pH, light, photoperiod, and temperature (Ahmad et al., 2020).

21.1.2 *Acer campestre* L. and *A. platanoides* L. (Field and Temperate Norway Maple)

The fungal endophytes from leaves (green and yellow) of field and temperate Norway maple were investigated using large-scale metabarcoding. The dominant endophytes belong to phyla: Ascomycota and Basidiomycota and classes: Dothideomycetes and Leotiomycetes. At the species level, abundant fungal species characterized were *Sawadaea bicornis*, *S. polyfida*, *Ramularia vallisumbrosae*, *Vishniacozyma victoriae*, and *Mycosphaerella harthensis* in the test tissue samples of *A. campestre* and *A. platanoides*. The study suggested that the endophytic fungi were responsible for the green islands of maple leaves (Wemheuer et al., 2019).

21.1.3 *Aerva javanica* Juss. Ex. Schult

The study of leaf endophytic fungus *Cercospora* sp. from *A. javanica* was reported for bioprocess optimization and biochemical characterization of the endo-metabolome with broad-spectrum antimicrobial (bacteria: *Acinetobacter baumannii*, *Bacillus subtilis*, *E. coli*, *E. coli* Kanr, *E. coli* Ampr, *Klebsiella pneumoniae*, *Mycobacterium smegmatis*, *Pseudomonas aeruginosa*, *P. fluorescens*, *Ralstonia solanacearum*, *Staphylococcus aureus*,

TABLE 21.1

Endophytic Fungi and Their Host Plants

Host Plant	Plant Tissues	Endophyte Isolated	Reference
Terminalia catappa, *T. mantaly*, and *Cananga odorata*	Leaves, stems, bark, roots, and flowers	*Fusarium*, *Phomopsis*, *Botryosphaeria*, *Penicillium*, *Nigrospora*, *Chaetomium*, *Cladosporium*, *Pestalotiopsis*, *Septoria*, *Lasiodiplodia*, *Fusicoccum*, *Corynespora*, *Phoma*, *Paraconiothyrium*, *Diaporthe*, *Guignardia*, *Mycosphaerella*, *Cercospora*, *Xylaria*, and *Colletotrichum*	Toghueo et al. (2017)
Taxuswallichiana Zucc.	Roots	*Aspergillus*, *Penicillium*	Adhikari and Pandey (2019)
Glycine max	Leaf, stem, and root	*Alternaria alternata*, *Arthrinium phaeospermun*, *Fusarium graminearum*, *F. oxysporum*, *F. equiseti*, *Scopulariopsis brevicaulis*, *Aspergillus terreus*, *A. flavus*, *Clonostachys rosea*, *Curvularia lunata*, *Macrophomina phaseolina*, and *Trichoderma saturnisporum*	Russo et al. (2016)
Evernia prunastri, *Ramalina fastigiata*, and *Pleurosticta acetabulum*	Lichen thalli	*Nemania serpens*, *Trametes versicolor*, *Trichoderma atroviride*, *T. caerulescens*, *Sordaria fimicola*, *Preussia persica*, *Plectania* sp., *Phoma herbarum*, *Periconia* sp., *Penicillium* sp., *Nemania aenea*, *Mucor racemosum*, *Leptosphaerulina chartarum*, *Epicoccum nigrum*, *Coniochaeta lignicola*, *Botrytis cinerea*, *Cladosporium* sp., *Biscogniauxia mediterranea*, *Aureobasidium melanogenum*, and *Aspergillus flavus*	Lagarde et al. (2018)
Panax ginseng	Roots	*Trichoderma citrinoviride*	Park et al. (2019)
Phoenix reclinata	Leaf	*Pestalotiopsis clavispora*	Alade et al. (2018)
Chenopodium quinoa	Roots	*Penicillium minioluteum*	Gonzalez-Teuber et al. (2018)
Populus deltoides and *P. trichocarpa*	Roots	*Atractiella rhizophila*	Velez et al. (2017)
Pelargonium sidoides	Roots	*Penicillium skrjabinii*	Aboobaker et al. (2019)
Trichoderma atroviride	Roots	*Salvia miltiorrhiza*	Zhou et al. (2019)
Aspergillus terreus and *Paecilomyces lilacinus*	Thalli	*Laurencia okamurai*	Li et al. (2020)
Phoma sp., *Nodulisporium* sp., and *Guignardia* sp.	Leaves	*Acanthus ilicifolius*	Chi et al. (2019)
Alternaria alternata, *Aspergillus terreus*, and *Alpestrisphaeria*	Roots, stems, leaves, and branches	*Vitex rotundifolia*	Yeh and Kirschner (2019)
Alternaria sp.	Leaves, bark, and xylem	*Citrus sinensis*	Juybari et al. (2019)
Alternaria sp.	Root	*Plumbago zeylanica* L.	Andhale et al. (2019)

TABLE 21.2

Diversity of Arbuscular Mycorrhizal Fungi (AMF) and Their Effects

AMF	Host Plant	Effects	Reference
Glomus, *Septoglomus*, *Rhizophagus*, *Kamienskia*, and *Sclerocystis*	*Sophoraflavescens*	Improved total nitrogen, available phosphorus concentration, invertase, soil organic matter, and urease	Song et al. (2019)
Glomus mosseae	Eveningprimrose	Alleviate water deficit stress	Mohammadi et al. (2019)
Rhizophagus intraradices and *Funneliformis mosseae*	*Thymus vulgaris*	Improvedmorphological traits, oil yield, and accumulation.	Arpanahi and Feizian (2019)
Glomus sp.	*Allium fistulosum* L.	Improved performance in Welsh onion	Akyol et al. (2019)
Acaulospora, *Funneliformis*, *Racocetra*, *Scutellospora*, *Claroideoglomus*, *Rhizophagus*, *Glomus*, *Entrophospora*, and *Diversispora*	Maize and wheat	Enhanced root colonization and soil alkaline phosphatase activity	Hu et al. (2015)

(*Continued*)

TABLE 21.2 (*Continued*)

Diversity of Arbuscular Mycorrhizal Fungi (AMF) and Their Effects

AMF	Host Plant	Effects	Reference
Diversispora sp., *Septoglomus constrictum*, *Funneliformis mosseae*, and *Rhizophagus irregularis*	*Allium scorodoprasum* and *Senecio fluviatilis*	Improved plant performance	Nobis et al. (2015)
Rhizophagus irregularis	*Panicum miliaceum*	Alleviate Cd toxicity in soybean	Cui et al. (2019)
Rhizophagus irregularis	Rice	Sustainable rice production by decreasing P loss	Zhang et al. (2020)
Glomus etunicatum	Maize	Carbon mineralization	Xu et al. (2019)
Funneliformis mosseae, *Rhizophagus irregularis*, and *Simiglomus hoi*	*Melilotus alba* Med. (white sweet clover)	Positive effects on phosphorus accumulation	Hack et al. (2019)
Rhizophagus intraradices	*Glycine max*	Soybean charcoal root rot control	Spagnoletti et al. (2020)
Acaulospora, *Claroideoglomus*, *Funneliformis*, *Glomus*, *Rhizophagus*, *Sclerocystis*, *Scutellospora*, and *Septoglomus*	*Zingiber montanum* and *Z. officinale*	Functional role in the growth and productivity of the gingers	Pandey et al. (2020)
Claroideoglomus claroideum, *Funneliformis mosseae*, *Gigaspora* sp., *Rhizophagus irregularis*, and *Scutellospora* sp.	Maize	Nutrients acquisition	Dias et al. (2018)
Acaulospora, *Archaeospora*, *Claroideoglomus*, *Diversispora*, *Glomus*, *Paraglomus*, and *Scutellospora*	*Digitaria macroblephara* and *Themeda triandra*	AMF decreased with grazing and precipitation and increased with soil phosphorus	Stevens et al. (2020)
Glomus intraradices	Watermelon	Improving chilling resistance	Bidabadi and Mehralian (2020)
Rhizophagus intraradices, *Glomus mosseae*, *G. aggregatum*, and *Claroideoglomus etunicatum*	*Thuja occidentalis*	Support cedar growth and nutrient supply	Anwar et al. (2020)
Funneliformis mosseae	Tomato	Increases the susceptibility of tomato plants to virus infection curly top disease	Ebrahimi et al. (2020)
Acaulospora, *Rhizoglomus*, *Entrophospora*, *Claroideoglomus*, *Funneliformis*, and *Gigaspora*	Rice	Increased plant growth and yield	Martins and Rodrigues (2020)
Funneliformis mosseae	Wheat	Plant tolerance to water deficit	Tarnabi et al. (2020)
Glomus irradicans	Green bean	Increased pod yield in salt stress	Motaleb et al. (2020)
Rhizophagus irregularis	*Medicago truncatula*	Emission of volatiles	Dreher et al. (2019)
Rhizophagus intraradices	Grapevine	Protect grapevine against GFLV	Hao et al. (2018)
Claroideoglomus etunicatum	Perennial ryegrass	Improve P uptake and growth of perennial ryegrass and to alleviate the damage caused by leaf spot incidence and drought stress	Deng et al. (2020)
Rhizophagus irregularis	*Andropogon gerardii*	Plant growth stimulation	Jansa et al. (2020)
Pisolithus tinctorius, *Rhizopogon villosuli*, *R. luteolus*, *R. amylopogon*, *R. fulvigleba*, *Scleroderma cepa*, *S. citrinum*, *Laccaria bicolor*, *L. laccata*, *Glomus intraradices*, *G. aggregatum*, *G. mosseae*, *G.brasilianum*, *G. monosporum*, *G. deserticola*, *G. clarum*, *G. etunicatum*, and *Gigaspora margarita*	*Cedrus libani*	Increased uptake of nutrients	Toprak (2020)
Acaulospora scrobiculata, *Glomus deserticola*, *G. intraradices* and *G. versiforme*	Tomato	Improved plant growth by phosphorus supply	El Maaloum et al. (2020)
Glomus sp.1, *Glomus* sp.2, *Septoglomus constrictum*, *Rhizophagus intraradices*, *Funneliformis mosseae*, and *Gigaspora margarita*	Olive	Improved plant growth	Chenchouni et al. (2020)
Glomus intraradices, *G. mosseae*, *G. aggregatum*, and *G. etunicatum*	*Festuca arundinacea*	Improved phytoremediation potential in tall fescue	Rostami and Rostami (2019)

TABLE 21.3

Potential of Endophytic Fungi in Alleviation of Plant Stress

Endophytic Fungi	Host Plant	Inoculated Plant	Response	Reference
A. japonicus	*Euphorbia indica* L.	Soybean and sunflower	High-temperature stress tolerance	Ismail et al. (2018)
Pleospora rosae, *Cochliobolus* sp., *Alternaria tenuissima*, and *Cladosporium macrocarpum*	*Kale and broccoli*	*A. thaliana*	Tolerance to water stress	Zahn and Amend (2019)
B. bassiana	*Parthenocissus quinquefolia* (L.) Planch.	Red oak (*Quercus rubra* L.)	Tolerance to drought stress	Ferus et al. (2019)
Meyerozyma caribbica	*Solanum xanthocarpum*	Maize	Tolerance to salt stress	Jan et al. (2019a)
Penicillium ruqueforti	*Solanum surattense*	Wheat	Phytostabilization of heavy metals	Ikram et al. (2018)
Trichoderma reesei	*S. surattense*	Wheat	Tolerance to salt stress	Ikram et al. (2019)
Exophiala pisciphila	*Arundinella bengalensis*	Maize	Tolerance to cadmium stress	Zhan et al. (2017)
A. flavus	*Chenopodium album*	Soybean	Tolerance to salt stress	Lubna et al. (2018b)
A. alternata	*Elymus dahuricus*	Wheat	Tolerance to water deficiency	Qiang et al. (2019)
Nectria haematococca	*Chrysanthemum indicum* L.	Tomato	Tolerance to drought stress	Valli and Muthukumar (2018)
Gaeumannomyces cylindrosporus	*Astragalusadsurgens* Pall.	Maize	Pb tolerance	Ban et al. (2017)
Yarrowia lipolytica	*Euphorbia milli* L.	Maize	alleviate salt stress	Jan et al. (2019b)
Thermomyces sp.	*Cullen plicata*	Cucumber	Heat stress tolerance	Ali et al. (2018)
Fusarium oxysporum	*Coriandrum sativum*	Tomato	Tolerance to salt stress	Rhouma et al. (2020)
Aureobasidium pullulans	*Boswellia sacra*	Cucumber	Tolerance to Cd and Pb stress	Ali et al. (2019a)
Alternaria alternata	*Elymus dahuricus*	Wheat	Tolerance to drought stress	Qiang et al. (2019)
Fusarium sp.	Pokkali rice	IR-64 rice	Tolerance to salinity stress	Sampangi-Ramaiah et al. (2019)
Fusarium sp.	*W. somnifera*	Tomato	Suppression of Fusarium crown and root rot	Nefzi et al. (2019)
Cochliobolus sp.	*Mirabilis jalapa* L.	Okra	Tolerance to Salt stress	Bibi et al. (2019)
Acrocalymma vagum and *Scytalidium lignicola*	*Ilex chinensis*	(*Medicago sativa* and *Ammopiptanthus mongolicus*)	Tolerance to Cd stress	Hou et al. (2020)
Curvularia sp.	*Suaeda salsa*	*Populus tomentosa*	Tolerance to salt stress	Pan et al. (2018)
Penicillium citrinum, *A. pullulans*, and *Dothideomycetes* sp.	*Citrus reticulate*	Mandarin	Tolerance to drought stress	Sadeghi et al. (2020)

and *X. campestris*. Fungi: *Candida albicans*, *Cryptococcus* sp., *Fusarium oxysporum*, *Kwoniella* sp., *Pseudozyma abaconensis*, *Saccharomyces cerevisiae*, *Sympodiomycopsis kandeliae*, and *Rhizoctonia solani*) and antioxidant activities. The findings proved that host plant botanicals changed the endophyte metabolomic profile and biocontrol (Mookherjee et al., 2020).

21.1.4 *Anoectochilus roxburghii*

Chaetomium globosum and *Colletotrichum gloeosporioides* from 277 strains of endophytic fungi isolated from roots, stems, leaves, and flowers of *Anoectochilus* and *Ludisia* orchids were reported for the growth and accumulation of active ingredients (flavonoids, kinsenoside, and polysaccharides) of *A. roxburghii*. The study concluded that the two endophyte strains were highly beneficial microbial resources with applied value in agriculture (Ye et al., 2020).

21.1.5 *Arabidopsis thaliana*

The seedlings of *A. thaliana* were inoculated with root endophytic fungus *Piriformospora indica* that resulted in decrease in H_2O_2, malondialdehyde, and freezing stress and increased levels of brassinolide (BR) and abscisic acid (ABA), soluble proteins, proline, and ascorbic acid. The results demonstrated that *P. indica* confers freezing tolerance and stimulated the expression of genes involved in the CBF-dependent pathway (Jiang et al., 2020).

21.1.6 *Arachis hypogaea* L. (Peanut)

The fungal endophyte *Phomopsis liquidambari* from the bark of *Bischofia polycarpa* was investigated for effects on iron (Fe) and molybdenum (Mo) uptake by peanut in monoculture soil. The results of endophyte inoculation in test plant improved the growth parameters, chlorophyll content, nitrate reductase (NR) activity, and Fe and Mo uptake (Su et al., 2019).

21.1.7 *Artemisia annua* L.

The findings revealed that dual (*Piriformospora indica* and *Azotobacter chroococcum*)-treated plants had better plant height and dry weights of shoot and root under salt stress than the un-inoculated *A. annua*. The treated plants reduced the oxidative damage in plants by decrease of MDA and H_2O_2 and increase of enzymatic (superoxide dismutase, catalase, ascorbate peroxidase, and glutathione reductase) and nonenzymatic (total flavonoids, phenolics, and carotenoids) antioxidants. The dual inoculation resulted in increase in artemisinin and proline content, and the findings demonstrate that the potentiality of symbiosis acts as a bio-ameliorator under salt stress (Arora et al., 2020).

21.1.8 *Camellia oleifera*

The endophytic fungi were from isolated leaves, barks, and fruits of *Camellia oleifera*, and their potential as biological control agents of *C. oleifera* anthracnose was evaluated. The isolates (81) were identified as 14 genera from two subdivisions (Deuteromycotina and Ascomycotina). The dominant species were *Pestalotiopsis* sp., *Penicillium* sp., and *Fusarium* sp. with high relative frequency. In the results of dual-culture experiments against the *C. oleifera* anthracnose pathogen, five strains among the isolates exhibited antifungal activity and endophytic *Oidium* sp. reported for the broad inhibition zones (Yu et al., 2018).

21.1.9 *Cannabis sativa*

The fungal endophytes from the root tissue segments of *C. sativa* were screened that support plant growth-promoting ability in rice and maize. The results showed eleven isolates of endophytes, among which the *Aspergillus niger* isolate was reported for the production of indole acetic acid (IAA), gibberellins (GA), siderophores, and phosphate solubilization. The culture filtrate and spore suspension of the *A. niger* isolate improved growth promotion in mutant *waito*-C rice (treated with uniconazole) and maize seedlings applied with uniconazole and yucasin. The endophyte isolate revealed to alleviate the effect of inhibitors of IAA and GA biosynthesis in test rice and maize (Lubna et al., 2018a).

21.1.10 *Castanea sativa* (Chestnut)

The study on foliar and gall tissues of fungal endophytes from chestnut and Asian chestnut gall wasp *Dryocosmus kuriphilus* (Hymenoptera: Cynipidae) was reported for 63 different fungal OTUs by ITS meta-barcoding. The fungal endophytes were rich in leaves than galls. The investigation on forest plant species and endophytes suggested that insect-induced galls provide a habitat for endophytic fungi and could be biocontrol agents of galling insects (Fernandez-Conradi et al., 2019).

21.1.11 *Cullen plicata*

The root thermophilic fungal endophyte characterized as *Thermomyces lanuginosus* was investigated for drought and heat stress in host plant *C. plicata*. The results of plant growth, metabolites, antioxidants, and photosynthetic characteristics after inoculation of test endophyte in *C. plicata* improved the tolerance to drought and heat stress (Ali et al., 2019b).

21.1.12 *Dendrobium exile* Schlechter (Orchidaceae)

The fungal isolates (40) from the roots (28) and seedlings (12) of *D. exile* were used in restoration-friendly cultivation by the *in situ/ex situ* seed-baiting technique to obtain germination-enhancing fungi for *Dendrobium* species. The isolated fungi were from seven different fungal species, i.e., three rhizoctonia fungi (*Tulasnella* sp. DerIV, DerV, and DesI), three non-*Rhizoctonia* fungi (*Nodulisporium* sp. DerI, *Xylaria plebeja* DerII, and *Colletotrichum* sp. DerIII) and one unidentified fungus DesII. Among the isolates, *Tulasnella* sp. DesI from seedlings was reported as more efficient that supported seed germination up to seedlings than the DerIV and DerV from adult roots. The results of the study indicated that orchids need different fungal partners for seed symbiotic germination vs. adult plant development. The diverse endophytic fungi resident of roots proved potential diverse functions in inoculated plants (Meng et al., 2019).

21.1.13 *Dicranum scoparium* (Bryophyta)

Chen et al. investigated fungi associated with adjacent living, senescing, and dead tissues of *D. scoparium* gametophytes by RNA metatranscriptomics (culture free) and inoculated about 900 surface-sterilized tissue fragments on malt extract agar (MEA). The results of metatranscriptomics detected 17 OTUs, i.e., Herpotrichiellaceae species, *Hyalocypha*, *Sistotrema*, *Epibryon*, *Mortierella*, *Cladosporium*, and fungi from order Helotiales, and 398 fungi (Sordariomycetes) isolated from surface-sterilized gametophytic tissues represented 61 OTUs. The findings revealed metatranscriptomics as a reliable approach to study the plant mycobiome (Chen et al., 2018).

21.1.14 *Dysosma versipellis* (Hance) M. Cheng ex Ying (Berberidaceae)

The effects of *Glomus mosseae* (AMF) inoculation in *D. versipellis* grown in copper (Cu) ion concentrations (0, 200, and 400 mg kg^{-1}) were investigated for different parameters (growth, lipid peroxidation (MDA and MRP), enzymatic (SOD, POD, and CAT) antioxidant activities, and active medicinal components) inpot experiments. The results of Cu stress with AMF inoculation enhanced the biomass and content of podophyllotoxin in roots and antioxidant capacity and reduced membrane lipid peroxidation in leaves of *D. versipellis*. The inoculum proved better adaptation, productivity, and quality of the test plant (Luo et al., 2020).

21.1.15 *Debregeasia salicifolia*

The study evaluated the diversity and bioactive potential of foliar endophytic fungi colonizing *D. salicifolia*. *Fusarium fujikuroi*, *Aspergillus tubingensis*, and *Rhizopus oryzae* were fungal endophytes that showed antibacterial activity (*S. aureus*, *B. spizizenii*, *Listeria monocytogenes*, *E. coli*, *Klebsiella pneumoniae*, *S. typhimrium*, and *Acinetobacter baumannii*) and antifungal activity against pathogenic *Aspergillus flavus* and *A. niger*. The fungal extracts indicated the presence of 21 compounds of diverse nature and structure through gas chromatography–mass spectrometry. The study highlighted the potential of *D. salicifolia* fungal endophytes, indicating that they were a source of potential therapeutic bioactive metabolites (Nisa et al., 2020).

21.1.16 *Euphorbia geniculata*

This study reported 22 isolates representing 21species from 15 genera of endophytic fungi from *E. geniculate*. Among the most common fungi was *Aspergillus* and *Isaria feline* isolated from both leaves and stem and *A. flavus*, *A. ochraceus*, *A. terreus* var. *terreus*, *Emercilla nidulans* var. *acristata*, and *Macrophomina phaseolina* colonized from both stem and root. The dual-culture assay of culture extracts from endophyte isolates against six strains of plant pathogens (*Eupenicillium brefeldianum*, *Penicillium echinulatum*, *Alternaria phragmospora*, *Fusarium oxysporum*, *F. verticilloid*, and *A. alternata*) proved antagonistic. The endophytic isolates such as *A. flavus*, *A. fumigatus*, and *F. lateritium* showed high antagonistic activity, and *Cladosporium herbarum*, *F. culomrum*, and *Sporotrichum thermophile* showed low activity. The secondary metabolites (terpenes and alkaloids) from culture extracts against three pathogenic fungi (*E. brefeldianum*, *P. echinulatum*, and *A. phragmospora*) from infected tomato plant were effectively inhibited and showed promising sensitivity. The study concluded that the leaf, stem, and root endophytic fungi from *E. geniculata* were effective antifungals (Kamel et al., 2020).

21.1.17 *Glycine max* L.

The nematicidal activity against soybean cyst nematode (SCN, *Heterodera glycines* Ichinohe) juveniles was tested from fungal endophytes from corn and soybean roots that were 401 morphotypes clustered into 108 operational taxonomic units (OTUs). The culture filtrates from isolates from order Hypocreales, *Fusarium* was the most commonly isolated nematicidal genus, and *Hirsutella rhossiliensis*, *Metacordyceps chlamydosporia*, and *Arthrobotrys iridis* were less common among the isolates (Strom et al., 2020).

The co-inoculation of *Paecilomyces formosus* LHL10 and *Penicillium funiculosum* LHL06 promoted plant growth attributes, photosynthetic activity, glutathione, catalase, and SOD activities, and macronutrient uptake and decreased lipid peroxidation and reduced endogenous abscisic acid and jasmonic acid levels in *G. max* under heavy metals (Ni, Cd, and Al) and high temperature and drought (HTD) stress. The results revealed that co-inoculation reduced metal accumulation and translocation in plants by downregulating heavy-metal ATPase and drought-related and heat shock protein 90 gene expression in co-inoculated plants that improved plant development under HTD stress. The test endophytes proved the possibility of cultivation of *G. max* in metal-contaminated soil in semiarid and high temperature conditions, promoting sustainable agriculture (Bilal et al., 2020).

The root endophytic *P. funiculosum* was investigated for the metal (Ni, Cu, Pb, Cr, and Al) stress-alleviation and regulation of stress-related proteins in *G. max*. The test plant was evaluated for physio-biochemical, molecular, and proteomic responses to combined heavy-metal toxicity. The endophyte-inoculated plants revealed tolerance to combined heavy metals and gibberellins and indole-3-acetic acid (IAA) production. This report proved that test endophyte could remediate the heavy metals in contaminated soils and final plant product safe from toxicity (Bilal et al., 2019).

The mycorrhizal fungi *Funneliformis mosseae* (syn. *Glomus mosseae*), *Rhizophagus intraradices* (syn. *G. intraradices*), and *Claroideoglomus etunicatum* (syn. *G. etunicatum*) from a salt marsh habitat were investigated for symbiotic performance of Clark (salt-tolerant) and Kint (salt-sensitive) soybean genotypes for salt stress tolerance and mutualistic interactions between AMF and the host plants. The AMF inoculation improved performance in both test genotypes by improved nodule formation, leghemoglobin content, nitrogenase activity, and IAA production. The colonization protected test plants under salt stress from membrane damage and reduced hydrogen peroxide, TBARS and lipid peroxidation. The study indicated that the AMF symbiosis of soybean genotypes improved the performance regardless of salt stress in plants (Hashem et al., 2019).

21.1.18 *Glycyrrhiza glabra* L. (liquorice; Leguminosae)

G. glabra from four different locations in the Northwestern Himalayas were reported for isolation of endophytic fungi from leaves, stems, and root (rhizome) tissue segments (1019). The most dominant class among the isolates from *G. glabra* was Dothideomycetes followed by Sordariomycetes, Mucoromycetes, Eurotiomycetes, Agaricomycetes, Euascomycetes, and Leotiomycetes. The endophytes (266) were grouped into 21 genera and 38 different taxa Ascomycota (18 genera), Zygomycota (*Mucor* and *Rhizopus*), and Basidiomycota (*Rhizoctonia*). The endophytic strains from underground plant tissues were *Fusarium incarnatum*, *Rhizoctonia* sp., *Mucor circinelloides*, *Lasiodiplodia theobromae*, *Macrophomina phaseolina*, *L. pseudotheobromae*, *Mucor hiemalis*, *Alternaria tenuissima*, *A. porri*, and *A. burnsii*and above-ground plant endophytic strains were *F. oxysporum*, *Talaromyces verruculosus*, *A. alternata*, *C. aeria*, *Didymella bryoniae*, *B. cinerea*, *Alternaria* sp., *A. brassicae*, *Cladosporium tenuissimum*, *Fusarium equiseti*, *Aspergillus flavus*, *Cladosporium cladosporioides*, *Stagonosporopsis cucurbitacearum*, *A. terreus*, *Xylaria* sp., and *Bionectria* sp. recovered from *G. glabra*. The isolates of endophytic *Phoma* sp. and *Fusarium* showed a strong affinity with the host plant. The high species richness was reported in host plants from the sub-tropical location than sub-temperate and the temperate locations. The isolated fungal taxa were positive for plant growth-promoting hormone, indole acetic acid (IAA), siderophore, phenolic and flavonoid contents, and hydrolytic enzymes (amylase, protease, and lipase) in varying concentrations and found no symptoms of disease in cocultivated plants. Increased root (rhizome) and shoot growth in the host was due to isolates, i.e., *S. cucurbitacearum*, *Bionectria* sp., and *A. terreus*. The extracts of fungal endophytes, i.e., *P. exigua*, *C. aeria*, *Phoma* sp., *B. dothidea*, *P. macrostoma*, *S. cucurbitacearum*, and *A. flavus* were active against *S. aureus*, and *Xylaria* sp. and *P. exigua* were potentially active against *E. coli*. The extracts of *S. cucurbitacearum* were active against both the Gram-positive pathogens (*S. aureus* and *B. cereus*) and *C. albicans*. The broad range of antimycotic activity (*C. albicans*, *F. oxysporum*, *Colletotrichum capsici*, *Geotrichum candidum*, *Sclerotinia* sp., and *A. fumigatus*) was proved with extracts from *Diaporthe terebinthifolii*, *D. cotoneastri*, and *T. verruculosus* (Arora et al., 2019).

21.1.19 Halophyte Plants (*Anabasis iranica, Seidlitzia rosmarinus, Salsola tomentos, Salsola yazdiana, Rubia tinctorum,* and *Artemisia annua*)

The endophytic fungi from test halophyte plants were 40 isolates: 23 roots, 15 stems, and 2 leaves. The root tissue isolates were 57%, and a high number of isolates were eleven fungi from *A. iranica* and low were two fungi from *S. rosmarinus* and *A. annua* among the halophytes. The endophytic *Aspergillus terreus* culture extracts showed production of enzymes (amylase, protease, cellulose, keratinase, and pectinase), antibacterial (*B. cereus*, *S. aureus*, *P. aeruginosa*, *Salmonella typhimurium* and *Candida albicans*) and antifungal against *A. fumigatus* (human pathogen) than other isolates *Acremonium*, *Paecilomyces*, *Microascus*, and *Monosorascus* (Jalili et al., 2020).

21.1.20 *Hordeum brevisubulatum* (Wild Barley)

The endophytic fungus *Epichloe bromicola* was investigated for its effect on polyamine metabolism in host (barley) plants under salt stress. The presence of *E. bromicola* infection resulted in amelioration of salt stress along with increase in dry weight, spermidine, and spermine and decrease of putrescine in total polyamines. The results of the study showed increase in the insoluble bound form, decrease-free form, and soluble conjugated forms of polyamines in the host plant under salt stress (Chen et al., 2019a).

The endophyte (*Epichloe bromicola*) -infected (E+) plants showed more tillers, higher biomass and yield, and higher chlorophyll content and superoxide dismutase activity than endophyte free (E–) plants under high salt stress in wild barley. The endophyte improved plant growth, physiological properties, and seed germination in 200 and 300 mM of NaCl. The results demonstrated that *E. bromicola* endophyte increased tolerance to salt stress by increased seed germination, growth, and altered plant physiology in wild barley (Wang et al., 2020).

21.1.21 *Helianthus annuus* L., *Solanum xanthocarpum*, *S. melongena*, and *Allium cepa* L.

The endophytic fungi isolated from stem, root, and leaf tissues from healthy plants (*H. annuus*, *S. xanthocarpum*, *S. melongena*, and *A. cepa*) were *Cephalosporium* sp., *Curvularia* sp., *Fusarium solani*, and *F. moniliforme*. The extracts from isolates were assayed for nematicidal potential (*Meloidogyne javanica*—root knot nematode), antibacterial (*Staphylococcus aureus*, *Pseudomonas aeruginosa*, *Salmonella typhimurium*, *Bacillus subtilis*, and *Escherichia coli*) and antifungal (*Fusarium solani*, *F. oxysporum*, *Macrophomina phaseolina*, and *Rhizoctonia solani*) activities against plant parasitic nematode, bacteria, and root-infecting fungi. The endophytic *Cephalosporium* sp. and *F. solani* were reported for strong nematicidal, antibacterial, and antifungal activity and new compounds with a wide range of novel antimicrobial action. The findings suggested *Cephalosporium* sp. and *F. solani* as a source for drug development and agrochemical production to protect agricultural crops from plant root diseases (Farhat et al., 2019).

21.1.22 *Jacaranda mimosifolia* D. Don

The biochar and five fungal isolates' (*Beauveria bassiana*, *Metarhizium anisopliae*, *Pochonia chlamydosporia*, *Purpureocillium lilacinum*, and *Trichoderma asperella*) spore suspension were sprayed on plant shoots (*J. mimosifolia*), and the soil surface was treated with Cu (200 mg dm–3), Mn (450 mg dm^{-3}), and Zn (450 mg dm^{-3}). The test fungal isolates increased Mn and Zn in shoot and root mass, and improved the translocation potential of Cu, Mn, and Zn from roots to shoots. The photochemical profile of the test plant based on chlorophyll a fluorescence showed improved tolerance to metal-contaminated soil. The tested fungal isolates and biochar studied were potential phytoremediators of Cu, Mn, and Zn that reduced the risk of metal content in plant tissues and leaching (Farias et al., 2020).

21.1.23 *Jatropha curcas* L. (Family: Euphorbiaceae)

The study explored the potential of laser biospeckle activity (LBSA) for the detection of endophytic colonization of leaves of *J. curcas*. *Alternaria*, *Aspergillus*, *Colletotrichum*, and *Nigrospora* genera of endophytes were obtained from E+ and E– leaves, and the increased water movements inside leaves were found to promote endophytic colonization. The results suggested LBSA as a tool to indirectly detect endophytic colonization (D'Jonsiles et al., 2020).

21.1.24 *Lactuca sativa* L.

The tissue segments of leaves, roots, and stems from *Beta vulgare*, *Parthenium argentatum*, *Zea mays*, *Saccharum officinarum*, *Colocasia esculenta*, *Cucurbita maxima*, and *Ficus carica* were reported to colonize 114 endophytic fungi. The colonization of endophytes was high in root tissues of *Z. mays* and leaves in other plants. Among the 24Cr-resistant endophytes, based on plant growth promotion, Cr tolerance, uptake, and detoxification abilities, *Aspergillus fumigatus*, *Fusarium proliferatum*, *Penicillium radicum*, and *Rhizopus* sp. removed the toxic Cr (culture media and soil) and supported the normal growth of Cr stressed *L. sativa* in the pot experiment. The resistant endophytes *Rhizopus* sp. Accumulated and detoxified Cr (intracellular), but *A. fumigatus* and *P. radicum* only detoxified 95%Cr (extracellular). The results of the study concluded that the selected endophytic fungi remediate Cr-contaminated soils for healthy and safe crop production free from toxic metals (Bibi et al., 2018).

21.1.25 *Lactuca serriola* L. (wild lettuce)

The effects of AMF)—*Rhizoglomus intraradices* and endophytic fungi, *Mucor* sp. or *Trichoderma asperellum* on plant growth, vitality, toxic metal accumulation, sesquiterpene lactone production, and flavonoid concentration in the presence of toxic metals by single and co-inoculation of *L. serriola* were evaluated. The AMF inoculation increased biomass yield of the plants grown on nonpolluted and polluted substrate and co-inoculation with the AMF and *Mucor* sp. found to increase biomass of plants grown on the polluted substrate. The co-inoculation with *T. asperellum* and the AMF increased plant biomass on the nonpolluted substrate. The co-inoculation of AMF and *Mucor* sp. increased Zn in leaves and roots. The sesquiterpene lactones in plant leaves were decreased by AMF inoculation in both substrates. The study showed that the double inoculation (AMF and endophytic fungi) was more beneficial than single fungus in unfavorable agricultural areas and toxic metal-polluted areas (Wazny et al., 2018).

21.1.26 *Lolium perenne* (Perennial Ryegrass; Poaceae)

The endophytic fungus *Epichloe festucae* var. *lolii* was investigated for hoverfly (Syrphidae) larvae and pupae abundance on endophyte-infected and uninfected *L. perenne* plants. The results proved that the endophyte in plants provided alkaloid production (direct defense) and improved the odor attractant for foraging aphid predators (indirect plant defense) (Fuchs and Krauss, 2019).

The infection rate of endophyte *Epichloefestucae*var. *lolii* on important agronomic traits (crown width, plant height, panicle number, green stage, over-wintering rate, etc.) was evaluated for the possible mechanisms of cold tolerance of the host subpopulations of *L. perenne*. The results after 3 years of screening, high endophyte infection rates in the tillers and seeds of plants, showed improved agronomic traits (increased root system and over-wintering rate) and enzymatic activities (SOD, POD, CAT, and APX) under low temperature stress (Chen et al., 2020).

21.1.27 *Medicago truncatula*

The root-endophytic fungal community of *M. truncatula* was described by amplicon mass sequencing and assessed plant performance. *Brachypodium pinnatum* induced higher and lower richness of Sordariomycetes and Glomeromycetes, whereas *Holcus mollis* decreased the OTU of the entire mycobiota that resulted in modifications in *M. truncatula* biomass. The results indicated that a given plant endophytic fungal community is determined by the neighboring plants (Vannier et al., 2020).

21.1.28 Mangrove Plants

The leaf endophytic fungi were characterized from the twenty mangrove plant species (*Acanthus ebracteatus* Vahl, *A. ilicifolius* L., *Aegiceras corniculatum* (L.) Blanco-, *Avicennia marina* (Forssk.) Vierh., *A. officinalis* L.-, *Bruguiera cylindrica* (L.) Blume, *B. gymnorhiza* (L.) Lam., *B. parviflora* (Roxb.) Wight & Arn.exGriff., *Ceriops tagal* (Perr.) C.B. Rob., *Rhizophora apiculate* Blume, *R. mucronata* Lam., *R. stylosa* Griff.-, *Excoecaria agallocha* L.- *mnitzera littorea* (Jack) Voigt, *L. racemosa* Willd.-, *Nypa fruticans* Wurmb, *Phoenix paludosa* Roxb.-, *Scyphiphora hydrophyllacea* C.F. Gaertn.-, *Sonneratia alba* Sm.-, *Xylocarpus granatum* J. Koenig-) elonging to ten different plant families (Acanthaceae, Myrsinaceae, Avicenniaceae, Rhizosphoraceae, Euphorbiaceae, Combretaceae, Arecaceae, Rubiaceae, Lythraceae and Meliaceae) from south Andaman Islands. From each plant, 40 leaf samples were prepared from apical, middle, and basal regions. The total isolates were 2,180 from all the test leaf tissue samples and *Phomopsis*, *Phyllosticta*, *Xylaria*, and *Colletotrichum* were most common isolates. This study exclusively reported the diversity and frequency of plant microbiomes from the mangrove ecosystem that supports the host plant fitness in a unique habitat (Rajamani et al., 2018).

21.1.29 *Mentha piperita*

The stem and leaf tissue segments (222) of *M. piperita* were evaluated for endophytic fungi and the 63 isolates (13 genera from Sordariomycetes (Hypocreaceae, Glomerellaceae, Nectriaceae, Diaporthaceae, and Chaetomiaceae), Dothideomycetes (Pleosporaceae and Botryosphaeriaceae), Saccharomycetes (Trichomonascaceae), and Eurotiomycetes (Trichocomaceae)) were screened for antifungal activity against chickpea rot pathogens (*Rhizoctonia solani*, *Botrytis cinerea*, *Fusarium oxysporum*, and *Sclerotinia sclerotiorum*) and antifungal metabolites. The results of the study characterized the endophytic *Acremonium* sp. effective biocontrol agent (BCA) against chickpea fungal phytopathogens (Chowdary and Kaushik 2018).

21.1.30 *Mitrephora wangii*

The tissue segments of leaves, stems, and flowers of *M. wangii* were evaluated for endophytic fungi, and the culture extracts were tested against five Gram-positive bacteria (*Staphylococcus aureus*, *S. epidermidis*, *S. agalactiae*, *Bacillus subtilis*, and *B. cereus*) and five Gram-negative bacteria (*Escherichia coli*, *Salmonella typhi*, *Klebsiella pneumoniae*, *Pseudomonas aeruginosa*, and *Shigella flexneri*) for antibacterial activity. The total isolates were 22 that were found to belong to genera:

Agrocybe, *Aspergillus*, *Colletotrichum*, *Nigrospora*, *Puccinia*, and *Ustilago*. Among the flower isolates, the *Aspergillus* sp. extract (β-thujaplicin ripropionin, callitrin, *cis*-thujopsenal, and *E*-ligustilide) was reported for antibacterial efficacy (Monggoot et al., 2018).

21.1.31 *Marchantia polymorpha* L.

The model liverwort *M. polymorpha* L. was explored for fungal endophyte community by both culturing and Illumina amplicon sequencing methods. Among endophytes characterized, 93 isolates from 50 species were from tested tissues. The endophyte isolates belonged to Ascomycota (class: Eurotiomycetes, Pezizomycetes, Saccharomycetes, Leotiomycetes, Dothideomycetes, and Sordariomycetes) and Basidiomycota. *Phoma herbarum* and *Colletotrichum* sp. were isolated from gametangiophores and *Xylaria cubensis* from rhizoids, gametangiophores, and thalli tissues. This study detected a diverse fungal community from *M. polymorpha* patches (Nelson and Jaw, 2019).

21.1.32 *Musa acuminata* (Banana)

The biocontrol effect of *Serendipita indica* (root-colonizing basidiomycete) on banana wilt disease by fungal pathogen *Fusarium oxysporum* f. sp. *cubense* (Foc) in Tianbaojiao' banana was studied by dual-culture experiments. The study examined the Foc resistance in *S. indica* colonized (S+) and noncolonized (CK) banana plants. The study indicated *S. indica* colonization on FocTR4 resistance of banana through regulation of antioxidant enzyme (superoxide dismutase (SOD), peroxidase (POD), catalase (CAT), and ascorbate peroxidase (APX)) activities (Cheng et al., 2020).

21.1.33 *Neurachne alopecuroidea*

The root endophyte *Drechslera* sp. strain from an Australian native grass *N. alopecuroidea* demonstrated efficacy against four plant pathogens (*Pythium ultimum*, *Rhizoctonia solani*, *Botrytis cinerea*, and *Alternaria alternata*) and was found capable of degrading the contaminant methyl tertiary-butyl ether (MtBE) used in gasoline. The metabolomic analysis revealed two major bioactive metabolites (monocerin and an alkynyl-substituted epoxycyclohexenone derivative) in culture filtrate resulted for antifungal activity. The *Drechslera* sp. strain and its compounds show promise in biocontrol and bioremediation for agriculture in MtBE-contaminated soil (d'Errico et al., 2020).

21.1.34 *Ocimum basilicum* (Sweet Basil)

The study investigated the growth of sweet basil on soil contaminated with lead and copper in a pot experiment inoculated with *Rhizophagus irregularis* (AMF) and *Serendipita indica* (beneficial endophyte) under greenhouse with defined conditions. The results of basil plants with inoculated fungi proved to increase in shoot and root dry weight under test conditions, but there was decrease of lead in shoots. The copper levels were reduced by *S. indica*, but the AM fungus affected the copper in the soil contaminated with both copper and lead. The AMF inoculation increased the essential oils (linalool and eucalyptol) in sweet basil grown on metal contaminated soils (Sabra et al., 2018).

21.1.35 *Oryza sativa* L. (Rice)

Fifteen plant species of Thar Desert, Rajasthan, India from family Fabaceae (8) and Poaceae (4) and one each from Bignoniaceae, Polygonaceae, and Salvadoraceae were evaluated for endophytic fungi with abilities to tolerate high temperature and drought stress in rice cultivar (IR-64). The colonization frequency was high in *Lasiurus scindicus*, *Polygonum glabrum*, and *Acacia* sp. and low in *Prosopis cineraria*. The endophytic fungi were 507 isolates from 82 operational taxonomic units (OTUs) and dominant were *Aspergillus*, *Alternaria*, *Chaetomium*, *Penicillium*, and *Nigrospora* sp., and among them, *Chaetomium* sp. was a thermotolerant endophyte that supported the plant from high-temperature tolerance. The drought tolerance in rice cultivar was reported at the early seedling stage by three OTUs of *Aspergillus* sp. (namely LAS-4, SAP-3, and SAP-6). The inoculated plants with *Chaetomium* sp. under temperature stress survived in high percentage with increased shoot and root growth compared to seedlings (noninoculated) (Sangamesh et al., 2018).

The study of stem bark endophytic fungus *Phomopsis liquidambaris* isolated from *Bischofia polycarpa* was investigated for biodegradability and remediation of phenanthrene *in vivo* of rice. The results showed that the endophyte established symbiosis with rice and degraded phenanthrene absorbed by the plant. The endophyte associated with rice resulted in high phenanthrene-degrading enzyme activities and gene expression levels in roots than in the shoot (Fu et al., 2020).

Acremonium strain D212 was isolated along with thirteen endophytic fungi from the buds of *P. notoginseng* seedlings. The isolate D212 was found to colonize the roots and promoted root growth, saponin biosynthesis, secrete indole-3-acetic acid (IAA), and jasmonic acid (JA) and developed resistance to root rot disease in *P. notoginseng*. The colonization of isolate D212 in the roots of the rice line Nipponbare was found to depend on methyl jasmonate (MeJA) and 1-naphthalenacetic acid (NAA) concentration. The degree of colonization of JA signaling-defective *coi1–18* mutant rice by D212 was found less than that of wild-type Nipponbare and miR393b-overexpressing lines of rice. The study concluded that D212 colonization needs MeJA but not NAA (Han et al., 2020).

The study demonstrated that salt-tolerant endophyte *Fusarium* sp. isolated from colonized salt-adapted Pokkali rice, promoted its growth and conferred salinity stress tolerance to the salt-sensitive rice variety IR-64. The colonized plants showed high assimilation rate and chlorophyll stability index, and comparative transcriptome analysis revealed 1,348 upregulated and 1,078 downregulated genes by MapMan and interaction network programs (Sampangi-Ramaiah et al., 2020).

The study of rice plants colonized by the beneficial root-colonizing endophytic fungus *P. indica* demonstrated that grain yield was higher in colonized plants grown in soil. The endophyte colonization promoted stomata closure, upregulation of antioxidant enzymes (catalase and glutathione reductase), and increase in the leaf surface temperature and diminished water stress-induced leaf wilting and impairments in photosynthetic efficiency in test plants under water stress. The colonized test plants lowered the malondialdehyde level (oxidative stress indicator) and enhanced the ratio of reduced-to-oxidized glutathione (Tsai et al. 2020).

21.1.36 *Oxalis corniculata*

The root endophytic fungi from *O. corniculata* were investigated for growth-promoting activities (secretion of indole acetic acid, gibberellins siderophore, and phosphate solubilization) in mutant dwarf rice Waito-C (GA-deficient). Among the 15 isolates, *A. fumigatus* and *Fusarium proliferatum* were reported for IAA and GA biosynthesis in culture filtrate and regulated GA in E+ plants. The study proves the potential test isolates for applied value growth-promoting ability of crop plants with stimulation of growth hormones (Bilal et al., 2018).

21.1.37 *Panax notoginseng*

The leaf fungal endophyte *Leptosphaeria* sp., (plant pathogen) from *P. notoginseng* was investigated for bioactive metabolites against fungi (*Botryosphaeria dothidea*, *Alternaria alternata f.* sp. *mali*, *Fusarium graminearum*, *Sclerotinia sclerotiorum*, *Verticillium dahliae* Kleb, *Bipolaris carbonum* Wilson, *Phytophthora parasitica*, *A. alternata*, *Rhizoctonia cerealis*, and *Botrytis cinerea* Pers.) and bacteria (*Micrococoecus lysodeikticus*, *Bacillus subtilis*, *B. cereus*, *M. luteus*, *Staphylococcus aureus*, *Proteus vulgaris*, *Salmonella typhimurium*, *Pseudomonas aeruginosa*, *Escherichia coli*, and *Enterobacter aerogenes*). The results detected seven compounds, among which three were newly invented compounds moderately effective against test bacteria and fungi (Chen et al., 2019b).

21.1.38 *Polystichum munitum* (Western Sword Fern; Dryopteridaceae)

The culture-independent approach in *P. munitum* for fungal endophytes was investigated directly by amplifying and sequencing the fungal DNA in the fern pinnae and blades. The isolates found were 264 operational taxonomic units (OTUs) that belong to families such as Helotiaceae, Nectriaceae, and Sebacinaceae in the 20 pinnae sampled in April, and the sample collected in May detected fungi from Helotiaceae and Sebacinaceae and a single member from Nectriaceae. This study reported *Flagellospora fusarioides* as common endophytes in both sampling periods and suggested that repeated sampling characterizes the fungal symbionts in the fern plant community (Younginger and Ballhorn, 2017).

21.1.39 *Pinus sylvestris* var. *mongolica*

The structure and distribution of endophyte fungal communities from wooden blocks of nine major coniferous species (*Pinus sylvestris* var. *mongolica*, *P. tabulaeformis*, *Pinus yunnanensis*, *Pinus massoniana*, *P. taeda*, *P. elliottii*, *P. koraiensis*, *Picea koraiensis*, and *Larix gmelinii*) were evaluated. Among the conifer trees, *P. sylvestris* host for European wood wasp, *Sirex noctilio* Fabricius (Hymenoptera: Siricidae), an invasive pest species that attack and damage the pine plantations. The endophyte isolation and colonization rates in *Pinus tabulaeformis*, followed by *P. sylvestris* var. *mongolica*, were lower than those in other test *Pinus* species. The study characterized the endophyte communities using internal transcribed spacer sequencing and morphological features in the host tree of *S. noctilio* and eight potential host tree species. The endophyte populations enumerated were 61 species, 34 genera, and 1,626 fungal strains from test conifer samples. The fungal endophyte community from each tree species harbored a unique structure, with the genus *Trichoderma*, *Penicillium*, *Aspergillus*, and *Fusarium* common in plant communities. The findings showed that the proportion of isolated endophytic fungi strongly inhibited parasitic wood wasp and its symbiotic fungus (*Amylostereum areolatum*) growth in *P. tabulaeformis*, *P. sylvestris* var. *mongolica*, and *P. yunnanensis* (Wang et al., 2019).

21.1.40 *Pinus tabulaeformis* Carr.

The seven ectomycorrhizal fungi (ECMF) isolated from sporocarps, i.e., *Suillus lactifluus* A.H. Sm. & Thiers. (Sl), *S. laricinus* (Sla), *S. bovinus* L. (Sb), *S. tomentosus* (Kauffman) Singer (St), *Handkea utriformis* Bull. (Hu), *Amanita vaginata* (Bull.) Lam. (Av), and *Schizophyllum* sp. Fr. (Ss) and dark septate endophytic fungi (DSE) species isolated from the roots of *Astragalus adsurgens* Pall from lead-zinc mine tailings, i.e., *Gaeumannomyces cylindrosporus* D. Hornby, Slope, Gutter. & Sivan (Gc), *Paraphoma chrysanthemicola* (Hollós) Gruyter (Pc), *Phialophora mustea* Neerg. (Pm), *Exophiala salmonis* Carmich. (Es), and *Cladosporium cladosporioides* Fres. (Cc) were evaluated for impacts on pine seedling growth and control of pine wilt disease by *Bursaphelenchus xylophilus*, a pinewood nematode (PWN) infection. The findings of pine inoculation with ECMF and DSE improved the seedling height by *S. laricinus* and *P. mustea*; increased the seedling diameter by *A. vaginata*, *C. cladosporioides*, and *G. cylindrosporus*; improved the root activity by *A. vaginata* and *P. mustea*; *A. vaginata*, *S. laricinus*, *C. cladosporioides*, and *P. chrysanthemicola*; and increased the seedlings' dry weight. The best inoculant strain that protected *P. tabulaeformis* seedlings against pine wilt disease in the study was *S. laricinus*. The findings of ECMF/DSE symbiosis proved it to enhance the resistance toward nematode infection and be a potential way for pine wilt disease prevention (Chu et al., 2019).

21.1.41 *Pisum sativum* L.

The benefits of plant endophytic fungi and bacteria on endophyte-free second-generation *Pisum sativum* L. under drought-stress were investigated for germination percentage, root and shoot length, reactive oxygen species (ROS) accumulation, antioxidant gene expression and protein content. The pea plants colonized by endophytic *Penicilium* SMCD2206, *Paraconiothyrium* SMCD2210, and *Streptomyces* sp. SMCD2215 strains improved seed germination and reduced

ROS accumulation levels in plant roots. The endophyte colonized plant leaves revealed downregulation of proline, superoxide dismutase (SOD), and manganese superoxide dismutase (MnSOD) genes under water deficiency (Kumari and Vujanovic, 2020).

21.1.42 *Pisum sativum* L. and *Triticum spelta* L.

The test plants were assessed in pot experiments using soils from conventional (CF) and organic farms (OF) for the indicator of plant microbiome such as AMF. The results showed that soil from organic farm has higher diversity and abundance of AMF (*Rhizophagus intraradices* and *Funneliformis mosseae*) with a colonization rate of 60% than 30% in conventional soil. The pea plant roots grown in organic soil showed both AMF sequences than *R. intraradices* in conventional soil. The significance of the study conveyed that farming practice plays a role in the microbiome with beneficial activities to plants (Gazdag et al., 2019).

21.1.43 *Phaseolus vulgaris* L. (Common Bean)

Beauveria bassiana (entomopathogenic fungus) was inoculated in the common bean by employing three methods (soil wetting, seed soaking, and leaf spraying) to evaluate the growth-promoting activity as an endophyte. The colonization of the inoculant was high in root tissue than leaves and stem. The effective method of inoculation for test fungi was soil wetting and leaf spraying. This study reports an effective approach of inoculation for plant growth improvement (Afandhi et al., 2019).

The endophytic *Trichoderma* (*T. harzianum* T8) from *P. vulgaris* root and AMF formulation (AMF2-*Gigaspora margarita*, *Glomus hoi*, and *Scutelospora gigantea*) from maizerhizosphere was investigated for nutrient uptake and resistance against pathogenic *Fusarium* strain (*F. solani*) for *Fusarium* root rot (FRR) in common bean. The combination of *T. harzianum* and AMF2 was reported for activation of mechanisms like competition, antibiosis, and induced plant defense (phenylalanine ammonia lyase and polyphenol oxidase activities). The results recorded increase in bean shoot biomass and chlorophyll pigment synthesis and biocontrol of FRR, respectively (Eke et al., 2019).

21.1.44 *Polygonum acuminatum* and *Aeschynomene fluminensis*

The root fragments were examined for endophytic fungal communities of *P. acuminatum* and *A. fluminensis* with soil contaminated with mercury (Hg). The findings of the study reported that 190 fungal isolates were distributed in two phyla, four classes, fifteen orders, twenty-seven families, and thirty-five genera from 480 root fragments. Two strains (*Aspergillus japonicas* and *Emericellopsis* sp.) from the contaminated area were positive for the production of five enzymes. The four isolates of endophyte communities (*Clonostachys rhizophaga*, *Aspergillus* sp., Glomeralleceae, and *Westerdykella* sp.) inhibited the growth of Gram-negative (*Escherichia coli*) and Gram-positive (*Staphylococcus saprophyticus*) bacteria. The Pb^{2+}sensitivity was reported in all isolates from *A. fluminensis*, and *Aspergillus* sp. was tolerant. The strains sensitive or inhibited by Zn^{2+} were *Bipolaris setariae* and *Phomopsis* sp., respectively. The endophyte *Falciformispora* sp. and *Trichoderma harzianum* showed resistance to cadmium. The findings suggested that phytoremediation of soils with toxic concentrations of mercury could be remediated by use of endophytes (Pietro-Souza et al., 2017).

The study used 32 root endophytic fungi from *A. fluminensis* and *P. acuminatum* for mercury bioremediation in *in vitro* and hostplants (*A. fluminensis* and *Zea mays*). The endophytic isolates *Aspergillus* sp. A31, *Curvularia geniculata* P1, Lindgomycetaceae P87, and *Westerdykella* sp. P71 were inoculated into the test host plant cultivated in the presence or absence of the metal. The test endophytes removed up to 100% of mercury from the culture medium in a species-dependent manner and promoted growth of inoculated plants in substrates with or without mercury (Pietro-Souza et al., 2020).

21.1.45 *Populus tremula* (Aspen)

Leaf samples with and without beetle damage (*Chrysomela tremula*) were evaluated for endophytic fungi from the aspen plant. The isolates of endophytes from leaves without beetle damage were *Cladosporium* sp., *C. cladosporioides*, and *Penicillium brevicompactum*, and *Arthrinium*, *Cryptococcus* sp., *Penicillium* sp., and *Penicillium expansum* were isolated from beetle damaged leaf samples, and isolates from beetle were *Rhodotorula* sp. and *Trichoderma*. The composition of salicinoid phenolic glycosides and competition between filamentous fungi (*P. brevicompactum* and *P. expansum*) and yeast-like forms (isolated from beetle-damaged leaf tissues) were demonstrated. The results showed that phenolics were dependent on plant genotypes, and the endophytic strain *P. brevicompactum* was stimulated by the yeast forms (Albrectsen et al., 2018).

21.1.46 *Phalaris arundinacea* and *Scirpus sylvaticus*

The study investigated the effects of four water regimes on AMF (*Rhizophagus irregularis*) root colonization of *Phalaris arundinacea* and *Scirpus sylvaticus* in constructed wetland. The results of AMF root colonization showed that two lower-water regimes were the most suitable. The root length, shoot height, biomass, and total phosphorus and chlorophyll contents of shoot of both wetland plants under the fluctuating water regimes were increased, and malondialdehyde (MDA) contents were decreased in both AMF-inoculated wetland plants (Hu et al., 2020).

21.1.47 *Securinega suffruticosa* (Pall.) Rehd

The distribution and diversity of culturable endophytic fungi in roots, stems, and leaves of *S. suffruticosa* and their antimicrobial activity against *Staphylococcus aureus*, *Escherichia coli*, *Pseudomonas aeruginosa*, *Enterococcus faecalis*, *Fusarium oxysporum*, *Phoma herbarum*, and *Colletotrichum siamense* were evaluated. The endophytic fungi isolates were 420, from which 20 genera and 35 species were identified through morphological and internal transcribed spacer (ITS) sequence analyses. The dominant genera were *Chaetomium*, *Fusarium*, *Cladosporium*, and *Ceratobasidium* (Du et al., 2020).

21.1.48 *Solanum lycopersicum*

The root endophyte *Fusarium solani* of tomato was investigated for colonization of legume plants (*Lotus japonicus* and *Medicago truncatula*). The endophytic isolate was found to colonize root tissues of *L. japonicus* and then translocated to the aerial parts. The results of the study revealed the mechanisms that endophytes utilize to enter and establish a relationship in the host plants (Skiada et al., 2019).

21.1.49 *Schedonorus arundinaceus* (Schreb.) Dumort.; syn. *Festuca arundinacea* (Tall Fescue)

The damage by native insect herbivore tiger moth *Paracles vulpina* (Arctiidae) on the invasive tall fescue and leaf quality traits, such as nutritional value and alkaloid contents of the endophyte *Epichloe coenophiala* (formerly *Neotyphodium coenophialum*), were evaluated. The results showed that two soil nutrients (N and P) increase the nutritional value of leaves and the concentration of fungal ergot alkaloids. The density of herbivores and herbivore damage on tall fescue leaves was low in plots fertilized with N and high in P-fertilized plots. The major finding established was the link between soil nutrients, endophyte-symbiotic plant invaders, and native insect herbivores in invasive tall fescue (Graff et al., 2020).

21.1.50 *Trisetum spicatum* (Poaceae)

The leaf endophytic fungi (*Epichloe*) were investigated for their role in acquisition of nitrogen (N) and phosphorus (P) in *T. spicatum*. The test plants were grouped as (with/without) endophyte under N and P fertilization. The results of the study proved that leaf endophytic fungi improved plant forage ability by acquisition of soil nutrients in *T. spicatum* (Buckley et al., 2019).

21.1.51 *Triticum aestivum* (Wheat)

The study was conducted to evaluate glyphosate residues that affect wheat growth, metabolite composition, and fungal endophyte colonization in plant root. The results proved that endophyte colonization was reduced to 10%, but there was no effect on shoot and root biomass. The residual soil borne glyphosate were found to alter wheat metabolism that impaired fungal root endophyte colonization. The low levels of glyphosate were sufficient to reduce fungal symbionts in the host plants (Claassens et al., 2019).

The structured screening approach explored two endophyte isolates, *Penicillium olsonii* ML37 and *Acremonium alternatum* ML38, as biocontrol agents (BCAs) against the wheat disease Septoria tritici blotch (STB) caused by ascomycete fungus *Zymoseptoria tritici* (syn. *Mycosphaerella graminicola*). The test endophytes proved both as fungicides and BCAs with potential for commercial use (Latz et al., 2020).

The composition of the fungal endophytic community was assessed inside wheat spikes at the flowering stage and during Fusarium head blight (FHB) disease of wheat heads. The isolated 69 OTUs were species from phylum Ascomycota and Basidiomycota. The most abundant species were *Fusarium graminearum* and *Cladosporium herbarum*, and other abundant species were *Sporobolomyces roseus*, *Vishniacozyma victoriae*, *Fusarium culmorum*, *Alternaria infectoria*, *Cryptococcus tephrensis*, *Aureobasidium pullulans*, and *Parastagonospora nodorum*. The endophytes belonging to the genera *Cladosporium*, *Itersonillia*, and *Holtermanniella* were naturally occurring endophytes that outcompete or prevent FHB and served as a source of potential biological control agents in wheat (Rojas et al., 2020).

21.1.52 *Vachellia farnesiana* (L.) Wight & Arn, (syn. *Acacia farnesiana* (L.) Willd)

The root endophytic fungi from *V. farnesiana* (heavy-metal hyper accumulator plant) were isolated and characterized for novel strategies for metal bioremediation. The morphological and phylogenetic analyses indicated that the fungal strains isolated with the ability to remove significant amounts of heavy metals from liquid cultures were from genera *Neocosmospora* and *Aspergillus*. *Neocosmospora* sp. on lead exposure secreted specific novel phenolic compounds and *Aspergillus* sp. decreased the pH in the medium due to malic and succinic acids produced. The test endophytic fungi showed the potential for bioremediation or metal-polluted areas' restoration (Salazar-Ramirez et al., 2020).

21.1.53 *Vitis vinifera* L.

The role of fungal endophytes (*Epicoccum nigrum* R2–21, *Alternaria alternate* XH-2) in the coloration, total anthocyanin concentrations, and phenylalanine ammonia-lyase (PAL) activities of grape berries was examined with the dual-culture system, and quantitative promotion of their total anthocyanidin concentrations was found. The test fungal strains modified the compositional patterns of grape cellular anthocyanidins that opened the avenue for application of endophytic fungi in grape plantations (Yu et al. 2020).

21.1.54 *Zea mays* L. (maize)

The study was conducted to establish endophytic *Beauveria bassiana* in maize plants and evaluate the level of endophyte establishment along with survival and fecundity test of the English grain aphid (*Sitobion avenae*) by spray of spore suspension on maize plants. The aphid-based assay was performed after two weeks of inoculation (*B. bassiana*).The colonization levels were 61% in inoculated leaves (old) and 19% in younger noninoculated leaves. The survival of aphid on plants was reduced up to 49% (inoculated) compared to control. The detection of endophyte in noninoculated younger leaves indicated the endophyte movement inside plants (Mahmood et al., 2019).

The effects of colonization by *B. bassiana* strains (GHA, PTG4, and PTG6) in maize crop production and drought tolerance were evaluated, and the results of the study showed 100% endophytic colonization in maize roots by the strains tested. The colonization rate was variable in shoots and leaves and showed tolerance to drought and flowered prior one to two weeks. The findings provide evidence that support endophyte application for seed treatments in agriculture and sustainable management (Kuzhuppillymyal-Prabhakarankutty et al., 2020).

21.2 Conclusion

The plant–fungi associations for beneficial purposes were reported by studies in different plants from diverse habitats. The fungi inhabiting the plant tissues were characterized based on culture-dependent and culture independent approaches. The culture morphology, microscopic observation, and DNA-based ITS investigations made the characterization of fungi to the species level. The identified species and their role in physiological processes of plants glorify the benefit of fungi in the applied field. The plants from cultivated, wild, and ornamental habitats were recorded to inhabit with fungi and support the growth and enhance the yields. The colonization of fungi is explored by molecular approaches for the species that are not traced in the culture-dependent techniques. The present studies of mycobiome in plants highlighted the revitalization of agroecosystems from conventional practices. The wide range of fungi-based products in agriculture need to popularize in the society to develop sustainable and safe production of plant products to fulfill the needs of the world population.

REFERENCES

Aboobaker, Z., A. Viljoen, W. Chen, P.W. Crous, V.J. Maharaj and S. van Vuuren. 2019. Endophytic fungi isolated from *Pelargonium sidoides* DC: Antimicrobialinteraction and isolation of a bioactive compound. *South African Journal of Botany* 122: 535–542.

Adhikari, P and A. Pandey. 2019. Phosphate solubilization of endophytic fungi isolated from *Taxus wallichiana* Zucc. roots. *Rhizosphere*. 9: 2–9.

Afandhi, A., T. Widjayanti, A.A.L. Emi, H. Tarno, M. Afiyanti and R.N.S. Handoko. 2019. Endophytic fungi *Beauveria bassiana* Balsamo accelerates growth of common bean (*Phaeseolus vulgaris* L.). *Chemical and Biological Technologies in Agriculture* 6: 11.

Ahmad R.Z., R. Khalid, M. Aqeel, F. Ameen and C.J. Li. 2020. Fungal endophytes trigger *Achnatherum inebrians* germination ability against environmental stresses. *South African Journal of Botany* Doi: 10.1016/j.sajb.2020.01.004.

Akyol, T.Y., R. Niwa, H. Hirakawa, H. Maruyama, T. Sato, T. Suzuki, A. Fukunaga, T. Sato, S. Yoshida, K. Tawaraya, M. Saito, T. Ezawa and S. Sato. 2019. Impact of introduction of arbuscular mycorrhizal fungi on the root microbial community in agricultural fields. *Microbes and Environments* 34(1): 23–32.

Alade, G.O., J.O. Moody, A.G. Bakare, O.R. Awotona, S. Adesanya, D. Lai, A. Debbab and P. Proksch. 2018. Metabolites from endophytic fungus; *Pestalotiopsis clavispora* isolated from *Phoenix reclinata* leaf. *Future Journal of Pharmaceutical Sciences* 4: 273–275.

Albrectsen, B.R., A.B. Siddique, V.H.G. Decker, M. Unterseher and K.M. Robinson. 2018. Both plant genotype and herbivory shape aspen endophyte communities. *Oecologia* 187: 535–545.

Ali, A.H., M. Abdelrahman, U. Radwan, S. El-Zayat, and M.A. El-Sayed. 2018. Effect of *Thermomyces* fungal endophyte isolated from extreme hot desert adapted plant on heat stress tolerance of cucumber. *Applied Soil Ecology* 124: 155–162.

Ali, A., S. Bilal, A. L. Khan, F. Mabood, A. Al-Harrasi and I-J. Lee. 2019a. Endophytic *Aureobasidiumpullulans* BSS6 assisted developments in phytoremediation potentials of Cucumissativus under Cd and Pb stress. *Journal of Plant Interactions* 14(1): 303–313.

Ali, A.H., U. Radwan, S. El-Zayat, and M.A. El-Sayed. 2019b. The role of the endophytic fungus, *Thermomyces lanuginosus*, on mitigation of heat stress to its host desert plant *Cullen plicata*. *Biologia Futura* 70: 1–7.

Andhale, N.B., M. Shahnawaz and A.B. Ade. 2019. Fungal endophytes of *Plumbago zeylanica* L. enhances plumbagin content. *Botanical Studies* 60: 21.

Anwar, G., E.A. Lilleskov and R.A. Chimner. 2020. Arbuscular mycorrhizal inoculation has similar benefits to fertilization for *Thuja occidentalis* L. seedling nutrition and growth on peat soil over a range of pH: Implications for restoration. *New Forests* 51: 297–311.

Arora, M., P. Saxena, M.Z. Abdin and A. Varma. 2020. Interaction between *Piriformospora indica* and Azotobacter chroococcum diminish the effect of salt stress in *Artemisia annua* L. by enhancing enzymatic and non-enzymatic antioxidants. *Symbiosis* 80: 61–73.

Arora, P., Z.A. Wani, T. Ahmad, P. Sultan, S. Gupta and S. Riyaz-Ul-Hassan. 2019. Community structure, spatial distribution, diversity and functional characterization of culturable endophytic fungi associated with *Glycyrrhiza glabra* L. *Fungal Biology* 123:373–383.

Arpanahi, A.A. and M. Feizian. 2019. Arbuscular mycorrhizae alleviate mild to moderate water stress and improve essential oil yield in thyme. *Rhizosphere* 9: 93–96.

Ban, Y.H., Z.Y. Xu, Y.R. Yang, H.H. Zhang, H. Chen and M. Tang. 2017. Effect of dark septate endophytic fungus *Gaeumannomyces cylindrosporus* on plant growth, photosynthesis and Pb tolerance of maize (*Zea mays* L.). *Pedosphere* 27(2): 283–292.

Bibi, S., A. Hussain, M. Hamayun, H. Rahman, A. Iqbal, M. Shah, M. Irshad, M. Qasimand and B. Islam. 2018. Bioremediation of hexavalent chromium by endophytic fungi; safe and improved production of *Lactuca sativa* L. *Chemosphere* 211: 653–663.

Bibi, S., G. Jan, F.G. Jan, M. Hamayun, A. Iqbal, A. Hussain, H. Rehman, A. Tawab and F. Kushdil. 2019. *Cochliobolus* sp. acts as a biochemical modulator to alleviate salinity stress in okra plants. *Plant Physiology and Biochemistry* 139: 459–469.

Bidabadi, S.S. and M. Mehralian. 2020. Arbuscular mycorrhizal fungi inoculation to enhance chilling stress tolerance of Watermelon. *Gesunde Pflanzen* Doi: 10.1007/s10343-020-00499-2.

Bilal, L., S. Asaf, M. Hamayun, H. Gul, A. Iqbal, I. Ullah, I.J. Lee and A. Hussain. 2018. Plant growth promoting endophytic fungi *Aspergillus fumigatus* TS1 and *Fusarium proliferatum* BRL1 produce gibberellins and regulates plant endogenous hormones. *Symbiosis* 76: 117–127.

Bilal, S., R. Shahzad, M. Imran, R. Jan and K.M. Kim. 2020. Synergistic association of endophytic fungi enhances *Glycine max* L. resilience to combined abiotic stresses. *Industrial Crops and Products* 143: 111931.

Bilal, S., R. Shahzad, A.L. Khan, A. Al-Harrasi, C.K. Kim and I.J. Lee. 2019. Phytohormones enabled endophytic *Penicillium funiculosum* LHL06 protects *Glycine max* L.from synergistic toxicity of heavy metals by hormonal and stress-responsive proteins modulation. *Journal of Hazardous Materials* 379: 120824.

Buckley, H., C.A. Young, N.D. Charlton, W.Q. Hendricks, B. Haley, P. Nagabhyru and J.A. Rudgers. 2019. Leaf endophytes mediate fertilizer effects on plant yield and traits in northern oat grass (*Trisetum spicatum*). *Plant Soil* 434: 425–440.

Chen, Z., C. Li, Z. Nan, J.F. White, Y. Jin, and X. Wei. 2020. Segregation of *Lolium perenne* into a subpopulation with high infection by endophyte *Epichloe festucae* var. *lolii* results in improved agronomic performance. *Plant Soil* 446: 595–612.

Chen, T., C. Li, J.F. White and Z. Nan. 2019a. Effect of the fungal endophyte *Epichloe bromicola* on polyamines in wild barley (*Hordeum brevisubulatum*) under salt stress. *Plant Soil* 436: 29–48.

Chen, K.H., H.L. Liao, A.E. Arnold, G. Bonito and F. Lutzoni. 2018. RNA-based analyses reveal fungal communities structured by a senescence gradient in the moss *Dicranum scoparium* and the presence of putative multi-trophic fungi. *New Phytologist* 218(4): 1597–1611.

Chen, H.Y., T.K. Liu, Q. Shi and X.L. Yang. 2019b. Sesquiterpenoids and diterpenes with antimicrobial activity from*Leptosphaeria* sp. XL026, an endophytic fungus in *Panax notoginseng*. *Fitoterapia* 137: 104243.

Chenchouni, H., M.N. Mekahlia and A. Beddiar. 2020. Effect of inoculation with native and commercial arbuscular mycorrhizal fungi on growth and mycorrhizal colonization of olive (*Olea europaea* L.). *Scientia Horticulturae* 261: 108969.

Cheng, C., D. Li, Q. Qi, X. Sun, M. R. Anue, B. M. David, Y. Zhang, X. Hao, Z. Zhang and Z. Lai. 2020. The root endophytic fungus *Serendipita* indica improves resistance of Banana to *Fusarium oxysporum* f. sp. *cubense* tropical race 4. *European Journal of Plant Pathology* 156: 87–100.

Chi, W.-C., K.-L. Pang, W.-L. Chen, G.-J. Wang and T.-H. Lee. 2019. Antimicrobial and iNOS inhibitory activities of the endophytic fungi isolated from the mangrove plant *Acanthus ilicifolius* var. *xiamenensis*. *Botanical Studies* 60: 4.

Chowdhary, K. and N. Kaushik. 2018. Biodiversity study and potential offungal endophytes of peppermint and effect of their extract on chickpea rot pathogens. *Archives of Phytopathology and Plant Protection* 51(3–4): 139–155.

Chu, H., C. Wang, Z. Li, H. Wang, Y. Xiao, J. Chen and M. Tang. 2019. The dark septate endophytes and ectomycorrhizal fungi effect on *Pinus tabulaeformis* Carr. Seedling growth and their potential effects to pine wilt disease resistance. *Forests* 10: 140.

Claassens, A., M.T. Rose, L.V. Zwieten, Z.(H). Weng and T.J. Rose. 2019. Soilborne glyphosate residue thresholds for wheat seedling metabolite profiles and fungal root endophyte colonization are lower than for biomass production in a sandy soil. *Plant Soil* 438: 393–404.

Cui, G., S. Ai, K. Chen and X. Wang. 2019. Arbuscular mycorrhiza augments cadmium tolerance in soybean by altering accumulation and partitioning of nutrient elements, and related gene expression. *Ecotoxicology and Environmental Safety* 171: 231–239.

Deng, J., F. Li and T.Y. Duan. 2020. *Claroideoglomus etunicatum* reduces leaf spot incidence and improves drought stress resistance in perennial ryegrass. *Australasian Plant Pathology* 49: 147–157.

d'Errico, G., V. Aloj, G.R. Flematti, K. Sivasithamparam, C.M. Worth, N. Lombardi, A. Ritieni, R. Marra, M. Lorito and F. Vinale. 2020. Metabolites of a *Drechslera* sp. endophyte with potential as biocontrol and bioremediation agent. *Natural Product Research*. DOI: 10.1080/14786419.2020.1737058.

Dias, T., P. Correia, L. Carvalho, J. Melo, A. de Varennes and C. Cruz. 2018. Arbuscular mycorrhizal fungal species differ in their capacity to overrule the soil's legacy from maize monocropping. *Applied Soil Ecology* 125 (2018) 177–183.

D'Jonsiles, M.F., G.E. Galizzi, A.E. Dolinko, M.V. Novas, E.C. Nakamurakare, and C.C. Carmaran. 2020. Optical study of laser biospeckle activity in leaves of *Jatropha curcas* L.: A non-invasive and indirect assessment of foliar endophyte colonization. *Mycological Progress* 19: 339–349.

Dreher, D., S. Baldermann, M. Schreiner and B. Hause. 2019. An arbuscular mycorrhizal fungus and a root pathogen induce different volatiles emitted by *Medicago truncatula* roots. *Journal of Advanced Research* 19: 85–90.

Du, W., Z. Yao, J. Li, C. Sun, J. Xia, B. Wang, D. Shi and L. Ren. 2020. Diversity and antimicrobial activity of endophytic fungi isolated from *Securinega suffruticosa* in the Yellow River Delta. *PLoS One* 15(3): e0229589.

Ebrahimi, S., O. Eini and D. Koolivand. 2020. Arbuscular mycorrhizal symbiosis enhances virus accumulation and attenuates resistancerelated gene expression in tomato plants infected with Beet curly top Iran virus. *Journal of Plant Diseases and Protection* Doi: 10.1007/s41348-020-00299-w.

El Maaloum, S., A. Elabed, Z. El Alaoui-Talibi, A. Meddich, A. Filali-Maltouf, A. Douira, S. Ibnsouda-Koraichi, S. Amir and C. El Modafar. 2020. Effect of arbuscular mycorrhizal fungi and phosphate-solubilizing bacteria consortia associated with phospho-compost on phosphorus solubilization and growth of Tomato seedlings (*Solanum lycopersicum* L.). *Communications in Soil Science and Plant Analysis* 51(5): 622–634.

Eke, P., L.N. Wakam, P.V.T. Fokou, T.V. Ekounda, K.P. Sahu, T.H.K. Wankeu and F.F. Boyom. 2019. Improved nutrient status and Fusarium root rot mitigation with an inoculant of two biocontrol fungi in the common bean (*Phaseolus vulgaris* L.). *Rhizopshere* 12: 100172.

Farias, C.P., G.S. Alves, D.C. Oliveira, E.I. de Melo and L.C.B. Azevedo. 2020. A consortium of fungal isolates and biochar improved the phytoremediation potential of *Jacaranda mimosifolia* D. Don and reduced copper, manganese, and zinc leaching. *Journal of Soils and Sediments* 20: 260–271.

Farhat, H., F. Urooj, A. Tariq, V. Sultana, M. Ansari, V.U. Ahmad and S. Ehteshamul-Haque. 2019. Evaluation of antimicrobial potential of endophytic fungi associated with healthy plants and characterization of compounds produced by endophytic *Cephalosporium* and *Fusarium solani*. *Biocatalysis and Agricultural Biotechnology* 18: 101043.

Fernandez-Conradi, P., T. Fort, B. Castagneyrol, H. Jactel and C. Robin. 2019. Fungal endophyte communities differ between chestnut galls and surrounding foliar tissues. *Fungal Biology* 42: 100876.

Ferus, P., M. Barta and J. Konopkova. 2019. Endophytic fungus *Beauveria bassiana* can enhance drought tolerance in red oak seedlings. *Trees* 33: 1179–1186.

Fu, W.-Q., M. Xu, K. Sun, X.-L. Chen, C.-C. Dai and Y. Jia. 2020. Remediation mechanism of endophytic fungus *Phomopsis liquidambaris* on phenanthrene *in vivo*. *Chemosphere* 243: 125305.

Fuchs, B. and J. Krauss. 2019. Can *Epichloe* endophytes enhance direct and indirect plant defence? *Fungal Ecology* 38: 98–103.

Gazdag, O., R. Kovacs, I. Paradi, A. Fuzy, L. Kodobocz, M. Mucsi, T. Szili-Kovacs, K. Inubuschi and T. Takacs. 2019. Density and diversity of microbial symbionts under organic and conventional agricultural management. *Microbes and Environments* 34(3):234–243.

Gonzalez-Teuber, M., A. Urzua, P. Plaza and L. Bascunan-Godoy. 2018. Effects of root endophytic fungi on response of *Chenopodium quinoa* to drought stress. *Plant Ecology* 219: 231–240.

Graff, P., P.E. Gundel, A. Salvat, D. Cristos and E.J. Chaneton. 2020. Protection offered by leaf fungal endophytes to an invasive species against native herbivores depends on soil nutrients. *Journal of Ecology* Doi: 10.1111/1365-2745.13371.

Hack, C.M., M. Porta, R. Schaufele and A.A. Grimoldi. 2019. Arbuscular mycorrhiza mediated effects on growth, mineral nutrition and biological nitrogen fixation of *Melilotus alba* Med. in a subtropical grassland soil. *Applied Soil Ecology* 134: 38–44.

Han, L., X. Zhou, Y. Zhao, S. Zhu, L. Wu, Y. He, X. Ping, X. Lu, W. Huang, J. Qian, L. Zhang, X. Jiang, D. Zhu, C. Luo, S. Li, Q. Dong, Q. Fu, K. Deng, X. Wang, L. Wang, S. Peng, J. Wu, W. Li, J. Friml, Y. Zhu, X. He and Y. Du. 2020. Colonization of endophyte *Acremonium* sp. D212 in *Panax notoginseng* and rice mediated by auxin and jasmonic acid. *Journal of Integrative Plant Biology* Doi: 10.1111/jipb.12905.

Hao, Z., D. van Tuinen, L. Fayolle, O. Chatagnier, X. Li, B. Chen, S. Gianinazzi and V. Gianinazzi-Pearson. 2018. Arbuscular mycorrhiza affects *grapevine fanleaf virus* transmission by the nematode vector *Xiphinema index*. *Applied Soil Ecology* 129: 107–111.

Hashem, A.E.F. Abd_Allah, A.A. Alqarawi, S. Wirth and D. Egamberdieva. 2019. Comparing symbiotic performance and physiological responses of two soybean cultivars to arbuscular mycorrhizal fungi under salt stress. *Saudi Journal of Biological Sciences* 26: 38–48.

Hou, L., J. Yu, L. Zhao and X. He. 2020. Dark septate endophytes improve the growth and the tolerance of *Medicago sativa* and *Ammopiptanthus mongolicus* under cadmium stress. *Frontiers in Microbiology* 10: 3061.

Hu, S., Z. Chen, M. Vosatka and J. Vymazal. 2020. Arbuscular mycorrhizal fungi colonization and physiological functions toward wetland plants under different water regimes. *Science of the Total Environment*. 716: 137040

Hu, J., A. Yang, A. Zhu, J. Wang, J. Dai, M.H. Wong and X. Lin. 2015. Arbuscular mycorrhizal fungal diversity, root colonization, and soil alkaline phosphatase activity in response to maize-wheat rotation and no-tillage in North China. *Journal of Microbiology*. 53(7): 454–461.

Ikram, M., N. Ali, G. Jan, A. Iqbal, M. Hamayun, F.G. Jan, A. Hussain and I.-J. Lee. 2019. *Trichoderma reesei* improved the nutrition status of wheat crop under salt stress. *Journal of Plant Interactions*. 14(1): 590–602.

Ikram, M., N. Ali, G. Jan, F.G. Jan, I.U. Rahman, A. Iqbal and M. Hamayun. 2018. IAA producing fungal endophyte *Penicillium roquefortiThom*., enhances stress tolerance and nutrients uptake in wheat plants grown on heavy metal contaminated soils. *PLoS One* 13(11): e0208150.

Ismail, M. Hamayun, A. Hussain, A. Iqbal, S.A. Khan and I.J. Lee. 2018. Endophytic fungus *Aspergillus japonicus* mediates host plant growth under normal and heat stress conditions. *BioMed Research International* Article ID 7696831.

Jalili, B., H. Bagheri, S. Azadi and J. Soltani. 2020. Identification and salt tolerance evaluation of endophyte fungi isolates from halophyte plants. *International Journal of Environmental Science and Technology* Doi: 10.1007/s13762-020-02626-y.

Jan, F.G., M. Hamayun, A. Hussain, A. Iqbal, G. Jan, S.A. Khan, H. Khan and I.J. Lee. 2019a. A promising growth promoting *Meyerozyma caribbica* from Solanum xanthocarpum alleviated stress in maize plants. *Bioscience Reports* 39: BSR20190290.

Jan, F.G., M. Hamayun, A. Hussain, G. Jan, A. Iqbal, A. Khan and I.J. Lee. 2019b. An endophytic isolate of the fungus *Yarrowia lipolytica*produces metabolites that ameliorate the negative impact of salt stress on the physiology of maize. *BMC Microbiology* 19: 3.

Jansa, J., P. Smilauer, J. Borovicka, H. Hrselova, S.T. Forczek, K. Slamova, T. Rezanka, M. Rozmos, P. Bukovska and M. Gryndler. 2020. Dead *Rhizophagus irregularis* biomass mysteriously stimulates plant growth. *Mycorrhiza* 30: 63–77.

Jiang, W., R. Pan, C. Wu, L. Xu, M.E. Abdelaziz, R. Oelmuller and W. Zhang. 2020. *Piriformospora indica* enhances freezing tolerance and post-thaw recovery in *Arabidopsis* by stimulating the expression of *CBF* genes. *Plant Signaling & Behavior* 15(4): 1745472.

Juybari, H.Z., M.A.T. Ghanbary, H. Rahimian, K. Karimi and M. Arzanlou. 2019. Seasonal, tissue and age influences on frequency and biodiversity of endophytic fungi of *Citrus sinensis* in Iran. *Forest Pathology* 49(6): e12559.

Kamel, N.M., F.F. Abdel-Motaal and S.A. El-Zayat. 2020. Endophytic fungi from the medicinal herb *Euphorbia geniculata* as a potential source for bioactive metabolites. *Archives of Microbiology* 202: 247–255.

Khan, A.A.H. 2019a. Plant-bacterial association and their role as growth promoters and biocontrol agents. In: Sayyed R. (eds) *Plant Growth Promoting Rhizobacteria for Sustainable Stress Management. Microorganisms for Sustainability*, vol 13. Springer, Singapore, pp. 389–419.

Khan, A.A.H. 2019b. Cytotoxic potential of plant nanoparticles. In: Abd-Elsalam K. and R. Prasad. (eds) *Nanobiotechnology Applications in Plant Protection. Nanotechnology in the Life Sciences*. Springer, Cham, pp. 241–265.

Khan, A.A.H. 2020. Endophytic fungi and their impact on agroecosystems. In: Khasim S., C. Long, K. Thammasiri and H. Lutken. (eds) *Medicinal Plants: Biodiversity, Sustainable Utilization and Conservation*. Springer, Singapore, pp. 443–499.

Khan, A.A.H., Naseem and B. Prathibha. 2011. Screening and potency evaluation of antifungal from soil isolates of *Bacillus subtilis* on selected fungi. *Advanced Biotech* 10(7): 35–37.

Khan, A.A.H., Naseem and S. Samreen. 2010. Antagonistic ability of *Streptomyces* species from composite soil against pathogenic bacteria. *Indian Journal of Biotechnology* 4(4): 222–225.

Khan, A.A.H., Naseem and B.V. Vardhini. 2015. Fungus mediated synthesis of metal nanoparticles. In: Sivaramaiah, G. (ed) *Proceedings of National Seminar on New Trends on Advanced Materials (NTAM)*. Paramount Publishing House, Hyderabad, pp. 82–88.

Khan, A.A.H., Naseem and B.V. Vardhini. 2016. Microorganisms and their role in sustainable environment. In: Prasad R. (ed) *Environmental Microbiology*. IK International Publishing House, New Delhi, pp 60–88.

Khan, A.A.H., Naseem and B.V. Vardhini. 2017. Resource-conserving agriculture and role of microbes. In: Prasad R. and N. Kumar. *Microbes & Sustainable Agriculture*. IK International Publishing House, New Delhi, pp 117–152.

Khan, A.A.H., A. Sadguna, V. Divya, S. Begum, Naseem and A.G. Siddiqui. 2014. Potential of microorganisms in clean-up the environment. *International Journal of Multidisciplinary and Current Research* 2: 271–285.

Kumari, V. and V. Vujanovic. 2020. Transgenerational benefits of endophytes on resilience and antioxidant genes expressions in pea (*Pisum sativum* L.) under osmotic stress. *Acta Physiologiae Plantarum* 42: 49.

KuzhuppillymyalPrabhakarankutty, L., P. TamezGuerra and R. GomezFlores. 2020. Endophytic *Beauveria bassiana* promotes drought tolerance and early flowering in corn. *World Journal of Microbiology and Biotechnology* 36: 47.

Lagarde, A., P. Jargeat, M. Roy, M. Girardot, C. Imbert, M. Millot and L. Mambu. 2018. Fungal communities associated with *Evernia prunastri*, *Ramalina fastigiata* and *Pleurosticta acetabulum:* -Three epiphytic lichens potentially active against *Candida* biofilms. *Microbiological Research* 211: 1–12.

Latz, M.A.C., B. Jensen, D.B. Collinge and H.J.L. Jorgensen. 2020. Identification of two endophytic fungi that control *Septoria tritici* blotch in the field, using a structured screening approach. *Biological Control* 141: 104128.

Li, H.-L., X.-M. Li, S.-Q. Yang, J. Cao, Y.-H. Li and B.-G. Wang. 2020. Induced terreins production from marine red algal-derived endophytic fungus *Aspergillus terreus* EN-539 co-cultured with symbiotic fungus *Paecilomyces lilacinus* EN-531. *The Journal of Antibiotics* 73: 108–111.

Lubna, A.S., M. Hamayun, H. Gul, I.-J. Lee, and A. Hussain. 2018a. *Aspergillus niger* CSR3 regulates plant endogenous hormones and secondary metabolites by producing gibberellins and indole acetic acid. *Journal of Plant Interactions* 13(1): 100–111.

Lubna, A.S., M. Hamayun, A.L. Khan, M. Waqas, M.A. Khan, R. Jan, I.-J. Lee and A. Hussain. 2018b. Salt tolerance of *Glycine max* L. induced by endophytic fungus *Aspergillus flavus* CSH1, via regulating its endogenous hormones and antioxidative system. *Plant Physiology and Biochemistry* 128: 13–23.

Luo, J., X. Li, Y. Jin, I. Traore, L. Dong, G. Yang and Y. Wang. 2020. Effects of arbuscular mycorrhizal fungi *Glomus mosseae* on the growth and medicinal components of *Dysosma versipellis* under copper stress. *Bulletin of Environmental Contamination and Toxicology* Doi: 10.1007/s00128-019-02780-1.

Mahmood, Z., T. Steenberga, K. Mahmood, R. Labouriau and M. Kristensen. 2019. Endophytic *Beauveria bassiana* in maize affects survival and fecundity of the aphid *Sitobion avenae*. *Biological Control* 137: 104017.

Martins, W.F.X and B.F. Rodrigues. 2020. Identification of dominant arbuscular mycorrhizal fungi in different rice ecosystems. *Agricultural Research* 9(1): 46–55.

Meng, Y.-Y., S.-C. Shao, S.-J. Liu and J.-Y. Gao. 2019. Do the fungi associated with roots of adult plants support seed germination? A case study on *Dendrobium exile* (Orchidaceae). *Global Ecology and Conservation* 17: e00582.

Mohammadi, M., S.A.M. Modarres-Sanavy, H. Pirdashti, B. Zand and Z. Tahmasebi-Sarvestani. 2019. Arbuscular mycorrhizae alleviate water deficit stress and improve antioxidant response, more than nitrogen fixing bacteria or chemical fertilizer in the evening primrose. *Rhizosphere* 9: 76–89.

Monggoot, S., T. Pichaitam, C. Tanapichatsakul and P. Pripdeevech. 2018. Antibacterial potential of secondary metabolites produced by *Aspergillus* sp., an endophyte of *Mitrephora wangii*. *Archives of Microbiology* 200(6): 951–959.

Mookherjee, A., M. Mitra, N.N. Kutty, A. Mitra and M.K. Maiti. 2020. Characterization of endo-metabolome exhibiting antimicrobial and antioxidant activities from endophytic fungus *Cercospora* sp. PM018. *South African Journal of Botany* Doi: 10.1016/j.sajb.2020.01.040.

Motaleb, N.A.A., S.A.A. Elhady and A.A. Ghoname. 2020. AMF and *Bacillus megaterium* neutralize the harmful effects of salt stress on bean plants. *Gesunde Pflanzen* 72: 29–39.

Nefzi, A., R.A.B. Abdallah, H. Jabnoun-Khiareddine, N. Ammar and M. Daami-Remadi. 2019. Ability of endophytic fungi associated with *Withania somnifera* L. to control *Fusarium* Crown and Root Rot and to promote growth in tomato. *Brazilian Journal of Microbiology* 50: 481–494.

Nelson, J. and A.J. Shaw. 2019. Exploring the natural microbiome of the model liverwort: Fungal endophyte diversity in *Marchantia polymorpha* L. *Symbiosis* 78: 45–59.

Nisa, S., N. Khan, W. Shah, M. Sabir, W. Khan, Y. Bibi, M. Jahangir, I.U. Haq, S. Alam and A. Qayyum. 2020. Identification and bioactivities of two endophytic fungi *Fusarium fujikuroi* and *Aspergillus tubingensis* from foliar parts of *Debregeasia salicifolia*. *Arabian Journal for Science and Engineering* Doi: 10.1007/s13369-020-04454-1.

Nobis, A., J. Błaszkowski and S. Zubek. 2015. Arbuscular mycorrhizal fungi associations of vascular plants confined to river valleys: Towards understanding the river corridor plant distribution. *Journal of Plant Research* 128:127–137.

Pan, X., Y. Qin and Z. Yuan. 2018. Potential of a halophyte-associated endophytic fungus for sustaining Chinese white poplar growth under salinity. *Symbiosis* 76: 109–116.

Pandey, R.R., S. Loushambam and A.K. Srivastava. 2020. Arbuscular mycorrhizal and dark septate endophyte fungal associations in two dominant ginger species of Northeast India. *Proceedings of the National Academy of Sciences, India Section B*. Doi: 10.1007/s40011-019-01159-w.

Park, Y.-H., R.C. Mishra, S. Yoon, H. Kim, C. Park, S.-T. Seo and H. Bae. 2019. Endophytic *Trichoderma citrinoviride* isolated from mountain-cultivated ginseng (*Panax ginseng*) has great potential as abiocontrol agent against ginseng pathogens. *Journal of Ginseng Research* 43: 408–420.

Pietro-Souza, W., I.S. Mello, S.J. Vendruscullo, G.F.d. Silva, C.N.d. Cunha, J.F. White and M.A. Soares. 2017. Endophytic fungal communities of *Polygonum acuminatum* and *Aeschynomene fluminensis* are influenced by soil mercury contamination. *PLoS One* 12(7): e0182017.

Pietro-Souza, W., F.d.C. Pereira, I.S. Mello, F.F.F. Stachack, A.J. Terezo, C.N.d. Cunha, J.F. White, H. Li and M.A. Soares. 2020. Mercury resistance and bioremediation mediated by endophytic fungi. *Chemosphere* 240: 124874.

Qiang, X., J. Ding, W. Lin, Q. Li, C. Xu, Q. Zheng and Y. Li. 2019. Alleviation of the detrimental effect of water deficit on wheat (*Triticum aestivum* L.) growth by an indole acetic acid-producing endophytic fungus. *Plant Soil* 439: 373–391.

Rajamani, T., T.S. Suryanarayanan, T.S. Murali and N. Thirunavukkarasu. 2018. Distribution and diversity of foliar endophytic fungi in the mangroves of Andaman Islands, India. *Fungal Ecology* 36: 109–116.

Rhouma, M.B., M. Kriaa, Y.B. Nasr, L. Mellouli and R. Kammoun. 2020. A new endophytic *Fusarium oxysporum* gibberellic acid: Optimization of production using combined strategies of experimental designs and potency on tomato growth under stress condition. *BioMed Research International* Mar 12; 2020:4587148.

Rojas, E.C., R. Sapkota, B. B. Jensen, H.J.L. Jorgensen, T. Henriksson, L.N. Jorgensen, M. Nicolaisen and D.B. Collinge. 2020. *Fusarium* head blight modifies fungal endophytic communities during infection of wheat spikes. *Microbial Ecology* 79: 397–408.

Rostami, M. and S. Rostami. 2019. Effect of salicylic acid and mycorrhizal symbiosis on improvement of fluoranthene phytoremediation using tall fescue (*Festuca arundinacea* Schreb). *Chemosphere* 232: 70–75.

Russo, M.L., A. Sebastian, Pelizza, M.N. Cabello, S.A. Stenglein, M.F. Vianna and A.C. Scorsetti. 2016. Endophytic fungi from selected varieties of soybean (*Glycine max* L. Merr.) and corn (*Zea mays* L.) grown in an agricultural area of Argentina. *Revista Argentina de Microbiología* 8(2): 154–160.

Sabra, M., A. Aboulnasr, P. Franken, E. Perreca, L.P. Wright and I. Camehl. 2018. Beneficial root endophytic fungi increase growth and quality parameters of sweet basil in heavy metal contaminated soil. *Frontiers in Plant Science* 9: 1726.

Sadeghi, F., D. Samsampour, M.A. Seyahooei, A. Bagheri and J. Soltani. 2020. Fungal endophytes alleviate drought-induced oxidative stress in mandarin (*Citrus reticulata* L.): Toward regulating the ascorbate–glutathione cycle. *Scientia Horticulturae* 261: 108991

Salazar-Ramirez, G., R.d.C. Flores-Vallejo, J.C. Rivera-Leyva, E. Tovar-Sanchez, A. Sanchez-Reyes, J. Mena-Portales, M.d.R. Sanchez-Carbente, M.F. Gaitan-Rodriguez, R.A. Batista-García, M.L. Villarreal, P. Mussali-Galante and J.L. Folch-Mallol. 2020. Characterization of fungal endophytes isolated from the metal hyperaccumulator plant *Vachellia farnesiana* growing in Mine tailings. *Microorganisms* 8(2): 226.

Sampangi-Ramaiah, M.H., Jagadheesh, P. Dey, S. Jambagi, M.M.V. Kumari, R. Oelmuller, K.N. Nataraja, K.V. Ravishankar, G. Ravikanth and R.U. Shaanker. 2020. An endophyte from salt-adapted Pokkali rice confers salt-tolerance to a salt-sensitive rice variety and targets a unique pattern of genes in its new host. *Scientific Reports* 10: 3237.

Sampangi-Ramaiah, M.H., K.V. Ravishankar, K.N. Nataraja and R.U. Shaanker. 2019. Endophytic fungus, *Fusarium* sp. reduces alternative splicing events in rice plants under salinity stress. *Plant Physiology Reports* 24: 487–495.

Sangamesh, M.B., S. Jambagi, M.M. Vasanthakumari, N.J. Shetty, H. Kolte, G. Ravikanth, K.N. Nataraja and R.U. Shaanker. 2018. Thermotolerance of fungal endophytes isolated from plants adapted to the Thar Desert, India. *Symbiosis* 75:135–147.

Skiada, V, A. Faccio, N. Kavroulakis, A. Genreb, P. Bonfante and K.K. Papadopoulou. 2019. Colonization of legumes by an endophytic *Fusarium solani* strain FsK revealscommon features to symbionts or pathogens. *Fungal Genetics and Biology* 127: 60–74.

Song J., Y. Hana, B. Baia, S. Jin, Q. He and J. Ren. 2019. Diversity of arbuscular mycorrhizal fungi in rhizosphere soils of the Chinese medicinal herb *Sophora flavescens* Ait. *Soil & Tillage Research* 195: 104423.

Spagnoletti F.N., M. Cornero, V. Chiocchio, R.S. Lavado and I.N. Roberts. 2020. Arbuscular mycorrhiza protects soybean plants against *Macrophomina phaseolina* even under nitrogen fertilization. *European Journal of Plant Pathology* 156: 839–849.

Stevens, B.M., J.R. Propster, M. Opik, G.W.T. Wilson, S.L. Alloway, E. Mayemba and N.C. Johnson. 2020. Arbuscular mycorrhizal fungi in roots and soil respond differently to biotic and abiotic factors in the Serengeti. *Mycorrhiza* 30: 79–95.

Strom, N., W. Hu, D. Haarith, S. Chen and K. Bushley. 2020. Corn and soybean host root endophytic fungi with toxicity towards the soybean cyst nematode. *Phytopathology* Doi: 10.1094/PHYTO-07-19-0243-R.

Su, C.-L., F.-M. Zhang, K. Sun, W. Zhang and C.-C. Dai. 2019. Fungal endophyte *Phomopsis liquidambari* improves Iron and Molybdenum nutrition uptake of peanut in consecutive monoculture soil. *Journal of Soil Science and Plant Nutrition* 19: 71–80.

Tarnabi, Z.M., A. Iranbakhsh, I. Mehregan and R. Ahmadvand. 2020. Impact of arbuscular mycorrhizal fungi (AMF) on gene expression of some cell wall and membrane elements of wheat (*Triticum aestivum* L.) under water deficit using transcriptome analysis. *Physiology and Molecular Biology of Plants* 26: 143–162.

Toghueo, R.M.K., I. Zabalgogeazcoa, B.R.V. de Aldana and F.F. Boyom. 2017. Enzymatic activity of endophytic fungi from the medicinal plants *Terminalia catappa*, *Terminalia mantaly* and *Cananga odorata*. *South African Journal of Botany* 109: 146–153.

Toprak, B. 2020. Early growth performance of mycorrhizae inoculated Taurus Cedar (*Cedrus libani* A. Rich.) seedlings in a nursery experiment conducted in inland part of Turkey. *Journal of Plant Nutrition* 43(2): 165–175.

Tsai, H-J., K.-H. Shao, M.-T. Chan, C.-P. Cheng, K.W. Yeh, R. Oelmuller and S.-J. Wang. 2020. *Piriformosporaindica* symbiosis improves water stress tolerance of rice through regulating stomata behavior and ROS scavenging systems. *Plant Signaling & Behavior* 15(2): 1722447.

Valli, P.P.S. and T. Muthukumar. 2018. Dark septate root endophytic fungus *Nectria haematococca* improves tomato growth under water limiting conditions. *Indian Journal of Medical Microbiology* 58(4): 489–495.

Vannier, N., A.-K. Bittebiere, C. Mony and P. Vandenkoornhuyse. 2020. Root endophytic fungi impact host plant biomass and respond to plant composition at varying spatio-temporal scales. *Fungal Ecology* 104: 100907.

Velez, J.M., T.J. Tschaplinski, R. Vilgalys, C.W. Schadt, G. Bonito, K. Hameed, N. Engle and C.E. Hamilton. 2017. Characterization of a novel, ubiquitous fungal endophyte from the rhizosphere and root endosphere of *Populus* trees. *Fungal Ecology* 27: 78–86.

Wang, Z., C. Li and J. White. 2020. Effects of *Epichloe* endophyte infection on growth, physiological properties and seed germination of wild barley under saline conditions. *Journal of Agronomy and Crop Science* 206: 43–51.

Wang, L., L. Ren, C. Li, C. Gao, X. Liu, M. Wang and Y. Luo. 2019. Effects of endophytic fungi diversity in different coniferous species on the colonization of *Sirex noctilio* (Hymenoptera: Siricidae). *Scientific Reports* 9: 5077.

Wazny, R, P. Rozpadek, R.J. Jedrzejczyk, M. Sliwa, A. Stojakowska, T. Anielskaand and K. Turnau. 2018. Does co-inoculation of *Lactuca serriola* with endophytic and arbuscular mycorrhizal fungi improve plant growth in a polluted environment? *Mycorrhiza* 28: 235–246.

Wemheuer, F., B. Wemheuer, R. Daniel and S. Vidal. 2019. Deciphering bacterial and fungal endophyte communities in leaves of two maple trees with green islands. *Scientific Reports* 9: 14183.

Xu, H., H. Shao and Y. Lu. 2019. Arbuscular mycorrhiza fungi and related soil microbial activity drive carbon mineralization in the maize rhizosphere. *Ecotoxicology and Environmental Safety* 182: 109476.

Ye, B., Y. Wu, X. Zhai, R. Zhang, J. Wu, C. Zhang, K. Rahman, L. Qin, T. Han and C. Zheng. 2020. Beneficial effects of endophytic fungi from the *Anoectochilus* and *Ludisiaspecies* on the growth and secondary metabolism of *Anoectochilus roxburghii*. *ACS Omega*. 5: 3487–3497.

Yeh, Y.-H. and R. Kirschner. 2019. Diversity of endophytic fungi of the coastal plant *Vitex rotundifolia* in Taiwan. *Microbes and Environments* 34(1): 59–63.

Younginger, B.S. and D.J. Ballhorn. 2017. Fungal endophyte communities in the temperate fern *Polystichum munitum* show early colonization and extensive temporal turnover. *American Journal of Botany* 104(8): 1188–1194.

Yu, M., J.-C. Chen, J.-Z. Qu, F. Liu, M. Zhou, Y.-M. Ma, S.-Y. Xiang, X.-X. Pan, H.-B. Zhang and M.-Z. Yang. 2020. Exposure to endophytic fungi quantitatively and compositionally alters anthocyanins in grape cells. *Plant Physiology and Biochemistry* 149: 144–152.

Yu, J., Y. Wu, Z. He, M. Li, K. Zhu and B. Gao. 2018. Diversity and antifungal activity of endophytic fungi associated with *Camellia oleifera*. *Mycobiology* 46(2): 85–91.

Zahn, G. and A.S. Amend. 2019. Foliar fungi alter reproductive timing and allocation in *Arabidopsis* under normal and water-stressed conditions. *Fungal Ecology* 41: 101–106.

Zhan, F., B. Li, M. Jiang, L. Qin, J. Wang, Y. He and Y. Li. 2017. Effects of a root-colonized dark septate endophyte on the glutathione metabolism in maize plants under cadmium stress. *Journal of Plant Interactions* 12(1): 421–428.

Zhang, S., X. Guo, W. Yun, Y. Xia, Z. You and M.C. Rillig. 2020. Arbuscular mycorrhiza contributes to the control of phosphorus loss in paddy fields. *Plant Soil* 447: 623–636.

Zhou, J., Z. Xu, H. Sun and H. Zhang. 2019. Smoke-isolated butenolide elicits tanshinone I production in endophytic fungus *Trichoderma atroviride* D16 from *Salvia miltiorrhiza*. *South African Journal of Botany* 124: 1–4.

22

Root Nodule Microbiome from Actinorhizal Casuarina *Plant*

Narayanasamy Marappa, D. Dhanasekaran, and Thajuddin Nooruddin
Bharathidasan University

CONTENTS

22.1 Introduction

Microorganisms are vital to the protection of life on the globe; nevertheless, we understand modest about the majority of microorganisms in the atmosphere such as soil, oceans, environment and even those alive on and in our own body. The plant microbiome plays a vital role in plant growth endorsement and soil productiveness for sustainable farming. Plants and soil are an important instinctive source harboring hotspots of microorganisms (Yadav, 2020). The soil microbiomes have significant responsibility in the prolongation of overall nutrient stability as well as bionetwork occupation. The microorganisms associated with plants such as endophytic, rhizospheric, and epiphytic with plant growth-promoting (PGP) aspects have been used as an essential and capable tool for sustainable farming. PGP microorganisms promote plant growth directly or obliquely, whichever, by liberate plant growth regulators; solubilization of phosphorus, zinc, and potassium; and production of siderophore, HCN, ammonia, and other minor metabolites, which are violent against pathogenic microorganisms. The plant growth-promoting microorganisms belonged to genera such as *Achromobacter*, *Arthrobacter*, *Rhizobium*, *Aspergillus*, *Azotobacter*, *Bacillus*, *Pantoea*, *Burkholderia*, *Frankia* *Methylobacterium*, *Pseudomonas*, *Paenibacillus*, *Piriformospora*, *Serratia*, *Penicillium*, *Azospirillum*, *Planomonospora*, *Streptomyces*, and *Gluconoacetobacter*.

These plant growth-promoting microorganisms might be used as biofertilizer alternatives for chemical fertilizers for appropriate farming (Yadav, 2020). Culture techniques have been allowed isolated microorganisms to be deliberate in detail, and molecular techniques such as metagenomics are gradually more allowing the identify caption of microorganisms in situ. The microbial communities, or microbiome, of diverse atmospheres have been deliberate in this approach, with the ambition of their perceptive biological function (Gilbert et al., 2010). The plant microbiome is a key determinant of plant healthiness and efficiency and has established significant awareness in current years (Turnbaugh et al., 2007). A demonstration of the significance of plant-microbe communications is the mycorrhizal fungi. Molecular proof recommends that their family with green algae were essential to the fruition of earth vegetation about 750 million years before (Berendsen et al., 2012). The majority plants, though remarkably not *Arabidopsis thaliana* and other Brassicaceae, have preserved this symbiosis, which support the root uptake of mineral nutrients such as phosphate. Besides, plant-associated microorganisms are key players in overall biogeochemical cycles.

In agricultural soils in specific, plants stimulate the microbial de-nitrification and methanogenesis, which provide emissions of N_2O and methane, correspondingly (Lebeis et al., 2012). These gases symbolize a loss of carbon and nitrogen from the system and provide to the conservatory outcome. Manipulation of the plant microbiome has the prospective to diminish the frequency of plant illness, enhance the agricultural fabrication, decrease the chemical effort, and decrease the emission of conservatory gases consequential in more sustainable agricultural practices. This ambition is seen essential for sustaining the world's mounting population. The rhizosphere is a region of rich, largely soil-derived, microbial variety, subjective to authentication of plant mucilage and root exudates (Bulgarelli et al., 2013). Microbial population of the rhizosphere and phyllosphere are called epiphytes, while microorganisms inhabiting within plant tissues, whether in leaves, roots, or stems, are called endophytes.

Microbes in this position protect favorable, unbiased, or harmful associations of unreliable familiarity with their host plants. Particular connections among microorganisms and replica plants such as in Rhizobium–legume symbioses are well implicit, but the popularity of the plant microbiome and its involvement in the comprehensive phenotype of the host is not hitherto well distinct. Prominently, the microbiome is powerfully prejudiced by the plant genome and possibly considered as a conservatory to form a subsequent genome or cooperatively to form a pan-genome. This chapter exclusively rewarded the prospect of root nodule microbiome from Actinorhizal *Casuarina* plant and its functional effects in abiotic stress tolerance for agricultural crop development (Bakker et al., 2012). The ultimate statement visualizes the prospect of the advantageous role of the plant microbiome in plant growth promotion as well as soil productiveness.

22.2 Actinorhizal *Casuarina* Root Nodule Microbiome Diversity Experimental Workflow

The 16S rRNA is composed of conserved and hypervariable regions. Whereas conserved regions are not significantly different across various microbial strains, the sequences of hypervariable regions are genus- or species-specific and differ in accordance with phylogenetic difference. Therefore, 16S rDNA serves as an identifier of biological species and is important for microbial phylogeny and taxonomic identification. 16S rDNA amplicon sequencing has become an important tool for the study of the composition of microbial communities in the environment. 16S rDNA amplicon sequencing includes the library construction using specific primers to amplify the variable region of prokaryotic 16S rDNA and data analysis of the 16S rDNA variable region sequence to identify the composition and abundance of prokaryotic microbes in the environment.

The proprietary workflow at GENEWIZ effectively amplifies the two variable regions of 16S rDNA (V3 and V4) and accurately identifies various species including archaea (Mahalingam et al., 2019). 18S / ITS rDNA amplicon sequencing includes the library construction using specific primers to amplify the variable region of eukaryotic 18S / ITS rDNA and data analysis to identify the composition and abundance of eukaryotic microorganisms in the environment. The Illumina MiSeq sequencing platform is widely used for 16S rDNA amplicon sequencing because of its deep sequencing depth, high throughput, short run-time, and high sequencing accuracy as well as reasonable cost. In recent years, pair-end chemistry has enabled the MiSeq sequencing platform to read longer, which further increased the accuracy of the results. 16S rDNA amplicon sequencing procedure includes genomic DNA extraction, quality control, rDNA variable region amplification, library construction, high-throughput sequencing, and data analysis. All the steps are important for data quality and quantity, which in turn affects the subsequent data analysis. In order to ensure data accuracy and reliability, every step has to pass strict quality control before pooling the library by adjusting the volume of each library according to the target data volume for Illumina MiSeq sequencing (Mahalingam et al., 2019; Egamberdieva, 2012). The workflow is shown in Figure 22.1.

22.3 Actinorhizal *Casuarina* Root Nodule Microbiome Statistical Data Analysis Workflow

First, adapters and low-quality data were filtered out from the original data. Then, the chimera sequences were removed to obtain the effective sequences for cluster analysis. Each cluster was called an operational taxonomic unit (OTU). The taxonomy analysis of the representative sequence of each OTU was then performed to obtain species distribution information. Based on the results of OTU analysis, α-diversity indices of each sample can be derived as well as the species richness and

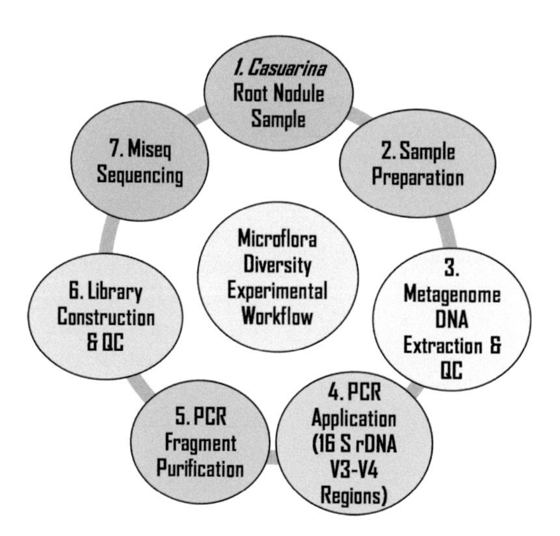

FIGURE 22.1 Root nodule microbiome diversity experimental workflow.

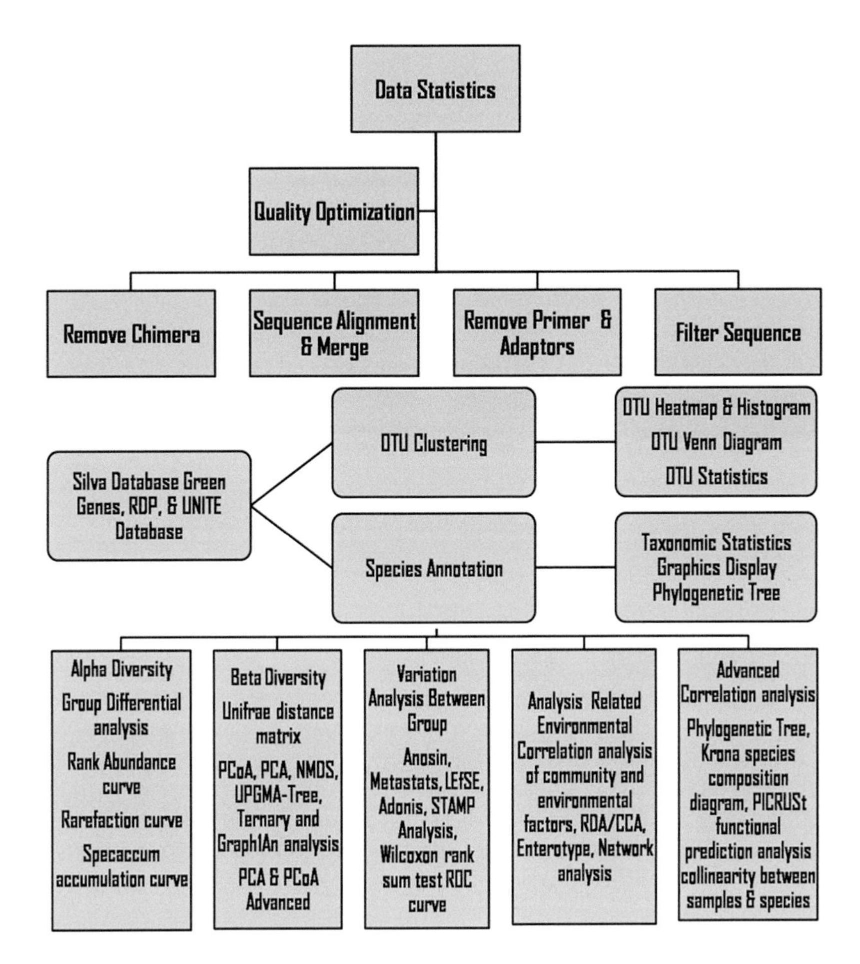

FIGURE 22.2 Root nodule microbiome statistical data analysis workflow.

evenness. Based on taxonomic information, statistical analysis of the community structure can be carried out at each classification level. UPGMA clustering tree and PCoA plots can be constructed based on Unifrac distance to illustrate the differences in the community structure between different samples or groups. Following the basic analysis above, a series of in-depth data mining can be carried out. For example, researchers can investigate the different community structure among different groups of samples using multiple statistical methods (Lundberg et al., 2012; Heyer et al., 2019). This could be further combined with the environmental factors and species diversity to discover the environmental factors important for community structure (Figure 22.2).

22.4 MetaVx™ Library Preparation and Illumina MiSeq Sequencing

Next-generation sequencing library preparations and Illumina MiSeq sequencing were conducted at GENEWIZ, Inc. (Suzhou, China). DNA samples were quantified using a Qubit 2.0 Fluor meter (Invitrogen, Carlsbad, CA, USA). The 30–50ng DNA was used to generate amplicons using a MetaVx™ Library Preparation kit (GENEWIZ, Inc., South Plainfield, NJ, USA). V3 and V4 hypervariable regions of prokaryotic 16S rDNA were selected for generating amplicons and the following taxonomy analysis. GENEWIZ designed a panel of proprietary primers aimed at relatively conserved regions bordering the V3 and V4 hypervariable regions of bacteria and Archaea16S rDNA. The v3 and v4 regions were amplified using forward primers containing the sequence "CCTACGGRRBGCASCAGKVRVGAAT" and reverse primers containing the sequence "GGACTACNVGGGTWTCTAATCC". First-round PCR products were used as templates for second-round amplicon enrichment PCR. At the same time, indexed adapters were added to the ends of the 16S rDNA amplicons to generate indexed libraries ready for downstream NGS sequencing on Illumina MiSeq (Hartman et al., 2017).

DNA libraries were validated by Agilent 2100 Bioanalyzer (Agilent Technologies, Palo Alto, CA, USA) and quantified by Qubit 2.0 Fluorometer. DNA libraries were multiplexed and loaded on an Illumina MiSeq instrument according to manufacturer's instructions (Illumina, San Diego, CA, USA). Sequencing was performed using a 2x300 paired-end (PE) configuration; image analysis and base calling were conducted by the MiSeq Control Software (MCS) embedded in the MiSeq instrument.

22.5 *Casuarina* Root Nodule Microbiome Data Analysis

The QIIME data analysis package was used for 16S rRNA data analysis. The forward and reverse reads were joined and assigned to samples based on barcode and truncated by cutting off the barcode and primer sequence. Quality filtering on joined sequences was performed, and sequences that did not fulfill the following criteria were discarded: sequence length <200bp, no ambiguous bases, and mean quality score≥20. Then, the sequences were compared with the reference database (RDP Gold database) using UCHIME algorithm to detect chimeric sequence, and then, the chimeric sequences were removed. The effective sequences were used in the final analysis. Sequences were grouped into operational taxonomic units (OTUs) using the clustering program VSEARCH (1.9.6) against the Silva 119 database preclustered at 97% sequence identity (Figure 22.3) (Wang et al., 2016).

The Ribosomal Database Program (RDP) classifier was used to assign the taxonomic category to all OTUs at a confidence threshold of 0.8. The RDP classifier uses the Silva 132 database, which has taxonomic categories predicted to the species level, and the heat map demonstrates the variability among the group and diversity (Figures 22.4 and 22.5). Sequences were rarefied prior to calculation of alpha and beta diversity statistics. Alpha diversity indexes were calculated in QIIME from rarefied samples using for diversity the Shannon index and for richness the Chao1 index. Beta diversity was calculated using weighted and unweighted UniFrac, and principal coordinate analysis (PCoA) was performed. The unweighted pair group method with arithmetic mean (UPGMA) tree from beta diversity distance matrix was built (Table 22.1).

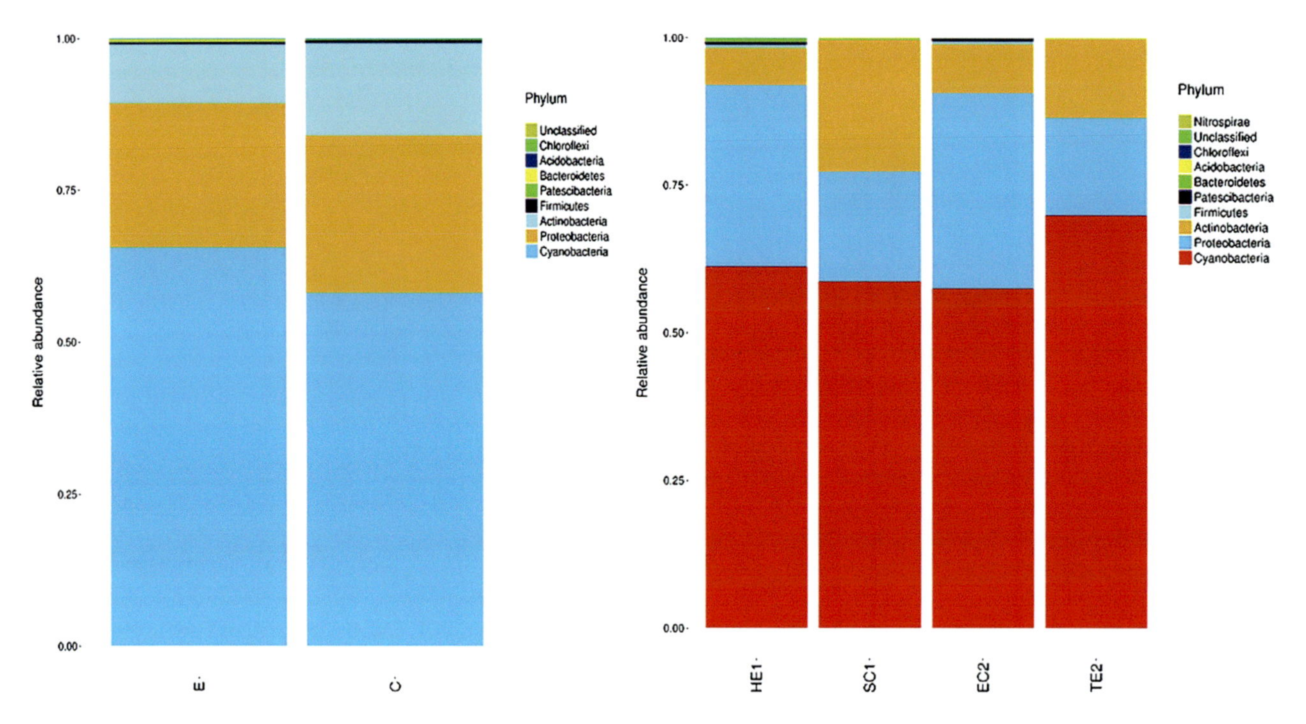

FIGURE 22.3 *Casuarina* root nodule microbiome in phylum level.

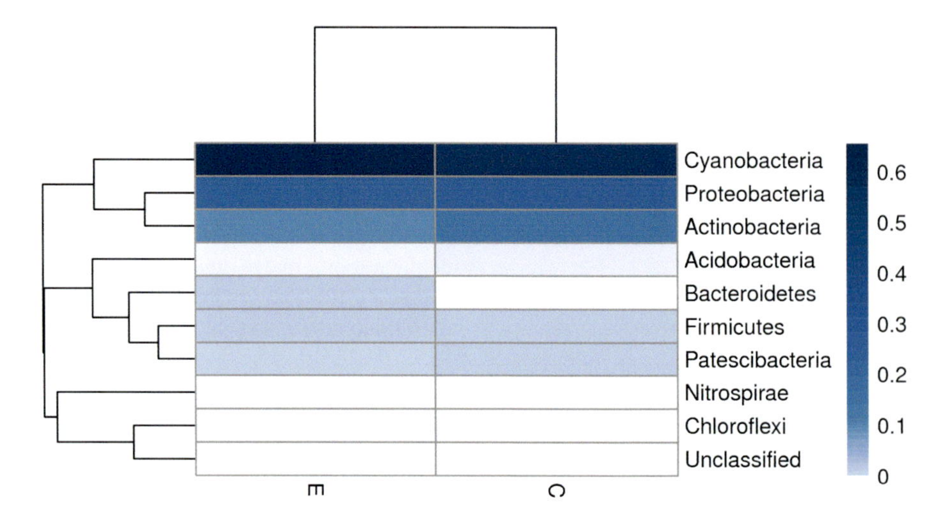

FIGURE 22.4 *Casuarina* root nodule microbiome heat map in group at phylum level.

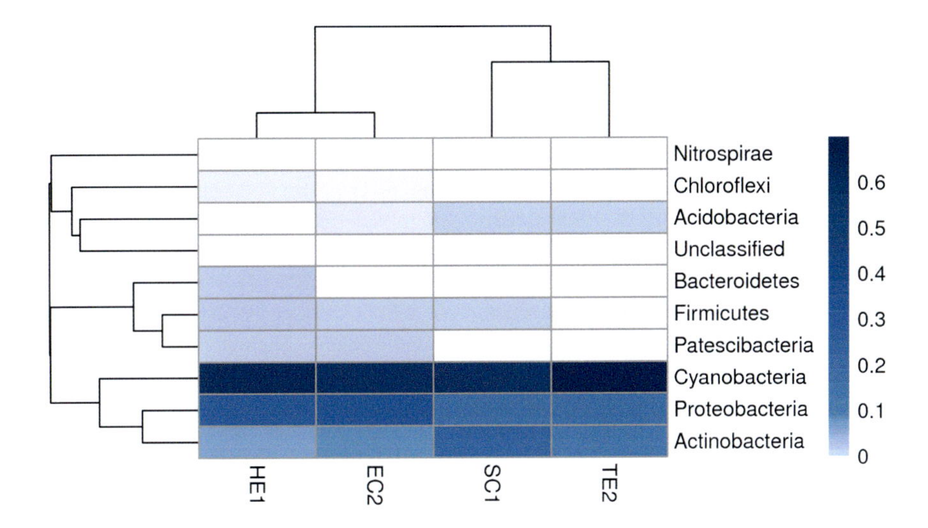

FIGURE 22.5 *Casuarina* root nodule microbiome heat map in different habitats at phylum level.

TABLE 22.1

List of Microbiome Sequence Analysis Software

Software	Version
Cutadapt	1.9.1
Vsearch	1.9.6
Qiime	1.9.1
RDP classifier	2.2
PyNAST	1.2
R	3.3.1
LEfSe	1.0
PICRUSt	1.0.0
STAMP	2.1.3
Cytoscape	3.5.1
GraPhlAn	1.0.0
Krona tools	2.7
Circos	0.69-

22.6 Plant Microbiome

Classic microbiology involves isolating and culturing microorganisms from an atmosphere using different nutrient media and growth situations depending on the target organisms. While acquiring a pure culture of an organism is a requisite for in-depth studies of its heredity and functioning, culture-dependent methods overlook the immense preponderance of microbial diversity in an atmosphere. Frequent culture-independent molecular methods are used in the microbial ecosystem. For studying prokaryotes, PCR amplification of the ever-present 16S ribosomal RNA (rRNA) genetic material is frequently used. Sequencing the changeable regions of this gene permits an accurate taxonomic classification. The use of high-throughput sequencing technology (Kent & Triplett, 2002) has been extensively approved since it allows taxonomy of thousands to millions of sequences in a model, illuminating the abundances of yet exceptional microbial variety. For studying eukaryotic microorganisms such as fungi, the corresponding rRNA gene (18S) might not provide adequate taxonomic favoritism, so the overexcited changeable internally transcribed spacer is frequently used. An inadequacy of PCR amplification of genomic DNA is intrinsically influenced by primer design (Hong et al., 2009) and usually merely identifies the goal organisms. Complex environments are inhabited by organisms from all domains of life. Eukaryotes, such as fungi, oomycetes, protozoa, and, nematodes, are omnipresent in soils and can be significant plant pathogens or symbionts, while others are bacterial grazers. The archaea carry out vital biochemical response, predominantly in agricultural soils, including ammonia oxidation and methanogenesis (Pinto & Raskin, 2012).

Viruses excessively are profuse and widespread and also influence the metabolism as well as population dynamics of their hosts (Kent & Triplett, 2002).

Microbes in a community interact with each other and the host plant, so it is significant to imprison as greatly of the diversity of a microbiome as potential. Toward, execute require the use of universal analyses including Metagenomics, metaproteomics, and Metatranscriptomics which allocate immediate evaluation and association of microbial populations transversely all province of life. Metagenomics are able to expose the purposeful impending of a microbiome, while metatranscriptomics and metaproteomics are making available to snapshots of community-wide gene appearance and protein profusion, correspondingly. Metatranscriptomics has exposed kingdom-level revolutionization in the formation of the crop-plant rhizosphere microbiome (Barea et al., 2005). The comparative profusion of eukaryotes in pea and oat rhizospheres was five-fold higher than in plant-free soil or the rhizosphere of contemporary hexaploid wheat. The pea rhizosphere in exacting was extremely enriched with fungi. Complementary molecular methods are able to complement such approaches. For example, constant isotope inquisitive allows organisms metabolizing an exacting labeled substrate to be well known (Turner et al., 2013). Unite these culture-based technique approaches should advance our consideration of plant-microbe communications at the intensity of the system.

22.7 The Rhizosphere Environment

The rhizosphere is the province of soil predisposed by plant roots through rhizodeposition of exudates, mucilage, and sloughed cells. Root exudates enclose a variety of compounds, predominately organic acids and sugars, except fatty acids, amino acids, vitamins, hormones, growth factors, and antimicrobial compounds (Bertin et al., 2003). Root exudates are key determinants of rhizosphere microbiome arrangement. The composition of root exudate preserves differs among the plant genus and cultivars with plant age in the developmental stage (Shi et al., 2011). Besides, the microbiome manipulates root exudates, as anemically mature plants have noticeably dissimilar exudate compositions from those predisposed by microorganisms. Various concurrences of *A. thaliana* have revealed the difference in root exudate compositions with correspondingly dissimilar rhizosphere bacterial communities, while the rhizosphere bacterial communities of other accessions have exposed high similarity. Root exudates are not the only constituent of rhizodeposition. The sloughing of root cells able to liberate mucilage deposits a large amount of substance into the rhizosphere, including plant cell wall polymers such as cellulose and pectin (Bais et al., 2006).

Cellulose degradation is wide among microbial inhabitants of high-organic-matter in soils. The disintegration of pectin releases methanol as a carbon source by other microorganisms, and active metabolism of methanol in the rhizosphere has been observed. Whereas a carbon source is provided to rhizosphere microorganisms, plant roots also provide a structure on which microorganisms can fasten. Behind this is the surveillance of significant overlie among microbes attaching to a root and to an inert timber structure (Broeckling et al., 2008).

Differences among the plant cultivars are noticeable while comparing microbial type and strains. The Proteobacteria habitually dominate samples, mostly those of α and β classes. Other main groups including Actinobacteria, Firmicutes, Bacteroidetes, Planctomycetes, Verrucomicrobia, and Acidobacteria of fastidious significance in the rhizosphere are plant growth-promoting rhizobacteria, which act throughout a variety of mechanisms (Micallef et al., 2009).

Nitrogen-fixing microorganisms, together with those that are free-living like *Azotobacter* spp. and symbiotic including *Frankia* and *Rhizobium*, etc., provide a resource of fixed nitrogen for the plant, and various bacteria can solubilize phosphorous- containing natural resources, escalating its bioavailability. Microbial exploitation of plant hormones, mainly auxins, ethylene, and gibberellins, leads to growth promotion or pressure tolerance. Various plant growth-promoting rhizobacteria are active violently toward plant pathogens via producing antimicrobials or by interfering through virulence factors via effectors conveyed by type 3 emission systems (T3SSs) (Mark et al., 2005). Actinobacteria, in particular, are recognized to generate a wide assortment of compounds with antibacterial, antiviral, antifungal, insecticidal, and nematicidal possessions. They frequently originate as one of the majority abundant bacterial classes in soil and rhizospheres and are predominantly enriched in the endophytic community.

Other bacteria act as illness antagonists, including *Pseudomonas fluorescens*, which fabricate the antifungal compound diacetylphloroglucinol (DAPG). *Pseudomonas* spp. producing DAPG has been exposed to modulate transcription in a different plant growth-promoting rhizobacterium, *Azospirillum brasilense*, escalating expression of genes concerned in wheat root migration and plant growth endorsement (DeAngelis et al., 2009). DAPG moreover affects other microbiota, including nematodes, wherever it originated to be harmful to various species but stimulatory to others. The occurrence of DAPG-producing *Pseudomonas spp.* in soils has been concerned in the occurrence of take-all refuse (Chaparro et al., 2013). Take-all is a cereal infection caused by the fungus *Gaeumannomyces graminis.*

In take-all refuse, illness harshness reduces with frequent cultivation of a plant like wheat. The soil becomes infection-oppressive as a result of the establishment of the aggressive microbial community (Cavaglieri et al., 2009). Other antagonists from the Proteobacteria, Firmicutes, and Actinobacteria contributed to soils suppressive toward the root-rotting fungus *Rhizoctonia.*

Antifungal-metabolite-producing *pseudomonads* were one of the major groups responsible for inhibition. Shifts in microbiomes have been associated with soils oppressive toward *Fusarium* and *Streptomyces scabies* (Lundberg et al., 2012). This proposes that an association of microorganisms contributes to suppressiveness, while the cause and upshot are habitually not discernible. An affluent and various microbiotas alone might be adequate to prevent disease by preventive access to roots and nutrients.

22.8 The Phyllosphere Atmosphere

The phyllosphere, or the aerial exterior of a plant, is measured be comparatively nutrient-poor in contrast with the rhizosphere. Microbial migration of plants is not homogenous but is pretentious by leaf structures such as veins, hairs, and stomata. Leaf surfaces are colonized by up to 107 microbes per cm^2 (Knief et al., 2012). The phyllosphere is a greatly more dynamic environment than the rhizosphere, with resident microbes subjected to great fluxes in temperature, humidity, and emission throughout the daytime and darkness. These abiotic factors also circuitously affect the phyllosphere microbiome during changes in plant metabolism. Precipitation and blustery weather in fastidious are thought to contribute to the chronological variability in inhabitant phyllosphere microbes (Mendes et al., 2011). Fascinatingly, the leaf metabolite summary of *A. thaliana* has been distorted by relevance of soil microorganisms to roots: improved absorption of several amino acids in the leaf metabolome was associated with amplified herbivore by insects, signifying cross-talk among the above- and below-ground parts of the plant. The bacterial and fungal communities in the phyllospheres of different plants have been profiled via PCR amplification of rRNA genes. Microbial prosperity seems to be superior in warmer, moister climates than in moderate ones. Proteobacteria (α and γ classes) are constantly the foremost bacterial phylum, with Bacteroidetes and Actinobacteria also usually found (Klein et al., 2013). The phyllospheres of more than a few plants in the Mediterranean were found to be subjugated by lactic acid bacteria (Firmicutes) throughout summer. Their type of metabolism was projected to allocate them to abide the warm and desiccated climate circumstances even though this was not evaluated between different periods.

At elevated microbial taxonomic levels, phyllosphere microbiomes of different plants can appear parallel; however, at the microbial species and strain levels, bleak differences are obvious, reflecting the delicately tuned metabolic adaptations obligatory to live in such an atmosphere (Badri et al., 2013). While rhizosphere microbiomes are analogous to soil, modest similarity has been found among phyllosphere microbiomes and those of the atmosphere. Proteogenomic analyses for a variety of phyllosphere microbiomes have exposed species that assimilate plant derivative ammonium, amino acids, and simple carbohydrates, implicating these compounds as main nitrogen and carbon sources in the phyllosphere (Vokou et al., 2012). Metagenomic analysis of taxonomically varied plant species has shown an abundance of different known and new microbial rhodopsins in the phyllosphere.

22.9 The Endosphere Atmosphere

Endophytic bacteria are those that inhabit within plant tissues. They are usually considered to be nonpathogenic, causing no noticeable symptoms, although they include dormant pathogens that, depending on ecological conditions and/or host genotype, can cause infections (Hallmann et al., 1997). Endophytes are reflection to be a sub-population of the rhizosphere microbiome; however, they also have uniqueness distinct from rhizospheric bacteria, signifying that not all rhizospheric bacteria can penetrate to plants, and/or that formerly within their hosts they revolutionize their metabolism and become tailored to their internal atmosphere (Compant et al., 2010). It is normally assumed that the bacteria that can be isolated from plant tissues subsequent to surface sterilization are "endophytic". In the most current updates, sonication was used to remove the exterior layers of plant tissue, and the residual tissue was used to identify the endophytic microbiome (Sessitsch et al., 2012).

Such study exposed that endophytic bacteria frequently reside in the intercellular apoplast and in deceased or dying cells and as nonetheless they have not been persuasively exposed to occupy living cells in the same prearranged manner as true endosymbioses. Once inside the roots, the bacteria inhabit the apoplast; however, their numbers appear to be prohibited, as they infrequently exceed 102–103 colony-forming units (cfu) per gram fresh weight and are habitually as low as 96 cfu per gram fresh weight, depending on plant age and genotype. Younger plants have superior bacterial concentration than established ones, and the concentrations of epiphytic bacteria are frequently greater than those of endophytes, such as by a factor of ten in the case of *Herbaspirillum* (James, 2000).

To put these endophyte records into viewpoint, symbiotically proficient legume nodules characteristically contain up to 1,015 cfu rhizobial bacteroids per gram fresh weight so it would appear that the numbers of endophytic bacteria are not so elevated as to necessitate the expansion of a dedicated organ, such as a nodule, to residence them. Most of our acquaintance regarding endophytic bacteria comes from work on a few well-deliberate 'model' organisms, such as *Azoarcus*, *Burkholderia*, *Gluconacetobacter*, *Herbaspirillum*, and *Klebsiella* spp., which were all isolated from nonlegumes, predominantly grasses (Reinhold-Hurek & Hurek, 2011). While these reports have specified much imminent into mechanisms of infection and migration, they notify us slightly on the true diversity of the bacteria in the endophytic microbiome. Culture-independent methods, such as analyses of 16S rRNA and nifH transcripts, and metagenome analyses have established a massive diversity of endophytes in the inexpensively significant crops such as sugarcane and rice. Fascinatingly, these reports recommend that rhizobia and other α-Proteobacteria are very frequent endophytes, as are β-Proteobacteria, γ-Proteobacteria, and Firmicutes.

22.10 Effect of Plant Host Microbiome

The communications among a plant and its microbiome are decidedly complex and forceful. The plant immune system (Box 1) in exacting is thought to have a key role in influential plant microbiome constitution. Mutants of *A. thaliana* deficient in systemic acquired resistance (SAR) have revealed the differences in rhizosphere bacterial community composition compared with wild variety (Hein et al., 2008), where chemical creation of SAR did not affect the considerable shifts in the rhizosphere bacterial community. In the phyllosphere of *A. thaliana*, orientation of salicylic acid-mediated protection reduced the diversity of endophytes, whereas plants scarce in jasmonate-mediated resistance illustrate superior epiphytic diversity (Kniskern et al., 2007). These findings propose that the effects of plant protection progression on the microbiome are erratic, and that SAR is dependable for controlling the populations of some bacteria. The construction of plant hormones such as indole-3-acetic acid (IAA) is extensive between plant-associated bacteria, predominantly the *rhizobia*, and some *Bacillus* spp. produce gibberellins (Ghosh et al., 2011). *Pseudomonas syringae* generate hormone analogs that obstruct with jasmonate and ethylene signaling, resulting in stomatal aperture and pathogen entry.

Some plant genes and pathways have roles in establishment of various interactions with unlike microorganisms; examples comprise the developmental pathways that are collective between mycorrhizal and rhizobial symbioses. It is not yet recognized whether and how these pathways act together with other members of the microbiome. Plants fabricate a wide selection of antimicrobial compounds both constitutively and in effect to pathogens (Melotto et al., 2006). Unexpectedly, yet, a recent inclusive study of the rhizosphere microbiome of these two genotypes establish a slight variation between the fungal community. This highlights that a little change in plant genotype can have intricate and surprising effects on the plant microbiome. With its complexity and vitality, chiefly in the natural atmosphere, it is vital not to ignore the plant microbiome when interpreting investigational data, particularly when it can lead to applications in the field. Hereditary alteration of plants, to resist infection for example, may have unforeseen consequences for the rest of the microbiome, which may or may not be physiologically significant (Bressan et al., 2009). The role of the microbiome and its association with plant health, effectiveness, and biogeochemical cycle obligation should be considered as much as the plant itself. An extension of this notion is that molecular breeding or genetic modification of plants could be used to modulate the microbiome deliberately, recruiting disease antagonists and plant-growth promoters to improve agricultural construction.

22.11 Various Applications of the Plant Microbiome

Rhizosphere complex microbial communities have potential which theater and vital position in nutrient cycling, decorative soil productivity, protect plant health and effectiveness. Particular microbiomes that are accruing secure to roots are considered to be the most composite ecosystem on the globe. Various microbial districts of rhizospheric microbiomes appreciably diverge by soil diversity, land use prototype, plant species, and host genotype. It is recognized that root exudates act as substrates and signaling molecules, which are fundamental for establishing plant–rhizobacterial interactions (Kour et al., 2019; Mendes et al., 2013). These survey priorities might facilitate us to manipulate the agricultural microbiome and thus to increase the association approach for enhanced fabrication and competence of global agriculture in a sustainable approach. The unique appearance for panorama study effort encloses enrichment and conservation of rhizosphere biodiversity and their potential application in agricultural soils. Endophytes are the microorganisms that are active in the internal tissues of plants. Endophytic microbes squeeze enormous implication for the roles that they play in organization with the host plants.

Endophytes are predictable to promote the acceleration of the host plants by a multiplicity of behavior such as detoxification of venomous composites, protection beside pathogens, and production of plant growth-promoting hormones (Rana et al., 2019; Suman et al., 2016). Various biotechnologically essential metabolites are moreover produced by the endophytes such as anticancer and antimicrobial compounds (Narayanasamy et al., 2020c). There are a prosperous variety of endophytes that require to be considered for biotechnological purposes (Narayanasamy et al., 2020c). Such endophytes contribute a considerable accountability in plant growth endorsement as these compose obtainable resistance to plant beside varied biological pressure and poisonous compounds, defend host plants beside abundant pathogens, and formulate various plant growth-promoting hormones. Endophytic microbes are also considerably essential as biotransformers of various chemicals and effortlessness in recycling of nutrients (Narayanasamy et al., 2020b). The endophytes besides expose a variety of industrial usages because they are well-known for the innovation of numerous fundamental enzymes and metabolites (Yadav et al., 2019).

Endophytes symbolize a vital ingredient of microbial diversity while 28 years of important evolution in the field expose the implication of endophytic microorganisms. Endophytic microbes are an unfamiliar group of organisms that have massive credibility for narrative pharmaceutical materials; they are renowned as antioxidant, antifungal, anticancer, and anti-inflammatory agents. Likewise, in recent years, an incredible perfection was made in their potential as favorable molecules beside various ailments. Current updates in further review are essential in bioprospecting narrative endophytic microbes and their function.

Bacterial and fungal endophytes are ubiquitously occupied in the interior tissue of living plants. Endophytic fungi discrete out while tropical districts to frosty province have enormous credibility in provisos of secondary metabolite production (Narayanasamy et al., 2018). It is appropriate to know that the assortment of bioactive fundamental compounds' estimation by these endophytic fungi is host-specific. They are dreadfully important to improve the suppleness of the endophyte and its host plants, for illustration, biotic and abiotic pressure forbearance (Rana et al., 2019; Yadav et al., 2019). The phyllosphere refers to the overall aerial plant exterior and acts as a habitat for microbes. Microbes constitute a compositionally composite community on the leaf surface. The microbiome of the phyllosphere is affluent in assortment of bacteria, fungi, cyanobacteria, actinobacteria, and viruses (Kumar et al., 2019; Muller et al., 2016).

Microorganisms are usually significance either epiphytic or endophytic approach of life cycle on phyllosphere atmosphere, which helps the host plant and efficient statement with the adjacent atmosphere. The phyllosphere is an exclusive environment occupied by an extensive diversity of microbes together with epiphytes, favorable and pathogenic, fungi, bacteria, and viruses (Bargabus et al., 2002). Sympathetic phyllosphere community arrangement, networking, and physiology are an enormous challenge.

Yet, extensive exploration on phyllosphere microbiome provides enormous prospects for the applications in profitable plant efficiency, exclusively farming and forestry, bio-network cleaning, and healthiness. Climate changeability has been, and persists to be, the major resource of fluctuations in worldwide food manufacture in developing countries (Oseni and Masarirambi, 2011). The significant threat of rising warming of world and premature rainfall proceedings, uneven winter seasons, more infection incidence, and yield failures is changeable (Adger et al., 2005). Tremendous environments symbolize exclusive bionetwork, which harbors narrative biodiversity. Microbial community is connected with plants upward in most varied circumstances, including the extreme of temperature, salinity, water deficiency and pH. To survive under such excessive circumstances, these organisms referred to as extremophiles, have residential adaptive features, which authorize them to cultivate optimally under one or more ecological boundaries, while polyextremophiles grow optimally under a manifold environment. These extremophiles are able to grow optimally in some of the earth's most antagonistic environment of temperature (−2°–20°C—psychrophiles; 60°– 115°C—thermophiles), salinity (2–5M NaCl—halophiles), and pH (9—alkaliphiles) (Yadav et al., 2019).

Microorganisms connected with crops are capable to endorse the plant augmentation. Numerous microorganisms have been reported that they can encourage plant growth directly or circuitously. Microorganisms have been exposed to endorse plant growth directly, e.g., *Frankia* by fixation of atmospheric nitrogen, solubilization of minerals such as phosphorus, potassium, and zinc; and creation of siderophores and plant growth hormones such as cytokinin, gibberellins, and auxin. Various bacteria carry plant growth obliquely, by construction of antagonistic substances by inducing resistance beside plant pathogens (Narayanasamy et al., 2020a; Tilak et al., 2005). High salinity in agricultural soil is the severe problem all over the globe, and it is also a vital ecological issue for decrease of growth and yield of farming crops.

For a solution, in the soils, the use of plant growth-promoting rhizobacteria, actinobacteria, and *Frankia* (PGPR) can reduce soil salinity and load of chemical fertilizers and pesticide in the agricultural field and recover soil healthiness, seed germination, crop growth, and efficiency under saline circumstances. PGPR is established as a probable microorganism that can withstand various impressive conditions like high temperature, pH, and salty soils (Yadav et al., 2019). Halophilic microorganisms are isolated from salty soils or rhizosphere of halophytic plants and demonstrate plant growth-promoting characters directly such as the construction of IAA, solubilization of phosphate, creation of siderophore, fixation of N_2, deaminase ACC action, or ultimately inhabits by controlling of phytopathogens under salty circumstances (Narayanasamy et al., 2020a,c; Verma et al., 2017).

Awareness of plant–microbe communications facilitates the strategy for the fortification of crops and salty soil remediation, and this kind of communication is also observed in the district for biological approving of microorganisms, which endorse halophyte to flexibility in a salinity-prosperous atmosphere. Deficiency is an obvious stress causing harmful consequences on plant growth and efficiency. In order to recompense the acquiesce loss due to deficiency, resourceful and sustainable strategies are obligatory for its organization. Drought anxiety forbearance is a composite attribute involving clusters of genes; therefore, genetic engineering to produce drought-resistant variety is a demanding mission. Under this circumstance, the purpose of plant growth- promoting microorganisms (PGPM) to moderate deficiency stress is gaining concentration as a gorgeous and gainful choice strategy (Kour et al., 2019).

Microbes' proficiency of coping with low temperatures is extensive in these normal atmospheres wherever they habitually symbolize the foremost flora, and they must, consequently, be regarded as the most victorious colonizers of our globe. Psychrophilic microbes are tailored to prosper well at low temperatures secure to the freezing point of water (Yadav et al., 2019). The microbial action of psychrophiles has yet been described at freezing temperatures. In general, psychrophilic microbes reveal superior growth yield and microbial activity at low temperatures evaluate to temperatures secure to the utmost temperature of growth and have more often been put forth as a clarification to victorious microbial version to the usual cold atmosphere. Panoramas of the cold habitats have led to the separation of an immense diversity of psychrotrophic microbes.

The cold-adapted microorganisms have probable biotechnological applications in crop growth, medicines, and industry. Biofertilizers are characteristically microbial formulations in organic transporter materials that develop soil fitness and crop growth and expansion. Microbial formulations might be organism-specific or a conglomerate of organisms. Various soil microbes are gifted with an assortment of capability ranging from fabrication of growth-enhancing substances to the discharge of substances, which improve the property of different abiotic stress circumstances such as deficiency, salinity, pH pressure, heat anxiety, impurity, and nutrient insufficiency. There has been a pointed augment in the world's inhabitants above the past few decades, which can be intimidating in stipulations of the food safety of the people. Thus, to provide to the enormous demand of food, agricultural manufacture must be augmented within a tiny span of time and with inadequate global agricultural land possessions. This condition has made the farmers all over the globe to rely greatly on the commercially obtainable chemical fertilizers for improved agricultural yield.

Nevertheless, there has been a major increase in the construction of crops, and these fertilizers have established to be harmful for our bionetwork as well as mammal and human healthiness. The deteriorative effect of the great chemical inputs in the agricultural scheme has not only defied the sustainability of crop construction but moreover the preservation of the environment superiority. Using biofertilizers is an innate, inexpensive, and eco-friendly approach to this difficulty (Verma et al., 2017). Biofertilizers encompass live microbes able to supply adequate nutrients to the plants, while maintaining lofty yield. With the ever-growing inhabitants, there appears no end to the stipulate of food, although with the accessibility of the chemical fertilizers, it was considered that the difficulty might be tackled.

REFERENCES

Adger, W. N., Arnell, N. W., & Tompkins, E. L. (2005). Successful adaptation to climate change across scales. *Global Environmental Change*, 15(2), 77–86.

Badri, D. V., Zolla, G., Bakker, M. G., Manter, D. K., & Vivanco, J. M. (2013). Potential impact of soil microbiomes on the leaf metabolome and on herbivore feeding behavior. *New Phytologist*, 198(1), 264–273.

Bais, H. P., Weir, T. L., Perry, L. G., Gilroy, S., & Vivanco, J. M. (2006). The role of root exudates in rhizosphere interactions with plants and other organisms. *Annual Review of Plant Biology*, 57, 233–266.

Bakker, M. G., Manter, D. K., Sheflin, A. M., Weir, T. L., & Vivanco, J. M. (2012). Harnessing the rhizosphere Microbiome through plant breeding and agricultural management. *Plant and Soil*, 360(1–2), 1–13.

Barea, J. M., Pozo, M. J., Azcon, R., & Azcon-Aguilar, C. (2005). Microbial co-operation in the rhizosphere. *Journal of Experimental Botany*, 56(417), 1761–1778.

Bargabus, R. L., Zidack, N. K., Sherwood, J. E., & Jacobsen, B. J. (2002). Characterisation of systemic resistance in sugar beet elicited by a non-pathogenic, phyllosphere-colonizing *Bacillus mycoides*, biological control agent. *Physiological and Molecular Plant Pathology*, 61(5), 289–298.

Berendsen, R. L., Pieterse, C. M., & Bakker, P. A. (2012). The rhizosphere microbiome and plant health. *Trends in Plant Science*, 17(8), 478–486.

Bertin, C., Yang, X., & Weston, L. A. (2003). The role of root exudates and allelochemicals in the rhizosphere. *Plant and Soil*, 256(1), 67–83.

Bressan, M., Roncato, M. A., Bellvert, F., Comte, G., el Zahar Haichar, F., Achouak, W., & Berge, O. (2009). Exogenous glucosinolate produced by Arabidopsis thaliana has an impact on microbes in the rhizosphere and plant roots. *The ISME Journal*, 3(11), 1243–1257.

Broeckling, C. D., Broz, A. K., Bergelson, J., Manter, D. K., & Vivanco, J. M. (2008). Root exudates regulate soil fungal community composition and diversity. *Applied and Environmental Microbiology*, 74(3), 738–744.

Bulgarelli, D., Schlaeppi, K., Spaepen, S., Van Themaat, E. V. L., & Schulze-Lefert, P. (2013). Structure and functions of the bacterial microbiota of plants. *Annual Review of Plant Biology*, 64, 807–838.

Cavaglieri, L., Orlando, J., & Etcheverry, M. (2009). Rhizosphere microbial community structure at different maize plant growth stages and root locations. *Microbiological Research*, 164(4), 391–399.

Chaparro, J. M., Badri, D. V., Bakker, M. G., Sugiyama, A., Manter, D. K., & Vivanco, J. M. (2013). Root exudation of phytochemicals in Arabidopsis follows specific patterns that are developmentally programmed and correlate with soil microbial functions. *PloS One*, 8(2), e55731.

Compant, S., Clément, C., & Sessitsch, A. (2010). Plant growth-promoting bacteria in the rhizo-and endosphere of plants: Their role, colonization, mechanisms involved and prospects for utilization. *Soil Biology and Biochemistry*, 42(5), 669–678.

DeAngelis, K. M., Brodie, E. L., DeSantis, T. Z., Andersen, G. L., Lindow, S. E., & Firestone, M. K. (2009). Selective progressive response of soil microbial community to wild oat roots. *The ISME Journal*, 3(2), 168–178.

Egamberdieva, D. (2012). Colonization of *Mycobacterium phlei* in the rhizosphere of wheat grown under saline conditions. *Turkish Journal of Biology*, 36(5), 487–492.

Ghosh, S., Ghosh, P., & Maiti, T. K. (2011). Production and metabolism of indole acetic acid (IAA) by root nodule bacteria (Rhizobium): A review. *Journal of Pure and Applied Microbiology*, 5, 523–540.

Gilbert, J. A., Meyer, F., Jansson, J., Gordon, J., Pace, N., Tiedje, J., & Glöckner, F. O. (2010). The earth Microbiome project: Meeting report of the "1st EMP meeting on sample selection and acquisition" at Argonne National Laboratory October 6th 2010. *Standards in Genomic Sciences*, 3(3), 249–253.

Hallmann, J., Quadt-Hallmann, A., Mahaffee, W. F., & Kloepper, J. W. (1997). Bacterial endophytes in agricultural crops. *Canadian Journal of Microbiology*, 43(10), 895–914.

Hartman, K., van der Heijden, M. G., Roussely-Provent, V., Walser, J. C., & Schlaeppi, K. (2017). Deciphering composition and function of the root Microbiome of a legume plant. *Microbiome*, 5(1), 2.

Hein, J. W., Wolfe, G. V., & Blee, K. A. (2008). Comparison of rhizosphere bacterial communities in *Arabidopsis thaliana* mutants for systemic acquired resistance. *Microbial Ecology*, 55(2), 333–343.

Heyer, R., Schallert, K., Siewert, C., Kohrs, F., Greve, J., Maus, I.,& Puttker, S. (2019). Metaproteome analysis reveals that syntrophy, competition, and phage-host interaction shape microbial communities in biogas plants. *Microbiome*, 7(1), 69.

Hong, S., Bunge, J., Leslin, C., Jeon, S., & Epstein, S. S. (2009). Polymerase chain reaction primers miss half of rRNA microbial diversity. *The ISME Journal*, 3(12), 1365–1373.

James, E. K. (2000). Nitrogen fixation in endophytic and associative symbiosis. *Field Crops Research*, 65(2–3), 197–209.

Kent, A. D., & Triplett, E. W. (2002). Microbial communities and their interactions in soil and rhizosphere ecosystems. *Annual Reviews in Microbiology*, 56(1), 211–236.

Klein, E., Ofek, M., Katan, J., Minz, D., & Gamliel, A. (2013). Soil suppressiveness to Fusarium disease: Shifts in root Microbiome associated with reduction of pathogen root colonization. *Phytopathology*, 103(1), 23–33.

Knief, C., Delmotte, N., Chaffron, S., Stark, M., Innerebner, G., Wassmann, R., & Vorholt, J. A. (2012). Metaproteogenomic analysis of microbial communities in the phyllosphere and rhizosphere of rice. *The ISME Journal*, 6(7), 1378–1390.

Kniskern, J. M., Traw, M. B., & Bergelson, J. (2007). Salicylic acid and jasmonic acid signaling defense pathways reduce natural bacterial diversity on Arabidopsis thaliana. *Molecular Plant-Microbe Interactions*, 20(12), 1512–1522.

Kour, D., Rana, K. L., Yadav, N., Yadav, A. N., Kumar, A., Meena, V. S.,& Saxena, A. K. (2019). Rhizospheric microbiomes: Biodiversity, mechanisms of plant growth promotion, and biotechnological applications for sustainable agriculture. In *Plant Growth Promoting Rhizobacteria for Agricultural Sustainability*, 19–65. Springer, Singapore.

Kumar, M., Kour, D., Yadav, A. N., Saxena, R., Rai, P. K., Jyoti, A., & Tomar, R. S. (2019). Biodiversity of methylotrophic microbial communities and their potential role in mitigation of abiotic stresses in plants. *Biologia*, 74(3), 287–308.

Lebeis, S. L., Rott, M., Dangl, J. L., & Schulze-Lefert, P. (2012). Culturing a plant Microbiome community at the cross-Rhodes. *New Phytologist*, 196(2), 341–344.

Lundberg, D. S., Lebeis, S. L., Paredes, S. H., Yourstone, S., Gehring, J., Malfatti, S.,& Edgar, R. C. (2012). Defining the core *Arabidopsis thaliana* root Microbiome. *Nature*, 488(7409), 86–90.

Mahalingam, S., Dharumadurai, D., & Archunan, G. (2019). Vaginal microbiome analysis of buffalo (Bubalus bubalis) during estrous cycle using high-throughput amplicon sequence of 16S rRNA gene. *Symbiosis*, 78(1), 97–106.

Mark, G. L., Dow, J. M., Kiely, P. D., Higgins, H., Haynes, J., Baysse, C., & O'Gara, F. (2005). Transcriptome profiling of bacterial responses to root exudates identifies genes involved in microbe-plant interactions. *Proceedings of the National Academy of Sciences*, 102(48), 17454–17459.

Melotto, M., Underwood, W., Koczan, J., Nomura, K., & He, S. Y. (2006). Plant stomata function in innate immunity against bacterial invasion. *Cell*, 126(5), 969–980.

Mendes, R., Garbeva, P., & Raaijmakers, J. M. (2013). The rhizosphere microbiome: Significance of plant beneficial, plant pathogenic, and human pathogenic microorganisms. *FEMS Microbiology Reviews*, 37(5), 634–663.

Mendes, R., Kruijt, M., De Bruijn, I., Dekkers, E., van der Voort, M., Schneider, J. H.,& Raaijmakers, J. M. (2011). Deciphering the rhizosphere microbiome for disease-suppressive bacteria. *Science*, 332(6033), 1097–1100.

Micallef, S. A., Shiaris, M. P., & Colon-Carmona, A. (2009). Influence of Arabidopsis thaliana accessions on rhizobacterial communities and natural variation in root exudates. *Journal of Experimental Botany*, 60(6), 1729–1742.

Muller, C. A., Obermeier, M. M., & Berg, G. (2016). Bioprospecting plant-associated microbiomes. *Journal of Biotechnology*, 235, 171–180.

Narayanasamy, M., Dhanasekaran, D. & Thajuddin, N. and MA. Akbarsha. (2020c). Morphological, molecular characterization and biofilm inhibition effect of endophytic *Frankia* sp. from root nodules of Actinorhizal plant *Casuarina* sp. *South African Journal of Botany*, 134, 72–83.

Narayanasamy, M., Dhanasekaran, D., & Thajuddin, N. (2020). Beneficial Microbes in Agro-Ecology. Chapter-11*Frankia*, 185–211. Doi:10.1016/b978-0-12-823414-3.00011-3. Academic Press is an imprint of Elsevier, UK, ISBN: 9780128172308.

Narayanasamy, M., Dhanasekaran, D., Vinothini, G., & Thajuddin, N. (2018). Extraction and recovery of precious metals from electronic waste printed circuit boards by bioleaching acidophilic fungi. *International Journal of Environmental Science and Technology*, 15(1), 119–132.

Narayanasamy, M., Lavania R.M., Dhanasekaran, D., & Thajuddin, N. (2020b). Recovery of gold and other precious metal resources from environmental polluted E-waste printed circuit board by bioleaching *Frankia*. *International Journal of Environmental Research*, 14, 165–176.

Narayanasamy, M., Lavania, R.M., Dhanasekaran, D., Thajuddin, N. (2020a) Plant growth-promoting active metabolites from *Frankia* spp. of actinorhizal *Casuarina* spp. *Applied Biochemistry and Biotechnology*, 5, 1–18.

Pinto, A. J., & Raskin, L. (2012). PCR biases distort bacterial and archaeal community structure in pyrosequencing datasets. *PloS one*, 7(8), e43093.

Rana, K. L., Kour, D., Sheikh, I., Yadav, N., Yadav, A. N., Kumar, V., & Saxena, A. K. (2019). Biodiversity of endophytic fungi from diverse niches and their biotechnological applications. In *Advances in Endophytic Fungal Research*, 105–144. Springer, Cham.

Reinhold-Hurek, B., & Hurek, T. (2011). Living inside plants: Bacterial endophytes. *Current Opinion in Plant Biology*, 14(4), 435–443.

Sessitsch, A., Hardoim, P., Döring, J., Weilharter, A., Krause, A., Woyke, T.,& Hurek, T. (2012). Functional characteristics of an endophyte community colonizing rice roots as revealed by metagenomic analysis. *Molecular Plant-Microbe Interactions*, 25(1), 28–36.

Shi, S., Richardson, A. E., O'Callaghan, M., DeAngelis, K. M., Jones, E. E., Stewart, A., & Condron, L. M. (2011). Effects of selected root exudate components on soil bacterial communities. *FEMS Microbiology Ecology*, 77(3), 600–610.

Suman, A., Yadav, A. N., & Verma, P. (2016). Endophytic microbes in crops: Diversity and beneficial impact for sustainable agriculture. In *Microbial Inoculants in Sustainable Agricultural Productivity*, 117–143. Springer, New Delhi.

Tilak, K. V. B. R., Ranganayaki, N., Pal, K. K., De, R., Saxena, A. K., Nautiyal, C. S., & Johri, B. N. (2005). Diversity of plant growth and soil health supporting bacteria. *Current Science*, 89(1)136–150.

Turnbaugh, P. J., Ley, R. E., Hamady, M., Fraser-Liggett, C. M., Knight, R., & Gordon, J. I. (2007). The human Microbiome project. *Nature*, 449(7164), 804–810.

Turner, T. R., Ramakrishnan, K., Walshaw, J., Heavens, D., Alston, M., Swarbreck, D., & Poole, P. S. (2013). Comparative metatranscriptomics reveals kingdom level changes in the rhizosphere microbiome of plants. *The ISME Journal*, 7(12), 2248–2258.

Verma, P., Yadav, A. N., Kumar, V., Singh, D. P., & Saxena, A. K. (2017). Beneficial plant-microbes interactions: Biodiversity of microbes from diverse extreme environments and its impact for crop improvement. In *Plant-Microbe Interactions in Agro-Ecological Perspectives*, pp. 543–580. Springer, Singapore.

Vokou, D., Vareli, K., Zarali, E., Karamanoli, K., Constantinidou, H. I. A., Monokrousos, N., & Sainis, I. (2012). Exploring biodiversity in the bacterial community of the Mediterranean phyllosphere and its relationship with airborne bacteria. *Microbial Ecology*, 64(3), 714–724.

Wang, W., Zhai, Y., Cao, L., Tan, H., & Zhang, R. (2016). Illumina-based analysis of core actinobacteriome in roots, stems, and grains of rice. *Microbiological Research*, 190, 12–18.

Yadav, A. N. (2020). Plant microbiomes for sustainable agriculture: Current research and future challenges. In *Plant Microbiomes for Sustainable Agriculture*, pp. 475–482. Springer, Cham.

Yadav, A. N., Yadav, N., Sachan, S. G., & Saxena, A. K. (2019). Biodiversity of psychrotrophic microbes and their biotechnological applications. *Journal of Applied Biology & Biotechnology*, 7, 99–108.

23

Growth Promotion Utility of the Plant Microbiome

S. Kalaiselvi
Government Arts and Science College (W)

A. Panneerselvam
A.V.V.M Sri Pushpam College Poondi (Auto)

CONTENTS

23.1 Introduction

Plant microbiomes can play an important role in protecting the plant from potential pathogens and at the same time improving growth, health, and production, thus being an advantage to plants (Berg et al., 2016; Haney et al., 2015). The bacterial and fungal microbiome, carrying out beneficial interactions, is important for supportable agriculture and has attracted more attention compared to particularly other groups of organisms. (Figure 23.1).

Plants live together with microbial communities to form interactions that are essential for the performance and survival of the host. Many studies have discovered a vast plant-associated microbial diversity. However, plants are a valuable part of a balanced diet including raw-eaten vegetables, fruits, and herbs. Edible plant microbiome and its diversity can be important for humans as (1) an additional aid to the diversity of our gut microbiome and (2) as a stimulus for the human immune system.

Reducing the use of chemical fertilizers and bacterial microbiome and the production of bioinoculants enable scientists to affect plant-beneficial activities such as limiting the action of phytopathogens and encouraging plant growth and health (Adesemoye and Kloepper, 2009). Rhizospheric or endophytic bacteria that increase plant growth are known as plant growth-promoting bacteria (PGPB) (Santoyo et al., 2016). PGPB may induce plant growth by direct or indirect mechanisms (Glick, 2012). Direct stimulation of plant growth occurs when a rhizobacterium facilitates the attainment of essential nutrients or regulates the level of hormones within a plant. Nutrient accomplishment facilitated by PGPB needed for plant growth usually includes elements such as nitrogen, phosphorus, and iron (Calvo et al., 2017).

Model system of field, transcriptomics, proteomics, metabolomics, and merge thereof were applied to plant microbiomes to inscription environmental accommodate at the community level for past decade. The collectivity of microbial genetic information serves as a vital basis for functional genomics approaches and may be deduced from shotgun metagenome sequencing, introduced by Handelsman et al. (1998) or by sequencing strain collections (Bai et al., 2015). In addition, the microbial metagenomes attained from a discrete plant section allow for the identification of habitat-specific gene improvement, potentially highlighting functional traits apropos for effective host colonization. Based on these analyses of bacterial communities, a number of characteristics are emerging as relevant under environmental conditions.

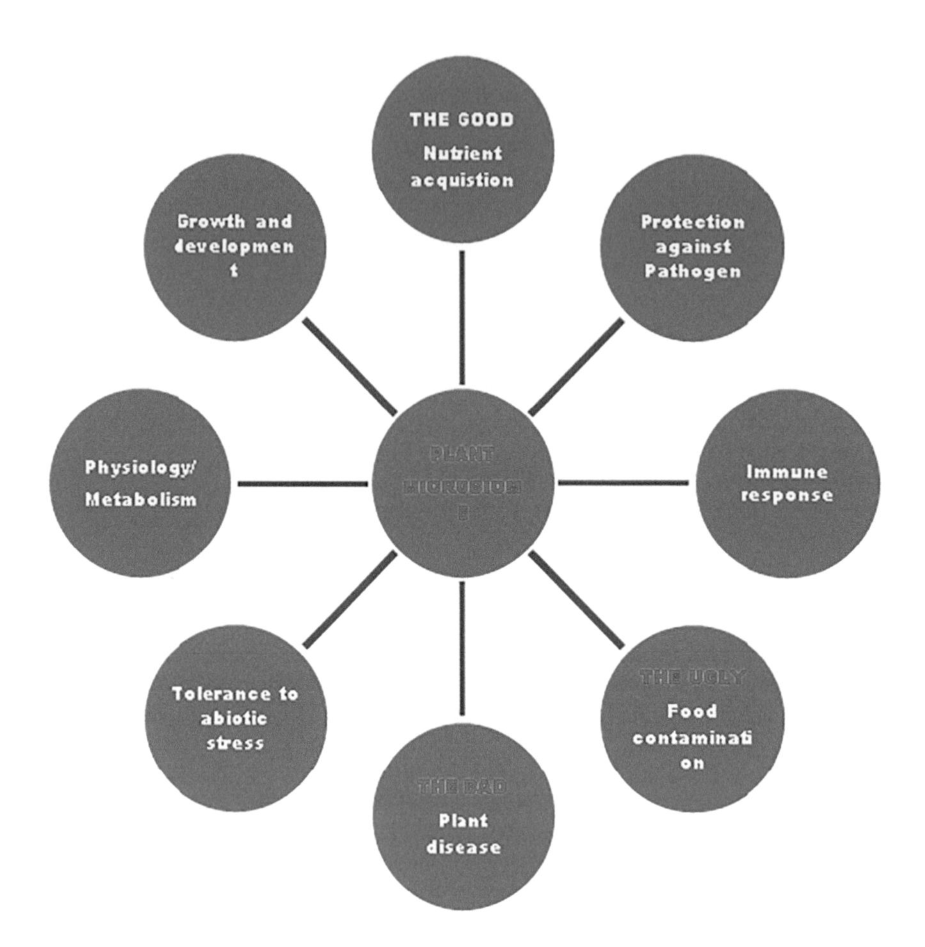

FIGURE 23.1 Overview of the functions and impact of plant-beneficial, plant-pathogenic, and human-pathogenic microorganisms on the host plant.

23.1.1 Mechanism of the Plant Microbiome

Regulation of hormone levels may imply PGPB in the synthesis of one or more phytohormones, auxins, cytokinins, and gibberellins (Bhattacharyya et al., 2015; Perez-Flores et al., 2017). Some PGPB can reduce the levels of the phytohormone ethylene by producing an enzyme 1-aminocyclopropane-1-carboxylate (ACC) deaminase that splits the compound ACC, which is the immediate pioneer of ethylene in all higher plants (Glick, 1995; Glick, 2012). Indirect plant growth development occurs when PGPB decrease plant damage due to infection with a plant pathogen, including harmful fungi and bacteria. This is usually due to the inhibition of the pathogens by PGPB (Ryan et al., 2008; Santoyo et al., 2012).

The mechanisms include the synthesis and excretion of antibiotics such as 2,4-diacetylphloroglucinol, proteases, chitinases, bacteriocins, siderophores, lipopeptides (such as iturin A, bacillomycin D, and mycosubtilin), and volatile organic compounds (Santoyo et al., 2012; Glick, 2012; Hernandez-Leon et al., 2015). The microbiome is an important component for the plant to carry out physiological functions, plant development or growth of essential organs such as the root, and improved acquisition of nutrients and water (Gutierrez-Luna et al., 2010).

In this review, we emphasize recent progress on supplementary aspects in plant microbiota research. Focus on bacteria that form structured communities in association with leaves, flowers, seeds, and roots are more interesting aspects of fungi and other eukaryotes are noticed. Microorganisms functionally diversify and survive in the process of occupying such niches within a certain habitat under competitive conditions, altogether resulting in the accompany of populations. Therefore, in order to achieve a systems-level understanding of the plant microbiota, there is a need to move from population inventories and phylogeny-inferred reputed functional traits to actual data on microbiota activities *in situ*.

> **Definitions:** Microorganisms (bacteria, fungi, viruses, archaebacteria, etc.) can be found everywhere in and on plants, animals, water, soil, food, and humans. Within each of those habitats, microorganisms live together in communities called *microbiomes*. Microbiomes have an effect on human health; therefore, scientists are researching how these communities of organisms coexist with each other, with us and the environment. The term "microbiome" was first coined in 2001 by Nobel Laureate and Microbiologist *Joshua Lederberg* (Marchesi and Ravel, 2015). Notably, a "History of Medicine" article in a recent Annals of Internal Medicine issue makes this same protest that Lederberg coined the term in 2001 (Podolsky, 2017).

The word "microbiome" is construed from the "omics" family of terminology; it should be used to describe the collective genomes of microbial species, while the collection of organisms themselves should be termed "*microbiota*." Whipps and colleagues used the term "microbiome" to explain the collection of microbes and their activities within a given environment. They state that:

> A convenient ecological framework in which to examine biocontrol systems is that of the microbiome. This may be defined as a characteristic microbial community occupying a reasonably well-defined habitat, which has distinct physiochemical properties. This term thus not only refers to the microorganisms involved but also encompasses their theatre of activity.

"Microbiome" is defined as combining of microbe and biome, describing the microbial ecosystem, inhabitants, and all, and not just genomes. Whereas the term is relatively new to our science vocabulary, the underlying concept and importance of microbiome work hearken back to the very beginning of microbial ecology and to Sergei Winogradsky in the 1800s.

Bulgarelli et al. (2013) refer to the microbiome as the set of genomes of the microorganisms in a particular habitat, *synonym of microbiota* to mean *microbial community*. Those microbial communities incorporated with the plant which can live, prevail, and interact with different tissues such as roots, shoots, leaves, flowers, and seeds (Haney and Ausubel, 2015; Haney et al., 2015; Mueller and Sachs, 2015; Nelson, 2017).

23.2 Diversity, Functional Potential, and Adaptation to the Plant Environment

Figure 23.2 denotes the regions or zones where the microbes can interact with the plant, live with root (rhizosphere), the aerial plant habitat sensibility or the leaf surface in interaction with the external environment (phyllosphere), the stem (caulosphere), the flowers (anthosphere), the fruits (carposphere), the endosphere (all inner parts), and the spermosphere (the exterior of germinated seed). While the discovery of specific microbiomes is initially associated with the rhizosphere, there are currently only a few other compartments where species-specific diversity was identified, e.g., the carposphere (Hardoim et al., 2015, Leff and Fierer, 2013). Given the estimated number of 370,000 species of higher plants, a great deal of work is still required before the details of global plant microbiome diversity will be fully cleared.

23.3 Rhizosphere

The rhizosphere is the soil portion persuaded by plant roots (Hartmann et al., 2008). Microecosystem is the major region where chemical communications and the exchange of compounds and nutrients occur between the plant and the soil microorganisms (Peiffer et al., 2013). A strategy to improve plant health and development includes the selection and modification of the rhizosphere microbiome (Chaparro et al., 2012).

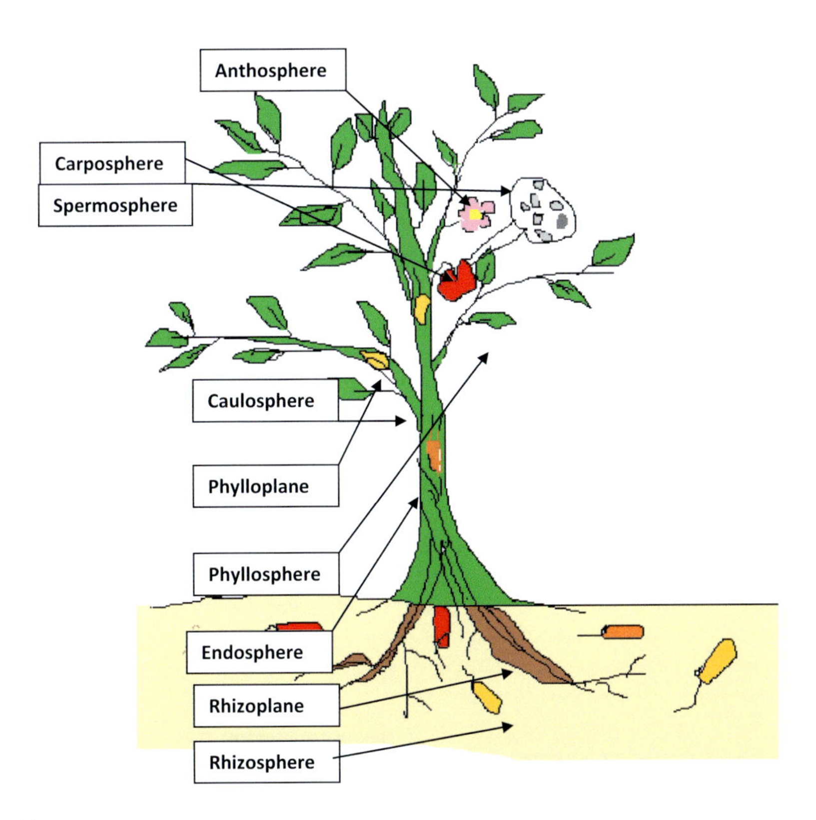

FIGURE 23.2 Plant microbiome.

23.4 Root Microbiota Transformation

Root microbiota are mostly horizontally transferred, i.e., they are derived from the soil environment which contains highly diverse microorganisms, dominated by Acidobacteria, Verrucomicrobia, Bacteroidetes, Proteobacteria, Planctomycetes, and Actinobacteria (Fierer, 2017).

Root microbiota may also be vertically transmitted via seeds. Seeds also represent an important source of microorganisms, which increase in the roots of the developing plant (Liu et al., 2012; Hardoim et al., 2012). Plants with their root system provide unique ecological niches to soil microbiota, which colonize the rhizosphere, roots, and to some extent above-ground parts (Hartmann, 2009). The narrow layer of soil under the direct influence of plant roots, i.e., the rhizosphere, is deliberate as a hot spot of microbial activity and represents one of the most complex ecosystems. Donn et al. (2015) recently analyzed root-driven changes in bacterial community structure of the wheat rhizosphere and found a 10-fold higher number of actinobacteria, pseudomonads, oligotrophs, and copiotrophs in the rhizosphere as assimilated to bulk soil. The authors also showed that rhizosphere and rhizoplane communities were altered over time, while the bulk soil population remained unaffected. Similarly, Kawasaki et al. (2016) urged that the *Brachypodium distachyon* (a model for wheat) rhizosphere was dominated by the Burkholderiales, Sphingobacteriales, and Xanthomonadales; moreover, the bulk soil was dominated by the order Bacillales. Rhizosphere effects are root-exuded, i.e., organic acids, amino acids, fatty acids, phenolics, plant growth regulators, nucleotides, sugars, putrescine, sterols, and vitamins are known to affect the microbial composition around roots (Hartmann et al., 2008). A group of protective secondary metabolites such as benzoxazinoids discharged by maize roots alter the composition of root-associated microbiota, and microorganisms belonging to Actinobacteria and Proteobacteria were found to be most affected by benzoxazinoids metabolites (Hu et al., 2018). Moreover, different plant species and genotypes, based on the type and composition of root exudates, influence the composition of rhizosphere microbiota.

For root or rhizosphere communities, genes relating to chemotaxis and motility have been identified as enhanced categories in the metagenomes of the particular microbiota of wheat and cucumber (Ofek-Lalzar et al., 2014) and metaproteome of rice (Knief et al., 2012). Different modes of motility occur and have been shown to be important for migration from soil toward roots and root colonization. For example, *Flavobacterium* (Bacteroidetes) possess unique gliding motility machinery that is functionally combined with a Bacteroidetes-specific type IX secretion system (Kniskern et al., 2007), and twitching motility via type IV pili is essential for migration of endophytic diazotrophic *Azoarcus* into rice roots (Bohm et al., 2007). This promotes that individual taxa have evolved specific strategies to successfully move to favorable locations and carry on in these environments. However, transcripts affiliated to motility were under render in a *Burkholderia* strain endophytic in potato plants; once cells have attached or entered a plant tissue, detecting that motility might no longer be needed (Sheibani-Tezerji et al., 2015).

23.5 Endosphere

Márquez-Santacruz et al. (2010) represent that endospheric microbiome can ensue as a result of the rhizosphere microbiome. Plant roots colonized inside the root are endosphere by a diverse range of bacterial endophytes. Entry of bacterial endophytes through internal root tissues often happens through root damage or emergence points of lateral roots as well as by active mechanisms. The colonization and transmission of endophytes within plants depend on many agents such as annuity of plant resources and endophytes able to colonize the plants.

Several ranges of bacterial taxa can enter the root tissues; examples of the most abundant phyla that repeatedly occurred in grapevine roots were Proteobacteria, Acidobacteria, Actinobacteria, Bacteroidetes, Verrucomicrobia, Planctomycetes, Chloroflexi, Firmicutes, and Gemmatimonadetes (Zarraonaindia et al., 2015; Samad et al., 2017; Faist et al., 2016). Most dominant families in the roots of rice are Rhizobiaceae, Comamonadaceae, Streptomycetaceae, and Bradyrhizobiaceae (Edwards et al., 2015).

Bacteria, archaebacteria, fungi, and viruses that inhabit, colonize, and retain inside the host without causing any harm to the plants are endophytes. Some reviews have investigated the colonization of internal plant tissues by endophytes. Santoyo et al. (2016) revealed available data about various entry points, the rhizosphere being an important region for microbe entry into the plant because of its nutrient-rich environment, produced by the root exudates, and its high concentration of microbes (Badri and Vivanco, 2009). There are other parts of entry, such as the lenticels, stomas, wounds, ruptures, and nodules. Santoyo et al. (2016) urged that endophytes can also be natural, by vertical transmission through seeds. Some researchers have explained that the internal microbiome of plants has an advantage over external microbiome (phyllosphere or rhizosphere) in that it is not affected by soil contingencies including the presence of bacterial exploiters. For example, siderophores of bacterial origin secreted in the rhizosphere can be captured and intercalated by other organisms such as fungi, which possess nonspecific transporters of the siderophore–Fe complexes (Philpott, 2006).

Various studies have exhibited that the composition of the rhizosphere is more diverse than that of the endosphere. In a study involving rhizospheric and endospheric 16S rRNA genes from Mexican husk tomato plants (*Physalis ixocarpa*), a higher number of operational taxonomic units (OTUs) occurred in a library of clones from the rhizosphere compared to that from the root endosphere, respectively (i.e., 86 vs. 17). It should be recorded that the number of clones obtained from the endosphere and rhizosphere was statistically related. The dominant genera found in the rhizosphere, such as *Stenotrophomonas*, *Burkholderia*, *Bacillus*, and *Pseudomonas*, were indicated in the internal tissues of the plant (Márquez-Santacruz et al., 2010).

23.6 Phyllosphere

Vorholt (2012) explained that the phyllosphere represents the surface and the apoplast of leaf tissue. The great importance of the phyllosphere microbiome on biocontrol, and the promotion of plant growth, has been reported for years. In fact, some authors have introduced that the foliar microbiota exert useful activities (Lindow and Brandl, 2003; Penuelas and Terradas, 2014), and many foliar microbiota can fix nitrogen, this being the main mechanism to provide nitrogen to plants growing in tropical humid ecosystems (Abril et al., 2005). Other beneficial roles of the microbiota that live on the plant surface include the indirect defense against pathogens and the production of plant hormones. Thus, the possibility exists that foliar microbiota could reduce the use of agrochemicals to control leaf pathogens and improve plant growth (Penuelas and Terradas, 2014).

Many endophytes spread systemically via the xylem to separate compartments of the plant such as stem, leaves, and fruits (Compant et al., 2010), although they can enter plant tissues through the meridional parts of the plant such as flowers and fruits (Compant et al., 2011). Depending on plant source allocation, different above-ground plant compartments host individual endophytic communities. It has been suggested that phyllosphere bacteria also derive from the soil environment and are propelled by the plant and environmental factors, although the latter having a more depth effect (Vorholt, 2012; Zarraonaindia et al., 2015; Wallace et al., 2018).

Structural analysis of phyllosphere or carposphere microbiota of the grapevine revealed that the dominant genera are *Pseudomonas*, *Sphingomonas*, *Frigoribacterium*, *Curtobacterium*, *Bacillus*, *Enterobacter*, *Acinetobacter*, *Erwinia*, *Citrobacter*, *Pantoea*, and *Methylobacterium* (Zarraonaindia et al., 2015; Kecskeméti et al., 2016). The analysis of endophytes of grape berries displayed a dominance of the genera *Ralstonia*, *Burkholderia*, *Pseudomonas*, *Staphylococcus*, *Mesorhizobium*, *Propionibacterium*, *Dyella*, and *Bacillus* (Campisano et al., 2014). Moreover, a recent work (Wallace et al., 2018) considered predominant taxa of the maize leaf microbiome across 300 diverse maize lines and found sphingomonads and methylobacteria. They also showed that the phyllosphere microbial composition was largely driven by environmental factors.

Analysis of the transcriptome of a leaf-pathogenic *Pseudomonas* strain exposed high expression of genes related to motility during epiphytic growth, whereas the expression strongly decreased when the bacteria were colonizing the apoplast. Ofek-Lalzar et al. (2014) revealed microbial adaptation to the rhizoplane by a combinatorial approach of metagenomics and metatranscriptomics. Those compared in the surrounding bulk soil, among other factors, genes, and transcripts tangled in lipopolysaccharide (LPS) biosynthesis were recognized (Ofek-Lalzar et al., 2014). Lipopolysaccharides are important for binding of bacteria to plant surface glycoproteins (Balsanelli et al., 2013). Delmotte et al. (2009) analyzed community proteogenomics, i.e., a combination of metagenome sequencing and metaproteome analysis to detect proteins high in the leaf microbiota of *A. thaliana*, clover, and soybean plants grown under environmental factors to decide the physiology of the colonizing microorganisms. Metagenome sequencing significantly increases protein identification, and proteins related to carbon metabolism and transport processes were determined to be high, revealing that they are important functional characters of the microbiota of all three plant species (Delmotte et al., 2009). Transport proteins comprised TonB-dependent receptors of different specificities, mainly assigned to *Sphingomonas*, β-barrel porins, and ABC transporters for amino acids, monosaccharides, and disaccharides. Complementary metabolomics approaches confirmed that glucose, fructose, and sucrose are obtainable on *A. thaliana* leaves, and sugars and amino acids (e.g., arginine) were diverse upon colonization by heterotrophic epiphytes (Ryffel et al., 2016). Diurnal cycles, i.e., methanol is a common substance available to leaf bacteria. Plants are produced in large quantities as a side output of pectin methylesterases during the cell wall remodeling essential for plant growth (Fall and Benson, 1996). Ubiquitously occurring methanol-consuming bacteria are mostly placed in the genus of *Methylobacterium* and gain a competitive advantage from methanol used during leaf colonization (Delmotte et al., 2009 and Sy et al., 2005). The phyllosphere is considered a tough environment with rapidly changing conditions and exposure to several stresses such as UV radiation, reactive oxygen species (ROS), and desiccation (Vorholt, 2012). Catalase and superoxide dismutase are enzymes essential for the detoxification of ROS, while pigmentation prevents and photolyase repairs UV-induced damage of nucleic acids. Extracellular polymeric substances and secretion of bioactive surfactants can increase water permeability and wettability of the plant cuticle, thereby developing epiphytic fitness during conditions of fluctuating humidity (Burch et al., 2014). Adaptation of bacteria were analyzed to the leaf surface by proteomics detected a response regulator (phyllosphere-induced regulator (PhyR)) important for epiphytic colonization of *Methylobacterium extorquens* (Gourion et al., 2006) and *Sphingomonas melonis* (Kaczmarczyk et al., 2011). This protein is a main regulator of the general stress response in Alphaproteobacteria (Francez-Charlot et al., 2015). As expected, transporters tangled in drug resistance were induced during phyllosphere colonization by *Arthrobacter*, and genes related to detoxification occurred enriched in the metagenome of the barley rhizosphere (Bulgarelli et al., 2015, Scheublin et al., 2014).

Moreover, the comparison of the metaproteomes in the phyllosphere and rhizosphere microbiota on rice plants (Knief et al., 2012) identified proteins involved in stress response, nutrient uptake, and one-carbon metabolism (Knief et al., 2012). Metagenome comparison represented the presence of nitrogenase genes both above and below ground, whereas the nitrogenase protein was indicated exclusively in the rhizosphere (Knief et al., 2012). Thus, beneath the extensive overlap of encoded functions in the microbiome of different plant parts, gene expression patterns may differ significantly and to a larger extent than one might deduce from gene enrichments.

Steven et al. (2018) identified *Pseudomonas* and Enterobacteriaceae as predominant taxa in apple flowers. Similarly, most abundant genus *Pseudomonas* were found in numerous studies on apple, almond, grapefruit, tobacco, and pumpkin flowers. Recently, seed-associated bacteria have been addressed and found to be compared to mostly Proteobacteria, Actinobacteria, Bacteroidetes, and Firmicutes (Liu et al., 2012; Barret et al., 2015; Johnston-Monje and Raizada, 2011; Rodriguez-Escobar et al., 2018).

Glassner et al. (2018) and Mitter et al. (2017) explained that seed microbiota are related to soil microbiota and also to those of flowers and fruits. Endophytes and above-ground microbiota are well known for their importance in promoting plant growth, improving disease resistance, and relieving stress tolerance (Hardoim et al., 2015).

Transcriptome, proteome, and metabolome analyses alone or combined with stable isotope probing hold great potential and together with spatially structured methods such as matrix-assisted laser desorption ionization (MALDI) imaging, nano-secondary ion mass spectrometry (SIMS), and single-cell Raman spectroscopy are a powerful tool in the rapidly improving field of microbial ecology (Eichorst et al., 2015; Hsu and Dorrestein, 2015; Huang et al., 2009; Ryffel et al., 2016).

23.7 Factors Affecting Plant Microbiota

In many plant organs, microbial composition is heft by a range of biotic and abiotic factors including soil pH, salinity, type, structure, moisture, organic matter, and exudates (Fierer, 2017), which are mostly related to root, whereas factors such as external environmental factors including climate, pathogen presence, and human practices (Hardoim et al., 2015) influence microbiota of above- and below-ground plant parts.

The plant species and genotype engage microorganisms from the soil environment where root morphology, exudates, and rhizodeposits play an important role in the recruitment of plant microbiota (Reinhold-Hurek et al., 2015; Hartmann et al., 2009; Ladygina and Hedlund, 2010; Chaparro et al., 2014). Plant species growing in the soil environment recruited significantly different microbial communities in both rhizosphere and root compartments (Hacquard, 2016; Samad et al., 2017; Aleklett et al., 2015). Using a 16S rRNA gene sequencing and shotgun metagenome approach, Bulgarelli et al. (2012) detected the root microbiota of different barley varieties and found that the host innate immune system and root metabolites mainly shaped the root microbial community structure.

Other host-related factors such as plant age and developmental stage, health, and fitness are also known to credit the plant bacterial community structure through affecting plant signaling (i.e., increased systemic resistance and systemic acquired resistance) and the composition of root exudates (Reinhold-Hurek et al., 2015; Aleklett et al., 2015). In the future, it will be exhilarating to study the impact of intensive agricultural practices on changes in the structure of plant microbiota.

23.8 Core and Satellite Microbiomes

Toju et al. (2018) explained that microorganisms that are closely associated with some plant species or genotype, independent of soil and environmental conditions are defined as the core plant microbiome. Pfeiffer et al. (2017) identified a core microbiome of potato (*Solanum tuberosum*) particularly comprising of *Bradyrhizobium*, *Sphingobium*, and *Microvirga*. Zarraonaindia et al. (2015) analyzed a grapevine core microbiome belonging to Pseudomonadaceae, Micrococcaceae, and Hyphomicrobiaceae independent of soil and climatic conditions. Edwards et al. (2015) mentioned bacteria particularly close to *Deltaproteobacteria*, *Alphaproteobacteria*, and *Actinobacteria* as a member of the rice core microbiome. The core plant microbiome is important to comprise keystone microbial taxa that are significant for plant fitness and acknowledge through evolutionary mechanisms of selection and enrichment of microbial taxa containing essential functions for the fitness of the plant holobiont.

Some microbial taxa that are found in low abundance in a less number of sites are called *satellite microbiomes* (Hanski, 1982; Magurran and Henderson, 2003). *Satellite microbiomes* can be defined based on the geographical range, local abundance, and habitat specificity (Jousset et al., 2017). The importance of satellite taxa is increasingly being comprehended as drivers of key functions for the ecosystem. Mallon et al. (2015) introduced that taxa found in low abundance are critical for reducing unwanted microbial invasions into soil communities. Low-abundance bacterial species have high contribution to the production of antifungal volatile compounds that defense the plant against soil-borne pathogens in a similar manner (Hol et al., 2015).

23.9 Functions of Plant Microbiota

The plant microbiome contains beneficial, neutral, or pathogenic microorganisms. Plant growth-promoting bacteria can increase plant growth by either direct or indirect mechanisms. Some PGPB produce phytohormones that affect plant growth through modulating endogenous hormone levels in association with a plant. Moreover, some PGPB can secrete an enzyme, 1-aminocyclopro pane-1-carboxylate (ACC) deaminase, which reduces the level of stress hormone ethylene in the plant.

Pseudomonas spp., *Arthrobacter* spp., and *Bacillus* spp. and other strains have been reported to enhance plant growth through the production of ACC deaminase. Rascovan et al. (2016) reported a diverse range of bacteria including *Pseudomonas* spp., *Paraburkholderia* spp., and *Pantoea* spp. found in wheat and soybean roots that exposed important plant growth promotion properties such as phosphate solubilization, nitrogen fixation, indole acetic acid and ACC deaminase production, and mechanisms involved in increased nutrient uptake, growth, and stress tolerance.

Some bacteria can cause disease symptoms through the production of phytotoxic compounds such as proteins and phytohormones. *Pseudomonas syringae* is an example for well-known plant pathogens having a very broad host range such as tomato, tobacco, olive, and green bean and *Erwinia amylovora* that causes fire blight disease of fruit trees and ornamentals plants. *Xanthomonas* species, *Ralstonia solanacearum*, and *Xylella fastidiosa* are also similar with many important diseases of crops like potato and banana (Mansfield et al., 2012). The severity of plant disease is based on the combination of multiple factors such as pathogen population size, host susceptibility, favorable environment, and biotic factors such as plant microbiota that collectively determine the outcome of plant pathogen interaction.

Both below- and above-ground plant-associated bacteria have been shown to increase host resistance against pathogen infection either through commensal pathogen interactions or through modulating plant defense (Rudrappa et al., 2008; De Vrieze et al., 2018). Antibiotics, lytic enzymes, pathogen-inhibiting volatile compounds, and siderophores are examples for biocontrol activities against pathogen invasion and disease (Hopkins et al., 2017; Berg et al., 2018).

Some bacteria protect the plant from pathogens through modulating plant hormone levels and inducing plant systemic resistance. The continuous use of agricultural soils can build pathogen pressure and can also produce disease-suppressive soils containing microorganisms mediating disease suppression (Santhanam et al., 2015; Duran et al., 2018). In particular, genera such as *Pseudomonas*, *Streptomyces*, *Bacillus*, *Paenibacillus*, *Enterobacter*, *Pantoea*, *Burkholderia*, and *Paraburkholderia* have been suggested for their role in pathogen suppression (Gomez Exposito et al., 2017; Schlatter et al., 2017). Trivedi et al. (2017) recently identified three bacterial taxa belonging to Acidobacteria, Actinobacteria, and Firmicutes that controlled the invasion of *Fusarium* wilt at a continental scale. Carrion et al. (2018) urged the disease-suppressive ability of *Paraburkholderia graminis* PHS1 against fungal root pathogen and combined soil suppressiveness with the synthesis of sulfurous volatile compounds such as dimethyl sulfoxide reductase and cysteine desulfurase. Endosphere bacterial community of all disease *Gaeumannomyces graminis* suppression and they identified endophytes belonging to Serratia and Enterobacter as most auspicious candidates against *Gaeumannomyces graminis* mentioned earlier Duran et al. (2018).

23.10 Application of Microbial Consortia

The microbial consortia application is an emerging study to give up lab to field balks (De Vrieze et al., 2018; Parnell et al., 2016). Moreover, for human food and health, microbial diversity is a main issue, and we should take care of plant-associated diversity and produce our food in a way that is optimal for this purpose. Biotechnological strategies can be promoted to contribute to this purpose. "Microbiome therapies" are an example of a promising method to maintain or improve plant-associated microbial diversity in combination with quality control (Gopal et al., 2013). Another example is the biocontrol agent *Bacillus amyloliquefaciens* FZB42, which was able to improve the overall plant-associated diversity (Erlacher et al., 2014). Berg et al. (2013) suggested that next-generation microbial inoculants took both the diversity and human health issues into consideration and in the future should significantly control plant diseases, generally enhance microbial diversity, and stimulate our immune system.

As earlier mentioned, plant-associated microbial communities can improve plant health and performance under many adverse conditions. Dangl et al. (2013) and Tkacz and Poole (2015) reported that the demand for a more sustainable agriculture and the exploration of these microbial functions have gained considerable interest from both academia and industry.

Several beneficial microorganisms have been commercialized (Berg, 2009); their efficacy has not always been coherent, but using a combination of strains could develop performance. A recent work reported by Wei et al. (2015) showed the ability of the plant pathogen *Ralstonia solanacearum* to invade different resident communities closely similar to *Ralstonia* strains on the basis of bacterial carbon source competition networks. Wei et al. (2015) reported that the resident communities with a clear niche overcome to the pathogen were best at reducing pathogen invasion in microcosms and in experimental plants. Meanwhile, the combination of biocontrol strains that depend on several mechanisms has been found to improve plant protection, provided that the strains were amicable (Stockwell et al., 2011) (Table 23.1).

23.11 Modern Tools to Explore PM Interactions

Basic mechanism was understanding the plant-specific microbiome, which is a good approach for its use in agriculture since

TABLE 23.1

Microbial Consortia Identified in the Host Plant

Plant	Approach	Main Findings Related to Rhizosphere Microbiome Composition	References
Oryza sativa L. in flooded plains	rDNA 16S amplicon analysis	Richest bacterial classes found in the vegetative stage were Gammaproteobacteria (276 OTUs), Alphaproteobacteria (237 OTUs), and Betaproteobacteria (211 OTUs) vegetative stage and Bacilli (46 OTUs) and Clostridia (18 OTUs) reproductive stage of rice plant.	King et al.(2009)
Oryza sativa L.	Bacterial 16S rRNA genes and the ITS2 regions of fungal rRNA	OTU were classified into the domains bacteria (99.96% of the total data set). *Acidovorax* (6.7%) and Caulobacter (5.7%) and fungi (100% of the total data set). Penicillium (30%) and *Wallemia* (7.9%).	Wang et al. (2016)
Erica andevalensis in a naturally metal-enriched and extremely acidic environment	16S rRNA gene clone library	Bacteria: of 101 sequenced clones, the majority were affiliated with the *Actinobacteria* (38 clones; 12 OTUs) followed by the *Acidobacteria* (21 clones; 10 OTUs), and *Proteobacteria* (18 clones; eight OTUs). Archaea: considering 27 clones, the community was composed by *Crenarchaeota* (21 clones; four OTUs) and *Euryarchaeota* (six clones; two OTUs)	Mirete et al. (2007)
Maize crop	*nif*H Cluster I clone library	*Azospirillum*, *Bradyrhizobium*, and *Ideonella* were abundant genera found in the rhizosphere, comprising *c*. 5%, 21% and 11% of the clones, respectively.27% of unidentified bacteria.	Roesch et al. (2007)

(*Continued*)

TABLE 23.1 (*Continued*)

Microbial Consortia Identified in the Host Plant

Plant	Approach	Main Findings Related to Rhizosphere Microbiome Composition	References
Oat microcosms	16S rRNA gene microarray	A total of 1,917 taxa were detected, and the community was predominated by *Proteobacteria* and *Firmicutes*. Less anticipated rhizosphere-competent phyla were also detected, including *Actinobacteria*, *Verrucomicrobia*, and *Nitrospira*	DeAngelis et al. (2009)
Tomato (Solanum lycopersicum)	the ITS2 regions of fungal rRNA	Fungal indicator taxa, which is *Aspergillus* (F_OTU364), *Trichoderma* (F_OTU4), and *Chrysosporium* (F_OTU31) with growth-promoting activity.	Zhang et al., (2016)
Deschampsia antarctica and *Colobanthus quitensis* in the Arctic	16S rRNA gene pyrosequencing	*Firmicutes* was the most abundant group, and *Acidobacteria* was rarely detected. The predominant genera found were *Bifidobacterium* (phylum *Actinobacteria*), *Arcobacter* (phylum *Proteobacteria*), and *Faecalibacterium* (phylum *Firmicutes*)	Teixeira et al. (2010)
Oak in a forest soil	16S rRNA gene pyrosequencing	In one of the rhizosphere samples, 5,619 OTUs were identified in the bacterial community. The predominant phyla were *Proteobacteria* (38%), *Acidobacteria* (24%), and *Actinobacteria* (11%). A large proportion of unclassified bacteria (20%) were observed	Uroz et al. (2010)
Sugar beet in agricultural soil	16S rRNA gene microarray	A total of 33,346 bacterial and archaeal OTUs were identified, and the community was dominated by *Proteobacteria* (39%), *Firmicutes* (20%), and *Actinobacteria* (9%). The *Gamma-* and *Betaproteobacteria* and *Firmicutes* were detected as the most dynamic taxa associated with disease suppression	Mendes et al. (2011)
Potato in field soil	16S rRNA gene microarray	A total of 2,432 OTUs were indicated in at least one of the samples. The highest number of OTUs belonged to the *Proteobacteria* (46%), followed by *Firmicutes* (18%), *Actinobacteria* (11%), *Bacteroidetes* (7%), and *Acidobacteria* (3%). The bacterial families *Streptomycetaceae*, *Micromonosporaceae*, and *Pseudomonadaceae* showed the strongest response at the potato cultivar level	Weinert et al. (2011)
Rhizophora mangle and *Laguncularia racemosa* in mangrove	Archaeal 16S rRNA gene pyrosequencing	About 300 archaeal OTUs were denoted. Four classes were found: *Halobacteria*, *Methanobacteria*, *Methanomicrobia*, and *Thermoprotei*	Pires et al. (2012)
Potato in field soil	Pyrosequencing	A total of 55,121 OTUs. *Actinobacteria* and *Alphaproteobacteria* were the most abundant groups, followed by *Gammaproteobacteria*, *Betaproteobacteria*, *Acidobacteria*, *Gemmatimonadetes*, *Firmicutes*, *Verrucomicrobia*, *Deltaproteobacteria*, *Cyanobacteria*, *Bacteroidetes*, and the TM7 group	Inceoglu et al. (2011)
Rhizophora mangle in mangrove	16S rRNA gene pyrosequencing	*Proteobacteria* was a large phylum in all samples covering 36%–40% of the total sequencing reads	Gomes et al. (2010)
Mammillaria carnea (cactus) in semi-arid environment	16S rRNA gene pyrosequencing	Dominant bacterial groups such as *Acidobacteria*, *Actinobacteria*, *Proteobacteria*, and *Bacteroidetes*	Torres-Cortes et al. (2012)
Arabidopsis thaliana in Cologne and Golm soils	16S rRNA gene pyrosequencing	About 1,000 OTUs were estimated in the rhizosphere and 1,000 OTUs in root compartments. The rhizosphere was dominated by *Acidobacteria*, *Proteobacteria*, *Planctomycetes*, and *Actinobacteria*. *Proteobacteria*, *Actinobacteria*, and *Bacteroidetes* were dominant phyla in root bacterial communities and significantly enriched related to soil and rhizosphere	Bulgarelli et al. (2012)
Arabidopsis thaliana in Mason farm and Clayton soils	16S rRNA gene pyrosequencing	18,783 bacterial OTUs were firstly identified, and 778 measurable OTUs were used for analysis. The rhizosphere microbiome was dominated by *Proteobacteria*, *Bacteroidetes*, *Actinobacteria*, and *Acidobacteria*. The endophytic compartment was dominated by *Actinobacteria*, *Proteobacteria*, and *Firmicutes* and was depleted of *Acidobacteria*, *Gemmatimonadetes*, and *Verrucomicrobia*	Lundberg et al. (2012)

OTUs, Operational taxonomic units.

the plant-associated microbiota greatly influences the host's phenotype, as mentioned earlier.

Research themes about PM application in agriculture deals with the use of microbial consortia, a group of species, whereas another research theme involves an exact genetic modification of either plant or microbe. The genetic modified methods, together with gene silencing, are widely used to study gene functions or trait improvement. Munoz et al. (2019) suggested that the transgenic technology is a promising approach to achieve a faster outcome, but the integration of a foreign genetic material limits its widespread use due to the regulatory effect. GE tools are of more interest that allow scientists to revamp genomic sequences in a more precise manner without the integration of a foreign gene (Knott and Doudna, 2018). Engineered endonucleases to create a double-strand break that endures DNA repair by endogenous mechanisms and generates different types of mutations were employed in genome editing technology (Gaj et al., 2013).

Brandt et al. (2019) reported that targeted genetic modifications can be accomplished through several ways, but three meganucleases or site-specific nucleases (SSNs) or site-directed nucleases (SDN) are most commonly used, i.e., transcription activator-like effector nucleases (TALENs), zinc finger nucleases (ZFNs), and CRISPR/Cas (Cas, CRISPR-associated) system. Genome editing by ZNFs and TALENs is based on the ability of DNA-binding domains that can particularly recognize almost any target DNA sequence. Therefore, the GE ability of ZNF/TALEN is mostly governed by the DNA-binding affinity and specificity of the assembled zinc-finger and TALE proteins (Gaj et al., 2013). The CRISPR/Cas system is supervened compared to ZFNs and TALENs in terms of simple designing, quality, cost-effective, higher efficiency, multiplexing, and specificity (Brandt et al., 2019). Thus, primarily CRISPR-mediated GE tools and their applications in PM studies were argued.

23.12 Clustered Regularly Interspaced Short Palindromic Repeats (CRISPR)/Cas Systems

The CRISPR/Cas is a type II bacterial immune system found in several prokaryotes including bacteria and archaea. To represent some components of CRISPR in *Escherichia coli*, but the function of these components was not known. The function and mode of action of CRISPR arrays as a programmed immune prokaryotic system against phages was characterized (Barrangou et al., 2007; Brouns et al., 2008). In this regard, the components of the CRISPR immune system were reprogrammed for CRISPR-mediated GE. The two main components of the CRISPR-based tool are the single gRNA (sgRNA) and Cas endonuclease (Jinek et al., 2012).

23.13 Conclusions and Future Challenges

In future, fundamentals of the PM interactions should be understood, and their engineering for suitable application in sustainable agriculture is the most suitable way to meet the food demand. In the past decade, research about PGP microbes and microbe-mediated plant protection was carried using only a few typical species. Therefore, in recent years, detailed molecular studies about microbe-mediated plant uses have been conducted to broaden the horizon of PM engineering for agriculture. The CRISPR/Cas technology has potential to help scientists to understand the basics of PM interactions and to develop ideal plant/microbes related to agricultural application. A higher number of plant species, more in-depth sequencing analysis of the plant microbiome, and metatranscriptomic data are peripheral to understanding the community-level molecular mechanisms under field conditions studied consequently. Identification of individual plant or microbial genes governing agronomic traits will facilitate CRISPR-based applications in sustainable agricultural practices.

REFERENCES

Abril, A.B., Torres, P.A., Bucher, E.H., 2005. The importance of phyllosphere microbial populations in nitrogen cycling in the Chaco semi-arid woodland. *Journal of Tropical Ecology* 21, 103–7.

Adesemoye, A.O., Kloepper J.W. (2009). Plant-microbes interactions in enhanced fertilizer-use efficiency. *Applied Microbiology and Biotechnology* 85, 1–12.

Aleklett, K., Leff, J.W., Fierer, N., Hart, M. (2015). Wild plant species growing closely connected in a subalpine meadow host distinct root-associated bacterial communities. *Peer Journal* 15, 3, e804.

Badri, D.V., Vivanco, J.M. (2009). Regulation and function of root exudates. *Plant Cell & Environment* 32, 666–81. Doi: 10.1111/j.1365-3040.2009.01926.x.

Bai, Y., Müller, D.B., Srinivas, G., Garrido-Oter, R., Potthoff, E., Rott, M., Hüttel, B. (2015). Functional overlap of the Arabidopsis leaf and root microbiota. *Nature* 528, 364.

Balsanelli, E, Tuleski, T.R, de Baura, V.A., Yates, M.G., Chubatsu, L.S, Pedrosa, F.D.O., Souza, E.M.D., Monteiro, R.A. (2013). Maize root lectins mediate the interaction with *Herbaspirillum seropedicae* via N-acetyl glucosamine residues of lipopolysaccharides. *PLOS One* 8, e77001.

Barrangou, R., Fremaux, C., Deveau, H., Richards, M., Boyaval, P., Moineau, S., Romero, D.A., Horvath, P. (2007). CRISPR provides acquired resistance against viruses in prokaryotes. *Science* 315, 1709–12.

Barret, M., Briand, M., Bonneau, S., Préveaux, A., Valière, S., Bouchez O, Hunault, G., Simoneau, P., Jacquesa, M.A. (2015). Emergence shapes the structure of the seed microbiota. *Applied and Environmental Microbiology* 81, 1257–66.

Berg, G. (2009). Plant-microbe interactions promoting plant growth and health: perspectives for controlled use of microorganisms in agriculture. *Applied Microbioogy and Biotechnology*. 84, 11–18.

Berg, M., Koskella, B. (2018). Nutrient- and dose-dependent microbiome-mediated protection against a plant pathogen. *Current Biology*. 28, 2487–92.

Berg, G., Rybakova, D., Grube, M., Köberl, M. (2016). The plant microbiome explored: implications for experimental botany. *The Journal of Experimental Botany* 67, 995–1002.

Berg, G., Zachow, C., Müller, H., Philipps, J., Tilcher, R. (2013) Next-generation bio-products sowing the seeds of success for sustainable agriculture. *Agronomy* 3, 648–56.

Bhattacharyya, D., Garladinne, M., Lee, Y.H. (2015). Volatile indole produced by rhizobacteriu *Proteus vulgaris* JBLS202 stimulates growth of *Arabidopsis thaliana* through auxin, cytokinin, and brassinosteroid pathways. *Journal of Plant Growth Regulation* 34, 158–68.

Bohm, M, Hurek, T, Reinhold-Hurek, B. (2007). Twitching motility is essential for endophytic rice colonization by the N2-fixing endophyte *Azoarcus* sp. strain BH72. *Molecular Plant –Microbe Interactions* 20, 526–33.

Brandt, M., Hiernaux, P., Rasmussen, K., Turker, C.J., Wigneron, J., Aziz Diouf, A., Herrmann, S.M., Zhang, W., Kergoat, L., Mbow, C., Abel, C., Auda, Y., Fensholt, R. (2019). Changes in rainfall distribution promote woody foliage production in the Sahel. *Communications Biology* 2, 133.

Brouns, S.J.J., Jore, M.M., Lundgren, M., Westra, E.R., Slijkhuis, R.J.H., Snijders, A.P.L., Dickman, M.J., Makarova, K.S., Koonin, E.V., van der Oost, J. (2008). Small CRISPR RNAs guide antiviral defense in prokaryotes. *Science*. 321, 960–64.

Bulgarelli, D., Garrido-Oter, R., Münch, P.C, Weiman, A., Dröge, J., Pan, Y., McHardy, A.C., Schulze-Lefert, P. (2015). Structure and function of the bacterial root microbiota in wild and domesticated barley. *Cell Host Microbe*, 17, 392–403.

Bulgarelli, D., Rott, M., Schlaeppi, K., Ver Loren van Themaat, E., Ahmadinejad, N., Assenza, F., Rauf, P., Huettel, B, Reinhardt, R., Schmelzer, E., Peplies, J., Gloeckner, F.O., Amann, R., Eickhorst, T., Schulze-Lefert, P. (2012). Revealing structure and assembly cues for *Arabidopsis* root-inhabiting bacterial microbiota. *Nature* 488, 91–5.

Bulgarelli, D., Schlaeppi, K., Spaepen, S., Ver Loren van Themaat, E., Schulze-Lefert, P. (2013). Structure and functions of the bacterial microbiota of plants. *Annual Review of Plant Biology* 64, 807–38.

Burch, A.Y., Zeisler, V., Yokota, K., Schreiber, L., Lindow, S.E. (2014). The hygroscopic biosurfactant syringafactin produced by *Pseudomonas syringae* enhances fitness on leaf surfaces during fluctuating humidity. *Environmental Microbiology* 16, 2086–98.

Calvo, P., Watts, D.B., Kloepper, J.W., Torbert, H.A. (2017). Effect of microbial-based inoculants on nutrient concentrations and early root morphology of corn (Zea mays). *Journal of Plant Nutrition* and *Soil Science* 180, 56–70.

Campisano, A., Antonielli, L., Pancher, M., Yousaf, S., Pindo, M., Pertot, I. (2014). Bacterial endophytic communities in the grapevine depend on pest management. *PLOS One* 9, e112763.

Carrion, V.J., Cordovez, V., Tyc, O., Etalo, D.W., Bruijn, I.D., Jager, C.L., de Jager Victor, V.C., Medema, H.M., Eberl, L. Raaijmakers, M.J. (2018). Involvement of Burkholderiaceae and sulfurous volatiles in disease suppressive soils. *ISME Journal* 12, 2307–21.

Chaparro, J.M., Badri, D.V., Vivanco, J.M. (2014). Rhizosphere microbiome assemblage is affected by plant development. *ISME Journal* 8, 790–803.

Chaparro, J.M., Sheflin, A.M., Manter, D.K., Vivanco, J.M. (2012). Manipulating the soil microbiome to increase soil health and plant fertility. *Biology and Fertility of Soils* 48, 489–99.

Compant, S., Clément, C., Sessitsch, A., 2010. Plant growth-promoting bacteria in the rhizo-and endosphere of plants: their role, colonization, mechanisms involved and prospects for utilization. *Soil Biology and Biochemistry* 42, 669–78.

Compant, S., Mitter, B., Colli-Mull, J.G., Gangl, H., Sessitsch, A. (2011). Endophytes of grapevine flowers, berries, and seeds: identification of cultivable bacteria, comparison with other plant parts, and visualization of niches of colonization. *Microbial Ecology* 62, 188–97.

Dangl, J.L., Horvath, D.M., Staskawicz, B.J. (2013). Pivoting the plant immune system from dissection to deployment. *Science* 341, 746–51.

DeAngelis, K.M., Brodie, E.L., DeSantis, T.Z., Andersen, G.L., Lindow, S.E., Firestone, M.K. (2009). Selective progressive response of soil microbial community to wild oat roots. *ISME Journal* 3, 168–178.

Delmotte, N., Knief, C., Chaffron, S., Innerebner, G., Roschitzki, B., Schlapbach, R., Von Mering, C., Vorholt, J.A. (2009). Community proteogenomics reveals insights into the physiology of phyllosphere bacteria. *PNAS* 106, 16428–33.

De Vrieze, M., Germanier, F., Vuille, N., Weisskopf, L. (2018). Combining different potato associated *Pseudomonas* strains for improved biocontrol of *Phytophthora infestans*. *Frontiers in Microbiology* 9, 2573.

Donn, S., Kirkegaard, J.A., Perera, G., Richardson, A.E., Watt, M. (2015). Evolution of bacterial communities in the wheat crop rhizosphere. *Environmental Microbiology* 17, 610–21.

Duran, P., Tortella, G., Viscardi, S., Barra, P.J., Carrion, V.J., Mora, M.D.L.L., Pozo, M.J. (2018). Microbial community composition in take-all suppressive soils. *Frontiers in Microbiology* 9, 2198.

Edwards, J., Johnson, C., Santos-Medellín, C., Lurie, E., Podishetty, N.K., Bhatnagar, S., Eisen, J.A., Sundaresan, V. (2015). Structure, variation, and assembly of the root-associated microbiomes of rice. *Proceedings of the National Academy of Sciences* 112, E911–20.

Eichorst, S.A., Strasser, F., Woyke, T., Schintlmeister, A., Wagner, M., Woebken, D. (2015). Advancements in the application of NanoSIMS and Raman microspectroscopy to investigate the activity of microbial cells in soils. *FEMS Microbiology Ecology* 91, fiv 106.

Erlacher, A., Cardinale, M., Grosch, R., Grube, M., Berg, G. (2014). The impact of the pathogen *Rhizoctonia solani* and its beneficial counterpart *Bacillus amyloliquefaciens* on the indigenous lettuce microbiome. *Frontiers in Microbiology* 5, 175. Doi: 10.3389/fmicb.00175.

Faist, H., Keller, A., Hentschel, U., Deeken, R. (2016). Grapevine (*Vitis vinifera*) crown galls host distinct microbiota. *Applied and Environmental Microbiology* 82, 5542–52.

Fall, R., Benson, A.A. (1996). Leaf methanol–the simplest natural product from plants. *Trends in Plant Science* 1, 296–301

Fierer, N. (2017). Embracing the unknown–disentangling the complexities of the soil microbiome. *Nature Reviews Microbiology* 15, 579–90.

Francez-Charlot, A., Kaczmarczyk, A., Fischer, H.M., Vorholt, J.A. (2015). The general stress response in Alphaproteobacteria. *Trends in Microbiology* 23, 164–71.

Gaj, T., Gersbach, C.A., Barbas, C.F. (2013) ZFN, TALEN, and CRISPR/Cas-based methods for genome engineering. *Trends in Biotechnology* 31(7), 397–405. Doi: 10.1016/j.tibtech.2013.04.004.

Glassner, H., Zchori-Fein, E., Yaron, S., Sessitsch, A., Sauer, U., Compant, S. (2018). Bacterial niches inside seeds of *Cucumis melo* L. *Plant Soil* 422, 101–13.

Glick, B.R. (1995). The enhancement of plant growth by free-living bacteria. *Canadian Journal of Microbiology* 41, 109–17.

Glick, B.R. (2012). Plant growth-promoting bacteria: mechanisms and applications. *Scientifica Article*. 963401.

Gomes, C.M., Cleary, D.F., Pinto, F.N., Egas, C., Almeida, A., Cunha, A., Mendonça-Hagler, L.C., Smalla, K. (2010). Taking root: enduring effect of rhizosphere bacterial colonization in Mangroves. Doi: 10.1371/journal.pone.0014065.

Gomez Exposito, R., Bruijn, I., de, Postma, J., Raaijmakers, J.M. (2017). Current insights into the role of rhizosphere bacteria in disease suppressive soils. *Frontiers in Microbiology* 8, 2529.

Gopal, M., Gupta, A., Thomas, G.V. (2013). Bespoke microbiome therapy to manage plant diseases. *Frontiers in Microbiology* 4, 355.

Gourion, B., Rossignol, M., Vorholt, J.A. (2006). A proteomic study of *Methylobacterium extorquens* reveals a response regulator essential for epiphytic growth. *PNAS* 103, 13186–91.

Gutierrez-Luna, F.M., Lopez-Bucio, J., Altamirano-Hernandez, J., Valencia-Cantero, E., de la Cruz, H., Macias-Rodríguez, L. (2010). Plant growth-promoting rhizobacteria modulate root-system architecture i: *Arabidopsis thaliana* through volatile organic compound emission. *Symbiosis* 51, 75–83.

Hacquard, S., Kracher, B., Hiruma, K., Munch, P.C., Garrido-Oter, R., Thon, M.R., Weimann, A., Damm, U., Dallery, J.F., Hainaut, M., Henrissat, B., Lespinet, O., Sacristan, S., Ver Loren van Themaat, E., Kemen, E., McHardy, A.C., Schulze-Lefert, P., O'Connell, R.J. (2016). Survival trade-offs in plant roots during colonization by closely related beneficial and pathogenic fungi. *Nature Communications* 7, 11362.

Handelsman, J., Rondon, M.R., Brady, S.F., Clardy, J., Goodman, R.M. (1998). Molecular biological access to the chemistry of unknown soil microbes: a new frontier for natural products. *Chemistry and Biology* 5, R245–49.

Haney, C.H., Ausubel, F.M. (2015). Plant microbiome blueprints. *Science* 349, 788–9.

Haney, C.H., Samuel, B.S., Bush, J., Ausubel, F.M. (2015). Associations with rhizosphere bacteria can confer an adaptive advantage to plants. *Nature Plants* 1, 15051.

Hanski, I. (1982). Dynamics of regional distribution: the core and satellite species hypothesis. *Oikos* 38, 210–21.

Hardoim, P.R., Hardoim, C.C., van Overbeek, L.S., van Elsas, J.D. (2012). Dynamics of seedborne rice endophytes on early plant growth stages. *PLOS One* 7, e30438.

Hardoim, P.R., van Overbeek, L.S., Berg, G., Pirttila, A.M., Compant, S., Campisano, A., Doring, M., Sessitsch, A. (2015). The hidden world within plants: ecological and evolutionary considerations for defining functioning of microbial endophytes. *Microbiology and Molecular Biology Reviews* 79, 293–320.

Hartmann, A., Rothballer, M., Schmid, M. (2008). Lorenz Hiltner, a pioneer in rhizosphere microbial ecology and soil bacteriology research. *Plant Soil* 312, 7–14.

Hartmann, A., Schmid, M., van Tuinen, D., Berg, G. (2009). Plant-driven selection of microbes. *Plant Soil* 321, 235–75.

Hernandez-Leon, R., Rojas-Solis, D., Contreras-Perez, M., del C. Orozco-Mosqueda, M., Macias-Rodríguez, L., Reyes-delaCruz, H., Valencia-Cantero, E., Santoyo, G. (2015). Characterization of the antifungal and plant growth-promoting effects of diffusible and volatile organic compounds produced by *Pseudomonas fluorescens* strains. *Biological Control* 81, 83–92.

Hol, W.H.G., Garbeva, P., Hordijk, C., Hundscheid, M.P.J., Gunnewiek, P.J.A.K., van Agtmaal, M., Kuramae, E.E., de Boer, W. (2015). Non-random species loss in bacterial communities reduces antifungal volatile production. *Ecology* 96, 2042–8.

Hopkins, S.R., Wojdak, J.M., Belden, L.K. (2017). Defensive symbionts mediate host– parasite interactions at multiple scales. *Trends* in *Parasitology* 33, 53–64.

Hsu, C.C., Dorrestein, P.C. (2015). Visualizing life with ambient mass spectrometry. *Current Opinion in Biotechnology* 31, 24–34.

Hu, L., Robert, C.A.M., Cadot, S., Zhang, X., Ye M., Li, B., Manzo, D., Chervet, N., Steinger, T, Van der Heiiden, M.G.A., Schlaeppi, K., Erb, M.(2018). Root exudate metabolites drive plant-soil feedbacks on growth and defense by shaping the rhizosphere microbiota. *Nature Communications* 9, 2738.

Huang, W.E., Ferguson, A., Singer, A.C., Lawson, K., Thompson, I.P, Kalin, R.M., Larkin, M.J., Bailey, M.J., Whiteley, A.S. (2009). Resolving genetic functions within microbial populations: in situ analyses using rRNA and mRNA stable isotope probing coupled with single-cell Raman-fluorescence in situ hybridization. *Applied* and *Environmental Microbiology* 75, 234–41.

Inceoglu, O.L., Al-Soud, W.A., Salles, J.F., Alexander, V.S., and Elsas, J.D.V. (2011). Comparative analysis of bacterial communities in a potato field as determined by pyrosequencing. *PLOS One* 6, e23321. Doi: 10.1371/journal.pone. 0023321.

Jinek, M., Chylinski, K., Fonfara, I., Hauer, M., Doudna, J.A., Charpentier, E. (2012). A programmable dual-RNA–guided DNA endonuclease in adaptive bacterial immunity. *Science* 337, 816–821.

Johnston-Monje, D., Raizada, M.N. (2011). Conservation and diversity of seed associated endophytes in Zea across boundaries of evolution, ethnography and ecology. *PLOS One* 6, e20396.

Jousset, A., Bienhold, C., Chatzinotas, A., Gallien, L., Gobet, A., Kurm, V., Kusel, K., Rillig, M.C., Rivett, D.W., Salles, J.F., Van der Heiiden, M.G., Youssef, N.H., Zhang, X., Hol, W.H. (2017). Where less may be more: how the rare biosphere pulls ecosystems strings. *ISME Journal* 11, 853–62.

Kaczmarczyk, A., Campagne, S., Danza, F., Metzger, L.C., Vorholt, J.A., Francez-Charlot, A. (2011). Role of *Sphingomonas* sp. strain Fr1 PhyR-NepR-σEcfG cascade in general stress response and identification of a negative regulator of PhyR. *Journal of Bacteriology* 193, 6629–38.

Kawasaki, A., Donn, S., Ryan, P.R., Mathesius, U., Devilla, R., Jones A, Watt. M. (2016). Microbiome and exudates of the root and rhizosphere of *Brachypodium distachyon*, a model for wheat. *PLOS One* 11, e0164533.

Kecskeméti, E., Berkelmann-Löhnertz, B., Reineke, A., Cantu, D. (2016). Are epiphytic microbial communities in the carposphere of ripening grape clusters (*Vitis vinifera* L.) different between conventional, organic, and biodynamic grapes? *PLOS One* 11, e0160852.

Knief, C., Delmotte, N., Chaffron, S., Stark, M., Innerebner, G, Wassmann, R., von Mering, C., Vorholt, J.A. (2012). Metaproteogenomic analysis of microbial communities in the phyllosphere and rhizosphere of rice. *ISME Journal* 6:1378–90.

Kniskern, J.M., Traw, M.B., Bergelson, J. (2007). Salicylic acid and jasmonic acid signaling defense pathways reduce natural bacterial diversity on *Arabidopsis thaliana*. *Molecular Plant-Microbe Interactions* 20, 1512–22.

Knott, G.J., Doudna, J.A. (2018). CRISPR-Cas guides the future of genetic engineering. *Science* 361, 866–869.

Ladygina, N., Hedlund, K. (2010). Plant species influence microbial diversity and carbon allocation in the rhizosphere. *Soil Biology* & *Biochemistry* 42, 162–8.

Leff, J.W., Fierer, N. (2013). Bacterial communities associated with the surfaces of fresh fruits and vegetables. *PLOS One* 8, e59310. Doi: 10.1371/journal.pone.0059310.

Lindow, S.E., Brandl, M.T. (2003). Microbiology of the phyllosphere. *Applied* and *Environmental Microbiology* 69, 1875–1883.

Liu Y, Zuo S, Xu L, Zou Y, Song W. (2012). Study on diversity of endophytic bacterial communities in seeds of hybrid maize and their parental lines. *Archives of Microbiology*. 194, 1001–12.

Lundberg, D.S., Lebeis, S.L., Paredes SH, Yourstone S, Gehring J, Malfatti S, Tremblay J, Engelbrektson, A., Kunin, V., Del Rio., T.G., Eickhorst, T., Ley, R.E., Hygenholtz, P., Tringe, S.G., Dangl, J.L. (2012). Defining the core Arabidopsis thaliana root microbiome. *Nature* 488, 86–90 Doi: 10.1038/nature11237.

Magurran, A.E., Henderson, P.A. (2003). Explaining the excess of rare species in natural species abundance distributions. *Nature* 422, 714–6.

Mallon, C.A., Poly, F., Le Roux, X., Marring, I., van Elsas, J.D., Salles, J.F. (2015). Resource pulses can alleviate the biodiversity–invasion relationship in soil microbial communities. *Ecology* 6, 915–26.

Mansfield, J., Genin, S., Magori, S., Citovsky, V., Sriariyanum, M., Ronald, P, Dow, M., Verdier, V., Beer, S.V., Machado, M.A., Toth, I., Salmond, G., Foster, G.D. (2012). Top 10 plant pathogenic bacteria in molecular plant pathology. *Molecular Plant Pathology* 13, 614–29.

Marchesi, J.R and Ravel, J. (2015). The vocabulary of microbiome research: a proposal. *Microbiome* 30(3), 31.

Mendes, R., Kruijt, M., de Bruijn, I., Dekkers, E., van der Voort, M., Schneider, J.H., Piceno, Y.M., Desantis, T.Z., Andersen, G.L., Bakker, P.A., Raajimakers, J.M. (2011). Deciphering the rhizosphere microbiome for disease-suppressive bacteria. *Science* 332, 1097–100.

Mitter, B.N., Pfaffenbichler, R., Flavell, S., Compant, L., Antonielli, A., Petric, A., Sessitsch, A. (2017). A new approach to modify plant microbiomes and traits by introducing beneficial bacteria at flowering into progeny seeds. *Frontiers in Microbiology* 8, 11. Doi: 10.3389/fmicb.2017.00011.

Mueller, U.G., Sachs, J.L. (2015). Engineering microbiomes to improve plant and animal health. *Trends in Microbiology* 23, 606–17.

Munoz, I.V., Sarrocco, S., Malfatti, L., Baroncelli, R., Vannacci, G. (2019). CRISPR-Cas for fungal genome editing: a new tool for the management of plant diseases. *Frontiers in Plant Science* 10, 1–5.

Ofek-Lalzar, M., Sela, N., Goldman-Voronov, M., Green, S.J., Hadar, Y., Minz, D. (2014). Niche and host associated functional signatures of the root surface microbiome. *Nature Communications* 5, 4950.

Parnell, J.J., Berka, R., Young, H.A., Sturino, J.M., Kang, Y., Barnhart, D.M., DiLeo, M.V. (2016). From the lab to the farm: an industrial perspective of plant beneficial microorganisms. *Frontiers in Plant Science* 7, 1110.

Peiffer, J.A., Spor, A., Koren, O., Jin, Z., Tringe, S.G., Dangl, J.L., Ley, R.E. (2013). Diversity and heritability of the maize rhizosphere microbiome under field conditions. *Proceedings of the National Academy of Sciences. U. S. A.* 110, 6548–6553.

Penuelas, J., Terradas, J. (2014). The foliar microbiome. *Trends in Plant Science* 19, 278–80.

Perez-Flores, P., Valencia-Cantero, E., Altamirano-Hernandez, J., Pelagio-Flores, R., Lopez-Bucio, J., Garcia-Juárez, P., Macias-Rodriguez, L. (2017). *Bacillus methylotrophicus* M4–96 isolated from maize (*Zea mays*) rhizoplane increases growth and auxin content in *Arabidopsis thaliana* via emission of volatiles. *Protoplasma* 254(6), 2201–2213. Doi: 10.1007/s00709-017-1109-9.

Pfeiffer, S., Mitter, B., Oswald, A., Schloter-Hai, B., Schloter, M., Declerck, S., Sessitsch, A. (2017). Rhizosphere microbiomes of potato cultivated in the High Andes show stable and dynamic core microbiomes with different responses to plant development. *FEMS Microbiology Ecology* 93(2), 1–12. https://doi.org/10.1093/ femsec/fiw242.

Philpott, C.C. (2006). Iron uptake in fungi: a system for every source. *Biochimica ET Biophysica Acta (bba) - Molecular Cell Research* 1763, 636–45.

Pires, A.C.C., Cleary, D.F.R., Almeida, A., Cunha, A., Dealtry, S., Mendonca-Hagler, L.C., Smalla, K., Gomes, N.C.M. (2012) Denaturing gradient gel electrophoresis and barcoded pyrosequencing reveal unprecedented archaeal diversity in mangrove sediment and rhizosphere samples. *Applied and Environmental Microbiology* 78, 5520–5528.

Podolsky, S.H. (2017). Historical perspective on the rise and fall and rise of antibiotics and human weight gain. *Annals of Internal Medicine* 166 (2), 133–1.

Rascovan, N., Carbonetto, B., Perrig, D., Diaz, M., Canciani, W., Abalo M., Alloati, J., Gonzalez-Anta, G., Vazquez, M.P. (2016). Integrated analysis of root microbiomes of soybean and wheat from agricultural fields. *Scientific Reports* 6, 28084.

Reinhold-Hurek, B., Bunger, W., Burbano, C.S., Sabale, M., Hurek, T. (2015). Roots shaping their microbiome: global hotspots for microbial activity. *The Annual Review of Phytopathology* 53, 403–24.

Rodriguez-Escobar, C., Mitter, B., Barret, M., Sessitsch, A., Compant, S. (2018). Commentary: seed bacterial inhabitants and their routes of colonization. *Plant Soil* 422, 129–34.

Roesch, L.F.W., Fulthorpe, R.R., Riva, A., Casella, G., Hadwin, A.K.M., Kent, A.D., Daroub, S.H., Camargo, F.A.O., Farmerie, W.G., Triplett, E.W. (2007). Pyrosequencing enumerates and contrasts soil microbial diversity. *ISME Journal* 1, 283–290.

Rudrappa, T., Czymmek, K.J., Pare, P.W., Bais, H.P. (2008). Root-secreted malic acid recruits beneficial soil bacteria. *Plant Physiology* 148, 1547–1556.

Ryan, R.P., Germaine, K., Franks, A., Ryan, D.J., Dowling, D.N. (2008). Bacterial endophytes: recent developments and applications. *FEMS Microbiology Letter* 278, 1–9.

Ryffel, F., Helfrich, E.J., Kiefer, P., Peyriga, L., Portais, J.C., Piel, J., Vorholt, J.A. (2016). Metabolic footprint of epiphytic bacteria on *Arabidopsis thaliana* leaves. *ISME Journal* 10, 632–43.

Samad, A., Trognitz, F., Compant, S., Antonielli, L., Sessitsch, A. (2017). Shared and hostspecific microbiome diversity and functioning of grapevine and accompanying weed plants. *Environmental Microbiology* 19, 1407–24.

Santhanam, R., van Luu, T., Weinhold, A., Goldberg, J., Oh, Y., Baldwin, I.T. (2015). Native root-associated bacteria rescue a plant from a sudden-wilt disease that emerged during continuous cropping. *Proceedings of the National Academy of Sciences USA* 112, E5013–20.

Santoyo, G., Moreno-Hagelsieb, G., Del Carmen Orozco-Mosqueda, M., Glick, B.R. (2016). Plant growth-promoting bacterial endophytes. *Microbiology Research* 183, 92–99.

Scheublin, T.R., Deusch, S., Moreno-Forero, S.K., Muller, J.A., van der Meer, J.R., Leveau, J.H.J. (2014). Transcriptional profiling of Gram-positive *Arthrobacter* in the phyllosphere: induction of pollutant degradation genes by natural plant phenolic compounds. *Environmental Microbiology* 16, 2212–25.

Schlatter, D., Kinkel, L., Thomashow, L., Weller, D., Paulitz, T. (2017). Disease suppressive soils: new insights from the soil microbiome. *Phytopathology* 107, 1284–97.

Sheibani-Tezerji, R., Rattei, T., Sessitsch, A., Trognitz, F., Mitter, B. (2015). Transcriptome profiling of the endophyte *Burkholderia phytofirmans* PsJN indicates sensing of the plant environment and drought stress. *mBio* 6, e00621–15.

Steven, B., Huntley, R.B., Zeng, Q. (2018).The influence of flower anatomy and apple cultivar on the apple flower phytobiome. *Phytobiomes Journal* 2, 171–9.

Stockwell, V.O., Johnson, K.B., Sugar, D., Loper, J.E. (2011). Mechanistically compatible mixtures of bacterial antagonists improve biological control of fire blight of pear. *Phytopathology* 101, 113–23.

Sy, A., Timmers, A.C., Knief, C., Vorholt, J.A. (2005). Methylotrophic metabolism is advantageous for *Methylobacterium extorquens* during colonization of *Medicago truncatula* under competitive conditions. *Applied and Environmental Microbiology* 71, 7245–52.

Teixeira, L.C.R.S., Peixoto, R.S., Cury, J.C., Sul, W.J., Pellizari, V.H., Tiedje, J., Rosado, A.S. (2010). Bacterial diversity in rhizosphere soil from Antarctic vascular plants of Admiralty Bay, maritime Antarctica. *ISME Journal* 4, 989–1001.

Tkacz, A., Poole, P. (2015). Role of root microbiota in plant productivity. *Journal of Experimental Botany* 66, 2167–75.

Toju, H., Peay, K.G., Yamamichi, M., Narisawa, K., Hiruma, K., Naito, K., Fukuda, S., Ushio, M., Nakaoka, S., Onoda, Y., Yoshida, K., Schlaeppi, K., Bai, Y., Sugiura, R., Ichihashi, Y., Minamisawa, K., Kiers, E.T. (2018). Core microbiomes for sustainable agroecosystems. *Nature Plants* 4, 247–57.

Torres-Cortes, G., Millan, V., Fernandez-Gonzalez, A.J., Fernandez-Lopez, M., Toro, N., Martinez-Abarca, F. (2012). Bacterial community in the rhizosphere of the cactus species *Mammillaria carnea* during dry and rainy seasons assessed by deep sequencing. *Plant Soil* 357, 275–288.

Trivedi, P., Delgado-Baquerizo, M., Trivedi, C., Hamonts, K., Anderson, I.C., Singh, B.K. (2017). Keystone microbial taxa regulate the invasion of a fungal pathogen in agroecosystems. *Soil Biology & Biochemistry* 111, 10–4.

Uroz, S., Buee, M., Murat, C., Frey-Klett, P., Martin, F. (2010). Pyrosequencing reveals a contrasted bacterial diversity between oak rhizosphere and surrounding soil. *Environmental Microbiology Reports* 2, 281–288.

Vorholt, J.A. (2012). Microbial life in the phyllosphere. *Nature Reviews Microbiology* 10, 828–840. Doi: 10.1038/nrmicro2910.

Wallace, J., Kremling, K.A., Kovar, L.L., Buckler, E.S. (2018). Quantitative genetics of the maize leaf microbiome. *Phytobiomes Journal*. Doi: 10.1094/PBIOMES-02-18-0008-R.

Wang, W., Zhai, Y., Cao, L., Zhang, R. (2016). Endophytic bacterial and fungal microbiota in sprouts, roots and stems of rice (*Oryza sativa* L.) *Microbiological Research* 188–189, 1–8.

Wei, Z., Yang, T., Friman, V.P, Xu, Y., Shen, Q., Jousset, A. (2015). Trophic network architecture of rootassociated bacterial communities determines pathogen invasion and plant health. *Nature Communications* 6, 8413.

Weinert, N., Piceno, Y., Ding, G.C., Meincke, R., Heuer, H., Berg, G., Schloter, M., Andersen, G., Smalla, K. (2011). PhyloChip hybridization uncovered an enormous bacterial diversity in the rhizosphere of different potato cultivars: many common and few cultivar-dependent taxa. *FEMS Microbiology Ecology* 75, 497–506.

Zarraonaindia, I., Owens, S.M., Weisenhorn, P., West, K., Hampton-Marcell, J., Lax, S., Bokulich, N.A., Mills, D.A., Martin, G., Taghavi, S., Van der Lelie, D., Gilbert, J.A. (2015). The soil microbiome influences grapevine-associated microbiota. *MBio Journal* 6(2), e02527–e2614.

Zhang S, Gan Y, Xu B. (2016). Application of plant-growth-promoting fungi *Trichoderma longibrachiatum* T6 enhances tolerance of wheat to salt stress through improvement of antioxidative defense system and gene expression. *Frontiers in Plant Science* 7, 1405.

Section V

Environmental Microbiome

24

Microbiome of Speleothems – Secondary Mineral Deposits

D. Mudgil
University of Delhi

CONTENTS

24.1 Introduction

Earth's subsurface has abundant active microbial biomass (Whitman et al., 1998). The biomass value is likely to be more than the value predicted for Earth's other environments (McMahon and Parnell, 2014). Life has been microscopic for a long time in the geological history of Earth, and microbes were the sole forms of life on Earth for about two billion years (Schopf and Walter, 1983). Microorganisms can be found in almost every environment on the planet's surface. Microorganism's metabolic activities have influenced shaping the global environment that we live including atmospheric oxygenation (Madigan et al., 2000). Caves are karstic landscapes, which represent unique subsurface ecosystems that are comparatively unexplored due to their location. Perpetual darkness, difficulty in access, flash floods, slippery surfaces, and loose rocks are some of the challenges that make these ecosystems hard to be accessible. In caves, there are three zones: entrance zone, twilight zone, and dark zone. The absence of sunlight inside the cave excludes the photosynthesis process. It is only at cave openings or entrances where photosynthesis occurs at low levels. Most cave ecosystems are heterotrophic and depend upon organic materials that fall in through cave openings, are carried by water, or are deposited by cave animals that travel to the surface.

24.2 Speleothems: Origin and Types

Speleothems are secondary mineral formations found in limestone caves, where stalagmites and stalactites, draperies, or flowstones are the most common type of speleothems. These are formed in caves by flowing, dripping, ponded, or seeping water. Speleothems are mainly composed of minerals such as calcite, aragonite, and gypsum. Some other minerals also have been found in speleothems in minor amounts. Speleothems are primarily composed of calcium carbonate, precipitated from groundwater that has percolated through the neighboring carbonate host rock. Water dynamics and the crystal growth behavior of the constituent minerals compete to dictate the shapes of speleothems (stalactites, stalagmites, flowstone, and other speleothems). Helictites, anthodites, gypsum flowers formed from seeping water, and various pool deposits take shapes dictated by the habit of crystal growth. Calcite speleothems are tan, orange, and brown in color, and their luminescence under ultraviolet light is attributed to the addition of humic and fulvic acid from overlying soils. The presence of certain trace elements gives a characteristic hue to some deposits.

Speleothems are deposited as a result of evaporation of water near cave entrances, while degassing of CO_2 from water droplets is mainly responsible for most speleothems from deep within caves. Water percolated through soil comes into contact with decaying organic matter usually, or water passes through limestone rock, dissolves away calcite, and carries it in solution. This enhances the partial pressure of CO_2 exceeding that of the cave atmosphere. Consequently, when this supersaturated water impacts the cave floor, degassing of carbon dioxide (evaporation) makes it supersaturated with calcite, which gets precipitated in formations called speleothems. (McDermott et al., 2006; Fairchild and Baker, 2012). These speleothems are of several types and are deposited on the roof, floors, and side walls in a cave. Speleothems are formed by a physicochemical reaction from primary minerals in a cave (Moore, 1952; Provencio and Polyak, 2001). *Stalagmites* grow upward from the floor of the cave and water dripping from cave roof formations generally feeds them. They are convex-shaped, and their size ranges from decimeter to meters. *Stalactites* hang from the cave ceiling and grow toward the cave floor. They are centimeters to meters in length and form as water flows down the formation and evaporates or with degassing of CO_2 leaving layers of calcite. Stalactites and stalagmites are the most common speleothem types. Dripping water control both of these speleothems, so they are considered as gravitational forms. When a stalactite and a stalagmite grow, they join to form a *column*. Generally, all ceiling formations are stalactites, but because there are so many distinctive types, there are specific terms designated to them. *Flowstones* are sheet-like deposits and found on cave floors and walls. *Moonmilk* is white, pasty, and cottage cheese-like deposits. *Draperies* are formed when calcite is deposited in thin sheets that hang in delicate folds like a drape. Very thin, long, and hollow stalactites known as *soda straws* are captivating. They grow as water runs down inside them having an elongated cylindrical shape (Figure 24.1). *Helictites* are twisted speleothems having a central canal with many spiral or branch-like projections. They project at all angles from ceiling walls, and the floor of caves seems to defy gravity. There are so many names given to these deposits based on their structure such as dogtooth spar, cave popcorn or coralloids, cave pearls, and snottites.

24.3 History of Speleothems' Microbiome

Cave ecosystems, despite shortage of nutrient and energy limitations due to permanent darkness, show a rich level of microbial biodiversity. Microorganisms are also brought with surface debris by streams feeding the caves. Cave entrance, sinkholes, underground hydrology, and drip waters serve as entry points for limited energy and nutrients in caves (Barton and Jurado, 2007). Microbes can also enter caves by gravity, air currents, or anthropogenic activities.

Earlier it was assumed that most of the microbes in caves are translocated soil heterotrophs or coliform bacteria brought into caves by surface water and detritus. Bacterial richness in cave sediments could be comparable to that in overlying soils (Ortiz et al., 2013), but the rock surfaces are normally populated by the lowest diversity natural microbial communities (Macalady et al., 2007; Yang et al., 2011). In the cave ecosystem, having an allochthonous energy source, microbes break down complex organic matter and serve as a food source for higher-level organisms. On the other hand, caves having little to zero allochthonous inputs harbor chemolithotrophic bacteria. A growing body of evidence suggests that chemolithoautotrophs may sustain cave ecosystems by ruling primary production and driving biogeochemical cycles (Sarbu et al., 1996; Desai et al., 2013; Ortiz et al., 2013). Chemolithoautotrophs' energy sources include dissolved reduced compounds such as CO_2 or H_2S present in groundwater. Oligotrophic conditions limit a single microorganism from performing all reactions essential for growth, which in turn promotes mutualistic associations (Barton and Jurado, 2007). The microbes in caves trail various pathways for their energy requirements including the formation of biofilms through mutualism (Tomczyk-Żak and Zielenkiewicz, 2016). The species composition of a cave is influenced by variation in the physicochemical factors, such as pH, nutrient availability, sunlight, moisture, sulfur, and other metal compounds. Today, almost every possible environment having even remote possibilities of presence of life has been explored for the occurrence of microorganisms, including cave environments (Barton and Luiszer, 2005; Cañaveras et al., 2006; Cañaveras et al., 2001; Chelius and Moore, 2004; Engel et al., 2004; Groth et al., 1999; Holmes et al., 2001; Northup and Lavoie, 2001).

Historically, microbial activity in geological environments was overlooked due to the incompetence of culturing these microorganisms from such environments (Amann et al., 1995). Cave biology got attention during the sixteenth century (Culver and Pipan, 2009). The first biological study in caves was started in the middle ages in Europe and China. Later, a few experts were involved in this field (Romero, 2009; Engel, 2010). In the mid-twentieth century, studies regarding the microbial life existing in cave ecosystems attracted scientific attention. In the late 1940s, microscopy and enrichment techniques were used, and many cave water and sediment studies showed the presence of bacteria and fungi (Caumartin, 1963). The microbial involvement in cave formations, such as deposits of saltpeter (Faust, 1949) and carbonate speleothems (Went, 1969), was suggested. Nevertheless, initial studies presumed the cave microorganisms as trans-located soil and groups of microbes. They were thought to be transported through air currents, drip waters, surface streams, and animals from the surface into caves. This made many researchers to develop the notion that the cave microbiological studies were of little scientific significance (Engel, 2010). The examination of deep-sea hydrothermal vent microbial systems (Jannasch and Mottl, 1985) and the Movile Cave, driven by chemoautotrophy (Sarbu et al., 1996), emphasized the captivating nature of severe dark environments. These findings marked the beginning of a new era of cave microbiology (Lee et al., 2012). The investigators were able to study the geologic environments by various molecular biology techniques independent of traditional microbial cultivation methods (Pace, 1997; Newman and Banfield, 2002). As a result, the collaborations of microbiologists and geologists were strengthened. This scientific union, known as geomicrobiology, includes geology and biology as two major branches. Today, this union includes several branches of geology and provides extensive information about the processes that had occurred in the past (Newman and Banfield, 2002). The significant role that microbial species play in these systems was inferred by the uncommon structures that were identified when a detailed study of the cave environment was undertaken (Cunningham et al., 1995; Høeg, 1946).

24.4 Methods to Study Speleothems' Microbiome

Numerous techniques have been used to confirm the microbial presence in spelean environments. Microbial features and biomass on mineral surfaces are examined by scanning electron microscopy, environmental scanning electron microscopy, and transmission electron microscopy (Siering, 1998). To study the intricate microbial diversity of cave ecosystems, numerous culture-dependent and culture-independent methods and techniques were used.

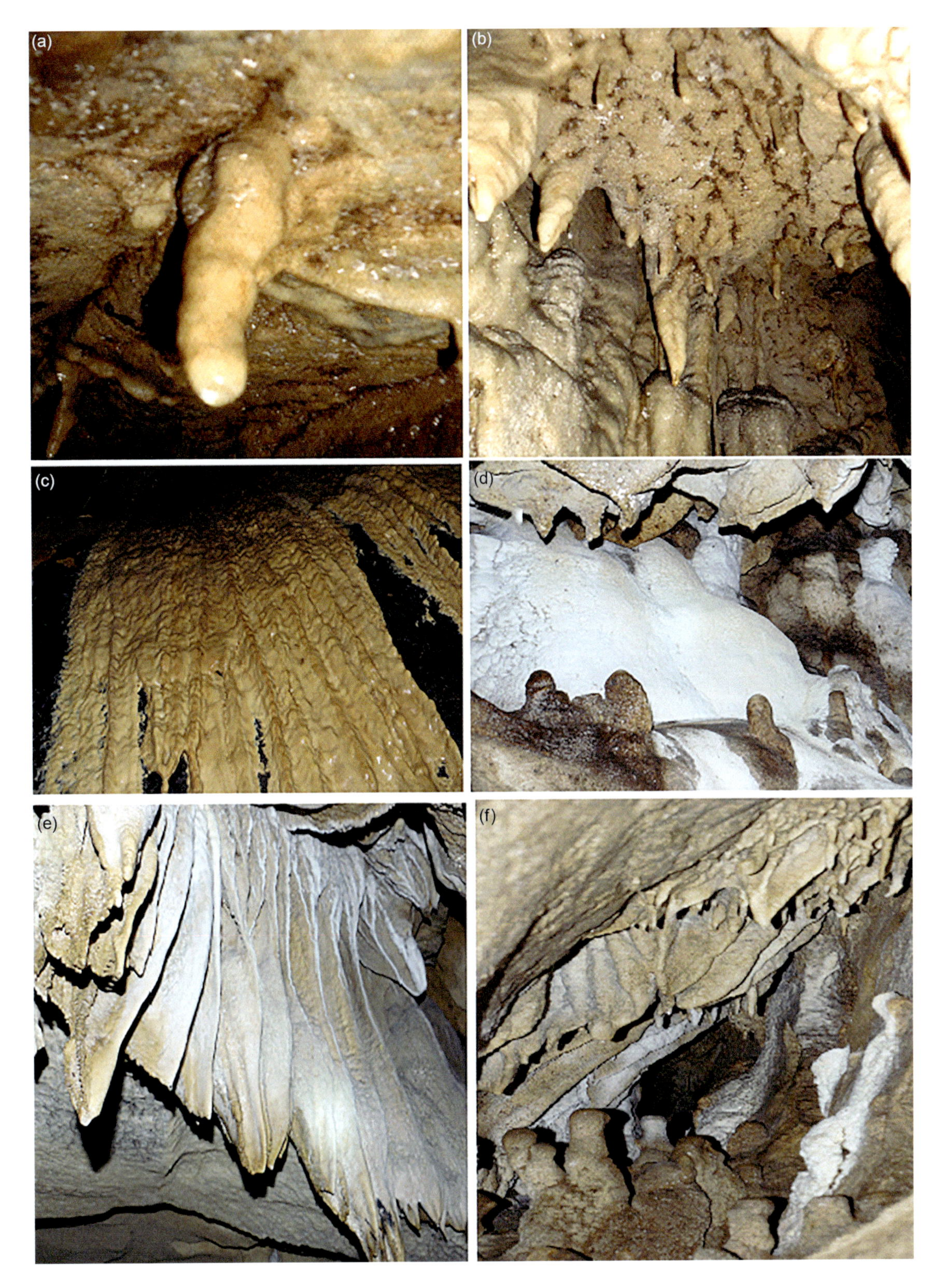

FIGURE 24.1 Typical forms of speleothems: (a) Stalactite, (b) soda straws, (c) flowstone, (d) moonmilk, (e) draperies, and (f) stalagmites.

24.4.1 Culture-Dependent Methods

Earlier cave microbiological studies used traditional culture-based methods and microscopy techniques (Engel, 2010). These studies discovered many microorganisms in caves, but the microbial diversity was much lower as compared to other terrestrial environments. Mostly, microbes were thought to be brought into caves from trans-located soil and/or aquatic microbes from the surface (Caumartin, 1963; Canganella et al., 2002; Engel, 2010). Although culture-dependent studies presented the leading insights in cave microbiology, they also underestimate the real cave microbial diversity. It has been predicted that these methods have successfully cultured only <1% of the microorganisms from environmental samples (Amann et al., 1995). Use of nutrient-rich culture media intended for medical microbiology poses another hurdle in earlier studies. Cave microorganisms adapted to nutrient-poor cave environments were unable to stop their metabolic pathways of scavenging nutrients, and mostly microbes die due to osmotic pressure (Koch, 1997; Barton, 2006). In such nutrient-rich media, omnipresent soil organisms, for example, *Pseudomonas* and *Bacillus*, were able to grow.

24.4.2 Culture-Independent Methods

Introduction of molecular genetics and culture-independent methods, especially rRNA-based techniques, has presented a unique microbial diversity in caves in addition to the existence of novel organisms (Barton et al., 2004; Gonzalez et al., 2006; Engel et al., 2010; Legatzki et al., 2011). Culture-independent methods allowed researchers to study bacterial communities in environmental samples by bypassing actually culturing them and with minimum interference with the environment under study, owing to small sample sizes. The study of the Movile Cave in Romania in the mid-1990s by Sabru et al. (1996) was a turning point in the history of cave microbial diversity. Present methods rely on molecular markers to study microbial diversity including 16S rRNA and 23S r RNA genes in addition to the examination of functional genes such as soxB, amoA, RuBisCO, rpoB, and recA or gyrB which are very necessary for cell function (Holmes et al. 2004). Genes are amplified from extracted DNA and microbial diversity and identification can be assessed by cloning of amplified genes and sequencing.

Environmental shotgun sequencing relies on the shredding of microbial genomic DNA by restriction enzymes into smaller fragments, afterwards introduced into appropriate vectors and sequenced individually. Fragment sequences are then reassembled based on, to create the original contiguous sequence, a contig. Stable isotope probing is an important technique advancing our knowledge about the interrelationship of environmental processes and microbial phylogeny. This is used to trace carbon from specific substrates into microbes that assimilate carbon from that substrate. Microbial populations are incubated with stable carbon isotope (^{13}C), and labeled DNA is isolated, which acts as a template for amplification of the 16S rRNA gene (Chen et al., 2009). Community fingerprinting methods such as denaturing gradient gel electrophoresis (DGGE) or temperature gradient gel electrophoresis (TGGE) are an important molecular tool to analyze community diversity or to follow general patterns in microbial communities. In DGGE/TGGE, 16S rRNA gene fragments are separated on the basis of differences in their melting behavior resulting in a pattern of bands in polyacrylamide gels containing either a linear gradient of DNA denaturants (a mixture of urea and formamide in DGGE) or a linear temperature gradient (TGGE). Theoretically, each band in gel symbolizes a unique sequence and consequently a unique species (Powell et al., 2003). In addition to this, various other genetic fingerprinting techniques such as amplified ribosomal DNA restriction analysis (ARDRA), random amplified polymorphic DNA (RAPD), restriction fragment length polymorphism (RFLP), and ribosomal intergenic spacer analysis (RISA) were used to characterize bacterial communities and differentiate bacterial isolates at the subspecies level, preferably even at the strain level. T-RFLP, however, is more advantageous over other fingerprinting techniques, because of its higher resolution, and direct reference can be made to the 16S rRNA gene sequence database (Marsh et al., 2000). FISH (fluorescent *in situ* hybridization) is a technique whereby fluorescently labeled oligonucleotide probes are bound to the targeted 16S rRNA in nucleic acids. This allows identifying cells at varying levels of taxonomic hierarchy in an environmental sample. FISH application in microbial systems offers identification and enumeration of microorganisms simultaneously without culturing them (Amann et al., 1990). Methods such as NanoSIMS (nanometer-scale secondary ion mass spectrometry) are used furthermore with FISH to reveal the metabolic processes. These methods help us to visualize the isotopic composition of cells. Recently, MALDI-TOF has been used to identify unknown bacteria at the species level by characterizing the components of cellular proteins. MALDI-TOF has several clear advantages: it is fast and accurate, and a large database of bacterial reference spectra is available (Sacchi et al., 2002). However, MALDI-TOF has some disadvantages. One important requirement for MALDI-TOF is that the bacterial cultures have to be pure. The resolving power of MALDI-TOF is lower than 16s rRNA analysis, and the technique is based on a chemical testing method that is not always 100% accurate for the identification of bacterial species.

Most recently, high-throughput DNA sequencing techniques are being applied for metagenomics analyses. This is advancing our knowledge of functional dynamics of microbial communities in environmental samples. "Next-generation" sequencing (NGS) technologies are better than traditional Sanger sequencing in terms of time and cost. These methods are based on different biochemistries. They dodge the cloning step before sequencing, essential in most Sanger sequencing. NGS studies resulted in the discovery of new microorganisms that were previously unreported due to cloning complications and prejudices. One major drawback of high-throughput next-generation sequencing technologies is that they require large amounts of DNA templates. Provided the oligotrophic conditions in caves, microbial biomass is very low, and the constituents of the rock surfaces (calcite) and secondary minerals occasionally constrain nucleic acid (DNA) extraction (Barton et al., 2006). However, overcoming these complications and applying these techniques to cave environments will significantly multiply our understanding about cave diversity and ecosystem function. In addition to these, biochemical analysis of genes, enzymes, and metabolites to figure out metabolic activities of cave microorganisms in many karst studies is also being used.

24.5 Microbes Reported from the Caves

Taxonomic comparison of the metagenomic results to earlier microscopic analyses from speleothem of the Tjuv-Ante's Cave, Northern Sweden confirmed the abundance of Actinobacteria and fungi. Genes related to photosynthesis, iron, and sulfur metabolism were detected suggesting the presence of chemoautotrophic bacteria. Furthermore, microbes known to cause calcium carbonate precipitation or biomineralization were also identified. The major group of bacteria and some rare groups reported from the caves are as follows:

a. *Proteobacteria:*

Proteobacteria are a diverse group of bacteria that are prevalent and abundant in caves. They were reported from Spanish Altamira Cave, famous for its paleolithic paintings; proteobacteria are dominant in the water dripping from the walls (Laiz et al., 1999; Schabereiter-Gurtner et al., 2002a); aggregations colonizing rocks (Portillo et al., 2008, 2009); rocks in caves Llonin, La Garma (Schabereiter-Gurtner et al, 2004); Tito Bustillo (Schabereiter-Gurtner et al., 2002b); Pajsarjevajama (Pašić et al., 2010), as well as in microbial mats on basalt walls of lava caves (Northup et al., 2011; Hathaway et al., 2014), on the stalactites of the Herrenberg Cave (Rusznyák et al., 2012), in the soil of the Niu Cave (Zhou et al. 2007), in the sediments of the Wind Cave (Chelius and Moore 2004), and in pools of the Barenschacht Cave (Shabarova and Pernthaler 2010). Microbiological studies revealed that Proteobacteria represented by far the most important bacterial group within the active microbial community in Altamira Cave colonies (Portillo et al., 2008). Members of the Gammaproteobacteria were identified from Cova des Pas De Vallgornera in the cave pools of anchialine waters using a culture-dependent method (Busquets et al., 2014). Microorganisms belonging to the Proteobacteria group dominate on the stalactites of the Herrenberg Cave representing more than 70% of the total 16S rRNA gene clones. Proteobacteria formed 22%–34% of the detected communities in fluvial sediments (Rusznyák et al., 2012). Proteobacteria is a versatile phylum, which was described as dominant in cave-dwelling communities exposed to high levels of human impact (Ikner et al., 2007). Lascaux Cave, which is a show cave and frequently visited by tourists, harbors colonies of Proteobacteria (Bastian et al., 2009). The alpha-, gamma-, and beta-proteobacteria are the most common classes of Proteobacteria that is identified as a dominant phylum in limestone caves. Proteobacteria have also been reported from extreme environments, such as sulfurous caves, e.g., Grotta Nuovadi Rio Garrafo (Jones et al., 2010), Parker (Angert et al., 1998), Cesspool (Engel et al., 2001), Lower Kane (Engel et al., 2003), and Movile (Sarbu, 2000) as the dominant phylum. In Grotta Sulfurea of the Frasassi cave system, *Proteobacteria* are involved in the sulfur cycle and represent >75% of the clone library created from the 16S rRNA genes of the water mat organisms (Macalady et al., 2006).

b. *Actinobacteria:*

They often inhabit the rock walls of caves, along with secondary mineral formations such as stalactites and stalagmites. This group were most frequently isolated from different caves including the walls of Pajsarjevajama Cave, Slovenia (Pašić et al., 2010), sediments collected from Wind Cave (Chelius and Moore, 2004), and paleolithic paintings of Altamira and Tito Bustillo Caves (Schabereiter-Gurtner et al., 2002a, 2002b). Culture-based studies from caves such as Cave of the Crystals in Chihuahua, Mexico (Quintana et al. 2013); hypogean environments (Groth and Saiz-Jimenez 1999); Altamira and Tito Bustillo, Spain (Groth et al., 1999); and Grotta dei Cervi, Porto Badisco, Italy (Groth et al., 2001) also reported Actinobacteria as one of the dominant groups of microbial cave populations. Culture-independent and molecular techniques-based species composition showed that Actinobacteria constituted (80% of the population) a main part of the microbial community in the Carlsbad Cavern (Barton et al., 2007). Actinobacteria were the second-largest taxonomic type in Grotta del Fiume of the Frasassi cave system; Llonin Cave (Schabereiter-Gurtner et al., 2004); Kartchner Cave (Ikner et al., 2007); Pajsarjevajama (Pašić et al., 2010); and vadose zone pools in the Barenschacht Cave (Shabarova and Pernthaler, 2010). Actinobacteria were supposed to take part in biomineralization processes (Barton et al., 2001; Cañveras et al., 2001; Groth et al., 2001; Jones 2001; Laiz et al., 1999) in the subsurface ecosystems.

c. *Firmicutes:*

Firmicutes are found frequently in microbial populations dwelling the surface of rock walls and sediments in caves. They can tolerate stress caused by oligotrophic conditions and water sacristy better than the Proteobacteria group. Dominant microorganisms identified in phototrophic biofilm from the Cave of Bats (Urzì et al., 2010) belonged to the genera Paenibacillus, Bacillus, and Staphylococcus, and approximately 50% of 16S rRNA sequences belonged to this group. This group dominated in the Kartchner Caverns (Ikner et al., 2007) in culture-dependent studies to assess the impact of tourism studied. Some of the microorganisms belonging to firmicutes from extremely acidic cave wall biofilms of the Frasassi cave system were able to reduce or oxidize sulfur, e.g., *Desulfotomaculum* spp. and *Sulfobacillus acidophilus* (Macalady et al., 2007). Firmicutes were part of the microbial communities from the following caves: Movile (Chen et al., 2009), Herrenberg Cave (Rusznyák et al., 2012), Lower Kane (Engel et al., 2010), Lolin, La Garma (Schabereiter-Gurtner et al. 2004), Altamira (Schabereiter-Gurtner et al., 2002a), Magura (Tomova et al., 2013), lava caves (Northup et al., 2011, Hathaway et al., 2014), Weebubbie (Tetu et al., 2013), and Barenschacht (Shabarova and Pernthaler, 2010).

d. *Bacteroidetes:*

Phylum formed the largest group in biofilms developing on ferromanganese deposits of the Carter Saltpeter Cave (Carmichael et al., 2013). *Bacteroidetes* were the second largest group of microorganisms in the white colonies inhabiting the rocks of the Altamira Cave. This type was represented by only a few metabolically active bacteria, mostly of the genus *Flavobacterium* (Portillo et al., 2009). In the Llonin Cave, *Bacteroidetes* were the third most dominant group (11.1% of 16S rRNA sequences) (Schabereiter-Gurtner et al., 2004). In the sediments of the Herrenberg Cave, *Bacteroidetes* were represented most frequently in the identified 16S rRNA sequences, constituting about 20% of all sequences. Although *Bacteroidetes* microorganisms are often found in caves, knowledge about their functional role in these ecosystems is limited. It is suggested that they are involved in the fermentation process and circulation of metal elements (Angert et al., 1998; Chelius and Moore 2004; Ikner et al., 2007; Macalady et al., 2006).

e. *Acidobacteria:*

The *Acidobacteria* group resides in various terrestrial ecosystems, in sub-surface ecosystems such as caves and acidic mine drainages, and a wide range of habitats (Baker and Banfield, 2003; Brofft et al., 2002, Kielak et al., 2016). In the study by Zimmermann and colleagues (2005a), a preliminary analysis of the species composition of *Acidobacteria* in the Altamira Cave showed that 25% of the microbes identified fit to different subgroups of Acidobacteria. Acidobacteria are ecologically important members of the microbial population of the Altamira Cave (Schabereiter-Gurtner et al., 2002a), Tito Bustillo (Schabereiter-Gurtner et al., 2002b), and La Garma (Schabereiter-Gurtner et al., 2004) and also in the soil of the Niu Cave (Zhou et al. 2007), Wind Cave sediments (Chelius and Moore, 2004), lava caves (Northup et al., 2011, Hathaway et al., 2014), and biofilm in the Roman catacombs (Zimmermann et al., 2005b). Meisinger and colleagues (2007) observed that *Acidobacteria* mats always appeared near the chemolithoautotrophs (dominant microbial groups from ε-and/or *γ-Proteobacteria*) (Engel et al., 2003, 2004). These organisms comprised a small group in the Grotta Sulfurea of the Frasassi cave system (Macalady et al., 2006), Herrenberg Cave (Rusznyák et al., 2012), Carter Saltpeter Cave (Carmichael et al., 2013), Weebubbie Cave (Tetu et al., 2013), and also the Barenschacht Cave (Shabarova and Pernthaler, 2010).

f. *Rarely found bacterial groups in caves:*

In addition to this, some bacterial groups are found rarely in the sub-surface ecosystem. The members of the *Gemmatimonadetes* group have been reported in the microbial population colonizing walls of the Pajsarjevajama (Pašić et al., 2010), Herrenberg Cave fluvial sediments (Rusznyák et al., 2012), soil of Niu Cave (Zhou et al., 2007), yellow and white mats of lava caves (Northup et al., 2011, Hathaway et al., 2014), and yellow colonies in the Altamira Cave (Portillo et al., 2008). *Spirochetes* formed a smaller fraction of bacterial population in the water mats of Grotta Nuova di Rio Garrafo (Jones et al., 2010), Movile (Chen et al., 2009), Lower Kane (Engel et al., 2010), and Wind (Chelius and Moore, 2004) Caves. Numerous *cyanobacteria* have been identified in phototrophic biofilm growing on the rocks at the entrance and exit of the Cave of Bats (Urzì et al., 2010). Artificial light installed for tourists in the Cave of Tito Bustillo induced strong growth and colonization of stalagmites and stalactites by *Scytonema julianum, Geitleria calcarea*, and *Pseudocapsa* sp. (Schabereiter-Gurtner et al., 2002b). These microorganisms were also found in lava caves (Northup et al., 2011, Hathaway et al., 2014) and slimes of Weebubbie Cave (Tetu et al., 2013). *Chlorobi* were identified only in small numbers in water mats of the Frasassi cave system at pH ~ 7.3 (Macalady et al., 2006), slimes of Weebubbie Cave (Tetu et al., 2013), and in ferromanganese deposits of the Carter Saltpeter Cave (Carmichael et al. 2013).

Microbiomes of caves also include archaea and fungi, but the majority of research groups are focused on bacterial diversity. In addition to these bacterial groups, the first demonstration of archaea in caves was by Northup et al. (2003), from the ferromanganese deposits of Lechuguilla Cave, USA. Other studies by Chelius and Moore, (2004), Barton et al. (2007), Barton et al. (2014), and Ortiz et al. (2014) also confirmed the presence of archaea in caves.

REFERENCES

Amann, R. I., Binder, B. J., Olson, R. J., Chisholm, S. W., Devereux, R., & Stahl, D. A. (1990). Combination of 16S rRNA-targeted oligonucleotide probes with flow cytometry for analyzing mixed microbial populations. *Applied and Environmental Microbiology, 56*(6), 1919–1925.

Amann, R. I., Ludwig, W., & Schleifer, K. H. (1995). Phylogenetic identification and in situ detection of individual microbial cells without cultivation. *Microbiological Reviews, 59*(1), 143–169.

Angert, E. R., Northup, D. E., Reysenbach, A. L., Peek, A. S., Goebel, B. M., & Pace, N. R. (1998). Molecular phylogentic analysis of a bacterial community in Sulphur River, Parker Cave, Kentucky. *American Mineralogist, 83* (11–12_Part_2), 1583–1592.

Baker, B. J., & Banfield, J. F. (2003). Microbial communities in acid mine drainage. *FEMS Microbiology Ecology, 44*(2), 139–152.

Banfield, J. F. (1997). Geomicrobiology: interactions between microbes and minerals. *Reviews in Mineralogy, 35*, 448.

Barton, H. A. (2006). Introduction to cave microbiology: a review for the non-specialist. *Journal of Cave and Karst Studies, 68*(2), 43–54.

Barton, H. A., Giarrizzo, J. G., Suarez, P., Robertson, C. E., Broering, M. J., Banks, E. D., … Venkateswaran, K. (2014). Microbial diversity in a Venezuelan orthoquartzite cave is dominated by the Chloroflexi (Class Ktedonobacterales) and Thaumarchaeota Group I. 1c. *Frontiers in Microbiology, 5*, 615.

Barton, H. A. (2006). Introduction to cave microbiology: a review for the non-specialist. *Journal of Cave and Karst Studies, 68*(2), 43–54.

Barton, H. A., & Luiszer, F. (2005). Microbial metabolic structure in a sulfidic cave hot spring: potential mechanisms of biospeleogenesis. *Journal of Cave and Karst Studies*, *67*(1), 28–38.

Barton, H. A., Spear, J. R., & Pace, N. R. (2001). Microbial life in the underworld: biogenicity in secondary mineral formations. *Geomicrobiology Journal*, *18*(3), 359–368.

Barton, H. A., Taylor, N. M., Kreate, M. P., Springer, A. C., Oehrle, S. A., & Bertog, J. L. (2007). The impact of host rock geochemistry on bacterial community structure in oligotrophic cave environments. *International Journal of Speleology*, *36*(2), 5.

Barton, H. A., Taylor, N. M., Lubbers, B. R., & Pemberton, A. C. (2006). DNA extraction from low-biomass carbonate rock: an improved method with reduced contamination and the low-biomass contaminant database. *Journal of Microbiological Methods*, *66*(1), 21–31.

Barton, H. A., Taylor, M. R., & Pace, N. R. (2004). Molecular phylogenetic analysis of a bacterial community in an oligotrophic cave environment. *Geomicrobiology Journal*, *21*(1), 11–20.

Bastian, F., & Alabouvette, C. (2009). Lights and shadows on the conservation of a rock art cave: the case of Lascaux Cave. *International Journal of Speleology*, *38*(1), 55–60.

Brofft, J. E., McArthur, J. V., & Shimkets, L. J. (2002). Recovery of novel bacterial diversity from a forested wetland impacted by reject coal. *Environmental Microbiology*, *4*(11), 764–769.

Busquets, A., Fornós, J. J., Zafra, F., Lalucat, J., & Merino, A. (2014). Microbial communities in a coastal cave: Cova des Pas de Vallgornera (Mallorca, Western Mediterranean). *International Journal of Speleology*, *43*(2), 8.

Cañaveras, J. C., Cuezva, S., Sanchez-Moral, S., Lario, J., Laiz, L., Gonzalez, J. M., & Saiz-Jimenez, C. (2006). On the origin of fiber calcite crystals in moonmilk deposits. *Naturwissenschaften*, *93*(1), 27–32.

Cañveras, J. C., Sanchez-Moral, S., Sloer, V., & Saiz-Jimenez, C. (2001). Microorganisms and microbially induced fabrics in cave walls. *Geomicrobiology Journal*, *18*(3), 223–240.

Canganella, F., Kuk, S. U., Morgan, H., & Wiegel, J. (2002). Clostridium thermobutyricum: growth studies and stimulation of butyrate formation by acetate supplementation. *Microbiological Research*, *157*(2), 149–156.

Carmichael, M. J., Carmichael, S. K., Santelli, C. M., Strom, A., & Bräuer, S. L. (2013). Mn (II)-oxidizing bacteria are abundant and environmentally relevant members of ferromanganese deposits in caves of the upper Tennessee River Basin. *Geomicrobiology Journal*, *30*(9), 779–800.

Caumartin, V. (1963). Review of the microbiology of underground environments. *Bulletin of the National Speleological Society*, *25*(Part 1), 1–14.

Chelius, M. K., & Moore, J. C. (2004). Molecular phylogenetic analysis of archaea and bacteria in Wind Cave, South Dakota. *Geomicrobiology Journal*, *21*(2), 123–134.

Chen, Y., Wu, L., Boden, R., Hillebrand, A., Kumaresan, D., Moussard, H., … Murrell, J. C. (2009). Life without light: microbial diversity and evidence of sulfur-and ammonium-based chemolithotrophy in Movile Cave. *The ISME Journal*, *3*(9), 1093–1104.

Culver, D. C., & Pipan, T. (2019). *The Biology of Caves and Other Subterranean Habitats*. Oxford University Press. United Kingdom, 336pp.

Cunningham, K. I., Northup, D. E., Pollastro, R. M., Wright, W. G., & LaRock, E. J. (1995). Bacteria, fungi and biokarst in Lechuguilla Cave, Carlsbad Caverns National Park, New Mexico. *Environmental Geology*, *25*(1), 2–8.

Desai, M. S., Assig, K., & Dattagupta, S. (2013). Nitrogen fixation in distinct microbial niches within a chemoautotrophy-driven cave ecosystem. *The ISME Journal*, *7*(12), 2411–2423.

Engel, A. S., Lee, N., Porter, M. L., Stern, L. A., Bennett, P. C., & Wagner, M. (2003). Filamentous "Epsilonproteobacteria" dominate microbial mats from sulfidic cave springs. *Applied and Environmental Microbiology*, *69*(9), 5503–5511.

Engel, A. S., Meisinger, D. B., Porter, M. L., Payn, R. A., Schmid, M., Stern, L. A., … Lee, N. M. (2010). Linking phylogenetic and functional diversity to nutrient spiraling in microbial mats from Lower Kane Cave (USA). *The ISME Journal*, *4*(1), 98–110.

Engel, S., Porter, M. L., Kinkle, B. K., & Kane, T. C. (2001). Ecological assessment and geological significance of microbial communities from Cesspool Cave, Virginia. *Geomicrobiology Journal*, *18*(3), 259–274.

Engel, A. S., Stern, L. A., & Bennett, P. C. (2004). Microbial contributions to cave formation: new insights into sulfuric acid speleogenesis. *Geology*, *32*(5), 369–372.

Fairchild, I. J., & Baker, A. (2012). *Speleothem Science: from Process to Past Environments* (Vol. 3). Wiley-Blackwell, Chichester, 432pp.

Faust, B. (1949). The formation of saltpeter in caves. *Bulletin of the National Speleological Society*, *11*, 17–23.

Gonzalez, J. M., Portillo, M. C., & Saiz-Jimenez, C. (2006). Metabolically active Crenarchaeota in Altamira cave. *Naturwissenschaften*, *93*(1), 42–45.

Groth, P. Schumann, L. Laiz, S. Sanchez-Moral, J. C., Cañveras, C., & Saiz-Jimenez, I. (2001). Geomicrobiological study of the grotta dei cervi, Porto Badisco, Italy. *Geomicrobiology Journal*, *18*(3), 241–258.

Groth, I., Vettermann, R., Schuetze, B., Schumann, P., & Sáiz-Jiménez, C. (1999). Actinomycetes in karstic caves of northern Spain (Altamira and Tito Bustillo). *Journal of microbiological methods*, *36*(1–2), 115–122.

Hathaway, J. J. M., Garcia, M. G., Balasch, M. M., Spilde, M. N., Stone, F. D., Dapkevicius, M. D. L. N., … Northup, D. E. (2014). Comparison of bacterial diversity in Azorean and Hawai'ian lava cave microbial mats. *Geomicrobiology Journal*, *31*(3), 205–220.

Høeg, O. A. (1946). Cyanophyceae and bacteria in calcareous sediments in the interior of limestone caves in Nord-Rana, Norway. *Nyatt Magasine for Naturvidenskapene*, *85*, 99–104.

Holmes, D. E., Nevin, K. P., & Lovley, D. R. (2004). Comparison of 16S rRNA, nifD, recA, gyrB, rpoB and fusA genes within the family Geobacteraceae fam. nov. *International Journal of Systematic and Evolutionary Microbiology*, *54*(5), 1591–1599.

Holmes, A. J., Tujula, N. A., Holley, M., Contos, A., James, J. M., Rogers, P., & Gillings, M. R. (2001). Phylogenetic structure of unusual aquatic microbial formations in Nullarbor caves, Australia. *Environmental Microbiology*, 3(4), 256–264.

Ikner, L. A., Toomey, R. S., Nolan, G., Neilson, J. W., Pryor, B. M., & Maier, R. M. (2007). Culturable microbial diversity and the impact of tourism in Kartchner Caverns, Arizona. *Microbial Ecology*, *53*(1), 30–42.

Jannasch, H. W., & Mottl, M. J. (1985). Geomicrobiology of deep-sea hydrothermal vents. *Science*, *229*(4715), 717–725.

Jones, B. (2001). Microbial activity in caves– a geological perspective. *Geomicrobiology Journal*, *18*(3), 345–357.

Jones, D. S., Tobler, D. J., Schaperdoth, I., Mainiero, M., & Macalady, J. L. (2010). Community structure of subsurface biofilms in the thermal sulfidic caves of Acquasanta Terme, Italy. *Applied and Environmental Microbiology*, *76*(17), 5902–5910.

Kielak, A. M., Barreto, C. C., Kowalchuk, G. A., van Veen, J. A., & Kuramae, E. E. (2016). The ecology of Acidobacteria: moving beyond genes and genomes. *Frontiers in Microbiology*, *7*, 744.

Koch, A. L. (1997). Microbial physiology and ecology of slow growth. *Microbiology and Molecular Biology Reviews*, *61*(3), 305–318.

Laiz, L., Groth, I., Gonzalez, I., & Saiz-Jimenez, C. (1999). Microbiological study of the dripping waters in Altamira cave (Santillana del Mar, Spain). *Journal of Microbiological Methods*, *36*(1–2), 129–138.

Lee, N. M., Meisinger, D. B., Aubrecht, R., Kovacik, L., Saiz-Jimenez, C., Baskar, S., ... Engel, A. S. (2012). 16 Caves and Karst Environments. *Life at Extremes: Environments, Organisms, and Strategies for Survival*, *1*, 320.

Legatzki, A., Ortiz, M., Neilson, J. W., Dominguez, S., Andersen, G. L., Toomey, R. S., ... Maier, R. M. (2011). Bacterial and archaeal community structure of two adjacent calcite speleothems in Kartchner Caverns, Arizona, USA. *Geomicrobiology Journal*, *28*(2), 99–117.

Macalady, J. L., Jones, D. S., & Lyon, E. H. (2007). Extremely acidic, pendulous cave wall biofilms from the Frasassi cave system, Italy. *Environmental Microbiology*, *9*(6), 1402–1414.

Macalady, J. L., Lyon, E. H., Koffman, B., Albertson, L. K., Meyer, K., Galdenzi, S., & Mariani, S. (2006). Dominant microbial populations in limestone-corroding stream biofilms, Frasassi cave system, Italy. *Applied and Environmental Microbiology*, *72*(8), 5596–5609.

Madigan, C. F., Lu, M. H., & Sturm, J. C. (2000). Improvement of output coupling efficiency of organic light-emitting diodes by backside substrate modification. *Applied Physics Letters*, *76*(13), 1650–1652.

Marsh, T. L., Saxman, P., Cole, J., & Tiedje, J. (2000). Terminal restriction fragment length polymorphism analysis program, a web-based research tool for microbial community analysis. *Applied and Environmental Microbiology*, *66*(8), 3616–3620.

McDermott, F. R. A. N. K., Schwarcz, H., & Rowe, P. J. (2006). Isotopes in speleothems. In *Isotopes in Palaeoenvironmental Research* (pp. 185–225). Springer, Dordrecht.

McMahon, S., & Parnell, J. (2014). Weighing the deep continental biosphere. *FEMS Microbiology Ecology*, *87*(1), 113–120.

Meisinger, D. B., Zimmermann, J., Ludwig, W., Schleifer, K. H., Wanner, G., Schmid, M., ... Lee, N. M. (2007). In situ detection of novel Acidobacteria in microbial mats from a chemolithoautotrophically based cave ecosystem (Lower Kane Cave, WY, USA). *Environmental Microbiology*, *9*(6), 1523–1534.

Moore, G. W. (1952). Speleothem—a new cave term. *National Speleological Society News*, *10*(6), 2.

Mudgil, D., Baskar, S., Baskar, R., Paul, D., & Shouche, Y. S. (2018). Biomineralization potential of Bacillus subtilis, Rummeliibacillus stabekisii and Staphylococcus epidermidis strains in vitro isolated from Speleothems, Khasi Hill Caves, Meghalaya, India. *Geomicrobiology Journal*, *35*(8), 675–694.

Newman, D. K., & Banfield, J. F. (2002). Geomicrobiology: how molecular-scale interactions underpin biogeochemical systems. *Science*, *296*(5570), 1071–1077.

Northup, D. E., Barns, S. M., Yu, L. E., Spilde, M. N., Schelble, R. T., Dano, K. E., ... Dahm, C. N. (2003). Diverse microbial communities inhabiting ferromanganese deposits in Lechuguilla and Spider Caves. *Environmental Microbiology*, *5*(11), 1071–1086.

Northup, E. K. H., & Lavoie, D. (2001). Geomicrobiology of caves: a review. *Geomicrobiology Journal*, *18*(3), 199–222.

Northup, D. E., Melim, L. A., Spilde, M. N., Hathaway, J. J. M., Garcia, M. G., Moya, M., ... Riquelme, C. (2011). Lava cave microbial communities within mats and secondary mineral deposits: implications for life detection on other planets. *Astrobiology*, *11*(7), 601–618.

Ortiz, M., Legatzki, A., Neilson, J. W., Fryslie, B., Nelson, W. M., Wing, R. A., ... Maier, R. M. (2014). Making a living while starving in the dark: metagenomic insights into the energy dynamics of a carbonate cave. *The ISME Journal*, *8*(2), 478–491.

Ortiz, M., Neilson, J. W., Nelson, W. M., Legatzki, A., Byrne, A., Yu, Y., ... Maier, R. M. (2013). Profiling bacterial diversity and taxonomic composition on speleothem surfaces in Kartchner Caverns, AZ. *Microbial Ecology*, *65*(2), 371–383.

Pace, N. R. (1997). A molecular view of microbial diversity and the biosphere. *Science*, *276*(5313), 734–740.

Pašić, L., Kovče, B., Sket, B., & Herzog-Velikonja, B. (2009). Diversity of microbial communities colonizing the walls of a Karstic cave in Slovenia. *FEMS Microbiology Ecology*, *71*(1), 50–60.

Portillo, M. D. C., Gonzalez, J. M., & Saiz-Jimenez, C. (2008). Metabolically active microbial communities of yellow and grey colonizations on the walls of Altamira Cave, Spain. *Journal of Applied Microbiology*, *104*(3), 681–691.

Portillo, M. C., Saiz-Jimenez, C., & Gonzalez, J. M. (2009). Molecular characterization of total and metabolically active bacterial communities of "white colonizations" in the Altamira Cave, Spain. *Research in Microbiology*, *160*(1), 41–47.

Powell, S. M., Bowman, J. P., Snape, I., & Stark, J. S. (2003). Microbial community variation in pristine and polluted nearshore Antarctic sediments. *FEMS Microbiology Ecology*, *45*(2), 135–145.

Provencio, P. V. J., & Polyak, P. (2001). Iron oxide-rich filaments: possible fossil bacteria in Lechuguilla Cave, New Mexico. *Geomicrobiology Journal*, *18*(3), 297–309.

Quintana, E. T., Badillo, R. F., & Maldonado, L. A. (2013). Characterisation of the first actinobacterial group isolated from a Mexican extremophile environment. *Antonie van Leeuwenhoek*, *104*(1), 63–70.

Romero, A. (2009). *Cave Biology: Life in Darkness*. Cambridge University Press. Cambridge, UK, 306 pp.

Rusznyák, A., Akob, D. M., Nietzsche, S., Eusterhues, K., Totsche, K. U., Neu, T. R., … Katzschmann, L. (2012). Calcite biomineralization by bacterial isolates from the recently discovered pristine karstic Herrenberg cave. *Applied and Environmental Microbiology*, *78*(4), 1157–1167.

Sacchi, C. T., Whitney, A. M., Reeves, M. W., Mayer, L. W., & Popovic, T. (2002). Sequence diversity of Neisseria meningitidis 16S rRNA genes and use of 16S rRNA gene sequencing as a molecular subtyping tool. *Journal of Clinical Microbiology*, *40*(12), 4520–4527.

Saiz-Jimenez, I. G. C. (1999). Actinomycetes in hypogean environments. *Geomicrobiology Journal*, *16*(1), 1–8.

Sarbu, S. M. (2000). Movile Cave: a chemoautotrophically based groundwater ecosystem. *Ecosystems of the world*, 319–344.

Sarbu, S. M., Kane, T. C., & Kinkle, B. K. (1996). A chemoautotrophically based cave ecosystem. *Science*, *272*(5270), 1953–1955.

Schabereiter-Gurtner, C., Saiz-Jimenez, C., Piñar, G., Lubitz, W., & Rölleke, S. (2002a). Altamira cave Paleolithic paintings harbor partly unknown bacterial communities. *FEMS Microbiology Letters*, *211*(1), 7–11.

Schabereiter-Gurtner, C., Saiz-Jimenez, C., Piñar, G., Lubitz, W., & Rölleke, S. (2002b). Phylogenetic 16S rRNA analysis reveals the presence of complex and partly unknown bacterial communities in Tito Bustillo cave, Spain, and on its Palaeolithic paintings. *Environmental Microbiology*, *4*(7), 392–400.

Schabereiter-Gurtner, C., Saiz-Jimenez, C., Piñar, G., Lubitz, W., & Rölleke, S. (2004). Phylogenetic diversity of bacteria associated with Paleolithic paintings and surrounding rock walls in two Spanish caves (Llonin and La Garma). *FEMS Microbiology Ecology*, *47*(2), 235–247.

Shabarova, T., & Pernthaler, J. (2010). Karst pools in subsurface environments: collectors of microbial diversity or temporary residence between habitat types. *Environmental Microbiology*, *12*(4), 1061–1074.

Siering, P. L. (1998). The double helix meets the crystal lattice: the power and pitfalls of nucleic acid approaches for biomineralogical investigations. *American Mineralogist*, *83*(11), 1593–1607.

Tetu, S. G., Breakwell, K., Elbourne, L. D., Holmes, A. J., Gillings, M. R., & Paulsen, I. T. (2013). Life in the dark: metagenomic evidence that a microbial slime community is driven by inorganic nitrogen metabolism. *The ISME Journal*, *7*(6), 1227–1236.

Tomczyk-Żak, K., & Zielenkiewicz, U. (2016). Microbial diversity in caves. *Geomicrobiology Journal*, *33*(1), 20–38.

Tomova, I., Lazarkevich, I., Tomova, A., Kambourova, M., & Vasileva-Tonkova, E. (2013). Diversity and biosynthetic potential of culturable aerobic heterotrophic bacteria isolated from Magura Cave, Bulgaria. *International Journal of Speleology*, *42*(1), 8.

Urzì, C., De Leo, F., Bruno, L., & Albertano, P. (2010). Microbial diversity in Paleolithic caves: a study case on the phototrophic biofilms of the Cave of Bats (Zuheros, Spain). *Microbial Ecology*, *60*(1), 116–129.

Went, F. W. (1969). Fungi associated with stalactite growth. *Science*, *166*(3903), 385–386.

Whitman, W. B., Coleman, D. C., & Wiebe, W. J. (1998). Prokaryotes: the unseen majority. *Proceedings of the National Academy of Sciences*, *95*(12), 6578–6583.

Yang, H., Ding, W., Zhang, C. L., Wu, X., Ma, X., He, G., … Xie, S. (2011). Occurrence of tetraether lipids in stalagmites Implications for sources and GDGT-based proxies. *Organic Geochemistry*, *42*(1), 108–115.

Zhou, J., Gu, Y., Zou, C., & Mo, M. (2007). Phylogenetic diversity of bacteria in an earth-cave in Guizhou Province, Southwest of China. *The Journal of Microbiology*, *45*(2), 105–112.

Zimmermann, J., Gonzalez, J. M., Sáiz-Jiménez, C., & Ludwig, W. (2005). Detection and phylogenetic relationships of highly diverse uncultured acidobacterial communities in Altamira Cave using 23S rRNA sequence analyses. *Geomicrobiology Journal*, *22*(7–8), 379–388.

Zimmermann, J., Gonzalez, J. M., & Saiz-Jimenez, C. (2006). Epilithic biofilms in Saint Callixtus Catacombs (Rome) harbour a broad spectrum of Acidobacteria. *Antonie Van Leeuwenhoek*, *89*(1), 203–208.

25

Microbiome of Marine Shallow-Water Hydrothermal Vents

Raju Rajasabapathy
Bharathidasan University

Chellandi Mohandass
CSIR – National Institute of Oceanography

Ana Colaço
Universidade dos Açores

Rathinam Arthur James
Bharathidasan University

CONTENTS

25.1 Introduction

Hydrothermal vents are associated with seafloor spreading zones, commonly occurring in mid-ocean ridges where two tectonic plates are moving apart and in basins near volcanic island arcs. Magma pockets are the energy engines that create volcanic activity. The molten rock (800°C–1200°C) beneath the magma discharges lavas onto the seafloor over periods ranging from <10 to >50,000 years between eruptions (Hammond 1997). Since the tectonic plate activities take place, the seawater sinks directly through the crevices in the crust and is exposed to the magma chambers. The cold seawater becomes heated up by the magma, then collects metals and minerals from the molten rocks, and shot out of an opening into the seawater (Brooks 2006), which forms chemical plume, sometimes called black smokers and white smokers based on the chemical compositions. Water emerging from the hot regions of some hydrothermal vents will be a supercritical fluid, which has physical properties between those of a gas and a liquid. In contrast to that of the ambient seawater, the temperature of superheated water in hydrothermal vents is above 400°C. The hottest ever-measured hydrothermal fluid temperature (464°C) was reported from the active venting Sisters Peak chimney on the Mid-Atlantic Ridge (Perner et al. 2014). Based on venting activity, the crevices in the venting ocean floor may enlarge in size and spread from 5 to 9 cm per year or as fast as 9–16 cm per year (Jones 1985).

Numerous hydrothermal venting sites and faunal assemblages at many mid-ocean ridges and back-arc basins have been explored since the discovery of hydrothermal vents along the Galapagos Ridge in 1977 (Corliss et al. 1979). Hydrothermal vent fluids are highly enriched with reduced inorganic chemicals, i.e., electron donors. Microbial communities take advantage of these compounds as energy sources and form the basis of the food chain (Lutz and Kennish 1993; Imhoff and Hugler 2009). The hydrothermal vent regions are not only limited to deeper areas of the ocean but also occur at shallow depths. This chapter will provide a brief overview of the microbiome and its diversity and functions in the shallow-water hydrothermal vent regions.

25.2 Difference between Deep-Sea and Shallow-Water Hydrothermal Vents

The accepted description for shallow hydrothermal vents is that they can occur at a depth of around 200 m (Tarasov et al. 2005) or up to 212 m (Price and Giovannelli 2017). The shallow-water vents differ from their deep-sea counterparts mainly by the presence of light, which penetrates up to 200 m depth in the ocean where photosynthetic organisms such as benthic microalgae and cyanobacteria are present (Sorokin 1991), and thus primary production takes place by photosynthetic organisms also (Dando et al. 1995). On the other hand, oxidized sulfur compounds are used by many heterotrophic members of the archaea and bacteria as electron acceptors for the anaerobic degradation of organic matter, although some can grow autotrophically. The environmental conditions in shallow hydrothermal vent regions vary from those in deep-sea hydrothermal systems and terrestrial hot springs concerning temperature, water pressure, pH, salinity, sunlight, etc. (Hirayama et al. 2007).

Shallow-water hydrothermal vent ecosystems are widespread that have been previously understudied when compared with deep-sea ecosystems. They are principally associated with heat generated by plate tectonics (associated with divergent, transform, and convergent boundaries). Most of the reported shallow vents are associated with Island arc volcanoes, and the remaining are evenly distributed across back-arc volcanoes, intra-plate volcanoes, and mid-ocean ridges (Price and Giovannelli 2017). Based on the InterRidge database, up to 200 m, there have been 45 confirmed and 5 inferred active reported hydrothermal vents (Figure 25.1). However, the literature suggests that the number of unexplored shallow vents may be much higher.

Deep-sea hydrothermal vents are highly productive ecosystems where chemolithoautotrophic microorganisms mediate the transfer of energy from the geothermal source to the higher trophic level. Relative to the majority of the deep sea, the areas around shallow submarine hydrothermal vents are biologically much higher productive, swarming complex communities with the energy gained from the chemicals dissolved in the vent fluids. The most commonly understood mode of metabolism thought to dominate the deep-sea hydrothermal vent microbial communities is chemolithoautotrophy, principally through the oxidation of iron compounds and reduced sulfur compounds (Jannasch and Mottl 1985). The habitat for these organisms is the hypoxic parts of the hydrothermal system, and correspondingly, many of them are thermophiles or hyperthermophiles (Karl 1995). Usually, there is high biomass of mostly endemic but species-poor fauna that depends on chemosynthesis-based production at deep-sea vent mid-ocean ridges (Tunnicliffe 1991). But contrastingly, shallow-water vents tend to have low biomass of a more diverse fauna with few endemic species (Kamenev et al. 1993; Dando et al. 1995; Morri et al. 1999).

Shallow hydrothermal vents discharge free gases from the sediment or from the crevices formed in the vent regions (Figure 25.2). The highest temperature was observed at 135.2°C with a pH ranging from 4 to 6 (Price et al. 2015). In Island arc-related shallow vents, these gases are dominated by carbon dioxide with trace amounts of sulfide and methane. Besides, the vent fluids are enriched with various metals and minerals, especially manganese and iron. Overall, the parameters such as the presence of light, terrestrial inputs, tidal cycles, and free gas phases are typical in shallow vents when compared to deep-sea counterparts. Based on the available information, Giovannelli and Price (2018) have divided the shallow hydrothermal vents into four environmental niches: (1) high-temperature sulfur-rich shallow-water vents, such as those found in the Hellenic Arc and the proximity of numerous active island volcanoes; (2) low-temperature sulfur-rich shallow vents, such as those present at Tor Caldara (Italy); (3) low-temperature carbon dioxide and iron-rich shallow vents, where sulfide and sulfur compounds are absent, such as those found in Ischia (Italy) and Faial Island, Azores (Portugal); and (4) low- and high-temperature alkaline shallow vents, characterized by high pH and the presence of appreciable concentrations

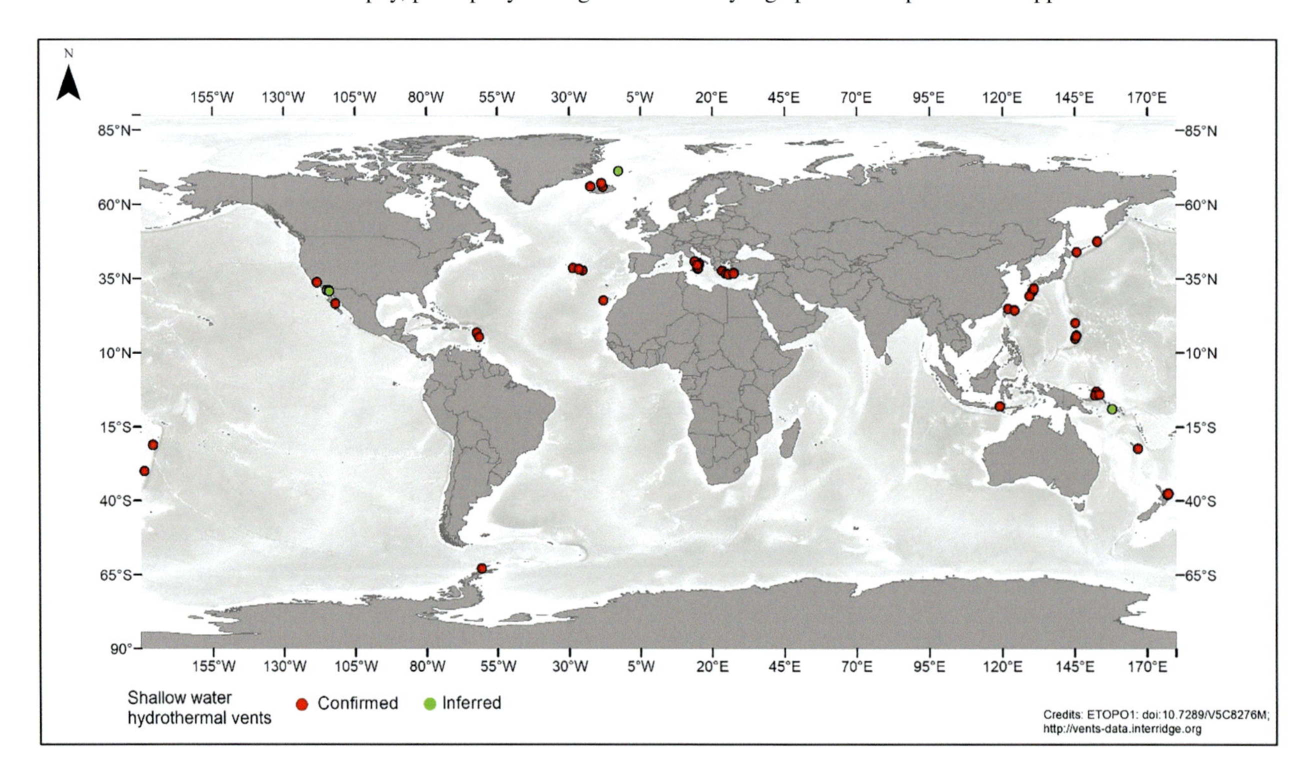

FIGURE 25.1 Map showing the location of confirmed and inferred shallow-water hydrothermal vents; locations have been obtained from InterRidge database ver. 3.4 (Beaulieu and Szafranski 2019).

FIGURE 25.2 Underwater photographs of a shallow-water hydrothermal vent (Espalamaca vent field in Azores, Portugal). (a) Presence of crevices, gas bubbles, and microbial mats; (b) gas bubbles without any crevice (Rajasabapathy et al. 2014).

of hydrogen and methane, such as those found in Prony Bay (New Caledonia) and Strytan Hydrothermal Field (Iceland). In 2018, scientists surprisingly discovered smoking ocean vents in shallow regions (Azores, Atlantic Ocean), which were thought to be typical in the deep-sea hydrothermal vents (Leahy 2018). Similar kinds of studies in the future will discover more detailed information about shallow vents and life in the hydrothermal region.

25.3 Life at Shallow-Water Hydrothermal Vents

The abundance of nutrients (silicate, phosphate, and nitrate), gases (CO_2, CH_4, H_2, and H_2S) and other reduced compounds (C_nH_n, S^0, $S_2O_3^{2-}$, and NH_4^+) in zones of shallow hydrothermal vents provides the condition for the use of two kinds of energy sources for primary production, i.e., sunlight (photosynthesis) and the oxidation of reduced compounds (chemosynthesis by bacteria). The chemosynthesis occurs both in the immediate surrounding area of venting fluid and in the surface layer of the water column, where it occurs together with intense photosynthesis. This surface photosynthesis is found below the layer of chemosynthesis, which is related to the circulation of hydrothermal fluids at the water surface. The contribution of each of these processes to total primary production depends on the physical and chemical conditions generated by the vents and on the range and adaptation potential of the organisms. The major and distinguishing biological components of shallow-water hydrothermal vents are an abundance of thermophilic bacteria and microbial mats undergoing a range of biogeochemical processes (Tarasov 2006). In general, the biological communities in the shallow vents can be divided into pelagic and benthic communities. The pelagic communities include phytoplankton, bacterioplankton, planktonic protozoa, and zooplankton. The benthic communities include thermophilic bacteria, microphytobenthos, meiobenthos, and macrobenthos.

25.4 Microbiome Studies in Shallow-Water Hydrothermal Vents

Previous studies have clearly shown the abundance of mesophilic, thermophilic, and hyperthermophilic archaea and bacteria in the shallow hydrothermal vent emissions (e.g. Dando et al. 1998; Sievert et al. 2000b; Hirayama et al. 2007; Rajasabapathy et al. 2014). Sunlight and high concentration of organic matter paved the way for the abundance of cyanobacteria, anoxygenic phototrophs, and plenty of heterotrophic microorganisms. For more than 10 years, next-generation sequencing technologies have been applied in several places including microbial diversity and richness in environmental samples. Concerning the shallow hydrothermal vents, only a

TABLE 25.1

Microbiome Studies in Shallow Hydrothermal Vents Using NGS Approaches

Study Area	Sequencing Method	Dominant Phyla/Class of Bacteria	Dominant Genera	Reference
Panarea Island, Italy	Illumina	Epsilon-proteobacteria	*Sulfurovum* and *Sulfurimonas*	Gugliandolo et al. (2015)
Kuei-shan Island, Taiwan	Pyrosequencing	Epsilon-proteobacteria	*Sulfurovum* and *Sulfurimonas*	Wang et al. (2015)
Black Point, off Panarea Island, Italy	Illumina	Alpha-proteobacteria and Gamma-proteobacteria	*Thiohalospira* and *Thiomicrospira*	Lentini et al. (2014)
NE Taiwan's coast (Yellow zone)	Pyrosequencing	Gamma-proteobacteria	*Thiomicrospira*	Zhang et al. (2012)
NE Taiwan's coast (White zone)		Epsilon-proteobacteria	*Nautilia*	
Kueishantao Island, Taiwan	Illumina	Gamma-proteobacteria	*Thiomicrorhabdus*	Tang et al. (2018)
Lesser Antilles, Dominica	Illumina	Gamma-proteobacteria	*Pseudomonas* and *Pseudoalteromonas*	Pop Ristova et al. (2017)

few studies used NGS approaches to study the microbial communities (Table 25.1).

Lentini et al. (2014) investigated the prokaryotic community structure and composition of a hydrothermal site, named Black Point, off Panarea Island (Eolian Islands, Italy). Proteobacteria (mainly consisting of the Alpha-, Gamma-, and Epsilon-proteobacteria) dominated in all the vent samples (high-temperature and low-temperature sampling points) followed by Actinobacteria and Bacteroidetes. The same research team (Gugliandolo et al. 2015) studied the prokaryotic community composition from a CO_2- and H_2S-rich Hot Lake thermal brine pool off Panarea Island (Eolian Islands, Italy) using Illumina sequencing technology. They studied the microbiome composition using V3 regions of the 16S rRNA gene at two time points that differed mainly concerning temperature conditions, high-temperature (94°C) and low-temperature (28.5°C). Both regions were dominated by members of the Epsilon-proteobacteria. Within this subphylum, bacteria of the genus *Sulfurimonas* were most frequently detected at the high-temperature region, while *Arcobacter* prevailed at the low-temperature point. He further revealed that hyperthermophilic and thermophilic groups of bacteria were dominant at the high-temperature region, whereas those related to nonthermophilic Bacteroidetes, Fusobacteria, and Actinobacteria were dominant at the low-temperature point. The unique feature of the shallow vent of the Hot Lake region was the co-occurrence of photosynthetic and chemolithotrophic microorganisms. Most recently, Gugliandolo and Maugeri (2019) reviewed the archaeal community especially in sediments collected from the Aeolian Islands. They revealed that hyperthermophilic members of Crenarchaeota (*Thermoprotei*) and Euryarchaeota (*Thermococci* and *Methanococci*) were found under the highest temperature condition, and the Mesophilic Euryarchaeota (*Halobacteria*, *Methanomicrobia*, and *Methanobacteria*) increased with decreasing temperatures.

Zhang et al. (2012) examined the bacterial and archaeal communities from the water column extending over a redoxocline gradient of a yellow and from a white hydrothermal vent, NE Taiwan's coast. Ribosomal tag pyrosequencing analysis showed that the bacterial and archaeal communities from the white hydrothermal plume were dominated by sulfur-reducing *Nautilia* and *Thermococcus*, whereas the yellow hydrothermal plume and the surface water were dominated by sulfide-oxidizing *Thiomicrospira* and Euryarchaeota Marine Group II, respectively. They concluded that sulfur-reducing and sulfide-oxidizing chemolithoautotrophs accounted for most of the primary biomass synthesis, and microbial sulfur metabolism fueled microbial energy flow and element cycling in the shallow hydrothermal systems off the coast of NE Taiwan.

Wang et al. (2015) studied microbial diversity in shallow-water hydrothermal sediments of Kuei-shan Island, Taiwan by the pyrosequencing method. Small-subunit ribosomal RNA gene-based high-throughput 454 pyrosequencing was used to characterize the assemblages of bacteria and archaea. The Epsilon-proteobacteria group was the most abundant in the vent sediment, but its abundance decreased with increasing distance from the vent area. Most of the Epsilon-proteobacteria belonged to the mesophilic chemolithoautotrophic genera *Sulfurovum* and *Sulfurimonas*. Further, they concluded that the dominant chemolithotrophic bacteria were potential sulfur-oxidizers, and microbial sulfur metabolism is most likely an important driving force for microbial energy flow and element cycling in Kuei-shan Island. In another shallow hydrothermal system (Kueishantao Island, Taiwan), fluorescence *in situ* hybridization, high-throughput 16S rRNA gene amplicon sequencing, and functional metagenomes were used to assess microbial communities (Tang et al. 2018). The results showed that the shallow-sea hydrothermal system was not only occupied with autotrophic bacteria but also abundant in heterotrophic bacteria.

Automated ribosomal intergenic spacer analysis (ARISA) and high-throughput Illumina sequencing combined with pore water geochemical analysis have been used to investigate microbial communities along geochemical gradients in two shallow-water hydrothermal systems off the island of Dominica (Lesser Antilles). The study results revealed that the shallow-water hydrothermal systems of Dominica harbored bacterial communities with high taxonomical and metabolic diversity, predominated by heterotrophic microorganisms associated with the Gamma-proteobacterial genera *Pseudomonas* and *Pseudoalteromonas*, indicating the importance of heterotrophic processes (Pop Ristova et al. 2017).

These metagenomics studies conducted in shallow hydrothermal vents are mainly from Taiwan and Italy. However, most of the microbiological studies from shallow vents have been performed using classical techniques, i.e., culture-based, clone library approach and denaturing gradient gel electrophoresis (Moyer et al. 1995; López-García et al. 2003; Hirayama et al. 2007; Zhou et al. 2009; Murdock et al. 2010; Mohandass et al. 2012; Giovannelli et al. 2013; Rajasabapathy et al. 2014; Rajasabapathy et al. 2018). The symbiosis between chemosynthetic bacteria and animals is prevalent across various ecosystems including shallow hydrothermal vents, cold seeps, and sulfide sediments. Hundreds of species from different phyla are known to harbor chemosynthetic symbioses, for hydrothermal vent macrofaunal (shrimps, crabs, mussels, and gastropods) and meiofaunal organisms (nematodes and copepods). In the shallow-water hydrothermal vent off Kuei-shan Island, known to be a species-poor habitat, symbiotic microbial diversity has been analyzed from the crab *Xenograpsus testudinatus*. The 16S r RNA gene amplicon pyrosequencing was used to investigate the diversity and composition of bacteria residing in various organs of *X. testudinatus*, as well as in surrounding seawater. Dominant bacteria were found to be the Gamma- and Epsilon-proteobacteria that might be capable of autotrophic growth by oxidizing reduced sulfur compounds (Yang et al. 2016). The microbial association in a nematode investigated by Bellec et al. (2019) revealed that the *Metoncholaimus albidus* (nematode) has a specific microbial community, distinct from its surrounding environment and characterized by high seasonal variability. Similar to environmental samples, the Epsilon- and Gamma-proteobacteria were dominant in the nematode with the members of *Campylobacter*, *Thiothrix*, and *Pseudoalteromonas*.

Overall, the existing shallow vent microbiology researches suggest that there are three major microbial assemblages appearing to be present worldwide, each mapping to the specific geochemical niches available at each of the shallow vent categories, i.e., (1) a community rich in the sulfur-oxidizing microbiome, generally consisting the members of the Epsilon-proteobacteria class, (2) a community dominated by iron oxidizers, normally represented by members of the Gamma- and Zeta-proteobacteria class, and (3) a community dominated by heterotrophic and mixotrophic members of the class Gamma-proteobacteria and/or Firmicutes (Giovannelli and Price 2018). Members of the Epsilon-proteobacteria have also been reported to represent a major part of microbial communities at deep-sea hydrothermal vents (e.g. Longnecker and Lopez-Garcia 2003).

Shallow hydrothermal vents offer a variety of habitats to metabolically diverse microbes. Though cultivation-based methods alone cannot explore the entire microbial community, they do elaborate their metabolic activities in biogeochemical cycles, which can be applied in environmental biotechnology. Stetter and colleagues isolated the first microorganism to grow optimally above 100°C named *Pyrodictium*, followed by *Pyrolobus*, which has a maximum growth temperature of 113°C, which encouraged the isolation of thermophiles, hyperthermophiles, and mesophiles from shallow-water hydrothermal vents (Stetter et al. 1983 and 1987). Since then, numerous mesophilic and thermophilic heterotrophs/chemolithoautotrophs have been isolated from several shallow-water hydrothermal vents (Table 25.2) and are available in various culture collections worldwide. It is noteworthy to mention here that there is no available pure culture of Epsilon-proteobacteria, which indicates a major gap in our knowledge of the ecology

TABLE 25.2

List of Bacteria and Archaea Isolated from Shallow Hydrothermal Vents and Available as Type Strains

Phylum	Species	Isolation Source	Relevant Genome Available in NCBI	Optimum Growth Temperature (°C)	Reference
Bacteria					
Aquificae	Aquifex pyrophilus†	Kolbeinsey Ridge (Iceland)		95	Huber et al. (1992)
	Hydrogenivirga caldilitoris†	Ibusuki (Japan)	RCCJ00000000	75	Nakagawa et al. (2004a)
	Hydrogenobacter *hydrogenophilus*†(Calderobacterium *hydrogenophilus*)	Vulcano island (Italy)	OBEN00000000	75	Kryukov et al. (1983), Stöhr et al. (2001)
	Thermovibrio ruber†	Lihir Island (Papua New Guinea)		75	Huber et al. (2002)
Bacteroidetes	Rhodothermus marinus	Isafjardardjup (Iceland)	CP001807	65	Alfredsson et al. (1988)
	Rhodothermus obamensis	Tachibana Bay (Japan		80	Sako et al. (1996b)
	Aequorivita nionensis (Vitellibacter nionensis)	Espalamaca Azores (Portugal)		30	Rajasabapathy et al. (2015b)
Calditrichaeota	Calorithrix insularis	Kunashir Island (Southern Kurils, Russia)		55	Kompantseva et al. (2017)
Chloroflexi	Ardenticatena maritima	Coastal hydrothermal field in Japan	BBZA00000000	65	Kawaichi et al. (2013)

(*Continued*)

TABLE 25.2 (*Continued*)

List of Bacteria and Archaea Isolated from Shallow Hydrothermal Vents and Available as Type Strains

Phylum	Species	Isolation Source	Relevant Genome Available in NCBI	Optimum Growth Temperature (°C)	Reference
Deferribacteres	Caldithrix palaeochoryensis	Milos Island (Greece)		60	Miroshnichenko et al. (2010)
Firmicutes	Acetoanaerobium pronyense	Prony Bay (southern New Caledonia)		35	Bes et al. (2015)
	Bacillus aeolius	Vulcano Island (Italy)		55	Gugliandolo et al. (2003)
	Filobacillus milensis	Milos Island (Greece)	SOPW00000000	37	Schlesner et al. (2001)
	Geobacillus vulcani (Bacillus vulcani)	Vulcano Island (Italy)	JPOI00000000	60	Caccamo et al. (2000)
	Serpentinicella alkaliphile	Prony Bay (New Caledonia)	SLYC00000000	37	Mei et al. (2016)
	Thermaerobacter litoralis	Satsuma Peninsula (Japan)		70	Tanaka et al. (2006)
	Thermaerobacter nagasakiensis	Tachibana Bay (Nagasaki Prefecture, Japan)		70	Nunoura et al. (2002)
	Vallitalea pronyensis	Prony Bay (southern New Caledonia)		30	Aissa et al. (2014)
Planctomycetes	*Blastopirellula retiformator*	Panarea Island (Italy)	SJPF00000000	30	Kallscheuer et al. (2020)
	Bremerella volcania	Panarea Island (Italy)	GCA_007748115	36	Rensink et al. (2020)
	Thermostilla marina	Vulcano Island (Italy)		55	Slobodkina et al. (2016c)
Alpha-proteobacteria	Citreicella manganoxidans	Espalamaca, Azores (Portugal)		30	Rajasabapathy et al. (2015a)
	Methyloceanibactercaenitepidi	Kyushu (Japan)	GCA_000828475	35	Takeuchi et al. (2014)
	Oceanicella actignis	Ribeira Quente, (Island of São Miguel, Azores)	GCA_008124525	50	Albuquerque et al. (2012)
	Paracoccus aurantiacus	Kueishantao Island in Taiwan (China)	VOPL00000000	28	Ye et al. (2020)
	Tepidicaulis marinus	Kagoshima Bay (Japan)	BBIO00000000	42	Takeuchi et al. (2014)
	Varunaivibrio sulfuroxidans	Tor Caldara, Tyrrhenian Sea (Italy)	SLZW00000000	30	Patwardhan and Vetriani (2016)
Beta-proteobacteria	Burkholderia insulsa	Ambitle Island (Papua New Guinea)	PVZM00000000	37	Rusch et al. (2015)
Delta-proteobacteria	Deferrisoma paleochoriense†	Milos Island (Greece)		60	Perez-Rodriguez et al. (2015)
	Desulfacinum hydrothermale†	Milos Island (Greece)	FWXF00000000	60	Sievert and Kuever (2000)
	Desulfurella multipotens†	Raoul Island (Kermadecarchipelago, New Zealand)	GCA_900101285	58	Miroshnichenko et al. (1994)
	Dissulfurirhabdusthermomarina†	Kuril Islands (Russia)	JAAGRR000000000	50	Slobodkina et al. (2016b)
	Hippea maritima	Bay of Plenty (New Zealand)	GCA_000194135	55	Miroshnichenko et al. (1999)
Gamma-proteobacteria	Galenea microaerophila†	Milos Island (Greece)		35	Giovannelli et al. (2012)
	Halothiobacillus kellyi†(Thiobacillus kellyi)	Milos Island (Greece)		40	Sievert et al. (2000a)
	Inmirania thermothiophila†	Kuril Islands (Russia)	RJVI00000000	65	Slobodkina et al. (2016a)

(*Continued*)

TABLE 25.2 (*Continued*)

List of Bacteria and Archaea Isolated from Shallow Hydrothermal Vents and Available as Type Strains

Phylum	Species	Isolation Source	Relevant Genome Available in NCBI	Optimum Growth Temperature (°C)	Reference
	Methylomarinovumcaldicuralii	Taketomi Island (Japan)		50	Hirayama et al. (2014)
	Methylomarinum vadi	Taketomi Island (Japan)	JPON00000000	37	Hirayama et al. (2013)
	Sulfurivirga caldicuralii†	Taketomi Island (Japan)	FSRE00000000	55	Takai et al. (2006)
	Thiobacillus prosperus†	Vulcano Island (Italy)	GCA_000754095	37	Huber and Stetter (1989)
Synergistetes	Dethiosulfovibrio marinus	White sea (Russia)		28	Surkov et al. (2001)
Thermotogae	Kosmotoga arenicorallina	Yaeyama Archipelago (Japan)	GCA_001636545	60	Nunoura et al. (2010)
	Marinitoga litoralis	Île Saint-Paul (Southern Indian Ocean)		60	Postec et al. (2010)
	Thermosipho africanus	Obock (Djibouti, Africa)	GCA_003351105	75	Huber et al. (1989)
	Thermotoga maritima	Vulcano island (Italy)	GCA_000230655	90	Huber et al. (1996)
	Thermotoga neapolitana	Lucrino (Italy)	GCA_000018945	90	Jannasch et al. (1988)
Archaea					
Crenarchaeota	*Acidianus infernus*†	Campi Flegrei and VulcanoIsland (Italy)	WFIY00000000	90	Segerer et al. (1986)
	Aeropyrum pernix	Kodakara-Jima Island (Japan)	GCA_000011125	90	Sako et al. (1996a)
	Ignicoccus hospitalis†	Kolbeinsey Ridge Iceland	GCA_000017945	90	Paper et al. (2007)
	Ignicoccus islandicus†	Kolbeinsey Ridge Iceland	GCA_001481685	90	Huber et al. (2000)
	Pyrobaculum aerophilum	Ischia (Italy)	GCA_000007225	100	Volkl et al. (1993)
	Pyrodictium abyssi	Shallow vents off Mexico and Iceland		97	Pley et al. (1991)
	Pyrodictium brockii	Vulcano Island (Italy)		105	Stetter et al. (1983)
	Pyrodictium occultum	Vulcano Island (Italy)	LNTB00000000	105	Stetter et al. (1983)
	Staphylothermus hellenicus	Milos Island (Greece)	GCA_000092465	90	Arab et al. (2000)
	Staphylothermus marinus	Vulcano Island (Italy)	GCA_000015945	90	Fiala et al. (1986)
	Stetteria hydrogenophila	Milos Island (Greece)		95	Jochimsen et al. (1997)
Euryarchaeota	*Archaeoglobus fulgidus*†	Vulcano Island and Stufe di Nerone (Italy)	GCA_000008665	83	Stetter (1988)
	Ferroglobus placidus†	Vulcano Island (Italy)	GCA_000025505	80	Hafenbradl et al. (1996)
	Methanococcus aeolicus†	Lipari Islands (near Sicily)	GCA_000017185	37	Kendall et al. (2006)
	Methanococcus thermolithotrophicus†	Naples (Italy)	AQXV00000000	65	Huber et al. (1982)
	Methanopyrus kandleri†	Kolbeinsey ridge (Iceland)	GCA_000007185	98	Kurr et al. (1991)
	Methanotorris igneus† (*Methanococcus igneus*)	Kolbeinsey ridge (Iceland)	GCA_000214415	88	Burggraf et al. (1990)
	Palaeococcus helgesonii	Vulcano Island (Italy)		80	Amend et al. (2003)
	Pyrococcus furiosus	Vulcano Island (Italy)	GCA_008245085	100	Fiala and Stetter (1986)
	Pyrococcus woesei	Vulcano Island (Italy)		95	Zillig et al. (1987)
	Thermococcus acidaminovorans	Vulcano Island (Italy)		85	Dirmeier et al. (1998)
	Thermococcus aegaeicus	Milos Island (Greece)		85	Arab et al. (2000)

(*Continued*)

TABLE 25.2 (*Continued*)

List of Bacteria and Archaea Isolated from Shallow Hydrothermal Vents and Available as Type Strains

Phylum	Species	Isolation Source	Relevant Genome Available in NCBI	Optimum Growth Temperature (°C)	Reference
	Thermococcus alcaliphilus	Vulcano Island (Italy)		85	Keller et al. (1995)
	Thermococcus gorgonarius	Bay of Plenty (New Zealand)	GCA_002214385	85	Miroshnichenko et al. (1998)
	Thermococcus litoralis	Lucrino, Bay of Naples (Italy)	GCA_000246985	85	Neuner et al. (1990)
	Thermococcus pacificus	Bay of Plenty (New Zealand)	GCA_002214485	85	Miroshnichenko et al. (1998)
	Thermococcus stetteri	Ushishir archipelago (Japan)		75	Miroshnichenko et al. (1989)

Modified from Giovannelli and Price (2018) and Updated.
Species with † indicates chemolithoautotrophs; remaining all are heterotrophs.

of one of the dominant members of shallow-water hydrothermal vent communities (Giovannelli and Price 2018). This is mainly because the available cultivation methods are not suitable for cultivating the Epsilon-proteobacteria members, which requires novel cultivation techniques such as iChip, FPMT, etc. (Nichols et al. 2010; Jung et al. 2018).

Studying individual genomes of microorganisms will give us a better understanding of their metabolic potential, their ability to cause diseases, and also their ability to survive in extreme environmental conditions. Several researchers have obtained partial or complete genomic information of shallow hydrothermal vent bacteria. Along with the pure culture details, in Table 25.2, we have also added relevant genome sequence information that is available in the GenBank database. In addition, researchers also sequenced bacterial genomes from various cultured isolates obtained from shallow-water vents (e.g. Handley et al. 2013; Filippidou et al. 2015; Lin et al. 2016; Pollo et al. 2016; Han et al. 2017). The complete genome of *Serinicoccus* sp. strain JLT9 isolated from Taiwan coast, possessed repertoire of genes responsible for the oxidation of reduced sulfur compounds. These genes encoded enzymes for the oxidation of reduced sulfur compounds including sulfide quinone oxidoreductase, mediating the oxidation of sulfide to elemental sulfur, rhodanese sulfur transferase for oxidation of thiosulfate to sulfite, and reverse dissimilatory sulfite reductase for oxidation of elemental sulfur to sulfite, adenosine 5-phosphosulfate reductase, and sulfate adenylyl transferase for oxidation of sulfite to sulfate (Han et al. 2017). The genome of another shallow vent bacteria, *Bacillus alveayuensis* strain 24KAM51, isolated from Milos (Greece) contains genes related to copper (copper-binding proteins and multicopper oxidase), manganese (manganese transporters and permeases), cadmium (cadmium transporter), zinc (zinc metalloproteases, proteases, and transporters), and arsenic (arsenic resistance protein, ArsB) resistance. Additionally, this bacterium also possesses genes encoding the NarH and NarZ proteins (nitrate reduction), as well as DsrE (sulfur reduction) (Filippidou et al. 2015). Much more studies on this aspect will obviously reveal the role of microbiome in such distinctive environments.

The diversity of functional genes and the rates of utilization of other substrates in energy-conserving reactions have rarely been investigated at shallow-hydrothermal systems (Akerman et al. 2011; Tang et al. 2013). The available few studies revealed that the role and contribution of alternative electron acceptors such as nitrate, nitrite, manganese, and iron (III) could be important in shallow-water hydrothermal vents. Zeta-proteobacteria have been reported as abundant in shallow-water hydrothermal vents where sulfide concentrations in the hydrothermal fluids are low and Fe(II) concentrations are high, suggesting the potential for biologically mediated iron oxidation. Considering these findings, the combination of metagenomic, metatranscriptomic, and metaproteomic studies along with *in situ* geochemistry and microbiological efforts will be needed to elucidate the importance of alternative electron acceptors and energetic metabolic pathways in shallow hydrothermal vent ecosystems (Price and Giovannelli 2017).

25.5 Role of Microorganisms in Shallow-Water Hydrothermal Vents

In the areas of hydrogen sulfide venting, the foremost role in biogeochemical processes belongs to the bacteria that oxidize or reduce a range of sulfur compounds during primary production (chemosynthesis and an oxygenic photosynthesis) or decomposition of organic matter. From the biogeochemical point of view, the applicable links in the sulfur cycle are groups of sulfur-reducing hyper thermophilic bacteria (including new species of archaea) and thermophilic bacteria. All these microbial members participate in the anaerobic degradation of organic matter and reduce sulfur to hydrogen sulfide. In shallow hydrothermal zones, the sulfur may have a deep (endogenic) origin or may be formed as a result of biological and chemical oxidation of hydrogen sulfide (Tarasov 2006). Studies have shown that Epsilon-proteobacteria members utilize reduced inorganic sulfur compounds (Inagaki et al. 2003) by sulfur-oxidation pathways and sulfur-reduction pathways. The detection of *sor* gene indicated the presence of direct sulfite oxidation, which was observed previously in the genome of *Sulfurovum* sp. NBC37-1 and *Sulfurimonas autotrophica* (Inagaki et al. 2003; Yamamoto et al. 2010).

The dominant CO_2 fixation mechanism of Epsilon-proteobacteria is by reductive tricarboxylic acid (rTCA) cycle (Huang 2012), which was originally discovered in green sulfur phototrophs, i.e., *Chlorobium* (Evans et al. 1966). Since then, it has been discovered in many chemoautotrophs including a sulfate-reducing Deltaproteobacterium (i.e., *Desulfobacter hydrogenophilus*), thermophilic *Aquificales* (e.g., *Hydrogenobacter* and *Aquifex*) and *Thermoproteales* (e.g., *Thermoproteus*). Most importantly, it has also been shown that deep-sea Epsilon-proteobacteria can utilize rTCA cycle as a carbon-fixation pathway (Nakagawa et al. 2004b; Hügler et al. 2010).

Peptides and polysaccharides (such as starch, glycogen, and pectin) are substrates for hyperthermophilic and mesophilic bacteria. However, full breakdown to CO_2 of high-molecular-weight compounds does not occur during bacterial fermentation. Further breakdown of the products of fermentation produced by extreme thermophilic bacteria (acetate, formate, etc.) is carried out by mesophilic bacteria at lower temperatures (<50°C–60°C) in the surface layers of the hydrothermal system, in the bottom sediments, or in the bacterial mats. The literature data demonstrate that such groups of bacteria develop in all hydrogen sulfide venting areas, and the rate of reduction of elemental sulfur is high. Such a pattern of anaerobic oxidation of organic substances is highly characteristic of shallow venting areas and has not been recorded previously in coastal ecosystems (Tarasov 2006).

Vent fluids at shallow-sea hydrothermal vents are usually enriched in H_2S, H_2, CH_4, Fe (II), and different trace elements and depleted in magnesium and sulfate compared to standard seawater concentration. Little is known about how microorganisms from marine hydrothermal environments interact with metals, but their interactions are generally described in one of three ways: the metals are toxic and elicit a response, they are oxidized or reduced to conserve energy in dissimilatory reactions, or they are taken up and utilized in assimilatory reactions (Holden and Adams 2003). Previous studies demonstrated that heterotrophic bacteria not only function as decomposers but also channel the dissolved organic and inorganic nutrients into higher trophic levels through microbial food-web (Azam et al. 1983; Azam 1998). The heterotrophic bacteria are highly abundant in the ocean and play a significant role in the biogeochemical cycle of carbon, nitrogen, and sulfur (Copley 2002; Karl 2002) in the hydrothermal vent ecosystem. Mn-oxidizing and Mn- and Fe-reducing bacteria can also play an important role in microbial systems at shallow-water vents, though the role of these bacteria in the balance of organic matter and the details of processes remain unknown (Tarasov et al. 2005).

25.6 Biotechnological Applications of Hydrothermal Vent Microorganisms

Though many microorganisms are already available in public culture collections, investigation of extremophilic organisms from various extreme environments like hydrothermal vents, cold seeps, and subterranean environments is required either with cultural methods or DNA-based molecular approaches to enhance the possibility of finding novel bioactive compounds. A high number of bacterial isolates belonging to Proteobacteria and Firmicutes isolated from Kolumbo submarine volcano of Santorini Island exhibited antimicrobial properties (Bourbouli et al. 2015). In another study, *Bacillus* spp. isolated from the shallow hydrothermal vent of Espalamaca (Azores, Portugal) also exhibited antibacterial activity (Ravindran et al. 2016).

Microorganisms inhabiting extreme environments often produce polymers and unusual enzymes to survive in high temperatures or high concentrations of H_2S and heavy metals (Maugeri et al. 2002). Bacterial exopolysaccharides (EPS) have also been isolated from shallow hydrothermal vents, especially those from mesophilic *Vibrio* and *Alteromonas* strains (Querellou 2003). The major areas concerned are cardiovascular diseases and tissue regeneration (proangiogenic effects/antithrombotic). Studies conducted by Guezennec (2002) on anticoagulant activities of the EPS showed that native EPS were deprived of effects, whereas sulfated derivatives were active.

Extremophiles are distinctive, which are adapted to thrive in ecological niches such as extreme pH, high or low temperatures, high salt concentrations, and high pressure. Therefore, biological systems and enzymes can function at temperatures between −5°C and 130°C, pH 0–12, salt concentrations of 3%–35%, and pressures up to1,000 bar (Bertoldo et al. 2002). Gugliandolo et al. (2012) characterized thermophilic bacilli from Panarea Island (Italy) to identify useful biomolecules for industrial purposes and environmental applications. The study revealed that many of the bacilli were thermophilic, alkalophilic, and haloalkaliphilic in nature. Most of the *Bacillus* spp. produced gelatinase, lipase, and amylase, and some of them were resistant to mercury. Erra-Pujada et al. (2001) isolated and purified type II pullulanase from *Thermococcus hydrothermalis* and characterized it for pullulanolytic and amylolytic activities. Undoubtedly, pullulanase can be used in tandem with other amylolytic enzymes for the conversion of starch to glucose, maltose, or fructose syrups (Saha and Zeikus 1989).

Cornec et al. (1998) investigated on thermostable esterases from hyperthermophilic archaeal and bacterial strains isolated from hydrothermal vents. The esterase activity exhibited a half-life of 22 h at 99°C and of 13 min at 120°C and retained its entire initial activity after incubation at 90°C for 8.5 h without any substrate and/or cofactor. Lipases from microbial origin have been used as an important biocatalyst in biomedical applications. Because of their tremendous catalytic action in a variety of organic solvents, they could be used for the synthesis of compounds of pharmaceutical concern. The majority of the isolates from the hydrothermal vent region in the Eolian Islands (Italy) showed lipolytic and amylolytic activities (Gugliandolo et al. 2012). Besides, a hemicellulolytic thermophilic bacterium, *Thermoanaerobacterium*, isolated from a shallow hydrothermal vent in Taiwan showed the highest levels of ethanol production from xylose or rice straw hemicellulosic hydrolysate at 70°C (Tsai et al. 2011). These findings confirm the potential of microbes originated from hydrothermal vents. Enzymes that have optimum activity at higher temperatures and pH are widely used in the household detergent, food, textile, pulp, paper, chemical, and leather-processing industries (Podar and Reysenbach 2006).

Apart from the production of thermostable enzymes and EPS, hydrothermal vent microbes play a vital role in metal recovery and detoxification. Further, they actively participate in the oxidation and reduction reaction. Rathgeber et al. (2002) isolated high numbers of tellurite- and selenite-reducing strains from the seawater samples near hydrothermal vents, bacterial films, and sulfide-rich rocks in Juan de Fuca Ridge in the Pacific Ocean. The growth of these bacterial members in K_2TeO_3- or Na_2SeO_3-amended media resulted in the accumulation of metallic tellurium or selenium. Around ten bacterial groups (most of them belong to *Pseudoalteromonas*) could tolerate up to 2500 μg mL^{-1} of K_2TeO_3 and up to 7000 μg mL^{-1} of Na_2SeO_3. Vetrini et al. (2005) isolated several bacteria resistant to mercury from hydrothermal fluids in EPR. Four moderate thermophiles (most of the isolates belonging to the genus *Alcanivorax*) and six mesophiles from the vent plume were resistant to >10 μM Hg(II) and reduced it to elemental mercury [$Hg_{(0)}$]. These heavy-metal-resistant and detoxifying hydrothermal vent bacteria may show promise in environmental applications. The low-temperature shallow venting area in the Espalamaca region is richly diversified with various metal-tolerant and metal-oxidizing heterotrophic bacteria. The maximum tolerable level for Mn, Pb, and Fe was 50 mM, 7.5 mM, and 1 mM, respectively. *Citreicella* sp. (VSW210) oxidized soluble Mn(II) quickly and survived in a high concentration of 50 mM Mn. Also, the majority of the bacterial phylotypes isolated from this vent were tolerant to multi-metals (Rajasabapathy 2015). From these studies, although limited, it is proposed that shallow hydrothermal systems may offer several biotechnological compounds.

25.7 Concluding Remarks

Shallow hydrothermal vents have considerable differences when compared to the deep-sea vents, mainly with the presence of sunlight, wave actions, low pressure, and terrestrial meteoric inputs. Knowledge of shallow hydrothermal vent microbial communities may offer significant information because they react quickly to changes in the concentrations and availability of chemicals within their environment. The available few metagenomic and classical microbiological studies revealed two types of bacterial communities, one is dominated by Gamma-proteobacteria members and another one dominated by Epsilon-proteobacteria. The microorganisms in the shallow-water hydrothermal vents represent a hot spot of potential biotechnological breakthroughs; still, we have only fewer reports. Since only a small percentage of microorganisms are cultivable (0.1%–1.0%), it is necessary to isolate much more novel microorganisms using new cultivation approaches. Most importantly, individual microbial genomes, together with wide-ranging metagenomic analysis, will provide us a complete picture of the hydrothermal vent microbiome.

Acknowledgments

RR acknowledges University Grants Commission, India for granting Dr. D. S. Kothari Postdoctoral Fellowship (BL/15–16/0370). This work was supported by Fundação para a Ciência e a Tecnologia (FCT) through IF/00029/2014/CP1230/CT0002 to A.C. and through the strategic projects UID/ 05634/2020. The authors thank Ricardo Medeiros from IMAR for technical support.

REFERENCES

Aissa FB, Postec A, Erauso G, et al. (2014) *Vallitalea pronyensis* sp. nov., isolated from a marine alkaline hydrothermal chimney. *Int. J. Syst. Evol. Microbiol.* 64: 1160–1165.

Akerman NH, Price RE, Pichler T, Amend JP (2011) Energy sources for chemolithotrophs in an arsenic- and iron- rich shallow-sea hydrothermal system. *Geobiol.* 9: 436–445.

Albuquerque L, Rainey FA, Nobre MF, da Costa MS (2012) *Oceanicella actignis* gen. nov., sp. nov., a halophilic slightly thermophilic member of the Alphaproteobacteria. *Syst. Appl. Microbiol.* 35(6): 385–389.

Alfredsson GA, Kristjansson JK, Hjorleifsdottir S, Stetter KO (1988) *Rhodothermus marinus*, gen. nov., sp. nov., a thermophilic, halophilic bacterium from submarine hot springs in Iceland. *J. Gen. Microbiol.* 134: 299–306.

Amend JP, Darcy R, Sheth SN, Zolotova N, Amend AC (2003) *Palaeococcus helgesonii* sp. nov., a facultatively anaerobic, hyperthermophilic archaeon from a geothermal well on Vulcano Island, Italy. *Arch. Microbiol.* 179: 394–401.

Arab H, Volker H, Thomm M (2000) *Thermococcus aegaeicus* sp. nov. and *Staphylothermus hellenicus* sp. nov., two novel hyperthermophilic Archaea isolated from geothermally heated vents off Palaeochori Bay, Milos, Greece. *Int. J. Syst. Evol. Microbiol.* 50: 2101–2108.

Azam F (1998) Microbial control of oceanic carbon flux: The plot thickens. *Science* 280: 694–696.

Azam F, Fenchel T, Gray JG, Meyer-Reil LA, Thingstad F (1983) The ecological role of water-column microbes in the sea. *Mar. Ecol. Prog. Ser.* 10: 257–263.

Beaulieu SE, Szafranski K (2019) InterRidge Global Database of Active Submarine Hydrothermal Vent Fields, Version 3.4. World Wide Web electronic publication available from http://vents-data.interridge.org Accessed on 2020-05-14.

Bellec L, Bonavita MC, Hourdez S (2019) Chemosynthetic ectosymbionts associated with a shallow-water marine nematode. *Sci. Rep.* 9: 7019. Doi: 10.1038/s41598-019-43517-8.

Bertoldo C, Grote R, Antranikian G (2002) Extremophiles: Life in extreme environments. *Encyclop. Environ. Microbiol.* 6: 1232–1237.

Bes M, Merrouch M, Joseph M, et al. (2015) *Acetoanaerobium pronyense* sp. nov., an anaerobic alkaliphilic bacterium isolated from a carbonate chimney of the Prony Hydrothermal Field (New Caledonia). *Int. J. Syst. Evol. Microbiol.* 65: 2574–2580.

Bourbouli M, Katsifas EA, Papathanassiou E, Karagauni AD (2015) The Kolumbo submarine volcano of Santorini Island is a large pool of bacterial strains with antimicrobial activity. *Arch. Microbiol.* 197: 539–552.

Brooks D (2006) Tropical Marine Ecology. Hays Cummings. June 5, 2006. Hydrothermal Vents.

Burggraf S, Fricke H, Neuner A, et al. (1990) *Methanococcus igneus* sp. nov., a novel hyperthermophilic methanogen from a shallow submarine hydrothermal system. *Syst. Appl. Microbiol.* 13: 263–269.

Caccamo D, Gugliandolo C, Stackebrandt E, Maugeri TL (2000) *Bacillus vulcani* sp. nov., a novel thermophilic species isolated from a shallow marine hydrothermal vent. *Int. J. Syst. Evol. Microbiol.* 50: 2009–2012.

Copley J (2002) All at sea. *Nature* 415: 572–574.

Corliss JB, Dymond J, Gordon LI, et al. (1979) Submarine thermal springs on the Galapagos Rift. *Science* 203: 1073–1083.

Cornec L, Robineau J, Rolland JL, Dietrich J, Barbier G (1998) Thermostable esterases screened on hyperthermophilic archaeal and bacterial strains isolated from deep-sea hydrothermal vents: Characterization of esterase activity of a hyperthermophilic archaeum, *Pyrococcus abyssi*. *J. Mar. Biotechnol.* 6: 104–110.

Dando P, Hughes J, Leahy Y, Niven S, Taylor L, Smith C (1995) Gas venting rates from submarine hydrothermal areas around the island of Milos, Hellenic Volcanic Arc. *Cont. Shelf. Res.* 15: 913–929.

Dando PR, Thomm M, Arab H, et al. (1998) Microbiology of shallow hydrothermal sites off Palaeochori Bay, Milos (Hellenic Volcanic Arc). *Cah. Biol. Mar.* 39: 369–372.

Dirmeier R, Keller M, Hafenbradl D, et al. (1998) *Thermococcus acidaminovorans* sp. nov., a new hyperthermophilic alkalophilic archaeon growing on amino acids. *Extremophiles* 2: 109–114.

Erra-Pujada M, Chang-Pi-Hin F, Debeire P, Duchiron F, O'Donohue MJ (2001) Purification and properties of the catalytic domain of the thermostable pullulanase type II from *Thermococcus hydrothermalis*. *Biotechnol. Lett.* 23: 1273–1277.

Evans M, Buchanan B, Arnon D (1966) A new ferredoxin-dependent carbon reduction cycle in a photosynthetic bacterium. *Proc. Natl. Acad. Sci. U.S.A.* 55 (4): 928–934.

Fiala G, Stetter KO (1986) *Pyrococcus furiosus* sp. nov. represents anovel genus of marine heterotrophic archaebacteria growing optimally at 100°C. *Arch. Microbiol.* 145: 56–61.

Fiala G, Stetter KO, Jannasch HW, Langworthy TA, Madon J (1986) *Staphylothermus marinus* sp. nov. represents a novel genus of extremely thermophilic submarine heterotrophic archaebacteria growing up to 98°C. *Syst. Appl. Microbiol.* 8: 106–113.

Filippidou S, Wunderlin T, Junier T, et al. (2015) Genome sequence of *Bacillus alveayuensis* strain 24KAM51, a halotolerant thermophile isolated from a hydrothermal vent. *Genome Announc* 3(4):e00982–15. Doi: 10.1128/genomeA.00982-15.

Giovannelli D, d'Errico G, Manini E, Yakimov M, Vetriani C (2013) Diversity and phylogenetic analyses of bacteria from a shallow-water hydrothermal vent in Milos island (Greece). *Front. Microbiol.* 4: 84. Doi: 10.3389/fmicb.2013.00184.

Giovannelli D, Grosche A, Starovoytov V, Yakimov M, Manini E, Vetriani C (2012) *Galenea microaerophila* gen. nov., sp. nov., a mesophilic, microaerophilic, chemosynthetic, thiosulfate-oxidizing bacterium isolated from a shallow-water hydrothermal vent. *Int. J. Syst. Evol. Microbiol.* 62(12): 3060–3066.

Giovannelli D, Price RE (2018) Marine shallow-water hydrothermal vents: Microbiology. *Encyclopedia of Ocean Sciences* (Third Edition) 4: 353–363.

Guezennec J (2002) Deep-sea hydrothermal vents: A new source of innovative bacterial exopolysaccharides of biotechnological interest? *J. Ind. Microbiol. Biotechnol.* 29: 204–208.

Gugliandolo C, Lentini V, Bunk B, Overmann J, Italiano F, Maugeri TL (2015) Changes in prokaryotic community composition accompanying a pronounced temperature shift of a shallow marine thermal brine pool (Panarea Island, Italy). *Extremophiles* 19: 547–559.

Gugliandolo C, Lentini V, Spanó A, Maugeri TL (2012) New bacilli from shallow hydrothermal vents of Panarea Island (Italy) and their biotechnological potential. *J. Appl. Microbiol.* 112: 1102–1112.

Gugliandolo C, Maugeri TL (2019) Phylogenetic diversity of archaea in shallow hydro-thermal vents of Eolian Islands, Italy. *Diversity* 11: 156. Doi: 10.3390/d11090156.

Gugliandolo C, Maugeri TL, Caccamo D, Stackebrandt E (2003) *Bacillus aeolius* sp. nov. a novel thermophilic, halophilic marine *Bacillus* species from Eolian Islands (Italy). *Syst. Appl. Microbiol.* 26: 172–176.

Hafenbradl D, Keller M, Dirmeier R, et al. (1996) *Ferroglobus placidus* gen. nov., sp. nov., a novel hyperthermophilic archaeum that oxidizes Fe^{2+} at neutral pH under anoxic conditions. *Arch. Microbiol.* 166: 308–314.

Hammond SR (1997) Offset caldera and crater collapse on Juan de Fuca ridge-flank volcanoes. *Bull. Volcano* 58: 617–627.

Han Y, Lin D, Yu L, Chen X, Sun J, Tang K (2017) Complete genome sequence of *Serinicoccus* sp. JLT9, an actinomycete isolated from the shallow-sea hydrothermal system. *Mar. Genomics* 32: 19–21.

Handley KM, Upton M, Beatson SA, Héry M, Lloyd JR (2013) Genome sequence of hydrothermal arsenic-respiring bacterium *Marinobacter santoriniensis* NKSG1T. *Genome Announc.* 1(3): e00231–13. Doi: 10.1128/genomeA.00231-13.

Hirayama H, Abe M, Miyazaki M, et al. (2014) *Methylomarinovum caldicuralii* gen. nov., sp. nov., a moderately thermophilic methanotroph isolated from a shallow submarine hydrothermal system, and proposal of the family Methylothermaceae fam. nov. *Int. J. Syst. Evol. Microbiol.* 64: 989–999.

Hirayama H, Fuse H, Abe M, et al. (2013) *Methylomarinum vadi* gen. nov., sp. nov., a methanotroph isolated from two distinct marine environments. *Int. J. Syst. Evol. Microbiol.* 63: 1073–1082.

Hirayama H, Sunamura M, Takai K, et al. (2007) Culture-dependent and independent characterization of microbial communities associated with a shallow submarine hydrothermal system occurring within a coral reef off Taketomi Island, Japan. *Appl. Environ. Microbiol.* 73(23): 7642–7656.

Holden JF, Adams MWW (2003) Microbe–metal interactions in marine hydrothermal environments. *Curr. Opin. Chem. Biol.* 7: 160–165.

Huang CI (2012) Molecular Ecology of Free-Living Chemoautotrophic Microbial Communities at a Shallow-Sea Hydrothermal Vent. Ph.D. dissertation, Universitaet Bremen.

Huber H, Burggraf S, Mayer T, WyschkonyI, Rachel R, Stetter KO (2000) *Ignicoccus* gen. nov., a novel genus of hyperthermophilic, chemolithoautotrophic Archaea, represented by two new species, *Ignicoccus islandicus* sp nov and *Ignicoccus pacificus* sp nov. *Int. J. Syst. Evol. Microbiol.* 50: 2093–2100.

Huber H, Diller S, Horn C, Rachel R (2002) *Thermovibrio ruber* gen. nov., sp. nov., an extremely thermophilic, chemolithoautotrophic, nitrate-reducing bacterium that forms a deep branch within the phylum Aquificae. *Int. J. Syst. Evol. Microbiol.* 52: 1859–1865.

Huber H, Stetter KO (1989) *Thiobacillus prosperus* sp. nov., represents a new group of halotolerant metal-mobilizing bacteria isolated from a marine geothermal field. *Arch. Microbiol.* 151: 479–485.

Huber H, Thomm M, Konig H, Thies G, Stetter KO (1982) *Methanococcus thermolithotrophicus*, a novel thermophilic lithotrophic methanogen. *Arch. Microbiol.* 132: 47–50.

Huber R, Langworthy TA, Konig H, et al. (1996) *Thermotoga maritime* sp. nov. represents a new genus of unique extremely thermophilic eubacteria growing up to 90°C. *Arch. Microbiol.* 144: 324–333.

Huber R, Wilharm T, Huber D, et al. (1992) *Aquifex pyrophilus* gen. nov. sp. nov., represents a novel group of marine hyperthermophilic hydrogen-oxidizing bacteria. *Syst. Appl. Microbiol.* 15: 340–351.

Huber R, Woese CR, Langworthy TA, Fricke H, Stetter KO (1989) *Thermosipho africanus* gen. nov., represents a new genus of thermophilic eubacteria within the "Thermotogales." *Syst. Appl. Microbiol.* 12: 32–37.

Imhoff I, Hugler M (2009) Life at deep sea hydrothermal vents - Oases under water. *The Int. J. Mar. Coast. Law* 24: 201–208.

Inagaki F, Takai K, Kobayashi H, Nealson KH, Horikoshi K (2003) *Sulfurimonas autotrophica* gen. nov., sp. nov., a novel sulfur-oxidizing ε-proteobacterium isolated from hydrothermal sediments in the Mid-OkinawaTrough. *Int. J. Syst. Evol. Microbiol.* 53: 1801–1805.

Jannasch HW, Huber R, Belkin S, Stetter KO (1988) *Thermotoga neapolitana* sp. nov. of the extremely thermophilic, eubacterial genus Thermotoga. *Arch. Microbiol.* 150: 103–104.

Jannasch HW, Mottl MJ (1985) Geomicrobiology of deep-sea hydrothermal vents. *Science* 229: 717–725.

Jochimsen B, Peinemann-Simon S, Volker H, et al. (1997) *Stetteria hydrogenophila*, gen. nov. and sp. nov., anovel mixotrophic sulfur-dependent crenarchaeote isolated from Milos, Greece. *Extremophiles* 1: 67–73.

Jones ML (1985) *Hydrothermal Vents of the Eastern Pacific: An Overview.* 6th ed. Vienna, VA: INFAX Corporation.

Jung D, Seo E-Y, Owen JS, et al. (2018) Application of the filter plate microbial trap (FPMT), for cultivating thermophilic bacteria from thermal springs in Barguzin area, eastern Baikal, Russia. *Biosci. Biotech. Biochem.* 82(9):1–9.

Kallscheuer N, Wiegand S, Heuer A, et al. (2020) *Blastopirellula retiformator* sp. nov. isolated from the shallow-sea hydrothermal vent system close to Panarea Island. *Antonie van Leeuwenhoek*, Doi: 10.1007/s10482-019-01377-2.

Kamenev GM, Fadeev VI, Selin NI, Tarasov VG, Maalakhov VV (1993) Composition and distribution of macro- and meiobenthos around sublittoral hydrothermal vents in the Bay of Plenty, New Zealand. *New Zeal. J. Mar. Fresh.* 27: 407–418.

Karl DM (1995) Ecology of free-living, hydrothermal vent microbial communities. In Karl DM (Ed.), Microbiology of deep-sea hydrothermal vents. CRC Press, Inc., Boca Raton, FL.

Karl DM (2002) Microbiological oceanography: Hidden in a sea of microbes. *Nature* 415: 590–591.

Kawaichi S, Ito N, Kamikawa R, Sugawara T, Yoshida T, Sako Y (2013) *Ardenticatena maritima* gen. nov., sp. nov., a ferric iron- and nitrate-reducing bacterium of the phylum "Chloroflexi" isolated from an iron-rich coastal hydrothermal field, and description of *Ardenticatenia classis* nov. *Int. J. Syst. Evol. Microbiol.* 63: 2992–3002.

Keller M, Braun F-J, Dirmeier R, et al. (1995) *Thermococcus alcaliphilus* sp. nov., a new hyperthermophilic archaeum growing on polysulfide at alkaline pH. *Arch. Microbiol.* 164: 390–395.

Kendall MM, Liu Y, Sieprawska-Lupa M, Stetter KO, Whitman WB, Boone DR (2006) *Methanococcus aeolicus* sp. nov., a mesophilic, methanogenic archaeon from shallow and deep marine sediments. *Int. J. Syst. Evol. Microbiol.* 56: 1525–1529.

Kompantseva EI, Kublanov IV, Perevalova AA et al. (2017) *Calorithrix insularis* gen. nov., sp. nov., a novel representative of the phylum Calditrichaeota. *Int. J. Syst. Evol. Microbiol.* 67(5): 1486–1490.

Kryukov VR, Savelyeva ND, Pusheva MA (1983) *Calderobacterium hydrogenophilum* nov. gen. nov. sp., an extremethermophilic hydrogen bacterium, and its hydrogenase activity. *Mikrobiologiya* 52: 781–788.

Kurr M, Huber R, Konig H, et al. (1991) *Methanopyrus kandleri*, gen. and sp. nov. represents a novel group of hyperthermophilic methanogens, growing at 110°C. *Arch. Microbiol.* 156: 239–247.

Leahy S (2018) https://www.nationalgeographic.com/news/2018/06/hydrothermal-vents-discovered-azores-science-environment/.

Lentini V, Gugliandolo C, Bunk B, Overmann J, Maugeri TL (2014) Diversity of prokaryotic community at a shallow marine hydrothermal site elucidated by Illumina sequencing technology. *Curr. Microbiol.* 69: 457–466.

Lin W, Chen H, Chen Q, Liu Y, Jiao N, Zheng Q (2016) Genome sequence of *Bacillus* sp. CHD6a, isolated from the shallow-sea hydrothermal vent, Mar. *Genomics* 25: 15–16.

LLópez-García P, Duperron S, Philippot P, Foriel J, Susini J, Moreira D (2003) Bacterial diversity in hydrothermal sediment and epsilonproteobacterial dominance in experimental microcolonizers at the Mid-Atlantic Ridge. *Environ. Microbiol.* 5(10): 961–976.

Lutz RA, Kennish M (1993) Ecology of deep-sea hydrothermal vent communities: A review. *Rev. Geophy.* 31: 211–242.

Maugeri TL, Gugliandolo C, Caccamo D, et al. (2002) A halophilic thermotolerant *Bacillus* isolated from a marine hot spring able to produce a new exopolysaccharide. *Biotechnol. Lett.* 24: 515–519.

Mei N, Postec A, Erauso G et al. (2016) *Serpentinicella alkaliphila* gen. nov., sp. nov., a novel alkaliphilic anaerobic bacterium isolated from the serpentinite-hosted Prony hydrothermal field, New Caledonia. *Int. J. Syst. Evol. Microbiol.* 66(11): 4464–4470.

Miroshnichenko ML, Bonch-Osmolovskaya EA, Neuner A, Kostrikina NA, Chernych NA, Aleksee VA (1989) *Thermococcus stetteri* sp. nov., a new extremely thermophilic marine sulfur-metabolizing Archaebacterium. *Syst. Appl. Microbiol.* 12(3): 257–262.

Miroshnichenko ML, Gongadze GA, Lysenko AM, Bonch-Osmolovskaya EA (1994) *Desulfurella multipotens* sp. nov., a new sulfur-respiring thermophilic eubacterium from Raoul Island (Kermadec archipelago, New Zealand). *Arch. Microbiol.* 161: 88–93.

Miroshnichenko ML, Gongadze GM, Rainey FA, et al. (1998) *Thermococcus gorgonarius* sp. nov. and *Thermococcus pacificus* sp. nov: Heterotrophic extremely thermophilic Archaea from New Zealand submarine hot vents. *Int. J. Syst. Evol. Microbiol.* 48: 23–29.

Miroshnichenko ML, Kolganova TV, Spring S, Chernyh N, Bonch-Osmolovskaya EA (2010) *Caldithrix palaeochoryensis* sp. nov., a thermophilic, anaerobic, chemo-organotrophic bacterium from a geothermally heated sediment, and emended description of the genus *Caldithrix*. *Int. J. Syst. Evol. Microbiol.* 60: 2120–2123.

Miroshnichenko ML, Rainey FA, Rhode M, Bonch-Osmolovskaya EA (1999) *Hippea maritima* gen. nov., sp. nov., a new genus of thermophilic, sulfur-reducing bacterium from submarine hot vents. *Int. J. Syst. Bacteriol.* 49: 1033–1038.

Mohandass C, Rajasabapathy R, Ravindran C, Colaco A, Santos RS, Meena RM (2012) Bacterial diversity and their adaptations in the shallow water hydrothermal vent at D Joao de Castro Seamount (DJCS), Azores, Portugal. *Cah. Biol. Mar.* 53: 65–76.

Morri C, Bianchi CN, Cocito S, et al. (1999) Biodiversity of marine sessile epifauna at an Aegean island subject to hydrothermal activity: Milos, Eastern Mediterranean Sea. *Mar. Biol.* 135: 729–739.

Moyer CL, Dobbs FC, Karl DM (1995) Phylogenetic diversity of the bacterial community from a microbial mat at an active, hydrothermal vent system, Loihi Seamount, Hawaii. *Appl. Environ. Microbiol.* 61(4): 1555–1562.

Murdock S, Johnson H, Forget N, Juniper SK (2010) Composition and diversity of microbial mats at shallow hydrothermal vents on Volcono 1, South Tonga Arc. *Cah. Biol. Mar.* 51: 407–413.

Nakagawa T, Nakagawa S, Inagaki F, Takai K, Horikoshi K (2004b) Phylogenetic diversity of sulfate-reducing prokaryotes in active deep-sea hydrothermal vent chimney structures. *FEMS Microbiol. Lett.* 232: 145–152.

Nakagawa S, Nakamura S, Inagaki F, Takai K, Shirai N, Sako Y (2004a) *Hydrogenivirga caldilitoris* gen. nov., sp. nov., anovel extremely thermophilic, hydrogen- and sulphur oxidizing bacterium from a coastal hydrothermal field. *Int. J. Syst. Evol. Microbiol.* 54: 2079–2084.

Neuner A, Jannasch HW, Belkin S, Stetter KO (1990) *Thermococcus litoralis* sp. Nov: A new species of extremely thermophilic marine archaebacteria. *Arch. Microbiol.* 153: 205–207.

Nichols D, Cahoon N, Trakhtenberg EM, et al. (2010) Use of Ichip for high-throughput *In Situ* cultivation of "uncultivable" microbial species. *Appl. Environ. Microbiol.* 76: 2445–2450.

Nunoura T, Akihara S, Takai K, Sako Y (2002) *Thermaerobacter nagasakiensis* sp. nov., a novel aerobic and extremely thermophillic marine bacterium. *Arch. Microbiol.* 177: 339–344.

Nunoura T, Hirai M, Imachi H, et al. (2010) *Kosmotoga arenicorallina* sp. nov. a thermophilic and obligately anaerobic heterotroph isolated from a shallow hydrothermal system occurring within a coral reef, southern part of the Yaeyama Archipelago, Japan, reclassification of *Thermococcoides shengliensis* as *Kosmotoga shengliensis* comb. nov., and emended description of the genus *Kosmotoga*. *Arch. Microbiol.* 192, 811–819.

Paper W, Jahn U, Hohn MJ (2007) *Ignicoccus hospitalis* sp. nov., the host of '*Nanoarchaeum equitans*'. *Int. J. Syst. Evol. Microbiol*, 57: 803–808.

Patwardhan S, Vetriani C (2016) *Varunaivibrio sulfuroxidans* gen. nov., sp. nov., a facultatively chemolithoautotrophic, mesophilic alphaproteobacterium from a shallow-water gas vent at Tor Caldara, Tyrrhenian Sea. *Int. J. Syst. Evol. Microbiol.* 66: 3579–3584.

Perez-Rodriguez I, Rawls M, Coykendall DK, Foustoukos DI (2015) *Deferrisoma paleochoriense* sp. nov., athermophilic, iron (III)-reducing bacterium from a shallow-water hydrothermal vent in the Mediterranean Sea. *Int. J. Syst. Evol. Microbiol.* 66: 830–836.

Perner M, Gonnella G, Kurtz S, LaRoche J (2014) Handling temperature bursts reaching 464°C: Different microbial strategies in the Sisters Peak hydrothermal chimney. *Appl. Environ. Microbiol.* 80(15): 4585–4598.

Pley U, Schipka J, Gambacorta A, et al. (1991) Pyrodictium abyssi sp. nov. represents a novel heterotrophic marine archaeal hyperthermophile growing at 110°C. *Syst. Appl. Microbiol.* 14(3): 245–253.

Podar M, Reysenbach AL (2006) New opportunities revealed by biotechnological explorations of extremophiles. *Curr. Opin. Biotechnol.* 17: 250–255.

Pollo SMJ, Charchuk R, Nesbø CL (2016) Draft genome sequences of *Kosmotoga* sp. strain DU53 and *Kosmotoga arenicorallina* S304. *Genome Announc.* 4(3):e00570–16. Doi: 10.1128/genomeA.00570-16.

Pop Ristova P, Pichler T, Friedrich MW, Bühring SI (2017) Bacterial diversity and biogeochemistry of two marine shallow-water hydrothermal systems off Dominica (Lesser Antilles). *Front. Microbiol.* 8: 2400. Doi: 10.3389/fmicb.2017.02400.

Postec A, Ciobanu M, Birrien J-L, Bienvenu N, Prieur D, Le Romancer, M (2010) *Marinitoga litoralis* sp. nov., a thermophilic, heterotrophic bacterium isolated from a coastal thermal spring on Ile Saint-Paul, Southern Indian Ocean. *Int. J. Syst. Evol. Microbiol.* 60: 1778–1782.

Price RE, Giovannelli D (2017) *A Review of the Geochemistry and Microbiology of Marine Shallow-Water Hydrothermal Vents, Reference Module in Earth Systems and Environmental Sciences.* Elsevier.

Price RE, LaRowe DE, Italiano F, Savov I, Pichler T, Amend JP (2015) Subsurface hydrothermal processes and the bioenergetics of chemolithoautotrophy at the shallow-sea vents off Panarea Island (Italy). *Chem. Geol.* 407: 21–45.

Querellou J (2003) Biotechnology of marine extremophiles. *Book of Abstracts, International Conference on the sustainable development of the Mediterranean and Black Sea environment.* Thessaloniki, Greece, 28 May-1 June, extended abstract.

Rajasabapathy R (2015) *Molecular Diversity of the Shallow Water Hydrothermal Vent (Azores) Bacteria, Their Adaptation and Biotechnological Potentials.* PhD thesis submitted to Goa University, India.

Rajasabapathy R, Mohandass C, Bettencourt R, Colaço A, Goulart J, Meena RM (2018) Bacterial diversity at a shallow-water hydrothermal vent (Espalamaca) in Azores Island. *Curr. Sci.* 115(11): 2110–2121.

Rajasabapathy R, Mohandass C, Colaço A, Dastager SG, Santos RS, Meena RM (2014) Culturable bacterial phylogeny from a shallow water hydrothermal vent of Espalamaca (Faial, Azores) reveals a variety of novel taxa. *Curr. Sci.* 106(1): 58–69.

Rajasabapathy R, Mohandass C, Dastager SG, Liu Q, Li W-J, Colaço A (2015a) *Citreicella manganoxidans* sp. nov., a novel manganese oxidizing bacterium isolated from a shallow water hydrothermal vent in Espalamaca (Azores). *Antonie van Leeuwenhoek* 108: 1433–1439.

Rajasabapathy R, Mohandass C, Yoon J-H, et al. (2015b) *Vitellibacter nionensis* sp. nov., isolated from a shallow water hydrothermal vent. *Int. J. Syst. Evol. Microbiol.* 65(2): 692–697.

Rathgeber C, Yurkova N, Stackebrandt E, Beatty T, Yurkov V (2002) Isolation of tellurite- and selenite-resistant bacteria from hydrothermal vents of the Juan de Fuca Ridge in the Pacific Ocean. *Appl. Environ. Microbiol.* 68(9): 4613–4622.

Ravindran C, Varatharajan GR, Rajasabapathy R, Sreepada RA (2016) Antibacterial activity of marine *Bacillus* substances against *Vibrio cholerae* and *Staphylococcus aureus* and *In vivo* evaluation using embryonic Zebrafish test system. *Ind. J. Pharm. Sci.* 78: 417–422.

Rensink S, Wiegand S, Kallscheuer N, et al. (2020). Description of the novel planctomycetal genus *Bremerella*, containing *Bremerella volcania* sp. nov., isolated from an active volcanic site, and reclassification of *Blastopirellula cremea* as *Bremerella cremea* comb. nov. *Antonie van Leeuwenhoek* . Doi: 10.1007/s10482-019-01378-1.

Rusch A, Islam S, Savalia P, Amend JP (2015) *Burkholderia insulsa* sp. nov., a facultatively chemolithotrophic bacterium isolated from an arsenic-rich shallow marine hydrothermal system. *Int. J. Syst. Evol. Microbiol.* 65: 189–194.

Saha BC, Zeikus JG (1989) Novel highly thermostable pullulanase from thermophiles. *TIBTECH* 7: 234–239.

Sako Y, Nomura N, Uchida A, et al. (1996a) *Aeropyrum pernix* gen. nov., sp. nov., a novel aerobic hyperthermophilic archaeon growing at temperatures up to 100°C. *Int. J. Syst. Bacteriol.* 46: 1070–1077.

Sako Y, Takai K, IshidaY, UchidaA, KatayamaY (1996b) *Rhodothermus obamensis* sp. nov., a modern lineage of extremely thermophilic marine bacteria. *Int. J. Syst. Bacteriol.* 46: 1099–1104.

Schlesner H, Lawson PA, Collins MD, et al. (2001) *Filobacillus milensis* gen. nov., sp. nov., a new halophilic spore-forming bacterium with Orn-D-Glu-typepeptidoglycan. *Int. J. Syst. Evol. Microbiol.* 51: 425–431.

Segerer A, Neuner A, Kristjansson JK, Stetter KO (1986) *Acidianus infernus* gen. nov., sp. nov., and *Acidianus brierleyi* comb. Nov: Facultatively aerobic, extremely acidophilic thermophilic sulfur-metabolizing archaebacteria. *Int. J. Syst. Bacteriol.* 36: 559–564.

Sievert SM, Heidorn T, Kuever J (2000a) *Halothiobacillus kellyi* sp. nov., a mesophilic, obligately chemolithoautotrophic, sulphur oxidizing bacterium isolated from a shallow water hydrothermal vent in the Aegean Sea, and emended description of the genus *Halothiobacillus*. *Int. J. Syst. Evol. Microbiol.* 50: 1229–1237.

Sievert SM, Kuever J (2000) *Desulfacinum hydrothermale* sp. nov., a thermophilic, sulfate-reducing bacterium from geothermally heated sediments near Milos Island (Greece). *Int. J. Syst. Evol. Microbiol.* 50: 1239–1246.

Sievert SM, Kuever J, Muyzer G (2000b) Identification of 16S ribosomal DNA-defined bacterial populations at a shallow submarine hydrothermal vent near Milos Island (Greece). *Appl. Environ. Microbiol.* 66(7): 3102–3109.

Slobodkina GB, Baslerov RV, Novikov AA, Viryasov MB, Bonch-Osmolovskaya EA, Slobodkin AI (2016a) *Inmirania thermothiophila* gen. nov., sp. nov., a thermophilic, facultatively autotrophic, sulfur-oxidizing gammaproteobacterium isolated from shallow-seahydrothermal vent. *Int. J. Syst. Evol. Microbiol.* 66: 701–706.

Slobodkina GB, Kolganova TV, Kopitsyn DS (2016b) *Dissulfurirhabdus thermomarina* gen. nov., sp. nov., a thermophilic, autotrophic, sulfite-reducing and disproportionating deltaproteobacterium isolated from a shallow-sea hydrothermal vent. *Int. J. Syst. Evol. Microbiol.* 66: 2515–2519.

Slobodkina GB, Panteleeva AN, Beskorovaynaya DA, Bonch-Osmolovskaya EA, Slobodkin AI (2016c) *Thermostilla marina* gen. nov., sp. nov., a novel thermophilic facultatively anaerobic planctomycete isolated from a shallow submarine hydrothermal vent. *Int. J. Syst. Evol. Microbiol.* 66: 633–638.

Sorokin DY (1991) Oxidation of reduced sulphur compounds in volcanic regions in the Bay of Plenty (New Zealand) and Matupy Harbour (New Britain, Papua-New Guinea). *Proc. USSR Acad. Sci. Ser. B3*: 376–387

Stetter KO (1988) *Archaeoglobus fulgidus* gen. nov., sp. nov.: A new taxon of extremely thermophilic Archaebacteria. *Syst. Appl. Microbiol.* 10: 172–173.

Stetter KO, Konig H, Stackebrandt E (1983) *Pyrodictium* gen. nov., a new genus of submarine disc-shaped sulphur reducing archaebacteria growing optimally at 105°C. *Syst. Appl. Microbiol.* 4: 535–551.

Stetter KO, Lauerer G, Thomm M, Neuner A (1987) Isolation of extremely thermophilic sulphate reducers: Evidence for a novel branch of archaebacteria. *Science* 236: 822–824.

Stohr R, Waberski A, Volker H, Tindall BJ, Thomm M (2001) *Hydrogenothermus marinus* gen. nov., sp. nov., a novel thermophilic hydrogen-oxidizing bacterium, recognition of *Calderobacterium hydrogenophilum* as a member of the genus *Hydrogenobacter* and proposal of the reclassification of *Hydrogenobacter* acidophilus as *Hydrogenobaculum acidophilum* gen. nov., comb. nov., in the phylum 'Hydrogenobacter/Aquifex'. *Int. J. Syst. Evol. Microbiol.* 51, 1853–1862.

Surkov AV, Dubinina GA, Lysenko AM, Glockner FO, Kuever J (2001) *Dethiosulfovibrio russensis* sp. nov., *Dethosulfovibrio marinus* sp. nov. and *Dethosulfovibrio acidaminovorans* sp. nov., novel anaerobic, thiosulfate-and sulfur-reducing bacteria isolated from 'Thiodendron' sulfur mats in different saline environments. *Int. J. Syst. Evol. Microbiol.* 51: 327–337.

Takai K, Miyazaki M, Nunoura T, et al. (2006) *Sulfurivirga caldicuralii* gen. nov., sp. nov., a novel microaerobic, thermophilic, thiosulfate-oxidizing chemolithoautotroph, isolated from a shallow marine hydrothermal system occurring in a coral reef. *Int. J. Syst. Evol. Microbiol.* 56: 1921–1929.

Takeuchi M, KatayamaT, Yamagishi T, et al. (2014) *Methyloceanibacter caenitepidi* gen. nov., sp. nov., a facultatively methylotrophic bacterium isolated from marine sediments near a hydrothermal vent. *Int. J. Syst. Evol. Microbiol.* 64: 462–468.

Tanaka R, Kawaichi S, Nishimura H, Sako Y (2006) *Thermaerobacter litoralis* sp. nov., a strictly aerobic and thermophilic bacterium isolated from a coastal hydrothermal field. *Int. J. Syst. Evol. Microbiol.* 56: 1531–1534.

Tang K, Liu K, Jiao N, et al. (2013) Functional metagenomic investigations of microbial communities in a shallow-sea hydrothermal system. *Plos One*, 8: e72958. Doi: 10.1371/journal.pone.0072958.

Tang K, Zhang Y, Lin D, et al. (2018) Cultivation-Independent and Cultivation-Dependent analysis of microbes in the shallow-sea hydrothermal system off Kueishantao Island, Taiwan: Unmasking heterotrophic bacterial diversity and functional capacity. *Front. Microbiol.* 9: 279. Doi: 10.3389/fmicb.2018.00279.

Tarasov VG (2006) Effects of shallow-water hydrothermal venting on biological communities of coastal marine ecosystems of the Western Pacific. *Adv. Marine Biol.* 50: 267–421.

Tarasov VG, Gebruk AV, Mironov AN, Moskalev LI (2005) Deep-sea and shallow-water hydrothermal vent communities: Two different phenomena? *Chem. Geol.* 224: 5–39.

Tsai T-L, Liu S-M, Lee S-C (2011) Ethanol production efficiency of an anaerobic hemicellulolytic thermophilic bacterium, strain NTOU1, isolated from a marine shallow hydrothermal vent in Taiwan. *Microbes Environ.* 26(4): 317–324.

Tunnicliffe V (1991) The biology of hydrothermal vents: Ecology and evolution. *Annu. Rev. Oceanogr. Mar. Biol.* 29: 3119–3207.

Volkl P, Huber R, Drobner E, et al. (1993) *Pyrobaculum aerophilum* sp.nov., a novel nitrate-reducing hyperthermophilic archaeum. *Appl. Environ. Microbiol.* 59: 2918–2926.

Wang L, Cheung MK, Kwan HS, Hwang J-S, Wong CK (2015) Microbial diversity in shallow-water hydrothermal sediments of Kueishan Island, Taiwan as revealed by pyrosequencing. J. *Basic Microbiol.* 55: 1308–1318.

Yamamoto M, Nakagawa S, Shimamura S, et al. (2010) Molecular characterization of inorganic sulfur-compound metabolism in the deep-sea epsilonproteobacterium *Sulfurovum* sp. NBC37-1. *Environ. Microbiol.* 12: 1144–1152.

Yang S-H, Chiang P-W, Hsu T-C, Kao S-J, Tang S-L (2016) Bacterial community associated with organs of Shallow Hydrothermal Vent Crab *Xenograpsus testudinatus* near Kuishan Island, Taiwan. *PLoS One* 11(3): e0150597. Doi: 10.1371/journal.pone.0150597.

Ye J, Lin D, Zhang M, et al. (2020) *Paracoccus aurantiacus* sp. nov., isolated from shallow-sea hydrothermal systems off Kueishantao Island. *Int. J. Syst. Evol. Microbiol.* 70: 2554–2559.

Zhang Y, Zhao Z, Chen C-TA, et al. (2012) Sulfur metabolizing microbes dominate microbial communities in Andesite-hosted shallow-sea hydrothermal systems. *PLoS One* 7(9): e44593. Doi: 10.1371/journal.pone.0044593.

Zhou H, Li J, Peng X, Meng J, Wang F, Ai Y (2009) Microbial diversity of a sulphide black smoker in main Endeavear hydrothermal vent field, Juan de Fuca Ridge. *J. Microbiol.* 47(3): 235–247.

Zillig W, Holz I, Klenk H-P, et al. (1987) *Pyrococcus woesei*, sp. nov., an ultra-thermophilic marine archaebacterium, representing a novel order, Thermococcales. *Syst. Appl. Microbiol.* 9: 62–70.

26

Diversity and Bioprospecting Potentials of Antarctic (Polar) Microbes

B. Abirami, K. Manigundan, M. Radhakrishnan, and V. Gopikrishnan
Sathyabama Institute of Science and Technology

P.V. Bhaskar
National Centre for Polar and Ocean Research

T. Shanmugasundaram
Bharathiar University

Syed G. Dastager
National Collection of Industrial Microorganisms

CONTENTS

26.1 Introduction

26.1.1 Microbial Diversity in Extreme Environments

Extreme environments have been considered as a *terra incognita* by scientists for a long time owing to the harsh conditions characterizing these habitats considered as incompatible with life. After the physical and chemical boundaries of life were established, it became evident that these environments, in addition to being copiously populated, also represent the habitats of the first living organisms on Earth (Casillo et al., 2019). Microorganisms that survive in extreme physiological and geochemical conditions, called extremophiles, have evolved themselves to survive in such conditions through the course of evolution. The boundaries under which extremophiles can thrive have been pushed in every direction comprised of extreme temperature, pH, salinity, pressure, and nutrient limitation. In addition, microbes can also survive the harsh conditions of space, an environment with extreme radiation, vacuum pressure, extremely variable temperature, and microgravity (Yamagishi et al., 2018). In the recent past, due to the availability of advanced sampling and analytical techniques, there are several novel extremophilic microbial species reported from different extreme environments. Each and every extremophilic group of microbes is blessed with an exciting survival mechanism to grow and reproduce in their respective extreme habitats. The success of mimicking the similar extreme conditions in the laboratory nowadays allows us to explore different extremophilic organisms for various bioprospecting applications. Four success stories of extremophilic microbes for bioprospecting are the thermostable DNA polymerases used in the polymerase chain reaction, various enzymes used in the process of making biofuels, organisms used in the mining process, and carotenoids used in the food and cosmetic industries. Making lactose-free milk; the production of antibiotics, anticancer, and antifungal drugs; and the production of electricity are some other potentials of extremophiles (Coker, 2016).

26.1.2 The Polar Environments (Cryosphere)

The regions of Earth surrounding its geographical poles (North Pole and South Pole) are known as polar regions. They are also known as frigid zones because these regions are dominated by Earth's polar ice caps, the northern resting on the Arctic Ocean and the southern on the continent of Antarctica. The climate of this region is characterized by cold winters and cool summers. Winters are characterized by continuous darkness, cold and stable weather conditions, and clear skies, whereas summers are characterized by continuous daylight, damp and foggy weather, and weak cyclones with rain or snow.

Some important terminologies related to cryosphere are described below:

Sea ice: Sea ice is frozen ocean water that floats on the surface of the sea. It forms, grows, and melts in the ocean. Sea ice occurs in both the Arctic and the Antarctic's winter, not completely disappearing but retreating in summer. Sea ice covers nearly 12% of the world's oceans.

Lake ice: Lake ice is a sheet or stretch of ice forming on the surface of lakes when the temperature drops below freezing (0°C). Formation of ice may be as simple as a floating layer that gradually thickens, or extremely complex, especially when the water is flowing rapidly.

River ice: The formation of ice in rivers is more complex than in lakes, mainly because of the impacts of water velocity and turbulence. River ice phenomena include formation, evolution, transport, accumulation, dissipation, and deterioration of different ice forms. The presence of ice on rivers is said to change their behavior and interfere with their use.

Snow cover: The area of land that is covered by accumulated snow at any given time is termed as snow cover. It helps in regulating the Earth's surface temperature when present and helps in filling rivers and reservoirs once it melts away. Snow cover reflects 80%–90% of sun's energy back into the atmosphere, thereby cooling the planet by regulating the exchange of heat between the Earth's surface and the atmosphere.

Glaciers: Glaciers are large, persistent accumulation of crystalline ice, snow, rock, sediment, and often liquid water that originates on land and moves downhill under its own weight. They form where the accumulation of snow exceeds its ablation over many years, even centuries.

Ice caps: Ice caps are mass of ice that covers less than 50,000 km^2. They are miniature ice sheets, usually found in the North and South Poles of Earth. They form primarily in polar and subpolar regions that are relatively flat and high in elevation.

Ice sheet: An ice sheet, also known as a continental glacier, is a mass of glacial ice that covers surrounding terrain and is greater than 50,000 km^2. It contains nearly 99% of freshwater. The only ice sheets present on the Earth are in Antarctica and Greenland.

Frozen grounds: Frozen ground occurs when the ground contains water, and the temperature of the ground goes down below 0°C. It can make a huge difference if the ground stays frozen all year or if the ground freezes and thaws.

26.1.3 Psychrophiles – Cryosphere Microbes

Low-temperature environments are numerous on Earth and have been successfully colonized by cold loving organisms termed a psychrophiles. Psychrophiles are most abundant in terms of biomass, diversity, and distribution among the various organisms thriving in extreme environments. Psychrophiles are reported from all three domains of life: bacteria, Archaea, and Eukarya. Extreme psychrophiles have been traditionally isolated from two important polar regions: Antarctic and Arctic (Margesin and Feller, 2010). Different categories of microbes especially bacteria, fungi, and algae are reported from all the above described polar samples. However, the diversity, distribution, and bioprospecting potentials of microbes from the different polar samples and regions may vary significantly.

More recently, permafrost representing more than 20% of terrestrial soils has revealed an unexpected biodiversity in cryopegs, i.e., saltwater pockets that have remained liquid for about 1,00,000 years at −10°C (Gilichinsky et al., 2005). In addition, high altitude mountains, glaciers, or natural caves are additional sources of cold-loving microorganisms. However, the largest psychrophilic reservoir is provided by oceans that have a constant temperature of 4°C below a depth of 1,000 m, irrespective of the latitude. In this chapter, we have described the microbial diversity especially bacteria, fungi, actinobacteria, and algae in the Antarctic and Arctic regions and their bioprospecting potentials.

26.2 Antarctica

Antarctica, one of the seven continents of the Earth, has a vast diversity of marine ecosystems covering from the circumpolar Southern Ocean's benthos, water source, and fluctuating sea-ice to vast lakes. They cover from ultra-oligotrophic proglacial lakes on Signy Island, the South Orkney Islands, to the permanently frozen lakes of McMurdo Dry Valleys to Vestfold Hills. Their salt content is high. Terrestrial ecosystems also include damp and brown soils on Signy Island Sea and endolithic cold desert on the edge of the polar plateau. Antarctica is the coldest continent and also termed as white desert. The Antarctic convergence and other circumpolar barriers are completely separated from the South Ocean itself from the northerly waters. The barrier between the Antarctic continent and northwards landscape acts as a geographical and climate barrier so that potential colonists will arrive via aerobiota, animal vectors, or wind spray and related aerosols. There are a fairly large variety of organisms within the maritime and coastal microbiota, but this becomes reduced further inland and south where life exists near its limits.

26.3 Microbial Diversity in Antarctica

Antarctic microorganisms, from the ecological point of view, must have evolved among environmental stressors like frequent freeze-thaw cycles, high UV exposure during summer, total absence of light during winter, nutrient starvation, and osmotic limitation. This may have aided the microorganisms in possessing specific biochemical adaptations that provide them selective advantages in such hostile environment. Secondary metabolite production, sporulation, conjugation, symbiosis, motility, bioluminescence, and biofilm formation are strategies developed by the microorganisms to survive in the environment (Wong et al., 2019). Despite the fact that microbes remain entombed and viable in glacier ice for hundreds of thousands of years, it is the metabolically active ones present in ecosystems such as subglacial lakes and desert soils of Antarctica that contribute to the global biogeochemical cycles (Campen et al., 2019).

26.3.1 Bacteria

The diversity of bacterial taxa in the Antarctic sea water corresponds to both the chemical and physical oceanographic variations. Reports on the bacterial community structure and composition in Antarctic waters are much less compared to the open ocean, especially the bacterial population (Giudice & Azzaro, 2019). Species-specific functional responses and shallow water community composition in Potter Cove, King George Island was studied by Abele et al. in 2017. Glacial melting has affected this shallow fjord by accelerating the microbial turnover rates. The dominant phyla found in the coastal waters of Potter Cove were *Proteobacteria* and *Bacteroidetes*, among which *Rhodobacteraceae* belonging to *Alphaproteobacteria* were in abundance. This bacterial community, under oxic and anoxic conditions, has the capability to switch between chemoheterotrophic and photosynthetic energy production. The composition of bacterial communities in the sea water surface at the coasts of Terra Nova Bay was reported to be highly influenced by both sea-ice and human activity. Investigations by Yakimov et al. in (2004) at Adeline Cove and Road Bay reported the abundance of *Gammaproteobacteria* at both sites, while the prevalence and relative abundance of *Alpha*- and *Betaproteobacteria* depended on the sampling site. Clone library sequencing found many *Gammaproteobacteria* related to chemoheterotrophic species and genera including *Pseudomonas, Pseudoalteromonas, Shewanella, Halomonas, Methylophaga, Marinobacter,* and *Colwellia*. These findings ponder a common bacterial community composition that could be observed in the Southern Ocean. Sequencing clone libraries from 16S rRNA and 16S rDNA analyses by Gentile et al. in (2006) from surface seawater of Evans Cove and Road Bay depicted a clear picture of composition of bacterioplankton assemblages dominated by *Gammaproteobacteria* and *Bacteroidetes*, showing low abundance of *Alphaproteobacteria*. *Gammaproteobacteria* were predominantly associated with the psychrophilic genera *Colwellia* and *Psychrobacter,* while *Bacteroidetes* were affiliated with *Polaribacter* and *Flectobacillus*. Sequences affiliated to *Betaproteobacteria*, especially the genera *Curvibacter* and *Aquaspirillum*, were found in abundance at Road Bay, but only in 16S rDNA searches, which probably implies that they might not be part of metabolically active community.

As a means to thrive in a harsh and seasonally fluctuant environment, bacterial communities trapped in ice must adopt unique adaptation strategies. Growth of such bacterial assemblages is nourished by a large supply of dissolved organic matter that mainly consists of carbohydrates exudates after the death and lysis of sea-ice organisms, in addition to the release of organic polymers by diatoms and bacteria. Copiotrophic bacteria, especially *Gammaproteobacteria*, *Flavobacteria*, and *Alphaproteobacteria*, dominate the Antarctic bacterial communities in both first-year ice and multiyear ice (Giudice & Azzaro, 2019). The possibility of bacterial phototrophy at the Antarctic sea-ice was recognized when Maas et al. in 2012 reported 16S rDNA sequences clustering with species encoding phototrophic genes. The *Roseobacter* is well known to produce bacteriochlorophyll-a and also to possess genes for aerobic anoxygenic photosynthesis, while the *Polaribacter* is photoheterotrophic containing the proteorhodopsin gene. This implies the abundance of photosynthetically competent bacteria in the Antarctic sea ice. Moreover, the presence of sulphate-reducing bacteria such as *Desulfofrigus, Sulfurospirillum,* and *Desulforhopalus* observed in a thick anoxic ice sheet implies the occurrence of reduction of sulphur compounds under suitable conditions in sea-ice and the development of anaerobic bacteria community.

Antarctic sediments are also the richest source for diverse bacterial community. Carr et al. (2013) reported the presence of 50 diverse bacterial phyla in the soft sediments collected from the surface of the sediment from Ross Ice Shelf (the world's largest ice shelf) using 16S rRNA gene pyrosequencing, which had diverse community structure than the deeper sediments. The members of phyla *Bacteroidetes, Firmicutes,* and *Chloroflexi* along with subphyla *Beta*-, *Delta*-, and *Gammaproteobacteria* dominate the bacterial community. The degree of diversity within the core changes according to the availability of the carbon source. Wölfl et al. (2014) reported high metabolically diverse and functionally versatile bacterial communities from Potter Cove sediments, implying their capability to respond to sudden environmental changes. *Alpha-* and *Gammaproteobacteria* representing *Proteobacteria* and *Bacteroidetes*, especially *Haliscomenobacter* spp. and *Flavobacteriaceae,* dominated the bacterial community followed by *Verrucomicrobia* and *Planctomycetes*. Some important bacterial genera reported from the Antarctic region are given in Table 26.1. Till date, approximately more than 50 bacterial genera are reported from Antarctic regions.

26.3.2 Actinobacteria

Actinobacteria are the largest phylum in the bacterial domain with the unparalleled ability to produce novel natural products, especially antibiotics. The relative abundance of Actinobacteria averages to 22%±4% making them the most abundant soil bacterial taxa across a wide range of ecological areas that include extreme environments. Actinobacteria

TABLE 26.1

Bacterial Diversity in Antarctic Habitats

Source	Genus/Species	Reference
Sea water	• *Actimicrobium antarcticum*, • *Aequorivita antarctica*, • *Antarcticimonas* flava, • *Bacillus* sp., • *Granulosicoccus antarcticus*, • *Halomonas* sp., • *Moraxella* sp., • *Oleispira Antarctica*, • *Polaribacter irgensii*, • *Pseudoalteromonas haloplanktis* • *Robiginitomaculum antarcticum*, • *Zhongshania antarctica*	Shivaji et al. (2019), Núñez-Montero and Barrientos (2018), Maiangwa et al. (2014), Murray and Grzymski (2007), Birolo et al. (2000)
Fast ice	• *Psychrobacter adeliensis* • *Psychrobacter salsus*	Shivaji et al. (2004)
Lake Water, Ponds	• *Antarctobacter heliothermus*, • *Lysobacter oligotrophicus*, • *Methylosphaera hansonii*, • *Psychrosinus fermentans*, • *Rhodoligotrophos appendicifer*, • *Roseibaca ekhonensis*, • *Roseisalinus antarcticus*, • *Roseovarius tolerans*, • *Saccharospirillum impatiens*, • *Shewanella* sp. • *Staleyagutti formis*	Shivaji et al. (2019), Núñez-Montero & Barrientos (2018)
Rock	• *Constrictibacter antarcticus*, • *Polymorphobacter fuscus*, • *Pseudomonas fragi*	Pershina et al. (2018), Núñez-Montero and Barrientos (2018), Shivaji et al. (2019)
Sea ice	• *Gelidibacter algens*, • *Lacinutrix copepodicola*, • *Polaribacter filamentus*, • *Polaromonas vacuolata*, • *Psychroflexus torques*, • *Psychroserpens burtonensis*, • *Subsaxibacter broadyi*, • *Subsaximicrobium wynnwilliamsii*	Shivaji et al. (2019)
Mats	• *Gillisia limnaea*, • *Janthinobacterium* sp., • *Leptolyngbya frigid*, • *Loktanella salsilacus*, • *Nodosilinea* sp., • *Plectolyngbya hodgsoni*, • *Pseudomonas* sp., • *Psychrobacter* sp., • Shewanella sp.	Shivaji et al. (2019), Núñez-Montero and Barrientos (2018), Rego et al. (2019)
Sediment	• *Pseudomonas* sp., • *Psychrobacter* sp. • *Psychromonas antarcticus*, • *Roseicitreum antarcticum*	Núñez-Montero and Barrientos (2018), Shivaji et al. (2019)

(*Continued*)

TABLE 26.1 (*Continued*)

Bacterial Diversity in Antarctic Habitats

Source	Genus/Species	Reference
Soil	• *Bradyrhizobium* sp., • *Enterococcus* sp., • *Hymenobacter glaciei*, • *Janthinobacterium* sp., • *Methylobacterium* sp., • *Paracoccus* sp., • *Pedobacter* sp., • *Phormidium autumnale*, • *Planococcus* sp., • *Pseudomonas* sp., • *Sejongia Antarctica*, • *Sphingomonas* sp.	Núñez-Montero and Barrientos (2018), Pershina et al. (2018)
Ornithogenic soil	• *Burkholderia* sp., • *Janthinobacterium* sp., • *Pseudomonas* sp., • *Sporosarcina* sp.	Núñez-Montero and Barrientos (2018)
Porifera	• *Colwellia* sp., • *Gillisia* sp. CAL575, • *Pseudomonas aeruginosa*, • *Psychrobacter* sp., • *Shewanella* sp., • *Winogradskyella* sp.	Lo Giudice et al. (2018)
Annelida	• *Flavobacterium* sp., • *Pseudomonas* sp.	Lo Giudice et al. (2018)

induce soil moisture uptake and improve microbial growth by the process of hydrolyzing complex polymeric substances and stabilizing clay particles and organic matter. Actinobacteria also provide various plant health and nutritional benefits, for example, *Streptomyces* species can ease plant biotic and abiotic stresses (Araujo et al., 2020). Actinobacteria are prolific producers of bioactive compounds essential for human and animal health. A study by Araujo et al. (2020) showed that soils with high biodiversity of Actinobacteria generally exhibit high biodiversity of total bacterial taxa. Till date, more than 425 genera are reported under this phylum including isolates from extreme ecosystems. There are several articles describing the diversity and bioprospecting potentials of actinobacteria (Berdy, 2012; Lam, 2006; Balagurunathan et al., 2010; Barka et al., 2016; Manikkam et al., 2019). They are widely distributed in various normal and extreme environments. There are several studies that have reported the distribution of actinobacteria in cold climatic conditions including the polar regions. Antarctica is well known for its diverse specific microbial ecosystems as a result of its extreme environment and varying local microenvironments. The actinobacterial populations that dominate specific locations of Northern Antarctica include *Nocardioides, Rubrobacter, Gaiellaceae,* and *Frankiaceae*, while *Actinoalloteichus, Euzebya, Streptomyces, Geodermatophilus, Actinomadura,* and *Conexibacter* appeared occasionally. When compared with samples collected from mainland, islands, or other parts of Antarctica, King Island and Northern Antarctica Casey station region were found to have a high proportion of 19% of similar actinobacterial OTUs (Araujo et al. 2020). This microbial similarity may be explained as the combination of persistence of actinobacteria in polar regions with slow evolution and shared geological history including the possibility of persistence of actinobacterial communities for millions of years (Altizer et al. 2013; Bajerski & Wagner, 2013; Ferrari et al., 2016; Shivlata & Tulasi, 2015; Qin et al., 2016). Based on molecular and phenotypic markers, Antony et al. in (2009) identified *Cellulosimicrobium cellulans* from Antarctic snow.

Millán-Aguiñaga et al. (2019) investigated the actinobacterial diversity and evolution in polar environments. Culture-dependent analysis revealed a total of 78% of the strains isolated to belong to the phylum *Actinobacteria* and all of it corresponding to rare actinomycete or non-*Streptomyces* genera. Almost all of the isolates belonged to five genera—*Agrococcus, Pseudonocardia, Microbacterium, Salinibacterium,* and *Rhodococcus*. Pyrosequencing data of sediment samples collected from Adelie Basin of Antarctica revealed that the sequence reads comprised nearly 10% actinobacteria (Carr et al., 2015). When compared to soil and sediments, actinobacterial diversity in surface waters and deeper zones such as McMurdo ice shelf and Victoria Land is less (Shivaji et al., 2019). Nearly 50 species of *Actinobacteria* were obtained from metagenomic analysis of basal ice from subglacial Lake Vostok, Antarctica. *Mycobacterium* spp. and *Streptomyces* spp. dominated other members of *Actinobacteria* in basal ice while *Frankia* spp., *Nocardia* spp., *Micromonospora* spp., and *Streptomyces* spp. predominate in accretion ice samples (Gura & Rogers, 2020). More than 25 actinobacterial genera are reported from the Antarctic samples (Table 26.2).

TABLE 26.2

Actinobacterial Genera Reported from Antarctic Habitats

Source	Genus/Species	Reference
Sea water	• *Arthrobacter* sp. • *Arthrobacter kerguelensis* • *Janibacter thuringiensis* • *Nesterenkonia* sp. • *Rhodococcus fascians*	Núñez-Montero and Barrientos (2018), Shivaji et al. (2017), Lo Giudice et al. (2007)
Soil	• *Arthrobacter* sp. • *Arthrobacter cryotolerans* • *Arthrobacter gangotriensis* • *Arthrobacter livingstonensis* • *Barrientosiimonas humi* • *Brevibacterium* sp. • *Cryobacterium psychrophilum* • *Curtobacterium psychrophilum* • *Demetria* sp. • *Friedmanniella antarctica* • *Gordonia* sp. • *Gordonia terrae* • *Janibacter* sp. • *Lapillicoccus* sp. • *Leifsonia psychrotolerans* • *Leifsonia* soli • *Micrococcus antarcticus* • *Micromonospora* sp. • *Modestobacter multiseptatus* • *Nocardioides* sp. • *Nocardioides glacieisoli* • *Nocardiopsis fildesensis* • *Rhodococcus* sp. • *Streptomyces* sp. • *Streptomyces fildesensis* • *Terrabacter lapilli*	Núñez-Montero and Barrientos (2018), Shivaji et al. (2017), Lee et al. (2012), Shivaji et al. (2019), Cheah et al. (2015), Yarzabal et al. (2016), Pershina et al. (2018), Lavin et al. (2016)
Ornithogenic soil	• *Arthrobacter* sp. • *Rhodococcus* sp.	Núñez-Montero and Barrientos (2018)
Sediment	• *Arthrobacter* sp. • *Arthrobacter antarcticus* • *Arthrobacter ardleyensis* • *Leifsonia antarctica* • *Marisediminicola antarctica* • *Nocardioides antarcticus*	Núñez-Montero and Barrientos (2018), Shivaji et al. (2017), Li et al. (2010), Shivaji et al. (2019)
Mats	• *Arthrobacter* sp. • *Arthrobacter roseus* • *Kocuria Polaris* • *Leifsonia aurea* • *Leifsonia rubra*	Núñez-Montero and Barrientos (2018), Shivaji et al. (2017)
Dry valley	• *Dermacoccus nishinomiyaensis* • *Flexivirga* sp. • *Kocuria* sp. • *Micrococcus* sp.	Rego et al. (2019)
Water	• *Friedmanniella lacustris* • *Nesterenkonia lacusekhoensis* • *Nocardioides aquaticus* • *Rhodoglobus vestalii*	Shivaji et al. (2017), Shivaji et al. (2019)
Sandstone, quartz, sea sand	• *Friedmanniella antarctica* • *Micromonospora endolithica* • *Sanguibacter antarcticus* • *Streptomyces hypolithicus*	Shivaji et al. (2017)
Glacial soil	• *Micrococcus roseus*	Maiangwa et al. (2015)
Moraine	• *Pseudonocardia antarctica*	Shivaji et al. (2017)
Penguin	• *Arthrobacter psychrochitiniphilus*	Shivaji et al. (2017)

26.3.3 Fungi

Generally, in marine ecosystems, fungi occur as spores, mycelium, or hypha fragments, either in dormant or active form. In oceans, marine fungi can occur as a mutualist, commensal, or parasite, in harmonic and disharmonic symbiosis. Similar to that of the land ecosystems, fungi in the ocean ecosystems play a vital ecological role in organic matter decomposition such as seaweeds, corals, vertebrates, and invertebrates. In Antarctica, marine fungi have been recovered from coastal waters, deep sea, macro algae, and marine mammals. Even though studies of the presence of fungi have been detected in the deep-sea sediments of Atlantic and Pacific oceans, studies of the Southern Ocean surrounding Antarctica are still in the initial stages (Rosa et al., 2019). The most represented fungal phyla in the soils of Antarctica include *Chytridiomycota, Glomeromycota, Zygomycota, Basidiomycota,* and *Ascomycota*. Due to global warming, it is estimated that the soils in the peninsula regions of the Antarctica are likely to be colonized by fungi with a 20%–27% increase in the richness of different species in the southernmost soils within the end of this century. Fungal genera including *Pseudogymnoascus, Mortierella, Antarctomyces, Penicillium, Rhodotorula,* and *Aspergillus* are reported to dominate other communities of different habitats of Antarctica (Oliveira et al., 2019).

Minimum information is available on fungal taxa present in marine sediments of the polar regions. The majority of the fungal communities identified in Antarctic marine sediments are shared with marine sediments of other oceans. For example, fungal genera *Penicillium, Cladosporium,* and *Rhodotorula* are commonly observed in marine sediments of all oceans. Several known taxa of fungi have been identified to be present in Antarctica by studies that isolated culturable fungi from the marine sediments (Rosa et al., 2019). The first study on cultivable fungi was represented by Gonçalves et al. in (2013) in the deep-sea sediments of Antarctica. 52 fungal isolates were obtained from marine sediments of depths 100, 500, 700, and 1,100 m. This study suggested that deep sea Antarctic sediments might represent an interesting microhabitat to recover and study the biology of psychrophilic fungi. Sediment samples of 5cm deep sea from Admiralty Bay, King George Island, when processed, contained 226 fungal isolates with species from 17 different genera (Wentzel et al., 2019). Yeast species identified from Antarctic shallow marine sediments include *Phenoliferia glacialis, Leucosporidiella muscorum, Nadsonia commutate, Vishniacozyma victoriae, Metschnikowia australis,* and *Candida glaebosa* (Vaz et al., 2011). Global fungal species such as *Candida, Aspergillus, Fusarium, Cladosporium, Rhodotorula,* and *Penicillium* have well adapted to the extreme conditions of the Antarctic seawater (Rosa et al., 2019). When compared with other areas, fungal species like *Glaciozyma, Holtermanniales,* and *Phenoliferia* were reported to occur frequently in the Antarctic marine sediments (Wentzel et al., 2019).

In order to simulate the limitations of the environment when culturing microorganisms from extreme environments, many unique isolation conditions have to be considered, which act as a major constraint in identifying marine Antarctic fungi. It is estimated that of the total species, only 20% have been isolated and grown in pure culture, explaining the reason for the lack of data in the literature regarding obligate marine fungi in the region. This has led to the uncultured analysis along with molecular studies that has reported a wide range of fungal taxa present in marine ecosystems, thereby suggesting that there might be more fungi still undiscovered in the Antarctic environment, though due to the complexity of the polar regions there is a knowledge gap regarding their ecological role in such habitat (Rosa et al., 2019). Novel studies using various culture media, conditions, and metagenomics techniques are needed to recover, better characterize, and understand the complexity, ecological role, and biotechnological potential of these polar fungi. There are several fungal genera reported from Antarctic samples (Table 26.3).

26.3.4 Algae

Algal diversity is relatively less resolved in Antarctica. While Southern Australia reported nearly 1,500 macroalgal species, Antarctica reported only 130 species of macroalgae. Infrequent sampling, logistic difficulties, difficulty of access, and the lack of exploration of Antarctica attribute to the low level of species richness (Ko et al., 2020). In the maritime Antarctic Peninsula, along eight islands of South Shetland Archipelago, a total of 104 benthic marine algae species that include 28 *Phaeophyceae*, 52 *Rhodophyta*, and 24 *Chlorophyta* were reported by Pellizzari et al. (2017). This represents nearly 82% of all the seaweed communities present in entire Antarctica. Six novel species were reported, of which four putative taxa were confirmed by their biogeographical distribution and the two new species were identified by their morphological and molecular characteristics (Rosa et al., 2019). A study by Das and Singh (2020) revealed a diverse algal population in Larsemann Hills, East Antarctica. This study observed that bryophytes such as *Bryum pseudotriquetrum* and *Bryoerythrophyllum recurvirostrum* act as suitable host plants for algal communities. As a whole, 16 algal species belonging to *Chlorophyta, Charophyta, Xanthophyta, Cyanophyta,* and *Bacillariophyta* were reported to be growing epiphytic on 7 bryophyte species sampled from different regions of Larsemann Hills. There are several algal cultures reported from Antarctic samples (Table 26.4).

26.4 Bioprospecting Potentials of Antarctic Microbes

Bioprospecting is a search for new or promising bioproducts from biological sources preferably from various applications. Molecules derived from natural products, particularly those produced by plants and microorganisms, have an excellent record of providing novel chemical structures for development as new pharmaceuticals. Microorganisms living under stress are the best sources for bioprospecting, and also, these organisms are the least explored. This chapter focuses on the bioprospecting and biotechnological potential of microorganisms isolated from the polar region of the Antarctic environment. In this context, let us have a glance through the major cold

TABLE 26.3

Fungal Genera Reported from Antarctic Habitats

Source	Genus/Species	Reference
Sea water	• *Acremonium* sp. • *Aspergillus pseudoglaucus* • *Candida spencermartinsiae* • *Candida zeylanoides* • *Cladosporium sphaerospermum* • *Cystobasidium slooffiae* • *Cystobasidium slooffiae* • *Exophiala xenobiotica* • *Glaciozyma antarctica* • *Graphium rubrum* • *Lecanicillium attenuatum* • *Leucosporidiella creatinivora* • *Leucosporidium scottii* • *Penicillium chrysogenum* • *Penicillium citreosulfuratum* • *Purpureocillium lilacinum*	Gonçalves et al. (2017), Vaz et al. (2011)
Shallow marine sediment	• *Cadophora* sp. • *Cystobasidium* sp. • *Holtermanniella* sp. • *Leucosporidium scottii* • *Metschnikowia* sp. • *Meyerozyma* sp. • *Paraconiothyrium* sp. • *Pestalotiopsis* sp. • *Phenoliferia* sp. • *Pseudocercosporella* sp. • *Toxicocladosporium* sp.	Wentzel et al. (2019), Vaz et al. (2011)
Deep marine sediment	• *Comospora* sp. • *Pleosporaceae* sp. • *Schizophyllum commune* • *Simplicillium lamellicola*	Gonçalves et al. (2015)
Shallow sediments	• *Cryptococcus* sp. • *Leucosporidiella muscorum* • *Rhodotorula* sp. • *Vishniacozyma victoriae*	Wentzel et al. (2019), Vaz et al. (2011)
Shallow to deep marine sediment	• *Penicillium solitum*	Gonçalves et al. (2013)

TABLE 26.4

Algal Cultures Isolated from Antarctic Habitats

Source	Genus/Species	Reference
Islet	• *Acanthococcus antarcticus* • *Adenocystis utricularis* • *Antarcticothamnion polysporum* • *Ascoseira mirabilis* • *Ballia callitricha* • *Callophyllis linguata* • *Curdiea racovitzae* • *Cystosphaera jacquinotii* • *Delisea pulchra* • *Desmarestia antarctica* • *Georgiella confluens* • *Gigartina skottsbergii* • *Halopteris obovata* • *Iridaea cordata* • *Monostroma hariotii* • *Myriogramme manginii* • *Neuroglossum ligulatum*	Ko et al. (2020)

(*Continued*)

TABLE 26.4 (*Continued*)

Algal Cultures Isolated from Antarctic Habitats

Source	Genus/Species	Reference
	• *Palmaria decipiens* • *Pantoneura plocamioides* • *Paraglossum lancifolium* • *Petroderma maculiforme* • *Phaeurus antarcticus* • *Phycodrys antarctica* • *Phyllophora ahnfeltioides* • *Picconiella plumose* • *Plocamium cartilagineum* • *Trematocarpus antarcticus* • *Ulothrix australis* • *Ulva bulbosa*	
Sublittoral zone	• *Desmarestia menziesii* • *Himantothallus grandifolius*	Ko et al. (2020)
Bryophyte	• *Coleochaete scutata* • *Gloeocapsopsis magma* • *Hormidiopsis crenulata* • *Luticola muticopsis* • *Nostoc commune* • *Nostoc fuscescens* • *Nostoc lichenoides* • *Nostoc punctiforme* • Nostoc sphaericum • *Oscillatoria sancta* • *Pinnularia borealis* • *Stigonema minutum* • *Xanthonema antarcticum*	Das and Singh (2020)

adaptations of psychrophiles and their application potentials with special reference to psychrophilic enzymes. Extreme environments provide microorganisms containing robust enzymes. Such microorganisms that flourish in extremes of temperature, pressure, acidity, or alkalinity are sources of extremophilic enzymes that are required by many industrial processes. Bioprospecting for such enzymes involves culturing and isolation of novel extremophiles from polar habitats. Microorganisms represent the largest reservoir of unexplored biodiversity and hence possess the greatest potential for the discovery of new natural products.

26.5 Bioactive Compounds/Products

Microorganisms from unexplored habitats such as the polar regions are a rich source of bioactive compounds from different sources. Polar regions, which refer to the regions of the Antarctic, are remote and challenging areas of Earth. A vast amount of new biological natural compounds with different activities, such as antimicrobial, antitumor, antiviral, and so on, have been isolated from microorganisms discovered from the polar ecosystem over the past few years. In 2013, Liu et al. examined a variety of new secondary metabolites with different activities derived from both Antarctic and Arctic microorganisms, while Skropeta and Wei in 2014 published a study of deep-sea isolated natural products that included several polar species.

Antarctic microorganisms have been reported as a potential source of bioactive compounds because of their survival in extreme environments. In order to develop different strategies to thrive in Antarctic environments, such as high UV radiation, lack of substrates, short-term intense heat during summer, and sustained low temperature, it becomes necessary for microorganisms to genetically adapt by producing secondary metabolites or bioactive compounds. Even though bacterial and fungal isolates produce metabolites that have antimicrobial activity, the bioactive compound production by actinobacterial genera *Streptomyces* dominates the rest. Various drugs from the bioactive compounds by Actinobacteria have been commercially manufactured such as Streptomycin, Tetracycline, Anthracycline, Neomycin, etc. Bioactive compounds such as Marinomycin, Penilactones, and Anthraquinones are reported to be potential antitumor agents. (Table 26.5)

26.5.1 Enzymes

The high activity of psychrophilic enzymes at low and moderate temperatures offers potential economic benefits through substantial energy savings in large-scale processes that would not require the expensive heating of reactors. A typical example is the industrial "peeling" of leather by proteases, which can be done at the temperature of tap water by cold-active enzymes instead of heating to 37°C for the process to be performed by mesophilic enzymes. Psychrophilic enzymes can also be useful in domestic processes. In the food industry, their properties allow the transformation or refinement of heat-sensitive products. The removal of lactose from milk by a psychrophilic β-galactosidase during cold storage has recently

TABLE 26.5

List of Bioactive Compounds Isolated from Antarctic Microbes

Activity	Bioactive Compound	Microbe	Reference
Antibacterial	Actinomycins	*Streptomyces anulatus*	Barka et al. (2015)
	Neomycin	*Streptomyces fradiae*	Barka et al. (2015)
	Streptomycin	*Streptomyces griseus*	Barka et al. (2015)
	Tetracycline	*Streptomyces aureofaciens*	Barka et al. (2015)
	Vancomycin	*Amycolatopsis orientalis*	Barka et al. (2015)
	Anthracyclin	*Micromonospora spp.*	Núñez-Montero and Barrientos (2018)
Antitumor agent	Anthraquinones	*Micromonospora spp.*	Barka et al. (2015)
	Marinomycin	*Marinospora spp.*	Barka et al. (2015)
	Salinosporamide	*Salinispora tropica*	Barka et al. (2015)
	Penilactones A and B	*Penicillium crustosum*	Wu et al. (2012)
Immunosuppressive agent	Brasilicardin	*Nocardia brasiliensis*	Barka et al. (2016)
Therapeutic enzyme	L-Asparaginase	*Streptomyces spp.*	Barka et al. (2016)
Immunostimulatory agent	Rubratin	*Nocardia rubra*	Barka et al. (2016)
Antiparasitic agent	Avermectins	*Streptomyces avermitilis*	Barka et al. (2016)
Antifungal		*Bradyrhizobium sp.*	Núñez-Montero and Barrientos (2018)
		Janthinobacterium sp.	Núñez-Montero and Barrientos (2018)
		Paracoccus sp.	Núñez-Montero and Barrientos (2018)
		Sphingomonas sp.	Núñez-Montero and Barrientos (2018)
		Pseudomonas sp.	Núñez-Montero and Barrientos (2018)
		Sporosarcina sp.	Núñez-Montero and Barrientos (2018)

been patented. As another example, cold-active pectinases can help to reduce viscosity and clarify fruit juices at low temperatures. The heat-lability of these enzymes also ensures their fast, efficient, and selective inactivation in complex mixtures. The use of a heat-labile alkaline phosphatase in molecular biology is probably the first biotechnological application proposed for a psychrophilic enzyme. Glycosidases are often used in the baking industry but can retain residual activity after cooking that alters the structure of the final product during storage; this can be avoided by the use of psychrophilic glycosidases. However, despite such powerful biotechnological potential, psychrophilic enzymes remain under-used, partly because the cost of production and processing at low temperatures is higher than the commercial enzymes that are presently in use.

Enzymes, lipase, and galactosidase were extensively isolated from Antarctic microbes. While lipases were found to be utilized in detergent and food industries along with bioremediation processes and biodiesel applications, galactosidases were exclusively utilized in the food industry (Parrilli et al., 2019). Applications of the α-amylase enzyme extend to the food, paper, detergent, fine-chemical, textile, and pharmaceutical industries (Wang et al., 2018). Multiple isolates of *Pseudoalteromonas* were found to produce cold-adapted proteases used in the cosmetic industry and in processing and preservation of foods (Sharma et al., 2018). Chitinase produced by *Pseudoalteromonas* sp. DL-6 isolated from Antarctic sea water was reported to aid in biocontrol of phytopathogens in cold environments in addition to biocontrol of microbial spoilage in refrigerated food (Parrilli et al., 2019) (Table 26.6).

26.5.2 Proteins

Transcription and translation are temperature-sensitive steps and psychrophiles have obviously adapted the process of protein synthesis to low temperatures. It is expected that the enzymatic activities that are involved in protein synthesis have the general traits of cold-adapted enzymes, high activity associated with low stability, but this aspect has not been consistently analyzed so far. However, low temperatures strengthen the interactions between DNA strands in the double helix and in the supercoiled state, therefore impairing unwinding and access to RNA polymerase. Low temperatures also promote unfavorable RNA secondary structures, which are likely to interfere with translation. Accordingly, it is expected that nucleic acid-binding proteins, which relieve the adverse effects of low temperatures, have a central role in the cold adaptation of psychrophiles. In this respect, it is significant that the five unique genes that have been detected in the genome of two cold-adapted Archaea are predicted to encode nucleic acid binding proteins. Moreover, increased posttranscriptional incorporation of dihydrouridine in tRNA from psychrophilic bacteria is thought to improve the conformational flexibility of RNA. Cold-acclimation proteins (CAPs) seem to be another important and general feature of cold-adapted microorganisms. A set of ~20 proteins is permanently synthesized during steady-state growth at low temperatures, but not at milder temperatures. Interestingly, some of the few CAPs that have been identified in cold-adapted bacteria are cold-shock proteins in mesophiles, such as the RNA chaperone *CspA*. It has been proposed that these CAPs are essential for the maintenance of both growth and the cell cycle at low temperatures, but their function is still poorly understood. These peptides and glycopeptides of various sizes decrease the freezing point of cellular water by binding to ice crystals during formation.

Extracellular ice crystal formation on tissues due to subzero temperatures can cause extensive damage resulting in cell membrane injury. Intracellular water passes to extracellular spaces due to persistent freezing, which leads to cell

TABLE 26.6

List of Enzymes Reported from Antarctic Microbes

Enzymes	Microbe	Source	Reference
α-amylase	*Pseudoalteromonas* sp. M175	Sea ice	Wang et al. (2018)
β-D-galactosidase	*Polaribacter irgensii* 23	Near shore marine waters	Murray and Grzymski, (2007)
β-galactosidases	*Pseudoalteromonas* sp.	Antarctic krill, sea water, deep sea sponges	Parrilli et al. (2019)
β-glucosidase	*Pseudoalteromonas* sp.	Deep sea sponges	Parrilli et al. (2019)
ATP-dependent RNA helicase	*Polaribacter irgensii* 23	Near shore marine waters	Murray and Grzymski (2007)
Chitinases	*Pseudoalteromonas* sp.	Sea water	Parrilli et al. (2019)
Lipase	*Halomonas* sp. BRI8	Sea	Jadhav et al. (2013)
	Moraxella TA144	Sea water	Feller et al. (1991)
	Pseudoalternomonas haloplanktis	Marine	Maiangwa et al. (2015)
	Pseudomonas antarctica		Reddy et al. (2004)
	Pseudomonas sp. strain AMS8	Soil	Ali et al. (2013)
	Psychrobacter sp.	Sea	Xuezheng et al. (2010)
	Psychrobacter salsus sp. nov.	Fast ice	Shivaji et al. (2004)
	Psychrobacter adeliensis sp. nov.	Fast ice	Shivaji et al. 2004
	Vibrio sp.		Lo Giudice et al. (2006)
DNA ligase	*Pseudoalteromonas haloplanktis*	Sea water	Parrilli et al. (2019)
Esterases	*Pseudoalteromonas* sp.	Sea water, Antarctic krill, sea-ice	Parrilli et al. (2019)
Proteases	*Glaciozyma antarctica, Pseudoalteromonas* sp.	Deep sea, sea-ice, sea water, deep sea sponges, deep sea sediment	Sharma et al. (2018), Parrilli et al. (2019)
Ribonuclease H1	*Polaribacter irgensii* 23	Near shore marine waters	Murray and Grzymski (2007)
Soluble acyl-ACP desaturase	*Polaribacter irgensii* 23	Near shore marine waters	Murray and Grzymski (2007)

TABLE 26.7

List of Psychrophilic Proteins Reported from Antarctic Microbes

Proteins	Microbe	Source	Reference
Cold shock protein	*Polaribacter irgensii* 23-P	Near shore marine waters	Murray and Grzymski (2007)
Antifreeze protein (AFP)	*Marinomonas primoryensis*	Lake	Garnham et. al. (2011)
Ice binding protein (FfIBP)	*Flavobacterium frigoris* PS1	Sea ice	Do et al. (2014)
ColAFP	*Collwellia* SLW05	Sea ice	Raymond et al. (2007)
AFP	*Williamsia* sp. strain D3	Soil	Guerrero et al. (2014)
AFP	*Nostoc* sp., *Phormidium* sp.	Fresh water	Raymond and Fritsen (2000)
AFP	Glaciozyma Antarctica PI12	Sea ice	Hashim et al. (2013)
AnpAFP	*Antarctomyces psychrotrophicus*	Mosses, soil, mats	Xiao et al. (2010)
TisAFP	*Typhula ishikariensis*	Mosses, soil, mats	Xiao et al. (2010)

dehydration and increased osmolarity. Organisms existing in environments where temperature is lower than the freezing point are found to survive with the help of unique adaptation proteins called antifreeze proteins (AFPs), sometimes referred to as ice-binding proteins (IBPs). Investigations have proved the applications of AFPs in medical and industrial fields owing to their distinct properties. The benefits of low-temperature storage and inhibition of ice crystallization contributed to its application in cryopreservation, cryosurgery, food storage, etc. (Tejo, 2020). Apart from bacterial isolates (Do et al., 2014) such as *Marinomonas primoryensis* and *Colwellia* sp., AFPs were also reported from actinobacteria, *Williamsia* sp. strain D3 (Guerrero et al., 2014), cyanobacteria, *Nostoc* sp., and *Phormidium* sp. (Raymond & Fritsen, 2000). Antarctic yeast *Glaciozyma Antarctica* PI12 along with fungal species including *Antarctomyces psychrotrophicus* and *Typhula ishikariensis* (Xiao et al., 2010) isolated from the terrestrial environments of Antarctica were found to possess AFPs. Some important psychrophilic proteins reported from Antarctic microbes are given in Table 26.7.

26.5.3 Pigments

The biota in polar regions is exposed to high irradiances, low temperatures, and periodic freeze-thaw cycles. In order to maintain the usual functions of the cell, the lipid membrane requires being in a fluid state, which can be achieved with a higher proportion of unsaturated fatty acids at lower temperature. The importance of pigments was studied in various bacteria grown at different temperatures and under freeze-thaw stress. It was documented that certain carotenoid pigments can give rigidifying effect to regulate the fluidity and stabilize the cell membrane (Hara et al., 1999). This stabilizing effect may further control the diffusion barrier for ions and molecular oxygen (Subczynski et al., 1994). For instance, carotenoids like zeaxanthin, β-cryptoxanthin, and β-carotene also showed subsequent reduction in fluidity (Jagannadham et al., 2000). Furthermore, carotenoid pigments can also efficiently reduce photo-damage (Jahns and Holzwarth, 2012) and influence osmoregulation, thereby preventing the effects of freeze-thaw and desiccation in a multistressor environment (Mueller et al., 2005).

Carotenoids maintain the food quality by protecting it from intense light and are also reported to be used as coloring agents for baked items, cooked sausages, and soft drinks (Chattopadhyay et al., 1997, Konuray & Erginkaya, 2015). Multiple endophytic fungal isolates isolated from Antarctica were reported to produce a dark blackish melanin pigment, which provided protection against environmental stresses in cold habitats (Rosa et al., 2009). Astaxanthin and Phoenicoxanthin produced from *Xanthophyllomyces dendrorhous* were reported to have a photoprotective role (Contreras et al., 2015). Dimitrova et al. (2013) reported the pigments torularhodin and torulene extracted from *Sporobolomyces salmonicolor* that has antioxidant activity. The multipurpose pigment, violacein, has the capability to dye both natural and synthetic fibers, which has led to its importance in the textile industry. Violacein were also found to display promising antibacterial and antifungal activities (Sajjad et al., 2020). Some important pigments reported from Antarctic microbes are given in Table 26.8.

26.5.4 Biosurfactants

Production of emulsifiers and biosurfactants by the extremophiles such as psychrophilic and psychrotrophic bacteria was reported in recent years (Giudice et al., 2010). Biosurfactants from cold-adapted organisms can interact with multiple physical phases—water, ice, hydrophobic compounds, and gases—at low and freezing temperatures and be used in sustainable (green) and low-energy-impact (cold) products and processes. One of the current most-valued biosurfactant-producing microorganisms *Moesziomyces antarcticus* (formerly known as *Pseudozyma/Candida antarctica*) was originally isolated from Lake Vanda in the Wright Valley, Antarctica. *M. antarcticus* is a yeast known for the production of industrially relevant molecules including extracellular enzymes (e.g., lipase) and glycolipid biosurfactants of the type mannosylerythritol lipids (MELs). Importantly, *M. antarcticus* can be produced at high yields (up to 100 g L^{-1}) from vegetable oils by intermittent feeding, which makes them attractive for industrial applications [21]. In addition, despite being synthesized under moderate conditions (around 25°C), MELs display remarkable performance also at subzero temperatures, for example, anti-agglomeration of ice crystals (Kitamoto et al., 2001) and freezing point depression (Madihalli et al., 2016). Perfumo et al. (2017) critically reviewed the biotechnological applications of biosurfactants produced by cold-loving microorganisms including those that are from Antarctica. Malavenda et al.

TABLE 26.8
Pigments Reported from Antarctic Microbes

Pigments	Microbe	Source	Reference
Astaxanthin	*Xanthophyllomyces dendrorhous*	Soil	Contreras et al. (2015)
β-carotene	*Cryptococcus albidus, Cryptococcus laurentii, Sporobolomyces salmonicolor*	Soil	Dimitrova et al. (2013)
β-carotene	*Thelebolus microsporus*	McLeod Island	Singh et al. (2014)
(carotenoid)	*Xanthophyllomyces Dendrorhous*	Soil	Contreras et al. (2015)
β-Carotenoid	*Cystofilobasidium capitatum, Sporobolomyces ruberrimus*	Antarctic zooplankton	Moliné et al. (2009)
Carotenoids	*Micrococcus roseus*	Antarctic soil	Chattopadhyay et al. (1997)
	Arthrobacter agilis	Sea ice	Fong et al. (2001)
	Dioszegia patagonica	Soil	Trochine et al. (2017)
	Hymenobacter actinosclerus	Soil	Órdenes-Aenishanslins et al. (2016)
	Chryseobacterium Chaponense	Soil	Órdenes-Aenishanslins et al. (2016)
	Kocuria polaris sp. Nov	McMurdo Dry Valley	Reddy et al. (2003)
	Pedobacter terrae	Doumer Island	Correa-Llantén et al. (2012)
Melanin	*Lysobacter oligotrophicus*	Microbial mats	Kimura et al. (2015)
	Nadsoniella nigra		Chyizhanska and Beregova (2009)
Mycosporine	*Cryptococcus* sp., *Torrubiella* sp.	Sedimentary rocks	Barahona et al. (2016)
Phoenicoxanthin	*Xanthophyllomyces dendrorhous*	Soil	Contreras et al. (2015)
Torularhodin	*Sporobolomyces salmonicolor*	Soil	Dimitrova et al. (2013)
Torulene	*Sporobolomyces salmonicolor*	Soil	Dimitrova et al. (2013)
Unidentified blue pigment	*Antarctomyces pellizariae* sp. nov.,	Coppermine Peninsula	de Menezes et al. (2017)
Violacein	*Janthinobacterium* sp. strain Ant5-2	Lake	Mojib et al. (2013)

(2015) also described in detail about screening and isolation of biosurfactant producing Antarctic and Arctic bacteria using hydrocarbon.

26.5.5 Polysaccharides

Many reviews have been already published about exopolysaccharides from marine bacteria and marine extremophiles (Casillo et al., 2018). Extracellular polymeric substances (EPSs) play a diversified ecological role, including cell adhesion to surfaces and cell protection, and are highly involved in the interactions between the bacterial cells and the bulk environments. Interestingly, EPSs find valuable applications in the industrial field, due to their chemical versatility. Antarctic bacteria have not been given the attention they deserve as producers of EPS molecules, and a very limited insight into their EPS production capabilities and biotechnological potential is available in the literature to date. Antarctic EPS-producing bacteria are mainly psychrophiles deriving from the marine environments (generally sea ice and seawater) around the continent (Lo Giudice et al., 2020). Exopolysaccharides (EPSs) may play an important role in the Antarctic marine environment, possibly acting as ligands for trace metal nutrients such as iron or providing cryoprotection for growth at low temperature and high salinity. Nichols et al. (2005) characterized the exopolysaccharides produced by Antarctic bacteria. The study revealed that the EPSs produced by the Antarctic bacteria were very diverse, even among six closely related *Pseudoalteromonas* isolates. Some strain produced unusually large polymers (molecular weight up to 5.7 MDa) including one strain in which EPS synthesis is stimulated by low temperature. In addition, Caruso et al. (2018) reported the production and biotechnological potential of extracellular polymeric substances from sponge-associated Antarctic bacteria *Winogradskyella* sp. strains CAL384 and CAL396, *Colwellia* sp. strain GW185, and *Shewanella* sp. strain CAL606.

26.5.6 Nanoparticle synthesis

Nowadays, nanotechnology has started playing a significant role in all domains. Biobased approaches play a vital role in the production of different nanomaterials for various applications. There are several articles published on the biosynthesis of nanoparticles using plants, microbes, marine organisms, or their products. However, there are very few reports on nanoparticle synthesis using psychrophilic microorganisms. Recently, Das et al. (2020) reported the biosynthesis of gold nanoparticles using the psychrotolerant Antarctic bacteria, *Bacillus* sp. GL 1.3 at all the incubation temperatures (4°C, 10°C, 25°C, 30°C, and 37°C) and its antibacterial activity against sulphate-reducing bacteria. The formed nanoparticles were to be of size 30–50 nm. Comet assay revealed that the genotoxic effect of GNP on SRB is responsible for the inhibition of its growth and sulfide production. Plaza et al. (2016) reported the biological synthesis of fluorescent nanoparticles by cadmium- and tellurite-resistant Antarctic bacteria belonging to the genera *Pseudomonas*, *Psychrobacter*, and *Shewanella*.

26.6 Polar Microbial Research: Opportunities and Challenges

Polar regions are the pristine environments for cold-loving microorganisms with diverse ecological and industrial significance. Antarctica is the coldest and driest among the seven continents of our earth. But they act as the richest source of genetic pool of psychrophilic and psychrotolerant microorganisms. There are several microbial genera including novel genera and species reported from Antarctic samples both at international and Indian levels. Outcomes of several studies also highlighted the bioprospecting potentials of Antarctic microbes. Hence, there is a great scope for working on polar microbes. However, our understanding and exploration of Antarctic microbes especially those with biotechnological significance is still in the stage of infancy. The known major challenges associated with polar microbial research are successful expedition to polar regions and collection, preservation and transportation of relevant samples, and on-board isolation of polar microbes and their long-term preservation of future bioprospecting applications. In India, the National Centre for Polar and Ocean Research (coordinating polar research in India), MoES, and other universities and research institutes are actively engaged in polar microbial research. However, the judicious and gainful utilization of microbial resources from polar environments at further level depend on the multidisciplinary collaborations.

Acknowledgment

The authors acknowledge the financial support for the project (MoES-NCPOR/R.No. NCPOR/2019/PACER-POP/BS-08) given by the ESSO-National Centre for Polar and Ocean Research, Ministry of Earth Sciences, under the PACER Outreach Programme (POP) Initiative. The authors also thank the authorities of the management of Sathyabama Institute of Science and Technology, Chennai for their encouragement and support.

REFERENCES

Abele D, Vazquez S, Buma AG, et al. 2017. Pelagic and benthic communities of the Antarctic ecosystem of Potter Cove: genomics and ecological implications. *Marine Geno.* 33, 1–1.

Ali M, Shukuri M, Mohd Fuzi SF, Ganasen M, Abdul Rahman RN, Basri M, Salleh AB. 2013. Structural adaptation of cold-active RTX lipase from *Pseudomonas* sp. strain AMS8 revealed via homology and molecular dynamics simulation approaches. *BioMed Res Int.* 2013, 2013.

Altizer S, Ostfeld RS, Johnson PT, Kutz S, Harvell CD. 2013. Climate change and infectious diseases: from evidence to a predictive framework. *Science.* 341(6145), 514–519.

Araujo R, Gupta VV, Reith F, Bissett A, Mele P, Franco CM. 2020. Biogeography and emerging significance of Actinobacteria in Australia and Northern Antarctica soils. *Soil Biol Biochem.* 146, 107805.

Bajerski F, Wagner D. 2013. Bacterial succession in Antarctic soils of two glacier forefields on Larsemann Hills, East Antarctica. *FEMS Microbiol Ecol.* 85(1), 128–142.

Balagurunathan R, Radhakrishnan M, Somasundram ST. 2010. L-glutaminase producing actinomycetes from marine sediments- selective isolation, semi-quantitative assay and characterization of potential strain. *Australian J Basic and Appl Sci*. 4(5), 698–705.

Barahona S, Yuivar Y, Socias G, Alcaíno J, Cifuentes V, Baeza M. 2016. Identification and characterization of yeasts isolated from sedimentary rocks of Union Glacier at the Antarctica. *Extremophiles*. 20(4), 479–491.

Barka EA, Vatsa P, Sanchez L, Gaveau-Vaillant N, Jacquard C, Klenk HP, Clément C, Ouhdouch Y, Van Wezel GP. 2016. Taxonomy, physiology and natural products of actinobacteria. *Microbiol Molecularbiol Rev*. 80(1), 1–43.

Berdy J. 2012. Thoughts and facts about antibiotics: where we are now and where we are heading. *J Antibiot*. 65(8), 385–395.

Birolo L, Tutino ML, Fontanella B, Gerday C, Mainolfi K, Pascarella S, Sannia G, Vinci F, Marino G. 2000. Aspartate aminotransferase from the Antarctic bacterium Pseudoalteromonas haloplanktis TAC 125: cloning, expression, properties, and molecular modelling. *Eur J. Biochem*. 267(9), 2790–2802.

Campen R, Kowalski J, Lyons WB, Tulaczyk S, Dachwald B, Pettit E, Welch KA, Mikucki JA. 2019, Microbial diversity of an Antarctic subglacial community and high-resolution replicate sampling inform hydrological connectivity in a polar desert. *Environ. Microbiol*. 21(7), 2290–2306.

Carr SA, Orcutt BN, Mandernack KW, Spear JR. 2015. Abundant Atribacteria in deep marine sediment from the Adélie Basin, Antarctica. *Front Microbiol*. 6, 872.

Carr SA, Vogel SW, Dunbar RB, Brandes J, Spear JR, Levy R, Naish TR, Powell RD, Wakeham SG, Mandernack KW. 2013. Bacterial abundance and composition in marine sediments beneath the R oss I ce S helf, A ntarctica. *Geobiology*. 11(4), 377–395.

Caruso C, Rizzo C, Mangano S, Poli A, Di Donato P, Finore I, Nicolaus B, Di Marco G, Michaud L, Lo Giudice A. 2018. Production and biotechnological potential of extracellular polymeric substances from sponge-associated Antarctic bacteria. *Appl Environ Microbiol*. 84, e01624–17.

Casillo A, Lanzetta R, Parrilli M, Corsaro MM. 2018. Exopolysaccharides from marine and marine extremophilic bacteria: structures, properties, ecological roles and applications. *Mar Drugs*. 16(2), 69.

Casillo A, Parrilli E, Tutino ML, Corsaro MM. 2019. The outer membrane glycolipids of bacteria from cold environments: isolation, characterization, and biological activity. *FEMS Microbiol. Ecol*. 95(7), fiz094.

Chattopadhyay MK, Jagannadham MV, Vairamani M, Shivaji S. 1997. Carotenoid pigments of an antarctic psychrotrophic bacterium *Micrococcus roseus:* temperature dependent biosynthesis, structure, and interaction with synthetic membranes. *Biochem Biophys Res Commun*. 239(1), 85–90.

Cheah Y, Lee L, Chieng C, Catherine Y, Wong V, Michael LC. 2015. Isolation, identification and screening of actinobacteria in volcanic soil of deception island (the Antarctic) for antimicrobial metabolites. *Polish Polar Res*. 36(1), 67–78.

Chyizhanska N, Beregova T. 2009. Effect of melanin isolated from Antarctic yeasts on preservation of pig livestock after ablactation. Український антарктичний журнал. 8, 382–385.

Coker JA. 2016. Extremophiles and biotechnology: current uses and prospects. *F1000Res*. 5, F1000.

Contreras G, Barahona S, Sepúlveda D, Baeza M, Cifuentes V, Alcaíno J. 2015. Identification and analysis of metabolite production with biotechnological potential in *Xanthophyllomyces dendrorhous* isolates. *World J. Microbiol Biotechnol*. 31(3), 517–526.

Correa-Llantén DN, Amenábar MJ, Blamey JM. 2012. Antioxidant capacity of novel pigments from an Antarctic bacterium. *J. Microbiol*. 50(3), 374–379.

Das SK, Singh D. 2020. Epiphytic Algae on the Bryophytes of Larsemann Hills, East Antarctica. *Nat. Acad. Sci Lett*. 1–5.

Das KR, Tiwari AK, Kerkar S. 2020. Psychrotolerant Antarctic bacteria biosynthesize gold nanoparticles active against sulphate reducing bacteria. *Prep Biochem Biotechnol*. 50(5), 438–444.

de Menezes GC, Godinho VM, Porto BA, Gonçalves VN, Rosa LH. 2017. Antarctomyces pellizariae sp. nov., a new, endemic, blue, snow resident psychrophilic ascomycete fungus from Antarctica. *Extremophiles*. 21(2), 259–269.

Dimitrova S, Pavlova K, Lukanov L, Korotkova E, Petrova E, Zagorchev P, Kuncheva M. 2013. Production of metabolites with antioxidant and emulsifying properties by Antarctic strain Sporobolomyces salmonicolor AL 1. *Appl. Biochem. Biotechnol*. 169(1), 301–311.

Do H, Kim SJ, Kim HJ, Lee JH. 2014. Structure-based characterization and antifreeze properties of a hyperactive ice-binding protein from the Antarctic bacterium Flavobacterium frigoris PS1. *Acta Crystallographica Sect. D: Biol. Crystall*. 70(4), 1061–1073.

Feller G, Thiry M, Arpignya JL, Gerday C. 1991. Cloning and expression in Escherichia antarctic strain Moraxellu TA144 c & i of three lipase-encoding genes from the psychrotrophic. *Gene*. 102(1), 111–115.

Ferrari BC, Bissett A, Snape I, van Dorst J, Palmer AS, Ji M, Siciliano SD, Stark JS, Winsley T, Brown MV. 2016. Geological connectivity drives microbial community structure and connectivity in polar, terrestrial ecosystems. *Environ Microbiol*. 18(6), 1834–1849.

Fong N, Burgess M, Barrow K, Glenn D. 2001. Carotenoid accumulation in the psychrotrophic bacterium Arthrobacter agilis in response to thermal and salt stress. *Appl Microbiol Biotechnol*. 56(5–6), 750–766.

Garnham CP, Campbell RL, Davies PL. 2011. Anchored clathrate waters bind antifreeze proteins to ice. *Proc. Nat. Acad. Sci*. 108(18), 7363–7367.

Gentile G, Giuliano L, D'Auria G, Smedile F, Azzaro M, De Domenico M, Yakimov MM. 2006. Study of bacterial communities in Antarctic coastal waters by a combination of 16S rRNA and 16S rDNA sequencing. *Environ. Microbiol*. 8(12), 2150–2161.

Gilichinsky D, Rivkina E, Bakermans C, et al. 2005. Biodiversity of cryopegs in permafrost. *FEMS Microbiol Ecol*. 53(1), 117–128.

Giudice AL, Azzaro M. 2019. Diversity and ecological roles of prokaryotes in the changing Antarctic marine environment. In Castro-Sowinski, Ed., *The Ecological Role of Micro-organisms in the Antarctic Environment*. Springer, Cham. 109–131.

Giudice, A.L., Bruni, V., De Domenico, M. and Michaud, L., 2010. Psychrophiles-cold-adapted hydrocarbon-degrading microorganisms. In *Handbook of Hydrocarbon and Lipid Microbiology*.

Gonçalves VN, Campos LS, Melo IS, Pellizari VH, Rosa CA, Rosa LH. 2013. *Penicillium solitum:* a mesophilic, psychrotolerant fungus present in marine sediments from Antarctica. *Polar Biol.* 36(12), 1823–1831.

Gonçalves VN, Carvalho CR, Johann S, et al. 2015. Antibacterial, antifungal and antiprotozoal activities of fungal communities present in different substrates from Antarctica. *Polar Biol.* 38(8), 1143–1152.

Gonçalves VN, Oliveira FS, Carvalho CR, Schaefer CE, Rosa CA, Rosa LH. 2017. Antarctic rocks from continental Antarctica as source of potential human opportunistic fungi. *Extremophiles.* 21(5), 851–860.

Guerrero LD, Makhalanyane TP, Aislabie JM, Cowan DA. 2014. Draft genome sequence of Williamsia sp. strain D3, isolated from the Darwin Mountains, Antarctica. *Genome Announcements.* 2(1), e01230-13.

Gura C, Rogers SO. 2020. Metatranscriptomic and metagenomic analysis of biological diversity in Subglacial Lake Vostok (Antarctica). *Biology.* 9(3), 55.

Hara M, Yuan H, Yang Q, et al. 1999. Stabilization of liposomal membranes by thermozeaxanthins: carotenoid-glucoside esters. *Biochim Biophys Acta Biomembr.* 1461, 147–154.

Jadhav VV, Pote SS, Yadav A, Shouche YS, Bhadekar RK. 2013. Extracellular cold active lipase from the psychrotrophic Halomonas sp. BRI 8 isolated from the Antarctic sea water. *Songklanakarin J. Sci. Technol.* 35(6), 623–630.

Jagannadham MV, Chattopadhyay MK, Subbalakshmi C, et al. 2000 Carotenoids of an Antarctic psychrotolerant bacterium, *Sphingobacterium antarcticus*, and a mesophilic bacterium, *Sphingobacterium multivorum*. *Arch Microbiol.* 173, 418–424.

Jahns P, Holzwarth AR. 2012. The role of the xanthophyll cycle and of lutein in photoprotection of photosystem II. *Biochim Biophys Acta Bioenerg.* 1817, 182–193.

Kimura T, Fukuda W, Sanada T, Imanaka T. 2015. Characterization of water-soluble dark-brown pigment from Antarctic bacterium, *Lysobacter oligotrophicus*. *J. Biosci. Bioeng.* 120(1), 58–61.

Kitamoto D, Yanagishita H, Endo A. 2001. Remarkable antiagglomeration effect of a yeast biosurfactant, diacylmannosylerythritol, on ice-water slurry for cold thermal storage. *Biotechnol Prog.* 17, 362–365.

Ko YW, Choi HG, Lee DS, Kim JH. 2020. 30 years revisit survey for long-term changes in the Antarctic subtidal algal assemblage. *Sci. Reports.* 10(1), 1–1.

Konuray G, Erginkaya Z. 2015 Antimicrobial and antioxidant properties of pigments synthesized from microorganisms. In A. Méndez-Vilas, Ed., *The Battle against Microbial Pathogens---Basic Science, Technological Advances and Educational Programs.* Formatex Research Center, Badajoz, 27–32.

Lam KS. 2006. Discovery of novel metabolites from marine actinomycetes. *Curr Opin Microbiol.* 9, 245–251.

Lavin PL, Yong ST, Wong CM, De Stefano M. 2016. Isolation and characterization of Antarctic psychrotroph Streptomyces sp. strain INACH3013. *Antarctic Sci.* 28(6), 433–442.

Lee CK, Barbier BA, Bottos EM, McDonald IR, Cary SC. 2012. The inter-valley soil comparative survey: the ecology of Dry Valley edaphic microbial communities. *The ISME J.* 6(5), 1046–1057.

Li HR, Yu Y, Luo W, Zeng YX. 2010. Marisediminicola antarctica gen. nov., sp. nov., an actinobacterium isolated from the Antarctic. *Int J Syst Evol Microbiol.* 60(11), 2535–2539.

Liu SB, Chen XL, He HL, et al. 2013. Structure and ecological roles of a novel exopolysaccharide from the Arctic Sea ice bacterium *Pseudoalteromonas* sp. strain SM20310. *Appl Environ Microbiol.* 79, 224–230.

Lo Giudice A, Brilli M, Bruni V, De Domenico M, Fani R, Michaud L. 2007. Bacterium–bacterium inhibitory interactions among psychrotrophic bacteria isolated from Antarctic seawater (Terra Nova Bay, Ross Sea). *FEMS Microbiol Ecol.* 60(3), 383–396.

Lo Giudice A, Michaud L, De Pascale D, De Domenico M, Di Prisco G, Fani R, Bruni V. 2006. Lipolytic activity of Antarctic cold-adapted marine bacteria (Terra Nova Bay, Ross Sea). *J Appl Microbiol.* 101(5), 1039–1048.

Lo Giudice A, Poli A, Finore I, et al. 2020. Peculiarities of extracellular polymeric substances produced by Antarctic bacteria and their possible applications. *Appl Microbiol Biotechnol.* 104, 2923–2934.

Lo Giudice A, Rizzo C. 2018. Bacteria associated with marine benthic invertebrates from polar environments: unexplored frontiers for biodiscovery? *Diversity.* 10(3), 80.

Maas EW, Simpson AM, Martin A, Thompson S, Koh EY, Davy SK, Ryan KG, O'Toole RF. 2012. Phylogenetic analyses of bacteria in sea ice at Cape Hallett, Antarctica. New Zealand *J Marine Freshwater Res.* 46(1), 3–12.

Madihalli C, Sudhakar, H, Doble, M. 2016. Mannosylerythritol lipid-A as a pour point depressant for enhancing the low-temperature fluidity of biodiesel and hydrocarbon fuels. *Energy Fuels.* 30, 4118–4125.

Maiangwa J, Ali MS, Salleh AB, Abd Rahman RN, Shariff FM, Leow TC. 2015. Adaptational properties and applications of cold-active lipases from psychrophilic bacteria. *Extremophiles.* 19(2), 235–247.

Malavenda R, Rizzo C, Michaud L, et al. 2015. Biosurfactant production by Arctic and Antarctic bacteria growing on hydrocarbons. *Polar Biol.* 38, 1565–1574.

Manikkam R, Pati P, Thangavel S, Venugopal G, Joseph J, Ramasamy B, Dastager SG. 2019. Distribution and Bioprospecting Potential of Actinobacteria from Indian Mangrove Ecosystems. In *Microbial Diversity in Ecosystem Sustainability and Biotechnological Applications* (pp. 319–353). Springer, Singapore.

Margesin R, Feller G. 2010. Biotechnological applications of psychrophiles. *Environ Technol.* 31(8–9), 835–844.

Millán-Aguiñaga N, Soldatou S, Brozio S, Munnoch JT, Howe J, Hoskisson PA, Duncan KR. 2019. Awakening ancient polar Actinobacteria: diversity, evolution and specialized metabolite potential. *Microbiology.* 165(11), 1169–1180.

Mojib N, Farhoomand A, Andersen DT, Bej AK. 2013. UV and cold tolerance of a pigment-producing Antarctic Janthinobacterium sp. Ant5-2. *Extremophiles.* 17(3), 367–378.

Moliné M, Libkind D, Diéguez MC, Van Broock M. 2009. Photoprotective role of carotenoid pigments in yeasts: experimental study contrasting naturally occurring pigmented and albino strains. *J Photoch Photobio B.* 95, 156–161.

Mueller DR, Vincent WF, Bonilla S, et al. 2005. Extremotrophs, extremophiles and broadband pigmentation strategies in a high arctic ice shelf ecosystem. *FEMS Microbiol Ecol.* 53, 73–87.

Murray AE, Grzymski JJ. 2007. Diversity and genomics of Antarctic marine micro-organisms. *Philosophical Trans Royal Soc B: Biol Sci*. 362(1488), 2259–2271.

Nichols CM, Lardiere SG, Bowman JP, Nichols PD, AE Gibson J, Guezennec J. 2005. Chemical characterization of exopolysaccharides from Antarctic marine bacteria. *Microb Ecol*. 49(4), 578–589.

Núñez-Montero K, Barrientos L. 2018. Advances in antarctic research for antimicrobial discovery: a comprehensive narrative review of bacteria from antarctic environments as potential sources of novel antibiotic compounds against human pathogens and microorganisms of industrial importance. *Antibiotics*. 7(4), 90.

Oliveira CE, Michel RF, Rosa CA, Rosa LH. 2019. *Fungi Present in Soils of Antarctica. Fungi of Antarctica---Diversity, Ecology and Biotechnological Applications*. Rosa, L.H., Ed., 1st ed., Springer, Cham, 43.

Órdenes-Aenishanslins N, Anziani-Ostuni G, Vargas-Reyes M, Alarcón J, Tello A, Pérez-Donoso JM. 2016. Pigments from UV-resistant Antarctic bacteria as photosensitizers in dye sensitized solar cells. *J Photochem Photobiol B: Biol*. 162, 707–714.

Parrilli E, Tedesco P, Fondi M, Tutino ML, Giudice AL, de Pascale D, Fani R. 2019. The art of adapting to extreme environments: the model system Pseudoalteromonas. *Phys Life Rev*.

Pellizzari F, Silva MC, Silva EM, Medeiros A, Oliveira MC, Yokoya NS, Pupo D, Rosa LH, Colepicolo P. 2017. Diversity and spatial distribution of seaweeds in the South Shetland Islands, Antarctica: an updated database for environmental monitoring under climate change scenarios. *Polar Biol*. 40(8), 1671–1685.

Perfumo A, Banat IM, Marchant R. 2018. Going green and cold: biosurfactants from low-temperature environments to biotechnology applications. *Trends Biotechnol*. 36, 277–289.

Pershina EV, Ivanova EA, Abakumov EV, Andronov EE. 2018. The impacts of deglaciation and human activity on the taxonomic structure of prokaryotic communities in Antarctic soils on King George Island. *Antarctic Sci*. 30(5), 278–288.

Plaza DO, Gallardo C, Straub YD, et al. 2016. Biological synthesis of fluorescent nanoparticles by cadmium and tellurite resistant Antarctic bacteria: exploring novel natural nanofactories. *Microb Cell Fact*. 15, 76.

Qin S, Li WJ, Dastager SG, Hozzein WN. 2016. Actinobacteria in special and extreme habitats: diversity, function roles, and environmental adaptations. *Front Microbiol*. 7, 1415.

Raymond JA, Fritsen CH. 2000. Ice-active substances associated with Antarctic freshwater and terrestrial photosynthetic organisms. *Antarctic Sci*. 12(4), 418–424.

Raymond JA, Fritsen C, Shen K. 2007. An ice-binding protein from an Antarctic sea ice bacterium. *FEMS Microbiol Ecol*. 61(2), 214–221.

Reddy GS, Matsumoto GI, Schumann P, Stackebrandt E, Shivaji S. 2004. Psychrophilic pseudomonads from Antarctica: *Pseudomonas antarctica* sp. nov., *Pseudomonas meridiana* sp. nov. and *Pseudomonas proteolytica* sp. nov. *Int J syst Evol Microbiol*. 54(3), 713–719.

Reddy GS, Prakash JS, Prabahar V, Matsumoto GI, Stackebrandt E, Shivaji S. 2003. *Kocuria polaris* sp. nov., an orange-pigmented psychrophilic bacterium isolated from an Antarctic cyanobacterial mat sample. *Int J syst Evol Microbiol*. 53(1), 183–187.

Rego A, Raio F, Martins TP, et al. 2019. Actinobacteria and cyanobacteria diversity in terrestrial antarctic microenvironments evaluated by culture-dependent and independent methods. *Front Microbiol*. 10, 1018.

Rosa LH, Pellizzari FM, Ogaki MB, de Paula MT, Mansilla A, Marambio J, Colepicolo P, Neto AA, Vieira R, Rosa CA. 2019. Sub-antarctic and antarctic marine ecosystems---an unexplored ecosystem of fungal diversity. In Rosa, L. H., Ed., *Fungi of Antarctica*. Springer, Cham. 221–242.

Rosa LH, Vaz AB, Caligiorne RB, Campolina S, Rosa CA. 2009. Endophytic fungi associated with the Antarctic grass Deschampsia antarctica Desv.(Poaceae). *Polar Biol*. 32(2), 161–167.

Sajjad W, Din G, Rafiq M, Iqbal A, Khan S, Zada S, Ali B, Kang S. 2020. Pigment production by cold-adapted bacteria and fungi: colorful tale of cryosphere with wide range applications. *Extremophiles*, 1.

Sharma M. 2019. Thermophiles vs. psychrophiles: cues from microbes for sustainable industries. In Sobti, R. C., Arora, N. K., Kothari, R., Eds., *Environmental Biotechnology: For Sustainable Future*, Springer, Singapore. 323–340.

Shivaji S. 2017. Bacterial biodiversity, cold adaptation and biotechnological importance of bacteria occurring in Antarctica. *Proc Indian Nat Sci Acad*. 83(2), 327–352.

Shivaji S, Chattopadhyay MK, Reddy GS. 2019. Diversity of Bacteria from Antarctica, Arctic, Himalayan Glaciers and Stratosphere. *Proc Indian Nat Sci Acad*. 85(4), 909–923.

Shivaji S, Reddy GS, Aduri RP, Kutty R, Ravenschlag K. 2004. Bacterial diversity of a soil sample from Schirmacher Oasis, Antarctica. *Cellular Mol Biol*. 50(5), 525–536.

Shivlata L, Tulasi S. 2015. Thermophilic and alkaliphilic Actinobacteria: biology and potential applications. *Front Microbiol*. 6, 1014.

Singh AK, Sad K, Singh SK, Shivaji S. 2014. Regulation of gene expression at low temperature: role of cold-inducible promoters. *Microbiology*. 160(7), 1291–1296.

Skropeta D, Wei L. 2014. Recent advances in deep-sea natural products. *Nat Prod Rep*. 31(8), 999–1025.

Subczynski WK, Wisniewska A, Yin JJ, et al. 1994. Hydrophobic barriers of lipid bilayer membranes formed by reduction of water penetration by alkyl chain unsaturation and cholesterol. *Biochemistry*. 33, 7670–7681.

Tejo BA, Asmawi AA, Rahman MB. 2020. Antifreeze proteins: characteristics and potential applications. *Makara J Sci*. 24(1), 8.

Trochine A, Turchetti B, Vaz AB, Brandao L, Rosa LH, Buzzini P, Rosa C, Libkind D. 2017. Description of Dioszegia patagonica sp. nov., a novel carotenogenic yeast isolated from cold environments. *Int J Syst Evol Microbiol*. 67(11), 4332–4339.

Vaz AB, Rosa LH, Vieira ML, Garcia VD, Brandão LR, Teixeira LC, Moliné M, Libkind D, Van Broock M, Rosa CA. 2011. The diversity, extracellular enzymatic activities and photoprotective compounds of yeasts isolated in Antarctica. *Brazilian J Microbiol*. 42(3), 937–947.

Wang X, Kan G, Ren X, Yu G, Shi C, Xie Q, Wen H, Betenbaugh M. 2018. Molecular cloning and characterization of a novel α-amylase from antarctic sea ice bacterium Pseudoalteromonas sp. M175 and its primary application in detergent. *BioMed Res Int*. 2018.

Wentzel LC, Inforsato FJ, Montoya QV, Rossin BG, Nascimento NR, Rodrigues A, Sette LD. 2019. Fungi from admiralty bay (King George Island, Antarctica) soils and marine sediments. *Microb Ecol*. 77(1), 12–24.

Wölfl AC, Lim CH, Hass HC, Lindhorst S, Tosonotto G, Lettmann KA, Kuhn G, Wolff JO, Abele D. 2014. Distribution and characteristics of marine habitats in a subpolar bay based on hydroacoustics and bed shear stress estimates—Potter Cove, King George Island, Antarctica. *Geo-Marine Lett*. 34(5), 435–446.

Wong SY, Charlesworth JC, Benaud N, Burns BP, Ferrari BC. 2019. Communication within East Antarctic Soil Bacteria. *Appl Environ Microbiol*. 86(1), e01968–19.

Wu G, Ma H, Zhu T, Li J, Gu Q, Li D. 2012. Penilactones A and B, two novel polyketides from Antarctic deep-sea derived fungus *Penicillium crustosum* PRB-2. *Tetrahedron*. 68(47), 9745–9749.

Xiao N, Suzuki K, Nishimiya Y, Kondo H, Miura A, Tsuda S, Hoshino T. 2010. Comparison of functional properties of two fungal antifreeze proteins from Antarctomyces psychrotrophicus and Typhula ishikariensis. *FEBS J*. 277(2), 394–403.

Xuezheng L, Shuoshuo C, Guoying X, Shuai W, Ning D, Jihong S. 2010. Cloning and heterologous expression of two cold-active lipases from the Antarctic bacterium Psychrobacter sp. G. *Polar Res*. 29(3), 421–429.

Yakimov MM, Gentile G, Bruni V, Cappello S, D'Auria G, Golyshin PN, Giuliano L. 2004. Crude oil-induced structural shift of coastal bacterial communities of rod bay (Terra Nova Bay, Ross Sea, Antarctica) and characterization of cultured cold-adapted hydrocarbonoclastic bacteria. *FEMS Microbiol Ecol*. 49(3), 419–432.

Yamagishi A, Kawaguchi Y, Hashimoto H, et al. 2018. Environmental data and survival data of *Deinococcus aetherius* from the exposure facility of the Japan experimental module of the international space station obtained by the tanpopo mission. *Astrobiology*.18, 1369–1374.

27

Alterations in Microbial Community Structure and Function in Response to Azo Dyes

Sandhya Nanjani and Haresh Kumar Keharia
Sardar Patel University

CONTENTS

27.1 Introduction

Humanity has evolved to use colors as a powerful and irreplaceable mode of communication. It is because the human brain is configured to perceive a message or an emotion by looking at colors more rapidly than speaking, reading, or writing any language. The implementation of colors is manipulated and exploited everywhere, whether it is marketing or regulations, entertainment or emotions, and philosophy or psychology. As a matter of fact, the fruit orange derives its name from the color it possesses, providing a logical link between taste and color. Thus, colors have eventually become an inevitable and integrated part of the lifestyle. So, over the decades, there has occurred a phenomenal increase in the fascination and craving for varied colors. The industrial revolution marked a renaissance from the application of natural pigments to synthetic dyes. In 1856, Sir William H. Perkin serendipitously discovered dye mauveine while synthesizing quinine, an antimalarial drug (Chengalroyen & Dabbs, 2013). Since then, the world has seen a sudden boom in the demands of colors and colored products. The escalated production of dyes, specifically in the textile industries, has threatened the health of the planet. As per World Bank, these industries contribute to 20% of the total industrial water contamination caused across the world (Wong et al., 2019). Such colossal pollution by dyeing industries has become a challenge demanding immediate attention of scientists globally to address the issue by developing a sustainable solution. The major concern is that the wastewater liberated from textile industries is heavily loaded with toxic and pernicious chemicals, including heavy metals, surfactants, dyes, pigments, phenolics, chlorinated organics, solvents, and high concentration of salts (Rather et al., 2019; Uysal et al., 2014). Moreover, the rate of anthropogenic pollution is very rapid as compared to the self-cleaning efficiency of nature. Due to the presence of dyes, almost 93% of the surface water available for consumption is contaminated by colored wastewaters (Gupta et al., 2015). The principal reason for this is the lack of complete fixation of dye onto substrates (i.e., acrylic fibers, cotton, polyester, etc.), which in turn is influenced by several parameters including dye shade, material to liquor ratio, pH, and method of application (Singh & Arora, 2011). Also, large quantities of wastewaters (~80%) are released from the textile processing and manufacturing units without following pollution control (Mani et al., 2019).

The effluent discharged from these industries has a devastating impact on the soil and water resources from the aesthetic viewpoint (Silveira et al., 2009). It is because the dyes are visible at the mere concentration of 1 mg L^{-1} (Pandey et al., 2007; Tony et al., 2009), and the treated effluent released carries approximately 20–30 mg L^{-1} dyes (Nilratnisakorn et al., 2007). Moreover, loads of sulfates in the wastewaters lead to acidification of natural water bodies, increase salinity, and contribute to H_2S (hydrogen sulfide) generation on anaerobic catabolism (Miran et al., 2018). Also, the dyes in the effluent are a long-standing menace to the lives at all the biological levels and are subject to biomagnification (Moopantakath & Kumavath, 2018). Based on the mode of dyeing, dyes are classified as direct, vat, reactive, mordant, and disperse. On the basis of charge, they are classified as acidic (anionic), basic (cationic), and (nonionic) (Marimuthu et al., 2013). However, depending on the type of chromophores, these are classified as nitro, nitroso, azo, triphenylmethane, indigo, and anthraquinone dyes (Ali, 2010). Besides, these can be classified as a synthetic or natural dye, depending on the source. China is the leading producer of dyes contributing to 40% of the total dyes manufactured globally(Wang et al., 2013), followed by India, which is the second-largest producer giving up one-third of the total dye exports (Sharma et al., 2007).

27.2 Azo Dyes and Their Hazards

Azo dyes embrace the most significant proportion, i.e., 3,000 different shades make up to 60%–70% of the total synthetic dyes applied in the textile industries (Sudha et al., 2018; Yaseen & Scholz, 2016; Yurtsever et al., 2015). The global annual discharge of dye-laden effluents solely from the textile plants accounts to approximately 2,80,000 tons (Franca et al., 2020a). Like textile processing units, these are also utilized in plastic, paper, leather, tannery, cosmetics, pulp, pharmaceuticals, and food industries (Ayed et al., 2019; Mishra & Maiti, 2018). These may be described as the class of refractory pollutants, which are comprised of one or several azo bridges (–N=N–) bound to the sp^2-hybridized carbon atom of substituted/nonsubstituted aromatic or heterocyclic rings (Pathak et al., 2014; Saratale et al., 2011). Based on the number of azo groups present, these are identified as mono-, di-, tris-, or poly-azo dyes (Chengalroyen & Dabbs, 2013). The benzene or naphthalene rings are often substituted by diverse groups viz., carboxyl (–COOH), hydroxyl (–OH), methyl ($–CH_3$), sulfonate ($–SO_3$), nitro ($–NO_2$), amine ($–NH_2$), and chloro (–Cl). (Chequer et al., 2013). The sulfonate groups confer water solubility to dyes (Sarayu & Sandhya, 2012), whereas the reactive functional groups (eg., –OH, –SH and –NH) are responsible for covalently bonding to the fibers viz., cotton, wool, nylon, and silk (Carmen & Daniela, 2012).

The presence of dyes/pigments blocks the penetration of visible light into surface water resulting in the reduction of photic or euphotic zones. The attenuated light intensities hinder the respiration and photosynthesis by phytoplankton and other primary producers, which in turn leads to oxygen-deficient conditions (Chacko & Subramaniam, 2011; Movafeghi et al., 2016). Thus, the presence of dyes disturbs the food chain by impeding the growth of several primary producers, thereby affecting the survival of other dependent organisms at all positions in the food web. Furthermore, the water properties like biological oxygen demand (BOD), salinity, pH, chemical oxygen demand (COD), and total organic carbon (TOC) are adversely affected (Saratale et al., 2009; Solís et al., 2012). Overall, the quality of the surface and groundwater gets highly compromised. Moreover, the removal of the reactive azo dyes has become a matter of utmost concern as these display toxicity merely at a concentration of 5.2 mg L^{-1} (Nilratnisakorn et al., 2007) and a have half-life time of several years (Mahajan et al., 2019). It is because the dyes can withstand debasement by heat, pH, visible and UV-light, and detergents (Jonstrup et al., 2013; Rane et al., 2015; Uysal et al., 2014).

The dye toxicity has been affirmed in several organisms like plants, mollusks, rats, microbes, fishes, and cultured mammalian cell lines (Chandanshive et al., 2017). In particular, the nonionic class of azo dyes is perceived as potentially toxic substances (Novotny et al., 2006). The exposure to dyes can be carcinogenic, mutagenic, estrogenic, and allergic resulting in eczema (Mahajan et al., 2019; Saba et al., 2015). Moreover, dyes and their products can prove to be injurious to the kidney, brain, reproductive system, central nervous system, and liver (Sarayu & Sandhya, 2012). Numerous endeavors are being made to comprehend holistic health risk by evaluating the acute and chronic toxicity and mutagenicity of a range of dyes on different classes of biological systems. Thus, various methods are being employed at a healthy pace to determine the direct or indirect effects of azo dyes on human and environmental health. Barathi et al. (2020) demonstrated that Reactive Blue 160 was toxic to human skin cell line CRL-1474 at very low IC_{50}, i.e., 19.77 μg mL^{-1} when exposed for 24 h. Moreover, they also illustrated that in zebrafish, the dye prompted deformities (such as eye defect, tail ulceration, and so forth) during embryo development, decreased hatching and survival rate and tempered cardiovascular functioning (Barathi et al., 2020). On entering the gastrointestinal tract, skin, or lungs in humans, it can lead to hemoglobin adducts and disturb blood formation (Wong et al., 2019). de Almeida et al. (2019) exemplified high mutagenicity of a binary dye solution containing Procion Red MX-5B and Acid Blue 161 employing Ames test, i.e., *Salmonella*/microcosm assay. Even if the dyes are relatively nontoxic, these may lead to the generation of toxic products if cleaved naturally. Rawat et al. (2018) scrutinized the environmental security of using Acid Orange 7 (AO7), which is otherwise applied as a nontoxic dye in the industries. The group ascertained that AO7 on treatment with a microbial consortium under saline conditions resulted in products that exhibited organism-level toxicity in *Vigna mungo* as well as cytotoxicity in root cells of *Allium cepa*. Such toxicological studies provide essential information for precautions and help environmental agencies in forming norms and laws for ecological administration (Sharma et al., 2007).

27.3 Mechanisms of Microbial Azo Dye Reduction

Bioremediation is a process of maneuvering organisms for degradation or the transformation of the sturdy xenobiotics. This approach of treating azo dye-laden effluents is being favored over other physicochemical methods (e.g., flocculation, ozonation, coagulation, advanced oxidation processes, adsorption, *etc.*) as it is economical and environmentally safe and brings about reduced sludge formation (Nachiyar et al., 2012). In particular, the physicochemical methods are incompatible with the spectrum of dyes, require more energy and heavy loads of additional chemicals, generate recalcitrant secondary pollutants, and are too complicated to perform (Movafeghi et al., 2016; Oliveira et al., 2020; Sivakumar, 2014; Vijayalakshmidevi & Muthukumar, 2014; Yaseen & Scholz, 2017). Biological degradation of dyes has been investigated employing various microorganisms (viz., bacteria, fungi, yeast, and algae), plants, or their associations (e.g., bacterial consortia, fungal-bacterial consortia, root-rhizobiome associations, mycorrhizae, and periphyton) (Gao et al., 2018; He et al., 2004; Nachiyar et al., 2012; Senan & Abraham, 2004; Shabbir et al., 2017; Wang et al., 2009). The response to a pollutant is contingent on the catabolic potential and metabolic versatility of the organism. Also, dye characteristics such as the number of –N=N– bonds, the type of reactive groups, and their arrangement is a pivotal parameter governing degradation efficiency (Sreedharan & Rao, 2019). The prospective microorganisms to be used for decolorization should not only possess catabolic potential but should also be highly resilient to the changes. The dye degradation can proceed in multifarious ways. Based on the oxygen availability, the dye can be either degraded by oxidation or reduction. Thus, in addition to the genetic potential of the organism being employed, the redox potential available at microenvironments forms a deciding factor of how the dye degradation proceeds, i.e., whether by oxidation or reduction.

27.3.1 Oxidative Mechanism of Decolorization

Under oxidative conditions, the decolorization is carried out by aerotolerant azoreductases, which are highly specific to the structure of dye (Padmavathy et al., 2003; Srisuwun et al., 2018). In order to express this enzyme, the organism needs to be acclimatized for a long time (Cui et al., 2015). Moreover, the degradation efficiency observed is low as the oxygen being an electrophile serves as the primary electron sink (Dos Santos et al., 2004). Also, the dyes are more likely to get adsorbed on the bacterial cells or fungal mycelia rather than undergoing oxidation (Cerboneschi et al., 2015; Robinson et al., 2001). Besides aerobic azoreductases, the other enzymes involved are laccases, phenoloxidase, lignin peroxidases (LiP), manganese peroxidases (MnP), veratryl alcohol oxidases, tyrosinases, N-demethylases, and cellobiose dehydrogenase (Solís et al., 2012).Another important class of enzymes is cytochrome P450 monooxygenases. These are diverse heme-complexed enzymes that can effectuate a range of reactions, including decarboxylation, demethylation, desaturation, aromatic dehalogenation, and carbon hydroxylation (Kelly & Kelly, 2013). Most of the bacteria are unable to degrade dye by the oxidative mechanism. The attempts have been made to enrich and cultivate organisms from various sources, including textile effluents, contaminated sediments, soil, fresh and marine water, animal excreta, food materials, and plants (i.e., endophytic bacterium). Some of the isolates proficient in oxidative degradation of the dye are *Bacillus* sp. (e.g., *B. stearothermophilus*), *Pseudomonas* sp., *Sphingomonas* sp., *Flavobacterium* sp., *Pigmentiphaga kullae* K24, and *Paenibacillus azoreducens* (Blumel & Stolz, 2003; Senan & Abraham, 2004). Earlier investigations have indicated that actinobacterial candidates also possess dye-oxidizing potential viz., *Streptomyces venezuelae* ATCC 10712, *Streptomyces griseus*, and *Streptomyces ipomoea.* (Moopantakath & Kumavath, 2018). Recently, the genome of soil actinobacterium, i.e., *Kocuria indica* DP-K7 was analyzed, and it was found that the methyl red (MR)-degrading attributes of the strain were due to the presence of two genes for flavin-independent azo reductase (Kumaran et al., 2020).

The oxidative decolorization has been well-studied in white-rot fungi (WRF), e.g., *Phanerochaete chrysosporium, Ceriporia lacerate, Phlebia tremellosa, Ganoderma lucidum, Hirschioporus larincinus, Coriolus versicolor, Inonotus hispidus, Pycnoporus sanguineus, etc.* (Annuar et al., 2009; Carmen & Daniela, 2012; Sreedharan & Rao, 2019). Like bacteria, the decolorization in WRF can be attributed to the production of nonspecific, free-radical-based, extracellular ligninolytic enzymes viz., peroxidases [i.e., lignin peroxidases (LiP) and manganese peroxidases (MnP)], laccases, and other oxidases such as veratryl alcohol oxidase (Tochhawng et al., 2019; Wang et al., 2013). These enzymes are produced during the primary phase of growth by some fungi, e.g., *Trametes versicolor*, whereas some express it under the secondary phase, e.g., *P. chrysosporium* (Keharia & Madamwar, 2002). Moreover, the white-rot and brown-rot fungi are reported to synthesize almost 150 types of cytochrome P450 (CYPs), most of which are functionally uncharacterized (Kelly & Kelly, 2013). Apart from these enzymes, WRF can also use the membrane-bound redox-active system for the decolorization (McMullan et al., 2001). An additional nonenzymatic mechanism of decolorization also prevails wherein the Fenton's reaction is supported by the biomolecules such as siderophores and organic acids in the presence of H_2O_2 produced by oxidases, e.g., glucose oxidase, cellobiose oxidase, and aryl alcohol oxidase (Gil et al., 2018). Other fungi reported for decolorization include *Neurospora crassa, Schizophyllum commune, Trichoderma sp.,* and *Cunninghamella elegans* (Ambrósio & Campos-Takaki, 2004; Banat et al., 1997). The chief clampdowns in the employment of fungi in the bioreactors are the necessity of complex nutrients, specified growth parameters, longer growth cycle, maintenance of functional phase, lower color removal efficiencies, and higher hydraulic retention time (HRT) to achieve complete decolorization (Banat et al., 1997; Kapdan et al., 2003;

Lade et al., 2012; Meerbergen et al., 2018; Shah, 2014). Also, through lab-scale studies, it has been proclaimed that under nonsterile conditions, the bacteria might out-grow the fungi and affect the secretion of fungal enzymes (Gao et al., 2008; Jonstrup et al., 2013; Zhou et al., 2014).

On a large scale, the dye reduction by oxidation is achieved by activated sludge or aerobic granular sludge (ABS), aerated lagoons, rotating biological contactors, stabilization ponds, trickling filters, oxidizing beds, and biological aerated filters (BAF) (Franca et al., 2020a; Ghaly et al., 2014; He et al., 2013; Lin et al., 2010; Wong et al., 2019). The activated carbon or bentonite can be supplemented to aeration tanks to improve the efficiency of aerobic treatment (Carmen & Daniela, 2012). As the dyes possess the electron-withdrawing chromophore (i.e., –N=N– bond) and sulfonate groups, these cannot be metabolized solitarily as a source of carbon and energy by microorganisms (Blumel & Stolz, 2003). Thus, the inclusion of a co-substrate (e.g., glucose, lactate, yeast extract, methanol, acetate, etc.) becomes indispensable in the bioprocess. However, the addition of co-substrates increases the cost of the remediation.

27.3.2 Reductive Mechanism of Decolorization

The academicians and researchers started showing interest in the development of anaerobic biosystems for dye reduction way back in the 1970s (Fatima et al., 2017). The higher molecular weight and presence of polar and bulky groups viz., sulfonate renders the dyes impermeable to the uptake by cell membranes, and thus, the bacterial reduction is assumed to proceed extracellularly (Solís et al., 2012). It is also well established through several shreds of evidence that decolorization under anoxic/anaerobic conditions is thoroughly nonspecific (Guo et al., 2010).The dyes behave as an oxidizing agent for the reduced molecules (e.g., flavin nucleotides), wherein the reaction can be either catalyzed by cytoplasmic reductases or electron transport molecules (Robinson et al., 2001; Tang et al., 2018). The process occurs in two steps: initially, on acquiring two electrons, the dye gets reduced to a hydrazine intermediate (eq. 1), which is further followed by addition of two more electrons (eq. 2) to generate aromatic amines (Eslami et al., 2016; Singh & Arora, 2011). The reaction is represented as follows:

$$\underset{\text{Azo dye}}{\left(R_1N = NR_2\right)} + 2H^+ + 2e^- \xrightarrow{\text{Reduction}} \underset{\text{Hydrazine intermediate}}{\left(R_1\text{-HN-NH-}R_2\right)} \tag{27.1}$$

$$\underset{\text{Hydrazine intermediate}}{\left(R_1\text{-HN-NH-}R_2\right)} + 2H^+ + 2e^- \xrightarrow{\text{Reduction}} \underset{\text{Aromatic amine}}{\left(R_1\text{-}NH_2\right)} + \underset{\text{Aromatic amine}}{\left(R_2\text{-}NH_2\right)} \tag{27.2}$$

Where R_1 and R_2 represent the heterocyclic/aryl amine substituted group.

These reactions result in the disappearance of the color from the solution and engender a plethora of aromatic amines such as amino-benzene, amino-benzidine, and amino-naphthalene (Khehra et al., 2005). Based on the type of electron donor used, azo reductases are categorized into two main classes. These are flavin-dependent and flavin-independent reductases that exhibit a preference for NADH and NAD(P)H, respectively (Mahmood et al., 2016). An additional type of flavin-dependent and NAD(P)H-preferring azo reductase has also been identified from the microbes (Chung, 2016). Most of these enzymes are recognized from facultative anaerobes (Albuquerque et al., 2005). In mammals, the azo (–N = N–) bond reduction is mediated by azoreductases produced by hepatic cells or the intestinal microflora (Puvaneswari et al., 2006; Xu et al., 2010). Several other nonspecific enzymes are reported to conduct reductive decolorization viz., NADH: DCIP reductases, riboflavin reductase, NADH/FMN-dependent AQS reductase, etc. (Chengalroyen & Dabbs, 2013; Imran et al., 2015a; Ling et al., 2009; Singh et al., 2015). Also, membrane-bound cytochromes, including OmcA, OmcB, and MtrF, can mediate electron transfer and are therefore known to function as a terminal azo reductase (Hong & Gu, 2010). Apart from the azo-reduction by nonspecific cytoplasmic and membrane-bound enzymes, it can also occur nonenzymatically via the reduced molecules, e.g., H_2S or Fe^+, which are released during anaerobic respiration of bacteria (Mishra & Maiti, 2018; Parshetti et al., 2006). Therefore, the reduction of dyes under anoxic/anaerobic conditions seems to be highly fortuitous (Chen et al., 2004).

The reductive decolorization has been reported from various anaerobes or microaerophiles viz., *Sphingomonas* sp strain BN6, *Sphinogmonas xenophaga* QYY, *Enterobacter aerogenes* PP002, *Citrobacter* sp. strain A1, *Pseudomonas* sp., *Bacillus* sp., and *Micrococcus glutamicus* NCIM-2168 (Chan et al., 2012; Gupta et al., 2015; Keck et al., 1997; Lin et al., 2010; Ling et al., 2009; Saratale et al., 2009; Sudha et al., 2018; Zablocka-Godlewska et al., 2012). Hitherto, several strains of *Shewanella* have been monitored for the decolorization of various azo dyes. Owing to the high metabolic and physiological diversity, the organisms of this genus can oxidize a variety of organic substrates and utilize metals as a terminal electron acceptor for survival and growth (Hong et al., 2007). Moreover, it has one of the wealthiest respiratory systems among the identified organisms (Zou et al., 2019). Some well-studied strains of this genera include *S. oneidensis* MR-1, *S. decolorationis* S12, *Shewanella* sp. strain IFN4, etc (Hong et al., 2007; Imran et al., 2016; Li et al., 2018; Xiao et al., 2012). In addition to this, some yeasts have also been reported to exhibit reductive decolorization. *Saccharomyces cerevisiae* MTCC 463 was shown to degrade 100 ppm of MR in distilled water within just 16 min by dint of enzymes such as azoreductase, tyrosinase, lignin peroxidase, and NADH-DCIP reductase (Jadhav et al., 2007). This is one of the most rapid known anoxic degradations. Some dye-reducing actinobacterial isolates are *Georgenia* sp. CC-NMPT-T3, *Arthrobacter aurescens* TC1, *Rhodococcus erythropolis* PR4, *Streptomyces hygroscopicus,* etc (Moopantakath & Kumavath, 2018).

Several anaerobic reactors implemented include anaerobic filters, thermophilic/mesophilic UASB (Up-flow anaerobic sludge blanket), or EGSB (expanded granular sludge bed), bio-electrochemical system (i.e., via anodic reduction), anaerobic sequence batch reactors (ASBRs), fixed-films, etc (Dos Santos et al., 2004; Lin et al., 2010; McHugh et al., 2003; Srisuwun et al., 2018; Zhang et al., 2019). The paramount advantage of performing degradation under such conditions is that higher dye removal efficiency is obtained, and the myriads of dyes can be treated as the reaction is nonspecifically mediated by the electron transfer chain (Mishra & Maiti, 2018). Besides, the anaerobes can generate biogas if the co-substrate is entirely mineralized to methane and carbon dioxide (Dhaouefi et al., 2019; Razo-Flores et al., 1997; Tan et al., 2000). The resulting biogas can be utilized to generate power and heat. Also, it is economical and environmentally safe as no aeration is required, and little sludge is generated (Wang et al., 2018). However, with due course of time, the degradation products such as aromatic amines get accumulated, which might be more carcinogenic, mutagenic, and/or toxic than the parent dye molecule (Feng et al., 2012; Franca et al., 2020b). Thus, the absolute degradation of intermediates and the lower rate owing to the shortfall of electron donors form the limiting factors during the anaerobic digestion (Cao et al., 2017). Additionally, the growth of the methanogens is inhibited due to the toxicity of the degradation products and competition of reducing equivalents by the degraders (Dai et al., 2016; Xie et al., 2016).

27.3.3 Alternative Oxidative-Reductive Mechanisms for Decolorization

Although the dye reduction under anoxic condition is promising, the complete mineralization cannot be easily achieved (McMullan et al., 2001), and the aromatic amines and other intermediates are presumably more carcinogenic and toxic than the parent dye molecule (Keharia & Madamwar, 2003). Also, if the aromatic amines are not well-metabolized, they are most likely to polymerize or oxidize to form a colored product on subsequent exposure to oxygen (Pandey et al., 2019; Kapdan et al., 2003). Thus, the anaerobic/anoxic decolorization is integrated into additional aerobic treatment either simultaneously or sequentially (Barragán et al., 2007; Mohanty et al., 2006). Such a strategy of treatment is concomitantly beneficial in the reduction of COD (Dafale et al., 2008a). Gadow and Li (2020) demonstrated > 98% color and COD removal with simultaneous methane production from 2-Naphthol Red laden wastewater utilizing a lab-scale continuous UASB/aerobic system (Gadow & Li, 2020). Various innovations have been made in integrated technology. Dhaouefi et al. (2019) proposed the utilization of an anaerobic-aerobic photobioreactor exploiting algal and bacterial symbiosis, which decolorized > 95% dyes in synthetic wastewater along with an adequate reduction in total nitrogen, organic carbon, and phosphorous at a HRT of eight days. Nonetheless, lately, the academic interest has shifted to search a single biological system, either organism or microbial community, which is befitting both sequential phases of this technology. Franciscon et al. (2009) isolated *Klebsiella* sp. strain VN-31 from activated sludge and demonstrated its implementation in detoxification of dye and its intermediates in subsequent microaerophilic and aerobic treatment in the single bioreactor. In order to attain a successful two-step application of technologies, the balance between oxidation and reduction reaction becomes a prerequisite. Thus, in contrast to the sequential application of technologies, Oliveira et al. (2020) recommended micro-aeration during anoxic/microaerophilic conditions to provide enough dissolved oxygen for microbial growth without disturbing the redox conditions. Sahinkaya et al. (2017) were the first group to demonstrate the application of intermittent aeration in dynamic membrane bioreactor technology for the treatment of synthetic textile wastewater containing heavy loads of dyes and chromium. Moreover, degradation of by-products formed as a result of biological treatment can be escalated by combining the bioremediation with various physicochemical approaches (e.g., Fenton's reaction) as posttreatment or pretreatment (Brindha et al., 2019).

27.3.4 Acceleration of Dye Reduction by Redox Mediators: Artificial and Biological

Due to the limited rate of dye decolorization, attempts have been made to accelerate the biological process by augmentation of several redox-active molecules, also referred to as electron shuttles (Chengalroyen & Dabbs, 2013). These are organic molecules capable of ferrying an electron from the donor (i.e., co-substrate) to the acceptor (i.e., azo dye) while increasing the reaction rate by several orders (Pereira et al., 2013; Sun et al., 2013). Ideally, for a compound to accelerate a nonmediated reaction and act as an effective electron shuttle, it should firstly lower the activation energy of the reaction and secondly possess standard redox potential (E'_0) falling between the half-reactions of azo bond reduction and primary electron donor oxidation (Van der Zee & Cervantes, 2009). The difference in the redox potential of the electron donors and acceptors is critical to electron transfer reaction (Yurtsever et al., 2015). The standard redox potential (E'_0) of azo dyes ranges from −180 to −450 mV, whereas that of primary electron donors lies from 290 to 430 mV (Hong & Gu, 2010). Moreover, the anaerobes/microaerophiles have to maintain a lower redox potential of ≤ -50 mV to spontaneously transfer an electron from the primary electron donor or redox intermediate to the chromophore (Dos Santos et al., 2004; Gao et al., 2018).

Some commonly used synthetic quinone-based mediators are anthraquinone-2-sulfonate (AQS), anthraquinone-2, 6-disulfonate (AQDS), and lawsone (2-hydroxy-1,4-naphthoquinone) (Ling et al., 2009; Van der Zee et al., 2001). Other molecules with redox-active properties are ethyl viologen, methyl viologen, and benzyl viologen (Kavita & Keharia, 2012). The cellular reducing equivalents performing electron transfer from respiratory chain to azo dyes are FMN, FAD, NADH, NADPH, and vitamin B_{12}(riboflavin) (Field & Brady, 2003; Imran et al., 2015a). Most of these could not diffuse across the cellular membrane except riboflavin (Puvaneswari et al., 2006). The presence of such biomolecules is sufficient to reduce the dyes nonspecifically. A thermophilic organism, *Novibacillus thermophiles* SG-1, was able to reduce the azo dye, Orange I anaerobically, despite being deficient in the azoreductase gene (Tang et al., 2018). Through the qPCR studies and genome inspection, it was found that the strain was producing riboflavin, which shuttled cellular electrons to the azo bond (Tang et al., 2018). In addition to these molecules, microorganisms produce heterogeneous molecules in the late log or stationary phase as a result of co-substrate consumption. These include cytochromes, cobalamins, porphyrins, pyridines, quinines, and phenazines (Guo et al., 2010). Several other metabolites synthesized by organisms can also mediate electron transfer. A metabolite, 3-hydroxyanthranilic acid from WRF *Picnoporus cinnabarinus,* was suggested to mediate electron transfer and reduce Fe^{3+} in a Fenton's reaction, thereby decolorizing several dyes (Santana & Aguiar, 2015). Besides, the application of "degradation intermediaries" as redox shuttling molecules in the azo bond breakdown has also been demonstrated by many groups. Through a series of classical experiments, Keck and his co-workers in 1997 were the first to propose the utilization of metabolites generated during oxidative degradation of naphthalene-2-sulfonate by *Sphingomonas xenophaga* BN6 in azo dye degradation under anaerobic conditions (Keck et al., 1997). Generally, the redox-active molecules are applicable during anaerobic decolorization. However, Cui et al. (2015) were the first group to demonstrate the application of quinone-based redox mediators such as menadione and lawsone in aerobic decolorization of azo dyes by *E. coli* strain CD-2 as well as activated sludge. In contrast to electron shuttles, the electrophilic molecules viz., oxygen, ferric ion, sulfate, and nitrate might interfere with the process due to their electron-withdrawing properties (Popli & Patel, 2015). Furthermore, the addition of synthetic soluble electron carriers during bio-reduction can prove to be highly expensive, and thus, attempts have been made to immobilize the shuttles by methods such as crosslinking and entrapment (Olivo-Alanis et al., 2018). Moreover, conductive materials such as graphene and ferroferric oxide (Fe_3O_4) can also be administered to improve the electron transfer during anaerobic degradation (Wang et al., 2018). Wang and co-workers demonstrated the improvement in Reactive Red 2 decolorization, methane production, and the activity of extracellular transport system on the addition of Fe_3O_4 in anaerobic sequence batch reactors (Wang et al., 2018).

27.4 Importance of Microbiome Interaction in Bioremediation

27.4.1 Plant-Microbe Interaction

Plant-based remediation, commonly known as phytoremediation, is one of the cost-effective *in situ* technologies used to remediate a gamut of pollutants from wastewaters, brownfields, constructed wetlands (CWs), and groundwater (Watharkar & Jadhav, 2014). It is accounted that 100 million miles of plant roots are present per acre of land, which makes them promising for pollutant removal in nature (Gerhardt et al., 2009). The process engages xenobiotic-tolerant plant species in conjunction with soil microbiome for the transformation of contaminants (Praveen et al., 2019). The foremost advantage of phytoremediation is that it is a solar-energy driven method that can be performed at meager cost without perturbing the contaminated site and requirement of any equipment (Bharathiraja et al., 2018). Also, it generates reduced sludge, irrespective of the nature of waste, i.e., chemical or biological (Sivakumar, 2014).The various techniques involved in phytoremediation are phytovolatilization, phytoextraction, phytomining, phytostabilization, phytotransformation (phytodegradation), and rhizoremediation (Lourenço et al., 2018; Saba et al., 2015). Of these, rhizoremediation is one of the strategies, which is entirely reliant on the microbial population residing in the rhizosphere of the plants. It exploits the symbiotic interaction of plant rhizosphere with its indigenous microbial population, i.e., either rhizospheric bacteria or mycorrhizae for the degradation of recalcitrant compounds (Hussain et al., 2018; Mishra et al., 2019; Tahir et al., 2016). Plants release a suite of substances from roots as exudates viz., amino acids, sugars, organic acids, vitamins, etc. which serve as a source of nutrition and energy for microbial growth, survival, and metabolism (Godheja et al., 2017; Kuiper et al., 2004). The secretions from the roots attract bacteria through chemotaxis, and thus, the type of secretions is crucial to the plant-microbe interaction (Drogue et al., 2012). A little variation in these exudates can affect the abundance, composition, and functionality of "rhizospheric microbial communities" or "rhizobiome" (Kotoky et al., 2018; Rajkumar et al., 2013). This process of microbiome enrichment, as moderated by the root secretions in the rhizosphere, is defined as the rhizospheric effect (Shaikh et al., 2018). Plant age is another factor affecting rhizobiome structure. The roots of young plants are swayed by the r-strategists, which are fast-growing organisms utilizing simple substrates, whereas the older plants are dominated by slow-growing k-strategists that can utilize complex substrates (Morgan et al., 2005). Other factors affecting the rhizobiome composition are the interplay of exometabolites, root morphology, plant genotype, and developmental stage (Sasse et al., 2018). The rhizobacteria, in reciprocation, benefit the plants in many ways. They not only mitigate the recalcitrant toxicity but also promote plant growth by fixing nitrogen, solubilizing minerals (e.g., P), sequestering Fe via siderophores, releasing antagonistic secondary metabolites, and inducing systemic resistance (Kumar et al., 2017; Rabbee et al., 2019). Moreover, rhizobacteria impart plants with tolerance against abiotic stress and boost the acquisition of nutrients (Ulrich et al., 2019). In addition to rhizobiome, the plants may also interact with endophytic, soil, and nodule-associated microbes (Morgan et al., 2005). The fungi produce copious amounts of diverse extracellular enzymes, which adds to the degradation potential of the plants (Otero-Blanca et al., 2018). The endophytic fungi impart plants with the tolerance to the metal stress while the soil fungi protect plant roots from directly interacting with metals (Deng & Cao, 2017).

The mode of dye curtailment is subject to the plant species employed. The various direct and indirect mechanisms of dye phytoremediation are portrayed in Figure 27.1. The direct mechanisms include the accumulation of dye in the various plant tissues. The water-soluble contaminants are quickly transported to plant tissues through the xylem much before they are acted upon by rhizobiome (Ijaz et al., 2016). Dye being highly water-soluble in nature is readily accumulated, which further demands *in planta* debasement by the endophytic bacterium. The plants may release several degradative oxidoreductases to gratify the process viz., lignin peroxidase, laccase, azo reductase, veratryl alcohol oxidase, tyrosinase, DCIP reductase, riboflavin reductase, superoxide dismutase (SOD), and catalase (Chandanshive et al., 2018; Watharkar & Jadhav, 2014). Plants also release several stress-related enzymes in response to dyes. The induction of the plant PODs (peroxidases) has been used as an indicator of the pollutant-driven stress in the plants (Davies et al., 2005). During decolorization, *Lemna minor* was found to release SOD, guaiacol peroxidase (GPX), and ascorbate peroxidase (APX) in the medium, and therefore, these enzymes were suggested to be used as bioindicators of dye-induced stress in plants (Imron et al., 2019). The indirect mechanisms of dye phytoremediation are dependent on the interaction of the plant with its associated microbiome. The plant boosts the growth of beneficial rhizospheric bacteria and mycorrhizae fungi, and as a payback, these organisms help plants survive in dye-contaminated environments. Moreover, the plants may enhance the co-metabolism of the pollutant by influencing the microbe-microbe interaction within the soil microbiome. Hitherto through many studies, it has been apparent that the plant-microbe interactions are capable of exacerbating the potency of phytoremediation (Dogra et al., 2018). Although a broad perspective of the mechanisms of dye phytoremediation has been gained, the cellular and metabolic intricacies behind the process remain uncovered (Rane et al., 2015).

Phytoremediation of various azo dyes, their mixtures, simulated wastewaters, and effluents discharged from the textile industry has been reported. A multitudinous variety of plants extending from wetland grasses such as *Phragmites australis* to floating plants such as *Lemna minor* have been exploited for the same either in CWs or simulated pond systems. However, duckweed such as *Lemna* is the first choice for ecotoxicologists as it is widespread and grows and reproduces at a faster rate (Sharma et al., 2007; Yaseen & Scholz, 2017). Table 27.1 provides the details of various nonedible ferns, grasses, and weeds used for azo dye removal by this process. Different operational variables, including dye reduction time, initial dye concentration, fresh weight of the plant, temperature, pH, and reusability of the plant were investigated for optimum dye removal (Khataee et al., 2013; Vafaei et al., 2012). Overall, the impact of dye adsorption and degradation on physiological responses of plants viz., growth, photosynthetic pigment, lipid peroxidation, phytochemical components, and antioxidant enzymes has been investigated. In the majority of the studies, the dye was found to exert chemical stress on the plant cells as a drastic upsurge in the activity of antioxidant enzymes such as catalase, peroxidase, and SOD during the process (Khataee et al., 2013; Vafaei et al., 2012). Altogether, the plant roots have proved to be central to the dye remediation (Kadam et al., 2018). Some groups also carried out the augmentation of the potent bacterial cultures (i.e., endophytic or root-associated bacteria) to foster the process (Kabra et al., 2013; Khandare & Govindwar, 2015; Shehzadi et al., 2014). In addition to the interaction of the plant with microbes, the other factors influencing the efficiency of phytoremediation are plant interaction with pollutants and substrates as well as biotic and abiotic constraints (Khandare & Govindwar, 2015). Although the majority of the studies were conducted in controlled experimental setups for a short time (i.e., up to a few weeks), the reports on the *in situ* treatment of dye-contaminated areas are scarce (Chandanshive et al., 2017). Thus, the current research of dye removal revolves around employing the CWs (Haddaji et al., 2019). The only stumbling block in the application of CWs is that it requires longer start time and is subject to seasonal changes (Shehzadi et al., 2014). Furthermore, if the effluent to be treated contains very high concentrations of pollutants, then it can inhibit the growth and survival of the phytoremediators leading to failure of the entire setup.

27.4.2 Metal-Microbe Interaction

The effluents discharged from the wool and textile industries may also lead to heavy metal contamination as these carry various metal-based mordants [usually chromium (Cr) complexes and dichromates] and metal-complexed dyes (Ghosh et al., 2016). The other heavy metals associated with the synthesis of such dyes are copper (Cu), cadmium (Cd), lead (Pb), and nickel (Ni) (Bhardwaj et al., 2014). Similar to dyes, heavy metals can contaminate water and soil and thereby threaten human and environmental health (Sobariu et al., 2017). The microbes can be strategically used to alleviate the toxicity of metal by its accumulation, precipitation, sorption, chelation, or transformation (i.e., via oxidation, reduction, or alkylation) (Brandl, 2002). The higher surface area to volume ratio bestows the microbial cells with a larger area for contact and thus helps in efficient metal sorption (Allam, 2017). Furthermore, the microbes can tolerate higher quantities of heavy metals by developing various intrinsic mechanisms such as changing the permeability of the cell wall, synthesizing extracellular polysaccharides, and remaining devoid of certain transport systems (Beveridge et al., 1996). The metals can additionally be methylated to form their respective volatile derivatives (Ehrlich, 1997). The microbes can also efflux the metal ions out of the cell by specific transporters (Joshi et al., 2014) or bind metals to the cell wall constituents or cytoplasmic metal-binding proteins/peptides such as phytochelatins and metallothioneins

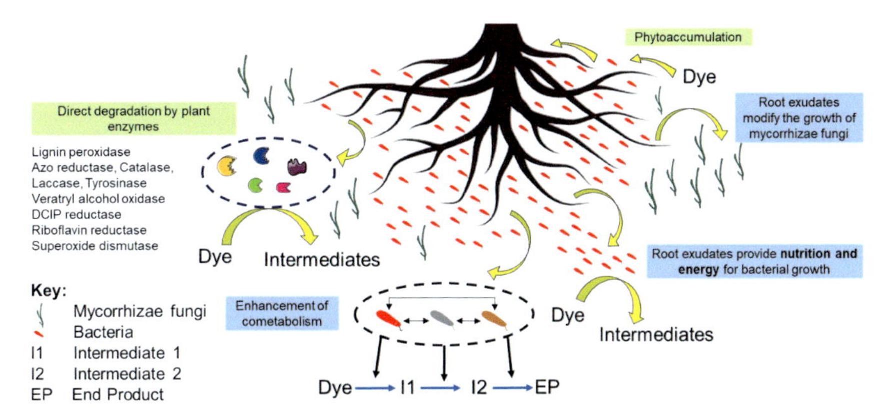

FIGURE 27.1 Plant-microbe interactions during dye phytoremediation. Direct removal by the plant is highlighted in green, whereas microbe-mediated removal is highlighted in blue.

TABLE 27.1

Phytoremediation of Diverse Azo Dyes Employing Various Plants with or without Co-inoculants

Plant Employed	Dye/s Treated	Significant Findings	References
Chara vulgaris	Textile effluent	The treatment of effluent (diluted to 10% concentration) with the macroalgae resulted in a drastic decrease in TDS, COD, BOD, and EC in 120h. The undiluted effluent was toxic to algae. Moreover, the algae accumulated the heavy metal cadmium from the effluent.	Mahajan et al. (2019)
Typha domingensis	Amarnath azo dye	The plant was employed in an RHFCW, and ~92% decolorization was attained. A reduction in other parameters such as COD, nitrate, and ammonia was also observed. A concomitant upsurge in the activity of enzymes such as SOD, CAT, GPX, GR, and APX with an increase in the dye concentrations indicated oxidative stress.	Haddaji et al. (2019)
Tagetes patula, Aster amellus, Portulaca grandiflora, and Gaillardia grandiflora	Textile wastewater	*In situ* remediation of dye-contaminated soil was performed employing high rate transpiration systems (HRTS) in 30days. The microbial count in the untreated soil was less as compared to planted soil. Photosynthetic pigments (such as carotenoid, chlorophyll a and chlorophyll b), and the oxidoreductase enzymes in the plants were induced in the presence of textile effluent. Also, dye accumulation was confirmed by anatomical studies.	Chandanshive et al. (2018)
Typha angustifolia and Paspalum scrobiculatum	Congo Red and textile effluent	Efficient *in situ* remediation of effluent was demonstrated by the co-plantation of plants in drenches (CWs). A significant decline in ADMI value, TDS, BOD, TSS, COD, and TDS was observed in 96h. Also, induction in azo reductase, laccase, lignin peroxidase, and veratryl alcohol oxidase activity in the roots of both plants was observed during decolorization of Congo Red.	Chandanshive et al. (2017)
Lemna minor	Reactive Blue 198 (RB 198), Acid Blue 113 (AB113), Direct Orange 46 (DO46), and Basic Red 46 (BR46)	BR46 reduction efficiency of constructed pond systems under semi-natural (outdoor) and controlled laboratory conditions was compared. It was found that the controlled setups exhibited 23% higher dye decolorization than semi-natural systems. Also, the plant growth was higher in laboratory setups. In both experiments, the growth rate of the plants was affected in the ponds containing dye.	Yaseen and Scholz (2017)
Spirodela polyrhiza	Direct Blue 129	Dyes impacted photosynthetic pigment production in plants. Also, the catalase and peroxidase activities were significantly elevated.	Movafeghi et al. (2016)
Ipomoea aquatic and Ipomoea hederifolia	Brown 3R and effluent	Soil-bed treatment (by *I. hederifolia*) when coupled with rhizofiltration (by *I. aquatic*) resulted in decolorization of>500 liters of effluent in 72h. Induction in plant enzymes and reduction in pigment was observed during decolorization of Brown 3R by *I. aquatic*. Further, the *in situ* remediations of effluents was upscaled to lagoons of 60,000 liters.	Rane et al. (2016)
Alternanthera philoxeroides	Remazol Red	The macrophyte was able to degrade the 70mg L^{-1} in 72h. Stem and root exhibited elevated riboflavin reductase and azo reductase activity. Also, there was a reduction in the plant pigments viz., chlorophyll a, chlorophyll b, and carotenoids. A rhizofiltration reactor of pilot-scale with *A. philoxeroides* was shown to decrease ADMI value, COD, BOD, TDS, and TSS from textile wastewater.	Rane et al. (2015)
Persicaria barbata	Reactive Black 5	Dye was efficiently removed in CWs utilizing rice agricultural waste as the substratum. Approximately 50% and 80% of dye were adsorbed in rice husk and its biochar, respectively. Removal of dye was improved to 90% by augmenting *Psychrobacter alimentarius* strain KS23.	Saba et al. (2015)
Typha domingensis (with endophytic bacteria)	Textile effluent	Efficient dye removal and a drastic reduction in COD, TDS, TSS, and BOD from the effluent were attained in 72h. The enhancement of degradation was achieved by the combined treatment of effluent with plant and endophytic bacterium (*Microbacterium arborescens* TYSI04 and *B. pumilus* PIRI30). The selected cultures were potential dye degraders with plant growth-promoting traits.	Shehzadi et al. (2014)
Consortium of *Petunia grandiflora* and *Gaillardia grandiflora*	Brilliant Blue G, Direct Blue GLL, and Rubin GFL, Scarlet RR, and Brown 3 REL and mixture	Induced veratryl alcohol oxidase and laccase activity in *P. grandiflora,* whereas increased activity in tyrosinase, riboflavin reductase, and lignin peroxidase of *G. grandiflora* roots were observed. Synergistic plant remediation exhibited improved decolorization. Also, the cyto-genotoxicity of the dye was decreased on treatment by the consortium.	Watharkar and Jadhav (2014)

(*Continued*)

TABLE 27.1 (*Continued*)

Phytoremediation of Diverse Azo Dyes Employing Various Plants with or without Co-inoculants

Plant Employed	Dye/s Treated	Significant Findings	References
Lemna minor	Textile wastewater and Acid Orange 10	The maximum color and COD reduction were attained in a CW in the four days of contact time. The pollutant removal was also performed for Acid Orange 10 under optimized process parameters (i.e., pH, dilution ratio, nutrient dosage, and contact time) to confirm the reproducibility of the process.	Sivakumar, (2014)
Azolla filiculoides	Acid Blue 92	The dye was efficiently removed, but the growth, as well as photosynthetic pigments in the frond, was affected. Also, an upsurge in the antioxidant enzymes &nd lipid peroxidation was detected.	Khataee et al. (2013)
Glandularia pulchella (Sweet) *Tronc.* (with *Pseudomonas monteilii* ANK)	Scarlet RR and mixture of dyes	The plant and bacterial consortium exhibited 100% decolorization of Scarlet RR in 48 h and 92 % decolorization of the dye mixture in 96 h. During dye removal, the plant displayed higher activities for DCIP reductase and lignin peroxidase, whereas the bacteria displayed higher activities for laccase, tyrosinase, and DCIP reductase. Four reactors demonstrating the application of soil, plant, bacteria, and consortium for decolorization of textile effluent were created. The consortium reactor exploiting plant-bacterial synergism proved to be most efficient in reducing ADMI value.	Kabra et al. (2013)
Azolla filiculoides	Basic Red 46 (BR46)	The aquatic fern removed the dye, but its growth was hindered. Moreover, a concurrent induction of antioxidant enzymes, i.e., SOD, peroxidase, and catalase, was observed in the plant cells.	Vafaei et al. (2012)
Lemna minor	Untreated effluent form textile industry	Maximum decolorization efficiencies of untreated textile effluent in a pond with a continuous-flow system were attained at HRT of three days.	Uysal et al. (2014)
Eichhornia crassipes (water hyacinth)	Red RB and Black B	Both the dye solutions were efficiently decolorized (> 95%) by the hydrophyte. An increase in n-hexadecanoic acid and a decrease in phytol content was observed during the process. Further, the used plants were used for vermicomposting with leaves and cow-dung.	Muthunarayanan et al. (2011)
Myriophyllum Spicatum and *Ceratophyllum demersum*	Basic Blue 41	Dye was reduced effectively in a lab-scale wetland system at HRTS ranging from 9 to18 days. The dye removal efficiency of both the plants was similar.	Keskinkan and Göksu (2007)
Typha angustifolia	Synthetic dye wastewater	Cattails cultured in glass bottles with or without soil were able to reduce ~ 60% of dyes and ~44% of sodium from textile wastewater in 14 days. Moreover, the dosage of 23.5 mg L^{-1} was found to be toxic.	Nilratnisakorn et al. (2007)
Phragmites australis	Acid Orange 7	Induction in peroxidase activity was observed during color removal in a vertical flow CW.	Davies et al. (2005)

(Ojuederie & Babalola, 2017). If the metal ion taken up by the microbial cell is not efficiently effluxed out by transporters, then an additional sequestration mechanism prevails in the cytoplasm wherein the metal ions are bound by inclusion bodies viz., polyphosphate granules (Haferburg & Kothe, 2007). The uptake of metals is influenced by several parameters, including environmental conditions, chemical characteristics of the pollutant, metal bioavailability, root zone (if the organism is rhizospheric), and the chelators (Mishra et al., 2019). A few gene clusters identified in bacteria that are dedicated to imparting resistance to metals such as cadmium, lead, mercury, chromium, and copper are *cadB*, *pbrA*, *merA*, *chrA*, and *copAB*, respectively (Das et al., 2016). Some bacterial and archaeal species can also conserve energy while oxidizing or reducing metals viz., manganese, iron, and cobalt at a large scale (Rajendran et al., 2003).

Scientific reports depicting exclusive transformation or bioaccumulation of dyes from the synthetic or industrial wastewater are available. However, fewer studies are present in which the dye removal has been proclaimed under the state of multiple stress, e.g., under concurrent higher metal concentrations. Ertuğrul et al. (2009) demonstrated the bioaccumulation of the dye and metals such as chromium(VI), copper(II), and nickel(II) by *Rhodotorula mucilaginosa*. The reduction of dyes and metals is subject to the electron transfer potential of the system, which in turn is reliant on the reduction efficiencies of the organism engaged and the reductive microenvironment available. Since a similar mechanism underlies the debasement of both the pollutants, their simultaneous removal appears to be effortless. Howbeit, this is not true as under anoxic/anaerobic conditions, the metal ions may act as the terminal electron acceptor and badly affect the dye removal efficiency. This shortcoming in concomitant dye and metal reduction can be overcome by supplementing electron transferring molecules. In a study conducted by Mahmood et al. (2015), the augmentation of electron shuttle (viz., hydroquinone, and uric acid) at lower concentrations was carried out for concurrent dye and metal reduction by *Pseudomonas putida* K1. Alternatively, a mixture of organisms can be utilized to withstand the stress of dyes and heavy metals, cooperatively.

Mishra and Malik (2014) illustrated efficient dye and metal removal from textile effluents deploying a fungal consortium consisting of *Aspergillus lentulus*, *Aspergillus terreus*, and *Rhizopus oryzae* (Mishra& Malik, 2014).

27.4.3 Soil-Microbe Interaction

Soil holds an extraordinary stature as the number of living organisms supported by fertile soil is more than the total number of humans that ever resided on Earth (Young & Crawford, 2019). No doubt why the first sample searched for the evidence of life from Mars is soil. Soils are nothing but microscale heterogeneous complexes serving as a porous habitat for the sustenance of diverse microbial communities (Crawford et al., 2012). These possess bountiful of water, gases, minerals, organics, humic substances, microorganisms, and plant roots adhered together by strong forces that are resistant to mechanical stress. The heterogeneity of a soil matrix at the micrometer scale is highly ordered as characterized by specific pore size and important in providing continuous transfer of air and water to flora and fauna (Young & Crawford, 2019). With the help of microbes, the soil performs several relevant functions such as supporting life in the biosphere of the planet, geochemical and nutrient cycling, maintaining fertility, storing carbon, and processing waste (Cai et al., 2019; Wilpiszeski et al., 2019). The interaction between soil and microorganisms can be biological, i.e., involving bacterial growth and secretion of enzymes, or it can be physical, i.e., associated with abiotic parameters such as water retention, cohesion to soil particles, and aggregate stabilization (Chenu & Stotzky, 2016). The soil texture and physical properties may affect the microbiome composition and activity significantly (Watteau & Villemin, 2018). In soils with low moisture levels, the air gaps are bridged by the filamentous bacteria, whereas in soil with high moisture content, the free-living motile microbes are prevalent (Traxler & Kolter, 2015).

Conversely, the composition of the microbial communities in the soil affects the soil structure, its aggregation, and nutrient retention in many ways. The exopolysaccharides (EPSs) produced by microbes are directly involved in soil aggregation and organo-mineral conglomeration (Costa et al., 2018). By trapping nutrients and water content, EPS serve as a reservoir of carbon source for the metabolism by the rhizospheric microbial community, which in turn mediates plant growth (Şengör, 2019). EPS can provide an excellent buffering capacity to the soil as they can lose a considerable amount of water with little changes in the water potential during transient water fluctuations (Chenu, 1995). Moreover, the EPS of microbial origin can influence bioleaching, biomineralization, and binding or adsorption of heavy metals (Lin et al., 2016). The commonly found genera known to stabilize and shape the soil structure by the EPS production and biofilm formation are *Bacillus, Paenibacillus,* and *Pseudomonas* (Şengör, 2019). Moreover, during soil formation, the enzymes released by the soil residents can be adsorbed to the humic and mineral colloids or can copolymerize with the phenolics (Huang et al., 2005). Besides interacting with soil, microbes might also interact with each other. Cooperation and competition are among the common inter- and intra-species interactions that prevail in the soil biofilm for resource utilization (Cai et al., 2019).

The soil-microbe complexes basically are self-organizing structures shaped over time as a result of the integrated impact of spatial interactions and the biological activities of life thriving there (Crawford et al., 2012). Wilpiszeski et al. (2019) apprehended the microbiome in the soil aggregates as "*microbial villages*" and discussed strategies for studying the interactions of microbes within themselves and with soil aggregates considering the ongoing wetting events and transfer of metabolites, genetic material, and viruses. It is proposed that a broader perspective about the genesis of soil aggregates could be gained by creating mock *microbial villages*, examining their biogeochemical properties, and then predicting the structure-function relationship based on interactions (Wilpiszeski et al., 2019). Several scientific advances are made by creating theoretical models and conducting experiments on the soil microcosms to unveil the beauty of microbe-microbe interaction within the soil and their impact on the soil architecture. But the influence of external factors on the physics and biology of the soil dynamics is underexplored. Although some reports related to soil-microbe-plant interaction in the plant roots and rhizospheres are available (Kushwaha et al., 2018; Mimmo et al., 2014), the details of alteration in the soil microstructure or micro-aggregates as driven by azo dye degraders are less addressed. Dyes can neither escape to the atmosphere by volatilization nor precipitate in soil or water due to their chemical nature. These are either accumulated, adsorbed, or degraded to varying extents by plant, soil, and water microbiome to simpler molecules.

27.4.4 Microbe-Microbe Interaction

A flurry of studies conducted to hunt an organism that serves as an all-embracing potential xenobiotic degrader has ended up without any success. However, it goes unsaid that it is quite an unpragmatic desire to have. Thus, for bioremediation, eventually, the scientific focus has shifted from the exploitation of single superbug to the mixture of a few to several microbes or an entire indigenous community. Within a community, the web of physiological and chemical tasks performed by the organisms is imperative to the survival of the cohort (Shong et al., 2012). Nevertheless, how do organisms with diverse metabolic traits and inherent growth rates complement each other to accomplish a community function of relevance? How is the community stability achieved? To manipulate the communities for remediation, fuel-production, or host-related associations, the consideration of underlying synergistic mechanisms becomes instrumental (Niehaus et al., 2019). It is because the synergistic interactions between the members are one of the leading causes of ameliorated community functions such as degradation (Allam, 2017; Holkar et al., 2016). Recently, under the umbrella of microbial community studies, promising strides have been made to decipher complex interactions viz., bacterial-bacterial, bacterial-fungal, or fungal-fungal interactions. However, a realistic approach to take benefit of the synergism requires a complete cognizance of the intricate microbial interactions. Moreover, the inter-microbial interactions are the actual drivers of the structure and dynamics of a community at spatial and temporal scale, and thus, decoding them is crucial for their successful application (Bernstein et al., 2019). One such meaningful interaction is cross-feeding, where one species produces a compound beneficial to the other isolate leading to the web of metabolic interdependencies (Adamowicz et al., 2018). The cross-feeding can be further classified into a metabolite-, substrate-, mutual-, or augmented cross-feeding depending on the type of biomolecules being released or exchanged (Smith et al., 2019). The transfer of diffusible molecules that referee the interactions and the metabolic networks within the microbes is affected by the physical proximity between the cells (Gupta et al., 2020). Thus, the spatial distribution of microbes within the community also needs to be well understood to modulate the function and the positive interactions. In addition to synergism, the other interactions such as competition, commensalism, mutualism, or parasitism are also a consequence of the unidirectional or bidirectional interplay of molecules (Xu et al., 2019).

Considering the benefits offered by the microbial communities, researchers are trying to make most of these in the niche segment of dye biodegradation. It is because during synergistic interactions, the different catabolic routes of the microorganisms are perceived to complement each other, which can lead to the complete mineralization of azo dyes (Chang et al., 2004). Also, the complexity of the consortium enables them to act on a variety of pollutants, making them worthy of biotransforming multiple pollutants within a time frame at a contaminated site (Senan & Abraham, 2004). Another advantage is that living in a community is thought to generate robustness to environmental fluctuations and promote stability to the members through time (Che & Men, 2019; Shanmugam et al., 2017). This makes them resistant to any invading species. The increased efficiency in a consortium or mixed community is a result of "division of labor" (DOL) or "functional compartmentalization" (Lindemann et al., 2016). DOL, which is the differentiation of the tasks within individuals or subpopulations in a community, is orchestrated via communication as mediated by signal molecules (Brenner et al., 2008). Contrastingly, a single redoubtable population or pure cultures may prove to be incompetent due to extensive metabolic load. Also, these may face competition repression in native populations, which are already well acclimatized to the existing environment (Sarayu & Sandhya, 2012). Moreover, pure cultures are specific to a dye and cannot be easily scaled up or maintained in large-scale operations typical of wastewater treatment systems (Dafale et al., 2008a, 2008b).

Delineating the role of individual members of the consortium or studying the interactions within them during degradation can aid in understanding the co-metabolism. The various methodologies used for the same include quantification of molecules interchanged by metabolomics, the use of microfluidics to detect inter-species synergism, computationally analyzing the co-occurrence of species through metagenomics, or strain-specific qPCR studies in a combinatorial form (Pacheco & Segrè, 2019; Ren et al., 2015). In contrast to the large complex communities with unknown microorganisms, the dynamics of communities with fewer culturable isolates can be easily deciphered. The role of each member can be understood by creating a "knockout community" wherein one of the members is eliminated, and its effect on the functioning of other could be inferred. This approach is the same as gene disruption studies in molecular biology, where the role of a gene in an organism is evaluated by constructing a knockout mutant in which the gene function of interest is eliminated. Alternatively, all the possible combinations and permutations of the members can be created, and the change in productivity can be used to categorize positive, negative, or neutral cohorts (Purswani et al., 2017). A similar approach was adopted by Nanjani et al. (2020), wherein through combinatorial studies, strong cooperation was demonstrated within the members of a ternary dye-degrading bacterial consortium SCP. Furthermore, the examination of the inter-microbial social behaviors and metabolic potencies revealed that *Cellulomonas* sp. APG4 was the functionally most significant isolate or an "actor," which was dependent on the *Stenotrophomonas acidaminiphila* APG1 and *Pseudomonas stutzeri* APG2 for complete decolorization (Nanjani et al., 2020). Additionally, the bio-calorimetric analysis of the isolates, along with their consortium, can also be carried out to entail synergism. Shanmugam et al. (2017) monitored dye degradation and metabolic heat-flux of monocultures (i.e., *Staphylococcus lentus*, *Bacillus flexus*, and *Pseudomonas aeruginosa*) as well as their consortium by a bioreaction calorimeter in order to reveal the metabolic roles of the isolates. Moreover, numerous reports are available on the application of the acclimatized mixed culture, designed consortium, or complex microbial communities for dye degradation. Zhang et al. (2019) developed Acid Orange 7 (AO7) degrading extremely thermophilic mixed culture in which Firmicutes and Proteobacteria were found to be dominating phyla accounting to>97% abundance (Zhang et al., 2019). The dominant genera identified in the same were *Caldanaerobacter*, *Pseudomonas*, *Xanthobacter*, *Azospirillum*, *Achromobacter*, and *Cellulomonas*. Various strategies have been used to improve the synergistic interactions within the complex communities. Oliveira et al. (2020) suggested the use of intermittent aeration during the anaerobic digestion in order to improve the growth of diverse microbes and boost synergistic interactions among them. Earlier studies have also shown that the augmentation of an appropriate substrate can strengthen the co-metabolism during treatment of dye-containing wastewater (Xie et al., 2018).

27.5 Dye-Induced Microbial Community Shifts and Adaptations

It is essential to develop an insight into the microbiome structure and function from the industrial effluent or dye-contaminated sources for successful *in situ* application or large-scale optimization during bioremediation. The diversity in a population is profoundly affected by environmental factors and the pollutants. In the presence of the pollutants, the autochthonous microbiome may face a selection pressure, and the population shifts toward the organisms degrading or tolerating xenobiotics (Khan & Malik, 2018). The effluent often decreases the microbial diversity and alters the biogeochemical cycling betiding in an ecosystem (Atashgahi et al., 2015; Drury et al., 2013). Moreover, the identification of the predominant species or genera from the polluted sources can aid in strategically improving the biodegradation by enriching the degraders.

The commonly applied methods for understanding community dynamics at multiple scales are polymerase chain reaction with denaturing gradient gel electrophoresis (PCR-DGGE), community-level physiological profiling (CLPP), phospholipid fatty acid (PLFA)/fatty acid methyl ester analysis (FAME), fluorescence *in situ* hybridization (FISH), metagenomics, metaproteomics, metagenomics, metatranscriptomics, and metabolomics studies (Aguiar-Pulido et al., 2016; Haines et al., 2002; Lacerda et al., 2007; Lehman et al., 1995; Lv et al., 2017; Punzi et al., 2015; Zhu et al., 2018). A few additional methods that can be employed for the same include plat counts, determining mol % G+C content, nucleic acid hybridization, microarrays, single-strand conformation polymorphism (SSCP), temperature gradient gel electrophoresis (TGGE), restriction fragment length polymorphism (RFLP), terminal restriction fragment length polymorphism (T-RFLP), and ribosomal intergenic spacer analysis (RISA)/amplified ribosomal DNA restriction analysis (ARDRA) (Fakruddin & Mannan, 2015).

27.5.1 In Soil

The soil is suffused with biotic and abiotic entities, which are highly balanced with each other. Any changes in the biotic characters of the soil can drastically affect the biogeochemical cycling in the ecosystem. Thus, studying the pollutant-driven repercussions on biological features of the soil is imperative. Like other pollutants, the dye might also lead to a decrease in the diversity of the microbial population. Through PCR-DGGE studies, Joshi et al. (2013) observed an overall reduction in the population diversity of the soil collected from the contaminated site when exposed to the diazo dye. Also, microbial diversity is impacted depending upon the kind of pollution. Patel et al. (2020) investigated microbial community diversity from six different dye- or oil-contaminated sludge samples collected from various sites (viz., CETP, port, ship-breaking yard, wastewater discharge site, etc.). It was revealed that unlike bacterial diversity, the fungal diversity was highly site-specific. Moreover, the species richness was dramatically affected in the soil containing heavy loads of dyes and their intermediates (Patel et al., 2020). The studies identifying dye ramifications on the soil populations based on the shifts in microbial diversity are limited (Table 27.2).

27.5.2 In Bioreactors

The microbiome of the seed sludge used in treatment plants might undergo many variations in the presence of the new wastewater in a way that the process can collapse (Yang et al., 2012). Thus, the supervision of the microbial population shift, in reactors, is critical as their overall performance is dependent on the microorganisms prevailing there. The investigation of population dynamics in response to doses of pollutants, co-substrates, redox mediators, or operating conditions can provide fundamental insights into the functioning of the reactors. A multitude of scientific research has been executed in this direction in various bioreactors (Table 27.2). Xie et al. (2018) investigated the influence of different carbon and nitrogen sources on the degradation of simulated wastewater in a hydrolysis acidification (HA) reactor. They found that a few genera viz., *Aeromonas, Acinetobacter,* and *Pseudomonas,* which are reported to decolorize dye, were present in all the samples. Moreover, such an inspection of the microbiome structure revealed the prevalence of interaction between the co-substrate and microbial community (Xie et al., 2018). Pan et al. (2017) observed a shift in the functionality of communities toward the carboxylic acid, polymer, and amine/amide utilization on augmentation of nanoscale zerovalent iron, NZVI (0.1, 0.2, and 0.5 g L^{-1}). The dominant phyla in three reactors (with 0, 0.2, and 0.5 g L^{-1} NZVI) identified were *Bacteroidetes, Firmicutes, Proteobacteria,* and *Verrucomicrobia,* but their abundances varied greatly in all of them (Pan et al., 2017). In the presence of NZVI, which works as an activator for the advanced oxidation process by persulfates in UASB, the *Bacteroidetes* and *Verrucomicrobia* were enriched (Pan et al., 2017). Based on the related research available, it can be deduced that the properties of the influents (e.g., dye type) and the additives (e.g., co-substrates, NZVI) govern the shifts in microbial community composition in the wastewater treatment process (Liu et al., 2017; Pan et al., 2017; Xie et al., 2018). From the various reports discussed in Table 27.1, it can be concluded that *Firmicutes, Proteobacteria* (*Alpha-, Beta-,* and *Gammaproteobacteria*), and *Bacteroidetes* were the most common predominant groups in the dye-degrading communities (He et al., 2017; Liu et al., 2017; Xie et al., 2016). Also, in most of the studies, the community structure was affected at all the taxonomic levels.

27.5.3 In the Human Intestine

A gamut of colors is added to the foodstuffs to make it look alluring to the human eye and stomach. The dyes used as nonnutritional additives might prove to be dangerous if introduced without following the USFDA (U.S. Food and Drug Administration) guidelines. For instance, Sudan I dye is illegally utilized in chili powder and curry powder, which was banned for its carcinogenic effects (Chung, 2016). All the Sudan dyes (i.e., Sudan I, II, III, and IV) have been classified as class 3 carcinogens by the "International Agency for Research on Cancer", IARC (Pan et al., 2012).

Once these dyes enter the human gut as a contaminant in the drugs, food, or cosmetics, they are degraded either enzymatically by the intestinal microflora or get abiotically reduced to aromatic amines in the reducing gut ecosystem (Handayani et al., 2007). The mammalian liver is also involved in the azo dye reduction with the help of several microsomal and cytosolic enzymes (Xu et al., 2010). The reduction of azo dyes generates the aromatic amines that can be further metabolized to form genotoxic metabolites owing to their DNA-binding abilities (Manganelli et al., 2016). Several members of human gut flora, specifically those belonging to the *Enterobacteriaceae* family, have been studied for azo dye decolorization. One of the most important intestinal dye degraders is *Klebsiella,* which can further enter wastewater or contaminate soil (Cui et al., 2012). The other essential intestinal microflora efficient in degradation may belong to the Gram-positive *Enterococcaceae* family. The cell-free extract of *Enterococcus faecalis* isolated from rat caecal samples exhibited high azo reductase activity (Chen et al., 2004).

TABLE 27.2
Details of Microbial Community Shifts in Response to Dye from Various Sources

S. No.	Studied in/on	Methodology Adapted	Dyes	Significant Finding	References
In soil					
1	Sludge from six different contaminated sites	Illumina MiSeq sequencing	Dyes form the effluent	• The microbial diversity was studied from all different sites. • Out of 20 phyla identified from samples, the abundance of *Proteobacteria, Firmicutes, Bacteroidetes, Nitrospirae, Cyanobacteria,* and *Euryarchaeota* ranged between 59% and 87%. • All the samples were mixed and subjected to *in situ* remediation. A significant reduction in dye and COD was observed by microbial consortium from site 1. • The fungal species identified belonged to *Ascomycota, Basidiomycota, Glomeromycota,* and *Blastocladiomycota*.	Patel et al. (2020)
2	Soil from dyeing pits	Plate counts	Methylene Blue, Congo Red, Violet, and Green dye	• The fungal diversity was analyzed from the three dye-contaminated soils. • Eight different fungal species were identified from the three sites. *Penicillium* sp., *A. niger* & *A. fumigatus* were common among all the sites. • *A. ochraceus* was identified as an active degrader.	Sani and Abdullahi (2018)
3	Soil form vegetable garden	PLFA profiling	Direct Red 81, Reactive Black 5, and Acid Yellow 19	• The structurally distinct dyes displayed different impacts on the microbial community. • Out of 29 PLFA, concentrations of 23 were affected whereas that of three (i.e., 22:0, 22:0 and 21:1 ω3c) were increased on azo dye augmentation. • As compared to gram-positive bacteria, the azo dyes were more toxic to gram-negative bacteria. • Reactive Black 5 led to a reduction of fungal PLFA, whereas Direct Red 81 led to a reduction in bacterial PLFA.	Imran et al. (2015b)
4	Soil collected from textile effluent disposal site	PCR-DGGE	Congo Red	• The soil microcosms were created to study the impact of diazo dye on the microbiome with an increase in time (i.e., up to 90 days). • On spiking dye, the initial population was retained up to 20 days, but later, an increase in the intensity of seven bands and a decrease in one band was observed. • Alpha- & gamma-proteobacteria were found to be dominant in the native community. Other members identified were *Sphingobacteria* and some uncultured bacteria.	Joshi et al. (2013)
In reactors					
1	Batch reactors (1L capacity)	PCR DGGE fingerprinting and Illumina HiSeq sequencing analysis	Direct Black 22	• The dye-mineralization and toxicity-reduction against Daphnia magna by recurrent aeration and glucose supplementation under anoxic conditions was demonstrated. • The microbial consortium used was developed from the soil and lignocellulosic material. • The abundance of *Klebsiella* increased, whereas *Enterococcus* disappeared and Streptomyces decreased on intermittent aeration.	Oliveira et al. (2020)

(Continued)

TABLE 27.2 (*Continued*)

Details of Microbial Community Shifts in Response to Dye from Various Sources

S. No.	Studied in/on	Methodology Adapted	Dyes	Significant Finding	References
				• In nonaerated experiments, the abundance of *Pseudomonas*, *Delftia*, *Lysinibacillus*, *Stenotrophomonas*, *Chryseobacterium*, *Acinetobacter*, *Pelosinus*, *Clostridium*, and other genera increased.	
2	HA reactor	Illumina MiSeq sequencing analysis	Reactive Black 5 (RB5) Remazol Brilliant Blue R (RBBR)	• The effect of different co-substrates on the reactor performance in terms of decolorization and COD removal was studied. • Highest decolorization & COD removal were observed on the addition of sucrose, which further led to the enrichment of *Raoultella*, *Desulfovibrio*, *Tolumonas*, and *Clostridium*. • The six groups, i.e., *Bacteroidetes*, *Firmicutes*, *Proteobacteria*, *Acidobacteria*, *Actinobacteria*, and *Spirochaetae*, accounted for >98% abundance in all samples.	Xie et al. (2018)
3	Anaerobic^3-Oxic^2-Sedimentation (A^3O^2S) reactor	Illumina MiSeq sequencing analysis	Acid Orange 7 (AO7), Methyl Orange (MO)	• A reactor with six compartments containing activated sludge was used to treat dyes. • *Comamonas*, Parabacteroides, Acinetobacter, *Acidaminococcus*, and Clostridium *sensu stricto* were identified in all the samples with a relative abundance of > 1%. • The LEfSe analysis signified that the microbial composition was drastically different in anaerobic and aerobic chambers and also was affected by the structure of dye & degradation intermediates. • The genera functional in azo bond reduction and further mineralization were identified.	Zhu et al. (2018)
4	Nanoscale zerovalent iron/ per-sulphate (NZVI/PS) enhanced UASB reactor	Biolog EcoPlates Illumina MiSeq sequencing technique	Brilliant Red X-3B	• The effects of NZVI/PS on anaerobic microbial communities were examined. • The abundance of the most abundant genus (*Lactococcus*) was found to decrease, while that of the genus *Akkermansia* increased significantly in the presence of 0.2 g L^{-1} NZVI during the biological treatment process. • Three strains were isolated from the sludge in the UASB reactors and identified. These were consistent with the results from the Illumina MiSeq high throughput sequencing. • The results indicated that Fe(0) was transformed into Fe(II)/Fe(III), which are beneficial for the microorganism growth, thus promoting their metabolic processes and microbial community.	Pan et al. (2017)
5	Hydrolysis acidification (HA) reactor	PCR-DGGE fingerprinting	Reactive Black 5 (RB5-azo) and Remazol Brilliant Blue R (RBBR-anthraquinone)	• The microbial composition in the HA reactor was conditioned by the type of dye (i.e., azo or anthraquinone) and not by dye concentrations. • All bacteria identified belonged to three groups, i.e., *Proteobacteria*, *Bacteroidetes*, and *Firmicutes*. • *Bacteroidetes* were able to degrade both dyes, whereas Bacilli preferred RBBR degradation.	Liu et al. (2017)

(*Continued*)

TABLE 27.2 (*Continued*)

Details of Microbial Community Shifts in Response to Dye from Various Sources

S. No.	Studied in/on	Methodology Adapted	Dyes	Significant Finding	References
6	Biological aerated filters (BAF)	Illumina MiSeq sequencing analysis	Acid Red B	• The shift in the bacterial and fungal population along with the stability of augmented yeast, i.e., *Magnusiomyces ingens* LH-F1 (a yeast) in BAF, was evaluated. • The bacterial diversity increased with an increase in influent dye concentration. • The *fungus Cosmospora* and *Guehomyces* were dominant in all the samples. • The augmented genera, i.e., *Magnusiomyces*, turned out to be dominant, indicating its survival during the process and successful bioaugmentation.	He et al. (2017)
7	HA reactor	Illumina MiSeq sequencing analysis	Simulated dyeing wastewater (RB5 and RBBR)	• Higher microbial richness and diversity were observed during RBBR degradation. • *Bacteroidetes, Firmicutes &Proteobacteria* were identified as dominant phyla. • The genera *Dysgonomonas* and *Klebsiella* were dominant in the RB5 sample, whereas *Prevotella* sp. and *Lactococcus* were dominant in the RBBR sample. • The common genera contributing to the degradation of both dyes were *Klebsiella* sp., *Desulfovibrio* sp., *Lactococcus*, etc.	Xie et al. (2016)
8	UASB-submerged aerated biofilter (SAB)	454-pyrosequencing analysis	Effluent of a denim factory (containing Direct Black 22)	• The anaerobic sludge was used as seed sludge for the treatment in the reactor. • Out of 8 identified phyla, UASB was downright dominated by *Firmicutesphylum* (> 95%) and that too by *Clostridium sensu stricto*. • The other identified phyla were the *Proteobacteria, Acidobacteria,* and *Verrucomicrobia*. • In SAB, out of 25 identified phyla, *Proteobacteria* (> 44%) SAB was predominant; the other phyla in SAB were *Firmicutes, Chloroflexi, Spirochetes,* and *Actinobacteria*. • The abundant genera in UASB were Clostridium and Pseudomonas, whereas in SAB were *Sulfuricurvum*.	Kochling et al. (2016)
In human intestine					
1	11 Prevalent human intestinal strains	Cell growth and viability assays	Sudan azo dyes (I, II, III & IV), Para Red and their reduction metabolites	• Sudan azo dyes & their metabolites affected human intestinal bacterial ecology as some species were selectively inhibited. More than cell viability, cell growth was affected. • The tested strains, i.e., *Clostridium perfringens, Lactobacillus rhamnosus, Enterococcus faecalis, Bifidobacterium catenulatum, E. coli,* and *Peptostreptococcus magnus* were differentially affected by different Sudan azo dyes. • 1-Amino-2-naphthol was one of the reduction products that were inhibitory.	Pan et al. (2012)

(*Continued*)

TABLE 27.2 (*Continued*)

Details of Microbial Community Shifts in Response to Dye from Various Sources

S. No.	Studied in/on	Methodology Adapted	Dyes	Significant Finding	References
2	35 prevalent human intestinal strains	Dye degradation assays	Sudan azo dyes (I, II, III, and IV) and Para Red	• *Clostridium indolis, Lactobacillus rhamnosus, Enterococcus faecalis*, and *Ruminococcus obeum* possessed dye reduction potential, whereas E. coli and *Peptostreptococcus magnus* were not able to reduce dyes greatly. • Bacteria from human colon were proficient in dye degradation as the degradation metabolites were identified by HPLC and LC/ESI-MS.	Xu et al. (2010)
In natural sources					
1	Textile dye polluted river sediment	16S qPCR assays and Illumina MiSeq sequencing analysis	Textile wastewater	• The potential of river sediment bacteria to degrade undetectable dyes and by-products was assessed. • The changes in degradation potential were consistent with changes in the sediment bacterial community over time. • The identified phylotypes related to unidentified *Xanthomonadaceae* member, *Acidithio bacillus, Thiomonas, Acidobacteriaceae* genera, and *Acidocella* might be involved in dye degradation. • After the closure of dye-house, there was a shift toward the enrichment of aniline degraders, i.e., related to *Desulfobulbaceae* genus *Desulfococcus, Halothiobacillus*, and *Rhodanobacter*.	Ito et al. (2016)

27.5.4 In Microbial Fuel Cells

Microbial fuel cells (MFCs) are the bioelectrochemical systems (BESs) wherein the anaerobes degrade organic matter and donate electrons to complete the electrical circuit, thereby generating electricity (Kiseleva et al., 2015). MFCs are being utilized for the bioelectricity generation concomitant with the decomposition of recalcitrants such as hydrocarbons, phenols, chlorophenols, and azo dyes (Nimje et al., 2011). These serve as a sustainable technology for bioremediation of industrial effluents for the fact that instead of consuming power, these generate power. Also, these lead to lower sludge generation and function at atmospheric temperatures and pressures (Mani et al., 2019). The textile effluents can be subjected to the treatment in anodic as well as the cathodic chamber. At the anode, the breakdown of dye to simpler products can be directly carried out by electroactive microbes under anaerobic conditions via nanowires or membrane-bound c-type cytochrome (Miran et al., 2018). A standard MFC has been depicted in Figure 27.2.

The transfer of the electrons to the electrophilic dye as well as the anode is the rate-limiting step which is overcome by supplementing synthetic (e.g., riboflavin and AQDS) or natural redox mediators (humic substances) or reduced inorganic molecules such as sulfides, etc. (Sun et al., 2013). Several organisms associated with electron transfer at anodes are *Shewanella* sp., *Geobacter metallireducens,* and *Geobacter sulfurreducens* (Nimje et al., 2011). These can carry out the transfer either by direct contact or by means of electron-ferrying compounds (Kiseleva et al., 2015). At the cathode, dye oxidation can be mediated by oxygen or enzymes such as laccase. Laccases (EC 1.10.3.2) are copper-complexed multinuclear oxidases produced by several bacteria, ligninolytic fungi, and higher plants (Zucca et al., 2016). Due to the non-specific mechanism of action, these can oxidize diverse phenolic and nonphenolic contaminants and thus hold potential applicability in the field of bioremediation (Bilal et al., 2019; Yang et al., 2017). Extensive research has been conducted to demonstrate the applicability of various anaerobes in different types of MFCs for the conversion of chemical energy into bioelectricity with simultaneous dye reduction. However, very scarce information is available on how the patterns of microbial population shift during or after dye reduction in MFCs (Table 27.3). Miran et al. (2018) analyzed the changes in the sulfate-reducing microbial communities during Acid Red 114 degradation in MFC suspension as well as on anode. Due to the enrichment of the sulfate-reducing community, a drastic shift in the dominant species from *Actinobacteria* in initial inoculum to *Proteobacteria* in MFC (suspension and anode) was reported. Also, at class level, *Alphaproteobacteria* were dominant in the initial inoculum and MFC suspension (i.e., 18.8% and 40.4%, respectively), whereas *Deltproteobateria* (52.7%) were dominant in biofilm formed at the anode (Miran et al., 2018). It is well-established that *Bacteroidetes* and *Firmicutes* are generally found to be involved in bioelectricity generation as well as the degradation of organic matter (Dai et al., 2020). The sustainable application and scale-up of MFCs are limited due to the requirement of costly cathodes (e.g., Pt) (Rojas et al., 2017).

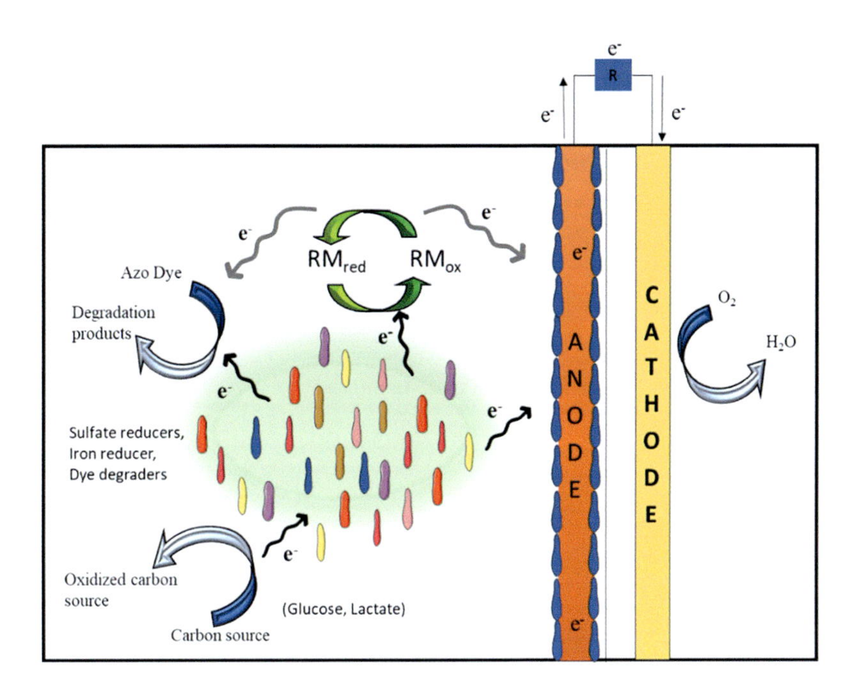

FIGURE 27.2 Mechanism of azo dye degradation and bioelectricity production in MFCs.

27.6 Role of Microbes in Natural Remediation of Dyes in Rivers and Sediments

When the textile effluents are disposed of in the natural resources like river and soil, the dye is bound to influence their indigenous communities. In response to the dye perturbation, these communities may adapt to either degrade or tolerate dyes and pigments. However, adaptation to dyes is not sufficient to safeguard these populations as the by-products generated on natural remediation lead to higher toxicity. Thus, an evaluation of how the diversity and degradation abilities of the microbiome alter with time becomes critical for the proper ecological survey. Research has been conducted in this direction concerning domestic and industrial pollution but not specific to dye pollution (de Jong et al., 2018; Köchling et al., 2017). However, Ito et al. (2016) studied the shift in microbial population and degradation of dye and intermediates (i.e., aromatic amine) from the river sediment, which was receiving treated but still colored effluent from the dye house. They demonstrated that during natural remediation, microbial communities degrade dyes followed by aromatic amines, which might take approximately two years or more. As the effluent discharge was ceased, dye degradation potential decreased, and aromatic amine-degrading potential initially increased, which eventually declined (Ito et al., 2016).

27.7 Prospective Implications of Microbial Cooperation in Dye Reduction

The success of any bioremediation process is controlled by the vigor of synergistic interactions present. Thus, in addition to understanding the evolution of microbial composition, diversity, and abundance, the co-occurrence of species during dye removal can be useful in improving the efficiency of the process. Furthermore, the interdependence of different genera in a community should be scrutinized meticulously. For instance, *Lactococcus* sp. is reported to degrade dye while producing lactate under an anoxic condition, whereas *Desulfovibrio* sp. is known to utilize lactate as an electron donor while reducing dyes under similar conditions. Thus, in a microbial community, if their abundances are shifting positively and concomitantly with degradation efficiency, it can be presumed that these might be involved in establishing a cycle of electron transfer through the interchange of biomolecules. Also, most of the work conducted lacks the analysis of archaeal diversity, which is equally essential as bacterial. It is well known that the uncultured archaea can perform several tedious tasks that bacteria cannot. Moreover, these are more resistant to extremely unfavorable conditions. These might not be found as a dominant group in the community but can still silently contribute to the function by manifesting the existing synergism. Such an integrated approach where the changes in taxa abundances are not looked upon individually but studied in correlation with each other up to the kingdom level can aid in delineating the mechanism in complex communities. Moreover, a holistic approach of the phylogenetic evaluation in association with the functional assignment to the genus level with respect to dye and intermediate degradation can be beneficial to recognize the structure of the active players during the process. For instance, Zhu et al. (2018) used multiple linear regression wherein the abundance of genera was considered as a response variable and azo bond reduction as an explanatory variable to understand functional communities. Several other in-depth statistical analyses, viz., principal component analysis (PCA), redundancy analysis (RDA), etc. along with the measurement of multiple parameters, can aid in the interpreting interaction of the dye with biotic and abiotic entities.

TABLE 27.3

Microbial Community Alterations in Various Microbial Fuel Cells

S. No.	Microbial Fuel Cell Type	Biodegrader / Enzyme	Dye Degraded	Significant Finding	References
1	Single chamber air cathode MFC	Anaerobic sludge from wastewater plant	Congo Red	• Along with bioelectricity production and decolorization, sulfide removal was also obtained. • In contrast to anodic biofilm, microbial diversity was lower on the cathode. • On acclimatization, a shift occurred with *Proteobacteria* to be most dominant in suspension as well as anodic and cathodic biofilm. • The functional analysis of communities revealed the pathways related to energy production, amino acid metabolism, intracellular trafficking, cell motility, vesicular transport, and secretion were enriched in suspension and anodic biofilm.	Dai et al. (2020)
2	MFC-anaerobic baffled reactor (MFC-ABR)	Sludge from municipal wastewater plant	Acid orange 7	• Microbial anode respiration dramatically reshaped the community structure and decreased the diversity of the microbial communities. • *Pseudomonas* was enriched by electricity generation. • Bacteria capable of methanogenesis and sulfate respiration were suppressed • Bacteria with versatile respiration (e.g., extracellular electron transfer and fumarate and nitrate respiration) were enriched, and biodegradation capabilities provided useful information to understand the enhanced degradation and detoxication of AO-7 by microbial anode respiration	Yang et al. (2018)
3	'H'-type reactors	*Shewanella oneidensis* and laccase	Acid Orange 7	• The anodic and cathodic degradation of dye was compared. • *Shewanella oneidensis* driven decolorization at anode was slow as compared to laccase-driven decolorization at cathode. • Also, the degradation products formed during anodic dye removal were unstable and more toxic than dye. • Moreover, the power generation was better in cathodic chamber.	Mani et al. (2019)
4	Dual compartment MFC	Sulfate-reducing microbial communities	Acid Red 114	• The microbial community was employed in anodic chamber to degrade dye, reduce sulfate, and decrease COD. • The parameters like COD/sulphate ratio and initial dye concentration were optimized. • The genus *Desulfovibrio* and phylum *Proteobacteria* were found to be dominant at anode.	Miran et al. (2018)
5	Single chamber air-cathode MFC	Anaerobic sludge	Congo Red	• Degradation enhancement was performed by addition of AQDS, humic acid and riboflavin. • Addition of electron shuttles altered the microbial community. • On electron shuttle augmentation, *Bacteroides nordii* strain JCM12987 disappeared, whereas uncultured *Chlorobi* clone T12, *Desulfuromonadaceae* bacterium, and uncultured *Endomicrobia* bacterium were enriched.	Sun et al. (2013)

27.8 Summary

Azo dyes are the refractory pollutants that have been muddling the health and stability of the ecosystem for years, either directly or indirectly. Soon after their discovery, the world witnessed a rapid commencement in the studies delineating their toxicity, which in turn called up for the development of a versatile and cost-effective technology for their remediation. Since then, many physicochemical and biological methods have been developed. Within the ambit of biological technologies, various organisms such as bacteria, fungi, and algae have been employed successfully in their pure forms. Nevertheless, since the dyes are designed to be oxidation-resistant, their degradation by oxidation is difficult to achieve. Contrastingly, technologies employing reduction result in higher degradation efficiencies. However, a daunting limitation associated with reductive methodologies is the appearance of the toxic aromatics formed as a result of the azo bond reduction. To overcome such limitations, scientific interests have switched from the application of pure cultures to microbial communities. In complex communities, understanding the interactions is critical for the manipulation of bioremediation. The literature in this chapter summarizes all plausible microbial interactions in view of dye remediation. It is apparent that the plant-microbe and inter-microbial interactions have been most widely studied for the dye remediation. The azo dyes, like any other pollutant, severely affect the microbial diversity of the microbiome. Many research groups have successfully unveiled the complexity of the microbial community structure from various dye-contaminated sources. However, the functional shifts in the communities have been undermined. Thus, a combinatorial approach that elaborates functional and structural dye-induced alterations in mixed communities should be adapted. Moreover, a strategic approach to identify the key degraders and their association with each other has become a prerequisite. There can be two main advantages of such investigations. Firstly, the large-scale processes in wastewater plants wherein the complex communities are applied can be manipulated in terms concerning synergistic and co-metabolic behavior. Secondly, these will pave the way for a new horizon in the ecological restoration of dye-contaminated sites (i.e., sediments, rivers, etc.). Moreover, the statistics published may not be sufficient to determine the precise risk and pressure of dyes on the ecosystem as dye-contaminated effluents are often associated with several other micropollutants whose ecotoxicological significance may be understated. So, alterations in the microbiome generated as a result of contamination of dye and other related pollutants demand a great deal of research.

Abbreviations

ABS: Aerobic Granular Sludge
ADMI: American Dye Manufacturer's Institute
AO7: Acid Orange 7
APX: Ascorbate Peroxidase
AQDS: Anthraquinone-2,6-Disulfonate
AQS: Anthraquinone-2-Sulfonate
ARDRA: Amplified Ribosomal DNA Restriction Analysis
ASBR: Anaerobic Sequence Batch Reactor
ATCC: American Type Culture Collection
BAF: Biological Aerated Filters
BOD: Biological Oxygen Demand
CAT: Catalase
CLPP: Community-Level Physiological Profiling
COD: Chemical Oxygen Demand
CWs: Constructed Wetlands
CYP: Cytochrome P450
DCIP: Dichlorophenol Indophenol
DGGE: Denaturing Gradient Gel Electrophoresis
DOL: Division of Labor
EC: Electrical Conductivity
EGSB: Expanded Granular Sludge Bed
EPSs: Exopolysaccharides
FAD: Flavin Adenine Dinucleotide
FDA: U.S. Food and Drug Administration
FISH: Fluorescence *in situ* Hybridization
FMN: Flavin Mononucleotide
GPX: Guaiacol Peroxidase
GR: Glutathione Reductase
HA: Hydrolysis Acidification
HRT: Hydraulic Retention Time
HRTS: High Rate Transpiration Systems
IARC: International Agency for Research on Cancer
LEfSe: Linear Discriminant Analysis Effect Size
LiP: Lignin Peroxidase
MBR: Membrane Bioreactor
MnP: Manganese Peroxidase
MR: Methyl Red
NAD(P)H: Nicotinamide Adenine Dinucleotide Phosphate
NADH: Nicotinamide Adenine Dinucleotide
NZVI: Nanoscale-Zerovalent Iron
PCR: Polymerase Chain Reaction
PLFA: Phospholipid Fatty Acid
POD: Peroxidases
PS: Persulfate
RFLP: Restriction Fragment Length Polymorphism
RHFCW: Re-circulating horizontal flow constructed wetland
RISA: Ribosomal Intergenic Spacer Analysis
SAB: Submerged Aerated Biofilter
SOD: Superoxide Dismutase
SSCP: Single-Strand Conformation Polymorphism
TDS: Total Dissolved Solids
TGGE: Temperature Gradient Gel Electrophoresis
TOC: Total Organic Carbon
T-RFLP: Terminal Restriction Fragment Length Polymorphism
TSS: Total Suspended Solids
UASB: Up-flow Anaerobic Sludge Blanket
UV: Ultra-Violet
WRF: White Rot Fungi

Declaration of interests: Authors declare no conflicts of interest.

REFERENCES

Adamowicz, E. M., Flynn, J., Hunter, R. C., & Harcombe, W. R. (2018). Cross-feeding modulates antibiotic tolerance in bacterial communities. *ISME Journal*, *12*(11), 2723–2735. Doi: 10.1038/s41396-018-0212-z.

Aguiar-Pulido, V., Huang, W., Suarez-Ulloa, V., Cickovski, T., Mathee, K., & Narasimhan, G. (2016). Metagenomics, metatranscriptomics, and metabolomics approaches for microbiome analysis. *Evolutionary Bioinformatics*, *12*, 5–16. Doi: 10.4137/EBO.S36436.

Albuquerque, M. G. E., Lopes, A. T., Serralheiro, M. L., Novais, J. M., & Pinheiro, H. M. (2005). Biological sulphate reduction and redox mediator effects on azo dye decolourisation in anaerobic-aerobic sequencing batch reactors. *Enzyme and Microbial Technology*, *36*(5–6), 790–799. Doi: 10.1016/j.enzmictec.2005.01.005.

Ali, H. (2010). Biodegradation of synthetic dyes - A review. *Water, Air, and Soil Pollution*, *213*(1–4), 251–273. Doi: 10.1007/s11270-010-0382-4.

Allam, N. (2017). Bioremediation efficiency of heavy metals and azo dyes by individual or consortium bacterial species either as free or immobilized cells: A comparative study. *Egyptian Journal of Botany*, *0*(0), 555–564. Doi: 10.21608/ejbo.2017.689.1040.

Ambrósio, S. T., & Campos-Takaki, G. M. (2004). Decolorization of reactive azo dyes by *Cunninghamella elegans* UCP 542 under co-metabolic conditions. *Bioresource Technology*, *91*(1), 69–75. Doi: 10.1016/S0960-8524(03)00153-6.

Annuar, M. S. M., Adnan, S., Vikineswary, S., & Chisti, Y. (2009). Kinetics and energetics of azo dye decolorization by *Pycnoporus sanguineus*. *Water, Air, and Soil Pollution*, *202*(1–4), 179–188. Doi: 10.1007/s11270-008-9968-5.

Atashgahi, S., Aydin, R., Dimitrov, M. R., Sipkema, D., Hamonts, K., Lahti, L., … Smidt, H. (2015). Impact of a wastewater treatment plant on microbial community composition and function in a hyporheic zone of a eutrophic river. *Scientific Reports*, *5*(November), 1–13. Doi: 10.1038/srep17284.

Ayed, L., Zmantar, T., Bayar, S., Charef, A., Achour, S., Mansour, H. B., & Mzoughi, R. E. (2019). Potential use of probiotic consortium isolated from kefir for textile azo dye decolorization. *Journal of Microbiology and Biotechnology*, *29*(10), 1629–1635. Doi: 10.4014/jmb.1906.06019.

Banat, I. M., Nigam, P., Singh, D., & Marchant, R. (1997). Microbial decolorization of textile-dye-containing effluents: A review. *Bioresource Technology*, *58*, 217–227.

Barathi, S., Karthik, C., S, N., & Padikasan, I. A. (2020). Biodegradation of textile dye Reactive Blue 160 by *Bacillus firmus* (Bacillaceae: Bacillales) and non-target toxicity screening of their degraded products. *Toxicology Reports*, *7*, 16–22. Doi: 10.1016/j.toxrep.2019.11.017.

Barragán, B. E., Costa, C., & Carmen Márquez, M. (2007). Biodegradation of azo dyes by bacteria inoculated on solid media. *Dyes and Pigments*, *75*(1), 73–81. Doi: 10.1016/j.dyepig.2006.05.014.

Bernstein, D. B., Dewhirst, F. E., & Segrè, D. (2019). Metabolic network percolation quantifies biosynthetic capabilities across the human oral microbiome. *ELife*, *8*, 1–33. Doi: 10.7554/eLife.39733.001.

Beveridge, T. J., Hughes, M. N., Lee, H., Leung, K. T., Poole, R. K., Savvaidis, I., … Trevors, J. T. (1996). Metal-Microbe Interactions: Contemporary Approaches. *Advances in Microbial Physiology*, *38*, 177–243. Doi::10.1016/s0065-2911(08)60158-7.

Bharathiraja, B., Jayamuthunagai, J., Praveenkumar, R., & Iyyappan, J. (2018). Phytoremediation techniques for the removal of dye in wastewater. In *Bioremediation: Applications for Environmental Protection and Management* (pp. 243–252). Springer, Singapore. Doi: 10.1007/978-981-10-7485-1_12.

Bhardwaj, V., Kumar, P., & Singhal, G. (2014). Toxicity of heavy metals pollutants in textile mills effluents. *International Journal of Scientific & Engineering Research*, *5*(7), 664–666.

Bilal, M., Rasheed, T., Nabeel, F., Iqbal, H. M. N., & Zhao, Y. (2019). Hazardous contaminants in the environment and their laccase-assisted degradation – A review. *Journal of Environmental Management*, *234*, 253–264. Doi: 10.1016/j.jenvman.2019.01.001.

Blumel, S., & A. Stolz. (2003). Cloning and characterization of the gene coding for the aerobic azoreductase from *Pigmentiphaga kullae* K24. *Applied Microbiology and Biotechnology*, *62*, 186–190. Doi: 10.1007/s00253-003-1316-5.

Brandl, H. (2002). Metal-microbe-interactions and their biotechnological applications for mineral waste treatment. *Recent Research Developments in Microbiology.*, *6*, 571–584.

Brenner, K., You, L., & Arnold, F. H. (2008). Engineering microbial consortia: A new frontier in synthetic biology. *Trends in Biotechnology*, *26*(9), 483–489. Doi: 10.1016/j.tibtech.2008.05.004.

Brindha, R., Santhosh, S., & Rajaguru, P. (2019). Integrated bio-chemo degradation of Mordant Yellow 10 using upflow anaerobic packed bed reactor (UAPBR) and tray type photo-Fenton reactor (TPFR). *Journal of Cleaner Production*, *208*, 602–611. Doi: 10.1016/j.jclepro.2018.10.158.

Cai, P., Sun, X., Wu, Y., Gao, C., Mortimer, M., Holden, P. A., … Huang, Q. (2019). Soil biofilms: Microbial interactions, challenges, and advanced techniques for ex-situ characterization. *Soil Ecology Letters*, *1*(3–4), 85–93. Doi: 10.1007/s42832-019-0017-7.

Cao, Z., Zhang, J., Zhang, J., & Zhang, H. (2017). Degradation pathway and mechanism of Reactive Brilliant Red X-3B in electro-assisted microbial system under anaerobic condition. *Journal of Hazardous Materials*, *329*, 159–165. Doi: 10.1016/j.jhazmat.2017.01.043.

Carmen, Z., & Daniela, S. (2012). Textile organic dyes – Characteristics, polluting effects and separation/elimination procedures from industrial effluents – A critical overview. In *In Organic Pollutants Ten Years After the Stockholm Convention-Environmental and Analytical Update* (Vol. 2741, pp. 55–86). Rijeka, Croatia: InTech.

Cerboneschi, M., Corsi, M., Bianchini, R., Bonanni, M., & Tegli, S. (2015). Decolorization of acid and basic dyes: Understanding the metabolic degradation and cell-induced adsorption/precipitation by Escherichia coli. *Applied Microbiology and Biotechnology*, *99*(19), 8235–8245. Doi: 10.1007/s00253-015-6648-4.

Chacko, J. T., & Subramaniam, K. (2011). Enzymatic degradation of azo dyes – A review. *International Journal of Environmental Sciences*, *1*(6), 1250–1260.

Chan, G. F., Gan, H. M., & Rashid, N. A. A. (2012). Genome sequence of *Citrobacter* sp. strain A1, a dye-degrading bacterium. *Journal of Bacteriology*, *194*(19), 5485–5486. Doi: 10.1128/JB.01285-12.

Chandanshive, V. V., Kadam, S. K., Khandare, R. V., Kurade, M. B., Jeon, B. H., Jadhav, J. P., & Govindwar, S. P. (2018). In situ phytoremediation of dyes from textile wastewater using garden ornamental plants, effect on soil quality and plant growth. *Chemosphere*, *210*, 968–976. Doi: 10.1016/j.chemosphere.2018.07.064.

Chandanshive, V. V., Rane, N. R., Tamboli, A. S., Gholave, A. R., Khandare, R. V., & Govindwar, S. P. (2017). Co-plantation of aquatic macrophytes *Typha angustifolia* and *Paspalum scrobiculatum* for effective treatment of textile industry effluent. *Journal of Hazardous Materials*, *338*, 47–56. Doi: 10.1016/j.jhazmat.2017.05.021.

Chang, J. S., Chen, B. Y., & Lin, Y. S. (2004). Stimulation of bacterial decolorization of an azo dye by extracellular metabolites from *Escherichia coli* strain NO_3. *Bioresource Technology*, *91*(3), 243–248. Doi: 10.1016/S0960-8524(03)00196-2.

Che, S., & Men, Y. (2019). Synthetic microbial consortia for biosynthesis and biodegradation: Promises and challenges. *Journal of Industrial Microbiology and Biotechnology*, *46*(9–10), 1343–1358. Doi: 10.1007/s10295-019-02211-4.

Chen, H., Wang, R., & Cerniglia, C. E. (2004). Molecular cloning, overexpression, purification, and characterization of an aerobic FMN-dependent azoreductase from *Enterococcus faecalis*. *Protein Expression & Purification*, *34*, 302–310. Doi: 10.1016/j.pep.2003.12.016.

Chengalroyen, M. D., & Dabbs, E. R. (2013). The microbial degradation of azo dyes : Minireview. *World Journal of Microbiology and Biotechnology*, *29*(3), 389–399. Doi: 10.1007/s11274-012-1198-8.

Chenu, C. (1995). Extracellular polysaccharides: An interface between microorganisms and soil constituents. In *Environmental Impacts of Soil Component Interactions: Land Quality, Natural and Anthropogenic Organics,* Volume 1 (p. 464). Retrieved from https://books.google.com/books?hl=en&lr=&id=qfTrg6iPwoEC&pgis=1.

Chenu, C., & Stotzky, G. (2016). Interactions between microorganisms and soil particles: An overview. In Huang P. M., Bollag J. M., Senesi N. (Eds.) *Interactions between Soil Particles and Microorganisms - Impact on the Terrestrial Ecosystem: An Overview*. (pp. 274–302). John Wiley & Sons, Chichester.

Chequer, F. D., de Oliveira, G. A. R., Ferraz, E. A., Cardoso, J. C., Zanoni, M. B., & de Oliveira, D. P., (2013). Textile dyes : Dyeing process and environmental impact. In *Eco-Friendly Textile Dyeing and Finishing* (Vol. 6, pp. 151–176). IntechOpen, London. Doi: 10.5772/53659.

Chung, K. T. (2016). Azo dyes and human health: A review. *Journal of Environmental Science and Health - Part C Environmental Carcinogenesis and Ecotoxicology Reviews*, *34*(4), 233–261. Doi: 10.1080/10590501.2016.1236602.

Costa, O. Y. A., Raaijmakers, J. M., & Kuramae, E. E. (2018). Microbial extracellular polymeric substances: Ecological function and impact on soil aggregation. *Frontiers in Microbiology*, *9*(JUL), 1–14. Doi: 10.3389/fmicb.2018.01636.

Crawford, J. W., Deacon, L., Grinev, D., Harris, J. A., Ritz, K., Singh, B. K., & Young, I. (2012). Microbial diversity affects self-organization of the soil - Microbe system with consequences for function. *Journal of the Royal Society Interface*, *9*(71), 1302–1310. Doi: 10.1098/rsif.2011.0679.

Cui, D., Li, G., Zhao, D., Gu, X., Wang, C., & Zhao, M. (2012). Microbial community structures in mixed bacterial consortia for azo dye treatment under aerobic and anaerobic conditions. *Journal of Hazardous Materials*, *221–222*, 185–192. Doi: 10.1016/j.jhazmat.2012.04.032.

Cui, D., Li, G., Zhao, D., & Zhao, M. (2015). Effect of quinoid redox mediators on the aerobic decolorization of azo dyes by cells and cell extracts from *Escherichia coli*. *Environmental Science and Pollution Research*, *22*, 4621–4630. Doi: 10.1007/s11356-014-3698-6.

Dafale, N., Wate, S., Meshram, S., & Nandy, T. (2008a). Kinetic study approach of remazol black-B use for the development of two-stage anoxic-oxic reactor for decolorization/biodegradation of azo dyes by activated bacterial consortium. *Journal of Hazardous Materials*, *159*, 319–328. Doi: 10.1016/j.jhazmat.2008.02.058.

Dafale, N., Rao, N. N., Meshram, S. U., & Wate, S. R. (2008b). Decolorization of azo dyes and simulated dye bath wastewater using acclimatized microbial consortium – Biostimulation and halo tolerance. *Bioresource Technology*, *99*(7), 2552–2558. Doi: 10.1016/j.biortech.2007.04.044.

Dai, R., Chen, X., Luo, Y., Ma, P., Ni, S., Xiang, X., & Li, G. (2016). Inhibitory effect and mechanism of azo dyes on anaerobic methanogenic wastewater treatment: Can redox mediator remediate the inhibition? *Water Research*, *104*, 408–417. Doi: 10.1016/j.watres.2016.08.046.

Dai, Q., Zhang, S., Liu, H., Huang, J., & Li, L. (2020). Sulfide-mediated azo dye degradation and microbial community analysis in a single-chamber air cathode microbial fuel cell. *Bioelectrochemistry*, *131*. Doi: 10.1016/j.bioelechem.2019.107349.

Das, S., Dash, H. R., & Chakraborty, J. (2016). Genetic basis and importance of metal resistant genes in bacteria for bioremediation of contaminated environments with toxic metal pollutants. *Applied Microbiology and Biotechnology*, *100*(7), 2967–2984. Doi: 10.1007/s00253-016-7364-4.

Davies, L. C., Carias, C. C., Novais, J. M., & Martins-Dias, S. (2005). Phytoremediation of textile effluents containing azo dye by using *Phragmites australis* in a vertical flow intermittent feeding constructed wetland. *Ecological Engineering*, *25*(5), 594–605. Doi: 10.1016/j.ecoleng.2005.07.003.

de Almeida, E. J. R., Christofoletti Mazzeo, D. E., Deroldo Sommaggio, L. R., Marin-Morales, M. A., de Andrade, A. R., & Corso, C. R. (2019). Azo dyes degradation and mutagenicity evaluation with a combination of microbiological and oxidative discoloration treatments. *Ecotoxicology and Environmental Safety*, *183*, 109484. Doi: 10.1016/j.ecoenv.2019.109484.

Deng, Z., & Cao, L. (2017). Fungal endophytes and their interactions with plants in phytoremediation: A review. *Chemosphere*, *168*(June), 1100–1106. Doi: 10.1016/j.chemosphere.2016.10.097.

Dhaouefi, Z., Toledo-Cervantes, A., Ghedira, K., Chekir-Ghedira, L., & Muñoz, R. (2019). Decolorization and phytotoxicity reduction in an innovative anaerobic/aerobic photobioreactor treating textile wastewater. *Chemosphere*, *234*, 356–364. Doi: 10.1016/j.chemosphere.2019.06.106.

Dogra, V., Kaur, G., Kumar, R., & Prakash, C. (2018). The importance of plant-microbe interaction for the bioremediation of dyes and heavy metals. In *Phytobiont and Ecosystem Restitution* (pp. 433–457). Springer, Singapore. Doi: 10.1007/978-981-13-1187-1_22.

Dos Santos, A. B., Bisschops, I. A. E., Cervantes, F. J., & Van Lier, J. B. (2004). Effect of different redox mediators during thermophilic azo dye reduction by anaerobic granular sludge and comparative study between mesophilic (30°C) and thermophilic (55°C) treatments for decolourisation of textile wastewaters. *Chemosphere*, *55*(9), 1149–1157. Doi: 10.1016/j.chemosphere.2004.01.031.

Drogue, B., Doré, H., Borland, S., Wisniewski-Dyé, F., & Prigent-Combaret, C. (2012). Which specificity in cooperation between phytostimulating rhizobacteria and plants? *Research in Microbiology*, *163*(8), 500–510. Doi: 10.1016/j.resmic.2012.08.006.

Drury, B., Rosi-Marshall, E., & Kelly, J. J. (2013). Wastewater treatment effluent reduces the abundance and diversity of benthic bacterial communities in urban and suburban rivers. *Applied and Environmental Microbiology*, *79*(6), 1897–1905. Doi: 10.1128/AEM.03527-12.

Ehrlich, H. L. (1997). Microbes and metals. *Applied Microbiology and Biotechnology*, *48*(6), 687–692. Doi: 10.1007/s002530051116.

Ertuğrul, S., San, N. O., & Dönmez, G. (2009). Treatment of dye (Remazol Blue) and heavy metals using yeast cells with the purpose of managing polluted textile wastewaters. *Ecological Engineering*, *35*(1), 128–134. Doi: 10.1016/j.ecoleng.2008.09.015.

Eslami, M., Ali, M., & Asad, S. (2016). Isolation, cloning and characterization of an azoreductase from the halophilic bacterium Halomonas elongata. *International Journal of Biological Macromolecules*, *85*, 111–116. Doi: 10.1016/j.ijbiomac.2015.12.065.

Fakruddin, M., & Mannan, K. S. Bin. (2015). Methods for analyzing diversity of microbial communities in natural environments. *Ceylon Journal of Science (Biological Sciences)*, *42*(1), 123–151. Doi: 10.4038/cjsbs.v42i1.5896.

Fatima, M., Farooq, R., Lindström, R. W., & Saeed, M. (2017). A review on biocatalytic decomposition of azo dyes and electrons recovery. *Journal of Molecular Liquids*, *246*, 275–281. Doi: 10.1016/j.molliq.2017.09.063.

Feng, J., Cerniglia, C. E., & Chen, H. (2012). Toxicological significance of azo dye metabolism by human intestinal microbiota. *Frontiers in Bioscience - Elite*, *4 E*(2), 568–586. Doi: 10.2741/e400.

Field, J. A., & Brady, J. (2003). Riboflavin as a redox mediator accelerating the reduction of the azo dye Mordant Yellow 10 by anaerobic granular sludge. *Water Science and Technology*, *48*(6), 187–193. Doi: 10.2166/wst.2003.0393.

Franca, R. D. G., Pinheiro, H. M., & Lourenço, N. D. (2020a). Recent developments in textile wastewater biotreatment: Dye metabolite fate, aerobic granular sludge systems and engineered nanoparticles. In *Reviews in Environmental Science and Biotechnology* (*Vol. 19*). Doi: 10.1007/s11157-020-09526-0.

Franca, R. D. G., Vieira, A., Carvalho, G., Oehmen, A., Pinheiro, H. M., Barreto Crespo, M. T., & Lourenço, N. D. (2020b). *Oerskovia paurometabola* can efficiently decolorize azo dye Acid Red 14 and remove its recalcitrant metabolite. *Ecotoxicology and Environmental Safety*, *191*, 110007. Doi: 10.1016/j.ecoenv.2019.110007.

Franciscon, E., Zille, A., Fantinatti-garboggini, F., Serrano, I., Cavaco-paulo, A., & Regina, L. (2009). Microaerophilic – aerobic sequential decolourization/ biodegradation of textile azo dyes by a facultative *Klebsiella* sp. strain VN-31. *Process Biochemistry*, *44*, 446–452. Doi: 10.1016/j.procbio.2008.12.009.

Gadow, S. I., & Li, Y. Y. (2020). Development of an integrated anaerobic/aerobic bioreactor for biodegradation of recalcitrant azo dye and bioenergy recovery: HRT effects and functional resilience. *Bioresource Technology Reports*, *9*, 100388. Doi: 10.1016/j.biteb.2020.100388.

Gao, Y., Yang, B., & Wang, Q. (2018). Biodegradation and decolorization of dye wastewater: A review. *IOP Conference Series: Earth and Environmental Science*, *178*(1), 1–5. Doi: 10.1088/1755-1315/178/1/012013.

Gao, D., Zeng, Y., Wen, X., & Qian, Y. (2008). Competition strategies for the incubation of white rot fungi under non-sterile conditions. *Process Biochemistry*, *43*(9), 937–944. Doi: 10.1016/j.procbio.2008.04.026.

Gerhardt, K. E., Huang, X. D., Glick, B. R., & Greenberg, B. M. (2009). Phytoremediation and rhizoremediation of organic soil contaminants: Potential and challenges. *Plant Science*, *176*(1), 20–30. Doi: 10.1016/j.plantsci.2008.09.014.

Ghaly, A. E., Ananthashankar, R., Alhattab, M., & Ramakrishnan, V. V. (2014). Production, characterization and treatment of textile effluents : A critical review. *Chemical Engineering & Process Technology*, *5*(1), 1–19. Doi: 10.4172/2157-7048.1000182.

Ghosh, A., Dastidar, M. G., & Sreekrishnan, T. R. (2016). Recent advances in bioremediation of heavy metals and metal complex dyes: Review. *Journal of Environmental Engineering (United States)*, *142*(9), C4015003. Doi: 10.1061/(ASCE)EE.1943-7870.0000965.

Gil, N. M. B., Pajot, H. F., Soro, M. d. M. R., de Figueroa, L. I. C., & Kurth, D. (2018). Genome-wide overview of Trichosporon akiyoshidainum HP-2023, new insights into its mechanism of dye discoloration. *3 Biotech*, *8*(10), 440. Doi: 10.1007/s13205-018-1465-y.

Godheja, J., Shekhar, S. K., & Modi, D. R. (2017). Bacterial rhizoremediation of petroleum hydrocarbons (PHC). *Plant-Microbe Interactions in Agro-Ecological Perspectives*, *2*, 495–519. Doi: 10.1007/978-981-10-6593-4_20.

Guo, J., Kang, L., Wang, X., & Yang, J. (2010). Decolorization and degradation of azo dyes by redox mediator system with bacteria. In *Biodegradation of azo dyes* (pp. 85–100). Springer, Berlin, Heidelberg. Doi: 10.1007/698_2009_46.

Gupta, V. K., Khamparia, S., Tyagi, I., Jaspal, D., Malviya, A., & Paper, R. (2015). Decolorization of mixture of dyes: A critical review. *Global Journal of Environmental Science and Management*, *1*(11), 71–94. Doi: 10.7508/gjesm.2015.01.007.

Gupta, S., Ross, T. D., Gomez, M. M., Grant, J. L., Romero, P. A., & Venturelli, O. S. (2020). Investigating the dynamics of microbial consortia in spatially structured environments. *Nature Communications*, *11*(1), 1–15. Doi: 10.1038/s41467-020-16200-0.

Haddaji, D., Ghrabi-Gammar, Z., Hamed, K. B., & Bousselmi, L. (2019). A re-circulating horizontal flow constructed wetland for the treatment of synthetic azo dye at high concentrations. *Environmental Science and Pollution Research*, *26*(13), 13489–13501. Doi: 10.1007/s11356-019-04704-2.

Haferburg, G., & Kothe, E. (2007). Microbes and metals: Interactions in the environment. *Journal of Basic Microbiology*, *47*(6), 453–467. Doi: 10.1002/jobm.200700275.

Haines, J. R., Herrmann, R., Lee, K., Cobanli, S., & Blaise, C. (2002). Microbial population analysis as a measure of ecosystem restoration. *Bioremediation Journal*, *6*(3), 283–296. Doi: 10.1080/10889860290777611.

Handayani, W., Meitiniarti, V. I., & Timotius, K. H. (2007). Decolorization of Acid Red 27 and Reactive Red 2 by *Enterococcus faecalis* under a batch system. *World Journal of Microbiology and Biotechnology*, 1239–1244. Doi: 10.1007/s11274-007-9355-1.

He, F., Hu, W., & Li, Y. (2004). Biodegradation mechanisms and kinetics of azo dye 4BS by a microbial consortium. *Chemosphere*, *57*, 293–301. Doi: 10.1016/j.chemosphere.2004.06.036.

He, M., Tan, L., Ning, S., Song, L., & Shi, S. (2017). Performance of the biological aerated filter bioaugmented by a yeast *Magnusiomyces ingens* LH-F1 for treatment of Acid Red B and microbial community dynamics. *World Journal of Microbiology and Biotechnology*, *33*(2), 1–11. Doi: 10.1007/s11274-017-2210-0.

He, Y., Wang, X., Xu, J., Yan, J., Ge, Q., Gu, X., & Jian, L. (2013). Application of integrated ozone biological aerated filters and membrane filtration in water reuse of textile effluents. *Bioresource Technology*, *133*, 150–157. Doi: 10.1016/j.biortech.2013.01.074.

Holkar, C. R., Jadhav, A. J., Pinjari, D. V, Mahamuni, N. M., & Pandit, A. B. (2016). A critical review on textile wastewater treatments : Possible approaches. *Journal of Environmental Management*, *182*, 351–366. Doi: 10.1016/j.jenvman.2016.07.090.

Hong, Y. G., & Gu, J. D. (2010). Physiology and biochemistry of reduction of azo compounds by *Shewanella* strains relevant to electron transport chain. *Applied Microbiology and Biotechnology*, *88*(3), 637–643. Doi: 10.1007/s00253-010-2820-z.

Hong, Y., Xu, M., Guo, J., Xu, Z., Chen, X., & Sun, G. (2007). Respiration and growth of *Shewanella decolorationis* S12 with an azo compound as the sole electron acceptor. *Applied and Environmental Microbiology*, *73*(1), 64–72. Doi: 10.1128/AEM.01415-06.

Huang, P. M., Wang, M. K., & Chiu, C. Y. (2005). Soil mineral-organic matter-microbe interactions: Impacts on biogeochemical processes and biodiversity in soils. *Pedobiologia*, *49*(6), 609–635. Doi: 10.1016/j.pedobi.2005.06.006.

Hussain, I., Puschenreiter, M., Gerhard, S., Schöftner, P., Yousaf, S., Wang, A., … Reichenauer, T. G. (2018). Rhizoremediation of petroleum hydrocarbon-contaminated soils: Improvement opportunities and field applications. *Environmental and Experimental Botany*, *147*, 202–219. Doi: 10.1016/j.envexpbot.2017.12.016.

Ijaz, A., Imran, A., Anwar ul Haq, M., Khan, Q. M., & Afzal, M. (2016). Phytoremediation: Recent advances in plant-endophytic synergistic interactions. *Plant and Soil*, *405*(1–2), 179–195. Doi: 10.1007/s11104-015-2606-2.

Imran, M., Crowley, D. E., Khalid, A., Hussain, S., Mumtaz, M. W., & Arshad, M. (2015a). Microbial biotechnology for decolorization of textile wastewaters. *Reviews in Environmental Science and Biotechnology*, *14*(1), 73–92. Doi: 10.1007/s11157-014-9344-4.

Imran, M., Negm, F., Hussain, S., Ashraf, M., Ashraf, M., Ahmad, Z., … Crowley, D. E. (2016). Characterization and purification of membrane-bound azoreductase from azo dye degrading *Shewanella* sp. strain IFN4. *CLEAN-Soil Air Water*, *44*(11), 1523–1530. Doi: 10.1002/clen.201501007.

Imran, M., Shaharoona, B., Crowley, D. E., Khalid, A., Hussain, S., & Arshad, M. (2015b). The stability of textile azo dyes in soil and their impact on microbial phospholipid fatty acid profiles. *Ecotoxicology and Environmental Safety*, *120*, 163–168. Doi: 10.1016/j.ecoenv.2015.06.004.

Imron, M. F., Kurniawan, S. B., Soegianto, A., & Wahyudianto, F. E. (2019). Phytoremediation of methylene blue using duckweed (*Lemna minor*). *Heliyon*, *5*(8), e02206. Doi: 10.1016/j.heliyon.2019.e02206.

Ito, T., Adachi, Y., Yamanashi, Y., & Shimada, Y. (2016). Long–term natural remediation process in textile dye–polluted river sediment driven by bacterial community changes. *Water Research*, *100*, 458–465. Doi: 10.1016/j.watres.2016.05.050.

Jadhav, J. P., Parshetti, G. K., Kalme, S. D., & Govindwar, S. P. (2007). Decolourization of azo dye methyl red by *Saccharomyces cerevisiae* MTCC 463. *Chemosphere*, *68*(2), 394–400. Doi: 10.1016/j.chemosphere.2006.12.087.

de Jong, A., van der Zaan, B., Geerlings, G., Roosmini, D., … Jetten, M. (2018). Decrease in microbial diversity along a pollution gradient in Citarum River Sediment. *BioRxiv*, 357111. Doi: 10.1101/357111.

Jonstrup, M., Kumar, N., Guieysse, B., Murto, M., & Mattiasson, B. (2013). Decolorization of textile dyes by *Bjerkandera* sp. BOL 13 using waste biomass as carbon source. *Journal of Chemical Technology and Biotechnology*, *88*(3), 388–394. Doi: 10.1002/jctb.3852.

Joshi, S. M., Inamdar, S. A., Patil, S. M., & Govindwar, S. P. (2013). Molecular assessment of shift in bacterial community in response to Congo red. *International Biodeterioration and Biodegradation*, *77*, 18–21. Doi: 10.1016/j.ibiod.2012.10.010.

Joshi, S. R., Kalita, D., Kumar, R., Nongkhlaw, M., & Swer, P. B. (2014). Metal-microbe interaction and bioremediation. *Radionuclide Contamination and Remediation Through Plants*, 235–251. Doi: 10.1007/978-3-319-07665-2.

Kabra, A. N., Khandare, R. V., & Govindwar, S. P. (2013). Development of a bioreactor for remediation of textile effluent and dye mixture: A plant-bacterial synergistic strategy. *Water Research*, *47*(3), 1035–1048. Doi: 10.1016/j.watres.2012.11.007.

Kadam, S. K., Chandanshive, V. V., Rane, N. R., Patil, S. M., Gholave, A. R., Khandare, R. V., … Govindwar, S. P. (2018). Phytobeds with *Fimbristylis dichotoma* and *Ammannia baccifera* for treatment of real textile effluent: An in situ treatment, anatomical studies and toxicity evaluation. *Environmental Research*, *160*(June 2017), 1–11. Doi: 10.1016/j.envres.2017.09.009.

Kapdan, I. K., Tekol, M., & Sengul, F. (2003). Decolorization of simulated textile wastewater in an anaerobic-aerobic sequential treatment system. *Process Biochemistry*, *38*(7), 1031–1037.

Kavita, B., & Keharia, H. (2012). Reduction of hexavalent chromium by *Ochrobactrum intermedium* BCR400 isolated from a chromium-contaminated soil. *3 Biotech*, *2*(1), 79–87. Doi: 10.1007/s13205-011-0038-0.

Keck, A., Klein, J., Kudlich, M., Stolz, A., Knackmuss, H. J., & Mattes, R. (1997). Reduction of azo dyes by redox mediators originating in the naphthalenesulfonic acid degradation pathway of *Sphingomonas* sp. strain BN6. *Applied and Environmental Microbiology*, *63*(9), 3684–3690. Doi: 10.1128/aem.63.9.3684-3690.1997.

Keharia, H., & Madamwar, D. (2002). Transformation of textile dyes by white-rot fungus *Trametes versicolor*. *Applied Biochemistry and Biotechnology*, *102–103*(3), 99–108.

Keharia, H., & Madamwar, D. (2003). Bioremediation concepts for treatment of dye containing wastewater : A review. *Indian Journal of Experimental Biology*, *4*, 1068–1075.

Kelly, S. L., & Kelly, D. E. (2013). Microbial cytochromes P450: Biodiversity and biotechnology. Where do cytochromes P450 come from, what do they do and what can they do for us? *Philosophical Transactions of the Royal Society B: Biological Sciences*, *368*(1612). Doi: 10.1098/rstb.2012.0476.

Keskinkan, O., & Göksu, M. Z. L. (2007). Assessment of the dye removal capability of submersed aquatic plants in a laboratory-scale wetland system using ANOVA. *Brazilian Journal of Chemical Engineering*, *24*(2), 193–202. Doi: 10.1590/S0104-66322007000200004.

Khan, S., & Malik, A. (2018). Toxicity evaluation of textile effluents and role of native soil bacterium in biodegradation of a textile dye. *Environmental Science and Pollution Research*, *25*(5), 4446–4458. Doi: 10.1007/s11356-017-0783-7.

Khandare, R. V., & Govindwar, S. P. (2015). Phytoremediation of textile dyes and effluents: Current scenario and future prospects. *Biotechnology Advances*, *33*(8), 1697–1714. Doi: 10.1016/j.biotechadv.2015.09.003.

Khataee, A. R., Movafeghi, A., Vafaei, F., Salehi Lisar, S. Y., & Zarei, M. (2013). Potential of the aquatic fern *Azolla filiculoides* in biodegradation of an azo dye: Modeling of experimental results by artificial neural networks. *International Journal of Phytoremediation*, *15*(8), 729–742. Doi: 10.1080/15226514.2012.735286.

Khehra, M. S., Saini, H. S., Sharma, D. K., Chadha, B. S., & Chimni, S. S. (2005). Comparative studies on potential of consortium and constituent pure bacterial isolates to decolorize azo dyes. *Water Research*, *39*(20), 5135–5141. Doi: 10.1016/j.watres.2005.09.033.

Kiseleva, L., Garushyants, S. K., Ma, H., Simpson, D. J. W., Fedorovich, V., Cohen, M. F., & Goryanin, I. (2015). Taxonomic and functional metagenomic analysis of anodic communities in two pilot-scale microbial fuel cells treating different industrial wastewaters. *Journal of Integrative Bioinformatics*, *12*(3), 273. Doi: 10.2390/biecoll-jib-2015-273.

Kochling, T., Djalma, A., Ferraz, N., Florencio, L., Kato, M. T., & Gavazza, S. (2016). 454-Pyrosequencing analysis of highly adapted azo dye-degrading microbial communities in a two- stage anaerobic – aerobic bioreactor treating textile effluent. *Environmental Technology*, *3330*(August), 0–7. Doi: 10.1080/09593330.2016.1208681.

Köchling, T., Sanz, J. L., Galdino, L., Florencio, L., & Kato, M. T. (2017). Impact of pollution on the microbial diversity of a tropical river in an urbanized region of northeastern Brazil. *International Microbiology*, *20*(1), 11–24. Doi: 10.2436/20.1501.01.281.

Kotoky, R., Rajkumari, J., & Pandey, P. (2018). The rhizosphere microbiome: Significance in rhizoremediation of polyaromatic hydrocarbon contaminated soil. *Journal of Environmental Management*, *217*, 858–870. Doi: 10.1016/j.jenvman.2018.04.022.

Kuiper, I., Lagendijk, E. L., Bloemberg, G. V, & Lugtenberg, B. J. J. (2004). Rhizoremediation : A beneficial plant-microbe interaction bioremediation : A natural method. *Molecular-Plant Microbe Interactions*, *17*(1), 6–15.

Kumar, S. S., Kadier, A., Malyan, S. K., Ahmad, A., & Bishnoi, N. R. (2017). Phytoremediation and rhizoremediation: Uptake, mobilization and sequestration of heavy metals by plants. *Plant-Microbe Interactions in Agro-Ecological Perspectives*, *2*, 367–394. Doi: 10.1007/978-981-10-6593-4_15.

Kumaran, S., Ngo, A. C. R., Schultes, F. P. J., & Tischler, D. (2020). Draft genome sequence of *Kocuria indica* DP-K7, a methyl red degrading actinobacterium. *3 Biotech*, *10*(4), 1–10. Doi: 10.1007/s13205-020-2136-3.

Kushwaha, A., Hans, N., Kumar, S., & Rani, R. (2018). A critical review on speciation, mobilization and toxicity of lead in soil-microbe-plant system and bioremediation strategies. *Ecotoxicology and Environmental Safety*, *147*, 1035–1045. Doi: 10.1016/j.ecoenv.2017.09.049.

Lacerda, C. M. R., Choe, L. H., & Reardon, K. F. (2007). Metaproteomic analysis of a bacterial community response to cadmium exposure. *Journal of Proteome Research*, *6*(3), 1145–1152. Doi: 10.1021/pr060477v.

Lade, H. S., Waghmode, T. R., Kadam, A. a., & Govindwar, S. P. (2012). Enhanced biodegradation and detoxification of disperse azo dye Rubine GFL and textile industry effluent by defined fungal-bacterial consortium. *International Biodeterioration & Biodegradation*, *72*, 94–107. Doi: 10.1016/j.ibiod.2012.06.001.

Lehman, R. M., Colwell, F. S., Ringelberg, D. B., & White, D. C. (1995). Combined microbial community-level analyses for quality assurance of terrestrial subsurface cores. *Journal of Microbiological Methods*, *22*(3), 263–281. Doi: 10.1016/0167-7012(95)00012-A.

Li, Q., Feng, X. L., Li, T. T., Lu, X. R., Liu, Q. Y., Han, X., … Xiao, X. (2018). Anaerobic decolorization and detoxification of cationic red X-GRL by *Shewanella oneidensis* MR-1. *Environmental Technology*, *39*(18), 2382–2389. Doi: 10.1080/09593330.2017.1355933.

Lin, D., Ma, W., Jin, Z., Wang, Y., Huang, Q., & Cai, P. (2016). Interactions of EPS with soil minerals: A combination study by ITC and CLSM. *Colloids and Surfaces B: Biointerfaces*, *138*, 10–16. Doi: 10.1016/j.colsurfb.2015.11.026.

Lin, J., Zhang, X., Li, Z., & Lei, L. (2010). Biodegradation of Reactive blue 13 in a two-stage anaerobic/ aerobic fluidized beds system with a *Pseudomonas* sp. isolate. *Bioresource Technology*, *101*(1), 34–40. Doi: 10.1016/j.biortech.2009.07.037.

Lindemann, S. R., Bernstein, H. C., Song, H. S., Fredrickson, J. K., Fields, M. W., Shou, W., … Beliaev, A. S. (2016). Engineering microbial consortia for controllable outputs. *ISME Journal*, *10*(9), 2077–2084. Doi: 10.1038/ismej.2016.26.

Ling, J., Hong, L., Jiti, Z., & Jing, W. (2009). Quinone-mediated decolorization of sulfonated azo dyes by cells and cell extracts from *Sphingomonas xenophaga*. *Journal of Environmental Sciences*, *21*(4), 503–508. Doi: 10.1016/S1001-0742(08)62299-8.

Liu, N., Xie, X., Yang, B., Zhang, Q., Yu, C., & Yang, B. (2017). Performance and microbial community structures of hydrolysis acidification process treating azo and anthraquinone dyes in different stages. *Environmental Science and Pollution Research*, *24*(1), 252–263. Doi: 10.1007/s11356-016-7705-y.

Lourenço, J., Mendo, S., & Pereira, R. (2018). Rehabilitation of radioactively contaminated soil: Use of bioremediation/ phytoremediation techniques. *In Remediation Measures for Radioactively Contaminated Areas*. Doi: 10.1007/978-3-319-73398-2_8.

Lv, T., Zhang, Y., Carvalho, P. N., Zhang, L., Button, M., Arias, C. A., … Brix, H. (2017). Microbial community metabolic function in constructed wetland mesocosms treating the pesticides imazalil and tebuconazole. *Ecological Engineering*, *98*, 378–387. Doi: 10.1016/j.ecoleng.2016.07.004.

Mahajan, P., Kaushal, J., Upmanyu, A., & Bhatti, J. (2019). Assessment of phytoremediation potential of *Chara vulgaris* to treat toxic pollutants of textile effluent. *Journal of Toxicology*, *2019*, 1–11. Doi: 10.1155/2019/8351272.

Mahmood, S., Khalid, A., Arshad, M., & Ahmad, R. (2015). Effect of trace metals and electron shuttle on simultaneous reduction of reactive black-5 azo dye and hexavalent chromium in liquid medium by *Pseudomonas* sp. *Chemosphere*, *138*, 895–900. Doi: 10.1016/j.chemosphere.2014.10.084.

Mahmood, S., Khalid, A., Arshad, M., Mahmood, T., & Crowley, D. E. (2016). Detoxification of azo dyes by bacterial oxidoreductase enzymes. *Critical Reviews in Biotechnology*, *36*(4), 639–651. Doi: 10.3109/07388551.2015.1004518.

Manganelli, S., Benfenati, E., Manganaro, A., Kulkarni, S., Barton-Maclaren, T. S., & Honma, M. (2016). New quantitative structure-activity relationship models improve predictability of ames mutagenicity for aromatic azo compounds. *Toxicological Sciences*, *153*(2), 316–326. Doi: 10.1093/toxsci/kfw125.

Mani, P., Fidal, V. T, Bowman, K., Breheny, M., Chandra, T. S., Keshavarz, T., & Kyazze, G. (2019). Degradation of azo dye (Acid orange 7) in a microbial fuel cell: Comparison between anodic microbial-mediated reduction and cathodic laccase-mediated oxidation. *Frontiers in Energy Research*, *7*, 101. Doi: 10.3389/FENRG.2019.00101.

Marimuthu, T., Rajendran, S., & Manivannan, M. (2013). A review on bacterial degradation of textile dyes. *Journal of Chemistry and Chemical Sciences*, *3*(3), 201–212.

McHugh, S., O'Reilly, C., Mahony, T., Colleran, E., & O'Flaherty, V. (2003). Anaerobic granular sludge bioreactor technology. *Reviews in Environmental Science and Biotechnology*, *2*(2–4), 225–245. Doi: 10.1023/B:RESB.0000040465.45300.97.

McMullan, G., Meehan, C., Conneely, A., Kirby, N., Robinson, T., Nigam, P., … Smyth, W. F. (2001). Microbial decolourisation and degradation of textile dyes. *Applied Microbiology and Biotechnology*, *56*, 81–87. Doi: 10.1007/s002530000587.

Meerbergen, K., Willems, K. A., Dewil, R., Impe, J. Van, Appels, L., & Lievens, B. (2018). Isolation and screening of bacterial isolates from wastewater treatment plants to decolorize azo dyes. *Journal of Bioscience and Bioengineering*, *125*(4), 448–456. Doi: 10.1016/j.jbiosc.2017.11.008.

Mimmo, T., Del Buono, D., Terzano, R., Tomasi, N., Vigani, G., Crecchio, C., … Cesco, S. (2014). Rhizospheric organic compounds in the soil-microorganism-plant system: Their role in iron availability. *European Journal of Soil Science*, *65*(5), 629–642. Doi: 10.1111/ejss.12158.

Miran, W., Jang, J., Nawaz, M., Shahzad, A., & Lee, D. S. (2018). Sulfate-reducing mixed communities with the ability to generate bioelectricity and degrade textile diazo dye in microbial fuel cells. *Journal of Hazardous Materials*, *352*, 70–79. Doi: 10.1016/j.jhazmat.2018.03.027.

Mishra, A., Bhattacharya, A., & Mishra, N. (2019). Mycorrhizal symbiosis: An effective tool for metal bioremediation. In *New and Future Developments in Microbial Biotechnology and Bioengineering* (pp. 113–128). Elsevier. Doi: 10.1016/b978-0-12-818258-1.00007-8.

Mishra, S., & Maiti, A. (2018). The efficacy of bacterial species to decolourise reactive azo, anthroquinone and triphenylmethane dyes from wastewater: A review. *Environmental Science and Pollution Research*, *25*(9), 8286–8314. Doi: 10.1007/s11356-018-1273-2.

Mishra, A., & Malik, A. (2014). Metal and dye removal using fungal consortium from mixed waste stream: Optimization and validation. *Ecological Engineering*, *69*, 226–231. Doi: 10.1016/j.ecoleng.2014.04.007.

Mohanty, S., Dafale, N., & Rao, N. N. (2006). Microbial decolorization of reactive black-5 in a two-stage anaerobic-aerobic reactor using acclimatized activated textile sludge. *Biodegradation*, *17*(5), 403–413. Doi: 10.1007/s10532-005-9011-0.

Moopantakath, J., & Kumavath, R. (2018). Bio-augmentation of Actinobacteria and their role in dye decolorization. In *New and Future Developments in Microbial Biotechnology and Bioengineering: Actinobacteria: Diversity and Biotechnological Applications* (pp. 297–304). Elsevier. Doi: 10.1016/B978-0-444-63994-3.00020-5.

Morgan, J. A. W., Bending, G. D., & White, P. J. (2005). Biological costs and benefits to plant-microbe interactions in the rhizosphere. *Journal of Experimental Botany*, *56*(417), 1729–1739. Doi: 10.1093/jxb/eri205.

Movafeghi, A., Khataee, A. R., Moradi, Z., & Vafaei, F. (2016). Biodegradation of Direct Blue 129 diazo dye by *Spirodela polyrrhiza*: An artificial neural networks modeling. *International Journal of Phytoremediation*, *18*(4), 337–347. Doi: 10.1080/15226514.2015.1109588.

Muthunarayanan, V., Santhiya, M., Swabna, V., & Geetha, A. (2011). Phytodegradation of textile dyes by Water Hyacinth (*Eichhornia crassipes*) from aqueous dye solutions. *International Journal of Environmental Sciences*, *1*(7), 1702–1717.

Nachiyar, C. V., Sunkar, S., Kumar, G. N., Karunya, A., Ananth, P. B., Prakash, P., & Jabasingh, S. A. (2012). Biodegradation of Acid Blue 113 containing textile effluent by constructed aerobic bacterial consortia : Optimization and mechanism. *Bioremediation and Biodegradation*, *3*(9), 1–9. Doi: 10.4172/2155-6199.1000162.

Nanjani, S., Rawal, K., & Keharia, H. (2020). Decoding social behaviors in a glycerol dependent bacterial consortium during Reactive Blue 28 degradation. *Brazilian Journal of Microbiology*. Doi: 10.1007/s42770-020-00303-3.

Niehaus, L., Boland, I., Liu, M., Chen, K., Fu, D., Henckel, C., … Momeni, B. (2019). Microbial coexistence through chemical-mediated interactions. *Nature Communications*, *10*(1), 1–12. Doi: 10.1038/s41467-019-10062-x.

Nilratnisakorn, S., Thiravetyan, P., & Nakbanpote, W. (2007). Synthetic reactive dye wastewater treatment by narrow-leaved cattails (*Typha angustifolia* Linn.): Effects of dye, salinity and metals. *Science of the Total Environment*, *384*(1–3), 67–76. Doi: 10.1016/j.scitotenv.2007.06.027.

Nimje, V. R., Chen, C. Y., Chen, C. C., Chen, H. R., Tseng, M. J., Jean, J. S., & Chang, Y. F. (2011). Glycerol degradation in single-chamber microbial fuel cells. *Bioresource Technology*, *102*(3), 2629–2634. Doi: 10.1016/j.biortech.2010.10.062.

Novotny, C., Dias, N., Kapanen, A., Malachov, K., Vandrovcova, M., Itavaara, M., & Lima, N. (2006). Comparative use of bacterial, algal and protozoan tests to study toxicity of azo- and anthraquinone dyes. *Chemosphere*, *63*(9), 1436–1442. Doi: 10.1016/j.chemosphere.2005.10.002.

Ojuederie, O. B., & Babalola, O. O. (2017). Microbial and plant-assisted bioremediation of heavy metal polluted environments: A review. *International Journal of Environmental Research and Public Health*, *14*(12). Doi: 10.3390/ijerph14121504.

Oliveira, J. M. S., de Lima e Silva, M. R., Issa, C. G., Corbi, J. J., Damianovic, M. H. R. Z., & Foresti, E. (2020). Intermittent aeration strategy for azo dye biodegradation: A suitable alternative to conventional biological treatments? *Journal of Hazardous Materials*, *385*, 121558. Doi: 10.1016/j.jhazmat.2019.121558.

Olivo-Alanis, D., Garcia-Reyes, R. B., Alvarez, L. H., & Garcia-Gonzalez, A. (2018). Mechanism of anaerobic bio-reduction of azo dye assisted with lawsone-immobilized activated carbon. *Journal of Hazardous Materials*, *347*, 423–430. Doi: 10.1016/j.jhazmat.2018.01.019.

Otero-Blanca, A., Folch-Mallol, J. L., Lira-Ruan, V., del Rayo Sánchez Carbente, M., & Batista-García, R. A. (2018). Phytoremediation and fungi: An underexplored binomial. In *Approaches in Bioremediation* (pp. 79–95). Springer, Cham. Doi: 10.1007/978-3-030-02369-0_5.

Pacheco, A. R., & Segrè, D. (2019). A multidimensional perspective on microbial interactions. *FEMS Microbiology Letters*, *366*(11), 1–11. Doi: 10.1093/femsle/fnz125.

Padmavathy, S., Sandhya, S., Swaminathan, K., Subrahmanyam, Y. V., Chakrabarti, T., & Kaul, S. N. (2003). Aerobic decolorization of Reactive azo dyes in presence of various cosubstrates. *Chemical and Biochemical Engineering Quarterly*, *17*(2), 147–151.

Pan, F., Zhong, X., Xia, D., Yin, X., Li, F., Zhao, D., & Ji, H. (2017). Nanoscale zero-valent iron / persulfate enhanced upflow anaerobic sludge blanket reactor for dye removal : Insight into microbial metabolism and microbial community. *Scientific Reports*, *7*(1), 1–12. Doi: 10.1038/srep44626.

Pan, H., Feng, J., He, G. X., Cerniglia, C. E., & Chen, H. (2012). Evaluation of impact of exposure of Sudan azo dyes and their metabolites on human intestinal bacteria. *Anaerobe*, *18*(4), 445–453. Doi: 10.1016/j.anaerobe.2012.05.002.

Pandey, K., Saha, P., & Rao, K. V. B. (2019). A study on the utility of immobilized cells of indigenous bacteria for biodegradation of reactive azo dyes. *Preparative Biochemistry and Biotechnology*, *0*(0), 1–13. Doi: 10.1080/10826068.2019.1692219.

Pandey, A., Singh, P., & Iyengar, L. (2007). Bacterial decolorization and degradation of azo dyes. *International Biodeterioration & Biodegradation*, *59*(2), 73–84. Doi: 10.1016/j.ibiod.2006.08.006.

Parshetti, G., Kalme, S., Saratale, G., & Govindwar, S. (2006). Biodegradation of Malachite Green by *Kocuria rosea* MTCC 1532. *Acta Chimica Slovenica*, *53*(4), 492–498.

Patel, V. R., Khan, R., & Bhatt, N. (2020). Cost-effective in-situ remediation technologies for complete mineralization of dyes contaminated soils. *Chemosphere*, *243*, 125253. Doi: 10.1016/j.chemosphere.2019.125253.

Pathak, H., Soni, D., & Chauhan, K. (2014). Evaluation of in vitro efficacy for decolorization and degradation of commercial azo dye RB-B by *Morganella* sp. HK-1 isolated from dye contaminated industrial landfill. *Chemosphere*, *105*, 126–132. Doi: 10.1016/j.chemosphere.2014.01.004.

Pereira, R. A., Pereira, M. F. R., Alves, M. M., & Pereira, L. (2013). Carbon based materials as novel redox mediators for dye wastewater biodegradation. *Applied Catalysis B, Environmental*, *144*, 713–720. Doi: 10.1016/j.apcatb.2013.07.009.

Popli, S., & Patel, U. D. (2015). Destruction of azo dyes by anaerobic–aerobic sequential biological treatment: A review. *International Journal of Environmental Science and Technology*, *12*(1), 405–420. Doi: 10.1007/s13762-014-0499-x.

Praveen, A., Pandey, V. C., Marwa, N., & Singh, D. P. (2019). Rhizoremediation of polluted sites: Harnessing plant-microbe interactions. In *Phytomanagement of Polluted Sites* (pp. 389–407). Elsevier. Doi: 10.1016/b978-0-12-813912-7.00015-6.

Punzi, M., Anbalagan, A., Börner, R. A., Jonstrup, M., & Mattiasson, B. (2015). Degradation of a textile azo dye using biological treatment followed by photo-Fenton oxidation: Evaluation of toxicity and microbial community structure. *Chemical Engineering Journal*, *270*, 290–299. Doi: 10.1016/j.cej.2015.02.042.

Purswani, J., Romero-Zaliz, R. C., Martín-Platero, A. M., Guisado, I. M., González-López, J., & Pozo, C. (2017). BSocial: Deciphering social behaviors within mixed microbial populations. *Frontiers in Microbiology*, *8*, 919. Doi: 10.3389/fmicb.2017.00919.

Puvaneswari, N., Muthukrishnan, J., & Gunasekaran, P. (2006). Toxicity assessment and microbial degradation of azo dyes. *Indian Journal of Experimental Biology*, *44*, 618–626.

Rabbee, M. F., Sarafat Ali, M., Choi, J., Hwang, B. S., Jeong, S. C., & B. K. hyun. (2019). *Bacillus velezensis*: A valuable member of bioactive molecules within plant microbiomes. *Molecules*, *24*(6), 1046. Doi: 10.3390/molecules24061046.

Rajendran, P., Muthukrishnan, J., & Gunasekaran, P. (2003). Microbes in heavy metal remediation. *Indian Journal of Experimental Biology*, *41*(9), 935–944.

Rajkumar, M., Prasad, M. N. V., Swaminathan, S., & Freitas, H. (2013). Climate change driven plant-metal-microbe interactions. *Environment International*, *53*, 74–86. Doi: 10.1016/j.envint.2012.12.009.

Rane, N. R., Chandanshive, V. V., Watharkar, A. D., Khandare, R. V., Patil, T. S., Pawar, P. K., & Govindwar, S. P. (2015). Phytoremediation of sulfonated Remazol Red dye and textile effluents by *Alternanthera philoxeroides*: An anatomical, enzymatic and pilot scale study. *Water Research*, *83*, 271–281. Doi: 10.1016/j.watres.2015.06.046.

Rane, N. R., Patil, S. M., Chandanshive, V. V., Kadam, S. K., Khandare, R. V., Jadhav, J. P., & Govindwar, S. P. (2016). *Ipomoea hederifolia* rooted soil bed and Ipomoea aquatica rhizofiltration coupled phytoreactors for efficient treatment of textile wastewater. *Water Research*, *96*, 1–11. Doi: 10.1016/j.watres.2016.03.029.

Rather, L. J., Jameel, S., Dar, O. A., Ganie, S. A., Bhat, K. A., & Mohammad, F. (2019). Advances in the sustainable technologies for water conservation in textile industries. In *Water in Textiles and Fashion* (pp. 175–194). Woodhead Publishing. Doi: 10.1016/b978-0-08-102633-5.00010-5.

Rawat, D., Sharma, R.S., Karmakar, S., Arora, L.S., Mishra, V. (2018). Ecotoxic potential of a presumably non-toxic azo dye. *Ecotoxicology and Environmental Safety*, 148, 528–537. Doi: 10.1016/j.ecoenv.2017.10.049

Razo-Flores, E., Luijten, M., Donlon, B., Lettinga, G., & Field, J. (1997). Biodegradation of selected azo dyes under methanogenic conditions. *Water Science and Technology*, *36*(6–7), 65–72. Doi: 10.1016/S0273-1223(97)00508-8.

Ren, D., Madsen, J. S., Sørensen, S. J., & Burmølle, M. (2015). High prevalence of biofilm synergy among bacterial soil isolates in cocultures indicates bacterial interspecific cooperation. *ISME Journal*, *9*(1), 81–89. Doi: 10.1038/ismej.2014.96.

Robinson, T., McMullan, G., Marchant, R., & Nigam, P. (2001). Remediation of dyes in textile effluent: A critical review on current treatment technologies with a proposed alternative. *Bioresource Technology*, *77*, 247–255. Doi: 10.1504/IJEP.2004.004190.

Rojas, C., Vargas, I. T., Bruns, M. A., & Regan, J. M. (2017). Electrochemically active microorganisms from an acid mine drainage-affected site promote cathode oxidation in microbial fuel cells. *Bioelectrochemistry*, *118*, 139–146. Doi: 10.1016/j.bioelechem.2017.07.013.

Saba, B., Jabeen, M., Khalid, A., Aziz, I., & Christy, A. D. (2015). Effectiveness of rice agricultural waste, microbes and wetland plants in the removal of Reactive Black-5 azo dye in microcosm constructed wetlands. *International Journal of Phytoremediation*, *17*(11), 1060–1067. Doi: 10.1080/15226514.2014.1003787.

Sahinkaya, E., Yurtsever, A., & Çınar, Ö. (2017). Treatment of textile industry wastewater using dynamic membrane bioreactor: Impact of intermittent aeration on process performance. *Separation and Purification Technology*, *174*, 445–454. Doi: 10.1016/j.seppur.2016.10.049.

Sani, Z. M., & Abdullahi, I. L. (2018). A preliminary study of soil fungal diversity of dye-contaminated soils in urban Kano. *Bayero Journal of Pure and Applied Sciences*, *10*(1), 336. Doi: 10.4314/bajopas.v10i1.67s.

Santana, C. S., & Aguiar, A. (2015). Effect of biological mediator, 3-hydroxyanthranilic acid, in dye decolorization by Fenton processes. *International Bio deterioration and Biodegradation*, *104*, 1–7. Doi: 10.1016/j.ibiod.2015.05.007.

Saratale, R. G., Saratale, G. D., Chang, J. S., & Govindwar, S. P. (2009). Ecofriendly degradation of sulfonated diazo dye C. I. Reactive Green 19A using *Micrococcus glutamicus* NCIM-2168. *Bioresource Technology*, *100*(17), 3897–3905. Doi: 10.1016/j.biortech.2009.03.051.

Saratale, R. G., Saratale, G. D., Chang, J. S., & Govindwar, S. P. (2011). Bacterial decolorization and degradation of azo dyes : A review. *Journal of the Taiwan Institute of Chemical Engineers*, *42*(1), 138–157. Doi: 10.1016/j.jtice.2010.06.006.

Sarayu, K., & Sandhya, S. (2012). Current technologies for biological treatment of textile wastewater – A review. *Applied Biochemistry and Biotechnology*, *167*(3), 645–661. Doi: 10.1007/s12010-012-9716-6.

Sasse, J., Martinoia, E., & Northen, T. (2018). Feed your friends: Do plant exudates shape the root microbiome? *Trends in Plant Science*, *23*(1), 25–41. Doi: 10.1016/j.tplants.2017.09.003.

Senan, R. C., & Abraham, T. E. (2004). Bioremediation of textile azo dyes by aerobic bacterial consortium. *Biodegradation*, *15*, 275–280.

Şengör, S. S. (2019). Review of Current Applications of Microbial Biopolymers in Soil and Future Perspectives [Chapter]. In *ACS Symposium Series* (Vol. *1323*, pp. 275–299). American Chemcial Society. Doi: 10.1021/bk-2019-1323.ch013.

Shabbir, S., Faheem, M., Ali, N., Kerr, P. G., & Wu, Y. (2017). Periphyton biofilms : A novel and natural biological system for the effective removal of sulphonated azo dye methyl orange by synergistic mechanism. *Chemosphere*, *167*, 236–246. Doi: 10.1016/j.chemosphere.2016.10.002.

Shah, M. P. (2014). An application of mixed consortium in microbial degradation of Reactive Red: Effective strategy of bioaugmentaiton. *Journal of Applied & Environmental Microbiology*, *2*(4), 143–154. Doi: 10.12691/jaem-2-4-8.

Shaikh S. S., Wani, S., & Sayyed, R. (2018). Impact of interactions between rhizosphere and rhizobacteria: A review. *Journal of Bacteriology & Mycology*, *5*(5), 1058–1.

Shanmugam, B. K., Easwaran, S. N., & Lakra, R. (2017). Metabolic pathway and role of individual species in the bacterial consortium for biodegradation of azo dye: A biocalorimetric investigation. *Chemosphere*, *188*, 81–89. Doi: 10.1016/j.chemosphere.2017.08.138.

Sharma, K. P., Sharma, S., Sharma, S., Singh, P. K., Kumar, S., Grover, R., & Sharma, P. K. (2007). A comparative study on characterization of textile wastewaters (untreated and treated) toxicity by chemical and biological tests. *Chemosphere*, *69*(1), 48–54. Doi: 10.1016/j.chemosphere.2007.04.086.

Shehzadi, M., Afzal, M., Khan, M. U., Islam, E., Mobin, A., Anwar, S., & Khan, Q. M. (2014). Enhanced degradation of textile effluent in constructed wetland system using *Typha domingensis* and textile effluent-degrading endophytic bacteria. *Water Research*, *58*, 152–159. Doi: 10.1016/j.watres.2014.03.064.

Shong, J., Jimenez Diaz, M. R., & Collins, C. H. (2012). Towards synthetic microbial consortia for bioprocessing. *Current Opinion in Biotechnology*, *23*(5), 798–802. Doi: 10.1016/j.copbio.2012.02.001.

Silveira, E., Marques, P. P., Silva, S. S., Lima-filho, J. L., Porto, A. L. F., & Tambourgi, E. B. (2009). Selection of *Pseudomonas* for industrial textile dyes decolourization. *International Biodeterioration & Biodegradation*, *63*(2), 230–235. Doi: 10.1016/j.ibiod.2008.09.007.

Singh, K., & Arora, S. (2011). Removal of synthetic textile dyes from wastewaters : A critical review on present treatment technologies. *Critical Reviews in Environmental Science and Technology*, *41*, 807–878. Doi: 10.1080/10643380903218376.

Singh, R. L., Singh, P. K., & Singh, R. P. (2015). Enzymatic decolorization and degradation of azo dyes - A review. *International Biodeterioration and Biodegradation*, *104*, 21–31. Doi: 10.1016/j.ibiod.2015.04.027.

Sivakumar, D. (2014). Role of Lemna minor Lin. in treating the textile industry wastewater. *International Journal of Environmental, Ecological, Geological Snd Mining Engineering*, *8*(3), 203–207.

Smith, N., Shorten, P. R., Altermann, E., Roy, N. C., & McNabb, W. C. (2019). The classification and evolution of bacterial cross-feeding. *Frontiers in Ecology and Evolution*, *7*, 153. Doi: 10.3389/fevo.2019.00153.

Sobariu, D. L., Fertu, D. I. T., Diaconu, M., Pavel, L. V., Hlihor, R. M., Drăgoi, E. N., … Gavrilescu, M. (2017). Rhizobacteria and plant symbiosis in heavy metal uptake and its implications for soil bioremediation. *New Biotechnology*, *39*, 125–134. Doi: 10.1016/j.nbt.2016.09.002.

Solís, M., Solís, A., Inés, H., Manjarrez, N., & Flores, M. (2012). Microbial decolouration of azo dyes : A review. *Process Biochemistry*, *47*(12), 1723–1748. Doi: 10.1016/j.procbio.2012.08.014.

Sreedharan, V., & Rao, K. V. B. (2019). Biodegradation of textile azo dyes. In *Nano science and Biotechnology for Environmental Applications* (pp. 115–139). Springer, Cham. Doi: 10.1007/978-3-319-97922-9_5.

Srisuwun, A., Tantiwa, N., Kuntiya, A., Kawee-ai, A., Manassa, A., Techapun, C., & Seesuriyachan, P. (2018). Decolorization of Reactive Red 159 by a consortium of photosynthetic bacteria using an anaerobic sequencing batch reactor (AnSBR). *Preparative Biochemistry and Biotechnology*, *48*(4), 303–311. Doi: 10.1080/10826068.2018.1431782.

Sudha, M., Bakiyaraj, G., Saranya, A., Sivakumar, N., & Selvakumar, G. (2018). Prospective assessment of the *Enterobacter aerogenes* PP002 in decolorization and degradation of azo dyes DB 71 and DG 28. *Journal of Environmental Chemical Engineering*, *6*(1), 95–109. Doi: 10.1016/j.jece.2017.11.050.

Sun, J., Li, W., Li, Y., Hu, Y., & Zhang, Y. (2013). Redox mediator enhanced simultaneous decolorization of azo dye and bioelectricity generation in air-cathode microbial fuel cell. *Bioresource Technology*, *142*, 407–414. Doi: 10.1016/j.biortech.2013.05.039.

Tahir, U., Yasmin, A., & Khan, U. H. (2016). Phytoremediation: Potential flora for synthetic dyestuff metabolism. *Journal of King Saud University - Science*, *28*(2), 119–130. Doi: 10.1016/j.jksus.2015.05.009.

Tan, N. C. G., Borger, A., Slenders, P., Svitelskaya, A., Lettinga, G., & Field, J. A. (2000). Degradation of azo dye Mordant Yellow 10 in a sequential anaerobic and bioaugmented aerobic bioreactor. *Water Science and Technology*, *42*(5–6), 337–344. Doi: 10.2166/wst.2000.0533.

Tang, J., Wang, Y., Yang, G., Luo, H., Zhuang, L., Yu, Z., & Zhou, S. (2018). Complete genome sequence of the dissimilatory azo reducing thermophilic bacterium *Novibacillus thermophiles* SG-1. *Journal of Biotechnology*, *284*, 6–10. Doi: 10.1016/j.jbiotec.2018.07.032.

Tochhawng, L., Mishra, V. K., Passari, A. K., & Singh, B. P. (2019). Endophytic Fungi: Role in Dye Decolorization. In *Advances in Endophytic Fungal Research* (pp. 1–15). Springer, Cham. Doi: 10.1007/978-3-030-03589-1_1.

Tony, B. D., Goyal, D., & Khanna, S. (2009). Decolorization of textile azo dyes by aerobic bacterial consortium. *International Biodeterioration & Biodegradation*, *63*(4), 462–469. Doi: 10.1016/j.ibiod.2009.01.003.

Traxler, M. F., & Kolter, R. (2015). Natural products in soil microbe interactions and evolution. *Natural Product Reports*, *32*(7), 956–970. Doi: 10.1039/c5np00013k.

Ulrich, D. E. M., Sevanto, S., Ryan, M., Albright, M. B. N., Johansen, R. B., & Dunbar, J. M. (2019). Plant-microbe interactions before drought influence plant physiological responses to subsequent severe drought. *Scientific Reports*, *9*(1), 1–10. Doi: 10.1038/s41598-018-36971-3.

Uysal, Y., Aktas, D., & Caglar, Y. (2014). Determination of colour removal efficiency of Lemna minor L. from industrial effluents. *Journal of Environmental Protection And*, *1726*(4), 1718–1726.

Vafaei, F., Khataee, A. R., Movafeghi, A., Salehi Lisar, S. Y., & Zarei, M. (2012). Bioremoval of an azo dye by Azolla filiculoides: Study of growth, photosynthetic pigments and antioxidant enzymes status. *International Biodeterioration and Biodegradation*, *75*, 194–200. Doi: 10.1016/j.ibiod.2012.09.008.

Van der Zee, F. P., Bouwman, R. H. M., Strik, D. P. B. T. B., Lettinga, G., & Field, J. A. (2001). Application of redox mediators to accelerate the transformation of reactive azo dyes in anaerobic bioreactors. *Biotechnology and Bioengineering*, *75*(6), 691–701. Doi: 10.1002/bit.10073.

Van der Zee, F. P., & Cervantes, F. J. (2009). Impact and application of electron shuttles on the redox (bio) transformation of contaminants : A review. *Biotechnology Advances*, *27*(3), 256–277. Doi: 10.1016/j.biotechadv.2009.01.004.

Vijayalakshmidevi, S. R., & Muthukumar, K. (2014). Phytoremediation of textile effluent pretreated with ultrasound and bacteria. *Journal of Environmental Chemical Engineering*, *2*(3), 1813–1820. Doi: 10.1016/j.jece.2014.07.017.

Wang, Z. W., Liang, J. S., & Liang, Y. (2013). Decolorization of Reactive Black 5 by a newly isolated bacterium *Bacillus* sp. YZU1. *International Biodeterioration & Biodegradation*, *76*, 41–48. Doi: 10.1016/j.ibiod.2012.06.023.

Wang, Z., Yin, Q., Gu, M., He, K., & Wu, G. (2018). Enhanced azo dye Reactive Red 2 degradation in anaerobic reactors by dosing conductive material of ferroferric oxide. *Journal of Hazardous Materials*, *357*, 226–234. Doi: 10.1016/j.jhazmat.2018.06.005.

Wang, H., Zheng, X.-W., Su, J.-Q., Tian, Y., Xiong, X.-J., & Zheng, T.-L. (2009). Biological decolorization of the reactive dyes Reactive Black 5 by a novel isolated bacterial strain *Enterobacter* sp. EC3. *Journal of Hazardous Materials*, *171*, 654–659. Doi: 10.1016/j.jhazmat.2009.06.050.

Watharkar, A. D., & Jadhav, J. P. (2014). Detoxification and decolorization of a simulated textile dye mixture by phytoremediation using *Petunia grandiflora* and, *Gailardia grandiflora*: A plant-plant consortial strategy. *Ecotoxicology and Environmental Safety*, *103*(1), 1–8. Doi: 10.1016/j.ecoenv.2014.01.033.

Watteau, F., & Villemin, G. (2018). Soil microstructures examined through transmission electron microscopy reveal soil-microorganisms interactions. *Frontiers in Environmental Science*, *6*(OCT), 1–10. Doi: 10.3389/fenvs.2018.00106.

Wilpiszeski, R. L., Aufrecht, J. A., Retterer, S. T., Sullivan, M. B., Graham, D. E., Pierce, E. M., … Elias, D. A. (2019). Soil aggregate microbial communities: Towards understanding microbiome interactions at biologically relevant scales. *Applied and Environmental Microbiology*, *85*(14), 1–18. Doi: 10.1128/AEM.00324-19.

Wong, J. K. H., Tan, H. K., Lau, S. Y., Yap, P. S., & Danquah, M. K. (2019). Potential and challenges of enzyme incorporated nanotechnology in dye wastewater treatment: A review. *Journal of Environmental Chemical Engineering*, *7*(4), 103261. Doi: 10.1016/j.jece.2019.103261.

Xiao, X., Xu, C., Wu, Y., Cai, P., Li, W., Du, D., & Yu, H. (2012). Biodecolorization of naphthol Green B dye by *Shewanella oneidensis* MR-1 under anaerobic conditions. *Bioresource Technology*, *110*, 86–90. Doi: 10.1016/j.biortech.2012.01.099.

Xie, X., Liu, N., Ping, J., Zhang, Q., Zheng, X., & Liu, J. (2018). Illumina MiSeq sequencing reveals microbial community in HA process for dyeing wastewater treatment fed with different co-substrates. *Chemosphere*, *201*, 578–585. Doi: 10.1016/j.chemosphere.2018.03.025.

Xie, X., Liu, N., Yang, B., Yu, C., & Zhang, Q. (2016). Comparison of microbial community in hydrolysis acidification reactor depending on different structure dyes by Illumina MiSeq sequencing. *International Biodeterioration & Biodegradation*, *111*, 14–21. Doi: 10.1016/j.ibiod.2016.04.004.

Xu, H., Heinze, T. M., Paine, D. D., Cerniglia, C. E., & Chen, H. (2010). Sudan azo dyes and Para Red degradation by prevalent bacteria of the human gastrointestinal tract. *Anaerobe*, *16*(2), 114–119. Doi: 10.1016/j.anaerobe.2009.06.007.

Xu, X., Zarecki, R., Medina, S., Ofaim, S., Liu, X., Chen, C., … Freilich, S. (2019). Modeling microbial communities from atrazine contaminated soils promotes the development of biostimulation solutions. *The ISME Journal*, *13*(2), 494–508. Doi: 10.1038/s41396-018-0288-5.

Yang, J., Li, W., Bun Ng, T., Deng, X., Lin, J., & Ye, X. (2017). Laccases: Production, expression regulation, and applications in pharmaceutical biodegradation. *Frontiers in Microbiology*, *8*, 832. Doi: 10.3389/fmicb.2017.00832.

Yang, Y., Luo, O., Kong, G., Wang, B., Li, X., Li, E., … Xu, M. (2018). Deciphering the anode-enhanced azo dye degradation in anaerobic baffled reactors integrating with microbial fuel cells. *Frontiers in Microbiology*, *9*, 1–10. Doi: 10.3389/fmicb.2018.02117.

Yang, Q., Wang, J., Wang, H., Chen, X., Ren, S., Li, X. , … Li, X. (2012). Evolution of the microbial community in a full-scale printing and dyeing wastewater treatment system. *Bioresource Technology*, *117*, 155–163. Doi: 10.1016/j.biortech.2012.04.059.

Yaseen, D. A., & Scholz, M. (2016). Shallow pond systems planted with *Lemna minor* treating azo dyes. *Ecological Engineering*, *94*, 295–305. Doi: 10.1016/j.ecoleng.2016.05.081.

Yaseen, D. A., & Scholz, M. (2017). Comparison of experimental ponds for the treatment of dye wastewater under controlled and semi-natural conditions. *Environmental Science and Pollution Research*, *24*(19), 16031–16040. Doi: 10.1007/s11356-017-9245-5.

Young, I. M., & Crawford, J. W. (2019). Interactions and self-organization in the soil-microbe complex. *Science*, *304*(5677), 1634–1637. Doi: 10.1126/science.1097394.

Yurtsever, A., Sahinkaya, E., Aktaş, Ö., Uçar, D., Çinar, Ö., & Wang, Z. (2015). Performances of anaerobic and aerobic membrane bioreactors for the treatment of synthetic textile wastewater. *Bioresource Technology*, *192*, 564–573. Doi: 10.1016/j.biortech.2015.06.024.

Zablocka-Godlewska, E., Przysta, W., & Grabińska-Sota, E. (2012). Decolourization of diazo Evans Blue by two strains of *Pseudomonas fluorescens* isolated from different wastewater treatment plants. *Water, Air and Soil Pollution*, *223*(8), 5259–5266. Doi: 10.1007/s11270-012-1276-4.

Zhang, F., Guo, X., Qian, D. K., Sun, T., Zhang, W., Dai, K., & Zeng, R. J. (2019). Decolorization of Acid Orange 7 by extreme-thermophilic mixed culture. *Bioresource Technology*, *291*(July), 121875. Doi: 10.1016/j.biortech.2019.121875.

Zhou, D., Zhang, X., Du, Y., Dong, S., Xu, Z., & Yan, L. (2014). Insights into the synergistic effect of fungi and bacteria for Reactive Red decolorization. *Journal of Spectroscopy*, *2014*, 1–4.

Zhu, Y., Cao, X., Cheng, Y., & Zhu, T. (2018). Performances and structures of functional microbial communities in the mono azo dye decolorization and mineralization stages. *Chemosphere*, *210*, 1051–1060. Doi: 10.1016/j.chemosphere.2018.07.083.

Zou, L., Huang, Y.H., Long, Z.E., & Qiao, Y. (2019). On-going applications of *Shewanella* species in microbial electrochemical system for bioenergy, bioremediation and biosensing. *World Journal of Microbiology and Biotechnology*, *35*(1), 1–9. Doi: 10.1007/s11274-018-2576-7.

Zucca, P., Cocco, G., Sollai, F., & Sanjust, E. (2016). Fungal laccases as tools for biodegradation of industrial dyes. *Biocatalysis*, *1*(1), 82–108. Doi: 10.1515/boca-2015-0007.

28

Soil Microbiome

Govindan Nadar Rajivgandhi
Sun Yat-Sen University

R.T.V. Vimala, Govindan Ramachandran, and Natesan Manoharan
Bharathidasan University

Wen Jun-Li
Sun Yat-Sen University

CONTENTS

28.1 Introduction

Microbiomes are microbial communities that live on and in plants, oceans, people, atmosphere, and soil. Soil microbiome represents the collective term of highly diverse ecosystems with interacting microbial groups of bacteria, archaea, viruses, fungi, and protozoa. The most favorable place for microbial growth is the soil abiotic environment, heterogeneous in nature with water-filled and disconnected air-filled pores and patchy resources (Mackelprang et al., 2016).

The presence of soil microbes depends on many environmental factors including geographic location, salinity, temperature, oxygen, and nutrients. Therefore, the soil environment is highly dynamic in nature under the influence of factors such as soil moisture, and temperature. Moreover, the stability and resilience of the soil microbiome is disturbed by the unknown consequences caused by climate change (Norby et al., 2016). Moreover, microbial characters that confer ecosystem resilience to climate change are required for predicting the ecosystem. The soil microbiome supports ecosystems that benefit humans such as exchange of plant growth-limiting nutrients;

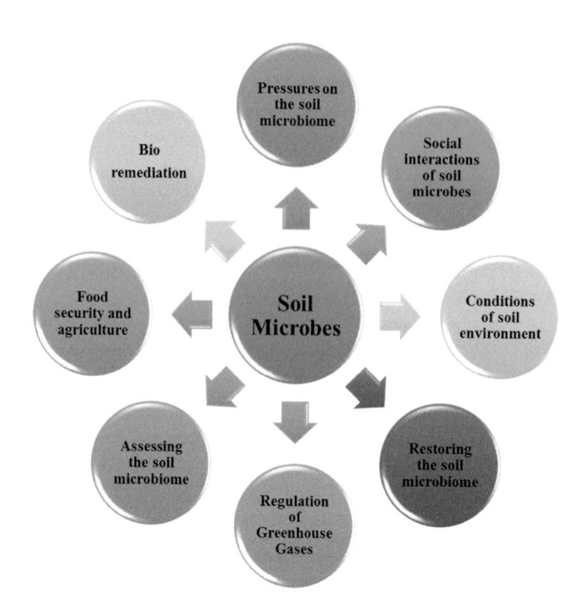

FIGURE 28.1 Entire study of this chapter.

decontamination of soils through bioremediation; changing the physical structure of soil; and a repository of undiscovered biochemicals including antibiotics that can be used to address antibiotic resistance.

In addition, soil microbial communities are very important to the soil fertility, which is useful for the success of agricultural crops. Moreover, the composition and diversity of these soil microbial communities are the key indicators of the soil fertility. However, a significant part of microbial life within the soil is not unexplored. Therefore, the chapter covers about the benefits provided by the soil microbiome, assessing the soil microbiome, soil indicator, pressures existing on the soil microbiome, land practices to restore the soil microbiome, and social interactions of soil microbes (Figure 28.1).

28.2 Benefits of the Soil Microbiome

28.2.1 Regulation of Greenhouse Gases

Soil microbes play a fundamental role in the cycling of the three major greenhouse gases: carbon dioxide (CO_2), methane (CH_4), and nitrous oxide (N_2O).

CO_2: Amount of carbon required for vegetation is stored in the soil. The disturbance of agricultural practices can stimulate the respiration, microbial decomposition, and increasing emissions (Smith, 2008).

CH_4: Bacteria decreases the amount of methane through the process of methanogenesis formed by archaea (Blake et al., 2015).

N_2O: Microbes can reduce N_2O into nitrogen gas, thereby turning the soil into fertile soil as N_2O is emitted predominantly due to the usage of nitrogen fertilizers (Hu et al., 2015).

28.2.2 Food Security and Agriculture

Soil microbes enhance plant growth by emitting chemicals that promote the formation of a relationship between plants and microbes (Hu et al., 2018). For example, a mutual beneficial relationship occurs in between mycorrhizal fungi and plants. Eighty percent of the plant species exhibit symbiotic relationship resulting in a wide connection between different species in the terrestrial ecosystem (Montesinos-Navarro et al., 2019). In addition, microbes in soil can break down organic matter in the soil into a mineral form which can be easily taken up by the plant (Schimel & Bennett, 2004). Therefore, a fertilizer is not required due to the presence of soil microbes.

Moreover, soil microbes are considered as an important component of disease-suppressive soils. Hyphae (long filamentous structures) can bind soil particles, through which bacteria and fungi leak out carbon-rich compounds (Hoorman et al., 2011). They are useful to provide a suitable medium for crop roots and to regulate water retention and drainage in dry/wet conditions. In addition, they protect plants with the help of microbes that include boosting the natural immune system of the plant (Van der Ent et al., 2009), competing with pathogen for nutrients or space in soil (Duijff et al., 1999), and secreting antibiotic compounds and lytic enzymes (Doornbos et al., 2012).

28.2.3 Assessing the Soil Microbiome

28.2.3.1 Technologies to Study the Soil Microbiome

Recently, advanced technologies have been developed to determine the types, proportions, and functions of the soil microbes. Apart from this, genomic approaches are being developed to identify the ecological role of microbes (Bender et al., 2016). Recently, techniques for growing microbes under lab conditions and screening cells for gene expression have been developed. Moreover, DNA sequences for microbes have been identified by bioinformaticians (Attwood et al., 2019).

28.2.3.2 Indicators for Soil Health

Generally, healthy soil depends on various factors such as soil type, location, and pH in which microbes are considered as a biological indicator to determine soil "health" (Ritz et al., 2009). Soil assessment predictive biomarkers are developed by combining sequencing data with other datasets (Trivedi et al., 2016).

28.2.4 Pressures on the Soil Microbiome

28.2.4.1 Farming

Conventional agricultural practices such as continuous cropping, intensive tillage, and use of chemical pesticides and fertilizers decrease soil fertility. They disturb microbes in soil such as fungi and bacteria interfering with ecological processes (Verbruggen et al., 2010). These practices affect reducing the pool of microbes and soil biodiversity. In addition, fertilizers and pesticides alter the communication between plants and soil microbes. They destroy the organic matter of the soil, which is essential for the soil microbiome, and reduce the soil capacity by reducing perturbations, such as drought (de Vries et al., 2012).

28.2.4.2 Urbanization

Soil is sealed due to the major impact of urbanization, in which impermeable materials such as a roads or buildings are covered. This leads to soil degradation and preventing the soil from their function (Scalenghe & Marsan, 2009). High levels of pollution and changes to urban climate are associated with detrimental effects on soil ecosystems. For example, soil pollution changes associated fungal communities, leading to adverse effects on tree health (Van Der Linde et al., 2018). Therefore, the structure and function of microbial communities are changed within urban areas. Soil microbes in urban areas are an important site of carbon storage, and networks of green space have been shown to be beneficial for pollinators. The reduction of soil microbial diversity has created a negative effect on the human immune system (Hanski et al., 2012).

28.2.4.3 Climate Change

Climate change and intensive agricultural practices affect the soil's nature. Climate change are caused due to changes of rising temperatures and increasing frequency of climate extremes such as flooding and drought (Field, 2014). Global changes lead to effects on increasing respiration rates, microbial metabolism, and CO_2 emissions from soils. Moreover, bacteria are not stable than fungal communities (de Vries et al., 2018). The soil structure protects the soil in flood mitigation.

Beneficial plant growth-promoting bacteria and fungi that live in the rhizosphere may balance the negative effect of drought by improving the plant growth even under stressful conditions. Plant growth-promoting bacterial microorganisms can be applied as seed coatings or as liquids to plants growing in the field. Plant growth-promoting bacterial strains, for example, *Rhizobium* sp. as inoculants are applied for biological nitrogen fixation, which is being associated with legumes. Presently, there is a growing interest for the application of inoculants as biofertilizers and biopesticides, which are used to mitigate the deleterious consequences of climate change (Compant et al., 2010) where some soil bacteria secrete the extracellular polymeric substances, resulting in producing biofilms that can protect plants from desiccation (Naylor & Coleman-Derr, 2018). Beneficial soil microorganisms could also be exploited to increase crop tolerance to drought stress through their production of phytohormones that stimulate plant growth, accumulation of osmolytes or other protective compounds, or detoxification of reactive oxygen species (Vurukonda et al., 2016). For example, some bacteria synthesize indole 3 acetic acid in the rhizosphere, resulting in increased root production that can help to alleviate water stress. Rhizosphere microorganisms secrete metabolites that can accumulate in plant cells to improve osmotic stress. Similarly, arbuscular mycorrhizal fungi secrete aquaporins that reduce water stress. Arbuscular mycorrhizal fungi are used to receive the water to the plant root by extending their mycelia into water-filled soil pores. In addition, soil microbes (arbuscular mycorrhizal fungi) as inoculants can be used to

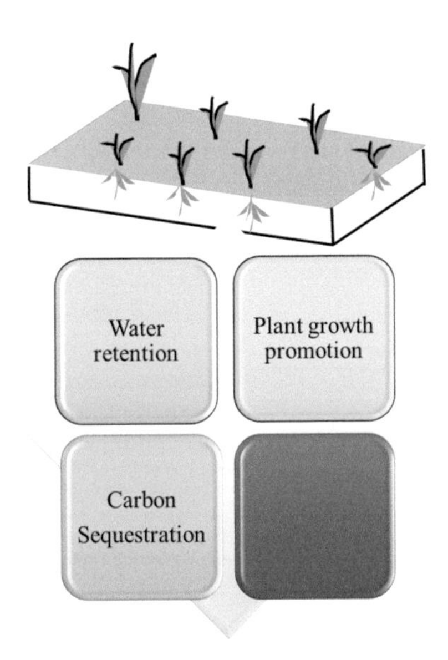

FIGURE 28.2 Negative consequences of climate change.

mitigate N_2O production. Thus, soil microorganisms were used to help and maintain ecosystem services in a changing climate (Figure 28.2).

28.2.5 Restoring the Soil Microbiome

Restoration approaches will need to be considered on a site-by-site basis. The best practice will differ depending on the factors such as soil type and land-management practice.

28.2.5.1 Incentivizing Good Practice

The Natural Capital Committee has suggested to pay the payments to farmers for improving soil health (Committee, 2017), and farmers can also suggest innovations to land practices. In addition, funding should be provided to improve the academic research to develop the soil health. Moreover, consultant support will be provided to enhance the soil health (Krzywoszynska, 2019) Factors such as short farm business tenancies can incentivize short-term productivity over soil health.

28.2.5.2 Targeted Approaches

Restoration of the ecosystem is used to transfer specific soil microbes into a degraded site where seeds are germinated when the soil microbiome is in contact with, allowing for greater crop yields without additional chemical inputs (Wubs et al., 2016).

28.2.5.3 Changing Land-Management Practices

Soil health will depend on its type or texture, soil management practices, and land use history. Sustainable land-management practices for improving food production decline soil productivity. Therefore, reducing input of fertilizers and pesticides may result in lower yields in the short term. But it may be beneficial for a long term, which is useful for maintaining soil health and microbiome. There are some methods that benefit the soil microbiome.

1. Soil organic matter supplies nutrients to the microbes, thereby improving the soil structure and enhancing the soil carbon storage capacity. It can be increased through the addition of cover crops and compost. The important source of soil organic matter is "digestate" obtained from anaerobic digesters, but it has a lower carbon to nitrogen ratio than conventional compost affecting soil organic matter (Johnson, 2018).
2. Chemical input reduction decreases the negative effect on the soil microbes
3. It increases the crop's diversity
4. Management strategies and sufficient organic matter inputs may offer the optimal approach.

28.2.5.4 Degradative Mechanisms of Persistent Organic Pollutants by Soil Microbes

Persistent organic pollutants (lipophilic in nature) cause harmful effects to the environment and human health with the help of biomagnification through the food chain (Chakraborty & Das, 2016). The presence of these diverse soil microorganisms has the ability to remove these toxic organic compounds such as polychlorinated biphenyls, polycyclic aromatic hydrocarbons, pesticides, plastics, etc. Gene mutation, gene rearrangement, and differential regulation in microbes are useful for their microbial survival even under unfavorable conditions (Thomas & Nielsen, 2005).

Bacterial genera such as *Burkholderia, Pseudomonas, Micrococcus, Stenotrophomonas, Corynebacterium, Staphylococcus, Mycobacterium, Desulfotomaculum, Rhodococcus, Sphingobium, Bacillus, Gordonia, Moraxella, Desulfovibrio, Aeromicrobium, Brevibacterium, Dietzia, Escherichia, Pelotomaculum,* and *Methanosaeta* are useful for degradation of persistent organic pollutants (Chakraborty & Das, 2016). Similarly, the fungal genera such as *Graphium, Neosartorya, Talaromyces, Amorphotheca, and Irpex* are the potential organisms for persistent organic pollutant degradation (Gupta et al., 2016b).

28.2.5.5 Plant Breeding and Microbiome Engineering

Certain beneficial soil microbial communities are made to breed the plants in which crops produce high levels of compounds thereby attracting beneficial soil microbes and increasing a symbiotic relationship between them. Combining techniques enhances the microbial diversity in the soil microbes (Bender et al., 2016). In addition, rapid exchange of the horizontal transmission of genes between microbes was carried out by broad-host-range plasmids (Pushpanathan et al., 2014). Elaborative research is yet to be carried out in order to identify the various microbial species from different environments with different functions.

28.2.5.6 Role of Microbial Enzymes and Encoding Genes

The degradation of persistent organic pollutants largely depends on their catabolic enzymes. Catabolic enzymes catalyze the complex degradation into simpler ones (Gupta et al., 2016b).Various genetic variations, particularly horizontal gene transfer, are important processes in adaptation and evolution of microbiomes and have involved in xenobiotic catabolic pathways (DeBruyn et al., 2012). Therefore, understanding catabolic pathways and their catabolic enzymes reveals an effective way to remediate the pollutants (Nzila, 2013).

Degradation of toxic organic compounds using various microbes through oxidative coupling is carried out by oxidoreductases. Microbes break down the chemical bonds by transferring the electrons from a reduced organic substrate to another chemical compound (Karigar and Rao, 2011). Oxidoreductases are divided into three types based on the reactions: monooxygenases, oxygenases, and dioxygenases.

Oxygenases involve in the oxidation of substrates by transferring oxygen from the reduced substrates such as flavin adenine dinucleotide (FAD)/nicotinamide adenine dinucleotide reduced (NAD)/nicotinamide adenine dinucleotide phosphate reduced (NADP) as the co-substrate. Oxygenases degrade many kinds of halogenated organic compounds (Karigar & Rao, 2011). Apart from this, there are other kinds of enzymes involving in the degradation of persistent organic pollutants. For example, laccases catalyze and involve in decarboxylation, demethylation, and demethylation of substrates such as polyphenols or polycyclic aromatic hydrocarbons (PAHs) (Zeng et al., 2016). Peroxidases are enzymes that oxidize lignin and other phenolic compounds in the presence of H_2O_2. These kinds of enzymes degrade the halogenated phenolic compounds, polycyclic aromatic compounds, and other aromatic compounds (Karigar & Rao, 2011).

In addition, several kinds of hydrolases such as celluloses, lipases, and proteases are involved in detoxifying the toxic molecules (Hiraishi, 2016). Aromatic dioxygenases degrade the aromatic compounds such as PAHs, biphenyls, and dioxins and are of considerable interest for bioremediation (Iwai et al., 2011). Naturally, bph and nah are the encoding genes for aromatic dioxygenases, which are used for the degradation of polychlorinated biphenyls (PCBs) and PAHs, respectively (Wang et al., 2014) (Table 28.1).

28.2.6 Social Interactions of Soil Microbes

28.2.6.1 *Quorum Sensing*

Quorum sensing is a process of cell-to-cell communication, which allows the bacteria to find out the information about cell density and species composition of the microbial community and adjust their gene expression profiles (Kylilis et al., 2018), in which microbes release signal molecules, in particular, Gram-positive and Gram-negative bacteria usually release autoinducing peptides as communication signals, which is useful for interspecies communication (Sedlmayer et al., 2018). Moreover, quorum sensing complicates the difference between prokaryotes and eukaryotes as it enables the bacteria to serve as multicellular organisms. Quorum sensing plays the major role in the process of soil microbiome such as biofilm formation, improving microbial stress resistance, and controlling virulence factor production.

28.2.6.2 *Biofilms*

A biofilm is a collection of microorganisms in which cells adhere to one other and stick on to the surface. These adherent cells are surrounded within a matrix of extracellular polymeric substances (Mah & O'Toole, 2001). For example, a coculture of *Pseudomonas protegens, Pseudomonas aeruginosa, and Klebsiella pneumoniae* forms a mixed species of biofilm that possesses the greater resistance to antimicrobials than individual species (Fredrickson, 2015). It has many advantages than free-living microorganisms: (1) ability to exchange genetic material and receive the nutrients from the environment; (2) safety from the surrounding environment; and (3) supplying different microenvironments for microorganisms (Sedlmayer et al., 2018). Indigenous bacterial communities have the ability to degrade persistent organic pollutants and transform the heavy metal contaminants. The low abundance and lack of access to contaminants and limitations in available nutrients of soil microbes are the major drawback of indigenous microbes (Edwards & Kjellerup, 2013), which causes insignificant transformation by these microorganisms. Indigenous biofilms constantly perform bioremediation in the environment, particularly in soil and sediment. Previous reports reported that mixed biofilms reach higher biodegradation efficiency on mixed PAH substrates due to increasing solubility of PAHs. In addition, previous reports showed that biofilms had an enhanced ability to degrade PCBs and dioxins. In addition, extracellular polymeric substances of biofilms play a main role in heavy metal biosorption. These substances form complexes with metal cations and promote the bioremediation of heavy metals.

TABLE 28.1

Catabolic Genes Found in Different Bacterial Genera for Organic Pollutant Degradation

Organic Pollutant	Catabolic Gene(s)	Representative Strain(s)	Reference(s)
Alkane	*alkB1B2*, P450, *almA*	*Alcanivoraxhongdengensis*A-11-3	Wang & Shao (2012)
4-chlorobenzoate	-	*Achromobacter sp. LBS1C*	Layton et al. (1992)
4-chlorobenzoate	-	*P. putida AC858*	Phillips et al. (2008)
	Alkane hydroxylase	-	Phillips et al. (2008)
	Catechol-2,3-dioxygenase	-	Phillips et al. (2008)
p-Toluenesulfonic acid	-	*Comamonastestosteroni* T-2	Tralau et al. (2001)
3-chloroaniline	-	*Delftiaacidovorans CA28*	Boon et al. (2001)
Chlorocatechol	-	*Pseudomonas sp. B13*	van der Meer et al. (2001)
Benzene	*tbc2ABCDEF*, *tomA012345*	*Burkholderia*sp. JS150, *Burkholderia, cepacia*G4	Hendrickx et al. (2006)
PAH	*nahAc*, *nagAc*, *nidA*, *phnABCHGF*, P450	*Pseudomonas putida*G7, *Burkholderia* sp. RP007, *Mycobacterium vanbaalenii* PYR-1, *Acidovorax* sp. NA3, *Irpexlacteus*	Kim et al. (2012); Singleton et al. (2009)
PCB	*bphABCD*, *rdhA*	*Pseudomonas putida* KF715, *Dehalococcoides* sp., *Rhodococcus* sp. R04	Pieper & Seeger (2008); Wang et al. (2014)
OPP	*opdA*	*Agrobacterium radiobacter* P230	Horne et al. (2002)
OCP	*linABCDE*	*Sphingobiumjaponicum* UT26	Okai et al. (2010)
Pyrethroid	*pytH*	*Sphingobium*sp. JZ-1	Wang et al. (2009)

PAH, polycyclic aromatic hydrocarbon; PCB, polychlorinated biphenyl; OPP, organic phosphorus pesticide; OCP, organochlorine pesticide.

28.2.7 Conditions of Soil Environment

The soil environments including soil type, temperature, bioavailability of nutrients, aeration status, soil moisture, water activity, presence of co-contaminants, and microbial competition are the major factors involving in the remedial system (Varjani & Upasani, 2017). Therefore, these factors need to be optimized to enhance the remedial efficiency of the system. Temperature is a major factor in bioremediation (Varjani et al., 2014). For example, 80% crude oil was biodegraded due to soil microbes where various environmental factors were included such as availability of optimal temperature, production of emulsion materials, and bacterial enzymes. They suggested that temperature and nitrogen demand were the important key factors that raised the bacterial efficiency for degrading crude oil components. Moreover, previous research reported that soil pH is the best predictor of bacterial and archaeal community composition. They suggested that soil pH would have a significant effect on the variety of soil microbiome structures which play the major role in soil function such as pollutants' biodegradation (Fierer, 2017).

28.2.8 Bioremediation

Soil microbes play the main role in removing the contaminants naturally (Dixit et al., 2015). It is a cheaper and more environmentally friendly method alternative to conventional methods. Moreover, microbes are successfully used to remediate the contaminated land filled with oil spills, heavy metals, and organic pollutants. Moreover, bioremediation depends on several factors such as soil pH, temperature, and availability of nutrients (Ojuederie & Babalola, 2017) (Figure 28.3).

28.2.8.1 Indigenous Microbes

To our knowledge, the survival ability of exogenous microbes and their growth in soil environments are difficult due their abiotic stressors (Varjani & Upasani, 2017). Previous study reported that indigenous microbes exert a high degree of competition with inoculated microbes (Perez-Garcia et al., 2016). Compared to single strains, microbial consortia showed advantage in that its high diversity of microbes could help the functional exogenous microorganisms to survive in new environments (Großkopf & Soyer, 2014). Researchers assessed the efficiency of microbial consortia and pure cultures for bioremediation. Moreover, crude oil biodegradation using bacterial consortium was composed of four strains such as *Pseudomonas* sp. BPS1-8 (69%), *Pseudomonas* sp. HPS2-5 (45%), *Bacillus* sp. IOS1-7 (64%), and *Corynebacterium* sp. BPS2-6 (41%) with the degradation rate of 77% (Sathishkumar et al., 2008). In addition, Festa et al. (2016) suggested that single strains as inoculants were the best options to remediate soil contamination than the consortium (Festa et al., 2016).

28.2.8.2 Bioavailability of Pollutants and Biosurfactants

The amount of substrate available to microbes is called bioavailability (Souza et al., 2014). The factors such as moisture, ageing, pH, texture, and composition are influenced by soil physicochemical properties. Low water solubility and low bioavailability are the main reasons for the existence of non-degradable organic pollutants (Chakraborty & Das, 2016). Previous reports suggested that different pollutants were degraded into different extents by microbes due to the bioavailability of the particular compound (Varjani & Upasani, 2017). Therefore, soil microorganisms have the ability to degrade different products (i.e., biosurfactants, biopolymers, solvents, and acids) to enhance remediation. Among all the products, biosurfactants play a critical role in improving the bioavailability of hydrocarbon pollutants. Therefore, the bioavailability of organic pollutants is enhanced using biosurfactants (Gupta et al., 2016a). Surfactant activity and hydrophobicity are the major factors involved in the interaction between microorganisms and insoluble substrates. Moreover, surfactants modify the cell surface by increasing hydrophilic and decreasing hydrophobic cultures (Owsianiak et al., 2009).

28.9 Conclusions

As soil pollution is attracting increasing attention, more research is focusing on the function of soil microbes. A variety of soil microorganisms were investigated to make contributions to detoxify the different environmental pollutants. Initially, microbiomes involve more metabolic flexibility to the complex pollutants. Diversity of microbes and their interrelated metabolic pathways are the special features to involve

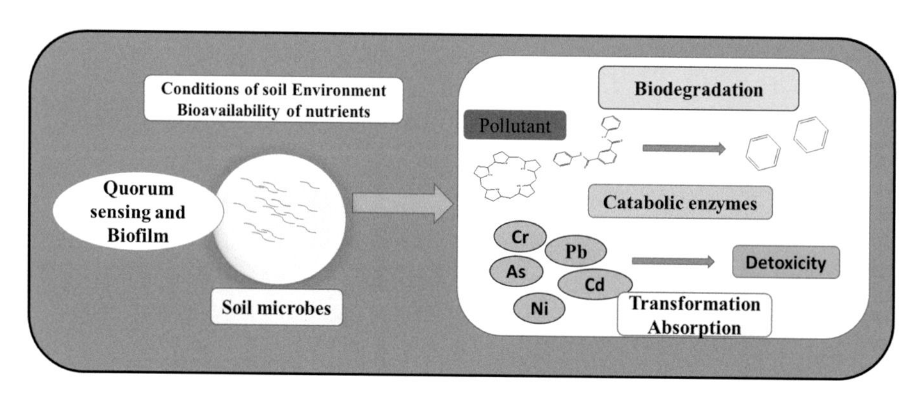

FIGURE 28.3 Bioremediation concept of soil microbiomes.

in different functions. Surfactants produced by microbes may affect the toxicity of pollutants to degrade and accelerate the process of bioremediation. Therefore, microbiomes were effective in degrading organic pollutants and were used to transform the heavy metals. DNA-, RNA-, and protein-based analyses about the soil microbiomes increase the knowledge about the taxonomic structure, interactive mechanisms, and functions of soil microbial communities.

Moreover, the networks of microorganisms among soil microbiome were the difficult processes such as competition, quorum sensing, induction, cooperation, and regulation. Degraders showed their capacity more stably and efficiently. In addition, various biotic and abiotic factors of the soil can influence the total amount of microbial biomass and its functions. Therefore, the chapter describes and provides the strategy of soil microbiomes toward various functions due to their nature such as environmental friendliness, sustainability, and low-cost remediating ability with high efficiency.

REFERENCES

Attwood, T.K., Blackford, S., Brazas, M.D., Davies, A., Schneider, M.V. 2019. A global perspective on evolving bioinformatics and data science training needs. *Briefings in Bioinformatics*, 20(2), 398–404.

Bender, S.F., Wagg, C., van der Heijden, M.G. 2016. An underground revolution: Biodiversity and soil ecological engineering for agricultural sustainability. *Trends in Ecology & Evolution*, 31(6), 440–452.

Blake, L.I., Tveit, A., Ovreas, L., Head, I.M., Gray, N.D. 2015. Response of methanogens in Arctic sediments to temperature and methanogenic substrate availability. *PLoS One*, 10(6), e0129733.

Boon, N., Goris, J., De Vos, P., Verstraete, W., Top, E.M. 2001. Genetic diversity among 3-chloroaniline-and aniline-degrading strains of the *Comamonadaceae*. *Applied and Environmental Microbiology*, 67(3), 1107–1115.

Chakraborty, J., Das, S. 2016. Molecular perspectives and recent advances in microbial remediation of persistent organic pollutants. *Environmental Science and Pollution Research*, 23(17), 16883–16903.

Committee, N.C. 2017. Improving natural capital: An assessment of progress.

Compant, S., Van Der Heijden, M.G., Sessitsch, A. 2010. Climate change effects on beneficial plant–microorganism interactions. *FEMS Microbiology Ecology*, 73(2), 197–214.

de Vries, F.T., Bloem, J., Quirk, H., Stevens, C.J., Bol, R., Bardgett, R.D. 2012. Extensive management promotes plant and microbial nitrogen retention in temperate grassland. *PLoS One*, 7(12), e51201.

de Vries, F.T., Griffiths, R.I., Bailey, M., Craig, H., Girlanda, M., Gweon, H.S., Hallin, S., Kaisermann, A., Keith, A.M., Kretzschmar, M. 2018. Soil bacterial networks are less stable under drought than fungal networks. *Nature Communications*, 9(1), 3033.

DeBruyn, J.M., Mead, T.J., Sayler, G.S. 2012. Horizontal transfer of PAH catabolism genes in Mycobacterium: Evidence from comparative genomics and isolated pyrene-degrading bacteria. *Environmental Science & Technology*, 46(1), 99–106.

Dixit, R., Malaviya, D., Pandiyan, K., Singh, U.B., Sahu, A., Shukla, R., Singh, B.P., Rai, J.P., Sharma, P.K., Lade, H. 2015. Bioremediation of heavy metals from soil and aquatic environment: An overview of principles and criteria of fundamental processes. *Sustainability*, 7(2), 2189–2212.

Doornbos, R.F., van Loon, L.C., Bakker, P.A. 2012. Impact of root exudates and plant defense signaling on bacterial communities in the rhizosphere. A review. *Agronomy for Sustainable Development*, 32(1), 227–243.

Duijff, B.J., Recorbet, G., Bakker, P.A., Loper, J.E., Lemanceau, P. 1999. Microbial antagonism at the root level is involved in the suppression of Fusarium wilt by the combination of non-pathogenic *Fusarium oxysporum* Fo47 and *Pseudomonas putida* WCS358. *Phytopathology*, 89(11), 1073–1079.

Edwards, S.J., Kjellerup, B.V. 2013. Applications of biofilms in bioremediation and biotransformation of persistent organic pollutants, pharmaceuticals/personal care products, and heavy metals. *Applied Microbiology and Biotechnology*, 97(23), 9909–9921.

Festa, S., Coppotelli, B.M., Morelli, I.S. 2016. Comparative bioaugmentation with a consortium and a single strain in a phenanthrene-contaminated soil: Impact on the bacterial community and biodegradation. *Applied Soil Ecology*, 98, 8–19.

Field, C.B. 2014. *Climate Change 2014–Impacts, Adaptation and Vulnerability: Regional Aspects*. Cambridge University Press.

Fierer, N. 2017. Embracing the unknown: Disentangling the complexities of the soil microbiome. *Nature Reviews Microbiology*, 15(10), 579–590.

Fredrickson, J.K. 2015. Ecological communities by design. *Science*, 348(6242), 1425–1427.

Großkopf, T., Soyer, O.S. 2014. Synthetic microbial communities. *Current Opinion in Microbiology*, 18, 72–77.

Gupta, A., Joia, J., Sood, A., Sood, R., Sidhu, C., Kaur, G. 2016a. Microbes as potential tool for remediation of heavy metals: A review. *Journal of Microbial and Biochemical Technology*, 8(4), 364–72.

Gupta, G., Kumar, V., Pal, A.K. 2016b. Biodegradation of polycyclic aromatic hydrocarbons by microbial consortium: A distinctive approach for decontamination of soil. *Soil and Sediment Contamination: An International Journal*, 25(6), 597–623.

Hanski, I., von Hertzen, L., Fyhrquist, N., Koskinen, K., Torppa, K., Laatikainen, T., Karisola, P., Auvinen, P., Paulin, L., Mäkelä, M.J. 2012. Environmental biodiversity, human microbiota, and allergy are interrelated. *Proceedings of the National Academy of Sciences*, 109(21), 8334–8339.

Hendrickx, B., Junca, H., Vosahlova, J., Lindner, A., Rüegg, I., Bucheli-Witschel, M., Faber, F., Egli, T., Mau, M., Schlömann, M. 2006. Alternative primer sets for PCR detection of genotypes involved in bacterial aerobic BTEX degradation: Distribution of the genes in BTEX degrading isolates and in subsurface soils of a BTEX contaminated industrial site. *Journal of Microbiological Methods*, 64(2), 250–265.

Hiraishi, T. 2016. Poly (aspartic acid)(PAA) hydrolases and PAA biodegradation:current knowledge and impact on applications. *Applied Microbiology and Biotechnology*, 100(4), 1623–1630.

Hoorman, J.J., Sa, J.C.d.M.J., Reeder, R. 2011. The biology of soil compaction. *Soil & Tillage Research*, 68, 49–57.

Horne, I., Sutherland, T.D., Harcourt, R.L., Russell, R.J., Oakeshott, J.G. 2002. Identification of an opd (organophosphate degradation) gene in an *Agrobacterium* isolate. *Applied and Environmental Microbiology,* 68(7), 3371–3376.

Hu, H.-W., Chen, D., He, J.-Z. 2015. Microbial regulation of terrestrial nitrous oxide formation: Understanding the biological pathways for prediction of emission rates. *FEMS Microbiology Reviews*, 39(5), 729–749.

Hu, L., Robert, C.A., Cadot, S., Zhang, X., Ye, M., Li, B., Manzo, D., Chervet, N., Steinger, T., Van Der Heijden, M.G. 2018. Root exudate metabolites drive plant-soil feedbacks on growth and defense by shaping the rhizosphere microbiota. *Nature Communications*, 9(1), 1–13.

Iwai, S., Johnson, T.A., Chai, B., Hashsham, S.A., Tiedje, J.M. 2011. Comparison of the specificities and efficacies of primers for aromatic dioxygenase gene analysis of environmental samples. *Applied and Environmental Microbiology*, 77(11), 3551–3557.

Johnson, K.L. 2018. Heat and soil vie for waste. *Nature*, 563(7733), 626–626.

Karigar, C.S., Rao, S.S. 2011. Role of microbial enzymes in the bioremediation of pollutants: A review. *Enzyme research*, 2011.

Kim, S.-J., Song, J., Kweon, O., Holland, R.D., Kim, D.-W., Kim, J., Yu, L.-R., Cerniglia, C.E. 2012. Functional robustness of a polycyclic aromatic hydrocarbon metabolic network examined in a nidA aromatic ring-hydroxylating oxygenase mutant of Mycobacterium vanbaalenii PYR-1. *Applied and Environmental Microbiology*, 78(10), 3715–3723.

Krzywoszynska, A. 2019. Making knowledge and meaning in communities of practice: What role may science play? The case of sustainable soil management in England. *Soil Use and Management*, 35(1), 160–168.

Kylilis, N., Tuza, Z.A., Stan, G.-B., Polizzi, K.M. 2018. Tools for engineering coordinated system behaviour in synthetic microbial consortia. *Nature Communications*, 9(1), 1–9.

Layton, A.C., Sanseverino, J., Wallace, W., Corcoran, C., Sayler, G.S. 1992. Evidence for 4-chlorobenzoic acid dehalogenation mediated by plasmids related to pSS50. *Applied and Environmental Microbiology,* 58(1), 399–402.

Mackelprang, R., Saleska, S.R., Jacobsen, C.S., Jansson, J.K., Taş, N. 2016. Permafrost meta-omics and climate change. *Annual Review of Earth and Planetary Sciences*, 44, 439–462.

Mah, T.-F.C., O'Toole, G.A. 2001. Mechanisms of biofilm resistance to antimicrobial agents. *Trends in Microbiology*, 9(1), 34–39.

Montesinos-Navarro, A., Valiente-Banuet, A., Verdú, M. 2019. Mycorrhizal symbiosis increases the benefits of plant facilitative interactions. *Ecography*, 42(3), 447–455.

Naylor, D., Coleman-Derr, D. 2018. Drought stress and root-associated bacterial communities. *Frontiers in Plant Science*, 8, 2223.

Norby, R.J., De Kauwe, M.G., Domingues, T.F., Duursma, R.A., Ellsworth, D.S., Goll, D.S., Lapola, D.M., Luus, K.A., MacKenzie, A.R., Medlyn, B.E. 2016. Model–data synthesis for the next generation of forest free-air CO_2 enrichment (FACE) experiments. *New Phytologist*, 209(1), 17–28.

Nzila, A. 2013. Update on the cometabolism of organic pollutants by bacteria. *Environmental Pollution*, 178, 474–482.

Ojuederie, O.B., Babalola, O.O. 2017. Microbial and plant-assisted bioremediation of heavy metal polluted environments: A review. *International Journal of Environmental Research and Public Health*, 14(12), 1504.

Okai, M., Kubota, K., Fukuda, M., Nagata, Y., Nagata, K., Tanokura, M. 2010. Crystal structure of γ-hexachlorocyclohexane dehydrochlorinase LinA from *Sphingobium japonicum* UT26. *Journal of Molecular Biology*, 403(2), 260–269.

Owsianiak, M., Szulc, A., Chrzanowski, L., Cyplik, P., Bogacki, M., Olejnik-Schmidt, A.K., Heipieper, H.J. 2009. Biodegradation and surfactant-mediated biodegradation of diesel fuel by 218 microbial consortia are not correlated to cell surface hydrophobicity. *Applied Microbiology and Biotechnology*, 84(3), 545–553.

Perez-Garcia, O., Lear, G., Singhal, N. 2016. Metabolic network modeling of microbial interactions in natural and engineered environmental systems. *Frontiers in Microbiology*, 7, 673.

Phillips, L.A., Germida, J.J., Farrell, R.E., Greer, C.W. 2008. Hydrocarbon degradation potential and activity of endophytic bacteria associated with prairie plants. *Soil Biology and Biochemistry*, 40(12), 3054–3064.

Pieper, D.H., Seeger, M. 2008. Bacterial metabolism of polychlorinated biphenyls. *Journal of Molecular Microbiology and Biotechnology*, 15(2–3), 121–138.

Pushpanathan, M., Jayashree, S., Gunasekaran, P., Rajendhran, J. 2014. Microbial bioremediation: A metagenomic approach. In *Microbial Biodegradation and Bioremediation*, *Elsevier*, pp. 407–419.

Ritz, K., Black, H.I., Campbell, C.D., Harris, J.A., Wood, C. 2009. Selecting biological indicators for monitoring soils: A framework for balancing scientific and technical opinion to assist policy development. *Ecological Indicators*, 9(6), 1212–1221.

Sathishkumar, M., Binupriya, A.R., Baik, S.H., Yun, S.E. 2008. Biodegradation of crude oil by individual bacterial strains and a mixed bacterial consortium isolated from hydrocarbon contaminated areas. *Clean-Soil, Air, Water*, 36(1), 92–96.

Scalenghe, R., Marsan, F.A. 2009. The anthropogenic sealing of soils in urban areas. *Landscape and Urban Planning*, 90(1–2), 1–10.

Schimel, J.P., Bennett, J. 2004. Nitrogen mineralization: Challenges of a changing paradigm. *Ecology*, 85(3), 591–602.

Sedlmayer, F., Hell, D., Müller, M., Ausländer, D., Fussenegger, M. 2018. Designer cells programming quorum-sensing interference with microbes. *Nature Communications*, 9(1), 1–13.

Singleton, D.R., Ramirez, L.G., Aitken, M.D. 2009. Characterization of a polycyclic aromatic hydrocarbon degradation gene cluster in a phenanthrene-degrading Acidovorax strain. *Applied and Environmental Microbiology*, 75(9), 2613–2620.

Smith, P. 2008. Land use change and soil organic carbon dynamics. *Nutrient Cycling in Agroecosystems*, 81(2), 169–178.

Souza, E.C., Vessoni-Penna, T.C., de Souza Oliveira, R.P. 2014. Biosurfactant-enhanced hydrocarbon bioremediation: An overview. *International Biodeterioration & Biodegradation*, 89, 88–94.

Thomas, C.M., Nielsen, K.M. 2005. Mechanisms of, and barriers to, horizontal gene transfer between bacteria. *Nature Reviews Microbiology*, 3(9), 711–721.

Tralau, T., Cook, A.M., Ruff, J. 2001. Map of the IncP1β plasmid pTSA encoding the widespread genes (tsa) forp-toluenesulfonate degradation in *Comamonas testosteroni* T-2. *Applied and Environmental Microbiology*, 67(4), 1508–1516.

Trivedi, P., Delgado-Baquerizo, M., Anderson, I.C., Singh, B.K. 2016. Response of soil properties and microbial communities to agriculture: Implications for primary productivity and soil health indicators. *Frontiers in Plant Science*, 7, 990.

Van der Ent, S., Van Hulten, M., Pozo, M.J., Czechowski, T., Udvardi, M.K., Pieterse, C.M., Ton, J. 2009. Priming of plant innate immunity by rhizobacteria and β-aminobutyric acid: Differences and similarities in regulation. *New Phytologist*, 183(2), 419–431.

Van Der Linde, S., Suz, L.M., Orme, C.D.L., Cox, F., Andreae, H., Asi, E., Atkinson, B., Benham, S., Carroll, C., Cools, N. 2018. Environment and host as large-scale controls of ectomycorrhizal fungi. *Nature*, 558(7709), 243–248.

van der Meer, J.R., Ravatn, R., Sentchilo, V. 2001. The clc element of Pseudomonas sp. strain B13 and other mobile degradative elements employing phage-like integrases. *Archives of Microbiology*, 175(2), 79–85.

Varjani, S., Thaker, M., Upasani, V. 2014. Optimization of growth conditions of native hydrocarbon utilizing bacterial consortium "HUBC" obtained from petroleum pollutant contaminated sites. *Indian Journal of Applied Research*, 4(10), 474–476.

Varjani, S.J., Upasani, V.N. 2017. A new look on factors affecting microbial degradation of petroleum hydrocarbon pollutants. *International Biodeterioration & Biodegradation*, 120, 71–83.

Verbruggen, E., Röling, W.F., Gamper, H.A., Kowalchuk, G.A., Verhoef, H.A., van der Heijden, M.G. 2010. Positive effects of organic farming on below-ground mutualists: Large-scale comparison of mycorrhizal fungal communities in agricultural soils. *New Phytologist*, 186(4), 968–979.

Vurukonda, S.S.K.P., Vardharajula, S., Shrivastava, M., SkZ, A. 2016. Enhancement of drought stress tolerance in crops by plant growth promoting rhizobacteria. *Microbiological Research*, 184, 13–24.

Wang, B.-z., Guo, P., Hang, B.-j., Li, L., He, J., Li, S.-p. 2009. Cloning of a novel pyrethroid-hydrolyzing carboxylesterase gene from *Sphingobium* sp. strain JZ-1 and characterization of the gene product. *Applied and Environmental Microbiology*, 75(17), 5496–5500.

Wang, S., Chng, K.R., Wilm, A., Zhao, S., Yang, K.-L., Nagarajan, N., He, J. 2014. Genomic characterization of three unique Dehalococcoides that respire on persistent polychlorinated biphenyls. *Proceedings of the National Academy of Sciences*, 111(33), 12103–12108.

Wubs, E.J., Van der Putten, W.H., Bosch, M., Bezemer, T.M. 2016. Soil inoculation steers restoration of terrestrial ecosystems. *Nature Plants*, 2(8), 1–5.

Zeng, J., Zhu, Q., Wu, Y., Lin, X. 2016. Oxidation of polycyclic aromatic hydrocarbons using Bacillus subtilis CotA with high laccase activity and copper independence. *Chemosphere*, 148, 1–7.

Index

Note: Page numbers in *italics* and **bold** refer to figures and tables.

D

E

F

G

H

N

Q

R

S

T

U

V

W

X

Y

Z